T0189058

Lecture Notes in Computer Science   10638

Commenced Publication in 1973
Founding and Former Series Editors:
Gerhard Goos, Juris Hartmanis, and Jan van Leeuwen

Editorial Board

More information about this series at http://www.springer.com/series/7407

Derong Liu · Shengli Xie
Yuanqing Li · Dongbin Zhao
El-Sayed M. El-Alfy (Eds.)

Neural
Information Processing

24th International Conference, ICONIP 2017
Guangzhou, China, November 14–18, 2017
Proceedings, Part V

 Springer

Editors
Derong Liu
Guangdong University of Technology
Guangzhou
China

Shengli Xie
Guangdong University of Technology
Guangzhou
China

Yuanqing Li
South China University of Technology
Guangzhou
China

Dongbin Zhao
Institute of Automation
Chinese Academy of Sciences
Beijing
China

El-Sayed M. El-Alfy
King Fahd University of Petroleum
  and Minerals
Dhahran
Saudi Arabia

ISSN 0302-9743   ISSN 1611-3349 (electronic)
Lecture Notes in Computer Science
ISBN 978-3-319-70138-7   ISBN 978-3-319-70139-4 (eBook)
https://doi.org/10.1007/978-3-319-70139-4

Library of Congress Control Number: 2017957558

LNCS Sublibrary: SL1 – Theoretical Computer Science and General Issues

Printed on acid-free paper

This Springer imprint is published by Springer Nature
The registered company is Springer International Publishing AG
The registered company address is: Gewerbestrasse 11, 6330 Cham, Switzerland

Preface

ICONIP 2017 – the 24th International Conference on Neural Information Processing – was held in Guangzhou, China, continuing the ICONIP conference series, which started in 1994 in Seoul, South Korea. Over the past 24 years, ICONIP has been held in Australia, China, India, Japan, Korea, Malaysia, New Zealand, Qatar, Singapore, Thailand, and Turkey. ICONIP has now become a well-established, popular and high-quality conference series on neural information processing in the region and around the world. With the growing popularity of neural networks in recent years, we have witnessed an increase in the number of submissions and in the quality of papers. Guangzhou, Romanized as Canton in the past, is the capital and largest city of southern China's Guangdong Province. It is also one of the five National Central Cities at the core of the Pearl River Delta. It is a key national transportation hub and trading port. November is the best month in the year to visit Guangzhou with comfortable weather. All participants of ICONIP 2017 had a technically rewarding experience as well as a memorable stay in this great city.

A neural network is an information processing structure inspired by biological nervous systems, such as the brain. It consists of a large number of highly interconnected processing elements, called neurons. It has the capability of learning from example. The field of neural networks has evolved rapidly in recent years. It has become a fusion of a number of research areas in engineering, computer science, mathematics, artificial intelligence, operations research, systems theory, biology, and neuroscience. Neural networks have been widely applied for control, optimization, pattern recognition, image processing, signal processing, etc.

ICONIP 2017 aimed to provide a high-level international forum for scientists, researchers, educators, industrial professionals, and students worldwide to present state-of-the-art research results, address new challenges, and discuss trends in neural information processing and applications. ICONIP 2017 invited scholars in all areas of neural network theory and applications, computational neuroscience, machine learning, and others.

The conference received 856 submissions from 3,255 authors in 56 countries and regions across all six continents. Based on rigorous reviews by the Program Committee members and reviewers, 563 high-quality papers were selected for publication in the conference proceedings. We would like to express our sincere gratitude to all the reviewers for the time and effort they generously gave to the conference. We are very grateful to the Institute of Automation of the Chinese Academy of Sciences, Guangdong University of Technology, South China University of Technology, Springer's *Lecture Notes in Computer Science* (LNCS), IEEE/CAA *Journal of Automatica Sinica* (JAS), and the Asia Pacific Neural Network Society (APNNS) for their financial support. We would also like to thank the publisher, Springer, for their cooperation in publishing the proceedings in the prestigious LNCS series and for sponsoring the best paper awards at ICONIP 2017.

September 2017

Derong Liu
Shengli Xie
Yuanqing Li
Dongbin Zhao
El-Sayed M. El-Alfy

ICONIP 2017 Organization

Asia Pacific Neural Network Society

General Chair

Derong Liu   Chinese Academy of Sciences and Guangdong University of Technology, China

Advisory Committee

Sabri Arik	Istanbul University, Turkey
Tamer Basar	University of Illinois, USA
Dimitri Bertsekas	Massachusetts Institute of Technology, USA
Jonathan Chan	King Mongkut's University of Technology, Thailand
C.L. Philip Chen	The University of Macau, SAR China
Kenji Doya	Okinawa Institute of Science and Technology, Japan
Minyue Fu	The University of Newcastle, Australia
Tom Gedeon	Australian National University, Australia
Akira Hirose	The University of Tokyo, Japan
Zeng-Guang Hou	Chinese Academy of Sciences, China
Nikola Kasabov	Auckland University of Technology, New Zealand
Irwin King	Chinese University of Hong Kong, SAR China
Robert Kozma	University of Memphis, USA
Soo-Young Lee	Korea Advanced Institute of Science and Technology, South Korea
Frank L. Lewis	University of Texas at Arlington, USA
Chu Kiong Loo	University of Malaya, Malaysia
Baoliang Lu	Shanghai Jiao Tong University, China
Seiichi Ozawa	Kobe University, Japan
Marios Polycarpou	University of Cyprus, Cyprus
Danil Prokhorov	Toyota Technical Center, USA
DeLiang Wang	The Ohio State University, USA
Jun Wang	City University of Hong Kong, SAR China
Jin Xu	Peking University, China
Gary G. Yen	Oklahoma State University, USA
Paul J. Werbos	Retired from the National Science Foundation, USA

Program Chairs

Shengli Xie	Guangdong University of Technology, China
Yuanqing Li	South China University of Technology, China
Dongbin Zhao	Chinese Academy of Sciences, China
El-Sayed M. El-Alfy	King Fahd University of Petroleum and Minerals, Saudi Arabia

Program Co-chairs

Shukai Duan	Southwest University, China
Kazushi Ikeda	Nara Institute of Science and Technology, Japan
Weng Kin Lai	Tunku Abdul Rahman University College, Malaysia
Shiliang Sun	East China Normal University, China
Qinglai Wei	Chinese Academy of Sciences, China
Wei Xing Zheng	University of Western Sydney, Australia

Regional Chairs

Cesare Alippi	Politecnico di Milano, Italy
Tingwen Huang	Texas A&M University at Qatar, Qatar
Dianhui Wang	La Trobe University, Australia

Invited Session Chairs

Wei He	University of Science and Technology Beijing, China
Dianwei Qian	North China Electric Power University, China
Manuel Roveri	Politecnico di Milano, Italy
Dong Yue	Nanjing University of Posts and Telecommunications, China

Poster Session Chairs

Sung Bae Cho	Yonsei University, South Korea
Ping Guo	Beijing Normal University, China
Yifei Pu	Sichuan University, China
Bin Xu	Northwestern Polytechnical University, China
Zhigang Zeng	Huazhong University of Science and Technology, China

Tutorial and Workshop Chairs

Long Cheng	Chinese Academy of Sciences, China
Kaizhu Huang	Xi'an Jiaotong-Liverpool University, China
Amir Hussain	University of Stirling, UK

James Kwok	Hong Kong University of Science and Technology, SAR China
Huajin Tang	Sichuan University, China

Panel Discussion Chairs

Lei Guo	Beihang University, China
Hongyi Li	Bohai University, China
Hye Young Park	Kyungpook National University, South Korea
Lipo Wang	Nanyang Technological University, Singapore

Award Committee Chairs

Haibo He	University of Rhode Island, USA
Zhong-Ping Jiang	New York University, USA
Minho Lee	Kyungpook National University, South Korea
Andrew Leung	City University of Hong Kong, SAR China
Tieshan Li	Dalian Maritime University, China
Lidan Wang	Southwest University, China
Jun Zhang	South China University of Technology, China

Publicity Chairs

Jun Fu	Northeastern University, China
Min Han	Dalian University of Technology, China
Yanjun Liu	Liaoning University of Technology, China
Stefano Squartini	Università Politecnica delle Marche, Italy
Kay Chen Tan	National University of Singapore, Singapore
Kevin Wong	Murdoch University, Australia
Simon X. Yang	University of Guelph, Canada

Local Arrangements Chair

Renquan Lu	Guangdong University of Technology, China

Publication Chairs

Ding Wang	Chinese Academy of Sciences, China
Jian Wang	China University of Petroleum, China

Finance Chair

Xinping Guan	Shanghai Jiao Tong University, China

Registration Chair

Qinmin Yang   Zhejiang University, China

Conference Secretariat

Biao Luo   Chinese Academy of Sciences, China
Bo Zhao   Chinese Academy of Sciences, China

Contents

Time Series Analysis

Social Networks

Bioinformatics, Information Security and Social Cognition

Data Mining

Low-Rank and Sparse Matrix Completion
for Recommendation

Zhi-Lin Zhao[1], Ling Huang[1], Chang-Dong Wang[1(✉)], Jian-Huang Lai[1],
and Philip S. Yu[2,3]

[1] School of Data and Computer Science, Sun Yat-sen University, Guangzhou, China
zhaozhl7@mail2.sysu.edu.cn, huanglinghl@hotmail.com,
changdongwang@hotmail.com, stsljh@mail.sysu.edu.cn
[2] Department of Computer Science, University of Illinois at Chicago,
Chicago, IL, USA
psyu@cs.uic.edu
[3] Institute for Data Science, Tsinghua University, Beijing, China

Abstract. Recently, recommendation algorithms have been widely used
to improve the benefit of businesses and the satisfaction of users in many
online platforms. However, most of the existing algorithms generate intermediate output when predicting ratings and the error of intermediate
output will be propagated to the final results. Besides, since most algorithms predict all the unrated items, some predicted ratings may be
unreliable and useless which will lower the efficiency and effectiveness of
recommendation. To this end, we propose a Low-rank and Sparse Matrix
Completion (LSMC) method which recovers rating matrix directly to
improve the quality of rating prediction. Following the common methodology, we assume the structure of the predicted rating matrix is low-rank
since rating is just connected with some factors of user and item. However, different from the existing methods, we assume the matrix is sparse
so some unreliable predictions will be removed and important results
will be retained. Besides, a slack variable will be used to prevent overfitting and weaken the influence of noisy data. Extensive experiments on
four real-world datasets have been conducted to verify that the proposed
method outperforms the state-of-the-art recommendation algorithms.

Keywords: Recommendation algorithms · Low-rank · Sparse

1 Introduction

Matrix completion is to recover a rectangular matrix from a sampling of its
entries. Among them, recovering a low-rank or approximately low-rank matrix
from the original matrix has been applied in many areas such as machine learning [1], data mining [7] and computer vision [17]. In particular, matrix factorization recovering an approximately low-rank matrix from a subset of ratings
is one of the most common collaborative filtering algorithms and has attracted
an increasing amount of attention. Many researches about matrix factorization

© Springer International Publishing AG 2017
D. Liu et al. (Eds.): ICONIP 2017, Part V, LNCS 10638, pp. 3–13, 2017.
https://doi.org/10.1007/978-3-319-70139-4_1

extract user and item latent feature vectors from the rating data or other auxiliary data first and then recover an approximately low-rank predicted rating
matrix according to those latent feature vectors. Normally, the dimensionality,
or the rank of latent feature vectors are low which means there are only few factors contribute to the preference of users and items. Those methods will generate
latent feature vectors as intermediate output, but in fact, we usually don't care
about those vectors but the final values of the predicted ratings. So there is a
gap between the input rating matrix and the predicted output rating matrix and
the error of the intermediate output will be propagated to the final prediction.

The common strategy of recommendation is to predict ratings for all items
that users have not rated, but we don't need to know all the predicted ratings because the number of items in recommendation list is limited [16,19,20].
And it will waste our time to sort all predicted ratings to select non-purchased
items with high predicted ratings. What's more, the original rating matrix is
quite sparse and the number of the predicted ratings is a lot larger. So, it may
produce some unreliable predictions which will reduce the effect of recommendation. Therefore, if we can filter out those unreliable predictions and generate
a sparse predicted rating matrix, we will improve both the efficiency and effectiveness of recommendation. Although sparsity has been considered in many
research domains such as classification [13], feature selection [18] and data compression [6], it has not been taken into account in predicting ratings in recommendation.

To address the above challenges, we propose a Low-rank and Sparse Matrix
Completion (LSMC) algorithm to get a low-rank and sparse predicted rating
matrix from the original rating matrix directly without any intermediate output
to avoid error propagation. As pointed out in [2], it's ill-posed if we complete a
matrix from a subset of its entries without imposing structural assumptions on
the matrix. Besides, the ratings depend on a small number of factors of users and
items as we usually assume in matrix factorization, so we expect the recovered
matrix is low-rank. But low-rank optimization is a nonconvex problem which is
difficult to solve, so we reasonably assume the matrix to be recovered from the
rating matrix is approximately low-rank. In order to get a sparse prediction, we
also assume that the matrix is sparse. When recovering the low-rank and sparse
matrix, we relax the error between the real ratings and the predicted ratings so
as to not only prevent overfitting but also reduce the impact of noisy data.

2   The Proposed Algorithm

In recommendation algorithm, the input user-item rating matrix $R \in \mathbb{R}^{m \times n}$
represents the rating relation between m users and n items which is quite sparse
normally. Each entry r_{ij} represents the rating of user i to item j and all the
ratings are in the interval $[R_{min}, R_{max}]$ which will vary on different datasets. If
a user has not rated an item, the corresponding value r_{ij} is equal to 0. I_{ij} is an
indicator function that is equal to 1 if user i has rated item j or 0 otherwise.
The task is to recover a low-rank and sparse predicted rating matrix $P \in \mathbb{R}^{m \times n}$
from the input user-item rating matrix R.

2.1 Objective Function

In matrix completion based recommendation, it is usually assumed that the predicted rating matrix P is low-rank. More concretely, relative to the number of users and items, the number of factors of users and items is small. So each rating relates to only a few factors and the recovered matrix P should be low-rank. Therefore we should minimize $rank(P)$. However, different from the common methodology of recommendation, we expect the recovered matrix P to be sparse so that it can take out unreliable predicted ratings automatically and drop less valuable ratings to enhance recommendation efficiency, i.e. reducing the number of ratings can decrease the computational complexity. Because l_0-norm counts the number of nonzero elements of a solution, we should minimize $\|P\|_0$ as well. P is recovered from R which means the elements in P are the combination of the elements in R with some other predicted ratings, so the nonzero ratings of R should be equal to the corresponding nonzero ratings in P. A strict equality constraint will be added to indicate this type of relation and the objective function is,

$$\min \quad rank(P) + \lambda\|P\|_0, \quad \text{s.t. } \|(R - P) \odot I\|_F = 0, \tag{1}$$

where λ is the sparsity coefficient to control the degree of sparsity and $\odot$ denotes the Hadamard product. However, the above nonconvex optimization problem is NP-hard. Fortunately, as pointed out in [5], if the rank of P is not too large and P is sparse, the tightest convex relaxation of $rank(P)$ and $\|P\|_0$ are $\|P\|_*$ and $\|P\|_1$ respectively where $\|\cdot\|_*$ and $\|\cdot\|_1$ denote the nuclear norm and the l_1-norm of a matrix, respectively. Another question is the constraint is too strict which may lead to overfitting problems. Besides, fitting noisy data perfectly will reduce the performance of prediction. So a slack variable ε is used to measure the upper error bound between the real ratings and the predicted ratings. Therefore, the relaxed objective function is,

$$\min \quad \|P\|_* + \lambda\|P\|_1, \quad \text{s.t. } \|(R - P) \odot I\|_F \le \varepsilon. \tag{2}$$

2.2 Optimization

In order to optimize the problem, we work with the approximate objective function. Applying the method used in [4], we rewrite the function in the conic form,

$$\min \quad \|P\|_* + \lambda\|P\|_1, \quad \text{s.t. } \begin{bmatrix} (R - P) \odot I \\ \varepsilon \end{bmatrix} \in \mathcal{K}, \tag{3}$$

where $\mathcal{K} = \{[x, t] : \|x\|_F \le t\}$ is the second-order cone and self-dual. To optimize the Problem (3), the objective function is separated by introducing an auxiliary variable Z firstly and rewritten as,

$$\min \quad \|P\|_* + \lambda\|Z\|_1, \quad \text{s.t. } \begin{bmatrix} (R - P) \odot I \\ \varepsilon \end{bmatrix} \in \mathcal{K}, P - Z = 0. \tag{4}$$

The Lagrangian is given by,

$$\mathcal{L} = \|P\|_* + \lambda\|Z\|_1 + <y, (R-P) \odot I> -s\varepsilon + <\Lambda, P-Z>, \quad (5)$$

where y, s and Λ are Lagrange multiplier and $< \cdot, \cdot >$ is inner product. We plan to use Proximal Algorithms [3] to optimize nuclear norm and l_1-norm problems, so the Frobenius norm terms of P and Z are required. Although we can add two terms $\frac{1}{\xi}\|P\|_F^2$ and $\frac{1}{\xi}\|Z\|_F^2$ with a large parameter ξ to the objective function, it will reduce the performance to some extent. Skillfully, we can introduce the two terms by forming the Augmented Lagrangian [3] without adding intentionally,

$$\mathcal{L} = \|P\|_* + \lambda\|Z\|_1 + <y, (R-P) \odot I> -s\varepsilon + \frac{\rho}{2}\|P - Z + u\|_F^2, \quad (6)$$

where u is scaled dual variable and ρ is a penalty parameter. According to Alternating Direction Method of Multipliers (ADMM) [3], we can optimize all the variables separately.

(1) Optimize P:

$$\|P\|_* + <y, (R-P) \odot I> + \frac{\rho}{2}\|P - Z + u\|_F^2 \Rightarrow \|P\|_* + \frac{\rho}{2}\|P - (\frac{y}{\rho} + Z - u)\|_F^2. \quad (7)$$

We assume $Q_k \Sigma_k W_k^T = \mathbf{SVD}(\frac{y_k}{\rho} + Z_k - u_k)$, where the operation $\mathbf{SVD}$ is Single Value Decomposition. According to Proximal Algorithms, we obtain,

$$P_{k+1} = \mathbf{prox}_{\frac{1}{\rho}}(\frac{y_k}{\rho} + Z_k - u_k) = Q_k \mathcal{S}_{\frac{1}{\rho}}(\Sigma_k)W_k^T, \quad (8)$$

where $\mathbf{prox}$ is proximal operator and $\mathcal{S}_\phi(x)$ is soft-thresholding (shrinkage) operator,

$$\mathcal{S}_\phi(x) = \begin{cases} x - \phi, & x > \phi \\ 0, & |x| \le \phi \\ x + \phi, & x < -\phi \end{cases}. \quad (9)$$

(2) Optimize y and s:
We can update the two Lagrange multiplier y and s by,

$$[y_{k+1}, s_{k+1}] = \mathcal{P}_\mathcal{K}\left([y_k, s_k] + \delta[(R - P_{k+1}) \odot I, -\varepsilon]\right), \quad (10)$$

where δ is step length and $\mathcal{P}_\mathcal{K}$ is the orthogonal projection onto $\mathcal{K}$ which is given by [4],

$$\mathcal{P}_\mathcal{K} = \begin{cases} [x, t], & \|x\|_F \le t \\ \frac{\|x\|_F + t}{2\|x\|_F}[x, \|x\|_F], & -\|x\|_F \le t \le \|x\|_F \\ [0, 0], & t \le -\|x\|_F \end{cases}. \quad (11)$$

(3) Optimize Z:

$$\frac{\lambda}{\rho}\|Z\|_1 + \frac{1}{2}\|Z - P - u\|_F^2, \tag{12}$$

so we can update the auxiliary variable Z by shrinkage operator,

$$Z_{k+1} = \mathbf{prox}_{\frac{\lambda}{\rho}}(P_{k+1} + u_k) = \mathcal{S}_{\frac{\lambda}{\rho}}(P_{k+1} + u_k). \tag{13}$$

(4) Optimize u: $u_{k+1} = u_k + P_{k+1} - Z_{k+1}$.

3 Theoretical Analysis

In this section, we analyze the relations between the parameters in the objective function in theory. When selecting parameters in implementation, the theorems in this section should be satisfied to guarantee the convergence and the availabilities of parameters.

Lemma 1 [10]. *Let $\partial\|x\|$ be the subgradient of norm $\|x\|$ and $y \in \partial\|x\|$. Then $\|y\|^* = 1$ if $x \neq 0$, and $\|y\|^* \leq 1$ if $x = 0$, where $\|\cdot\|^*$ is the dual norm of $\|\cdot\|$.*

Lemma 2. *During each iteration, the following inequality holds $\|y_k\|_F \leq \sqrt{\mathcal{R}}(1 + \lambda\sqrt{m})$ where $\mathcal{R} = \min(m, n)$.*

Proof. If we get the optimal P_{k+1}^* and Z_{k+1}^* in the $(k+1)^{th}$ iteration, we can get,

$$0 \in \partial_P \mathcal{L}(P_{k+1}^*, Z_{k+1}^*, y_k, s_k, u_k), 0 \in \partial_Z \mathcal{L}(P_{k+1}^*, Z_{k+1}^*, y_k, s_k, u_k). \tag{14}$$

Combining with the update of u, the result is,

$$y_k - \rho u_{k+1} \in \partial\|P_{k+1}\|_*, \frac{\rho}{\lambda} u_{k+1} \in \partial\|Z_{k+1}\|_1. \tag{15}$$

The dual norms of $\|\cdot\|_*$ and $\|\cdot\|_1$ are $\|\cdot\|_2$ and $\|\cdot\|_\infty$ respectively. According to Lemma 1, we can get,

$$\|y_k - \rho u_{k+1}\|_2 \in \{0, 1\}, \quad \frac{\rho}{\lambda}\|u_{k+1}\|_\infty \in \{0, 1\}. \tag{16}$$

Using the property of matrix norm equivalence, we obtain,

$$\frac{\rho}{\lambda\sqrt{m}\sqrt{\mathcal{R}}}\|u_{k+1}\|_F \leq \frac{\rho}{\lambda\sqrt{m}}\|u_{k+1}\|_2 \leq \frac{\rho}{\lambda}\|u_{k+1}\|_\infty \leq 1, \tag{17}$$

and

$$\frac{1}{\sqrt{\mathcal{R}}}\|y_k - \rho u_{k+1}\|_F \leq \|y_k - \rho u_{k+1}\|_2 \leq 1. \tag{18}$$

Using Minkowski inequality for Eq. (18), we can obtain,

$$\|y_k\|_F - \rho\|u_{k+1}\|_F \leq \|y_k - \rho u_{k+1}\|_F \leq \sqrt{\mathcal{R}} \Rightarrow \|y_k\|_F \leq \sqrt{\mathcal{R}}(1 + \lambda\sqrt{m}). \tag{19}$$

Theorem 1. *In order to ensure $P = Z$, we should keep $\lambda < \rho$.*

Proof. According to Eq. (16), we can get $\rho\|u_{k+1}\|_\infty \leq \lambda$ and $\|u_{k+1}\|_\infty \in \{0, \frac{\lambda}{\rho}\}$. From the update method of u, we know that the scaled dual variable u_k is the sum of the residuals $u_{k+1} = u_0 + \sum_{i=1}^{k}(P_i - Z_i)$ and we want $u^* = 0$ eventually to ensure $P = Z$. If $\rho \leq \lambda$, $\|u_{k+1}\|_\infty$ can be 0 or $\frac{\lambda}{\rho}$ which may cause $P \neq Z$. But if $\lambda < \rho$, $\|u_{k+1}\|_\infty$ can be only equal to 0 which means $P = Z$.

Theorem 2. *To assure the slack variable ε is valid, the following inequality should be satisfied $\varepsilon \leq \frac{3\sqrt{\mathcal{R}}(1+\lambda\sqrt{m})}{\delta} + \frac{s_0}{2^{k-1}\delta}$.*

Proof. Using Minkowski inequality, we obtain,

$$\delta\|(R - P) \odot I\|_F - \|y_k\|_F \leq \|y_k + \delta(R - P) \odot I\|_F. \tag{20}$$

We use the constraint in Eq. (2) and adjust Eq. (20),

$$\begin{aligned}
\varepsilon &\leq \frac{1}{\delta}(\|y_k + \delta(R - P) \odot I\|_F + \|y_k\|_F) \leq \frac{1}{\delta}(2s_{k+1} + \|y_k\|_F) \\
&\leq \frac{1}{\delta}((2 - (\frac{1}{2})^k)\sqrt{\mathcal{R}}(1 + \lambda\sqrt{m}) + \frac{s_0}{2^{k-1}} + \|y_k\|_F) \\
&\leq \frac{1}{\delta}((3 - (\frac{1}{2})^k)\sqrt{\mathcal{R}}(1 + \lambda\sqrt{m}) + \frac{s_0}{2^{k-1}}) \leq \frac{3\sqrt{\mathcal{R}}(1 + \lambda\sqrt{m})}{\delta} + \frac{s_0}{2^{k-1}\delta}.
\end{aligned} \tag{21}$$

The second and the third inequalities use Eqs. (10) and (11). The fourth inequalities use Lamma 2. So $\frac{3\sqrt{\mathcal{R}}(1+\lambda\sqrt{m})}{\delta} + \frac{s_0}{2^{k-1}\delta}$ is the upper bound of ε. When training the model, ε is used to restrict the error of prediction. But if ε is larger than the upper bound, the parameter will be useless.

4   Experiments

In this section, we analyze the influence of the sparsity coefficient λ and the slack variable ε on the proposed LSMC algorithm and compare the performance with seven state-of-the-art recommendation algorithms on four real-world datasets.

4.1   Dataset and Evaluation Methodology

The four datasets used in our experiments are BaiduMovie[1], FindFoods[2], Jester[3] and MovieLens[4]. In order to evaluate the quality of the recommendation algorithms, two widely used evaluation measures, namely Mean Absolute Error (MAE) and Root Mean Square Error (RMSE), will be used to measure the accuracy of the predicted ratings, which are defined as follows:

$$MAE = \frac{1}{T}\sum_{i,j}|r_{ij} - p_{ij}|, \quad RMSE = \sqrt{\frac{1}{T}\sum_{i,j}(r_{ij} - p_{ij})^2},$$

where T denotes the number of tested ratings.

---

[1] http://openresearch.baidu.com.
[2] http://snap.stanford.edu/data/web-FineFoods.html.
[3] http://eigentaste.berkeley.edu/dataset.
[4] http://grouplens.org/datasets/movielens.

4.2   Parameter Analysis

We analyze the effect of the sparsity coefficient λ and the slack variable ε. Without loss of generality, we assume $\varepsilon = mn\gamma$ where γ can be viewed as the average slack of each rating in the case sparsity has not been taken into account. We vary the value of λ from 0 to 1 continually with $\gamma = 0, 0.001$ and 0.01. The results are shown in Figs. 1 and 2.

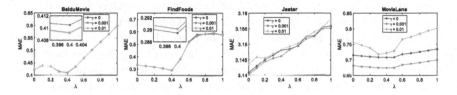

Fig. 1. The values of MAE with different λ and γ on the four datasets.

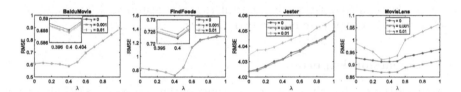

Fig. 2. The values of RMSE with different λ and γ on the four datasets.

Generally speaking, as λ increases, the performance on the BaiduMovie, Find-Foods and MovieLens datasets will increase to the optimal value first and then decrease. Because the three datasets are quite sparse, recommendation algorithms will get some unreliable predictions on those datasets. If we select a suitable λ which is around 0.4, the LSMC algorithm can improve the performance by removing some unreliable predictions. But if λ is too large, we may take out some valuable predictions which will reduce the performance. To illustrate the point better, we plot the sub-matrixes of the predicted rating matrix on the FindFoods dataset in Fig. 3 where black points and white points represent those rating that *can* and *can not* be predicted respectively. As λ increases, the number of predictions we can get will decrease. The LSMC algorithm generates the best result on the three datasets when $\lambda = 0.4$. If λ is too large, the performance

Fig. 3. The predicted rating matrix on the FindFoods dataset. The values of λ are from 0 to 1 with 0.1 increment from left to right.

is terrible because we miss many useful predictions. On the other hand, the ratio of the degree of low-rank will be smaller which will fail to capture the important information from the rating matrix better. On the Jester dataset, the performance will approximately linearly decrease as λ increases. The reason is that the dataset is quite dense and the number of unknown ratings is smaller than the known ratings. So almost all the predicted ratings are reliable and sparsity is needless in this situation.

The parameter γ has little effect on the BaiduMovie and FindFoods datasets. The reason might be that the amount of noisy data is relatively small on the two datasets. But on the BaiduMovie dataset, the performance of the LSMC algorithm will be slightly better if we can relax the error ($\gamma > 0$). On the FindFoods dataset, when λ is around 0.4, the result will be the best in the situation $\gamma = 0$. If we increase λ to around 0.8, the strict constraint will lead to the worst result. On the other two datasets, the amount of noisy data is relatively larger and the influence of γ is obvious. When $\gamma = 0.001$, the performance will be the best. If the constraint is strict, the noisy data will reduce the effect and the phenomenon of overfitting may appear. On the contrary, if the constraint is too loose, the predicted ratings can not fit the real ratings well. So the result is less-than-ideal in either case on the two datasets.

4.3   Comparsion Experiments

We compare the results of the predicted ratings of the proposed LSMC algorithm with seven state-of-the-art recommendation algorithms, i.e., User-Based Collaborative Filtering (**UBCF**) [15], Item-Based Collaborative Filtering (**IBCF**) [15], Slope One (**SO**) [9], Bayesian Similarity (**BS**) [8], Improving Regularized Singular Value Decomposition (**IRSVD**) [11], Sparse Covariance Matrix Factorization (**SCMF**) [14], Matrix Factorization to asymmetric user similarities (**MF-AMSD**) [12]. For the those matrix completion methods, we set the dimensionality of latent feature vector $d = 5$. For the proposed LSMC algorithm, we take the best results from Figs. 1 and 2, so the corresponding parameters can be obtained from the two figures.

The comparison results are shown in Table 1. The accuracies generated by the SO algorithm are quite poor on all the datasets and the proposed LSMC algorithm can improve the performance by at least 10%. The performances of UBCF, IBCF, SCMF and MF-AMSD are similar on the Jester and MovieLens datasets. Compared with these four algorithms on the two datasets, the proposed LSMC algorithm can make about 15% improvement and about 10% on the other two datasets. SCMF is better than MF-AMSD on the Jester datatset but worse on the other three datasets, and the proposed LSMC algorithm can achieve the improvement from 3% to 45%. Compared with BS, the proposed LSMC algorithm can make a noticeable improvement on the FindFoods dataset and improve about 5% on the other three datasets. Although the performance of the IRSVD algorithm is impressive on all the datasets, the proposed LSMC algorithm still can achieve improvement from 0.5% to 14%. Each item

Table 1. The values of MAE and RMSE on the four datasets of the eight recommendation algorithms. The best results are highlighted in bold.

Dataset	Measures	UBCF	IBCF	SO	BS	IRSVD	SCMF	MF-AMSD	LSMC
BaiduMovie	MAE	0.4431	0.4789	0.5283	0.4274	0.4112	0.5090	0.4418	**0.4091**
	RMSE	0.6156	0.6646	0.6989	0.5971	0.5908	0.6905	0.6108	**0.5876**
FindFoods	MAE	0.4150	0.3117	0.4976	0.4524	0.3461	0.3364	0.5322	**0.2968**
	RMSE	0.8049	**0.6516**	0.8218	0.8055	0.7518	0.7571	1.1107	0.6816
Jester	MAE	3.3265	3.3081	3.5515	3.3531	3.1931	3.3335	3.3423	**3.1403**
	RMSE	4.2177	4.1898	4.5205	4.2752	4.0820	4.2158	4.2381	**4.0237**
MovieLens	MAE	0.7654	0.7744	0.8744	0.7147	0.7139	0.7489	0.7739	**0.6742**
	RMSE	0.9726	0.9833	1.0618	0.9266	0.9150	0.9807	1.0389	**0.8690**

on FindFoods has many ratings so the item similarities can be calculated accurately which makes IBCF work well. The measure of RMSE is sensitive to large errors while MAE is sensitive to the accumulation of small errors. Because there are some cold users and items on the dataset leading to some large prediction errors, IBCF outperforms LSMC in terms of RMSE on FindFoods. Overall, the proposed LSMC algorithm can make a better rating prediction than most of the existing state-of-the-art recommendation algorithms on real-world datasets. The main reason is that LSMC recovers a predicted rating matrix from the original rating matrix directly without any intermediate output, which can eliminate the propagation of error of the prediction model. Besides, the sparsity can remove some unreliable predicted ratings automatically, and a slack variable is used to not only prevent overfitting but also weaken the influence of noisy data.

5   Conclusion

In this paper, we have proposed a novel matrix completion method termed LSMC for recommendation to improve the quality of rating prediction. We assume the structure of the matrix is low-rank. Sparsity is used to remove some unreliable predictions and retain the important results. In order to prevent overfitting and weaken the impact of noisy data, a slack variable is utilized to relax the prediction error. The objective function is optimized by ADMM and theoretical analysis has shown some important relations between the parameters. Extensive experiments have been conducted on four real-world datasets to confirm the effectiveness of the proposed algorithm. The results have indicated that our algorithm can get a sparse prediction and outperforms state-of-the-art recommendation algorithms.

Acknowledgments. This work was supported by the Fundamental Research Funds for the Central Universities (16lgzd15) and Tip-top Scientific and Technical Innovative Youth Talents of Guangdong special support program (No. 2016TQ03X542).

References

1. Argyriou, A., Evgeniou, T., Pontil, M.: Multi-task feature learning. In: Advances in Neural Information Processing Systems, pp. 41–48 (2006)
2. Bhaskar, S.A.: Probabilistic low-rank matrix recovery from quantized measurements: application to image denoising. In: 2015 49th Asilomar Conference on Signals, Systems and Computers, pp. 541–545 (2015)
3. Boyd, S.P., Parikh, N., Chu, E., Peleato, B., Eckstein, J.: Distributed optimization and statistical learning via the alternating direction method of multipliers. Found. Trends Mach. Learn. **3**(1), 1–122 (2011)
4. Cai, J., Candès, E.J., Shen, Z.: A singular value thresholding algorithm for matrix completion. SIAM J. Optim. **20**(4), 1956–1982 (2010)
5. Candès, E.J., Li, X., Ma, Y., Wright, J.: Robust principal component analysis. J. ACM **58**(3), 1–39 (2011)
6. Chao, T., Lin, Y., Kuo, Y., Hsu, W.H.: Scalable object detection by filter compression with regularized sparse coding. In: Proceedings of the IEEE Conference on Computer Vision and Pattern Recognition, pp. 3900–3907 (2015)
7. Cheng, Y., Yin, L., Yu, Y.: LorSLIM: low rank sparse linear methods for Top-N recommendations. In: 2014 IEEE International Conference on Data Mining, pp. 90–99 (2014)
8. Guo, G., Zhang, J., Yorke-Smith, N.: A novel Bayesian similarity measure for recommender systems. In: Twenty-Third International Joint Conference on Artificial Intelligence, pp. 2619–2625 (2013)
9. Lemire, D., Maclachlan, A.: Slope one predictors for online rating-based collaborative filtering. In: Proceedings of the 2005 SIAM International Conference on Data Mining, pp. 471–475 (2005)
10. Lin, Z., Liu, R., Su, Z.: Linearized alternating direction method with adaptive penalty for low rank representation. In: Advances in Neural Information Processing Systems, pp. 612–620 (2011)
11. Paterek, A.: Improving regularized singular value decomposition for collaborative filtering. In: Kdd Cup & Workshop, pp. 39–42 (2007)
12. Pirasteh, P., Hwang, D., Jung, J.J.: Exploiting matrix factorization to asymmetric user similarities in recommendation systems. Knowl.-Based Syst. **83**, 51–57 (2015)
13. Roberge, J., Rispal, S., Wong, T., Duchaine, V.: Unsupervised feature learning for classifying dynamic tactile events using sparse coding. In: 2016 IEEE International Conference on Robotics and Automation, pp. 2675–2681 (2016)
14. Shi, J., Wang, N., Xia, Y., Yeung, D.Y., King, I., Jia, J.: SCMF: sparse covariance matrix factorization for collaborative filtering. In: Twenty-Third International Joint Conference on Artificial Intelligence, pp. 2705–2711 (2013)
15. Wang, J., de Vries, A.P., Reinders, M.J.: Unifying user-based and item-based collaborative filtering approaches by similarity fusion. In: Proceedings of the 29th Annual International ACM SIGIR Conference on Research and Development in Information Retrieval, pp. 501–508 (2006)
16. Zhang, D.C., Li, M., Wang, C.D.: Point of interest recommendation with social and geographical influence. In: IEEE International Conference on Big Data, pp. 1070–1075 (2016)
17. Zhang, Y., Jiang, Z., Davis, L.S.: Learning structured low-rank representations for image classification. In: Proceedings of the IEEE Conference on Computer Vision and Pattern Recognition, pp. 676–683 (2013)

18. Zhang, Z., Bai, L., Liang, Y., Hancock, E.R.: Joint hypergraph learning and sparse regression for feature selection. Pattern Recogn. **63**, 291–309 (2017)
19. Zhao, Z.L., Wang, C.D., Lai, J.H.: AUI&GIV: recommendation with asymmetric user influence and global importance value. PLoS ONE **11**(2), e0147944 (2016)
20. Zhao, Z.L., Wang, C.D., Wan, Y.Y., Lai, J.H., Huang, D.: FTMF: recommendation in social network with feature transfer and probabilistic matrix factorization. In: 2016 International Joint Conference on Neural Networks, pp. 847–854 (2016)

A Brain Network Inspired Algorithm: Pre-trained Extreme Learning Machine

Yongshan Zhang[1], Jia Wu[2], Zhihua Cai[1(✉)], and Siwei Jiang[1]

[1] Department of Computer Science, China University of Geosciences,
Wuhan 430074, China
{yszhang,zhcai}@cug.edu.cn
[2] Department of Computing, Faculty of Science and Engineering,
Macquarie University, Sydney, NSW 2109, Australia
jia.wu@mq.edu.au

Abstract. Extreme learning machine (ELM) is a promising learning method for training "generalized" single hidden layer feedforward neural networks (SLFNs), which has attracted significant interest recently for its fast learning speed, good generalization ability and ease of implementation. However, due to its manually selected network parameters (e.g., the input weights and hidden biases), the performance of ELM may be easily deteriorated. In this paper, we propose a novel pre-trained extreme learning machine (P-ELM for short) for classification problems. In P-ELM, the superior network parameters are pre-trained by an ELM-based autoencoder (ELM-AE) and embedded with the underlying data information, which can improve the performance of the proposed method. Experiments and comparisons on face image recognition and handwritten image annotation applications demonstrate that P-ELM is promising and achieves superior results compared to the original ELM algorithm and other ELM-based algorithms.

Keywords: Extreme learning machine · ELM-based autoencoder · Pre-trained parameter · Classification

1 Introduction

Extreme learning machine (ELM) [1] is a useful learning method for training "generalized" single hidden layer feedforward neural networks (SLFNs), which shows its good performance in various research studies [2]. Compared with traditional neural networks which adjust the network parameters iteratively, in ELM, the input weights and hidden layer biases are randomly generated, while the output weights are analytically determined by using Moore-Penrose (MP) generalized inverse. Due to its extremely fast learning speed, good generalization

---

The original version of this chapter was revised. The title of the paper has been corrected. The erratum to this chapter is available at https://doi.org/10.1007/978-3-319-70139-4_94

© Springer International Publishing AG 2017
D. Liu et al. (Eds.): ICONIP 2017, Part V, LNCS 10638, pp. 14–23, 2017.
https://doi.org/10.1007/978-3-319-70139-4_2

ability and ease of implementation, ELM has drawn great attention in academia [3,4]. However, the manually assigned network parameters often degrade the performance of ELM.

In order to enhance the performance of ELM, researchers have proposed a number of improved methods from different perspectives, such as ensemble learning [5], voting scheme [6], weighting method [7] and instance cloning [4]. Liu and Wang [5] embedded ensemble learning into the training phase of ELM to mitigate the overfitting problem and improve the predictive stability. Cao et al. [6] incorporated multiple independent ELM models into a unified framework to enhance the performance in a voting manner. Zong et al. [7] proposed a weighting scheme method for ELM by assigning different weights for each example. The aforementioned methods for ELM have achieved good performance in some specific problems. However, they do not solve the primary problem in ELM (i.e., the random generation of the network parameters). Therefore, the performance of the above-mentioned methods may be compromised. How to select suitable network parameters for ELM is still an opening problem.

In reality, the original data can provide valuable information according to its different representations. Therefore, it is imperative for ELM to determine the network parameters based on the original data. A straightforward approach to solve the above problem is to use the idea of autoencoder. Autoencoder [8] is a special case of artificial neural network usually used for unsupervised learning, where the output layer are with the same neurons as the input layer. In autoencoder, the learning procedure can be divided into the processes of encoding and decoding [9,10]. The input data is mapped to a high-level representation in the encoding stage, while the high-level representation is mapped back to the original input data in the decoding stage. By doing so, autoencoder can explore the underlying data information and encode these information into the output weights.

Based on the above observations, in this paper, we propose a novel pre-trained extreme learning machine (P-ELM for short), where an ELM-based autoencoder (ELM-AE) is adopted to pre-train the suitable network parameters. The proposed P-ELM encodes the data information into the learned network parameters, which can achieve satisfactory performance for further learning. Experiments on face image recognition and handwritten image annotation applications demonstrate that the proposed P-ELM consistently outperforms other state-of-the-art ELM algorithms. The advantages of P-ELM can be summarized as follows:

- P-ELM falls into the category of data-driven methods, which can successfully find the proper network parameters for further learning.
- P-ELM is simple in both theory and implementation, which inherits the advantages of the original ELM.
- P-ELM is a nonlinear learning model and flexible in modeling different complex real-world relationships.

The remainder of the paper is structured as follows. Section 2 surveys the related work. Section 3 presents the proposed P-ELM method. The experiments are demonstrated in Sect. 4. Finally, we conclude the paper in Sect. 5.

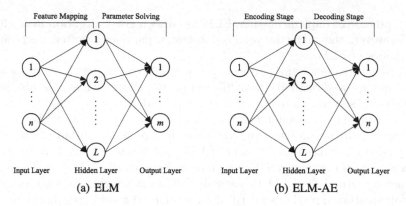

Fig. 1. Illustration of network structures for (a) ELM and (b) ELM-AE.

2   Related Work

Extreme learning machine (ELM) is an elegant learning method, which was originally proposed for SLFNs and then extended to "generalized" SLFNs [1]. In ELM, the hidden neurons need not be neuron alike and the networks parameters are without iterative tunning. The network structure of ELM is shown in Fig. 1(a). The basic ELM can fundamentally be regarded as a two-stage learning system, which can be spilt into feature mapping and parameter solving [11,12]. In the feature mapping stage, ELM randomly selects the input weights and hidden biases to calculate the hidden layer output matrix via an activation function. In the parameter solving stage, the output weights are analytically determined according to the Moore-Penrose (MP) generalized inverse and the smallest norm least-squares solution of general linear system. To accelerate the learning speed, Huang et al. [13] presented a constrained-optimization-based ELM and provided two effective solution for different size of training data. The learning theories and real-world applications of ELM are well-developed in the literature [2].

Apart from ELM-based SLFNs, the ELM theories can be also applied to built an ELM-based autoencoder (ELM-AE) [14]. Autoencoder is always used to be a feature extractor and usually functions as a basic unit in a multilayer learning model [15]. In recent years, autoencoder has been widely used for tackling numerous real-world applications, e.g., cross-language learning problem and domain adaption problem. Similar to the ELM, an ELM-AE can be also regarded as a two-stage process, where the input data is first mapped to a high-level representation, and then the high-level latent representation is mapped back to the original input data [16]. The network structure of ELM-AE is shown in Fig. 1(b). The main difference between ELM and ELM-AE is the output layer. In ELM, the output layer is to predict the target value for given data. By contrast, in ELM-AE, the output layer is to reconstruct the original input data. Due to the unique learning mechanism, ELM-AE extracts the informative features through the hidden layer and encodes the underlying data information into the output weights. Motivated by these, we propose to employ an ELM-AE to pre-train the network parameters for P-ELM in this paper.

3   Proposed Method

In this section, we present the proposed pre-trained extreme learning machine (P-ELM). Specifically, P-ELM is achieved through the following steps: (1) Employ an ELM-AE for network parameter learning; (2) Train the P-ELM model with the learned network parameters; and (3) Predict the class labels of the testing instances. Algorithm 1 reports the learning process of the proposed P-ELM.

3.1   Parameter Learning

In P-ELM, the most important aspect is to choose the suitable network parameters based on the original data. To this end, we use an ELM-AE to learn the network parameters. Given N distinct training examples $\mathcal{D} = \{(\boldsymbol{x}_i, \boldsymbol{t}_i)\}_{i=1}^N$, where $\boldsymbol{x}_i \in \mathbb{R}^n$ is the input data and $\boldsymbol{t}_i \in \mathbb{R}^m$ is the expectation output, the encoding process in ELM-AE with L hidden neurons can be presented as the following equation:

$$h(\boldsymbol{x}_i) = g(\boldsymbol{\alpha} \cdot \boldsymbol{x}_i + \boldsymbol{b}), \quad i = 1, 2, ..., N; \tag{1}$$

where $\boldsymbol{\alpha} \in \mathbb{R}^{L \times n}$ is the input weight matrix, $\boldsymbol{b} \in \mathbb{R}^{L \times 1}$ is the hidden neuron bias vector, $g(\cdot)$ is an activation function, and $h(\boldsymbol{x}_i)$ is the high-level latent representation for the input data $\boldsymbol{x}_i$. By contrast, the decoding process in ELM-AE can be formulated as follows:

$$h(\boldsymbol{x}_i)\boldsymbol{\varpi} = \boldsymbol{x}_i, \quad i = 1, 2, ..., N; \tag{2}$$

where $\boldsymbol{\varpi} \in \mathbb{R}^{L \times n}$ is the output weight matrix. Equation (2) can be also rewritten as the compacted form based on the whole dataset:

$$\mathbf{H}\boldsymbol{\varpi} = \mathbf{X}. \tag{3}$$

To enhance the performance of ELM-AE, the output weight matrix $\boldsymbol{\varpi}$ can be updated by minimizing the objective fuction: $L(\boldsymbol{\varpi}) = \frac{1}{2}||\boldsymbol{\varpi}||^2 + \frac{C}{2}||\mathbf{X} - \mathbf{H}\boldsymbol{\varpi}||^2$. The calculation of the output weight matrix $\boldsymbol{\varpi}$ can be solved by Eq. (4) according to the relationship between the number of training samples N and the number of hidden neurons L.

$$\boldsymbol{\varpi} = \begin{cases} \left(\dfrac{\mathbf{I}}{C} + \mathbf{H}^T\mathbf{H} \right)^{-1} \mathbf{H}^T\mathbf{X}, & \text{if } N \geq L \\ \mathbf{H}^T \left(\dfrac{\mathbf{I}}{C} + \mathbf{H}\mathbf{H}^T \right)^{-1} \mathbf{X}, & \text{if } N < L \end{cases} \tag{4}$$

The unique parameter learning mechanism enables ELM-AE to encode the underlying information of the original data into the output weights, which can be used as the input weights for the P-ELM model to achieve better performance. This is a data-driven method, which can adaptively search the suitable network parameters based on the specific data.

Algorithm 1. Pre-trained Extreme Learning Machine (P-ELM)

Input:
   Training dataset $\mathcal{D} = \{(\boldsymbol{x}_i, \boldsymbol{t}_i)\}_{i=1}^{N}$; Any testing instance $\boldsymbol{x}^{test} \in \mathcal{D}^{test}$;
   Activation function $g(\cdot)$; Number of hidden neurons L; Parameter C;
Output:
   The predicted class label $c(\boldsymbol{x}^{test})$ of testing instance $\boldsymbol{x}^{test}$;
      //P-trained Parameter Learning:
   1: Randomly assign the input weights $\boldsymbol{\alpha}$ and hidden biases $\boldsymbol{b}$ for ELM-AE;
   2: Calculate the hidden layer output $\mathbf{H}$ in ELM-AE by Eq. (1);
   3: Calculate the output weights $\boldsymbol{\varpi}$ in ELM-AE by Eq. (4);
      //P-ELM Model Training:
   4: Compute the input weights as $\boldsymbol{\varpi}^{T}$ and the hidden biases as $\boldsymbol{b}'$, where the ith hidden
      layer bias $b'_i = (\sum_{j=1}^{n} \varpi_{ij})/n, i = 1, 2, ..., L$ in P-ELM;
   5: Calculate the hidden layer output matrix $\mathbf{H}'$ in P-ELM by Eq. (5);
   6: Calculate the output weights $\boldsymbol{\beta}$ in P-ELM by Eq. (7);
      //Instance Label Prediction:
   7: Predict the underlying class label $c(\boldsymbol{x}^{test})$ for testing instance $\boldsymbol{x}^{test}$;
   8: Return the class label $c(\boldsymbol{x}^{test})$.

3.2   Model Training

In this section, we aim to formulate the learning model of the proposed pre-trained extreme learning machine (P-ELM). As described in the previous section, we use the output weights $\boldsymbol{\varpi}$ learned by ELM-AE as the input weights for P-ELM. In P-ELM, the input weights can be represented as $\boldsymbol{\varpi}^{T}$, and the hidden layer biases can be expressed as $\boldsymbol{b}'$, where the ith hidden layer bias is $b'_i = (\sum_{j=1}^{n} \varpi_{ij})/n, i = 1, 2, ..., L$. Therefore, the proposed P-ELM with L hidden neurons can be formulated as:

$$
\begin{aligned}
t_i &= \sum_{j=1}^{L} \beta_j g(\boldsymbol{\varpi}_j^{T} \cdot \boldsymbol{x}_i + b'_j) \\
&= g(\boldsymbol{\varpi}^{T} \cdot \boldsymbol{x}_i + \boldsymbol{b}')\boldsymbol{\beta} \qquad , \qquad i = 1, 2, ..., N; \\
&= \boldsymbol{h}'(\boldsymbol{x}_i)\boldsymbol{\beta}
\end{aligned}
\tag{5}
$$

where $\boldsymbol{h}'(\boldsymbol{x}_i) = g(\boldsymbol{\varpi}^{T} \cdot \boldsymbol{x}_i + \boldsymbol{b}')$ is the hidden layer output for the input data $\boldsymbol{x}_i$ and $\boldsymbol{\beta} \in \mathbb{R}^{L \times m}$ is the output weight matrix of the proposed P-ELM. Mathematically, Eq. (5) can be rewritten as the following compacted form:

$$
\mathbf{H}'\boldsymbol{\beta} = \mathbf{T}.
\tag{6}
$$

   To calculate the output weight matrix $\boldsymbol{\beta}$, Eq. (6) can be solved by minimizing the objective function: $L(\boldsymbol{\beta}) = \frac{1}{2}||\boldsymbol{\beta}||^2 + \frac{C}{2}||\mathbf{T} - \mathbf{H}'\boldsymbol{\beta}||^2$. Similar to Eq. (4), the output weight matrix $\boldsymbol{\beta}$ can be calculated as the following equation according to the relationship between the number of training samples N and the number of hidden neurons L.

$$\beta = \begin{cases} \left(\dfrac{\mathbf{I}}{C} + \mathbf{H}'^{T}\mathbf{H}'\right)^{-1} \mathbf{H}'^{T}\mathbf{T}, & \text{if } N \geq L \\ \mathbf{H}'^{T}\left(\dfrac{\mathbf{I}}{C} + \mathbf{H}'\mathbf{H}'^{T}\right)^{-1} \mathbf{T}, & \text{if } N < L \end{cases} \tag{7}$$

The training process of P-ELM is determined by Eq. (5). Different from the traditional ELM with randomly generated network parameters, the proposed P-ELM uses the network parameters pre-trained by ELM-AE for model training. By doing so, the performance of P-ELM can be improved. This is the major difference between P-ELM and the original ELM.

3.3   Label Prediction

In the testing phase, the class labels of each testing instance is predicted by the trained P-ELM model. The testing instances are used to calculate the output of hidden layer based on the pre-trained input weights and hidden layer biases. Then, the class labels of the testing instances can be determined by Eq. (6). Indeed, instance label prediction in the proposed P-ELM is similar to the prediction process in ELM.

4   Experimental Results

To validate the performance of the proposed method, the experiments are conducted on face image recognition [17] and handwritten image annotation [14] respectively. Classification accuracy [18,19] and running time [20] are used as the evaluation metrics. The reported results are based on 10-fold cross validation (CV). In P-ELM, the parameter C is tuned by a grid-search strategy from $\{0.01, 0.1, 1, 10, 100, 1000\}$, the sigmoid function is applied as the activation function for the hidden layer, and the setting of the number of hidden neurons depends on specific applications. For comparison purposes, we use four ELM-based methods compared to P-ELM, including a faster ELM method (ELM) [13], ensemble based ELM (EN-ELM) [5], voting based ELM (V-ELM) [6] and weighting based ELM (W-ELM) [7].

4.1   Face Image Recognition

In this section, we report the performance of P-ELM on face image recognition real-world application. The corresponding datasets used in the experiments are the ORL and Yale face image recognition datasets[1]. The ORL dataset contains 400 face images with the size of 32×32, which belongs to 10 different people. These images were taken at different times, varying the lighting, facial expressions and facial details. The Yale dataset has 165 face images with the size of 32×32 of different facial expressions conducted by 10 different people (Fig. 2).

---

[1] http://www.cad.zju.edu.cn/home/dengcai/Data/FaceData.html.

(a) ORL   (b) Yale

Fig. 2. Example images from different face image databases: (a) ORL and (b) Yale.

In Table 1, we report the experimental results of P-ELM and other baselines with 50 hidden neurons on two different face image datasets. The results indicate that P-ELM are with high testing accuracy and low standard deviation compared to other baselines. P-ELM achieves 74.50% testing accuracy with 4.06% standard deviation on the ORL dataset, and 60.63% testing accuracy with 8.04% standard deviation on the Yale dataset. In terms of both training time and testing time, P-ELM is superior to EN-ELM and V-ELM, and slightly inferior to ELM and W-ELM. Besides, the experimental results for all compared methods with different numbers of hidden neurons are given in Fig. 3. From Fig. 3, we can observe that P-ELM always significantly outperforms other baselines on both the ORL and Yale datasets. P-ELM's remarkable performance on face image recognition owes to the unique of parameter learning mechanism, which guarantees that P-ELM can achieve superior performance.

Table 1. Performance comparison on face image recognition.

Dataset	Measure	Algorithm				
		ELM	EN-ELM	V-ELM	W-ELM	P-ELM
ORL	Accuracy (%)	69.00	69.75	72.5	57.25	74.50
	Acc. Std. (%)	5.92	6.92	5.14	6.58	4.06
	Training time (s)	0.0109	0.7472	0.0905	0.0106	0.0328
	Testing time (s)	0.0042	0.4992	0.0094	0.0041	0.0047
Yale	Accuracy (%)	51.25	50.63	53.75	52.58	60.63
	Acc. Std. (%)	13.76	11.58	10.29	11.49	8.04
	Training time (s)	0.0078	0.2590	0.0406	0.0086	0.0312
	Testing time (s)	0.0047	0.1888	0.0187	0.0042	0.0062

4.2   Handwritten Image Annotation

For handwritten image annotation application, we report the performance of P-ELM in this section. In the experiments, we use the USPS and MNIST handwritten image annotation datasets[2]. The USPS dataset contains 9298 different gray-scale handwritten digit images with the size of 16×16. The MNIST dataset

---

[2] http://www.cad.zju.edu.cn/home/dengcai/Data/MLData.html.

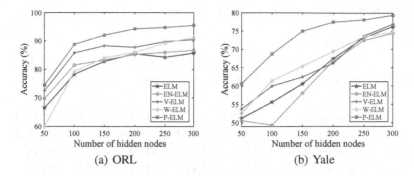

Fig. 3. Performance comparison with respect to the number of hidden neurons on face image recognition: (a) ORL and (b) Yale.

used in the experiments consists of 10000 images of handwritten numbers with the size of 28 × 28, where each digital number consists of 1000 images. For the USPS and MNIST datasets, they are both associated with 10 different categories of "0" through "9" (Fig. 4).

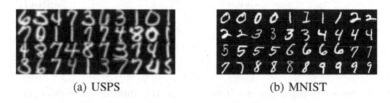

Fig. 4. Example images from different handwritten image databases: (a) USPS and (b) MNIST.

In Table 2, the results on handwritten image datasets show the performance of P-ELM and other baselines with 100 hidden neurons. P-ELM achieves 91.86% testing accuracy with 0.79% standard deviation on the USPS dataset, and 88.44% testing accuracy with 0.94% standard deviation on the MNIST dataset, which shows its superiority compared to other baselines. In terms of training time, P-ELM needs a little more running time than ELM, achieves slightly superior performance than W-ELM, and runs much faster than EN-ELM and V-ELM. In terms of testing time, P-ELM is slightly inferior to ELM and W-ELM, and significantly superior to EN-ELM and V-ELM. In addition, the simulation results for P-ELM and other baseline methods with various numbers of hidden neurons are presented in Fig. 5. As can be observed from Fig. 5, P-ELM is always superior to the baselines on the USPS dataset, and achieves better or comparable performance compared to other baselines on the MNIST dataset. The above observation suggests that P-ELM is also effective on handwritten image annotation, mainly because that it uses an ELM-AE to learn the suitable network parameters for P-ELM.

Table 2. Performance comparison on handwritten image annotation.

Dataset	Measure	Algorithm				
		ELM	EN-ELM	V-ELM	W-ELM	P-ELM
USPS	Accuracy (%)	89.44	89.61	90.32	87.91	91.86
	Acc. Std. (%)	1.02	1.07	1.11	0.95	0.79
	Training time (s)	0.2309	6.9748	1.6357	0.6257	0.4212
	Testing time (s)	0.0156	4.7471	0.0796	0.0152	0.0172
MNIST	Accuracy (%)	81.66	81.81	84.29	77.22	88.44
	Acc. Std. (%)	1.24	1.45	1.17	1.74	0.94
	Training time (s)	0.5647	12.9094	1.7023	0.8375	0.7192
	Testing time (s)	0.0172	8.4287	0.0858	0.0203	0.0265

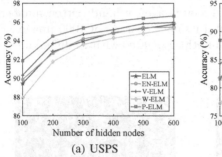
(a) USPS

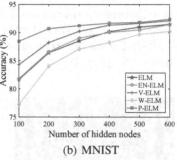
(b) MNIST

Fig. 5. Performance comparison with respect to the number of hidden neurons on handwritten image annotation: (a) USPS and (b) MNIST.

5   Conclusion

In this paper, we proposed a novel method called pre-trained extreme learning machine (P-ELM for short). The proposed P-ELM is a data-driven method, which uses an ELM-AE to intelligently determine the suitable network parameters for diverse learning tasks. The unique parameter learning mechanism, including the processes of encoding and decoding, ensures that P-ELM can encode the underlying information of the original data into the network parameters. Experiments and comparisons on face image recognition and handwritten image annotation (each application contains two datasets) demonstrate the superior performance of the proposed P-ELM compared to baseline methods.

Acknowledgments. This work is supported in part by the National Nature Science Foundation of China (Grant Nos. 61403351 and 61773355), the Key Project of the Natural Science Foundation of Hubei Province, China (Grant No. 2013CFA004), the National Scholarship for Building High Level Universities, China Scholarship Council (No. 201706410005), and the Self-Determined and Innovative Research Founds of CUG (No. 1610491T05).

References

1. Huang, G.B., Zhu, Q.Y., Siew, C.K.: Extreme learning machine: theory and applications. Neurocomputing **70**(1–3), 489–501 (2006)
2. Huang, G., Huang, G.B., Song, S., You, K.: Trends in extreme learning machines: a review. Neural Netw. **61**, 32–48 (2015)
3. Zhang, Y., Wu, J., Cai, Z., Zhang, P., Chen, L.: Memetic extreme learning machine. Pattern Recogn. **58**, 135–148 (2016)
4. Zhang, Y., Wu, J., Zhou, C., Cai, Z.: Instance cloned extreme learning machine. Pattern Recogn. **68**, 52–65 (2017)
5. Liu, N., Wang, H.: Ensemble based extreme learning machine. IEEE Signal Process. Lett. **17**(8), 754–757 (2010)
6. Cao, J., Lin, Z., Huang, G.B., Liu, N.: Voting based extreme learning machine. Inf. Sci. **185**(1), 66–77 (2012)
7. Zong, W., Huang, G.B., Chen, Y.: Weighted extreme learning machine for imbalance learning. Neurocomputing **101**(3), 229–242 (2013)
8. Ap, S.C., Lauly, S., Larochelle, H., Khapra, M., Ravindran, B., Raykar, V.C., Saha, A.: An autoencoder approach to learning bilingual word representations. In: Advances in Neural Information Processing Systems, pp. 1853–1861 (2014)
9. Vincent, P., Larochelle, H., Bengio, Y., Manzagol, P.A.: Extracting and composing robust features with denoising autoencoders. In: 25th International Conference on Machine Learning, pp. 1096–1103 (2008)
10. Wang, H., Shi, X., Yeung, D.Y.: Relational stacked denoising autoencoder for tag recommendation. In: 29th AAAI Conference on Artificial Intelligence, pp. 3052–3058 (2015)
11. Bai, Z., Huang, G.B., Wang, D., Wang, H., Westover, M.B.: Sparse extreme learning machine for classification. IEEE Trans. Cybern. **44**(10), 1858–1870 (2014)
12. Zhang, R., Lan, Y., Huang, G.B., Xu, Z.B.: Universal approximation of extreme learning machine with adaptive growth of hidden nodes. IEEE Trans. Neural Netw. Learn. Syst. **23**(2), 365–371 (2012)
13. Huang, G.B., Zhou, H., Ding, X., Zhang, R.: Extreme learning machine for regression and multiclass classification. IEEE Trans. Syst. Man Cybern. Part B Cybern. **42**(2), 513–529 (2012)
14. Kasun, L.L.C., Zhou, H., Huang, G.B., Chi, M.V.: Representational learning with elms for big data. IEEE Intell. Syst. **28**(6), 31–34 (2013)
15. Hinton, G.E., Salakhutdinov, R.R.: Reducing the dimensionality of data with neural networks. Science **313**(5786), 504–507 (2006)
16. Tang, J., Deng, C., Huang, G.B.: Extreme learning machine for multilayer perceptron. IEEE Trans. Neural Netw. Learn. Syst. **27**(4), 809–821 (2015)
17. Yang, Y., Wu, Q.J.: Multilayer extreme learning machine with subnetwork nodes for representation learning. IEEE Trans. Cybern. **46**(11), 2570–2583 (2016)
18. Wu, J., Cai, Z., Zeng, S., Zhu, X.: Artificial immune system for attribute weighted naive bayes classification. In: IEEE International Joint Conference on Neural Networks, pp. 1–8 (2013)
19. Wu, J., Hong, Z., Pan, S., Zhu, X., Cai, Z., Zhang, C.: Multi-graph-view learning for graph classification. In: 14th IEEE International Conference on Data Mining, pp. 590–599 (2014)
20. Wu, J., Pan, S., Zhu, X., Zhang, C., Wu, X.: Positive and unlabeled multi-graph learning. IEEE Trans. Cybern. **47**(4), 818–829 (2017)

K-Hop Community Search Based on Local Distance Dynamics

Lijun Cai[1], Tao Meng[1(✉)], Tingqin He[1], Lei Chen[1], and Ziyun Deng[2]

[1] College of Information Science and Engineering, Hunan University,
Changsha 410082, China
{ljcai,mengtao,hetingqin,chenleixyz123}@hnu.edu.cn
[2] Changsha Commerce and Tourism College, Changsha 410082, China
dengziyun@126.com

Abstract. Community search aims at finding a meaningful community that contains the query node and also maximizes (minimizes) a goodness metric, which has attracted a lot of attention in recent years. However, most of existing metric-based algorithms either tend to include the irrelevant subgraphs in the identified community or have computational bottleneck. Contrary to the user-defined metric algorithm, how can we search the natural community that the query node belongs to? In this paper, we propose a novel community search algorithm based on the concept of k-hop and local distance dynamics model, which can natural capture a community that contains the query node. Extensive experiments on large real-world networks with ground-truth demonstrate the effectiveness and efficiency of our community search algorithm and has good performance compared to state-of-the-art algorithm.

Keywords: Community search · Interaction model · Complex network

1 Introduction

Most complex networks in nature and human society, such as social networks and communication networks, contain community structures. The goal of community detection is to identify all communities in the entire network, which is a fundamental graph mining task which has been well-studied in the literature [1–3]. Recently, a different but related problem called community search have studied, which is to find the most likely community that contains the query node [4]. It has a wide range of applications in complex networks analysis, such as social contagion modeling and social circle detection [5].

In all the previous studies on these problems, a goodness metric is usually used to identify whether a subgraph forms a community. Many approaches have been proposed to find a subgraph contains the query node and the goodness metric is maximized or minimized, such as k-core [6, 7], k-truss [8, 9] and densest graph [10]. However, most of the existing goodness metrics do not address the "free rider effect" issue, that is, nodes irrelevant to query node or far away from it are included in the identified community [9, 10]. Moreover, real-world applications often generate massive-scale graphs and require efficient processing. Therefore, achieving strong scalability together with high-quality community search is still a challenging, open research problem.

© Springer International Publishing AG 2017
D. Liu et al. (Eds.): ICONIP 2017, Part V, LNCS 10638, pp. 24–34, 2017.
https://doi.org/10.1007/978-3-319-70139-4_3

In this paper, instead of introducing a new goodness metric for community search like k-core or k-truss, we consider the problem of community search from a new point view: local distance dynamics. The basic idea is to envision the nodes which k-hop away from a query node as an adaptive local dynamical system, where each node only interacts its local topological structure. Relying on a proposed local distance dynamics model, the distances among node will change over time, where the nodes sharing the same community with the query node tend to gradually move together while other nodes will keep far away from each other. Such interplay eventually leads to a steady distribution of distances and a meaningful community is naturally found.

The remainder of the paper is organization as follow: Sect. 2 gives related preliminary with our work. The details of our proposed algorithm are described in Sect. 3. Extensive experimental evaluation is presented in the Sect. 4. Section 5 provides our brief conclusion.

2 Preliminary

For the purpose of community search, some necessary definitions are first introduced. In this paper, we focus on an undirected and unweighted simple graph $G = (V, E)$, where V and E are a set of nodes and edges, respectively. Other type of graphs, such as directed and weighted can be handled with only slight modifications.

The structure of a node can be described its neighbors, and the distance between two nodes always according to how they share neighbors. The neighbors of a node is a node set of composed of all its adjacent nodes and the node itself.

Definition 1 (neighbors of node u). Given an undirected graph $G = (V, E)$, the neighbors of node u, denoted by $N(u)$, and is defined as follows:

$$N(u) = \{v \in V | \{u, v\} \in E\} \cup \{u\} \tag{1}$$

In order to discover local community for a given node, the method first need to initialize the distance for each edge. We use the popular Jaccard Distance [5] to measure the initial distance between two adjacent nodes.

Definition 2 (Jaccard Distance). Given an undirected graph $G = (V, E)$, the Jaccard Distance between node u and node v is defined as:

$$d(u, v) = 1 - \frac{|N(u) \cap N(v)|}{|N(u) \cup N(v)|} \tag{2}$$

3 Our Algorithm

3.1 Problem Definition

In order to efficiently identify the scope of the community that the query node belongs to, we use an observation of real-world graphs: the best community for a given node is in the neighborhood of the node. This observation is based on a well-known property of real-world graphs: small world effect, which is the name given to the finding that the average path (hop) between vertices in a network is small, and the local structure of network still has obvious grouping characteristics [11, 12].

Based on this property, we can prunes the distance evaluation for the nodes that are more than k-hop away from the query node. Specifically, our algorithm first roughly detects a k-hop subgraph by searching the nodes are k-hop away from the query node. It then refines the k-hop subgraph to find a best community that the query node belongs to by the local distance dynamics model. Before proceeding further, we given the formal definition of k-hop subgraph is as follows.

Definition 3 (K-Hop Subgraph). Given a graph $G = (V, E)$, a query node $q \in V$ and an integer k > 0. G' is a k-hop subgraph if and only if G' is connected, and each node u in G' has distance at most k-hop away from the node q.

On the basis of the definitions of k-hop subgraph, we give the definition of the k-hop community as follows, where the parameter k controls the scope of the community.

Definition 4 (K-Hop Community). Given a graph $G = (V, E)$, a query node $q \in V$ and an integer k > 0. C is a k-hop community, if C satisfies the following constrains.

- **Connectivity.** C is connected k-hop subgraph and contained node q.
- **Cohesiveness.** Employing the proposed local distance dynamics model on k-hop subgraph, where nodes in the target community will move together while other nodes will keep far away from the query node q.

Clearly, the *connectivity* constrain requires that the k-hop community containing the query node q be connected. In addition, the *cohesiveness* constrain makes sure that each node is as close as possible to the query node in the k-hop community. With the connectivity and cohesiveness constrains, we can ensure that the k-hop community is a connected and cohesive subgraph. The following example illustrate the definition of k-hop community.

Let us consider the graph show in Fig. 1. Assume that k = 1 and q is the query node. By Definition 3, we can see that the 1-hop subgraph included by node set $\{q, h_{11}, h_{12}, h_{13}, h_{14}, h_{15}, h_{16}\}$. However, it includes node h_{13} which is intuitively not relevant

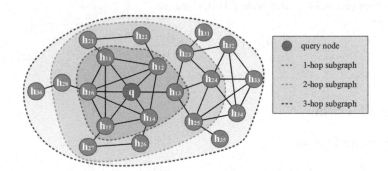

Fig. 1. An example graph for k-hop community.

to the query node. Relying on a proposed local distance dynamics model, the nodes q, h_{11}, h_{12}, h_{14}, h_{15} and h_{16} are moving together, while the node h_{13} keeping far away from them. As a result, the target community for node q is $C = \{q, h_{11}, h_{12}, h_{14}, h_{15}, h_{16}\}$.

Problem Definition. The problem of k-hop community search studied in this paper is defined as follows. Given a graph $G = (V, E)$, a query node $q \in V$ and an integer k > 0, find a k-hop community containing q.

3.2 Local Distance Dynamics Model

After specifying the scope of the target community, the next crucial step is to determine the interaction model among nodes in k-hop subgraph to simulate the distance dynamics. In the following, we will elaborate how the distance changes in local distance dynamics model.

Definition 5 (Core Edge). Given a k-hop subgraph $G'(V', E') \subseteq G(V, E)$, the edge $e = \{u, v\} \in E$ is core edge *iff* $e = \{u, v\} \in E'$.

As shown in Fig. 2(a). In this example network, node $q \in V$ is the query node and k = 1, and the circled by dotted line denote 1-hop subgraph. According to Definition 5, the edges $e(u,v)$, $e(u,a)$, $e(u,b)$, $e(u,q)$, $e(v,a)$, $e(v,b)$, $e(v,q)$, $e(a,b)$, $e(a,q)$ and $e(b,q)$ are core edges since them contained in 1-hop subgraph.

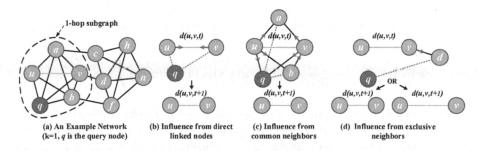

(a) An Example Network
(k=1, q is the query node)

(b) Influence from direct
linked nodes

(c) Influence from
common neighbors

(d) Influence from exclusive
neighbors

Fig. 2. Distance dynamics of one core edge influencing by three distinct interaction patterns.

Formally, let $e = \{u, v\} \in E$ be a core edge between two adjacent nodes u and v, and the $d(u,v)$ is its initial distance. Obviously, any change of the distance $d(u,v)$ actually results from the variation of node u and node v. Relying on its complete local topological structure (see Fig. 2), there are three distinct interaction patterns that allows influencing the distance $d(u,v)$.

Local Pattern 1. Here we consider the first interaction pattern: influence from direct linked nodes u and v (see Fig. 2(b)). Through mutual interactions, one node attracts another to move towards itself, and thus leads to the decrease of distance $d(u,v)$. Formally, we define the change of $d(u,v)$ from the influence of the direct linked nodes, DI, as follows:

$$DI = -\left(\frac{f(1-d(u,v)).(1-d(u,q))}{deg(u)} + \frac{f(1-d(u,v)).(1-d(v,q))}{deg(v)}\right) \quad (3)$$

In pattern DI, $f(\cdot)$ is a coupling function and $f(.) = sin(.)$ is used in this study. The term $1-d(.,.)$ implies the similarity between two direct linked nodes u and v, the more similar the two node have, the higher influence they will have. The term $1/deg(.)$ is a normalized factor which is used to consider the different influences between linked nodes with diverse degrees.

Take friendship network as an example. In general, each people affects their know people, and tends to increase their cohesiveness gradually. Moreover, the more similar the two people are, the higher influence between each other they will have; the more similar to the query people, the more likely to share the same community with the query people; the people with more friends are harder to be influenced comparing to the people with less friends.

Local Pattern 2. The second interaction pattern happens when there exists some common neighbors between nodes u and v (see Fig. 2(c)). The common neighbors between node u and v, denoted by $CN = (N(u)-u) \cap (N(v)-v)$. As the common neighbors have both links with node u and v, they attract the two nodes to move towards itself, and thus result in the decrease of distance $d(u,v)$. Formally, to characterize the change of the distance $d(u,v)$, we define the CI, indicating the influence from the interactions of common neighbors, as follows:

$$CI = -\sum_{x \in CN} \left(\begin{array}{c} \frac{1}{deg(u)} \cdot f(1-d(x,u)) \cdot (1-d(x,v)) \\ + \frac{1}{deg(v)} \cdot f(1-d(x,v)) \cdot (1-d(x,u)) \end{array} \right) \cdot (1-d(x,q)) \quad (4)$$

In pattern CI, the two terms $1 - d(x,u)$ and $1 - d(x,v)$ indicate the similarity of common neighbor x with node u and v, respectively. If the x is more similar to u, the influence from x on v is more similar to the influence from u. The term $1 - d(x,q)$ implies the similarity between common neighbor x and the query node q, the more similar to the query node, the higher influence the common node x will have.

Let us reconsider the friendship network as an example. Obviously, when two people share many common friends, their similar degree becomes large, and tends to increase their cohesiveness gradually. Furthermore, if their common friends are close to the query people, they tend to share the same community with the query people.

Local Pattern 3. The influence from exclusive neighbors is the third interaction pattern (see Fig. 2(d)). The exclusive neighbors only belongs to node u or v, and donated by $EN(u) = N(u) - (N(u) \cap N(v))$ and $EN(v) = N(v) - (N(u) \cap N(v))$, respectively. In this pattern, each exclusive neighbor may have the positive or negative influence to the distance $d(u,v)$. To determine the positive or negative influence of exclusive neighbors on the distance, a similarity-based heuristic strategy is proposed. The basic idea is to investigate whether each exclusive neighbor of node u is similar with the query node q, and vice versa. If the exclusive neighbor of node u is similar with the query node q, the movement of node u towards the exclusive neighbor results in the decrease of distance $d(u,v)$. Formally, we define the degree of positive or negative influence on the distance $d(u,v)$ from the exclusive neighbor as follows:

$$\sigma(x,q) = \begin{cases} (1 - d(x,q)) & (1 - d(x,q)) \geq \lambda \\ (1 - d(x,q)) - \lambda & \text{otherwise} \end{cases} \tag{5}$$

In the above Eq. 5, the term λ is a cohesive parameter, and will be further discussed in Sect. 4.2. Then, we define the change of $d(u,v)$ from the influence of exclusive neighbors, EI, as follows:

$$EI = \begin{pmatrix} -\sum_{x \in EN(u)} \left(\dfrac{1}{deg(u)} \cdot f(1 - d(x,u)) \cdot \sigma(x,q) \right) \\ -\sum_{y \in EN(v)} \left(\dfrac{1}{deg(v)} \cdot f(1 - d(y,v)) \cdot \sigma(y,q) \right) \end{pmatrix} \tag{6}$$

Finally, by considering three interaction patterns together, the dynamics of the distance $d(u,v)$ on core edge $e(u,v)$ over time is govern by:

$$d(u,v,t+1) = d(u,v,t) + DI(t) + CI(t) + EI(t) \tag{7}$$

In the above Eq. 7, the term $d(u,v,t+1)$ is the new distance at time step $t+1$. $DI(t)$, $CI(t)$ and $EI(t)$ are three different influence from the directed nodes, common neighbors, and exclusive neighbors on the distance $d(u,v,t)$ at time step t.

4 Experiments

4.1 Experiments Setup

In our experiments, we compare our algorithm with two representative community search algorithms: *K-Core* [6] and *K-Truss* [8]. We implemented all algorithms in Python and ran the experiments on a Windows Server with 2 Intel Xeon E5-2600 series processors and 176 GB main memory. For all experiments, without further statement, *K-Core* and *K-Truss* specify the default value of k to 6.

We used six large real-world networks in our experiments: *Amazon, DBLP, Youtube, LiveJournal, Orkut* and *Friendster*. These networks are provided with ground-truth community memberships and publicly available at https://snap.stanford.edu/data. We test the performance of the three algorithms to search local community by *Relative Density, Diameter* and *F-socre*, which are widely adopted by other community search methods [6, 8, 10].

4.2 Influence of Parameters

Our *K-Hop* algorithm uses two parameters: k and λ. In this subsection, we investigate the influence of k and λ on result of community. We are interested in the changes of community size and the accuracies with the different value of k and λ.

We first give our analysis about the influence of k on community size and accuracies. For *K-Hop* algorithm, the parameter k is used to determine the scope of local interaction. In general, it is expect that the community size is small changes with k. To verify this conjecture, we studied the sensitive of parameter k in a LFR network. Similar results were obtained on other networks.

The results are show in Fig. 3 and they verify our conjecture. From the Fig. 3(a), we can clearly see that the community size is very small when k = 1. When k = 2 and larger, the community size is almost stable. From the Fig. 3(b), we can clearly see that k = 1 is the critical point on which the minimum F-score is found. After this, the F-score is almost stable.

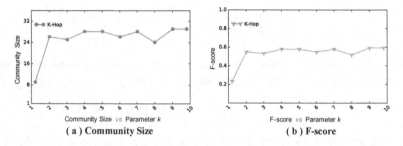

Fig. 3. Sensitive of parameter k.

The value of λ is the most important parameter for local distance dynamics model. For the *K-Hop* algorithm, the λ is used to determine the negative or positive interaction influence on the distances from exclusive neighbors. In general, it is expected that the community size monotonically decreases with λ.

Figure 4 shows our results on a synthetic network. From Fig. 4(a), we can clearly see that the community size monotonically decreases with λ. From the Fig. 4(b), we can clearly see that λ = 0.3 is critical point on which the maximal value of F-score is found. Before this, most of nodes in k-hop subgraph are move together to the query

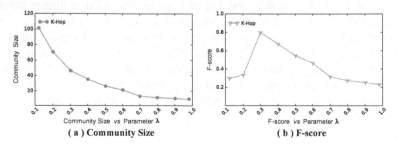

Fig. 4. Sensitive of parameter λ.

node when λ increase. After this, irrelevant nodes quickly keep far away from the query node due to the strong constraint on the closeness of a community.

From the results, we found that *K-Hop* is not sensitive to the searching results. In general, a λ value between 0.3 and 0.6 is normally sufficient to achieve a good result. We recommend a value for k, of 2. For *K-Hop*, we set the k = 2 and λ = 0.5 as default parameters.

4.3 Evaluation on Real Networks

We first evaluate the effectiveness of the selected methods on real networks. For each networks, we randomly select 100 query nodes with degree ranging from 10 to 100. The query node is selected from a random ground-truth community.

The Fig. 5(a) shows the relative density of the selected algorithms on different networks. It can be see that the *K-Truss* method better than other methods on most networks. Focus on the *K-Core* and *K-Hop* algorithms, it is not difficult to find that, the performance of *K-Hop* algorithm is exceeded to the *K-Core* method. In addition, the *K-Hop* algorithm is very close to the *K-Truss* algorithm on some real-world networks.

Figure 5(b) discusses the diameter of community search of various algorithms on real-world networks. From the Fig. 5(b), we can get the following observations. (1) For *K-Core*, the value of the diameters on six networks are very uneven, which imply the performance of the *K-Core* algorithm is very unstable on real-world networks. (2) For *K-Truss*, we can find that, *K-Truss* has better result and stability than *K-Core* algorithm on real-world networks. (3) For *K-Hop*, six real-world networks have the good results, the average value of the diameter is lesser than 4. (4) Focus on *K-Core* and *K-Truss* two algorithms, we can observe that these algorithms have larger diameter, this may be caused by the free rider effect.

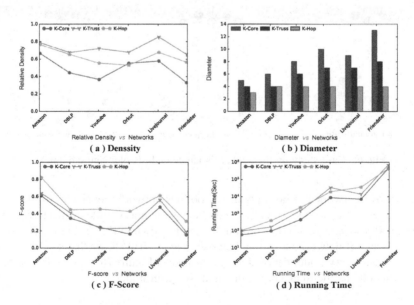

Fig. 5. Performance of community search of different algorithms on real-world networks.

Figure 5(c) shows the F-score of the identified community using different algorithms. We can see that the F-score value of *K-Hop* is 5% to 10% higher than those of other methods. If the nodes in the irrelevant community are selected as the identified community, these algorithms will identified irrelevant communities and will causes the low F-score value.

From the Fig. 5(d), when the scale of the network is small, the running time of all algorithms are small; Along with the increase of the scale of network, the running time are increase gradually. It is important to note that the *K-Core* is run faster than *K-Truss* and *K-Hop*, but it accuracies is low. The *K-Truss* and *K-Hop* algorithms have similar performance.

4.4 Case Study

To validate the effectiveness of our local distance dynamics model, we select two well-known UCI real-world networks with ground truth, use *K-Hop* algorithm to search the community structure with different query node.

The first network is the Zachary's karate club network, consisting of 34 vertices and 78 undirected edges. In this case study, we use node "1" and "34" as the query node, respectively. After set k = 2 and λ = 0.5, we got the community result shown in Fig. 6. Figure 6(a) shows the ground truth of karate club network, which covers 2 classes. Figure 6(b) shows the detection results with "1" as the query node, denoted by green nodes. Figure 6(c) shows the detection results with "34" as the query node, denoted by green nodes.

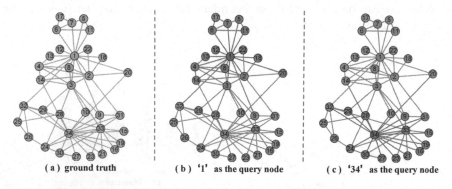

    (a) **ground truth**   (b) **'1' as the query node**   (c) **'34' as the query node**

Fig. 6. Case study on the Zachary's karate club. (Color figure online)

The second network is Books about US politics network, consisting of 105 nodes and 441 edges. Here, we use node "8" and "66" as the query node, respectively. After set k = 2 and λ = 0.5, we got the community result shown in Fig. 7. Figure 7(a) shows the ground truth of network, covering 3 classes. Figure 7(b) shows the detection results with "8" as the query node, denoted by green nodes. Figure 7(c) shows the detection results with "66" as the query node, denoted by green nodes.

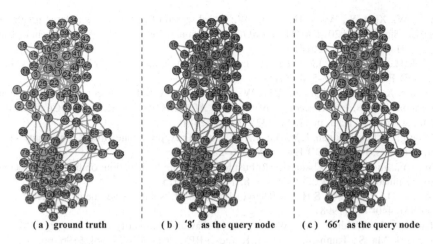

(a) ground truth (b) '8' as the query node (c) '66' as the query node

Fig. 7. Case study on the Books about US politics. (Color figure online)

5 Conclusions

In this paper, we introduce a new community search algorithm, called K-Hop, to automatically find the best community containing a query node in networks based on local distance dynamics. Extensive experimental is executed on six real-world networks show the effectiveness and efficiency of local distance dynamics model and search algorithm. Our future work will consider the community search on heterogeneous network based on the intuitive local distance dynamic model.

Acknowledgements. This work was supported by the National Natural Science Foundation of China (61174140, 61472127, 61272395); China Postdoctoral Science Foundation (2013M540628, 2014T70767); Natural Science Foundation of Hunan Province (14JJ3107); Excellent Youth Scholars Project of Hunan Province (15B087).

References

1. Xu, X., Yuruk, N., Feng, Z., Schweiger, T.A.: Scan: a structural clustering algorithm for networks. In: Proceedings of the 13th ACM SIGKDD International Conference on Knowledge Discovery and Data Mining, pp. 824–833. ACM (2007)
2. Shao, J., Han, Z., Yang, Q., Zhou, T.: Community detection based on distance dynamics. In: Proceedings of the 21th ACM SIGKDD International Conference on Knowledge Discovery and Data Mining, pp. 1075–1084. ACM (2015)
3. Newman, M.E.: Modularity and community structure in networks. Proc. Nat. Acad. Sci. **103**, 8577–8582 (2006)
4. Sozio, M., Gionis, A.: The community-search problem and how to plan a successful cocktail party. In: Proceedings of the 16th ACM SIGKDD International Conference on Knowledge Discovery and Data Mining, pp. 939–948. ACM (2010)
5. Ugander, J., Backstrom, L., Marlow, C., Kleinberg, J.: Structural diversity in social contagion. Proc. Nat. Acad. Sci. **109**, 5962–5966 (2012)

6. Cui, W., Xiao, Y., Wang, H., Wang, W.: Local search of communities in large graphs. In: Proceedings of the 2014 ACM SIGMOD International Conference on Management of Data, pp. 991–1002. ACM (2014)
7. Li, R.H., Qin, L., Yu, J.X., Mao, R.: Influential community search in large networks. Proc. VLDB Endow. **8**, 509–520 (2015)
8. Huang, X., Cheng, H., Qin, L., Tian, W., Yu, J.X.: Querying K-truss community in large and dynamic graphs. In: Proceedings of the 2014 ACM SIGMOD International Conference on Management of Data, pp. 1311–1322. ACM (2014)
9. Huang, X., Lakshmanan, L.V., Yu, J.X., Cheng, H.: Approximate closest community search in networks. Proc. VLDB Endow. **9**, 276–287 (2015)
10. Wu, Y., Jin, R., Li, J., Zhang, X.: Robust local community detection: on free rider effect and its elimination. Proc. VLDB Endow. **8**, 798–809 (2015)
11. Watts, D.J., Strogatz, S.H.: Collective dynamics of 'small-world' networks. Nature **393** (6684), 440–442 (1998)
12. Kunze, M., Weidlich, M., Weske, M.: Behavioral similarity – a proper metric. In: Rinderle-Ma, S., Toumani, F., Wolf, K. (eds.) BPM 2011. LNCS, vol. 6896, pp. 166–181. Springer, Heidelberg (2011). doi:10.1007/978-3-642-23059-2_15

An Improved Feedback Wavelet Neural Network for Short-Term Passenger Entrance Flow Prediction in Shanghai Subway System

Bo Zhang[1], Shuqiu Li[2], Liping Huang[2(✉)], and Yongjian Yang[1]

[1] College of Software, Jilin University, Changchun 130012, China
dazhangbo_01@163.com, yyj@jlu.edu.cn
[2] College of Computer Science and Technology, Jilin University,
Changchun 130012, China
shuqiu@jlu.edu.cn, huangliping5727@163.com

Abstract. Subway traffic prediction is of great significance for scheduling and anomalies detection. A novel model of multi-scale mixture feedback wavelet neural network(MMFWNN) is proposed to predict the short-term entrance flow of Shanghai subway stations. Firstly, passengers are classified into two categories of commuter and non-commuter by mining the travel pattern and identifying the travel pattern stability, which finds that the non-commuters travel is more susceptible to the meteorology status. The proposed prediction model adds a transitional layer to adapt the feedback mechanism, thus to improve the robustness with associative memorizing and optimization calculation. Thus MMFWNN is advantageous to the nonlinear time-varying short-term traffic flow prediction. We evaluate our model in the Shanghai subway system. The experimental results show that the MMFWNN model is more accurate in predicting the short-term passenger entrance flow in subway stations.

Keywords: Wavelet neural network · Subway flow prediction · Travel pattern · Data mining

1 Introduction

The urban rail transit is the main public transportation in large cities. With the overwhelming rail transit passenger volume, subway station presents a supersaturation state especially at peak periods [1]. To alleviate this problem, accurate short-term flow prediction is essential for transportation management.

Early research assessed the historical average, time-series, neural network, and proved that the nonparametric regression model has outperformed other models. Some models, such as ARIMA (Autoregressive integrated moving average model) [2], Seasonal-ARIMA [3], Kalman filter [4], BP (Back propagation) [5], RBF (Radial basis function) [6], have stable and self-adaptive ability. Zhang and Yang [7] proposed spline function to apply to flow prediction. Because of the irregular and stochastic characteristics, these prediction models always result

© Springer International Publishing AG 2017
D. Liu et al. (Eds.): ICONIP 2017, Part V, LNCS 10638, pp. 35–45, 2017.
https://doi.org/10.1007/978-3-319-70139-4_4

in unsatisfactory prediction. RNN (Recurrent neural network) [8] was proposed to cope with these temporal dependencies by the virtue of a short-term memory. On the basis of RNN, LSTM (Long short term memory networks) [9]and CW_RNN (Clockwork recurrent neural network) [10] are able to identify complex dependencies between temporally distant inputs more accurately.

Specifically, the emerging wavelet neural network (WNN) with strong capability of self-learning and high precision also has strong approximation ability and fault-tolerant performance. Sun et al. [11] combined the wavelet and SVM to predict Beijing subway passenger flow. Doucoure et al. [12] used artificial WNN to achieve an intelligent management system. Many studies have proved that WNN has a good performance in short-term prediction [13]. Meanwhile, these researches do not take into account of the travel patterns of passengers. Researches [14]on travel patterns were mentioned in other fields. In a summary, our main contributions in this paper are as follows:

(1) The passengers are classified into commuter and non-commuter according to the travel pattern stability.
(2) We proposed a novel structure of wavelet neural network with feedback which is suitable for forecasting short-term passenger flow. By adding a feedback of Elman networks into the wavelet neural network, the model incorporates the high-precision of wavelet networks and the memorizing ability of RNN. The prediction model as a real-time system combines the time series and historical results of time delay which makes the network having a strong associative memory and optimized calculation ability.
(3) We evaluate our model on the real dataset of smart card records and meteorology condition dataset in the Shanghai subway system. Experiment results identify the effectiveness and efficiency advantage over baselines.

This paper is organized as follows: Sect. 2 illustrates the framework of online forecasting system. Section 3 describes the proposed MMFWNN prediction method. Section 4 clarifies the evaluation methods and compares the results obtained by the proposed model with original wavelet neural network and other neural networks. Section 5 finally concludes this paper.

2   Preliminaries and Framework

Data Set Description: Our data were collected in Shanghai subway system and meteorological department from April 1st to April 30th,2015. The details of the cleaned data set are shown in Table 1. The meteorology <tem rain AQI wind> is a vector corresponding to the same time period. <tem rain AQI wind> represents temperature, rainfall, air quality index and wind speed respectively.

Definition (Commuter): Passengers with fixed travel patterns are defined as commuter. There are five patterns of commuter as shown in Fig. 1. Passengers are identified through the smart card ID. Combination of travel pattern and

Table 1. Details of the datasets.

Data sources			Shanghai
Time span			1st–30th Apr 2015
Subway data	# Records		About 2.8 million
	Proportion of commuter		25%
Meteorology data	Rainfall (#hours)		81
	Aqi (# hours)	Heavy pollution	9
		Moderate pollution	8
		Light pollution	158
		Others	545
	Temperature span		5.0°C–30°C
	Wind speed span		0–5.5 m/s

travel time period is adopted to distinguish a passenger is a commuter or a random traveller. Specific to each passengers travel records, we label him or her as a commuter if more than 80% of his or her travel records are fixed travel pattern during the same time period. Contrary to commuter, other passenger are labelled as non-commuter.

As shown in Fig. 2, the offline process of MMFWNN model includes two parts: (1) Classification of travel patterns: According to definition, the total flow is classified into commuter and non-commuter. For commuter, a simple neural network or a historical mean estimation will be used considering the travel pattern hereinafter would be consistent. For non-commuter, feedback wavelet neural network will be used to predict the traffic flow. (2) Feedback wavelet neural network learning: A transitional layer is added in order to join the feedback mechanism into the wavelet neural network. The input of hidden layer is a linear superposition of the input layer and the transitional layer. The input layer loads short time series and meteorological factors. The transitional layer returns further optimized delayed history information. The excitation function of hidden layer is wavelet function.

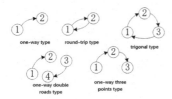

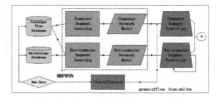

Fig. 1. Travel patterns.   **Fig. 2.** Framework of our model.

As shown in Fig. 2, the online process consists of 3 steps, classification, prediction and mixing. New data will be stored in the database, and it will be classified into commuter or non-commuter. Then the process data after and historical data from database will be input into the prediction model. The non-commuter output will be finally returned as a feedback value.

3   Multi-scale Mixture Feedback Wavelet Neural Network

3.1   Commuter and Non-commuter

The solid line in Fig. 3 is the mean of passenger entrance traffic of xujiahui in every time slot, and the shadow is the floating range in all days which can be regarded as confidence interval. Figure 3 illustrates that either in working days or holidays, the commuter is much more stable and less than the non-commuter. In order to represent stability, we define the log value of mean standard deviation as

$$\Gamma = \ln(\frac{\sum_{i=1}^{n} \delta_i}{n} + 1). \tag{1}$$

where n represents n time slots of a day and δ_i represents the standard deviation of passengers in the ith time slot.

$$\delta = \frac{1}{m} \sum_{i=1}^{m} (x_i - \overline{x})^2. \tag{2}$$

Γ reflects the stability of short-term flow. The larger value indicates that the unstable flow is more easily influenced by external factors, whereas the smaller value indicates that traffic flow is only related to time. A day is divided into 144 time slots of 10 min. As shown in Table 2, commuters owns small value of Γ than non-commuters. The fluctuation magnitude between commuter and non-commuter is different, therefore the influencing factors are exploited respectively.

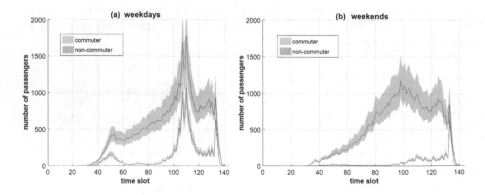

Fig. 3. The comparison of stability.

Table 2. The value of Γ

Time	Commuter	Non-commuter
Weekdays	4.6900	5.9321
Weekends	4.6752	7.0351

3.2 Multi-scale Features

Time Features. Intuitively, the travel pattern on workdays is different from non-workdays. Figure 4 exhibites the passenger entrance flow of Xujiahui. The area around Xujiahui is a bustling entertainment and business district, so passenger flow in the morning is less than both the evening and the night peak. It is obvious that traffic flow on weekdays is similar, consisting of morning peak, daytime period, evening peak, and evening period. Therefore, work days and non-work days are separated, and times slots are extracted as time feature.

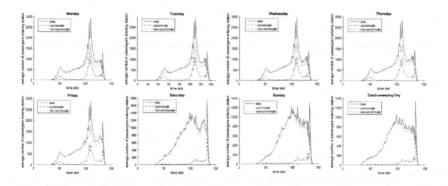

Fig. 4. Station entrance traffic.

Meteorological Features. Intuitively, people tend to choose a convenient way of transportation or stay home under extreme weather conditions, which leads to traffic decreasing. Figure 5 shows the influence of rainfall. It is obvious in Fig. 5a that commuter flow is not affected by rainfall, while in Fig. 5b, the non-commuter flow is reduced due to rainfall. At the same time, rainfall has little influence on the flow during morning and evening peak periods while other periods are obviously affected. In doing so, peak period is labeled as 1, and other periods is labeled 0 to represent the time feature.

In Fig. 6, three non-work days are chosen. On 19th Apr, the non-commuter flow of Xujiahui is smaller than the other two days as shown in Fig. 6(b, d), which is because of the poor AQI condition on 19th. While the commuter flow is fluctuated tranquilly in Fig. 6(a, c). In summary, we do not infer significant effect on subway flow with temperature and wind speed, so two dimensional meteorological features < rain AQI > are used.

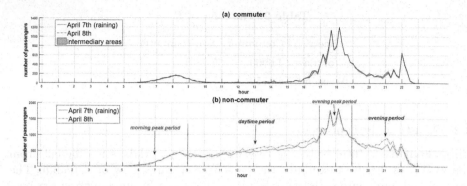

Fig. 5. Influence of rainfall.

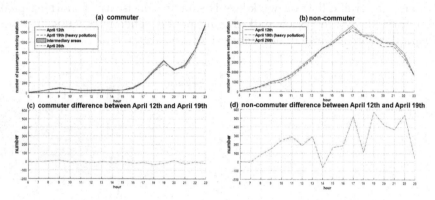

Fig. 6. Influence of the air quality index.

Time Series. A time series is a sequence in chronological order. The data is described either by a possible linear or non-linear auto-regressive process of the form:

$$x(k) = f[x(k-1), x(k-2), ..., x(k-n)].\tag{3}$$

where f is a function describing the relationship between the n past values and the present value. Currently, one-step-ahead time series prediction model is applied. In order to solve the value of n, the Spearman correlation coefficient was used to measure the correlation between the present and the past flow. The Spearman correlation coefficient is

$$\rho = 1 - \frac{6\sum_{i-1}^{N} d_i^2}{N(N^2-1)}.\tag{4}$$

where N stands for the amount of data and d denotes a collection of difference. We compared the correlation of previous time with current time and observed that it decreases as time goes forward. To finalize our analysis, the historical passenger flow whose correlation coefficient is greater than 0.85 are chosen as the input vector. There are four flow features greater than 0.85.

3.3   Feedback Wavelet Neural Network Model

WNN shown in Fig. 7 is a functional interlinkage neural network based on the wavelet function. $X_1, X_2, ..., X_k$ denote input parameters of wavelet neural network. $Y_1, Y_2, ..., Y_k$ are the expressions of the predicted output. ω_{jk} and ω_{ij} represent the link coefficient. When the input sequence is $x_i(i = 1, 2, 3...)$, The output of the hidden layer can be obtained as

$$S(j) = f_{(j)}(\frac{\sum_{i=1}^{k}\omega_{jk}x_i - b_j}{a_j}) \quad j = 1, 2, ..., l. \tag{5}$$

b_j stands for the translation factor of the wavelet basis function $f_{(j)}$, a_j stands for the scaling factor of the wavelet basis function $f_{(j)}$. In our work, the wavelet basis function is selected to represent the transfer function of hidden layer nodes, i.e.: $f(x) = \cos(1.75x)\exp(-x^2/2)$. The calculating formula for the network output layer can be expressed as

$$y(k) = \sum_{j=1}^{l}\omega_{ij}S(j) \quad k = 1, 2, ..., m. \tag{6}$$

where ω_{ij} is the link coefficient between the hidden layer and output layer. The hidden layer has l nodes and the output layer has m nodes.

Feedback mechanism improves the network to become a feedback dynamical system macroscopically. As shown in Fig. 8, a transitional layer is added into WNN as a delay operator. The input of the transitional layer is the output of the output layer. The output of the transitional layer is fed into the hidden layer. The activation function of the transitional layer is a simple linear function. The new output of hidden layer at time t can be obtained as

$$S(j)^t = f_{(j)}(\frac{\sum_{i=1}^{k}\omega_{jk}x_i - b_j + \omega'\sum_{i=1}^{m}Y_i^{(t-1)}}{a_j}) \quad j = 1, 2, ..., l. \tag{7}$$

where $Y_1^{(t-1)}, Y_2^{(t-1)}, ..., Y_m^{(t-1)}$ represent the output of the output layer at time $t-1$, $Y^{(0)}$ is 0 by default. Similarly, error reverse transmission is applied to optimize the parameters of feedback wavelet neural network. Parameters modification process of FWNN are as follows: (1) Calculating network error, i.e.,

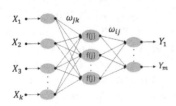

Fig. 7. WNN topology.

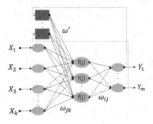

Fig. 8. FWNN topology.

$e = \sum_{k=1}^{k} yn(k) - y(k)$ where $yn(k)$ represents the desired output and $y(k)$ denotes the prediction output. (2) Changing network weights according to the error e, i.e., $\omega_{n,k}^{(t)} = \omega_{n,k}^{(t)} + \Delta\omega_{n,k}^{(t)}$. (3) $\Delta\omega_{n,k}^{(t)} = -\eta\frac{\partial e}{\partial\omega_{n,k}^{(t-1)}}$.

4 Validation and Results

4.1 Metrics and Baselines

Three representative subway stations are shown in Table 3: Xujiahui (XJH), ShangHai Railway Station (SRS) and Century Avenue (CA). The XJH station situated in the business center has heavy traffic flow, conversely, the CA station has low and stable traffic flow. The SRS station has much larger flow and strong fluctuations. We randomly select 70% of the data as training data and set the remainder of data as testing data. Root Mean Squared Error (RMSE) and Mean Relative Error (MRE) are considered to be the metrics for measuring forecasting accuracy. The following results are all the average values of 20 tests.

Table 3. Prediction error of passenger entrance flow.

	Xujiahui(XJH)				ShangHai Railway Station(SRS)				Century Avenue(CA)			
	Weekday		Weekend &Holiday		Weekday		Weekend &Holiday		Weekday		Weekend &Holiday	
	RMSE	MRE	RMSE	MRE	RMSE	MRE	RMSE	MRE	RMSE	MRE	RMSE	MRE
BP	135.088	0.138	120.768	0.110	160.000	0.141	132.996	0.145	35.867	0.130	22.800	0.133
WNN	129.993	0.122	116.291	0.106	153.559	0.136	121.362	0.143	33.731	0.131	22.337	0.123
LSTM	117.888	0.095	105.915	0.099	139.609	0.125	111.434	0.130	30.550	0.117	22.267	0.126
FWNN	120.833	0.090	105.237	0.089	140.604	0.122	110.699	0.131	32.630	0.127	22.552	0.120
MFWNN	110.974	0.087	98.887	0.087	134.083	0.120	100.743	0.127	25.764	0.110	20.656	0.110
MMFWNN	93.1952	0.074	90.769	0.085	122.280	0.115	98.7545	0.124	24.925	0.107	20.163	0.108

In order to confirm the performance of our model MMFWNN, five baselines BP (Back Propagation Neural Network), WNN (Wavelet Neural Network), LSTM (Long Short Term Memory Networks), FWNN (Feedback Wavelet Neural Network), MFWNN (Multi-scale Feedback Wavelet Neural Network) are conducted as shown below. The input features of four baselines (BP, WNN, LSTM, FWNN) are four high correlation traffic flow values before predicted time. The number of the hidden layer node is set to 10 based on experience and the number of the output layer node is only one. In MFWNN, 7 input layer nodes (seven features) and 15 hidden layer nodes are chosen, while pattern classification is not added into this model. Seven features include four traffic flow, two meteorological data, and one time correction feature. In MMFWNN, the non-commuter section uses FWNN (7 input layer nodes, 15 hidden layer nodes, 1 output layer node) and the commuter section uses simple historical mean estimation to avoid overkill. Forward feedback gradient descent are chosen for all optimization parameters in the training.

4.2 Results

As shown in Table 3, LSTM and FWNN in XJH station and SRS station is superior to other neural networks, which shows the feedback neural network and the recurrent neural network have good performance for time series data depending on the historical information, due to their potential memory structure. Nevertheless, the performance of all networks are almost the same in CA station. This is because the low and smooth traffic flow of the CA station is difficult to reflect the advantages of FWNN and LSTM (RNN). The purpose of subway traffic flow prediction is to monitor and avoid the situation of overload traffic, therefore stations whose flow is small are not within the scope of the investigation. Overall, the performance of LSTM and FWNN are similar. As shown in Table 4, the training time of FWNN is less than LSTM because of the complex structure of LSTM. The performance of MFWNN including meteorological information is generally better than FWNN, especially at weekends. It illustrates that passengers are influence by weather condition at weekends. Considering the last row in Table 3, the prediction errors by MMFWNN in XJH station on weekdays and at weekends decrease by about 16.02% and 8.21% compared with the model of MFWNN, respectively. This is because the commuter proportion of passengers on weekdays is so high that the introduction of pattern classification is better to improve the performance, while the weekends is the opposite. The reduction of RMSE proves the effectiveness of this method for short-term subway traffic forecasting.

Table 4. Average running time of subway traffic prediction model.

	WNN	LSTM	FWNN	MFWNN	MMFWNN
AVERAGE TIME	50.84s	184.54s	59.35s	64.76s	69.05s

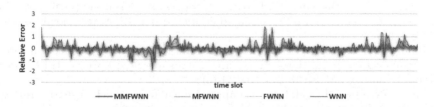

Fig. 9. Traffic prediction relative error.

As shown in Fig. 9, as much as 70% interception from the test set illustrates that MMFWNN has a lower relative error than other models. It is intuitive that the prediction performance of neural network is remarkable due to the ability of fitting nonlinear peculiarity and accurately describing the irregular change of short-term traffic flow. In summary, the results demonstrate the classification of commuter and the involvement of the meteorological characteristics improve the prediction performance of real-time subway system.

5 Conclusions

To predict the short-term passenger entrance flow in a subway system, we improve the wavelet neural network by adding feedback layer and incorporating travel pattern classification and the influencing factor of meteorology feature. The subway passengers are classified into two categories of commuters and non-commuters with the travel pattern identifying. And the meteorology features of aqi and rainfall and time feature are respectively extracted to identify the correlation between them with passenger flow. Experiment results show that the improved wavelet neural network with the extracted influencing factors achieves more accurate prediction result compared to original wavelet neural network with a little additional computing time. When compared to the LSTM, the feedback wavelet neural network has a similar performance, but the latter one presents a better efficiency. The data set size is not large enough to consider the seasonal traffic trend, and this will be added in the future research with more factors, such as special events to prevent the abnormal flow. In the future, the statistical test will be used to demonstrate the validity of the experimental results. Additionally, the model will be testified with other data set in other research area.

Acknowledgement. This research is partly supported by the National Nature Science Foundation of China under Grand no. 61272412 and Jilin Province Science and Technology Development Program under Grant no. 20160204021GX.

References

1. Si, B., Fu, L., Liu, J., Shiravi, S., Gao, Z.: A multi-class transit assignment model for estimating transit passenger flowsa case study of Beijing subway network. J. Adv. Transp. **50**(1), 50–68 (2015)
2. Calheiros, R.N., Masoumi, E., Ranjan, R., Buyya, R.: Workload prediction using arima model and its impact on cloud applicationsqos. IEEE Trans. Cloud Comput. **3**(4), 449–458 (2015)
3. Lippi, M., Bertini, M., Frasconi, P.: Short-term traffic flow forecasting: an experimental comparison of time-series analysis and supervised learning. IEEE Trans. Intell. Transp. Syst. **14**(2), 871–882 (2013)
4. Sun, J.: Examples of validating an adaptive kalman filter model for short-term traffic flow prediction. In: Twelfth COTA International Conference of Transportation Professionals, pp. 912–922 (2015)
5. Wei, Y., Chen, M.C.: Forecasting the short-term metro passenger flow with empirical mode decomposition and neural networks. Transp. Res. Part C Emerg. Technol. **21**(1), 148–162 (2012)
6. Niu, D., Lu, Y., Xu, X., Li, B.: Short-term power load point prediction based on the sharp degree and chaotic RBF neural network. Math. Prob. Eng. **2015**(3), 1–8 (2015)
7. Zhang, D.Y., Yang, H.N.: Passenger flow analysis in subway using a kind of neural network. Appl. Mech. Mater. **713–715**, 2284–2287 (2015)

8. Kim, Y.H., Abdallah, C.T., Lewis, F.L.: A dynamic recurrent neural-network-based adaptive observer for a class of nonlinear systems. Automatica **33**(8), 1539–1543 (1997)
9. Gers, F.A., Schmidhuber, J., Cummins, F.: Learning to forget: continual prediction with LSTM. Neural Comput. **12**(10), 2451 (2000)
10. Koutnik, J., Greff, K., Gomez, F., Schmidhuber, J.: A clockwork RNN. In: Proceedings of the 31st International Conference on Machine Learning (ICML 2014) (2014)
11. Sun, Y., Leng, B., Guan, W.: A novel wavelet-SVM short-time passenger flow prediction in Beijing subway system. Neurocomputing **166**(C), 109–121 (2015)
12. Doucoure, B., Agbossou, K., Cardenas, A.: Time series prediction using artificial wavelet neural network and multi-resolution analysis: application to wind speed data. Renewable Energy **92**, 202–211 (2016)
13. Wickerhauser, M.V.: Adapted Wavelet Analysis from Theory to Software, p. 160. A.K. Peters (1994)
14. Bhat, C.: Modeling the commute activity-travel pattern of workers: formulation and empirical analysis. Transp. Sci. **35**(35), 61–79 (2001)

Social and Content Based Collaborative Filtering for Point-of-Interest Recommendations

Yi-Ning Xu, Lei Xu, Ling Huang, and Chang-Dong Wang(✉)

School of Data and Computer Science, Sun Yat-Sen University, Guangzhou, China
{xuyn7,xulei28}@mail2.sysu.edu.cn, huanglinghl@hotmail.com,
changdongwang@hotmail.com

Abstract. The rapid development of Location-based Social Networks (LBSNs) has led to the great demand of personalized Point-of-interests (POIs) recommendation. Although previous researches have presented a variety of methods to recommend POIs by utilizing social relation, geographical mobility data and user content profile, they fail to address user/location's cold-start problem with high-dimensional sparse data, and overlook the compatibility of social relation, content based methodology and collaborative filtering. To cope with these challenges, we analyze user's check-in preference and find that it may be influenced in two spaces, namely Social Propagation Influence Space and Individual Attribute Influence Space. To this end, we propose a Social and Content based Collaborative Filtering Model (SCCF), which consists of a Social Relation Preference based Model (SRPB) considering social friends' preference and a User Location Content-based Model (ULCB) matching the user attributes with location features. Extensive experiments on real-world datasets firmly demonstrate that the proposed SCCF model outperforms the state-of-the-art approaches while addressing cold-start problems in POI recommendation.

Keywords: POI recommendation · Collaborative filtering · Social network · Content profile · Cold-start

1 Introduction

The appearance of location-based social networks (LBSNs), such as Yelp, Foursquare, and Facebook Places, have offered users abundant Point-of-Interests (POIs) information through social networks [1]. LBSNs enable users to receive check-in data, establish connections with each other, and share location preference among social circle. Although they are beneficial for users to explore interesting and favorite locations, there is an urgent need for personalized recommendation.

Recently, various methods have been proposed for personalized POI recommendation, but most of them still suffer severe cold-start problems, leading to poor performance for new user and new location. For example, when a new location is appended into a network, it's time-consuming and difficult to label its

© Springer International Publishing AG 2017
D. Liu et al. (Eds.): ICONIP 2017, Part V, LNCS 10638, pp. 46–56, 2017.
https://doi.org/10.1007/978-3-319-70139-4_5

tags or figure out its target population. Additionally, in spite of many methods building POI recommender system with geographical mobility data [2–4], or collaborative preferences of users' friends and check-in history [5–8], few of them seamlessly combine both of the two resources for further improving performance. For instance, Li et al. [9] incorporate a set of potential locations where each user's social friends, location friends, and neighboring friends have checked-in before into matrix factorization. But it fails to take user's personal attributes into account, so that the recommendation may deviate from user's actual preference. Moreover, due to the sparsity of users' check-in data, visit frequency matrix consists of tremendous uncertain elements, leading to degenerated prediction accuracy. For example, Lian et al. [10] vary the confidence with check-in frequency and incorporate content-aware methodology into collaborative filtering. But only utilizing check-in frequency as user behavior, it still suffers the problem of unknown frequency and redundant matrix dimension, which results in low prediction accuracy and high complexity.

To address the above issues, we combine the social aspect and user aspect together as a new benchmark of judgement. Since the influential friends' suggestions and users' personal preference affect a user's check-in decisions in many different ways, the influence space is separated into two parts [11], namely Social Propagation Influence Space and Individual Attribute Influence Space. Accordingly, Social and Content based Collaborative Filtering Model (SCCF) is proposed, which consists of a Social Relation Preference based Model (SRPB) considering social friends' preference and a User Location Content-based Model (ULCB) matching the user content profiles with location profiles. We only utilize the user/location profiles as features to reduce the matrix dimensionality in content-based collaborate filtering of ULCB, and take the advantage of the top-Z influential friends' preference in SRPB to solve the cold-start problem. The previous grading data reflects explicit information about users' preference towards POIs, and the star level of POIs directly indicates the popularity and reputation of these locations. As a result, some useful explicit behaviors of users such as users' grading data, and the influence level of social relation are considered in our POI recommendation model. Finally, to evaluate the effectiveness of the proposed models, extensive experiments are conducted on the Yelp LBSN datasets using various evaluation metrics. The results have confirmed the effectiveness of the proposed models.

2 Preliminaries

2.1 Notation

Suppose there are M users and N locations, $\mathbf{x}_u \in \mathcal{R}^P$ and $\mathbf{y}_i \in \mathcal{R}^Q$ are the attribute vectors of user u and location i respectively, where P and Q represent the number of features for user and location respectively. Given a user-friend matrix $\mathbf{S} \in \mathcal{R}^{M \times Z}$, $\mathbf{s}_u$ is the set of user u's friends, where each friend is denoted by f. In addition, $\mathbf{c}_i$ indicates the popularity of location i, $g_{u,i}$ shows the preference degree of a user u to location i, and $w_{u,i}$ is the confidence level on prediction for user u to location i.

2.2 Definition

Since users' choices on check-ins are often affected by both friends' suggestion and the extent to which locations' features meet users' preference, the influence space is separated into two parts as follows, which are regarded as two factors influencing user's check-in preference [11].

Definition 1 (Social Propagation Influence Space). *The social space of user u affects location recommendation for user u by propagating influential friends' preference, which consists of user u's friend list, and each friend f's influence level $\widetilde{s_{u,f}}$ towards user u and his preference probability vector $\mathbf{g}_f$ to location set.*

Definition 2 (Individual Attribute Influence Space). *The individual space of user u is generated by user u's attributes vector $\mathbf{x}_u$, locations' features in $\mathbf{y}$ and the latent factors among them, which affects location recommendation by matching the user content profiles with location profiles according to attribute similarity.*

3 The Proposed Method

3.1 Learning Confidence

In the ICCF (Implicit-feedback based Content-aware Collaborative Filtering) [10], location recommendation is based on a user-location matrix $\mathbf{C} \in \mathcal{R}^{M \times N}$, where each element $c_{u,i}$ indicates the visit frequency of user u to location i. It treats all unvisited locations as negative samples [12], ensuring that the confidence for negative samples is assigned equally. So the overall confidence $w_{u,i}$ is set as,

$$w_{u,i} = \begin{cases} \alpha(c_{u,i}) + 1, & \text{if } c_{u,i} > 0 \\ 1, & \text{otherwise} \end{cases}$$

where $\alpha(c_{u,i})$ is a monotonically increasing function of visit frequency.

However, due to the complexity of users' check-in preference, only considering implicit behavior for POI recommendation is not enough. To address this defect, user's explicit behaviors are appended to promote similarity between prediction and user's preference, such as users' grading data, reviews, and votes.

Our model first introduces two line vectors $\bar{\mathbf{h}} \in \mathcal{R}^N$ and $\widetilde{\mathbf{h}} \in \mathcal{R}^N$, with each element $\bar{h}_i$ representing the total check-in frequency of location i among all time intervals and $\widetilde{h}_i$ indicating the average rating in user review for location i. To enable the positive confidence increasing with total visit frequency, $\bar{h}_i$ may require further data preprocessing by applying a monotonically increasing function $\alpha(\bar{h}_i)$. Thus, denoted by $\mathbf{c} \in \mathcal{R}^N$, the location's total check-ins and average rating marked by users are combined, with each element c_i indicating the popularity of location i as $c_i = \gamma \bar{h}_i + (1 - \gamma)\tilde{h}_i$, where γ is the tuning parameter.

Subsequently, a user-location matrix $\mathbf{G} \in \mathcal{R}^{M \times N}$ where $g_{u,i}$ denotes the preference grade of user u to location i, is obtained by two sparse user-location matrix $\mathbf{T} \in \mathcal{R}^{M \times N}$ and $\mathbf{V} \in \mathcal{R}^{M \times N}$, where $t_{u,i}$ represents the star level in review given by user u to location i and $v_{u,i}$ indicates the corresponding votes given by other users towards this review. Furthermore, considering the reliability of this grade, votes are utilized as weight for star level, and the general grade matrix $\mathbf{G}$ is defined as follows,

$$y_{u,i} = \begin{cases} \text{sigmoid}(v_{u,i})t_{u,i} & \text{, if } (t_{u,i}) > 3 \\ \text{sigmoid}(-v_{u,i})t_{u,i} & \text{, otherwise} \end{cases} \tag{1}$$

where sigmoid function is used to transform the codomain of $v_{u,i}$ from $[0, +\infty)$ to $(0, 1)$, and controls the weight in $(0, 0.5]$ if user's review $t_{u,i}$ implies negative preference to location i, otherwise in $(0.5, 1)$ for positive preference. To avoid data sparsity, the unknown elements in $\mathbf{G} \in (0, 3)$ are supplemented with $g_{u,i} = 1$, which signifies a general preference of user to specific locations.

In order to better estimate user's preference, by combining locations' popularity in $\mathbf{c}$ and user's preference degree in $\mathbf{G}$, the confidence matrix $\mathbf{W}$ is modified as $w_{u,i} = g_{u,i}\alpha(c_i) + 1$, where $w_{u,i}$ is the confidence level for user u to location i, and an initialization value $w_{u,i} = 1$ is assigned for the unknown elements to avoid invalid confidence.

3.2   Recommendation Models

The Social Relation Preference-Based Model. Let $\mathbf{S} \in \mathcal{R}^{M \times Z}$ be a user-friend matrix indicating popularity and public influence level with Z referring to the number of top-Z friends whose social impact outperform others and f representing the index of the friend's user ID, where each element $s_{u,f}$ is evaluated by the fan number of friend f. Moreover, grade $g_{f,i}$ is appended as weight to assess friend f's preference influence level towards location i, where the linear weighted average of these top-Z friends' influence level denotes the total social influence. On the other hand, if a user has less than Z friends, the SRPB model will consider all relevant friends's preference, since social prediction is obtained by the average level of friends' preference.

Although social friends' preference has restricted candidate POIs, prediction may far deviate from user's preference if social relation is only considered. Hence, in order to connect social friends' influence with users' own preference, top-Z friends' total social influence is multiplied by the confidence level $w_{u,i}$. So the element in social prediction matrix $\bar{\mathbf{D}} \in \mathcal{R}^{M \times N}$ is defined as follows,

$$\bar{d}_{u,i} = \frac{\sum_f \widetilde{s_{u,f}} g_{f,i}}{Z} w_{u,i} \tag{2}$$

where $\widetilde{s_{u,f}}$ is normalized by the fan number of friend f, since $\mathbf{S}$'s codomain varies in $[0, +\infty)$ which is too extensive to express friends' public influence.

The User-Location Content-Based Model. Although being influenced by social relation, users still tend to follow their personal preference, which can be revealed by user attributes and location features.

For addressing the cold-start problem in LBSN network, a general solution is to incorporate content-based methodology into collaborative filtering [13]. As mentioned in the previous study [12], confidence matrix in implicit-feedback collaborative filtering steers clear and efficient for sampling negative items. However, state-of-the-art content-aware CF algorithm ICCF [10] whose loss function given as follows suffers from the problem of redundant matrix dimension, encapsulating the IDs of both users and locations as additional features,

$$\mathcal{L} = \frac{1}{2}\sum_{u,i} w_{u,i}(r_{u,i} - \widetilde{p}_u'\widetilde{q}_i)^2 + \frac{\lambda}{2}\Big(\sum_u \|\widetilde{p}_u - \mathbf{U}'x_u\|^2 \\ + \sum_i \|\widetilde{q}_i - \mathbf{V}'y_i\|^2 + (\|\mathbf{U}\|_F^2 + \|\mathbf{V}\|_F^2)\Big) \tag{3}$$

where $r_{u,i} = \mathcal{I}(c_{u,i} > 0)$ is in 0/1 rating matrix $\mathbf{R}$, indicating whether user u has visited location i, and $\widetilde{p}_u \in \mathcal{R}^L$, $\widetilde{q}_i \in \mathcal{R}^L$ are integrated latent factor vectors of user u's/location i's ID and user/location attribute value.

The exceeded dimensionality is actually useless in improving performance, but increases algorithm complexity and the average execution time. To improve this defect, the ULCB model only utilizes user/location profiles as features and reduces the dimension of transform matrices $\mathbf{U}$ and $\mathbf{V}$.

Let $\mathbf{X} \in \mathcal{R}^{M \times P}$ be a user-user attribute matrix, where each row vector $\mathbf{x}_u$ indicates the user attribute value of user u and P is the number of user attributes. Similarly, location features are encapsulated into a location-location feature matrix $\mathbf{Y} \in \mathcal{R}^{N \times Q}$, where the row vector $\mathbf{y}_i$ represents location feature value of location i and Q is the number of location features. To constitute the ULCB model, the transform matrices $\mathbf{U} \in \mathcal{R}^{P \times L}$ and $\mathbf{V} \in \mathcal{R}^{Q \times L}$ obtained by UV matrix decomposition, are also utilized to estimate user's personal preference, where L is the number of latent factors. Accordingly, these transform matrices can match users from user attributes, latent factors and location features to locations, and element in individual prediction matrix $\hat{\mathbf{D}}$ is defined as below,

$$\hat{d}_{u,i} = (\mathbf{x_u}\mathbf{U}\mathbf{V}'\mathbf{y_i'})w_{u,i} \tag{4}$$

which connects user's implicit behavior and explicit behavior, by multiplying the composite confidence level $w_{u,i}$ to prediction $\mathbf{x_u}\mathbf{U}\mathbf{V}'\mathbf{y_i'}$. Thus, the objective function is generated as follows, with a regularized term to avoid over-fitting.

$$\mathcal{L} = \frac{1}{2}\sum_{u,i} w_{u,i}(r_{u,i} - \mathbf{x_u}\mathbf{U}\mathbf{V}'\mathbf{y_i'})^2 + \frac{\lambda}{2}(\|\mathbf{U}\|_F^2 + \|\mathbf{V}\|_F^2) \tag{5}$$

Integrated Recommendation. Accordingly, being divided into two aspects that complement each other, both social relation based model SRPB and individual attributes based model ULCB in two separate influence space are able to improve performance in solving cold-start problem. Since user's check-in preference is so complex that it may be influenced by time, geographical distance and so on, the most influential space of these two differs in diverse situations. As a result, the best method to integrally improve cold-start problem is incorporating the two sub-models, where the tuning parameter controls the proportion of these two prediction components.

To approach user's genuine preference, the integrated model SCCF combines both user friends' influence level and user's personal preference modeled in two sub-models through linear weighting, denoted by $\mathbf{D}$ as follows,

$$d_{u,i} = \mu d_{u,i} + (1 - \mu)\hat{d}_{u,i} \tag{6}$$

where μ is the tuning parameter.

3.3 Parameter Optimization

In the proposed ULCB Model, based on the personal preference prediction formula in Eq. (4), transform matrices $\mathbf{U}$ and $\mathbf{V}$ are learned by minimizing the objective function in regularized optimization problem Eq. (5) by using the gradient decent method.

$$\frac{\partial \mathcal{L}}{\partial \mathbf{u}_j} = \sum_{u=1}^{m} \sum_{i=1}^{n} w_{u,i} x_j^u (\mathbf{x}^u \mathbf{U} \mathbf{V}'(\mathbf{y}^i)' - r_{u,i}) \cdot (\mathbf{y}^i \cdot \mathbf{V}) + \lambda \mathbf{u}_j$$

$$\frac{\partial \mathcal{L}}{\partial \mathbf{U}} = \sum_{u=1}^{m} \sum_{i=1}^{n} w_{u,i} (\mathbf{x}^u)' (\mathbf{x}^u \mathbf{U} \mathbf{V}'(\mathbf{y}^i)' - r_{u,i}) \cdot (\mathbf{y}^i \cdot \mathbf{V}) + \lambda \mathbf{U}$$

where $\mathbf{u}_j$ is the j-th row of $\mathbf{U}$, and x_j^u is the same as $x_{u,j}$ representing the attribute value of user u to user attribute j. Hence, the iteration equation to update transform matrix $\mathbf{U}$ by using gradient descent method is given as follows,

$$\mathbf{U}^{t+1} = \mathbf{U}^t - \eta [\sum_{u=1}^{m} \sum_{i=1}^{n} w_{u,i} \mathbf{x}_u' (\mathbf{x}_u \mathbf{U} \mathbf{V}' \mathbf{y}_i' - r_{u,i}) \cdot (\mathbf{y}_i \cdot \mathbf{V}) + \lambda \mathbf{U}^t]$$

Similar for the update of $\mathbf{V}$.

To simplify the optimization process, Alternating Least Squares (ALS) method is adopted with variable substitution, decomposing rating matrix into two matrices, the user latent factor matrix $\mathbf{P} \in \mathcal{R}^{M \times L}$ and location latent factor matrix $\mathbf{Q} \in \mathcal{R}^{N \times L}$. During the process of matrix factorization, the unknown elements will be estimated. Setting the gradient with respect to $\mathbf{P}_u$ and $\mathbf{Q}_i$ to zero, the update principles are obtained as follows,

$$\mathbf{p}_u = (\mathbf{Q}' \mathbf{W}^u \mathbf{Q} + \lambda \mathbf{I}_L)^{-1} \mathbf{Q}' \mathbf{W}^u \mathbf{r}_u \tag{7}$$

$$\mathbf{q}_i = (\mathbf{P}'\mathbf{W}^i\mathbf{P} + \lambda\mathbf{I}_L)^{-1}\mathbf{P}'\mathbf{W}^i\mathbf{r}_i \qquad (8)$$

where $\mathbf{r}_u$ is the u-th row, $\mathbf{r}_i$ is the i-th column in rating matrix $\mathbf{R}$, and $\mathbf{W}^u$ is a $N \times N$ diagonal matrix, $\mathbf{W}^i$ is a $M \times M$ diagonal matrix with $W_{ii}^u = W_{uu}^i = w_{u,i}$.

After initializing the two latent factor matrices randomly, fixing $\mathbf{P}$ and $\mathbf{Q}$ alternately, the update principles are utilized to update $\mathbf{p}_u$ and $\mathbf{q}_i$ until convergence and optimize the integrated latent factor matrices.

4 Experiment

4.1 The Experimental Setup

Dataset. A series of Yelp datasets containing over 100 million reviews of businesses worldwide is utilized in our experiments. Before adopting these datasets to our model, the ratio of active users and sociable users are analyzed and shown in Fig. 1. The statics indicate that approximately 72% users comment less than 15 reviews, and around 56% users have no friend in this network. For experimental purpose, since our algorithm depends on social information, only users with strongly-connected social relation and enough activity during recent period are considered. Constraining selected users with friend relationship and at least 15 check-in records, we eventually obtain 14,142 users and 2,780 POIs.

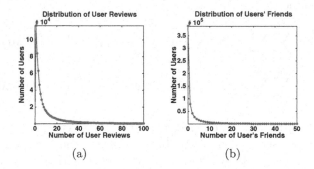

(a)   (b)

Fig. 1. Distribution of user reviews and friends.

Moreover, taken geographical barriers of locations into account, locations in diverse states are inappropriate to recommend simultaneously. Accordingly, the original location dataset is split into four subsets based on locations' geographical distance. And the data quantity and sparsity in each subset are shown in Table 1.

Experiment Settings and Evaluation Metrics. The tuning parameter in the proposed SCCF model, μ and λ are set as 0.01 respectively. When $\tilde{H}_i > 3$, which indicates the visited users have positive feedback, γ is set to 0.05, otherwise γ is set to -0.05. Also, according to the average number of user friends, $Z = 10$

Table 1. Statistics of location dataset.

#State	Number of users	Number of locations	Sparsity of rating
#AZ	14142	301	0.1589%
#IL and #WI	14142	678	0.0544%
#PA	14142	739	0.0878%
#SC and #NC	14142	840	0.0833%

is used for Top-Z friends selection, and number of latent factor is $L = 50$. The manipulation function $\alpha(\bar{H}_i) = 1 + \log(1 + \bar{H}_i \times (10^\epsilon))$ where $\epsilon = -1$, is used to generate steady monotone increasing value. For evaluation metrics, three widely used metrics, namely recall@K, precision@K and F1-Measure@K are used for top-K recommendation [8,14].

4.2   Comparison Results

Two types of cold-start experiments, new user and new POI recommendation, are conducted to evaluate models' performance.

Comparison on New User Recommendation. To accomplish new user recommendation, elements in rating dataset are initialized to one if visited, otherwise to zero. And the baseline ICF's recommendation is based on matrix factorization of rating matrix, and also varies the confidence with check-in frequency as ICCF. Observing the metrics results of ICF, ICCF and the proposed SCCF model on four sub-datasets shown in Fig. 2, we find that our model far outperforms the state-of-the-art content aware collaborative filtering. From the results, since ICF tends to approach zero for the unknown elements, it performs much poorer than ICCF and SCCF in the cold-start situation. On the other hand, F1-Measure of SCCF outperforms ICCF's over 30% as shown in Table 2, especially when K is no larger than 3, which implies a high data sparsity problem. Meanwhile, SCCF outperforms ICCF over 35% in terms of both recall and precision. In other words, the proposed SCCF model is more accurate than common CF models, and more stable during new user recommendation. Additionally, the independent sub-models' performance are also shown in Table 2, which indicates that SRPB and ULCB alternately have better performance on the four datasets, but still not as good as SCCF's overall.

Comparison on New POI Recommendation. Similar to new user recommendation, the new POI recommendation results are shown in Table 3. And it can be seen that the proposed SCCF model has remarkably achieves triple performance over ICCF in terms F1-measure, and performs better than independent sub-models SRPB and ULCB. Furthermore, since ICCF assigns unvisited locations a lower confidence, it outperforms explicit-feedback content-aware

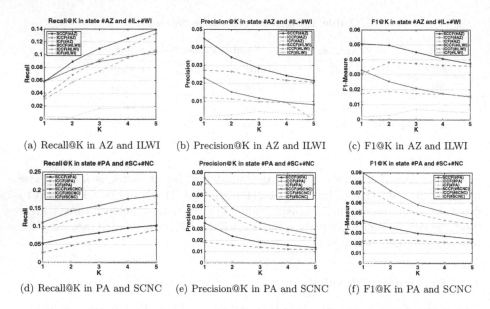

(a) Recall@K in AZ and ILWI (b) Precision@K in AZ and ILWI (c) F1@K in AZ and ILWI

(d) Recall@K in PA and SCNC (e) Precision@K in PA and SCNC (f) F1@K in PA and SCNC

Fig. 2. Comparison results on the four datasets in the user cold-start situation.

Table 2. Comparison on new user in terms of F1-Measure.

#State	Parameter	SCCF	ICCF	Improvement	SRPB	ULCB
#AZ	$k = 1$	**0.0506**	0.0307	64.82%	0.0418	0.0311
	$k = 2$	0.0497	0.0382	30.10%	0.0354	0.0359
	$k = 3$	0.0450	0.0375	20.00%	0.0315	0.0345
#IL and #WI	$k = 1$	**0.0330**	0.0172	91.86%	0.0279	0.0117
	$k = 2$	0.0254	0.0189	34.39%	0.0219	0.0097
	$k = 3$	0.0208	0.0173	20.23%	0.0175	0.0081
#PA	$k = 1$	**0.0425**	0.0221	92.31%	0.0395	0.0081
	$k = 2$	0.0355	0.0234	51.71%	0.0342	0.0109
	$k = 3$	0.0299	0.0228	31.14%	0.0276	0.0131
#SC and #NC	$k = 1$	**0.0898**	0.0754	19.10%	0.0523	0.0754
	$k = 2$	0.0724	0.0605	19.67%	0.0425	0.0555
	$k = 3$	0.0583	0.0490	18.98%	0.0358	0.0454

CF model, such as LibFM, of the best configuration by a significant margin [10]. In other words, performing more effective than state-of-the-art implicit and explicit feedback content-aware CF models, the proposed SCCF model is much more appropriate to be conducted in complex LBSNs and cold-start situation.

Table 3. Comparison on new POI in terms of F1-Measure.

#State	Parameter	SCCF	ICCF	Improvement	SRPB	ULCB
#AZ	$k = 1$	**0.0293**	0.0067	337.3%	0.0282	0.0013
	$k = 2$	0.0190	0.0068	179.4%	0.0184	0.0013
	$k = 3$	0.0161	0.0057	182.5%	0.0133	0.0042
#IL and #WI	$k = 1$	**0.0086**	0.0020	330.0%	0.0009	0.0085
	$k = 2$	0.0061	0.0027	125.9%	0.0014	0.0055
	$k = 3$	0.0051	0.0023	121.7%	0.0017	0.0048
#PA	$k = 1$	**0.0169**	0.0053	218.9%	0.0162	0.0052
	$k = 2$	0.0122	0.0051	139.2%	0.0115	0.0032
	$k = 3$	0.0100	0.0047	112.8%	0.0094	0.0027
#SC and #NC	$k = 1$	**0.0200**	0.0053	277.4%	0.0194	0.0012
	$k = 2$	0.0137	0.0051	168.6%	0.0123	0.0008
	$k = 3$	0.0106	0.0046	130.4%	0.0091	0.0006

5   Conclusions

In this paper, we have proposed a novel integrated SCCF model for POI recommendation which consists of two sub-models, namely SRPB and ULCB. In SCCF, user's social relations are exploited to model social friends' preference influence level to user, and user/location profiles contribute to generate latent factor matrices in approaching user's preference. Experimental results have confirmed that the proposed SCCF model outperforms the baseline models, and is more appropriate to reduce algorithm complexity and improve performance for cold-start and sparsity issues.

Acknowledgment. This work was supported by the Fundamental Research Funds for the Central Universities (16lgzd15) and Tip-top Scientific and Technical Innovative Youth Talents of Guangdong special support program (No. 2016TQ03X542).

References

1. Zhang, D.-C., Li, M., Wang, C.-D.: Point of interest recommendation with social and geographical influence. In: IEEE International Conference on Big Data (Big Data), pp. 1070–1075 (2016)
2. Cheng, C., Yang, H., King, I., Lyu, M.R.: Fused matrix factorization with geographical and social influence in location-based social networks. In: AAAI, vol. 12, p. 1 (2012)
3. Liu, Y., Wei, W., Sun, A., Miao, C.: Exploiting geographical neighborhood characteristics for location recommendation. In: CIKM, pp. 739–748 (2014)
4. Ye, M., Yin, P., Lee, W.-C.: Location recommendation for location-based social networks. In: SIGSPATIAL, pp. 458–461 (2010)

5. Konstas, I., Stathopoulos, V., Jose, J.M.: On social networks and collaborative recommendation. In: SIGIR, pp. 195–202 (2009)
6. Ma, H., Zhou, D., Liu, C., Lyu, M.R., King, I.: Recommender systems with social regularization. In: WSDM, pp. 287–296 (2011)
7. Wang, H., Terrovitis, M., Mamoulis, N.: Location recommendation in location-based social networks using user check-in data. In: SIGSPATIAL, pp. 374–383 (2013)
8. Ye, M., Yin, P., Lee, W.-C., Lee, D.-L.: Exploiting geographical influence for collaborative point-of-interest recommendation. In: SIGIR, pp. 325–334 (2011)
9. Li, H., Ge, Y., Zhu, H.: Point-of-interest recommendations: learning potential check-ins from friends. In: KDD, pp. 975–984 (2016)
10. Lian, D., Ge, Y., Zhang, F., Yuan, N.J., Xie, X., Zhou, T., Rui, Y.: Content-aware collaborative filtering for location recommendation based on human mobility data. In: ICDM, pp. 261–270 (2015)
11. Li, H., Hong, R., Zhu, S., Ge, Y.: Point-of-interest recommender systems: a separate-space perspective. In: ICDM, pp. 231–240 (2015)
12. Hu, Y., Koren, Y., Volinsky, C.: Collaborative filtering for implicit feedback datasets. In: ICDM, pp. 263–272 (2008)
13. Pazzani, M.J.: A framework for collaborative, content-based and demographic filtering. Artif. Intell. Rev. 13(5–6), 393–408 (1999)
14. Liu, B., Fu, Y., Yao, Z., Xiong, H.: Learning geographical preferences for point-of-interest recommendation. In: KDD, pp. 1043–1051 (2013)

Modeling Server Workloads for Campus Email Traffic Using Recurrent Neural Networks

Spyros Boukoros[1], Anupiya Nugaliyadde[2], Angelos Marnerides[3],
Costas Vassilakis[4], Polychronis Koutsakis[2], and Kok Wai Wong[2(✉)]

[1] Department of Computer Science,
Technische Universität Darmstadt, Darmstadt, Germany
sboukoros@gmail.com
[2] School of Engineering and Information Technology,
Murdoch University, Perth, Australia
{a.nugaliyadde,p.koutsakis,k.wong}@murdoch.edu.au
[3] School of Computing and Communications,
Lancaster University, Lancaster, UK
angelos.marnerides@lancaster.ac.uk
[4] Department of Informatics and Telecommunications,
University of Peloponnese, Tripoli, Greece
costas@uop.gr

Abstract. As email workloads keep rising, email servers need to handle this explosive growth while offering good quality of service to users. In this work, we focus on modeling the workload of the email servers of four universities (2 from Greece, 1 from the UK, 1 from Australia). We model all types of email traffic, including user and system emails, as well as spam. We initially tested some of the most popular distributions for workload characterization and used statistical tests to evaluate our findings. The significant differences in the prediction accuracy results for the four datasets led us to investigate the use of a Recurrent Neural Network (RNN) as time series modeling to model the server workload, which is a first for such a problem. Our results show that the use of RNN modeling leads in most cases to high modeling accuracy for all four campus email traffic datasets.

Keywords: Email traffic · Model server workload · Recurrent Neural Network · Time series modeling

1 Introduction

The inherently quick way of email communication, together with the ability it offers to attach files and multimedia content to messages have led to its worldwide acceptance both for personal and for corporate use. Employees tend to view emails within 6 s from the time they arrive [1]. Misuse of this powerful tool is something that naturally occurs, as with every kind of technology. Irresponsible parties use its ability to carry files and/or reach numerous customers for their own, sometimes not legal, actions (spam email). According to [2], Japan's Gross Domestic Product was reduced by 0.1% due to the spam traffic. Spam emails can also break the trust in a corporation by forcing

© Springer International Publishing AG 2017
D. Liu et al. (Eds.): ICONIP 2017, Part V, LNCS 10638, pp. 57–66, 2017.
https://doi.org/10.1007/978-3-319-70139-4_6

infected computers to spam as well and causing worldwide servers to block that corporation's servers, hence isolating the corporation temporarily. Spam traffic accounted for 66% of the worldwide email traffic in 2013 [3]. Consequently, Internet Service Providers (ISPs), corporations and universities have to deal with millions of spam emails every day. Both spam and regular emails arrive at such great volumes that it becomes a matter of crucial importance that servers can cope with the heavy workload and do not crash or exhibit degraded email delivery performance.

All of the above facts, regarding regular and spam email traffic show the urgent need for accurate email traffic prediction, which will help system administrators to take actions to optimize the way they allocate the storage space, processing resources or the bandwidth that they have at their disposal. By doing so, they will be able to avoid system crashes and failures and offer users a better quality of service. Gomez et al. [4] found that message sizes could be represented by lognormal distributions at the body and the tail. Their measurement period was one week. They also modeled the arrival process and the popularity of various email receivers. The Poisson arrival process was shown to fit their workload. The popularity of objects was modeled with a Zipf-like distribution. Bertolotti and Calzarossa [5] also collected the SMTP logs from email servers and modeled the workload. They modeled the message sizes, interarrival times and the number of recipients. The lognormal distribution was found to be the best fit for the message sizes. The interarrival times were shown to fit Weibull and Pareto distributions, in contrast with the conclusions in [4]. In [6] Shah and Noble present a large-scale study on an email server. They model various parameters from the message sizes to the number of words emails consist of. Their measurement period lasted more than 7 months. Regarding the modeling of message sizes, which is the focus of this study, they noticed that the cumulative distribution function (CDF) is symmetric under log scale. Hence, they concluded in this empirical way that their data must be distributed with a lognormal distribution. The main body was modeled with a lognormal distribution while the tail was modeled with a Pareto, following the lead of [4]. In this way, the workload was modeled with high accuracy. While [4] found that spam emails have smaller sizes than regular ones, [6] claims the opposite. However, both of the above studies concluded that spam traffic is distributed with a lognormal distribution. Paxson modeled wide–area transport Layer Protocol (TCP) connections [7]. SMTP connections are TCP connections for transferring emails. Unlike the previously mentioned studies and in accordance to our work he used goodness of fit tests to back up his findings. Regarding the SMTP connections, he found that the empirical distribution was bimodal and justified that from the fact that users sent either simple text mail or files. He decided to model it with two lognormal distributions, breaking the data in two populations, one below the 80th percentile and the other above.

In our previous work in [13] we modeled the email traffic data collected over nine weeks from the Technical University of Crete (TUC). We evaluated various well-known distributions from the relevant literature on workload characterization, in terms of their fitting accuracy to our data. By using leave-one-out cross validation in order to predict the incoming and outgoing traffic, we achieved in certain cases high accuracy in our email message size predictions, with the exception of some outliers which could not be predicted. In contrast with previous work in the field, we found that the lognormal distribution does not provide the best fit for any of the categories that we divided our traffic into. Instead, the best fit is provided by the log-logistic and Generalized Extreme Value distributions.

However, as it will be explained in the following sections, when evaluating the same methodology over three new datasets from other universities, we found that the prediction accuracy was smaller. For this reason, we decided to use a Recurrent Neural Network (RNN) and treat this as a time series problem.

To the best of our knowledge from the literature review, this is the first time that RNN is used for modeling email traffic, and the use of time series modeling for email size prediction is proposed for the first time as well. We anticipated that time series modeling could be an option given that the email traffic can be viewed as a series of events over a time period. The email traffic prediction accuracy with the use of the RNN time series prediction was found to be substantially higher, for all four datasets, than the probabilistic modeling approach.

The only work in the literature that is slightly relevant to our work is the study in [14], where the authors try to extract information from individual email histories, focusing on understanding how an individual communicates over time with recipients in their social network.

2   Methodology

2.1   Data Collection and Processing

With the invaluable help of academic colleagues and of the technical staff of four universities, we have collected a vast amount of email logs. The four universities were the Technical University of Crete (TUC), Greece, the University of the Peloponnese (UoP), Greece, Murdoch University, Australia and Liverpool John Moores University (LJMU), UK.

We got two separate kinds of logs, for the non-spam and the spam emails. The non-spam emails are the emails that arrived at the server and were not stopped by the filter or classified later as spam. The spam traffic that is blocked from the anti-spam filter is not recorded because the connection is closed before the email actually arrives. The emails that arrive at the servers but are classified as spam are saved into folders with their whole body.

We decided to break our data into 4 categories depending on whether they represent system or users' emails, and whether they are incoming or outgoing. The system emails consist mainly of server to server communication or diagnostic emails as well as no-reply messages sent to various users. The decision to break the data into categories was based on the fact that the system emails are sent out in bulk, usually to the whole university to inform everyone about events. Therefore, these emails are of a different nature, so we decided to consider them as a different category and model them separately, to achieve higher accuracy.

Summarizing, the following features were used to describe an e-mail: category (user, spam, system), direction (incoming, outgoing), timestamp (processed at week granularity) and size. More information on the data we collected from each university is presented in Sect. 4.

2.2 Modeling with Probability Distributions

We wanted to study whether our servers' workloads could be modeled with any of the well-known, from the literature, distributions for workload characterization and modeling. This approach serves as an implicit comparison of our conclusions with those of the previous works in the literature, on email traffic modeling. We should mention that the email datasets for those works were not available, to the best of our knowledge, for a direct comparison. The only email corpuses that are available on the web contain actual text from emails, to be used for linguistic analysis. Hence, we relied on our four datasets, which were significantly large.

We used the maximum likelihood estimation method to obtain the parameters which lead the distributions to produce size populations with the same mean and standard deviation as those for each week of our study. The distributions used were the uniform, exponential, gamma, weibull, log-logistic, lognormal and Generalized Extreme Value (GEV). The maximum likelihood estimation method returns a vector with the estimated parameters at the 95% significance level. Simulations were run in Matlab.

2.3 Time Series Modeling Using Recurrent Neural Network

As explained in Sect. 1, we used Recurrent Neural Network (RNN) for creating the model and treated the problem as a time series prediction. We combined all the weeks, except one, for each category separately and tried to predict the last week's email traffic sizes. We assumed that the data in different weeks have time series patterns that can be used to predict the remaining part of the dataset.

A RNN simulates a discrete dynamic system that has input (Xt), output (Yt) and hidden layers [15]. In general, a RNN takes the input sequence to the hidden layers to work out the information about the history of all the past elements. As a result, the output of the hidden layers can have some form of discrete time series similar to the output of the deep multilayer networks. The idea of RNN is that it can connect prior information to the present task, such as using previous data sequences to inform the understanding of the present data sequence to predict future data sequences.

Längkvist et al. [16] discuss various techniques using deep learning for time series as well as the recursive strategies performance in time series. Rather et al. [17] show the use of a traditional Recurrent Neural Network (RNN) with a hybrid model to achieve high time series prediction. The time series predictions are based on previous data, where memory networks and recurrent networks have a higher efficiency than any other deep learning method [18].

In this work, a RNN with 2 hidden layers was implemented with a sigmoid function. We used a batch size of 25 and ran the RNN for 10000 epochs. All these parameters were decided based on a trial and error approach. We used the leave-one-out cross validation technique when assessing the model. For each dataset, the data from all weeks except one were used for training, and this procedure was repeated as many times as the number of weeks of each dataset. The established model from the training is then used to predict the data in the last week of the dataset for testing.

3   Statistical Tests

We used five statistical tests to evaluate the accuracy of our two main approaches (probabilistic and RNN time series).

The first test is the Q-Q plot, a powerful goodness-of-fit test [9] which graphically compares two datasets in order to determine whether the datasets come from populations with a common distribution (if they do, the points of the plot should fall approximately along a 45° reference line). More specifically, a Q-Q plot is a plot of the quantiles of the data versus the quantiles of the fitted distribution. A z-quantile of X is any value x such that $P((X \leq x) = z$. We have plotted the quantiles of the real data with the respective quantiles of the various distribution fits.

The second test is the Kolmogorov–Smirnov (KS) test [10], which tries to determine if two datasets differ significantly. The KS-test has the advantage of making no assumption about the distribution of data, i.e., it is non-parametric and distribution-free. The KS-test uses the maximum vertical deviation between the two curves as its statistic D.

The third test is the Anderson-Darling (AD) test [8], which is a modification of the Kolmogorov-Smirnov test. It places more weight to the tails in comparison to the K-S Test. The test statistic belongs, like the Kolmogorov-Smirnov test, to the family of quadratic empirical distribution function statistics, which measure the distance between the hypothesized and the empirical CDF.

The fourth test is the Kullback-Leibler (KL) Divergence test [11] which measures the information loss between two distributions. It indicates how many extra bits we are going to need if we code samples using the Q probability distribution function instead of P. The test is non-symmetric meaning that if we reverse the P and Q (probability distributions functions) we get different results.

The fifth test is the Relative Percentage Error (RPE) [12], which gives a metric on how different one population is from another. By measuring the absolute difference between the two populations, we do not discriminate which one is bigger or smaller. Of course, we wish to achieve results as close to 0% as possible in order to find a modeling approach that has high accuracy.

RPE is defined as:

$$RPE = |Y - X|/X * 100\%$$

where Y is the predicted value and X the real observation.

4   Results

4.1   Incoming Traffic for Users

This section focuses on our modeling results for the incoming users' traffic. The range of the total number of incoming emails per week and the total number of bytes contained in the emails is presented for each university's dataset in Table 1, together with the number of weeks during which the data was collected.

Table 1. Incoming users emails' numbers and total size

	Min #emails	Max #emails	Min Gbytes	Max Gbytes	# of weeks
TUC	72379	134864	4.34	23.87	9
LJMU	142884	359020	11.60	31.3	4
Murdoch	1132	410672	0.03	56.2	52
UoP	37735	51726	2.00	4.7	4

Our statistical tests agreed that the log-logistic distribution provides the closest fit to our data for Murdoch and TUC, while GEV provided the closest fit to our data for UoP and LJMU. The fact that the KS test and AD test agreed with each other confirmed that the log-logistic/GEV distribution, respectively, is closest to both the tails and the main body of the distribution.

For the probabilistic approach, we found that the RPE results are significantly different between the 98% and the 100% of the quantiles because of the outliers, which tend to have extremely large sizes, something that the distribution methods cannot predict. Therefore, these outliers, usually amounting to 1–2% of our traffic in terms of bytes, cause very large errors. The results presented in Table 2 and for the rest of the paper for the probabilistic modeling approach in this section correspond to 98% of the traffic, excluding the outliers. On the contrary, the results presented throughout the paper for RNN have been derived without removing any outliers from the training or testing datasets, since RNN is resilient to the existence of outliers.

Table 2. Prediction error for incoming users' traffic

	RPE (%)	
	RNN	Probabilistic
TUC	13.9	21.5
LJMU	4.2	8.4
Murdoch	14.2	32.7
UoP	9.2	16.7

As shown in Table 2, despite modeling the whole dataset (including the outliers) RNN is able to largely outperform the probabilistic approach for all datasets, based on the RPE metric.

4.2   Incoming System Traffic

This section focuses on our modeling results for the incoming system traffic. The range of the total number of incoming emails per week and the total number of bytes contained in the emails is presented for each university's dataset in Table 3. It should be noted that our dataset from LJMU contained data only for incoming users' email and for spam traffic, therefore we had no data for incoming system traffic.

As shown in Table 4, RNN again largely outperforms the probabilistic approach for all datasets, based on the RPE metric.

Table 3. Incoming system emails' numbers and total size

	Min #emails	Max #emails	Min Gbytes	Max Gbytes	# of weeks
TUC	49985	318944	0.2	2.1	9
LJMU	–	–	–	–	–
Murdoch	3149	77686	0.03	6	52
UoP	838	7166	0.003	0.06	4

Table 4. Prediction error for users' incoming system traffic

	RPE (%)	
	RNN	Probabilistic
TUC	2.1	20.8
Murdoch	4.2	9.3
UoP	7	23

4.3   Outgoing Users' Traffic

This section focuses on our modeling results for the outgoing users' traffic. The range of the total number of outgoing users' emails per week and the total number of bytes contained in the emails is presented for each university's dataset in Table 5.

Table 5. Outgoing users emails' numbers and total size

	Min #emails	Max #emails	Min Gbytes	Max Gbytes	# of weeks
TUC	16611	74222	2.3	11.3	9
LJMU	–	–	–	–	–
Murdoch	573	103205	0.01	13.6	52
UoP	4730	102396	0.2	3.8	4

As shown in Table 6, RNN again clearly outperforms the probabilistic approach for all datasets.

Table 6. Prediction error for outgoing users' traffic

	RPE (%)	
	RNN	Probabilistic
TUC	9.4	14.7
Murdoch	25.3	40.8
UoP	13.7	29.6

4.4   Outgoing Traffic for System Emails

This section presents our modeling results for the outgoing system traffic. The range of the total number of outgoing system emails per week for each university's dataset is presented in Table 7. The vast majority (almost 99%) of these emails have a size smaller than 6 Kbytes. This means that the servers rarely send attachments. Instead, they send short plain messages.

Table 7. Outgoing system emails' numbers and total size

	Min #emails	Max #emails	Min Gbytes	Max Gbytes	# of weeks
TUC	50480	233653	0.22	2.7	9
LJMU	–	–	–	–	–
Murdoch	770	305947	0.007	1.31	52
UoP	2630	8996	0.01	0.05	4

As shown in Table 8, RNN once again largely outperforms the probabilistic approach for the TUC and UoP datasets. For the Murdoch dataset the probabilistic approach is shown to have a marginally smaller error, however the results for the probabilistic approach refer to 98% of the traffic, excluding the outliers. If the outliers are included, as they are for RNN, the RPE for the probabilistic approach becomes very high for all types of traffic of all datasets.

Table 8. Prediction error for outgoing system traffic

	RPE (%)	
	RNN	Probabilistic
TUC	5.3	10.0
Murdoch	23.3	22.6
UoP	4.4	20.4

4.5   Spam Traffic

This section focuses on our modeling results for the spam traffic. The range of the total number of outgoing system emails per week for each university's dataset is presented in Table 9. It should be noted that our dataset from Murdoch University did not contain data for spam traffic.

Table 9. Spam emails' numbers and total size

	Min #emails	Max #emails	Min Gbytes	Max Gbytes	# of weeks
TUC	1577	2372	0.029	0.089	9
LJMU	27116	77110	0.94	2.68	4
Murdoch	–	–	–	–	–
UoP	5469	8182	0.07	0.26	4

This is the only case where for one of the datasets (UoP) the probabilistic modeling outperforms RNN, as shown in Table 10 (for the TUC and LJMU datasets again RNN excels). One reason for this different result is that the outliers of the specific spam dataset make the accurate prediction over the whole dataset difficult, whereas the result presented for the probabilistic approach, as explained earlier, focuses on 98% of the traffic, excluding the outliers. The effect of DNS black list settings (which eliminate spam messages at the initial handshake phase, before any data are received, examined and stored), as well as the efficiency of spam detection filters in UoP will be investigated further, to gain insight on the reasons behind the high errors of RNN for the UoP dataset.

Table 10. Prediction error for spam traffic

	RPE (%)	
	RNN	Probabilistic
TUC	17.1	17.7
LJMU	18.7	36.9
UoP	57.1	25

5 Conclusions

In this work, we model the workload of the email servers of four universities. We initially evaluated various well-known distributions from the relevant literature on workload characterization, in terms of their fitting accuracy to our data. We found that the accuracy varied, depending on the email traffic category (incoming/outgoing, users/system email or spam) and that even in the cases where a significant accuracy was achieved for the vast majority of the email traffic sizes, there were outliers which could not be accurately predicted.

For this reason, we implemented a Recurrent Neural Network using time series prediction, as an alternative method, and we found that it was able to achieve a significantly higher accuracy for all types of email traffic, by treating the datasets as time series. The impressive result with the use of the RNN is that it outperforms the probabilistic approach in terms of accuracy although the RNN models the entirety of the datasets, including outliers, whereas the probabilistic approach is not used for the outliers, where it fails completely in their modeling.

We believe that our results offer a solid basis for larger scale future work on email traffic modeling and prediction, which will acquire data from a much larger pool of servers. In our view, it will be very interesting for ISPs to clarify whether these new results are associated with the current nature of emails in general, or if they are limited by the type of the dataset, i.e., if campus email traffic has different characteristics than that of a private Internet Service Provider.

Acknowledgements. We would like to sincerely thank Mr. Panagiotis Kontogiannis, Head of the Educational Computational Infrastructure at the Technical University of Crete, Mr. Martin Connell, Senior Systems Engineer at LJMU and Mr. Mario Pinelli, Manager of Computer Services and IT at Murdoch University. Without their help with collecting the datasets this research would not have been possible.

References

1. Jackson, T., Dawson, R., Wilson, D.: The cost of email interruption. J. Syst. Inf. Technol. **5**, 81–92 (2001)
2. Takemura, T., Ebara, H.: Spam mail reduces economic effects. In: Proceedings of the 2nd IEEE International Conference on the Digital Society (2008)
3. Kashyap, A., et al.: Internet Security Threat report (2014). http://www.symantec.com/content/en/us/enterprise/other_resources/b-istr_main_report_v19_21291018.en-us.pdf. Accessed 15 June 2017
4. Gomez, L.H., Cazita, C., Almeida, J.M., Almeida, V., Meira Jr., W.: Workload models of spam and legitimate e-mails. Perform. Eval. **64**(7–8), 690–741 (2007)
5. Bertolotti, L., Calzarossa, M.C.: Workload characterization of email servers. In: Proceedings of SPECTS (2000)
6. Shah, S., Noble, B.D.: A study of e-mail patterns. Softw. – Pract. Exp. **37**(14), 1515–1538 (2007)
7. Paxson, V.: Empirically-derived analytic models of wide-area TCP connections. IEEE/ACM Trans. Netw. **2**(4), 316–336 (1994)
8. Anderson, T.W., Darling, D.A.: Asymptotic theory of certain "goodness of fit" criteria based on stochastic processes. Ann. Math. Stat. **23**(2), 193–212 (1952)
9. Law, A.M., Kelton, W.D.: Simulation Modeling and Analysis, 2nd edn. McGraw-Hill, New York City (1991)
10. Massey, F.J.: The Kolmogorov-Smirnov test for goodness of fit. J. Am. Stat. Assoc. **46**(253), 68–78 (1951)
11. Kullback, S., Leibler, R.A.: On information and sufficiency. Ann. Math. Stat. **22**(1), 79–86 (1951)
12. Lanfranchi, L.I., Bing, B.K.: MPEG-4 bandwidth prediction for broadband cable networks. IEEE Trans. Broadcast. **54**(4), 741–751 (2008)
13. Boukoros, S., Kalampogia, A., Koutsakis, P.: A new highly accurate workload model for campus email traffic. In: Proceedings of the International Conference on Computing, Networking and Communications (ICNC), pp. 1–7 (2016)
14. Navaroli, N., DuBois, C., Smyth, P.: Statistical models for exploring individual email communication behavior. In: Proceedings of the Asian Conference on Machine Learning (2012)
15. Hüsken, M., Stagge, P.: Recurrent neural networks for time series classication. Neurocomputing **50**(C), 223–235 (2013)
16. Längkvist, M., Karlsson, L., Loutfi, A.: A review of unsupervised feature learning and deep learning for time-series modeling. Pattern Recogn. Lett. **42**(1), 11–24 (2014)
17. Rather, A.M., Agarwal, A., Sastry, V.: Recurrent neural network and a hybrid model for prediction of stock returns. Expert Syst. Appl. **42**(6), 3234–3241 (2015)
18. Bontempi, G., Ben Taieb, S., Le Borgne, Y.-A.: Machine learning strategies for time series forecasting. In: Aufaure, M.-A., Zimányi, E. (eds.) eBISS 2012. LNBIP, vol. 138, pp. 62–77. Springer, Heidelberg (2013). doi:10.1007/978-3-642-36318-4_3

Multiclass Imbalanced Classification Using Fuzzy C-Mean and SMOTE with Fuzzy Support Vector Machine

Ratchakoon Pruengkarn[✉], Kok Wai Wong, and Chun Che Fung

School of Engineering and Information Technology,
Murdoch University, Perth, Australia
{r.pruengkarn,k.wong,l.fung}@murdoch.edu.au

Abstract. A hybrid sampling technique is proposed by combining Fuzzy C-Mean Clustering and Synthetic Minority Oversampling Technique (FCMSMT) for tackling the imbalanced multiclass classification problem. The mean number of classes is used as the number of instances for applying undersampling and oversampling. Using the mean as the fixed number of the required instances for each class can prevent the within-class imbalance data from being eliminated erroneously during undersampling. This technique can decrease both within-class and between-class errors, and thus can increase the classification performance. The study was conducted using eight benchmark datasets from KEEL and UCI repositories and the results were compared against three major classifiers based on G-mean and AUC measurements. The results reveal that the proposed technique could handle most of the multiclass imbalanced datasets used in the experiments for all classifiers and retain the integrity of the original data.

Keywords: FCM · SMOTE · FSVM · Imbalanced data

1 Introduction

Imbalanced data classification is a challenging problem and has drawn significant attention from researchers in data mining and pattern recognition disciplines [1]. In binary imbalanced classification, the instances in one class (minority class) are much less in comparison to the other class (majority class). However, in case of multiclass problems with imbalanced data, there may be multiple majority and minority classes which cause skewed distributions [2]. Most existing techniques convert multiclass imbalanced dataset into binary class imbalanced datasets and then apply one-against-all or one-versus-one approaches directly to the binary imbalanced datasets [3–5]. Unfortunately, combining the results from classifiers learned from different binary class datasets may cause potential classification error [6].

This study focused on applying Fuzzy C-Mean (FCM) clustering method and Synthetic Minority Oversampling Technique (SMOTE) to address the multiclass imbalanced classification problem, without the need to convert dataset into a binary class problem while retaining a similar number of instances to the original dataset.

© Springer International Publishing AG 2017
D. Liu et al. (Eds.): ICONIP 2017, Part V, LNCS 10638, pp. 67–75, 2017.
https://doi.org/10.1007/978-3-319-70139-4_7

Three classifiers were implemented to compare the classification performance of the proposed method. The performance was evaluated in terms of G-mean and AUC.

The structure of this paper is as follows. Section 2 reviews previous research on handling multiclass imbalanced data. Section 3 describes the proposed method. The data used in the experiments and experimental setup are explained in Sect. 4 while Sect. 5 contains results of the classification performance from the benchmark datasets. Section 6 presents the conclusion of this study.

2 Related Work

Several solutions of imbalanced problem have been proposed in data mining and pattern recognition areas, including resampling of the data. Sampling methods consists of undersampling and oversampling methods. For undersampling method, the majority instances are removed from the majority class. Cluster based undersampling [7] is implemented for reducing the gap between the number of majority class instances and the number of minority class instances, with the objective to generate good quality training dataset. The clustering based undersampling using the nearest neighbors of the cluster centres has presented better results when compared to UnderBagging, SMOTEBagging and RUSBoost algorithms [8]. The ensemble Fuzzy C-Mean Clustering [9] has also been used for balancing the size of the classes and the results from FCM method has generated stable classification performance and high prediction accuracy.

In the case of oversampling, the new instances are created from the existing instances of minority class. The Synthetic Minority Oversampling Technique (SMOTE) [10] was proposed by generating a synthetic instances rather than replication the minority instances in order to tackle with the overfitting problem. The Mahalanobis Distance-based Over-sampling technique (MDO) [11] applies Mahalanobis distance instead of Euclidean distance for generating synthetic instances and reduces the risk of overlapping between different class regions in the multiclass problem.

In additional, several studies have combined undersampling and oversampling techniques and they dealt with multiclass problem to ensure between-class imbalance is reduced without excessive use of undersampling and oversampling. SCUT [2] was proposed to handle both within-class and between-class imbalance by implementing the Exception Maximization and SMOTE algorithms. The Different Contribution Sampling (DCS) method [12] was proposed based on the contribution of the support vector and the nonsupport vector to obtain the diverse training sets without either losing important information or adding trivial information.

3 Proposed Methodology

Combining FCM and SMOTE techniques called FCMSMT is proposed in this paper in order to handle the issue of imbalance between classes in multiclass problem. By applying FCMSMT algorithm, the number of instances in a new balanced dataset should be almost similar to the original dataset. This will decrease the impact of within-class and between-class errors, and it will also reduce excessive oversampling.

Regarding to undersampling, FCM is used to put the data into different clusters. The number of clusters used in this paper is five. SMOTE is also applied to create artificial examples in the minority class. First of all, the original dataset is divided into individual target classes. The number of instances for each class is calculated in order to compute the mean number of instances of all classes. For each individual target class, the number of instances is compared to the mean number of instances. If it is greater than the mean number, the FCM undersampling is applied to the class instances in order to reduce the number of instances to be equal to the mean number, and at least one instance is selected from each cluster. For example, the Autos dataset used in this study consists of 6 classes with 3, 20, 48, 46, 29 and 13 respectively (see Table 1). The mean number for this dataset is 27. The number of instances in class 3 is 48 which is greater than 27; therefore the FCM undersampling is performed. The instances in class 3 is clustered into five clusters using FCM algorithm. Then, twenty-seven instances are selected randomly from the clusters and stored in the new FCM dataset. FSM algorithm is then applied to class 4 and class 5 in order to get the number of instances to be the same as the mean number.

Table 1. Summary of the imbalanced datasets in the experiments

Datasets	Sources	# instances	# classes	Class distribution	Imbalanced ratio
Autos	KEEL	159	6	3/20/48/46/29/13	16.00
Dermatology	KEEL	358	6	111/60/71/48/48/20	5.55
Ecoli	KEEL	336	8	143/77/52/35/20/5/2/2	71.50
Lymphography	KEEL	148	4	2/81/61/4	40.50
Pageblocks	KEEL	548	5	492/33/3/8/12	164.00
Thyroid	KEEL	720	3	17/37/666	39.18
Wine Quality	UCI	6497	7	30/216/2138/2836/1079/193/5	567.20
Yeast	KEEL	1484	10	244/429/463/44/51/163/35/30/20/5	92.60

By contrast, the SMOTE algorithm is applied when the number of class instances is less than the mean. For instance, oversampling of the instances in class 1, class 2 and class 6 are created using SMOTE in order to get the same amount of the mean number and stored in the new SMOTE datasets. However, in case of the number of class instances is equal to the mean number, the instances remain the same. Finally, the new dataset is combined after applying FCMSMT techniques.

4 Experimental Design

This section describes the dataset characteristics and the experimental steps used in this study.

4.1 Data Characteristics

This study focused on multiclass imbalanced classification problem. Seven benchmark datasets from KEEL [13] and the Wine Quality dataset UCI [14] repositories are used in experiments. An overview of the datasets used in the experiments is presented in Table 1.

4.2 Experimental Processes

The experiments were conducted to handle multiclass imbalanced data classification and the results were compared between conventional techniques and the proposed techniques. Regarding to conventional techniques, the original dataset was divided in to training set and testing set. Sampling techniques were then applied on training set only [3–6]. As a result, the number of new dataset instances could be more or less than the number of original dataset instances. Subsequently, the proposed techniques were applied for data balancing on the original dataset, followed by dividing the data into training set and testing set as shown in Fig. 1. The number of new dataset instances would then become almost the same as the original dataset.

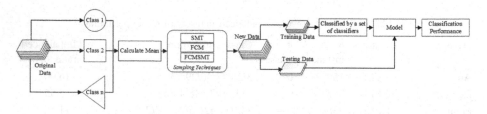

Fig. 1. Process of the proposed method.

The new balancing dataset was assessed using 10-fold cross-validation strategy. Each dataset consisted of ten folds with the same number of instances with 80% training dataset and 20% testing dataset. The instances are rearranged in each fold to obtain reliable results. A set of classifiers used in the experiments was defined at the first phase of the research.

- *Neural Network (NN)* can be used to extract complex patterns and detect trends [15]. The feed forward neural network was created with a single hidden layer, 18 nodes in the hidden layer, Scaled Conjugate Gradient as a training function and cross-entropy as performance function.
- *Support Vector Machine (SVM)* is primarily set to maximise hyperplane which will guarantee the input patterns are classified correctly [16]. Radial Basis Function (RBF) was used with one standardise data.
- *Fuzzy Support Vector Machine (FSVM)* assigns different fuzzy membership values to training instances based on their importance in order to handle the problem of outliers and noise [17]. Fuzzy memberships or weights were computed using exponential decaying. The optimal memberships were selected using highest G-mean value [18].

The three major sampling techniques were applied to the dataset: (1) undersampling, (2) oversampling, and (3) combining of both techniques. The details of each sampling technique are described as follows.

- *Oversampling*: for classes which have a number of instances less than the mean number, the SMOTE technique was used to replicate the number of instances until they were equal to the mean number. T-SMT and SMT are used to denote the application of the SMOTE technique in a conventional technique and the proposed technique respectively.
- *Undersampling*: the FCM technique was utilised to decrease the number of class instances to be the same as the mean number, where the number of class instances was greater than the mean number. T-FCM and FCM are acronyms for applying the FCM technique in a conventional technique and the proposed technique respectively.
- *Combination of both techniques*: FCMSMT is a technique wherein undersampling with the FCM technique and oversampling with SMOTE technique, were utilised as applicable to balance the classes to match the mean number. T-FCMSMT and FCMSMT are abbreviations for applying the FCMSMT technique in the conventional technique and the proposed technique respectively.

5 Result and Discussion

In this section, the results from the experiments are used to evaluate the classification performance based on G-mean and AUC measures. The proposed techniques are applied to eight commonly used datasets from both KEEL and UCI repositories and compared with three conventional classifiers: NN, SVM and FSVM. The results obtained through the proposed methods are presented from Tables 2, 3, 4, 5, 6, 7, 8 and 9.

Table 2. The number of instances in each dataset

Datasets	Original	T-FCM	T-SMT	T-FCMSMT	FCM	SMT	FCMSMT
Autos	*159*	123	194	*158*	117	204	*162*
Dermatology	*358*	308	410	*360*	296	422	*360*
Ecoli	*336*	219	456	*339*	190	482	*336*
Lymphography	*148*	94	204	*150*	80	216	*148*
Pageblocks	*548*	242	856	*550*	166	935	*550*
Thyroid	*720*	380	1060	*720*	294	1146	*720*
Wine Quality	*6497*	3883	9114	*6500*	3228	9765	*6496*
Yeast	*1484*	921	2050	*1487*	777	2187	*1480*

Table 2 shows the number of instances in the original datasets and the sampling datasets. It can be seen that the proposed technique, FCMSMT, retained similar number of instances as in the original datasets.

The results of this study showed that the proposed FCMSMT presented the best classification performance when compared to the conventional methods by using

Table 3. The G-mean of the datasets from NN classifier

Datasets	Original	T-FCM	T-SMT	T-FCMSMT	FCM	SMT	FCMSMT
Autos	0.6101	0.5093	*0.6707*	0.6254	0.5766	*0.7577*	0.7224
Dermatology	*0.9818*	0.9802	0.9772	0.9796	0.9758	*0.9843*	0.9802
Ecoli	*0.6819*	0.6582	0.6747	0.6663	0.6149	*0.8990*	0.8977
Lymphography	0.5098	0.6478	0.7110	*0.9051*	0.6656	0.9387	*0.9459*
Pageblocks	0.4874	0.6146	0.8766	*0.9256*	0.6450	0.9695	*0.9785*
Thyroid	0.1733	0.2292	0.7354	*0.7613*	0.3239	0.8518	*0.8676*
Wine Quality	0.2580	0.2975	0.4209	*0.4774*	0.2902	0.6178	*0.6589*
Yeast	0.6271	0.6504	0.6866	*0.6852*	0.6288	0.7441	*0.7590*

Table 4. The G-mean of the datasets from SVM classifier

Datasets	Original	T-FCM	T-SMT	T-FCMSMT	FCM	SMT	FCMSMT
Autos	0.4147	0.3805	0.5162	*0.5493*	0.4097	0.5642	*0.5966*
Dermatology	0.0000	*0.2477*	0.0076	0.2120	0.0493	0.0000	*0.3767*
Ecoli	0.4895	0.5403	0.5896	*0.6085*	0.5421	0.8814	*0.8958*
Lymphography	0.0556	*0.2574*	0.0614	0.2289	0.2683	0.5297	*0.6242*
Pageblocks	0.3284	0.5043	*0.7127*	0.7055	0.5053	*0.9729*	0.9498
Thyroid	0.3496	0.0694	0.0463	*0.6167*	0.0926	0.7993	*0.8616*
Wine Quality	0.3669	0.4113	0.4591	*0.5709*	0.3902	0.6837	*0.7375*
Yeast	0.4643	0.4931	0.5201	*0.5653*	0.4752	0.7301	*0.7770*

Table 5. The G-mean of the datasets from FSVM classifier

Datasets	Original	T-FCM	T-SMT	T-FCMSMT	FCM	SMT	FCMSMT
Autos	0.7238	0.6627	*0.7973*	0.7390	0.6520	*0.9006*	0.8502
Dermatology	0.9810	0.9805	*0.9819*	0.9802	0.9763	*0.9849*	0.9819
Ecoli	0.6829	*0.6885*	0.6807	0.6754	0.6680	0.8962	*0.9051*
Lymphography	0.5383	0.5393	0.5383	*0.5633*	0.5300	0.9296	*0.9634*
Pageblocks	0.7412	0.7668	*0.9040*	0.8950	0.7962	0.9724	*0.9825*
Thyroid	0.3802	0.4267	0.5500	*0.7337*	0.4042	0.8858	*0.8919*
Wine Quality	0.2230	0.3046	0.3801	*0.4810*	0.2868	0.5931	*0.6596*
Yeast	0.6802	0.6911	0.6980	*0.7110*	0.6843	0.7501	*0.7736*

Table 6. The AUC of the datasets with NN classifier

Datasets	Original	T-FCM	T-SMT	T-FCMSMT	FCM	SMT	FCMSMT
Autos	0.7104	0.6597	*0.7540*	0.7157	0.7039	*0.8010*	0.7798
Dermatology	*0.9822*	0.9808	0.9780	0.9804	0.9770	*0.9848*	0.9808
Ecoli	*0.8112*	0.7947	0.8043	0.7946	0.7715	*0.9078*	0.9064
Lymphography	0.7009	0.7740	0.8126	*0.9067*	0.7918	0.9397	*0.9487*
Pageblocks	0.6943	0.7590	0.9059	*0.9405*	0.7696	0.9704	*0.9790*
Thyroid	0.5481	0.5684	*0.7771*	0.7715	0.6137	0.8647	*0.8766*
Wine Quality	0.5573	0.5704	*0.5865*	0.5955	0.5595	0.6880	*0.7059*
Yeast	0.7272	0.7379	*0.7490*	0.7380	0.7361	0.7873	*0.7927*

Table 7. The AUC of the datasets with SVM classifier

Datasets	Original	T-FCM	T-SMT	T-FCMSMT	FCM	SMT	FCMSMT
Autos	0.6379	0.6127	0.6888	*0.6977*	0.6260	0.7045	*0.7130*
Dermatology	0.5000	*0.5612*	0.5010	0.5386	0.5069	0.5000	*0.5850*
Ecoli	0.6991	0.7301	0.7539	*0.7649*	0.7262	0.8938	*0.9036*
Lymphography	0.5077	*0.5686*	0.5097	0.5581	0.5730	0.6935	*0.7473*
Pageblocks	0.6086	0.6966	*0.8222*	0.8144	0.7052	*0.9735*	0.9517
Thyroid	0.6014	0.5121	0.5081	*0.6868*	0.5155	0.8202	*0.8703*
Wine Quality	0.6046	0.6140	0.6361	*0.6690*	0.5973	0.7328	*0.7674*
Yeast	0.6478	0.6650	0.6660	*0.6826*	0.6527	0.7683	*0.8008*

Table 8. The AUC of the datasets with FSVM classifier

Datasets	Original	T-FCM	T-SMT	T-FCMSMT	FCM	SMT	FCMSMT
Autos	0.8120	0.7558	*0.8467*	0.7918	0.7556	*0.9072*	0.8729
Dermatology	0.9818	0.9813	*0.9826*	0.9809	0.9774	*0.9854*	0.9825
Ecoli	0.8122	*0.8162*	0.8087	0.8031	0.7995	0.9061	*0.9117*
Lymphography	0.7287	0.7278	0.7287	*0.7396*	0.7333	0.9319	*0.9651*
Pageblocks	0.8256	0.8401	*0.9290*	0.9206	0.8627	0.9729	*0.9830*
Thyroid	0.6198	0.6419	0.6710	*0.7557*	0.6277	0.8960	*0.9021*
Wine Quality	0.5486	0.5767	0.5783	*0.5994*	0.5639	0.6889	*0.7083*
Yeast	0.7570	*0.7630*	0.7648	0.7531	0.7619	0.7957	*0.8038*

Table 9. The percentage of performance improvement between original data and FCSMT technique with FSVM classifier

Datasets	G-mean	AUC	Datasets	G-mean	AUC
Autos	12.64	6.09	Pageblocks	24.12	15.74
Dermatology	0.09	0.07	Thyroid	51.15	28.22
Ecoli	22.23	9.95	Wine quality	43.67	15.98
Lymphography	42.51	23.64	Yeast	9.33	4.68

G-mean and AUC measurements. In addition, when the proposed FCMSMT was used with the FSVM classifier, they provided better results for almost all datasets. This is mainly due to the exponential decay memberships in FSVM, which are robust to the imbalanced datasets. As can be seen in Table 9, the proposed FCMSMT improved the classification performance from the original datasets with approximately 25.72% of G-mean and 13.05% of AUC. However, the classification results for Autos, Dermatology and Yeast in this study did not improve much. This is because the imbalance ratios for Autos and Dermatology datasets were not high and the class distribution had almost similar number of instances in each class. This means that the original classifiers were able to use the original dataset to classify all the classes and therefore has no impact on the performance of Autos and Dermatology. For the classifiers used for the Yeast dataset, they could recognise most minority classes in the original dataset. It was observed that the classifiers for Yeast dataset can classify more minority instances when the proposed technique was implemented.

6 Conclusion

In this study, a hybrid FCM and SMOTE (FCMSMT) approach is introduced to handling multiclass imbalanced problem. The classification performance of eight benchmark datasets from KEEL and UCI repositories were evaluated by G-mean and AUC. The results reveal that the proposed technique provided the best result on most of the datasets on both G-mean and AUC for all classifiers, and it retained the class distribution of the datasets. Considering conventional method and proposed method, it can be seen that the proposed method performed better than the conventional method, and the proposed technique presented good results on both conventional and proposed methods. In addition, the average percentage of classification performance improvement between original datasets and after implementing FCMSMT in term of G-mean and AUC are 25.72 and 13.05, respectively.

References

1. López, V., Fernández, A., Herrera, F.: On the importance of the validation technique for classification with imbalanced datasets: addressing covariate shift when data is skewed. Inf. Sci. **257**, 1–13 (2014)
2. Agrawal, A., Viktor, H.L., Paquet, E.: SCUT: multi-class imbalanced data classification using SMOTE and cluster-based undersampling. In: 7th International Joint Conference on Knowledge Discovery, Knowledge Engineering and Knowledge Management (IC3 K), pp. 226–234. Lisbon (2015)
3. Jeatrakul, P., Wong, K.W., Fung, C.C.: Classification of imbalanced data by combining the complementary neural network and SMOTE algorithm. In: Wong, K.W., Mendis, B.S.U., Bouzerdoum, A. (eds.) ICONIP 2010. LNCS, vol. 6444, pp. 152–159. Springer, Heidelberg (2010). doi:10.1007/978-3-642-17534-3_19
4. Ou, G., Murphey, Y.L.: Multi-class pattern classification using neural networks. Pattern Recogn. **40**(1), 4–18 (2007)
5. Fernández, A., del Jesus, M.J., Herrera, F.: Multi-class imbalanced data-sets with linguistic fuzzy rule based classification systems based on pairwise learning. In: Hüllermeier, E., Kruse, R., Hoffmann, F. (eds.) IPMU 2010. LNCS, vol. 6178, pp. 89–98. Springer, Heidelberg (2010). doi:10.1007/978-3-642-14049-5_10
6. Wang, S., Yao, X.: Multiclass imbalance problems: analysis and potential solutions. IEEE Trans. Syst. Man Cybern. Part B Cybern. **42**(4), 1119–1130 (2012)
7. Rahman, M., Davis, D.N.: Addressing the class imbalance problem in medical datasets. Int. J. Mach. Learn. Comput. **3**(2), 224–228 (2013)
8. Lin, W.C., Tsai, C.F., Hu, Y.H., Jhang, J.S.: Clustering-based undersampling in class-imbalanced data. Inf. Sci. **409–410**, 17–26 (2017)
9. Kocyigit, Y., Seker, H.: Imbalanced data classifier by using ensemble fuzzy c-means clustering. In: The IEEE-EMBS International Conference on Biomedical and Health Informatics (BHI 2012), pp. 952–955. Hong Kong (2012)
10. Chawla, N., Bowyer, K., Hall, L., Kegelmeyer, P.: SMOTE: synthetic minority over-sampling technique. Artif. Intell. Res. **16**(1), 321–357 (2002)
11. Abdi, L., Hashemi, S.: To combat multi-class imbalanced problems by means of over-sampling techniques. IEEE Trans. Knowl. Data Eng. **28**(1), 238–251 (2016)

12. Jian, C., Gao, J., Ao, Y.: A new sampling method for classifying imbalanced data based on support vector machine ensemble. Neurocomputing **193**(1), 115–122 (2016)
13. KEEL Data-Mining Software Tool: Data Set Repository. http://sci2s.ugr.es/keel/imbalanced.php. Accessed 30 May 2017
14. Lichman, M.: UCI Machine Learning Repository. http://archive.ics.uci.edu/ml. Accessed 30 May 2017
15. Dumitru, C., Maria, V.: Advantages and disadvantages of using neural networks for predictions. Ovidius University Ann. Econ. Sci. Ser. **13**(1), 444–449 (2013)
16. Karamizadeh, S., Abdullah, S.M., Halimi, M., Shayan, J., Rajabi, M.J.: Advantage and drawback of support vector machine functionality. In: International Conference on Computer, Communications, and Control Technology (I4CT 2014), pp. 63–65. Langkawi (2014)
17. Batuwita, R., Palade, V.: FSVM-CIL: fuzzy support vector machines for class imbalance learning. IEEE Trans. Fuzzy Syst. **18**(3), 558–571 (2010)
18. Pruengkarn, R., Wong, K.W., Fung, C.C.: Data cleaning using complementary fuzzy support vector machine technique. In: Hirose, A., Ozawa, S., Doya, K., Ikeda, K., Lee, M., Liu, D. (eds.) ICONIP 2016. LNCS, vol. 9948, pp. 160–167. Springer, Cham (2016). doi:10.1007/978-3-319-46672-9_19

Incremental Matrix Reordering
for Similarity-Based Dynamic Data Sets

Parisa Rastin[(✉)] and Basarab Matei

LIPN-CNRS, UMR 7030, Université Paris 13,
99 Avenue J-B. Clément, 93430 Villetaneuse, France
{rastin,matei}@lipn.univ-paris13.fr

Abstract. Visualization methods are important to describe the underlying structure of a data set. When the data is not described as a vector
of numerical values, a visualization can be obtained through the reordering of the corresponding similarity matrix. Although several methods of
reordering exist, they all need the complete similarity matrix in memory.
However, this is not possible for the analysis of dynamic data sets. The
goal of this paper is to propose an original algorithm for the incremental reordering of a similarity matrix adapted to dynamic data sets. The
proposed method is compared with state-of-the-art algorithms for static
data-sets and applied to a dynamic data-set in order to demonstrate its
efficiency.

Keywords: Matrix reordering · Incremental · Relational data ·
Dynamic data sets

1 Introduction

Data visualization is the presentation of data in a graphical format. It is an
important step for exploratory data analysis, as it enables analysts to grasp
difficult concepts or identify new patterns, and greatly helps the choice of data
mining tools to apply. We focus here on the construction of useful visualization
of the data-set's structure. Such visualization gives a lot of information about
the data-set, and can be used "as it" or be the basis of a choice of parameter
values for an automatic analysis, especially for unsupervised analysis (clustering)
when no *a priori* knowledge is available about the data structure (see [7] for a
review).

These visualization methods are especially important when the data is not
described as a vector of numerical values. Indeed, most data mining algorithms
can only be applied to vectorial data. Non-vectorial datasets (such as text,
images, sequences, etc.) are usually defined by their relations or their similarities
with each other. Such similarities are represented by a matrix, which can be easily
visualised using different colors for different values of the cells. This representation has been used for almost a century in many domains: biology, neurology,
social science, supply management, transportation and artificial intelligence [3].

© Springer International Publishing AG 2017
D. Liu et al. (Eds.): ICONIP 2017, Part V, LNCS 10638, pp. 76–84, 2017.
https://doi.org/10.1007/978-3-319-70139-4_8

However, the object in the matrix must be organized in an order that reflect the data structure: objects that are similar to each other should be positioned close to each other in the matrix. To find the optimal order of objects in a similarity matrix, several authors proposed different reordering algorithms optimizing different cost functions: Optimal Leaf Ordering algorithm [1] ("OLO") and Grauvaeus and Wainer algorithm [11] ("GW") minimize the Hamiltonian Path Length. Spectral seriation [2] ("Spect") optimizes the 2-SUM problem whereas Multidimensional Scaling [5] ("MDS") optimizes the Least Square Criterion. Finally, Visual Assessment of clustering Tendency [4] ("VAT") tries to find a Minimum Spanning Tree (MST) in a weighted connected graph representing the distance matrix and Hierarchical Clustering [12] ("HC") minimizes the dissimilarity of adjacent objects. See [14] for a review.

In this paper, we wish to propose an incremental reordering algorithm adapted to dynamic non-vectorial data-sets. In a dynamic data-set, new object are added constantly, increasing the size of the data-set without a fixed limit. This type of dataset is increasingly frequent with the development of Internet and other communication networks or with the development of sensor networks. The analysis of a dynamic data stream is challenging because its structure can change over time ("concept drift") and it is not possible to keep all of the information in memory [10]. The method must treat the objects "on the fly", in an unpredictable order. In addition, the objects composing the stream of information in such data-sets are often non-vectorial (for example tweets, videos, images, etc.). Here we propose the first method able to create incrementally, from a dynamic non-vectorial data set, a similarity matrix in an ordered way that allows the visualization of the data underlying structure. We propose first a "static" version of the algorithm adapted to non-dynamic data sets, in order to be able to compare our proposal with the state-of-the-art reordering algorithms. Then, we present a version fully adapted to dynamic data sets, allowing a visualization of the "current" structure of the data-set and being able to forget "outdated" information in a reasonable computation time and memory consumption.

The paper is organized as follows. In Sect. 2 we describe our new algorithms. In Sect. 3 we show an experimental validation. A conclusion and future work perspectives are given in Sect. 4.

2 The Proposed Incremental Reordering Algorithm

In this section we propose two versions of an incremental algorithm for matrix reordering. The first version is adapted to static data sets: the reordered matrix must describe the full structure of the data. The second version is adapted to dynamic data sets: in that case the reordered matrix must represent the current structure of the data set, and must be able to "forget" outdated information.

2.1 Algorithm for Static Data Sets

The main idea of this approach is to incrementally optimise the Hamiltonian Path Length [6] of the ordered relational data. We consider that the order of the objects in a dissimilarity matrix corresponds to a path through a graph where each node represents an object and is visited exactly once: a Hamilton path. By minimizing this path length, we globally minimize the similarities between adjacent objects. The same criteria is used in algorithms such as "GW" or "OLO". However, we propose here to incrementally optimize this criteria. The full algorithm is described in Algorithm 1. The algorithm complexity is in $O(n(n+1)/2)$.

Algorithm 1. Static algorithm

Require: $\mathcal{O}, d$
1: $D = [0], L = [o^1]$
2: **for** each new object o^k **do**
3:     Compute the similarities $d_{kp} = d(o^k, o^p), o^p \in L$
4:     Compute λ_p^k by using Eq. 1
5:     Compute $p_{opt} = argmin_p(\lambda_p)$
6:     Update L by inserting o^k
7:     Update D by inserting d_{kp} at the position p_{opt}
8: **end for**

Let's introduce some notations. We call D the $n \times n$ dissimilarity matrix to reorder, or to construct in an ordered way, n being the total number of objects. In our method, we do not need a dissimilarity matrix a priori. A set of objects $\mathcal{O} = \{o^1, \ldots, o^n\}$ and a dissimilarity measure d are sufficient, as the objects are treated one by one to construct the ordered matrix. Let k be an integer $1 \leq k \leq n$ representing the number of objects already processed. Also let $L(k) = \{o^1, \ldots, o^k\}$ be the list of objects already treated and $D(k)$ be the matrix under construction.

The first object, which is arbitrarily fixed, forms the initial list $L(1)$. Then, for each new object o^k, the algorithm adds o^k to the list $L(k)$ and compares the best position of o^k in the matrix $D(k-1)$, by minimizing the Hamilton path length. At each step, the path length is increased by a value λ_p, which depends on the chosen position p of the new object and the dissimilarities $d(o^k, o^p)$ and $d(o^k, o^{p+1})$. For each possible position p, we compute λ_p according to Eq. 1 and we choose $p_{opt} = argmin_p(\lambda_p)$. This solution is inspired by a solution from the travelling salesman problem proposed in [17]. The matrix $D(k-1)$ is then updated with the addition of the dissimilarities between o^k and the object in $L(k-1)$, at the p_{opt} position. At the end of the process, the algorithm produces an ordered matrix with a minimized Hamiltonian Path Length.

$$\lambda_p^k = \begin{cases} d(o^k, o^p), \text{ for } p = 1 \\ \\ d(o^k, o^{p-1}) + d(o^k, o^p) - d(o^{p-1}, o^p), \\ \quad \text{ for } 2 \leq p \leq k-1 \\ \\ d(o^k, o^{p-1}), \text{ for } p = k \end{cases} \tag{1}$$

2.2   Algorithm for Dynamic Data Sets

In this section, we aim to present an algorithm to reorder dynamic relational data (Algorithm 2). The difference here is that new objects in $\mathcal{O} = \{(o^1, t^1), \ldots, (o^n, t^n)\}$ are dynamically presented to the system, so each object o^k is now associated with a time stamp t^k. The total number of objects in the stream is potentially infinite, and we wish to propose a visualization of the current data's structure at any time. To this end, the constructed matrix is constrained to keep a fixed size, in order to remove outdated information and to assure a reasonable memory and computation cost. In this case we define a parameter s which indicates the maximum size of the matrix. When a new object is presented to the system, if the updated matrix is bigger than s_{max}, the row and column corresponding to the older object is removed. An alternative possibility could be to keep a fixed time window instead of a fixed size. In that case, we define a minimum time stamp t_{min} and, at the end of each step, objects older than t_{min} are removed from the matrix. This algorithm is incremental by construction and is adapted to visualize the temporal variations of the data structure. The algorithm complexity is in $O(n)$.

Algorithm 2. Dynamic algorithm

Require: s_{max} or t_{min}, $\mathcal{O}$, d
1:   $D = [0], L = [(o^1, t^1)]$
2: **for** each new object (o^k, t^k) **do**
3:      Compute the similarities $d_{kp} = d(o^k, o^p), o^p \in L$
4:      Compute λ_p^k by using Eq. 1
5:      Compute $p_{opt} = argmin_p(\lambda_p)$
6:      Update L by inserting (o^k, t^k)
7:      Update D by inserting d_{kp} at the position p_{opt}
8:      Remove from L and D any data o^i with $t^i < t_{min}$ or the oldest object if $size(D)$
       $> s_{max}$
9: **end for**

---

3   Results

In this section we present the experimental protocol we used to validate the proposed algorithm. We first tested the quality of the reordered matrix and compared the results with the quality of the state-of-the-art algorithms. The aim is to demonstrate that the performance of the proposed method is at least similar to its competitors. Then, we present an application of our dynamic reordering approach to show that it is able to produce a dynamic visualization of the data structure.

3.1   Evaluation of the Proposed Algorithm Quality

The quality of the proposed algorithm was tested using seven classical indexes for matrix reordering (BAR, Moore Stress, Least Square, Weighted Gradient, Path Length, Inertia and 2SUM, see [14]). Its quality is compared with the quality of six existing methods: "OLO" [1], "GW" [11], "HC" [12], "VAT" [4], "Spectral" [9] and "MDS" [5]. The algorithms and the quality indices were computed using the "seriation" package [12] in R 3.3.2. The performances of the algorithms were tested on ten different data-sets, including 3 artificial and 4 real vectorial data-sets. The 3 last data-sets are real non-vectorial data built from a set of protein sequences, whose similarity is computed using scores of both mismatch and gap in the two sequences of amino acids [15]. Table 1 summarizes the data-sets' properties. See [16] for a detailed description.

Table 1. Description of the experimental data-sets

Data-set	Size	Dimension	#Clusters	Note
Art1	1000	5	4	Artificial with Gaussian clusters
Art2	1000	2	6	Artificial with Gaussian clusters
Art3	1000	2	7	Artificial with non Gaussian clusters
Iris	150	4	3	Real data
Digits	537	64	3	Real data
Glass	214	10	3	Real data
Wine	178	13	3	Real data
Prot1	50	Unknown	3	Non-vectorial real data
Prot2	129	Unknown	5	Non-vectorial real data
Prot3	400	Unknown	2	Non-vectorial real data

Figure 1 shows the mean values (± standard error) of the different quality indices over the ten data sets for each tested algorithm. In order to test the statistical significativity of the differences between the proposed approach and the other algorithms, we performed a Friedman test with post-hoc analysis [8]. To improve the readability of the tables, the indexes' values were normalized using a Min-Max normalization [13] to fit in the $[0, 1]$ interval. In this manner, for each data-set and each index, the algorithm with the "best" quality has a value of 1, whereas the algorithm with the "worst" quality has a value of 0. In addition, the detailed values are given for the "Path Length" index (for which the proposed approach is significantly better) and the "Weighed gradient" (for which the proposed approach is not significantly better) (Tables 2 and 3).

It is clear from these results that the proposed algorithm performs very well in comparison to the state-of-the-art algorithms. The main difference is that our algorithm works in an incremental way from a random presentation of the data, while the other algorithms can use any information about the similarity matrix

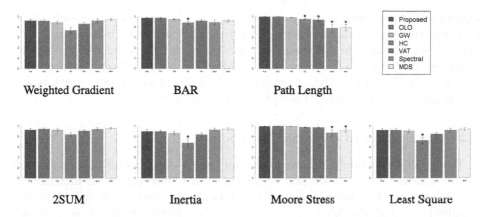

Fig. 1. Mean values (± standard error) of the different quality indexes over the ten data sets for each tested algorithm. "*" denotes a significant difference between an existing approach and the proposed algorithm (Friedman test with post-hoc analysis).

Table 2. Path length values (normalized).

Algorithm	Art1	Art2	Art3	Iris	Digits	Glass	Wine	Prot1	Prot2	Prot3
Prop	1.00	1.00	1.00	1.00	0.99	1.00	1.00	1.00	1.00	1.00
OLO	1.00	1.00	1.00	1.00	1.00	0.99	1.00	1.00	1.00	0.99
GW	0.99	1.00	1.00	0.98	0.98	0.95	1.00	1.00	1.00	0.94
HC	0.98	0.99	0.98	0.95	0.92	0.88	0.94	0.99	1.00	0.90
VAT	0.94	0.98	0.99	0.90	0.81	0.83	0.98	0.99	1.00	0.97
Spect	0.89	0.90	0.85	0.88	0.32	0.45	0.99	0.98	0.99	0.52
MDS	0.89	0.86	0.78	0.88	0.38	0.66	0.99	0.96	0.94	0.58

Table 3. Least square values (normalized).

Algorithm	Art1	Art2	Art3	Iris	Digits	Glass	Wine	Prot1	Prot2	Prot3
Prop	0.96	0.99	0.89	0.95	0.70	0.87	1.00	1.00	0.99	0.86
OLO	0.94	0.95	0.88	0.94	0.77	0.92	1.00	1.00	0.99	0.81
GW	0.94	0.93	0.76	0.93	0.80	0.83	1.00	1.00	0.99	0.87
HC	0.95	0.59	0.56	0.48	0.75	0.70	0.44	0.99	0.97	0.78
VAT	0.86	0.86	0.78	0.81	0.72	0.71	0.94	1.00	0.97	0.81
Spect	0.99	0.99	0.95	1.00	0.71	0.79	1.00	1.00	1.00	0.76
MDS	0.99	0.98	0.94	1.00	0.85	0.88	1.00	1.00	0.96	0.79

at any time. Despite this limitation, although the obtained results are not always better than the competitors depending on the indexes and data sets, its quality remains comparatively very high. The statistical comparisons, based on our experimental data-sets, show that the quality of the proposed approach is not significantly lower than the existing algorithms for the seven indexes tested. Is actually significantly better than "HC" for five of the seven indexes and significantly better than "VAT", "Spectral" or "MDS" for a few indexes. The proposed algorithm is the best method to minimize the Path Length index, which is not surprising as it was conceived to optimize this criteria.

3.2   Application to Dynamic Data Sets

In the previous section we showed that our algorithm performs at least as well as the existing methods despite computing the reordering in an incremental way from a random presentation of the data. In this section we will show the capability of our algorithm to visualize the dynamic structure of an evolving data-set.

We generated two data-sets of 10, 000 two-dimensional vectors with a predefined temporal organization (Figs. 2 and 3). The number, the shape and the position of clusters varies over time. Figures 2 and 3 shows a visualization of the data-set's dynamic and the corresponding visualization of the data structure with the proposed algorithm. All similarities are computed using the Euclidean distance. We have chosen to keep the matrix size to 1000 data points, in order to have a representation of the "current" structure of the data for ten periods of time. Note that the ten matrix visualizations presented in Figs. 2 and 3 are "snapshots" of the same matrix at different moment of the process. Each of the 10000 data points is presented once to the algorithm, in an order randomly chosen to follow the data-set's distributions dynamic (i.e. from distributions that vary over time).

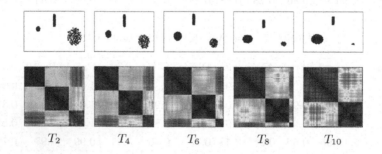

T_2 T_4 T_6 T_8 T_{10}

Fig. 2. Example of visualization obtained using the proposed incremental reordering method.

One can observe that the clusters are perfectly represented in the matrix. The emergence and disappearance of clusters is clearly visible, as well as the graduate shift in size and positions. In addition, the progressive split of one cluster into two new clusters in Fig. 3 can be seen in the matrix from one big cluster up to two well separated smaller clusters.

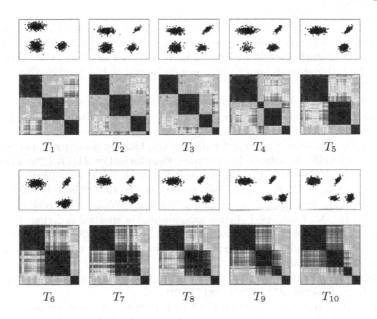

Fig. 3. Example of visualization obtained using the proposed incremental reordering method.

4   Conclusion

In this paper, we propose an incremental approach to matrix reordering to visualize the structure of static and dynamic data sets. Our algorithm performs very well on static data sets in comparison to the state-of-the-art algorithms: its quality remains comparable or higher, despite working in an incremental way from a random presentation of the data, while the other algorithms can use any information about the similarity matrix at any time. In addition, the algorithm assures a reasonable memory and computation cost for dynamic data sets, and the visualizations show that our approach is perfectly suitable to detecting temporal variations in the data structure, such as changes in density, appearance and disappearance of clusters of data or changes in similarities between clusters. Our future work will be to test this approach on real applications in order to visualize the dynamics of complex data stream. We are currently working on Tweet analysis and we wish to produce useful visualizations on the dynamic structure of Tweets over time.

References

1. Bar-Joseph, Z., Gifford, D.K., Jaakkola, T.S.: Fast optimal leaf ordering for hierarchical clustering. Bioinformatics **17**(Suppl. 1), S22–S29 (2001)
2. Barnard, S.T., Pothen, A., Simon, H.D.: A spectral algorithm for envelope reduction of sparse matrices. In: Proceedings of the 1993 ACM/IEEE Conference on Supercomputing 1993, pp. 493–502. ACM, New York (1993)

3. Behrisch, M., Bach, B., Riche, N.H., Schreck, T., Fekete, J.D.: Matrix reordering methods for table and network visualization. In: Computer Graphics Forum (2016)
4. Bezdek, J.C., Hathaway, R.J.: VAT: a tool for visual assessment of (cluster) tendency. In: International Joint Conference on Neural Networks (IJCNN), vol. 3, pp. 2225–2230 (2002)
5. Buja, A., Swayne, D.F., Littman, M.L., Dean, N., Hofmann, H., Chen, L.: Data visualization with multidimensional scaling. J. Comput. Graph. Stat. **17**(2), 444–472 (2008)
6. Caraux, G., Pinloche, S.: Permutmatrix: a graphical environment to arrange gene expression profiles in optimal linear order. Bioinformatics **21**(7), 1280–1281 (2005)
7. Chen, C.H., Hrdle, W., Unwin, A.: Handbook of Data Visualization, 1st edn. Springer-Verlag TELOS, Santa Clara (2008). doi:10.1007/978-3-540-33037-0
8. Conover, W.J.: Practical Nonparametric Statistics. Wiley, New York (1981)
9. Ding, C., He, X.: Linearized cluster assignment via spectral ordering. In: International Conference on Machine Learning, New York, NY, USA, p. 30 (2004)
10. Gama, J.: Knowledge Discovery from Data Streams, 1st edn. Chapman & Hall/CRC, Boca Raton (2010)
11. Gruvaeus, G., Wainer, H.: Two additions to hierarchical cluster analysis. Br. J. Math. Stat. Psychol. **25**(2), 200–206 (1972)
12. Hahsler, M., Hornik, K., Buchta, C.: Getting things in order: an introduction to the R package seriation. J. Stat. Softw. **25**(3), 1–34 (2008)
13. Han, J., Kamber, M.: Data Mining: Concepts and Techniques, 2nd edn., USA (2006)
14. Liiv, I.: Seriation and matrix reordering methods: an historical overview. Stat. Anal. Data Mining **3**(2), 70–91 (2010)
15. Mount, D.W.: Sequence and genome analysis. Bioinformatics: Cold Spring Harbour Laboratory Press: Cold Spring Harbour 2 (2004)
16. Rastin, P., Matei, B., Cabanes, G., El Baghdadi, I.: Signal-based autonomous clustering for relational data. In: International Joint Conference on Neural Networks, IJCNN 2017 (2017)
17. Rosenkrantz, D.J., Stearns, R.E., Lewis II, P.M.: An analysis of several heuristics for the traveling salesman problem. SIAM J. Comput. **6**(3), 563–581 (1977)

Power Consumption Prediction for Dynamic Adjustment in Hydrocracking Process Based on State Transition Algorithm and Support Vector Machine

Xiao-Fang Chen, Ying-Can Qian, and Ya-Lin Wang[✉]

School of Information Science and Engineering,
Central South University, Changsha 410083, China
ylwang@csu.edu.cn

Abstract. Power consumption is an important part of energy consumption in hydrocracking, which occupies about 43%–47% of the total energy consumption. In the daily production management, the real-time power consumption is manually recorded from the voltmeter. However, it is difficult to collect the power consumption especially in the dynamic adjustment. In this paper, a power consumption prediction model is proposed for dynamic adjustment in the hydrocracking process, which is based on state transition algorithm (STA) and support vector machine (SVM). A SVM regression model is developed to map the complex nonlinear relationship between power parameters and the power consumption in the dynamic adjustment of hydrocracking, and the state transition algorithm is used to optimize the parameters of SVM regression model. The experimental results demonstrate that the prediction accuracy of the model is close to the fitting accuracy and the modeling time is reduced.

Keywords: Hydrocracking · Dynamic adjustment · Power consumption prediction model · State transition algorithm-support vector machine (STA-SVM)

1 Introduction

Hydrocracking is one of the most important modern technology for deep processing of heavy oil refining, which can deal with numerous of raw materials and produce products of high quality.

To construct an intelligent decision system for dynamic adjustment in hydrocracking process, it is very important to evaluate the dynamic adjustment from historical dataset with performance of adjustment time, stability and energy consumption. Power consumption is an important part of energy consumption in hydrocracking, which occupy about 43%–47% of the total energy consumption. However, it is very difficult to measure and record the power consumption, especially during the dynamic adjustment of the hydrocracking process. This is

© Springer International Publishing AG 2017
D. Liu et al. (Eds.): ICONIP 2017, Part V, LNCS 10638, pp. 85–94, 2017.
https://doi.org/10.1007/978-3-319-70139-4_9

mainly because that the power consumption should be manually recorded from the voltmeter. Hence, it is necessary to establish a power consumption prediction model for dynamic adjustment.

So far, some soft computing techniques, such as artificial neural network (ANN), genetic algorithm and art colony optimization, autoregressive integrated moving average model (AIMAM), generalized autoregressive conditional heteroskedasticity (GARCH) and its extended models, have been used to estimate the energy consumption. Garshasbi [1] presented a hybrid genetic algorithm and Monte Carlo simulation approach to predict hourly energy consumption. Kalogirou [2] used the artificial neural networks to predict the energy consumption of a passive solar building. Hu [3] predicted electricity consumption by using a neural-network-based grey forecasting approach. The artificial neural network method is suitable for modeling of complex processes, especially when the data samples are sufficient. The model can describe the nonlinearity of the solving problem. However, the training time of this kind of model is long and it is prone to get to local optimum value in the space due to the phenomenon of "over fitting" or "under fitting". Meanwhile, the generalization ability of the model will be reduced.

In the last decades, SVM, developed by Vapnik et al. [4], has been applied to solve some pattern recognition and regression problems. These SVM algorithms have a solid theoretical foundation rooted in statistical learning theory and has been applied to many fields, including energy consumption prediction. Xing [5] established a prediction model of energy consumption on beer enterprise based on support vector machine. Zhang et al. [6] used hybrid algorithm of SVM and PSO on energy consumption prediction in iron making process.

Because of the limited data samples for modeling, in this paper, a kind of power consumption estimation model was established based on support vector machine (SVM) method. Then, we proposed a method for searching the optimal parameters of SVM based on the state transition algorithm which is an efficient approach for parameters selection of SVM. At the same time, we select the main factors, which will affect the energy consumption, by correlation analysis. It can offer high capability in choosing inputs that are relevant in predicting dependent variables. And the sampled data of power consumption parameters were regrouped and optimized through K-fold cross-validation method. The experimental results demonstrate that the prediction accuracy of the model is close to the fitting accuracy. Meanwhile, the generalization ability and stability of the model are improved, and the learning time is reduced.

2 STA-SVM Model

2.1 State Transition Algorithm: Background [7, 8]

State transition algorithm is a kind of new intelligent optimization algorithm based on the idea of state transition space, which was proposed by Zhou et al. 2011. The detailed form of the state transition process is shown as follows:

$$\begin{cases} x_{k+1} = A_k x_k + B_k u_k \\ y_k = f(x_{k+1}) \end{cases} \tag{1}$$

where $x_k \in R^n$ denotes a state, corresponding to a solution to the optimization problem, A_k and B_k are the state transition matrixes, u_k is a function related to state x_k and historical state, and f is a fitness function. To solve the continuous optimization problems, four special transformation operators are defined.

(1) Rotation transformation

$$x_{k+1} = x_k + \alpha \frac{1}{n \, \|x_k\|_2} R_r x_k \tag{2}$$

where α is a positive constant called the rotation factor; $R_r \in R^{n \times n}$ is a random matrix with its entries to the range of $[-1, 1]$; and $\|\cdot\|_2$ is the 2-norm of a vector.

(2) Translation transformation

$$x_{k+1} = x_k + \beta R_t \frac{x_k - x_{k-1}}{n \, \|x_k - x_{k-1}\|_2} \tag{3}$$

where β is a positive constant called the translation factor and $R_t \in R^{n \times n}$ is a random variable with its components in the range of $[0, 1]$.

(3) Expansion transformation

$$x_{k+1} = x_k + \gamma R_r x_k \tag{4}$$

where γ is a positive constant called the translation factor and $R_e \in R^{n \times n}$ is a random diagonal matrix with its elements obeying Gaussian distribution.

(4) Axesion transformation

$$x_{k+1} = x_k + \delta R_a x_k \tag{5}$$

where δ is a positive constant called the translation factor and $R_a \in R^{n \times n}$ is a random diagonal matrix with its entries obeying Gaussian distribution and only one random index having a nonzero value.

2.2   STA-SVM Based Predictive Framework

The optimization of parameters is an indispensable part of the modeling. Based on the given training set, testing set and kernel function, we should select the appropriate parameters to train the model. The main parameters of the model is the penalty coefficient C and the width of the kernel function σ. In this paper, we use the state transition algorithm to select parameters of the forecasting SVM model. To get the optimal parameters, the root-mean-square error (RMSE) criterion is applied on the cross validation set, which is defined as

$$RMSE_{CV} = \sqrt{\frac{\sum_{i=1}^{N_{CV}} (\hat{y}_i - y_i)^2}{N_{CV}}} \tag{6}$$

where N_{CV} is the number of samples in the cross validation set, $\hat{y}_i$ and y_i are estimated values and real values of the i_{th} training sample.

The operation procedure of the SVM parameters optimization based on STA are as follows

Step 1. Initialize the parameters of the algorithm, the number of search individuals of state transition algorithm SE and the value of four operating factor α, β, γ, and δ, set the number of iterations N and search range, and make the current iteration number $nm = 1$.

Step 2. According to the SVM model, the optimization objective function is established, and then calculate the fitness value of each individual.

Step 3. The optimal individual (best) of the initial population are obtained.

Step 4. Carry out the state transition of the relevant operations.

Step 5. Determine whether the iteration number reaches the maximum value, if it is out of circulation, output results; otherwise go back to **Step 2**.

The state transition algorithm integrated in this approach is depicted in the Fig. 1.

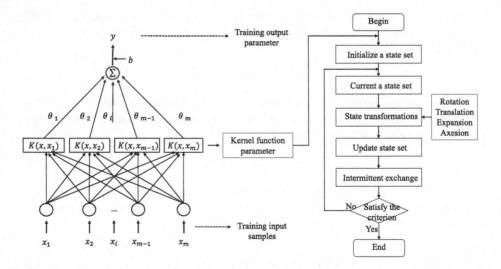

Fig. 1. Using STA to select parameters for building up the forecasting SVM model

3   Power Consumption Estimation

The power consumption during the dynamic adjustment of hydrocracking is mainly in hydrogen compressors, pumps and air cooler. Generally the power consumption of the raw oil pump and the new hydrogen compressor occupies a large proportion of the total power consumption.

In this paper, we take the power consumption during the dynamic adjustment of hydrocracking as the object of study. There are ten factors of the power consumption, including cumulative inlet amount during dynamic adjustment, the Cold hydrogen consumption of each bed in the reactor, circulating hydrogen consumption in the reaction system, hydrogen production of new hydrogen compressor and the purity of hydrogen. The output of this model is power consumption during dynamic adjustment hydrocracking. Based on the data set: $\{x_k, y_k\}_{k=1}^{N}$, we establish the prediction model $y = f(x)$ and make it have good generalization ability.

3.1 Select the Main Influence Parameters of Power Consumption

In the adjustment process, there are many factors will affect the energy consumption, but for modeling, there are not all parameters must be taken into consideration. In fact, some parameters are not the main cause of qualitative analysis. Therefore, analysing these parameters will be good to further simplify the model structure and highlight the main influencing factors.

In this paper, we select the main factors, which will affect the energy consumption, by Grey Relational Analysis [9]. The Table 1 shows the power parameters in the power consumption. The results of grey correlation matrix are displayed in Table 2. From Table 2, we can see that there are two input variables of the prediction model, including cumulative inlet amount in dynamic adjustment (x_1) and hydrogen production of new hydrogen compressor in dynamic adjustment (x_8).

Table 1. The power parameters in the power consumption

Variables	Variable description
x_1	Cumulative inlet amount
x_2	The Cold hydrogen consumption in the second bed in hydrofining reactor
x_3	The Cold hydrogen consumption in the third bed in hydrofining reactor
x_4	The Cold hydrogen consumption in the first bed in hydrocracking reactor
x_5	The Cold hydrogen consumption in the second bed in hydrocracking reactor
x_6	The Cold hydrogen consumption in the third bed in hydrocracking reactor
x_7	The Cold hydrogen consumption in the last bed in hydrocracking reactor
x_8	Hydrogen production of new hydrogen compressor

Table 2. The result of grey relation analysis for the power parameters

Variables	x_1	x_2	x_3	x_4	x_5	x_6	x_7	x_8
Correlation	0.8170	0.6565	0.7419	0.7583	0.7321	0.7126	0.7353	0.8297

3.2   Selection of Kernel Function

The radial basic function (RBF) kernel formulated is used in this study due to its efficiency compared with other kernel functions (linear, polynomial, and sigmoid). The linear kernel is outperformed by the remaining kernels because of its linearly limited capability. Polynomial kernels are acceptable but if a high degree is used, numerical difficulties tend to happen [10]. The sigmoid kernel matrix may not be positive definite and in general its accuracy is not better than RBF [11].

3.3   Estimate Generalization Error by Cross Validation

In order to extract the valid information of the sample set, in this paper, K-fold cross-validation [12] was used to determine the modeling samples. For model training, parameters optimization and model testing, the data set is divided into two parts: training data set and testing data set. First, we divided the training sample set into K parts randomly, which are mutually disjoint subsets and have equal size. Then, take turns to use k−1 parts of them to train the model, and the remainder is used for the model testing. At the same time, the training set is applied to the fitness function calculation of STA. In this process, there are k training sessions of STA evolution and k classifiers will be established. At last, the test recognition rate of K-CV is the average value of the K classifiers recognition rate of the K group test sets corresponding to the STA training.

4   Case Study

In this section, power consumption in dynamic adjustment of hydrocracking is provided to compare the prediction of the three regression methods.

4.1   Description of the Hydrocracking

Hydrocracking process is a chemical process, which can converts the heavy oil to light oil (e.g. gasoline, kerosene, diesel or the raw material of olefin by catalytic cracking and cracking) by hydrogenation, cracking and isomerization reactions under the condition of high temperature, high pressure and catalyst. A flow-chart of this process is given in Fig. 2. For this prediction model, the output of this model is power consumption during dynamic adjustment hydrocracking. There are two input variables of the prediction model, including cumulative inlet amount during dynamic adjustment and hydrogen production of new hydrogen compressor during dynamic adjustment.

A total of 107 data samples have been collected in this process. For model training, parameter optimization and model testing, the data set is divided into two parts: 80 data for historical data set, 27 data for testing data set. In order to examine the data characteristic, Fig. 3 shows the trend plots of the normalized input and output variables for the training data. The two figure, which located

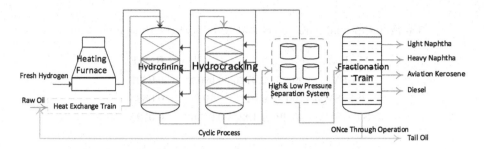

Fig. 2. Flowchart of the hydrocracking model

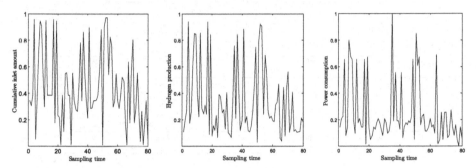

Fig. 3. The trend plots of the normalized input and output variables for the training data

in left and middle, show the trends of input variables, and the right one is the characteristic of the output variable. From these subfigures, it is easy to see the nonlinear relations between the output variable and input variables.

4.2 Results of the Prediction Model

First, the prediction models are built by using different combination of parameters involving optimization in each algorithm (e.g. GA, PSO, STA). Then the RMSEs are calculated and the optimal parameters that minimize the RMSEs are obtained. At last, the models with optimal parameter are used online to predict the output variable of query samples.

For the modeling results, Table 3 shows the selected optimal parameters and the corresponding RMSEs on the training set and the testing set.

According to the RMSE criteria, the RMSE value of the STA-SVM is the smallest and that of the PSO-SVM is the largest both in the cross training set and testing set. More detailed comparison of the quality prediction results in the testing set of the three prediction methods is shown in Fig. 4. Furthermore, the prediction errors of these three methods are depicted in Fig. 5. It is easy to see that the prediction of the STA-SVM matches well with the actual measurement of the power consumption during dynamic adjustment hydrocracking

Table 3. Optimal parameters and RMSEs of the three algorithm

Method	c	g	RMSE$_{CV}$	RMSE$_T$
GA-SVM	99.38	2.3720	0.0963	0.0801
PSO-SVM	100	0.01	0.1036	0.0868
STA-SVM	95.40	0.1925	0.0836	0.0795

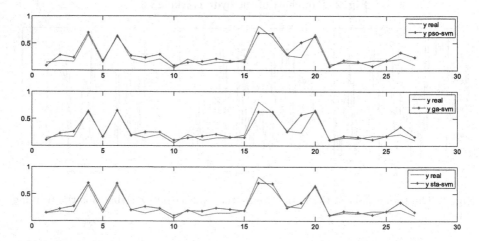

Fig. 4. The trend plots of the normalized input and output variables for the training data

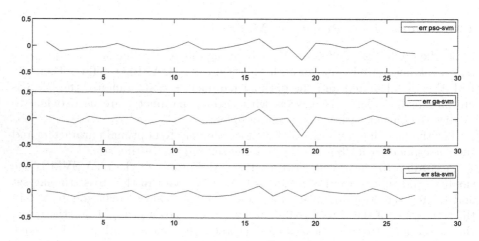

Fig. 5. The prediction error of these three methods

while the PSO-SVM leads to poor prediction results. There are significant deviations between the actual and predicted values across the whole process using GA-SVM. All the results indicate the superiority of the presented STA-SVM method in power consumption during dynamic adjustment hydrocracking.

5 Conclusion

In this paper, the analysis of power consumption in dynamic adjustment of hydrocracking is provided so as to modeling the prediction problem. The STA algorithm is introduced to improve the performance of SVM by providing optimal kernel parameters. We select the main factors, which will affect the energy consumption, by Grey Relational Analysis, and the sampled data of power consumption parameters were regrouped and optimized through K-fold cross-validation method. The experimental results of the practical data from the dynamic adjustment of hydrocracking process shows that the STA-SVM approach is of higher prediction accuracy and reliability than GA-SVM and PSO-SVM. Thus, the presented prediction model of power consumption can be implemented for energy consumption prediction in other processes.

Acknowledgments. This work is supported by the Major Program of the National Natural Science Foundation of China (61590921); the Program of the National Natural Science Foundation of China (61374156); the Fundamental Research Funds for the Central Universities of Central South University (2017zzts707).

References

1. Garshasbi, S., Kurnitski, J., Mohammadi, Y.: A hybrid genetic algorithm and monte carlo simulation approach to predict hourly energy consumption and generation by a cluster of net zero energy buildings. J. Appl. Energy **179**, 626–637 (2016)
2. Kalogirou, S.A., Bojic, M.: Artificial neural networks for the prediction of the energy consumption of a passive solar building. J. Energy **25**(5), 479–491 (2000)
3. Hu, Y.C.: Electricity consumption prediction using a neural-network-based grey forecasting approach. J. Oper. Res. Soc. **68**(10), 1–6 (2016)
4. Vapnik, V., Golowich, S.E., Smola, A.: Support vector method for function approximation, regression estimation, and signal processing. J. Adv. Neural Inf. Process. Syst. **9**, 281–287 (1996)
5. Xing, J., Haiwei, W.U.: Prediction model of energy consumption on beer enterprise based on support vector machine. J. Jilin Univ. (2014)
6. Zhang, Y., Zhang, X., Tang, L.: Energy consumption prediction in ironmaking process using hybrid algorithm of SVM and PSO. In: Wang, J., Yen, G.G., Polycarpou, M.M. (eds.) ISNN 2012. LNCS, vol. 7368, pp. 594–600. Springer, Heidelberg (2012). doi:10.1007/978-3-642-31362-2_65
7. Zhou, X., Yang, C., Gui, W.: State transition algorithm. J. Ind. Manag. Optim. **8**(4), 1039–1056 (2012)
8. Zhou, X., Yang, C., Gui, W.: A new transformation into state transition algorithm for finding the global minimum. In: International Conference on Intelligent Control and Information Processing, vol. 170, pp. 674–678. IEEE (2012)

9. Hashemi, S.H., Karimi, A., Tavana, M.: An integrated green supplier selection approach with analytic network process and improved grey relational analysis. Int. J. Prod. Econ. **159**(159), 178–191 (2015)
10. Keerthi, S.S., Lin, C.J.: Asymptotic behaviors of support vector machines with Gaussian kernel. J. Neural Comput. **15**(7), 1667–1689 (2003)
11. Lin, H.T., Lin, C.J.: A study on sigmoid kernels for SVM and the training of non-PSD kernels by SMO-type methods. J. Neural Comput. **27**(1), 15–23 (2005)
12. Moreno-Torres, J.G., Sez, J.A., Herrera, F.: Study on the impact of partition-induced dataset shift on k-fold cross-validation. IEEE Trans. Neural Netw. Learn. Syst. **23**(8), 1304–1312 (2012)

Learning with Partially Shared Features for Multi-Task Learning

Cheng Liu, Wen-Ming Cao, Chu-Tao Zheng, and Hau-San Wong[✉]

Department of Computer Science, City University of Hong Kong,
Kowloon Tong, Hong Kong
{cliu272-c,wenmincao2-c,ctzheng2-c}@my.cityu.edu.hk,
cshswong@cityu.edu.hk

Abstract. The objective of Multi-Task Learning (MTL) is to boost learning performance by simultaneously learning multiple relevant tasks. Identifying and modeling the task relationship is essential for multi-task learning. Most previous works assume that related tasks have common shared structure. However, this assumption is too restrictive. In some real-world applications, relevant tasks are partially sharing knowledge at the feature level. In other words, the relevant features of related tasks can partially overlap. In this paper, we propose a new MTL approach to exploit this partial relationship of tasks, which is able to selectively exploit shared information across the tasks while produce a task-specific sparse pattern for each task. Therefore, this increased flexibility is able to model the complex structure among tasks. An efficient alternating optimization has been developed to optimize the model. We perform experimental studies on real world data and the results demonstrate that the proposed method significantly improves learning performance by simultaneously exploiting the partial relationship across tasks at the feature level.

Keywords: Multi-Task Learning · Partially task relationship

1 Introduction

Different from standard machine learning methods in which each task is learnt individually, the objective of multi-task learning (MTL) [2] is to boost learning performance by discovering the relationships among tasks. This learning framework has been successfully applied to many domains: bioinformatics [8, 10], computer vision [12, 13], speech recognition [16] and so on.

Various MTL techniques have been developed to jointly learn multiple tasks. A large number of MTL methods assume that all tasks share common structures and the related parameters are close to each other. For example, Obozinski et al. [15] proposed $\ell_{2,1}$ norm to select a common set of features for all tasks, and Regularized MTL [4] encourages the model parameters of all tasks to be close to each other. In addition, the algorithms in [1] assume that the parameters of all tasks lie in a low dimensional space. Such strong assumptions do not hold in some real-world

© Springer International Publishing AG 2017
D. Liu et al. (Eds.): ICONIP 2017, Part V, LNCS 10638, pp. 95–104, 2017.
https://doi.org/10.1007/978-3-319-70139-4_10

applications, and sharing information with unrelated tasks may degrade the learning performance. To address this issue, some MTL methods assume that there exist outlier tasks which are irrelevant to other tasks. For example, robust multi-task learning (RMTL) [3] is able to simultaneously identify the low-dimensional structure and outliers among multiple tasks. Robust multi-task feature learning (rMTFL) [5] is able to simultaneously identify outlier tasks and select a common feature space. Other widely studied MTL algorithms are the task clustering methods. Jacob et al. [6] proposed a clustered multi-task learning method (CMTL) which assigns all tasks into different disjoint groups based on a regularization function. The approach proposed in [7] assigned all tasks into different disjoint groups based on integer programming, while assumed that the tasks lie in a low dimensional subspace within each group.

Fig. 1. Some examples of MNIST data.

However, relevant tasks are partially sharing information in some real-world applications. For example, in handwritten digits data application, there exist some similarities among the shapes of different digits. As an example of MNIST data shown in Fig. 1, the pairs of digits {6, 8}, {8, 9}, {1, 7} and {1, 4} have similar structures, and the tasks of learning the characteristics of these digit pairs are partially related. This indicates that the discriminative features of related tasks can partially overlap. The aforementioned methods are difficult to capture the flexible structure among tasks. Instead of assuming that the groups of tasks are disjoint, GOMTL [9] and SCMTL [14] allow partial overlap between the task clusters. Specifically, these models decompose the coefficient matrix into two matrices: one matrix represents the bases of latent tasks, and another one includes the associated linear combination coefficients for each task. Figure 2 provides an illustration of GOMTL and SCMTL. Therefore, any two tasks are allowed to overlap in one or more bases, and the partially shared information between two tasks can be exploited through their combination coefficients.

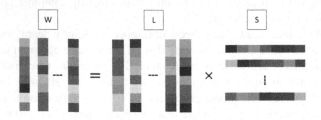

Fig. 2. Illustration of GOMTL and SCMTL

However, both GOMTL and SCMTL exploit these relationships *at the task level* and cannot capture more complex task structures. Moreover, these models cannot yield sparse solution and this may be too restrictive in some real-world applications, especially in a high-dimensional setting. In this study, we develop a new MTL approach to exploit partial tasks relationship *at the feature level*, which is able to selectively exploit shared information across the tasks while produce a task-specific sparse solution for each task. In particular, this new formulation assumes that the coefficients of each task can be decomposed into two components, one for capturing shared knowledge among tasks, and the other for generating a sparse pattern and selectively exploiting the shared information. Figure 3(a) provides an illustration of the proposed method. The coefficients U are able to capture shared information across tasks, while the coefficients V are for producing a sparse solution for each task.

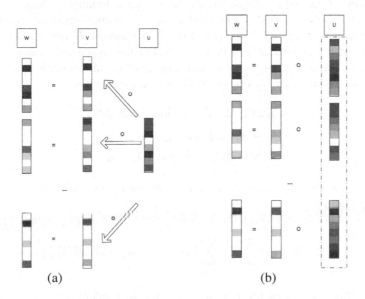

(a)   (b)

Fig. 3. Illustration of the proposed method

2   The Proposed Model

Suppose we have m tasks and the corresponding datasets are $\mathcal{D} = \{(X^1, Y^1), (X^2, Y^2), ..., (X^m, Y^m)\}$, where $X^i = \left[x_1^i, x_2^i, ..., x_{n_i}^i\right] \in \mathbb{R}^{n_i \times d}$ is the data matrix for the ith task with n_i samples and d features, while the corresponding vector $Y^i = \left[y_1^i, y_2^i, ..., y_{n_i}^i\right] \in \{-1, 1\}$ is for binary classification problem and $Y^i = \left[y_1^i, y_2^i, ..., y_{n_i}^i\right] \in \mathbb{R}$ is for regression task. Defining the coefficient matrix as $= \{w^1, w^2, ..., w^m\} \in \mathbb{R}^{d \times m}$, the standard regularized multi-task learning model can be formulated as:

$$min_{W,b}\, \mathcal{L}(\mathcal{D}, W, c) + \lambda \mathcal{R}(W) \tag{1}$$

where $\mathcal{L}(W, \mathcal{D})$ is the loss function and $\mathcal{R}(W)$ is a regularization term. As shown in Fig. 3(a), we expect that the coefficient vector of each task can simultaneously identify the common structure for all tasks and the individual sparse solution for each task. Thus, we can decompose the coefficient matrix W into two components: V (sparse set of coefficients for each task) and u (common shared coefficients for all tasks), and each element of the coefficient matrix W can be expressed as follows:

$$w_{kj} = u_j v_{kj}, k = 1, 2, \dots, m; j = 1, 2, \dots, d$$

However, utilizing a single common vector to capture the shared information across all tasks may too restrictive in some applications. Instead of exploiting shared information across all tasks by a single vector, we use an individual coefficient vector for each task, and all the vectors are expected to share a common structure. Therefore, the coefficient matrix W can be decomposed into two matrices: U and V. We apply the trace norm on the coefficient matrix U to identify the shared information across all tasks, and we use the l1 norm penalty to characterize the task-specific sparse solution for each task. Formally, the proposed MTL can be modeled as:

$$\min_{W,b} \mathcal{L}(\mathcal{D}, W, C) + \lambda_1 \|U\|_* + \lambda_2 \|V\|_{1,1}$$
$$s.t.\ W = U \circ V : w_{kj} = u_{kj} v_{kj} \tag{2}$$
$$k = 1, 2, \dots, m; j = 1, 2, \dots, d$$

where λ_1, λ_2 are tuning parameters, and the following loss function $\mathcal{L}(\mathcal{D}, W, C)$ is adopted in this study:

$$\mathcal{L}(\mathcal{D}, W, c) = \sum_{k=1}^{m} \frac{1}{n_k} \sum_{i=1}^{n_k} \log \left(1 + \exp\left(-y_i^k \left(x_i^k W_k\right)\right)\right) \tag{3}$$

Figure 3(b) provides an illustration of the proposed model (2).

3   Optimization

The proposed model (2) is non-convex, and we develop an alternating convex optimization process to solve the problem, where we iteratively optimize one of the two components, while fixing the other. Specifically, the proximal method was employed to estimate each component. The details of the optimization procedure are as follows:

Optimizing U with fixed L: Fixing the component L, the objective function (2) can be expressed with respect to component U as follows:

$$min_U \, \mathcal{Q}(U) + \lambda_1 \mathcal{R}(U) \tag{4}$$

where $\mathcal{Q}(U) = \mathcal{L}(U \circ V(t-1), X)$ and $\mathcal{R}(U) = \|U\|_*$ .The optimization problem (4) can be solved by the PG (proximal gradient descent) method, based on the following quadratic approximation in each iteration:

$$
\begin{aligned}
U(t+1) = min_U \mathcal{Q}(U(t)) &+ \langle (U(t) - U(t+1)), \nabla \mathcal{Q}(U(t)) \rangle \\
&+ \frac{\gamma}{2} \|U(t) - U(t+1)\|_F^2 + \lambda_1 \mathcal{R}(U)
\end{aligned} \tag{5}
$$

where γ denotes the step length (estimated by the line search method). Let $\Theta = \left(U(t) - \frac{1}{\gamma} \nabla \mathcal{Q}(U(t)) \right)$, and Eq. (5) can be further expressed as:

$$U(t+1) = min_P \frac{1}{2} \|U - \Theta\|_2^2 + \frac{\lambda_1}{\gamma} \|U\|_* \tag{6}$$

The following theorem [3] provides the closed form solution for $U(t+1)$:

Theorem 1. *The singular value decomposition of Θ can be defined as follows:*

$$\Theta = \mathcal{J} \Sigma \mathcal{K}^T \text{ and } \Sigma = \text{diag}(\{\sigma_i\}_{i=1}^r)\text{-}$$

where $r = rank(\Theta)$, $\mathcal{J} = \mathbb{R}^{d \times r}$, $\mathcal{K} = \mathbb{R}^{m \times r}$ and $\{\sigma_i\}_{i=1}^r$ denotes the singular values. Therefore the closed form solution for (6) is given by:
0

$$U(t+1) = \mathcal{J} \text{diag}\left(\left\{ \sigma_i - \frac{\lambda_1}{2\gamma} \right\}_+ \right) \mathcal{K}^T \tag{7}$$

where $\{a\}_+ = max(a, 0)$.

Optimizing L with Fixed U. For fixed U, we can also use PG to estimate L. Define $\mathcal{Q}(V) = \mathcal{L}(U(t) \circ V, X)$ and $\mathcal{R}(V) = \|V\|_{1,1}$, the optimization problem for component L can be modeled as:

$$V(t+1) = min_P \frac{1}{2} \|V - \Phi\|_2^2 + \frac{\lambda_2}{\gamma} \|V\|_{1,1} \tag{8}$$

where $= V(t) - \frac{1}{\gamma} \nabla \mathcal{Q}(V(t))$. We consider the following closed form function to update (8):

$$v_{kj} = \text{sign}(\varphi_{kj}) \left(\varphi_{kj} - \frac{\lambda_2}{\gamma} \right)_+ \quad j = 1, 2, \ldots, d; k = 1, 2, \ldots, m$$

The overall alternating optimization procedure for problem (2) is given in Algorithm 1.

Algorithm 1: Alternating Optimization Algorithm

Input: $X, Y, \lambda_1, \lambda_2$
Initialize: $t = 0, q_{kj}(0) = 1, p_{kj}(0) = 0.$
$(k = 1,2, \ldots, m; j = 1,2, \ldots, d)$

Repeat: $t = 0,1, \ldots,$
Step 1(estimate U): optimize $U(t)$ with fixed $V(t-1)$ as follows:

$$U(t+1) = \mathcal{J} \text{diag}\left(\left\{\sigma_i - \frac{\lambda_1}{2\gamma}\right\}_+\right) \mathcal{K}^T$$

Step 2 (solve for V): optimize $V(t)$ with fixed $U(t)$ as follows:

$$v_{kj} = \text{sign}(\varphi_{kj})\left(\varphi_{kj} - \frac{\lambda_2}{\gamma}\right)_+ \quad j = 1,2, \ldots, d; k = 1,2, \ldots, m$$

Step 3: set $W(t) = U(t) \circ V(t)$
Until the convergence of W
Output: $W = W(t+1)$

---

4   Experiments

In this section, we evaluate the performance of the proposed MTL model on a number of real-world image data sets. The competing multi-task learning models used in this study include: the $\ell_{1,1}$ norm [17], the $\ell_{2,1}$ norm [15], MTFL [1], CMTL [6], GOMTL [9] and AMTL [11] algorithms. We use 5-fold cross-validation (CV) to select the values for λ_1 and λ_2, and compute solutions based on the parameter range from [10^{-5}, 10]. In addition, for CMTL, the number of task clusters is pre-defined. For all experiments, we perform cross-validation over the values $\{1, 2, \ldots, m\}$ to select the optimal number of clusters.

In this study, we focus on multi-task classification problems based on the logistic loss (3), and several image classification data sets are used in our experimental studies: JAFFE, ORL, Yale, MNIST, USPS, AR and PIE. Following previous studies [7], we transform the multi-class data into a multi-task form. Specifically, we consider each one-vs-all classification problem as a single task. In this experiment, we use 30% of the data for training, and the rest for testing. For each setting, we perform the experiments 30 times. To evaluate the performance of the multi-task classification model, we use AUC and the Youden index (YI).

Table 1. The learning performances (± standard deviation) of 7 different MTL methods for the image data sets (the best performance is indicated in bold).

		$\ell_{1,1}$	$\ell_{2,1}$	MTFL	CMTL	GOMTL	AMTL	Proposed
AR	YI	0.719 ± 0.04	0.635 ± 0.06	0.559 ± 0.03	0.601 ± 0.06	0.617 ± 0.06	0.629 ± 0.05	**0.741 ± 0.04**
	AUC	0.861 ± 0.03	0.821 ± 0.04	0.788 ± 0.02	0.804 ± 0.04	0.811 ± 0.03	0.819 ± 0.04	**0.882 ± 0.03**
PIE	YI	0.943 ± 0.04	0.920 ± 0.05	0.847 ± 0.05	0.874 ± 0.05	0.883 ± 0.05	0.891 ± 0.05	**0.948 ± 0.03**
	AUC	0.972 ± 0.02	0.958 ± 0.03	0.929 ± 0.03	0.937 ± 0.03	0.944 ± 0.03	0.955 ± 0.03	**0.976 ± 0.01**
JAFFE	YI	0.406 ± 0.07	0.439 ± 0.07	0.284 ± 0.07	0.204 ± 0.05	0.278 ± 0.07	0.281 ± 0.07	**0.464 ± 0.07**
	AUC	0.703 ± 0.05	0.722 ± 0.04	0.643 ± 0.04	0.601 ± 0.04	0.638 ± 0.04	0.642 ± 0.04	**0.735 ± 0.05**
ORL	YI	0.823 ± 0.03	0.786 ± 0.04	0.713 ± 0.05	0.666 ± 0.03	0.679 ± 0.04	0.693 ± 0.03	**0.831 ± 0.03**
	AUC	0.895 ± 0.02	0.883 ± 0.02	0.856 ± 0.03	0.815 ± 0.01	0.829 ± 0.02	0.847 ± 0.02	**0.907 ± 0.02**
Yale	YI	0.620 ± 0.02	0.558 ± 0.02	0.367 ± 0.03	0.494 ± 0.03	0.477 ± 0.03	0.489 ± 0.03	**0.652 ± 0.02**
	AUC	0.823 ± 0.02	0.787 ± 0.02	0.683 ± 0.03	0.724 ± 0.02	0.706 ± 0.02	0.711 ± 0.02	**0.837 ± 0.02**
MNIST	YI	0.639 ± 0.02	0.651 ± 0.02	0.640 ± 0.02	0.601 ± 0.02	0.632 ± 0.03	0.648 ± 0.02	**0.672 ± 0.02**
	AUC	0.817 ± 0.02	0.841 ± 0.02	0.831 ± 0.02	0.816 ± 0.01	0.822 ± 0.02	0.837 ± 0.02	**0.847 ± 0.02**
USPS	YI	0.692 ± 0.02	0.681 ± 0.01	0.634 ± 0.01	0.682 ± 0.02	0.673 ± 0.02	0.688 ± 0.02	**0.705 ± 0.02**
	AUC	0.840 ± 0.01	0.832 ± 0.01	0.806 ± 0.01	0.829 ± 0.02	0.812 ± 0.02	0.835 ± 0.01	**0.851 ± 0.01**

Table 1 shows the learning performances of the different methods on the seven image data sets in terms of AUC and Youden Index value. From this table, we have the following observations:

1. The proposed method achieves the best learning performance among various methods. In these image applications, the related categories only have a small number of shared features. Exploring the partial relationship of tasks at the feature level will improve the learning performance.
2. MTFL assumes that all tasks share a common structure, which is too restrictive and fails to capture the correct structure in these applications. The learning performance of MTFL is even less satisfactory than the single task method (the $\ell_{1,1}$ norm).
3. The performances of sparse models (the proposed model, the $\ell_{1,1}$ norm and the $\ell_{2,1}$ norm) are better than those of the other four MTL methods. This may be due to the fact that only a small subset of features are relevant to the target. The noise features (e.g. background features) may degrade the learning performance.
4. While GOMTL and AMTL are able to capture the partial relationship at the task level, the performance of the proposed approach is better than those of these methods. The reason could be that our method is able to capture the task structure in a more flexible way at the feature level.

The algorithm may converge to a local minimum since the objective function is biconvex (the optimization problem is convex for one of the two components when the other is fixed). We have performed experimental studies to show the convergence capability of our alternating iterative procedure. As seen in Fig. 4, the value of objective function is decreasing and the algorithm converges to a fixed value. In addition, the convergence rate is very fast. In particular, for JAFFE, AR and Yale, the number of iterations is less than 15.

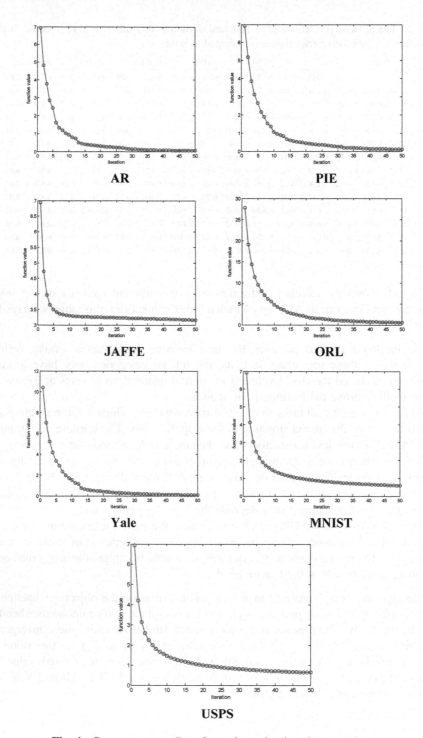

Fig. 4. Convergence studies of our alternating iterative procedure.

5 Conclusion

In this paper, we propose a new MTL model to exploit partial task relationships at the feature level. In particular, the coefficient matrix can be decomposed into two components: one for discovering shared information across the tasks, and one for producing a task-specific sparse pattern for each task. Comparing with the GOMTL model, which exploits partial relationship through the sharing of some of the latent tasks, this flexible formulation allows different tasks to partially share features, such that we can more accurately model the complex structure among the tasks. An efficient alternating optimization procedure has been proposed to solve the biconvex optimization problem. We perform experimental studies on several image data sets, and the experimental results demonstrate that the proposed method improve learning performance by simultaneously exploiting partial relationship across tasks at the feature level.

Acknowledgments. The work described in this paper was partially supported by a grant from the Research Grants Council of the Hong Kong Special Administrative Region, China [Project No. CityU 11300715], and a grant from City University of Hong Kong [Project No. 7004674].

References

1. Argyriou, A., Evgeniou, T., Pontil, M.: Multi-task feature learning. In: Advances in Neural Information Processing Systems, pp. 19–41 (2007)
2. Caruana, R.: Multitask learning. Mach. Learn. **28**(1), 41–75 (1997)
3. Chen, J., Zhou, J., Ye, J.: Integrating low-rank and group-sparse structures for robust multi-task learning. In: Proceedings of the 17th ACM SIGKDD International Conference on Knowledge Discovery and Data Mining, pp. 42–50 (2011)
4. Evgeniou, T., Pontil, M.: Regularized multi-task learning. In: Proceedings of the Tenth ACM SIGKDD International Conference on Knowledge Discovery and Data Mining, pp. 109–117 (2004)
5. Gong, P., Ye, J., Zhang, C.: Robust multi-task feature learning. In: Proceedings of the 18th ACM SIGKDD International conference on Knowledge Discovery and Data Mining, pp. 895–903 (2012)
6. Jacob, L., Vert, J.P., Bach, F.R.: Clustered multi-task learning: a convex formulation. In: Advances in Neural Information Processing Systems, pp. 745–752 (2009)
7. Kang, Z., Grauman, K., Sha, F.: Learning with whom to share in multi-task feature learning. In: Proceedings of the 28th International Conference on Machine Learning, pp. 521–528 (2011)
8. Kim, S., Xing, E.P.: Tree-guided group lasso for multi-response regression with structured sparsity, with an application to eQTL mapping. Ann. Appl. Stat. **6**(3), 1095–1117 (2012)
9. Kumar, A., Daume III, H.: Learning task grouping and overlap in multi-task learning. arXiv preprint arXiv:1206.6417 (2012)
10. Lee, S., Zhu, J., Xing, E.P.: Adaptive multi-task lasso: with application to eQTL detection. In: Advances in Neural Information Processing Systems, pp. 1306–1314 (2010)
11. Lee, G., Yang, E., Hwang, S.J.: Asymmetric multi-task learning based on task relatedness and loss. In: Proceedings of the 33rd International Conference on Machine Learning, pp. 230–238 (2016)

12. Liu, A.A., Su, Y.T., Nie, W.Z., Kankanhalli, M.: Hierarchical clustering multi-task learning for joint human action grouping and recognition. IEEE Trans. Pattern Anal. Mach. Intell. **39**(1), 102–114 (2017)
13. Lu, X., Li, X., Mou, L.: Semi-supervised multitask learning for scene recognition. IEEE Trans. Cybern. **45**(9), 1967–1976 (2015)
14. Maurer, A., Pontil, M., Romera-Paredes, B.: Sparse coding for multitask and transfer learning. In: Proceedings of the 30th International Conference on Machine Learning, pp. 343–351 (2013)
15. Obozinski, G., Taskar, B., Jordan, M.: Multi-task feature selection. Statistics Department, UC Berkeley, Technical report 17 (2006)
16. Parameswaran, S., Weinberger, K.Q.: Large margin multi-task metric learning. In: Advances in Neural Information Processing Systems, pp. 1867–1875 (2010)
17. Tibshirani, R.: Regression shrinkage and selection via the lasso. J. R. Stat. Soc. **73**(3), 267–288 (1996)

Power Users Behavior Analysis
and Application Based on Large Data

Xiaoya Ren[✉], Guotao Hui, Yanhong Luo, Yingchun Wang,
Dongsheng Yang, and Ge Qi

Northeastern University, Shenyang, China
nmg_renxy@163.com

Abstract. In this paper, a persona and users' segmentation model are established by analyzing the power users' data. In order to further complete the historical database, the paper adopts the method of questionnaire to collect information. Then according to the characteristics of power users, the index system is established, and the index is selected. Different construction methods are adopted for different models. Here, the K-means algorithm is used to cluster the second level indicators in the users' behavior attribute, and the users' label is extracted according to the clustering results. Finally, power users' persona is implemented. It can be proved that the model is effective in dealing with massive data, and provides reliable data support for decision making.

Keywords: Big data · Persona · Users' segmentation · Index selection

1 Introduction

With the large increase in the number of power customers and electricity consumption, the sale of electricity has also continued to increase dramatically, which objectively increased the intensity and complexity of the power marketing work. When faced with a large number of marketing business data in power system, how to use the existing methods and technology to identify potential value of big data from the electric power marketing system has attracted more and more attention.

In the literature [1–3], the clustering algorithm is applied to the customer segmentation of bank, and different marketing plans are implemented to different clients according to the clients' attributes. The clustering algorithm is applied to the study of the efficiency evaluation model of commercial banks, and the evaluation of bank efficiency is realized through the establishment of the relevant evaluation system [4, 5]. In the literature [6, 7], the K-means clustering algorithm is used to realize the customer loss prediction analysis, and different retention measures against the loss of customer are implemented according to clients' different attributes. In order to realize the power users' segmentation, this paper uses K-means clustering method to analyze a historical payment data and investigative payment data. Due to the properties of power users are different, their payment behavior will also have obvious differences. Through users' segmentation, the correlation analysis between the natural attributes and the payment method is realized. In the end, it can be more accurate and more targeted to build and plan the way of payment.

© Springer International Publishing AG 2017
D. Liu et al. (Eds.): ICONIP 2017, Part V, LNCS 10638, pp. 105–114, 2017.
https://doi.org/10.1007/978-3-319-70139-4_11

In the big data platform, the characteristics of the behavior of the customer payment are formed by processing and analyzing the characteristics data of the customers' electricity payment. On this basis, combined with the existing customer base information and feature information, the power persona is formed by the deeply fusion data and cross analysis. Then use the persona to analyze the clustering results, which can clearly shows the payment preferences, characteristics and needs of users. In the end, different user groups have adopted different methods of electricity payment planning.

2   The Establishment of Marketing System of Electric Power Users

Because users' information (users' name, home address, payment and collection records etc.) of the power department contained in the database cannot accurately describe the natural characteristics and behavior characteristics of the users' payment. Therefore, in order to establish a more accurate picture of the power users, this paper adopts a questionnaire survey method to obtain more users' information and payment preferences. Then the K-means clustering algorithm is adopted to carry out data mining on two aspects of users' natural attributes and payment behavior. The clustering of user groups is described with the feature labels, and they can be located by these labels. It needs to notice that labels need to have the characteristics of semantic and short text, which can give people intuitive understanding and convenient computer processing data. After the user is subdivided according to the label, using the user groups of the various features and labels of the persona, this method can achieve more accurate marketing and decision-making. The precision marketing system model is shown in Fig. 1.

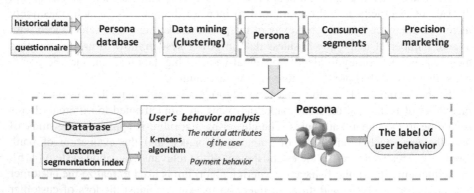

Fig. 1. Precision marketing system model.

2.1   The Definition and Characteristics of the Portrait of the User

Document [6] definition: the persona is the abstract data abstraction for a virtual character image process, but also the basis to achieve users' accurate subdivision and better find out the contents of the data. The feature of the persona is that it contains more three-dimensional, more comprehensive users' information. Each persona has a label that makes it easier to process or filter data. When planning and constructing diversified payment methods, there is an important principle that the users are as the basis of the study. Here, we use the persona to describe a class of user groups that represent each of the databases of different actual users. Through the persona can better analyze users' preferences to determine the users' needs. It provides a scientific decision-making basis for the implementation of fine marketing for electric power enterprises, so as to improve the payment services and achieve efficient payment.

2.2 Power Persona

This is the basic users' information provided by the city (home address, users' ID etc.) and payment history data, and questionnaire data provided as basic data support. In this paper, labels are used to describe the various features of the user, and a comprehensive review of the literature has been constructed in the framework of the power persona shown in Fig. 2.

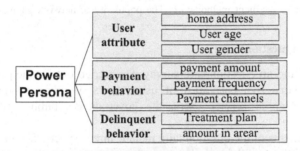

Fig. 2. Power persona.

3 Survey on Natural Attributes and Payment Behavior of Electric Power Users

3.1 Objective of Questionnaire Design

In order to more accurately for the persona, we need to expand the data base on the historical database. Here is a survey of the way to collect information. Through the questionnaire survey, we can according to users' attributes and payment preferences (payment methods, tendency of payment, the payment amount and arrears situation) further subdivided user, understand and master the user can maintain the preferences, and actively respond to users' needs. According to the users' demand for construction payment methods and form of payment, the existing problems can be improved. In [8] pointed out that a crucial impact questionnaire survey design of survey data and investigation results, so the design of a scientific and effective questionnaire is an important integral part of.

3.2   Establishment of Index System

Literature [9] pointed out that in order to better achieve the survey and evaluation, a scientific index system should be established. The goal of this survey is to understand the users' payment preference and users' expectations, which belong to the hidden data cannot be obtained directly. So the index system should be gradual developed, forming a series of secondary indexes which conclude specific problems in the questionnaire. These two levels of evaluation index constitute the index system of the power users' preference evaluation, as shown in Table 1.

Table 1. Power users' preference evaluation index.

1. Users' natural attributes
(1) The gender of the user; (2) The age of the user; (3) Home address
2. Payment preference
(1) Payment methods: a total of 10 species, divided into three categories: business payment, online payment, online payment; (2) Payment frequency; (3) Payment amount
3. User expectations:
(1) Diversification of payment methods; (2) The popularity of network payment

3.3   Design of Questionnaire

In this paper, a random questionnaire survey was carried out for the power users in a city, which contains five categories of users: users of urban residents, industrial enterprises, commercial service users, the government's public utility users and important users.

The main contents of the questionnaire include three aspects. The first is the natural attribute of the user to fill in, such as gender, age of user; the second part is the investigation of payment behavior of users, including payment frequency and payment methods; the third part mainly collect opinions and suggestions of users, such as whether the new payment networks, whether the expansion of network payment methods, enhance mutual understanding expectations and needs of users. It is feasible to deepen the understanding of the users' expectations and requirements by means of questionnaire.

3.4   Sample Size Distribution

The social survey is based on the principle to analyze clustered power users, which is generally adopted in the world of "95% confidence interval, 4% relative error level". This paper uses questionnaire to a random sample of 5000 users conducted a sample survey in the data cleaning stage excluding 101 invalid questionnaires, valid 4899 questionnaires. The 4899 questionnaires were used as the sample space, and each questionnaire was used as a sample. According to the classification of the age group, the corresponding numbers of each category are shown in Table 2.

Table 2. Sample distribution of questionnaire.

Users' age	Sample's number	The total sample proportion
Under the age of 25	545	11.12%
At the age of 25–35	1557	31.78%
At the age of 36–50	1866	38.09%
Over the age of 50	931	19.01%

4 Method Design of Power User Segmentation Based on K-Means Algorithm

After building the framework of the persona, the third chapter begins data mining on historical data and survey questionnaire data, using K-means algorithm to correlate the seemingly unrelated data and classification. At the end of the cluster, the user group is the difference persona.

4.1 K-Means Algorithm Principle

K-means clustering analysis technology is basic and most wide used cluster analysis algorithm. The basic idea is first randomly select k samples as the initial centers of the K class, and the remaining samples allocated to its nearest points according to the principle of minimum distance which is also called resemble degree. Then iterative calculation average of each class as a new cluster center, until the squared error criterion function converges to optimal approximate square error criterion function, which is:

$$E = \sum_{i=1}^{k} \sum_{x \in C_i} |x - m_i|^2 \qquad (1)$$

In the formula, E is the sample space of all objects squared error and x is to sample data in the database, and m_i is the average of all samples. It is reusing the mean value of the objects in each cluster a "center" of the object.

4.2 Power Users's Clustering Based on K-Means Algorithm

The principle of the algorithm is combined with the power users' clustering. Accepted input k in k-means algorithm is the class number a clustered power user. Generally, the value of the optimal number of clusters needs experimentally determination; the amount of each sample in database containing is to be clustered data object. Clustering based on K-means algorithm is the data of image divided into a certain value of the clustering center process. The processing procedure is as follows:

(1) Data import: Import each users' information and payment data;
(2) Data preprocessing: Digitize the text information in the data. It's easy to cross analyze data by standardizing the different dimensions of the data;

(3) Selection of segmentation variables: According to different requirements, choose the corresponding clustering evaluation index, which is a description of each label user group;

(4) Enter the initial clustering value: According to the general literature and several tests to determine the initial cluster optimal clustering effect value;

(5) The data mining process: According to the different needs of the cluster, the corresponding information of each user is distributed according to the principle of shortest distance from the nearest cluster center and the cluster centers are computed newly when each class set add new data. Iterate calculation until every user is complete assigned according to the principle of the algorithm and then end point clustering. Clustering flow chart is as shown in Fig. 3.

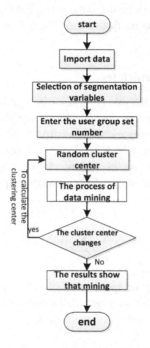

Fig. 3. Users' segmentation based on "persona".

4.3   Users' Segmentation Based on "Persona"

Based on the data mining and the precision marketing system model, this paper build the subdivision model from the two angles of the user attributes and the users' payment behavior, as well as considering the breakdown factors. The subdivision model is shown in Table 3. Before clustering, the specific clustering variables need to be identified, and then use the labels in the subdivision model to K-means clustering. After a certain number of iterations, each entity in the database is classified into a cluster. Characteristics and the label of each user group can be selected to complete the portrait, in order to determine the marketing decisions.

Table 3. Segmentation model of electric power users.

Basis of subdivision	Factor	
Users' attribute	Gender	
	Age (old, middle-aged and young)	
	Home address	
Payment behavior	Payment frequency: The number of segments	
	Payment methods: business hall, online payment, online collection	
	Payment amount	
	Delinquent behavior	Amount
		Repayment behavior

On the one hand, the model label the power user, realize persona, and explore the difference of different groups of users; on the other hand, the model reviews the group of each user from the value creation angle, providing the basis of user management and marketing service.

5 An Example Application of Power User Clustering Based on K-Means Algorithm

The comprehensive statistical analysis software SPSS is chosen to analyze data. The screened questionnaire data and city-provided payment history data are as the database. The data were normalized (pretreatment) before importing data, and the questionnaire options or text should be replaced by the corresponding figures, for the convenience of clustering in SPSS when they can be used as the numerical treatment options. Here, this paper will proceed from the two aspects of user attributes and payment behavior of cluster analysis.

5.1 Clustering Based on Users' Natural Attributes

According to the power users' accurate portrait segmentation model, when clustering analysis in users' natural attributes is done, the input variables are the users' age, occupation, and cultural degree. According to the literature review and experimental debugging, clustering number is tentatively scheduled for 5 when users' classification is being done. The deviations of 5 kinds of users' center point iteration are presented in Table 4. After the fifth iteration, the deviations of 5 kinds of users' center point are less than stopping criterion 0.02, and when the sixth iteration is finished, centers of clustering don't change any more, namely clustering analysis is completed.

Table 5 shows the final cluster center, which is the average value of each category in the various variables, which can help to give the actual meaning to the classification. Portrait of five types of user groups can be constructed according to the corresponding indicators, and in Table 5, sample distribution of the five groups of users is showed.

Table 4. The iterative process of clustering by users' natural attributes.

Iteration	Change of cluster center				
	1	2	3	4	5
1	1.568	1.288	1.295	1.490	1.624
2	.387	.132	.291	.192	.493
3	.279	.148	.089	.088	.408
4	.056	.131	.027	.119	.205
5	.000	.142	.114	.078	.007
6	.000	.000	.000	.000	.000

According to Table 5, five categories of users can be subdivided out, and each variable is regarded as a label for each user group portrait. The second category and third category users account for large percentage. The second users are young, low education, freelancer and likely to be a student; obviously the third belonging to the elderly user group. The number of the two categories of users exceeded 50%, being regarded as key marketing objects. Therefore, reasonable building and planning of the business hall and network should be considered.

Table 5. The final cluster center of the users' natural attributes.

	Class set				
	1	2	3	4	5
Age	2	2	4	2	3
Degree of education	2	2	1	4	3
Profession	2	4	4	2	4
Distribution of user groups	769.00	1290.0	1369.0	613.00	834.00

Note: age: 1: $\leq$ 25; 2: 25-35;3: 36-50;4: $\geq$ 50. Education level: Lower than Senior school, senior school, Junior college degree, university degree, Master degree or above. Occupation: 1, administrative organs 2, public institutions 3, enterprise 4, free professionals 5, others

5.2   Clustering Based on the Behavior of the Payment

According to the subdivision model, users' payment behavior variables contains: payment habits, arrears processing, payment method and payment tendency (i.e., users' expectations). Clustering the data which has been standardized processed, the number of cluster test is 3–6. Through the test, it is found that the clustering effect is best when the K value is 4. The iteration process is shown in Table 6. The clustering has finished during the sixth iteration process.

Table 6 shows the final cluster center which formed through 5 iterations. Which shows the distribution of samples in accordance with the clustering of the payment behavior. The power users can be divided into four categories referred to the payment behavior indicators. According to Table 7, whatever the kind of users' payment habit is, they all trend to pay off after arrearage. Fortunately, every user can timely paid in arrears.

Table 6. The iteration process of clustering by users' payment behavior.

Iteration	Change of cluster center			
	1	2	3	4
1	1.526	3.161	2.028	2.405
2	.143	.248	.157	.287
3	.090	.190	.000	.461
4	.009	.156	.000	.366
5	.000	.034	.000	.075
6	.000	.000	.000	.000

After the users' segmentation, the users' payment characteristics are not same, each user class has its different payment preferences. First class users' paying fee way is the collecting bank, and their anticipation is self-service terminals; while the second category of users obviously prefer the new payment methods, the daily fee for online payment and are willing to try mobile payment; the third type of users' payment habit is mobile phone fee, but tends to choose the payment business office, which probably that the user is not within the scope covered by the business hall, it is recommended to reestablish the business hall payment points according to the users' distribution; the fourth category of users belong to the sustain development type, with no need for the construction and planning of new payment methods.

Synthesizing Table 7, we can raise two questions from all user group distributions, one is the city's business hall and payment networks distribution cannot meet the needs of users, most users still tend to payment business offices to ensure the electricity is accurate, timely, safe arrival, it is recommended to re-plan the distribution of business offices and outlets according to users' distribution, or increase the propaganda of other modes of payment, reduce the cost of investment; second, it can be seen from the distribution of secondary users, new payment method is gradually being accepted and recognized. So if we strengthen the propaganda, it can effectively reduce the congestion at the peak period of payment and reduce the pressure of line charges.

Table 7. The final clustering center of users' payment behavior.

	Class set			
	1	2	3	4
Payment habits	2	2	2	2
Treatment plan	1	1	1	1
Payment methods	2	4	5	2
Preference	2	7	1	5
Distribution of user groups	1544.000	1288.000	1143.000	693.000

Note: payment habits: 1:Pre deposit;2:Arrears payment;3: Until received a reminder. Treatment: 1:Payment as soon as possible;2: Payment at month's end;3:Payment in next month
Payment method: 1:Self-service terminal;2:Bank;3:Community payment;4:Online payment;5:WeChat, Alipay;6:Others. Payment preference: 1:Business hall;2: Self-service terminal;3: 95588 online;4: Bank card;5:Bank;6: Online payment;7: Mobile payment;8: Community payment;9: Others

6 Conclusions

First of all, this paper adopts a questionnaire survey to further supplement the "persona" of the database; according to the index system, the users' segmentation model is constructed, and the variables are obtained. Then based on the users' natural attributes and data of payment behavior, data analysis is performed using K-means clustering algorithm and SPSS software, which can realize the clustering in two aspects of the users' attributes and payment behavior. Finally, the specific labels for each category of users are extracted. Accurately Persona and precise segmentation for all kinds of users are well achieved, providing the corresponding basis for the development of accurate marketing strategy.

Acknowledgements. This work is supported by the National Natural Science Foundation of China (61403073).

References

1. Xiaoning, Q.: K mean clustering algorithm in commercial bank customer classification in the application. Comput. Simul. **28**(6) (2011)
2. Wei, D., Chunrong, Z., Weijian, H.: Application of improved K-means clustering algorithm in customer segmentation. J. Hebei Univ. Econ. Bus. **35**(1) (2014)
3. Zhandong, X.: Research on efficiency evaluation model of commercial banks based on clustering analysis. Finance and Econ. (4) (2015)
4. Xiuxian, Z.: Based on the Data Mining Commercial Bank CRM System Research and Design. Nanjing University of Aeronautics and Astronautics, Nanjing (2014)
5. Mengjie, Y.: Data modeling — user product development in painting from the concrete to the abstract design. Technol. Res. (6) (2014)
6. Kang, Z.: Mobile phone user portrait in the scheme. Inf. Commun. Platf. Big Data **3**, 266–267 (2014)
7. Mingiun, Z., Huaguang, Z.: The neural network PID controller with BP optimized. Control Instrum. Chem. Ind. **37**(4), 5–9 (2010)
8. Zhang, X., Hui, G., Luo, Y.: ADP approach to solve unknown nonlinear zero-sum game. J. Northeast. Univ. (Nat. Sci.) **33**(12), 1673–1676 (2012)
9. Mingjun, Z., Huaguang, Z.: RBF neural network controller optimized, by genetic algorithm. Electr. Mach. Control **11**(2), 183–187 (2007)
10. Shuxian, L., Huaguang, Z.: Genetic algorithm approach to mixed fixed H2/H ∞ optimal adaptive noise cancellation. Instrum. Tech. Sens. **11**, 16–18 (2003)
11. Mingjun, Z., Huaguang, Z.: Neural network PID controller optimized by GA. J. Jilin Univ. Eng. Technol. Ed. **35**(1), 91–96 (2005)
12. Xiangui, W., Anna, W., Guotao, H.: An improved envelope fitting algorithm for the empirical mode decomposition. J. Northeast. Univ. Nat. Sci. **11**, 1535–1538 (2015)
13. Yang, L.: Power supply enterprises rely on big data marketing difference expand payment channels. Operations **10**, 74–75 (2014)
14. Peng, Z., Yijing, L.: Observation for consumers portrait. Internet Mark. **11**, 30–32 (2013)

Accelerated Matrix Factorisation Method for Fuzzy Clustering

Mingjun Zhan and Bo Li[✉]

School of Electronic and Information Engineering,
South China University of Technology, Guangzhou 510640, China
zmj.scut@gmail.com, leebo@scut.edu.cn

Abstract. Factorised fuzzy c-means (F-FCM) based on semi nonnegative matrix factorization is a new approach for fuzzy clustering. It does not need the weighting exponent parameter compared with traditional fuzzy c-means, and not sensitive to initial conditions. However, F-FCM does not propose an efficient method to solve the constrained problem, and just suggests to use a *lsqlin()* function in MATLAB which lead to slow convergence rate and nonconvergence. In this paper, we propose a method to accelerate the convergence rate of F-FCM combining with a non-monotone accelerate proximal gradient (nmAPG) method. We also propose an efficient method to solve the proximal mapping problem when implementing nmAPG. Finally, the experiment results on synthetic and real-world datasets show the performances and feasibility of our method.

Keywords: Nonnegative matrix factorization · Factorised fuzzy c-means · Non-monotone accelerate proximal gradient

1   Introduction

Nonnegative Matrix Factorization (NMF) [8,9] as a linear dimensionality reduction technique can be used in various data mining applications such as hyperspectral imaging [11], document clustering [17] and signal processing [1], etc. By decomposing the original nonnegative data matrix, NMF can generate two low-rank nonnegative matrix factors which are sparse and easily interpretable factors [12,21,22].

Some of previous works have demonstrated that NMF can be used in data clustering. Ding et al. [3] has proved that symmetric NMF, which imposes a near-orthogonality restriction on the factors, is identical with nonnegative relaxation of kernel k-means clustering, as well as the laplacian-based spectral clustering [20]. Furthermore, Pompili et al. [16] demonstrated that NMF with orthogonality constraints (ONMF) is equivalent to spherical k-means with weighted variant. Then Ding et al. [4] developed several various methods on the theme of NMF named Semi-NMF and Convex-NMF, which extended the range of applications of NMF into mixed signs data. Based on Semi-NMF's property, Trigeorg et al. [19] presented a deep Semi-NMF model which can extract the hidden features of the

© Springer International Publishing AG 2017
D. Liu et al. (Eds.): ICONIP 2017, Part V, LNCS 10638, pp. 115–123, 2017.
https://doi.org/10.1007/978-3-319-70139-4_12

given data. Recently, Suleman [18] came up with an alternative method for fuzzy clustering named factorized fuzzy c-means (F-FCM) which is based on a convex and semi-nonnegative matrix factorization. In Suleman's work, the weighting exponent parameter of conventional fuzzy c-means (FCM) is eliminated. And this method is not sensitive to the initialization. However, F-FCM suffers from slow convergence rate and nonconvergence problems [18].

F-FCM algorithm updates one matrix factor while another factor is fixed. After several times iterations, it outputs matrices of cluster centers V and membership degrees U. However, the F-FCM algorithm need to optimize U by solving N constrained least squared problems with both equality and inequality constraints:

$$min_{\boldsymbol{\mu}_k} f(\boldsymbol{\mu}_k) = \|\mathbf{x}_k - V\boldsymbol{\mu}_k\|_2^2, \quad 1 \leq k \leq N$$

$$s.t. \quad \sum_{i=1}^{c} \mu_{ik} = 1, \quad 0 \leq \mu_{ik} \leq 1, \quad 1 \leq i \leq C \tag{1}$$

where N is the number of data points on dataset X, and C is the number of clusters. The kth data in X is denoted as $\mathbf{x}_k$ and μ_{ik} is the membership degree of $\mathbf{x}_k$ belonging to the ith cluster. In [18], F-FCM just utilizes $lsqlin()$ function in MATLAB to solve the problem (1). In fact, $lsqlin()$ function is sort of the active-set algorithm which is similar to the methods described in [5,6]. The active-set algorithm is time-consuming and difficult to seek the solution. Therefore, F-FCM has a quite slow convergence rate, and, sometimes, failed to converge within a certain number of iterations in practice.

In this paper, we aim to accelerate the convergence speed of F-FCM algorithm by employing a non-monotone accelerate proximal gradient (nmAPG) method proposed in [10] to solve the problem (1). We also propose a fast method to deal with the proximal mapping problems when we implement nmAPG. This accelerated F-FCM method has a faster convergence speed, in the meanwhile, it maintains the original advantages and performance of F-FCM.

2   Accelerated Factorised-FCM

From the constraints in (1), we know that $\boldsymbol{\mu}_k$ belongs to the unit simplex S_c:

$$S_c = \{\boldsymbol{\alpha} = (\alpha_1, \alpha_2, \cdots, \alpha_c) : \alpha_i \geq 0 \wedge \sum_{i=1}^{c} \alpha_i = 1\} \tag{2}$$

Here, S_c is a convex set. Inspired by NeNMF [7], we infer that the gradient of cost function $f(\boldsymbol{\mu}_k)$ is Lipschitz continuous and has a Lipschitz constant $L = \|V^T V\|_2$. Therefore, for accelerating the convergence speed of F-FCM, we take full advantage of this particular convex structure of problem (1). Moreover, S_c is the domain of the function $f(\boldsymbol{\mu}_k)$, then we can obtain its indicator function $I_C(\boldsymbol{\mu}_k)$ easily:

$$I_C(\boldsymbol{\mu}_k) = \begin{cases} 0, & \text{if } \boldsymbol{\mu}_k \in C \\ +\infty, & \text{if } \boldsymbol{\mu}_k \notin C \end{cases} \tag{3}$$

So the constrained least squared problem (1) can be rewritten as:

$$min_{\boldsymbol{\mu}_k} g(\boldsymbol{\mu}_k) = \|\mathbf{x}_k - V\boldsymbol{\mu}_k\|_F^2 + I_C(\boldsymbol{\mu}_k) \tag{4}$$

The first term of this equation is convex, and its gradient is Lipschitz continuous. The second term is non-smooth, but convex. Hence, according to the theorem in [10], our problem satisfies the necessary pre-condition of nmAPG. The convergence rate in our situation is $O(1/k^2)$.

2.1 The Proximal Mapping Problem

The type of nmAPG is similar to the accelerated proximal gradient (APG) method. APG can extrapolate the next point by combining the current point and the previous point, and then solves the proximal mapping problem. nmAPG solves the problem in similar way.

The proximal mapping of the function I_C is Euclidean projection [2] onto C:

$$prox_{I_C}(\boldsymbol{\mu}_k) = argmin_{\boldsymbol{\alpha} \in C} \|\boldsymbol{\alpha} - \boldsymbol{\mu}_k\|^2 = P_C(\boldsymbol{\mu}_k) \tag{5}$$

The subset C can be viewed as the intersection of a hyperplane and a hypercube:

$$C = \{\boldsymbol{\alpha} \mid \mathbf{1}^T\boldsymbol{\alpha} = 1, 0 \preceq \boldsymbol{\alpha} \preceq 1\} \tag{6}$$

Given any $\mathbf{x} \in \mathbb{R}^{n \times 1}$, we define a projection operator $P_{[0,1]}(\mathbf{x})$ as below:

$$P_{[0,1]}(\mathbf{x})_i = \begin{cases} 0, & \text{if } x_i \leq 0 \\ x_i, & \text{if } 0 < x_i < 1 \\ 1, & \text{if } x_i \geq 1 \end{cases} \tag{7}$$

Accoring to [15], we can solve the proximal mapping problem (5) using

$$prox_{I_C}(\boldsymbol{\mu}_k) = P_C(\boldsymbol{\mu}_k) = P_{[0,1]}(\boldsymbol{\mu}_k - \lambda\mathbf{1}) \tag{8}$$

where λ is the solution of the followed equation:

$$\mathbf{1}^T P_{[0,1]}(\boldsymbol{\mu}_k - \lambda\mathbf{1}) = 1 \tag{9}$$

2.2 The Proximal Mapping Algorithm

Equation (9) is piecewise linear function and also decreases monotonously while λ increases. Its minimum value is 0. Taking advantage of this property, the function can decrease to one gradually by increasing λ from an initial value. First, we initialize λ by assuming all elements in $\boldsymbol{\mu}_k$ satisfy:

$$P_{[0,1]}(\boldsymbol{\mu}_k - \lambda)_i = \mu_{ik} - \lambda \tag{10}$$

So λ can be obtained by

$$\sum_{i=1}^{c}(\mu_{ik} - \lambda) = 1 \Rightarrow \lambda = (\sum_{i=1}^{c} \mu_{ik} - 1)/c \tag{11}$$

Then we put λ into the Eq. (9) to check whether our assumption is correct for each μ_{ik}. If λ satisfies Eq. (9), the current λ is the final solution. If not, it means that our assumption is not correct for all elements. There are three cases for those elements in vector $\boldsymbol{\mu}_k$ which do not satisfy assumption (10): (i) $\exists \mu_{mk} \in \boldsymbol{\mu}_k$, $\mu_{mk} - \lambda < 0$; (ii) $\exists \mu_{nk} \in \boldsymbol{\mu}_k$, $\mu_{nk} - \lambda > 1$; (iii) both case 1 and 2.

In case 2, because of the restriction of Eq. (11), there must exist elements μ_{mk} making $\mu_{mk} - \lambda < 0$, so as to make sure the sum to be 1. So, actually, there are only two situations, in which Eq. (9) dose not hold: (i) $\exists \mu_{mk} \in \boldsymbol{\mu}_k$, $\mu_{mk} - \lambda < 0$; (ii) $\exists \mu_{mk}, \mu_{nk} \in \boldsymbol{\mu}_k$, $\mu_{mk} - \lambda < 0$, $\mu_{nk} - \lambda > 1$.

In these two situations, the function $\mathbf{1}^T P_{[0,1]}(\boldsymbol{\mu}_k - \lambda\mathbf{1})$ is bigger than 1. Then we can increase λ to make Eq. (9) to be satisfied. Actually, some μ_{mk} are less than λ and have no contribution to the function. Then, we can erase the elements, and the rest satisfy Eq. (10). So, λ can be computed by:

$$\lambda_{new} = (\sum_{i=1}^{c} \mu_{ik} - 1)/c' \tag{12}$$

where c' represents the number of elements. In Algorithm 1, we demonstrate the proximal mapping algorithm.

Algorithm 1. Algorithm for the proximal mapping operator

Input: vector $\boldsymbol{\mu}_k$, the number of elements in $\boldsymbol{\mu}_k$
Output: the proximal mapping of vector $\boldsymbol{\mu}_k$
calculate the initial lambda $\lambda = (\sum_{i=1}^{c} \mu_{ik} - 1)/c$
compute $prox_{I_C}(\boldsymbol{\mu}_k) = P_{[0,1]}(\boldsymbol{\mu}_k - \lambda)$
While $sum(prox_{I_C}(\boldsymbol{\mu}_k)) \neq 1$
   a) set all $u_{ik} < \lambda$ to 0, set the number of nonzero elements in $\boldsymbol{\mu}_k$ to c';
   b) calculate the new lambda using $\lambda = (\sum_{i=1}^{c} \mu_{ik} - 1)/c'$;
   c) compute $prox_{I_C}(\boldsymbol{\mu}_k) = P_{[0,1]}(\boldsymbol{\mu}_k - \lambda)$ again;
End while
Return $prox_{I_C}(\boldsymbol{\mu}_k)$;

2.3   Accelerated F-FCM

Then, we can apply nmAPG to solve the problem (1) easily in this subsection. The method is presented in Algorithm 2. This algorithm can obtain the membership degree U. The cluster center V can be obtained by following:

$$\mathbf{v}_i = \frac{\sum_{k=1}^{N} \mu_{ik}\mathbf{x}_k}{\sum_{k=1}^{N} \mu_{ik}}, \quad 1 \leq i \leq c \tag{13}$$

In addition, nmAPG allows larger step size, instead of a fixed step size. We can also calculate the step size by backtracking line search [13] to further accelerate the convergence speed.

Algorithm 2. nmAPG for constrained least squared problem in F-FCM

Initiate $Z_1 = U_1 = U_0, t_1 = 1, t_0 = 0, \eta \in [0, 1)$,
$\delta > 0, c_1 = F(U_1), q_1 = 1, \alpha_x < \frac{1}{L}, \alpha_y < \frac{1}{L}$
for $k = 1, 2, 3, \cdots$ do
    $Y_k = U_k + \frac{t_{k-1}}{t_k}(Z_k - U_k) + \frac{t_{k-1}-1}{t_k}(U_k - U_{k-1})$
    $Z_{k+1} = P_C(Y_k - \alpha_x \nabla f(Y_k))$
    if $F(Z_{k+1}) \le c_k - \delta \|Z_{k+1} - Y_k\|^2$ then
        $U_{k+1} = Z_{k+1}$
    else
        $V_k = P_C(U_k - \alpha_y \nabla f(U_k))$
        $U_{k+1} = \begin{cases} Z_{k+1}, & \text{if } F(Z_{k+1}) \le F(V_{k+1}) \\ V_{k+1}, & \text{otherwise} \end{cases}$
    end if
    $t_{k+1} = \frac{\sqrt{4(t_k)^2+1}+1}{2}$
    $q_{k+1} = \eta q_k + 1$
    $c_{k+1} = \frac{\eta q_k c_k + F(U_{k+1})}{q_{k+1}}$
end for

---

3   Experiments

In this section, two types of dataset are employed, synthetic dataset and real-life dataset. The former one is a Normal-4 dataset [14], while the later one contains the same 12 real-life datasets used in [18]. The experiments are conducted on Windows 10 with an Intel Core i5 3.20 GHz CPU and 8 GB RAM. The parameter ε is set to 10^{-3}, and the maximum of iterations is 1000.

To keep the comparison fair, we follow the strategies in [18]. When measuring the clustering accuracy, the clustering result is given by the matrix $U^* = [\mu_{ik}^*]$:

$$\mu_{ik}^* = \begin{cases} 1 & \text{if } \mu_{ik} > 0.5 \\ 0 & \text{otherwise} \end{cases} \tag{14}$$

Using the index v_S in [18] we can obtain the optimal number of clusters c^* by varying c from 2 to $c^* + 2$ (never low than 6) in a step of 1.

1. Experiments on Normal-4 Dataset. At first, we compare the speed of clustering between AF-FCM and F-FCM. The two algorithms run 100 times on each number of clusters (ranged from 2 to 6) to obtain the average running time. Figures 1 and 3 show the experiment result, from which we can see that F-FCM takes 390 s per trial on average while AF-FCM only takes 132 s per trial. AF-FCM reduces the objective function much faster than F-FCM at the same time and need much less CPU time than F-FCM. Table 1 shows the comparison of clustering accuracy for FCM, F-FCM, AF-FCM and Semi-NMF on Normal-4. We can also see that F-FCM and AF-FCM perform best. And AF-FCM has the same clustering accuracy as F-FCM algorithm and the lowest standard deviation (Std).

2. Experiments on Real-life Dataset. We compare the running time between AF-FCM and F-FCM on the 12 real-life datasets. The running time is measured in second. Table 2 recorded the running time's average and standard deviation on 12 real-life datasets. It can be observed that the speed of AF-FCM is faster than F-FCM on real datasets, especially on the big datasets like Waveform Database (the last two rows in Table 2). In terms of clustering accuracy, Fig. 2 demonstrates 4 clustering algorithms' performances on the 12 real-life datasets. The number

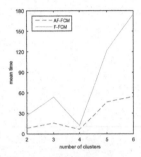
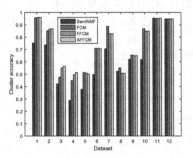

Fig. 1. Average running time of AF-FCM and F-FCM on Normal-4 dataset.

Fig. 2. The average clustering accuracy on 12 real-life datasets.

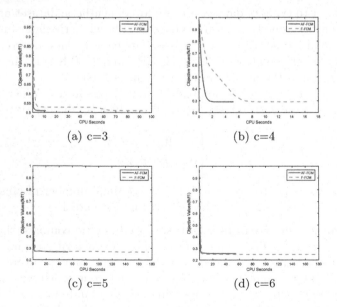

(a) c=3

(b) c=4

(c) c=5

(d) c=6

Fig. 3. CPU time on various classes

Table 1. Clustering accuracy on Normal-4 dataset.

Algorithm	Min	Max	Mean	Std
FCM	0.8575	0.9125	0.8877	0.0102
F-FCM	0.9175	0.9587	0.9358	0.0082
AF-FCM	0.9175	0.9587	0.9358	0.0082
Semi-NMF	0.8104	0.5637	0.9313	0.0100

Table 2. Running time's average and standard deviation on 12 real-life datasets.

Dataset	#Cluster	Time		Std	
		F-FCM	Ours	F-FCM	Ours
Breast cancer wisconsin (Original)	2	5.2	1.9	0.4	0.6
Breast cancer wisconsin (Diagnostic)	2	5.3	1.0	0.3	0.1
Ecoli	8	46.8	20.6	14.8	9.5
Glass identification	6	17.4	10.9	4.1	7.2
Image segmentation	7	20.1	6.6	7.2	2.0
Ionosphere	2	3.5	0.9	0.4	0.1
Iris	3	2.6	1.0	0.6	0.2
Liver disorders	2	4.3	0.5	0.4	0.1
Pima Indians diabetes	2	7.5	1.6	0.4	0.2
Seeds	3	4.8	2.1	1.4	0.8
Waveform database generator (Ver.1)	2	39.3	5.1	9.2	0.6
Waveform database generator (Ver.2)	2	36.2	7.2	7.9	1.0

of horizontal axis of Fig. 2 refer to the order of the 12 real-life datasets listed in Table 2. On the most of real-life datasets, the performances of AF-FCM and F-FCM are similar, but AF-FCM performs better than F-FCM on the No. 3 and 4 dataset. Except the No. 5, 7, 8 and 10 dataset, AF-FCM and F-FCM have better performance than FCM. Except No. 8 dataset, the performance of Semi-NMF is left behind of the rest three algorithms on others 11 datasets.

4   Conclusion

In this paper, we have proposed an accelerated factorized fuzzy c-means (AF-FCM) method for fuzzy clustering task. AF-FCM uses a non-monotone accelerate proximal gradient method to solve the constrained least squared problem occurred in F-FCM. It has a special strategy for the proximal mapping operator. All of these approaches have helped AF-FCM obtain a faster convergence speed. The numerical experiments on synthetic and real-world datasets showed that the clustering performance of AF-FCM is almost same with F-FCM, but the convergence speed of AF-FCM is much faster than F-FCM.

Acknowledgement. This research was supported by the National Natural Science Foundation of China (Grant Nos. 11627802, 51678249), by the Science and Technology Projects of Guangdong (2013A011403003), and by the Science and Technology Projects of Guangzhou (201508010023).

References

1. Cichocki, A., Zdunek, R., Amari, S.I.: New algorithms for non-negative matrix factorization in applications to blind source separation. In: Proceedings of IEEE International Conference Acoustics, Speech, Signal Processing, vol. 5, pp. 621–624 (2006)
2. Combettes, P.L., Pesquet, J.C.: Proximal splitting methods in signal processing. In: Bauschke, H., Burachik, R., Combettes, P., Elser, V., Luke, D., Wolkowicz, H. (eds.) Fixed-Point Algorithms for Inverse Problems in Science and Engineering, pp. 185–212. Springer, New York (2011). doi:10.1007/978-1-4419-9569-8_10
3. Ding, C., He, X., Simon, H.: On the equivalence of nonnegative matrix factorization and spectral clustering. In: SDM, vol. 5, pp. 606–610 (2005)
4. Ding, C., Li, T., Jordan, M.: Convex and semi-nonnegative matrix factorizations. IEEE Trans. Pattern Anal. Mach. Intell. **32**(1), 45–55 (2010)
5. Gill, P.E., Murray, W., Saunders, M.A., Wright, M.H.: Procedures for optimization problems with a mixture of bounds and general linear constraints. ACM Trans. Math. Softw. (TOMS) **10**(3), 282–298 (1984)
6. Gill, P.E., Murray, W., Wright, M.H.: Practical Optimization (1981)
7. Guan, N., Tao, D., Luo, Z., Yuan, B.: NeNMF: an optimal gradient method for non-negative matrix factorization. IEEE Trans. Sig. Process. **60**(6), 2882–2898 (2012)
8. Lee, D.D., Seung, H.S.: Learning the parts of objects by non-negativ matrix factorization. Nature **401**(6755), 788–91 (1999)
9. Lee, D.D., Seung, H.S.: Algorithms for non-negative matrix factorization. In: Advances in Neural Information Processing Systems, pp. 556–562 (2001)
10. Li, H., Lin, Z.: Accelerated proximal gradient methods for nonconvex programming. In: Advances in Neural Information Processing Systems, pp. 379–387 (2015)
11. Ma, W.K., et al.: A signal processing perspective on hyperspectral unmixing: insights from remote sensing. IEEE Sig. Process. Mag. **31**(1), 67–81 (2014)
12. Nicolas, G.: The why and how of nonnegative matrix factorization. In: Regularization, Optimization, Kernels, and Support Vector Machines, vol. 12, no. 257 (2014)
13. Nocedal, J., Wright, S.: Numerical Optimization, pp. 185–212 (2006)
14. Pal, N.R., Bezdek, J.C.: On cluster validity for the fuzzy c-means model. IEEE Trans. Fuzzy Syst. **3**(3), 370–379 (1995)
15. Parikh, N., Boyd, S.: Proximal algorithms. Found. Trends Optim. **1**(3), 127–239 (2013)
16. Pompili, F., Gillis, N., Absil, P., Glineur, F.: Two algorithms for orthogonal non-negative matrix factorization with application to clustering. Neurocomputing **141**, 15–25 (2014)
17. Shahnaz, F., Berry, M., Pauca, V., Plemmons, R.: Document clustering using non-negative matrix factorization. Inf. Process. Manag. **42**(2), 373–386 (2006)

18. Suleman, A.: A convex semi-nonnegative matrix factorisation approach to fuzzy c-means clustering. Fuzzy Sets Syst. **270**, 90–110 (2015)
19. Trigeorgis, G., Bousmalis, K., Zafeiriou, S., Schuller, B.: A deep semi-NMF model for learning hidden representations. In: ICML, pp. 1692–1700 (2014)
20. Zha, H., He, X., Ding, C., Gu, M., Simon, H.D.: Spectral relaxation for k-means clustering. In: Advances in Neural Information Processing Systems, pp. 1057–1064 (2001)
21. Zhou, G., Cichocki, A., Zhang, Y., Mandic, D.P.: Group component analysis for multiblock data: common and individual feature extraction. IEEE Trans. Neural Netw. Learn. Syst. **27**(11), 2426–2439 (2015)
22. Zhou, G., Zhao, Q., Zhang, Y., Adali, T., Xie, S., Cichocki, A.: Linked component analysis from matrices to high-order tensors: applications to biomedical data. Proc. IEEE **104**(2), 310–331 (2015)

Mining Mobile Phone Base Station Data Based on Clustering Algorithms with Application to Public Traffic Route Design

When Shen, Zhihua Wei$^{(\boxtimes)}$, and Zhiyuan Zhou

Department of Computer Science and Technology,
Tongji University, Shanghai 201804, China
zhihua_wei@tongji.edu.cn

Abstract. It attracts a lot of attention that how to use mobile phone base station data to predict user behavior and design the public traffic route. In this paper, we extend the classic algorithms to design the shuttle bus route. The contribution of this paper is mainly manifested on (1) we integrate the classical machine learning methods DBSCAN and GMM to complete mobile phone base station data modeling, so that to learn the residents' spatial travel pattern and temporal habits; (2) we apply the Public Route Scale Estimation Model to design the shuttle bus routes and departure intervals based on the modeling results of (1). Experimental results show that our model based on DBSCAN and GMM can effectively mine the significance of historical data of mobile phone base station and can successfully be applied to real-world problems like public traffic route design.

Keywords: Phone base station data · DBSCAN · GMM · Travelling behavior analysis · Public traffic route design

1 Introduction

With the increasing popularity of mobile phones, we can get a lot of mobile phone base station data, which is able to well reflect users travel behavior. The mobile phone base station data provides unprecedented insight into urban dynamics and human activity, providing a great opportunity for research and real-world applications. For instance, we can use mobile phone base station data to learn the residents' travelling track and then guide the residents travelling.

A great number of researches have been done to the analysis of such data. Calabrese et al. [1] point out that mobile phone data will ultimately provide both micro- and macroscopic views of cities and help understand citizens behaviors and patterns. Li et al. [2] use mobile phone base station data to study residents travelling track. They use GIS data analysis method to analyze the residents' travelling track and get residents travelling patterns. Wu et al. [3] propose a traffic semantic framework. They analyze the cell detail record (CDR) data in Beijing, and extract four features of base stations to tag the traffic semantic

© Springer International Publishing AG 2017
D. Liu et al. (Eds.): ICONIP 2017, Part V, LNCS 10638, pp. 124–133, 2017.
https://doi.org/10.1007/978-3-319-70139-4_13

attribute of the base stations. Dong et al. [4] use a K-means clustering method to divide the commuting traffic zones based on the CDR data and propose traffic zone attribute-index to indicate the preference on working or residential.

In previously mentioned mobile phone data researching, researches focus on the trajectory characteristics extraction of spatial distribution, but ignoring the effects of time distribution. The time distribution of citizens travelling is also important, such as, for bus time table design. In this paper, we use unsupervised classify methods to learn both trajectory characteristics and time distribution of residents travelling.

We are invited to work with Administration and Services Center (hereinafter referred to as ANS Center) of Xuhui District, Shanghai, to design the shuttle bus route within one kilometer of the ANS Center. Shanghai Telecom provides mobile data of citizens who going to ANS Center during investigations. The shuttle bus route design includes (1) determining the shuttle bus stops, and (2) inferring the most possible rush hour of the route, and (3) determining departure intervals.

For the first task above, we use DBSCAN to focus on spatial travel pattern analysis. Ma et al. [5] use DBSCAN algorithm to analyze the identified trip chains of riders and then learn their historical travel patterns. Le et al. [6] adopt DBSCAN algorithm to mine the travel pattern of each smart card user and then segment transit passengers into four identifiable types, which is helpful for transit operators to understand their passengers and provide oriented services. They point out that, compare to other approaches, DBSCAN provides flexibility in defining the group of stops that the passenger repeatedly choose [7]. In this paper, we use DBSCAN to gather residents' travelling trajectory which are close to each other, and choose the cluster center as shuttle bus stops.

For the second task above, we use GMM to learn the temporal habits of residents. The GMM is one of the most widely used models in cluster analysis [8]. Many researches have been done that use GMM to learn travel patterns. Cui et al. [9] propose a concept of Extreme Index (EI) based on the mixture Gaussian model to depict the extreme level of the passengers' travel pattern. Briand et al. [10] describe the temporal habits of the passenger by using a mixture of Gaussians, thus the different times of typical use as well as the variances around these peaks can be extracted. Lee and Sohn [11] introduce a formulation of Gaussian mixture model that could recognize route-use patterns. Qiao et al. [12] propose a trajectory prediction model based on GMM named GMTP. GMTP uses GMM to model complex motion patterns and calculates the probability distribution of different patterns so that the trajectory data are divided into different components, and then predicts the trajectories of moving objects. In this paper, we use GMM to learn the distribution of residents' travelling time, and mine the travelling rush hours.

In the end, we adopt the method proposed by Wu et al. [13] to finish the third task of calculating departure intervals, which will be introduced in Sect. 3.3.

Different from the previous studies that only use DBSCAN or GMM to do travel pattern analysis, this paper combines these two classic clustering methods and then respectively cluster the geographical location and travelling time. On the one hand, we use DBSCAN to cluster geographically similar trajectories and find the appropriate shuttle bus stops. On the other hand, we use GMM to learn the time distribution of residents' travelling (mainly the peak of travelling time).

The structure of this paper is as follows: in Sect. 2, we introduce the experiment approaches we adopt; next, in Sect. 3, we will show the experiments of shuttle bus route design; in the end, in Sect. 4 we summary our work.

2   Model

In this Section, we introduce our experiment model. In Sect. 2.1, we show the travelling trajectory modeling based on DBSCAN algorithm. Next, in Sect. 2.2 we introduce how to use GMM to model travelling time.

2.1   Travelling Trajectory Modeling Based on DBSCAN

Density-Based Spatial Clustering of Applications with Noise(DBSCAN), originally proposed by Ester et al. [14], is a data clustering algorithm based on the notion of density, where clusters are considered as the sets of points that lie inside or on the border of high-density regions in spatial databases [14–16]. That means DBSCAN gathers points which are close to each other and marks noise points which lie alone in low-density regions. With clusters of arbitrary shape, DBSCAN can easily find out expected clusters. The key idea of DBSCAN is that for each point of a cluster the neighborhood of a given radius has to contain at least a minimum number of points. Below are definitions of some symbols.

Our experimental data is provided by Shanghai Telecom, which recording the trajectory information of residents going to ANS center during investigations(about one week). But we only focus on the trajectory that is within one kilometer from ANS center.

The trajectory records within one kilometer from ANS center is referred as $Track = \{Tr_1, Tr_2, \cdots, Tr_i, \cdots, Tr_N\}$, where N is the records number and $Tr_i = \{t_i, l_i\}$, t_i is tracking time, l_i is trajectory point including longitude and latitude, $l_i = (lo_i, la_i)$.

In order to select shuttle bus stops, we extract the geographic information from the original dataset, the new dataset is referred as L, $L = \{l_1, l_2, \cdots, l_i, \cdots, l_N\}$.

Below is the modeling process.

ε-neighborhood, for a given trajectory point l_i and radius ε, ε-neighborhood is the area that with center l_i and radius ε. We use $N_\varepsilon(l_i)$ to express the ε-neighborhood of trajectory point l_i.

MinPts, the minimum number of trajectory points in ε-neighborhood of Core trajectory point.

Core trajectory point, for a given trajectory point l_i, if the number of trajectory points in its ε-neighborhood is not less than MinPts, then l_i is a Core trajectory point.

Border trajectory point, for a given trajectory point l_i, if its ε-*neighborhood* contains points less than *MinPts*, then l_i is a *Border trajectory point*.

Directly density-reachable, if a trajectory point $l_j \in N_\varepsilon(l_i)$ and $|N_\varepsilon(l_i)| >$ *MinPts*, we say l_j is *directly density-reachable* from trajectory point l_i.

Density-reachable, if there is a chain of trajectory points $l_1, l_2, \cdots, l_m$ and l_{i+1} is *directly density-reachable* from $l_i (0 < i < m)$, then we say point l_1 is *density-reachable* from point l_m.

Density-connected, if there is a trajectory point l_t, point l_t is *density-reachable* from point l_i and l_j and then we say point l_i and l_j are *density-connected*. To include two border trajectory points into the same cluster, the two points required to be *density-connected*.

The aim of DBSCAN modeling is to find the collections of *density-connected* trajectory points in L. After modeling, we get the cluster centers $Avg = \{a_1, ..., a_k, ..., a_{Kdbs}\}$ as the candidates of bus stops, where $Kdbs$ is the number of collections and $a_k = (lo_k, la_k)$.

2.2   Travel Time Distribution Modeling Based on GMM

In order to model the travel time distribution, we extract the time information from $Track$, the new dataset is referred as T, $T = \{t_1, t_2, \cdots, t_j, \cdots, t_N\}$. Because the shuttle buses work within a small range(within one kilometer from the center), and the distance between two stops is very short, so we can roughly think that the rush hour of all stops are the same.

GMM is a probabilistic model composed of K Gaussian distributions. The probability of each observation is the result of K Gaussian mixture [17].

Below is the modeling process.

K, number of distributions.

N, number of travel time records.

$\mu = \{\mu_1, \mu_2, \ldots, \mu_i, \ldots, \mu_K\}$, K-dimensional mean vector, where μ_i is the mean of distribution i, which means the i-th rush hour.

$\Sigma = \{\Sigma_1, \Sigma_2, \ldots, \Sigma_i, \ldots, \Sigma_K\}$, K-dimensional variance vector, where Σ_i is the variance of distribution i, which means the range of the i-th rush hour.

$W = \{W_1, W_2, \ldots, W_i, \ldots, W_K\}$, K-dimensional weight vector, where W_i is the weight of distribution i, which means the probability of residents travelling around the i-th rush hour.

$Likeli_f(T, \mu, \Sigma, W)$, likelihood function.

$N(t_{j=1\cdots N} \mid \mu_i, \Sigma_i)$, probability density function of distribution i.

$F(t_{j=1\cdots N} \mid \mu, \Sigma, W)$, probability distribution of travel time record j.

Suppose t_j is generated by GMM. The generation process includes, first, randomly selecting a distribution i from K distributions with probability W_i, then selecting t_j from this distribution with probability $N(t_j \mid \mu_i, \Sigma_i)$. Therefore, the probability of t_j generated by GMM is $\sum_{i=1}^{K} W_i N(t_j \mid \mu_i, \Sigma_i)$.

For a given set of travel time records $T = \{t_1, t_2, \cdots, t_j, \cdots, t_N\}$, suppose we know it is consistent with Gaussian mixture distributions, so the probability density function of T is

$$F(t_j \mid \mu, \Sigma, W) = \sum_{i=1}^{K} W_i N(t_j \mid \mu_i, \Sigma_i), j = 1, 2, \ldots, N \qquad (1)$$

wherein,

$$N(t_j \mid \mu_i, \Sigma_i) = \frac{1}{\sqrt{2\pi \Sigma_i}} \exp\{-\frac{1}{2\Sigma_i}(t_j - \mu_i)^2\} \qquad (2)$$

After modeling, we use EM algorithm to fix (locally) maximum likelihood parameters of GMM. The likelihood function is

$$Likeli_f(T, \mu, \Sigma, W) = \sum_{j=1}^{N} \log F(t_j \mid \mu, \Sigma, W) \qquad (3)$$

The algorithm will not stop until the likelihood function converged.

EM algorithm is to find the maximum likelihood or maximum a posteriori of parameters in statistical models depending on unobserved latent variables. EM algorithm is an iteration method involving two steps which are expectation step (E-step) and maximization step (M-step) [17–19]. Since we use EM algorithm to estimate the parameters of GMM, we will explain E-step and M-step by showing how E-step and M-step work on GMM.

E-step: For a given set of $T = \{t_1, t_2, \cdots, t_j, \cdots, t_N\}$, $w_j(i)$ means the probability of t_j being generated by distribution i. We suppose the algorithm knows all the parameters of GMM (initialization or the results of last iteration), then estimate the $w_j(i)$ with formula 4.

$$w_j(i) = \frac{W_i N(t_j \mid \mu_i, \Sigma_i)}{\sum_{k=1}^{K} W_k N(t_j \mid \mu_k, \Sigma_k)} \qquad (4)$$

M-step: on the basis of the E-step, updating parameters with formula 5, formula 6 and formula 7.

$$\mu_i = \frac{1}{\sum_{j=1}^{N} w_j(i)} \sum_{j=1}^{N} w_j(i) t_j \qquad (5)$$

$$\Sigma_i = \frac{1}{\sum_{j=1}^{N} w_j(i)} \sum_{j=1}^{N} w_j(i)(t_j - \mu_i)^2 \qquad (6)$$

$$W_i = \frac{\sum_{j=1}^{N} w_j(i)}{N} \qquad (7)$$

The parameters are updated with every iteration until the likelihood function converged.

It is obvious that the rush hours of going to the ANS Center and leaving the ANS Center are different, thus we group the dataset T into two subsets TG and TL, indicating the tracking time of going to the ANS center and leaving the ANS center. $TG = \{tg_1, tg_2, \cdots, tg_i, \cdots\}$ and $TL = \{tl_1, tl_2, \cdots, tl_i, \cdots\}$. Then, we model the two datasets with the above method to get the rush hours of going to the ANS center and leaving the ANS center.

3   Experiments

3.1   Shuttle Bus Stops Selecting and Driving Route Design

Figure 1 shows the transport situation around ANS Center, including 13 main bus stations (8 halfway stations, 2 terminal stations and 3 hub stations). Theoretically, the clustering centers of DBSCAN algorithm should near these main stations.

We run DBSCAN on dataset L to cluster the places where people are intensive. Table 1 shows the clustering result $Avg = \{a_1, a_2, a_3, a_4\}$, here $Kdbs$ is 4. It indicates that there are 4 people-intensive places that can be selected as bus stops, see Fig. 2. We can see from Fig. 2 that station B, C and D are all near those main stations in Fig. 1. The nearest station from B is Humin Road, Liuzhou Road station, the nearest station from C is Qinzhou Road, Guansheng Road station and The nearest station from D is Sanjiang Road, Longcao Road station. A is near the Shilong Road station, which is a metro station and does not been mapped out in Fig. 1.

The result indicates that our experimental result is fit of the actual situation and discovers the hidden important stations at the same time.

According to the experimental results and customer demand, we select the final stations from Avg. The selected stations are (A) Shilong Road station,

Table 1. DBSCAN clustering result.

Station	Longitude and latitude
A	(121.44749999999964, 31.161944440000052)
B	(121.43376999999975, 31.162949999999038)
C	(121.43419720000017, 31.172344439999968)
D	(121.44504729999986, 31.173680860000008)

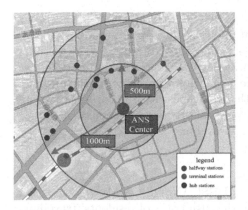

Fig. 1. Graphic presentation of ANS center's transport situation.

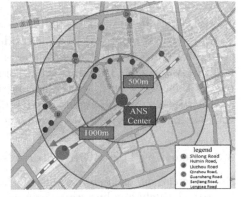

Fig. 2. Station fixing result based on DBSCAN.

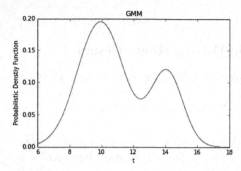
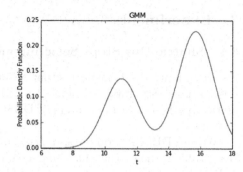

Fig. 3. GMM result of *TrackGoto*.
Rush hour¹: [9:55 a.m., 2:04 p.m.];
Weight¹: [0.706, 0.294].

Fig. 4. GMM result of *TrackLeave*.
Rush hour²: [3:41 p.m., 11:02 a.m.];
Weight¹: [0.641, 0.359].

(B) Humin Road, Liuzhou Road station and (D) Sanjiang Road, Longcao Road station.

Then we design 3 routes considering the actual road conditions.

Route 1, the shuttle bus departs from station D and gets through ANS Center.

Route 2, the shuttle bus departs from station D, then passes station B and gets through ANS Center.

Route 3, the shuttle bus departs from station A and gets through ANS Center.

3.2 Learning Rush Hours

In order to design bus timetable, we need to know the rush hours of shuttle bus routes. We run GMM on dataset TG and TL. Taking the actual situation into account, we set parameter $K = 2$ for both TG and TL. Figures 3 and 4 shows the results of GMM.

We can see that, residents usually go to ANS center at about 9:55 a.m. or about 2:04 p.m., and leave ANS center at about 11:02 a.m. or around 3:41 p.m. $Weight^1$ and $Weight^2$ tell us that residents going to ANS center in the morning are more than afternoon, but residents leaving ANS center in the morning are less than afternoon. This shows that some people go to the center in the morning but will not leave until afternoon, and this is in line with the actual situation. This is because some people spend a day to handle a great amount of things, such as enterprise registration, contract approval and so on.

3.3 Learning Departure Intervals

By using classic machine learning methods DBSCAN and GMM, we can design the routes (including bus stops) and learn the rush hours of residents' travelling. Now we apply these information to the Public Route Scale Estimation Model [13] and calculate the departure intervals.

The calculation formula is

$$C_{od} = 60Rr/t \tag{8}$$

where C_{od} is the carrying capacity of shuttle route (the unit is people-time per rush hour), R is rated passenger number, r is load factor of route per rush hour, t is departure interval (the unit is minute) and the coefficient 60 makes the unit of t to be minute. In our study, the rated passenger number (R) of shuttle buses is 25 and load factor (r) adopted is 0.4 according to [13].

Table 2 shows the statistic results of person-time, which is the carrying capacity (C_{od}) we expected. Since we only have the data of China Telecom, but do not have data of China Mobile and China Unicom, we augment the statistic value with the same percentage of the market share of China Telecom (about 14% at that time).

Table 2. Statistic results of person-time.

Station	Rush hour of TG	Person-time	Rush hour of TL	Person-time
A	9:55 a.m.	92	11:02 a.m.	29
	2:04 p.m.	71	3:41 p.m.	79
D	9:55 a.m.	14	11:02 a.m.	29
	2:04 p.m.	29	3:41 p.m.	43
B	9:55 a.m	214	11:02 a.m.	165
	2:04 p.m.	129	3:41 p.m.	171

Compute departure intervals with formula 8, Table 3 shows the departure intervals of different routes.

Table 3. Departure intervals of rush hours.

Route (station)	Rush hour of TG	Interval (min)	Rush hour of TL	Interval (min)
Route 1(D)	9:55 a.m.	46.86	11:02 a.m.	20.69
	2:04 p.m.	20.69	3:41 p.m.	13.95
Route 2(D+B)	9:55 a.m.	2.63	11:02 a.m.	3.09
	2:04 p.m.	3.80	3:41 p.m.	2.80
Route 3(A)	9:55 a.m.	6.52	11:02 a.m.	20.69
	2:04 p.m.	8.45	3:41 p.m.	7.59

It should be noted that our main task is to use the data analysis methods to provide options for the Xuhui District government and then the government makes the final decision. We are proud that the government staff affirm the work we have done. They will design a questionnaire to collect social opinions about our three routes and then, in conjunction with real-word road conditions and budgetary considerations, they will establish the final design on the basis of our project.

4   Conclusion

In this paper, we use the classic machine learning methods to dig out the important hidden information of the mobile phone base station data, including the residents' spatial travel pattern and temporal habits. Then we apply the information to traffic model and successfully design the shuttle bus routes (including bus stops and rush hours) and calculate departure intervals. The experiment results show that our modeling methods combined with traffic knowledge can effectively solve bus route planning problems. However, it is important to note that the experiment is made on small data set, which may lead to error. But, the research in this paper provides a valuable reference on traffic planning. In the future, we will do experiments on larger data set and the results can get closer to the reality.

Acknowledgments. The work is partially supported by the National Nature Science Foundation of China (Nos. 61573259 and 61673301), the program of Further Accelerating the Development of Chinese Medicine Three Year Action of Shanghai (No. ZY3-CCCX-3-6002), and the National Science Foundation of Shanghai (No. 15ZR1443800).

References

1. Calabrese, F., Ferrari, L., Blondel, V.D.: Urban sensing using mobile phone network data: a survey of research. ACM Comput. Surv. **47**(2), 1–20 (2014)
2. Li, P., Gao, Y.W., Wu, J.W., Li, X., Wu, B.B.: Residents traveling track and analysis methods based on mobile phone data. Adv. Mater. Res. **926**, 2730–2734 (2014)
3. Wu, M., Dong, H., Ding, X., Shan, Q., Chu, L., Jia, L.: Traffic semantic analysis based on mobile phone base station data. In: International Conference on Intelligent Transportation Systems, pp. 617–622. IEEE (2014)
4. Dong, H., Wu, M., Ding, X., Chu, L., Jia, L., Qin, Y., Zhou, X.: Traffic zone division based on big data from mobile phone base stations. Transp. Res. Part C: Emerg. Technol. **58**, 278–291 (2015)
5. Ma, X., Wu, Y.J., Wang, Y., Chen, F., Liu, J.: Mining smart card data for transit riders travel patterns. Transp. Res. Part C Emerg. Technol. **36**, 1–12 (2013)
6. Le, M.K., Bhaskar, A., Chung, E.: Passenger segmentation using smart card data. IEEE Trans. Intell. Transp. Syst. **16**(3), 1537–1548 (2015)
7. Le, M.K., Bhaskar, A., Chung, E.: A modified density-based scanning algorithm with noise for spatial travel pattern analysis from Smart Card AFC data. Transp. Res. Part C: Emerg. Technol. **6** (2015, in press). Corrected Proof, April 2015
8. Fraley, C., Raftery, A.E.: Model-based clustering, discriminant analysis, and density estimation. J. Am. Stat. Assoc. **97**(458), 611–631 (2002)

9. Cui, Z., Long, Y., Ke, R., Wang, Y.: Characterizing evolution of extreme public transit behavior using smart card data. In: 2015 IEEE 1st International Smart Cities Conference (ISC2), pp. 1–6 (2015)
10. Briand, A.S., Côme, E., Mohamed, K., Oukhellou, L.: A mixture model clustering approach for temporal passenger pattern characterization in public transport. Int. J. Data Sci. Anal. 1(1), 37–50 (2016)
11. Lee, M., Sohn, K.: Inferring the route-use patterns of metro passengers based only on travel-time data within a Bayesian framework using a reversible-jump Markov chain Monte Carlo (MCMC) simulation. Transp. Res. Part B: Methodol. 81, 1–17 (2015)
12. Qiao, S.J., Jin, K., Han, N., Tang, C.J., Gesangduoji, G.L.A.: Trajectory prediction algorithm based on Gaussian mixture model. J. Softw. 26(5), 1048–1063 (2015). Jian, R., BaoX. (eds.)
13. Wu, J., Zheng, Y., Chen, X.: Approaches to planning of subway station transfer facility in urban areas. J. Tongji Univ. (Nat. Sci.) 36(11), 1501–1506 (2008)
14. Ester, M., Kriegel, H.P., Sander, J., Xu, X.: A density-based algorithm for discovering clusters a density-based algorithm for discovering clusters in large spatial databases with noise. In: International Conference on Knowledge Discovery and Data Mining, pp. 226–231. AAAI Press (1996)
15. Campello, R.J.G.B., Moulavi, D., Sander, J.: Density-based clustering based on hierarchical density estimates. In: Pei, J., Tseng, V.S., Cao, L., Motoda, H., Xu, G. (eds.) PAKDD 2013. LNCS (LNAI), vol. 7819, pp. 160–172. Springer, Heidelberg (2013). doi:10.1007/978-3-642-37456-2_14
16. Sander, J., Ester, M., Kriegel, H.P., Xu, X.: Density-based clustering in spatial databases: the algorithm gdbscan and its applications. Data Min. Knowl. Disc. 2(2), 169–194 (1998)
17. Yu, G., Sapiro, G., Mallat, S.: Solving inverse problems with piecewise linear estimators: from Gaussian mixture models to structured sparsity. IEEE Trans. Image Process. 21(5), 2481–2499 (2012)
18. Dempster, A.P., Laird, N.M., Rubin, D.B.: Maximum likelihood from incomplete data via the EM algorithm. J. Roy. Stat. Soc. Ser. B (Methodol.) 1–38 (1977)
19. Kehtarnavaz, N., Nakamura, E.: Generalization of the EM algorithm for mixture density estimation. Pattern Recogn. Lett. 19(2), 133–140 (1998)

Extracting Deep Semantic Information for Intelligent Recommendation

Wang Chen, Hai-Tao Zheng$^{(\boxtimes)}$, and Xiao-Xi Mao

Tsinghua-Southampton Web Science Laboratory, Graduate School at Shenzhen,
Tsinghua University, Shenzhen, China
{chen-w16,mxx14}@mails.tsinghua.edu.cn, zheng.haitao@sz.tsinghua.edu.cn

Abstract. In recent years, there have been many works focusing on combing ratings and reviews to improve the performance of recommender system. Comparing with the rating based algorithms, these methods can be used to alleviate the data sparsity problem in a certain extent. However, they lack the ability to extract the deep semantic information from plaintext reviews. In addition, they do not take the consistence of the latent semantic space of user profiles and item representations into account. To address these problems, we propose a novel method named as Deep Semantic Hybrid Recommendation Method (DSHRM). We utilize deep learning technologies to extract user profiles and item representations from reviews and make sure both of them are in a consistent latent semantic space. We combine ratings and reviews to generate better recommendations. Extensive experiments on real-world datasets show that our method significantly outperforms other six state-of-the-art methods, including LFM, SVD++, CTR, RMR, BoWLF and LMLF methods.

Keywords: Recommender system · Deep learning · Text mining

1   Introduction

With the rapid growth of the Internet, we have entered into an information overload era. Today recommender systems have been incorporated into many application areas. The main purpose of the recommender system is to predict the user's preference for a particular item.

Although traditional rating based algorithms seem to achieve significant success, such as Latent Factor Model (LFM) [1] and SVD++ [2]. These methods are still suffering from the data sparsity problem. Along with an integer-valued score, users often post reviews which generally describe why they like or dislike the items. Some hybrid methods combing ratings and reviews have been proposed to alleviate the data sparsity problem, such as Hidden Factors as Topics (HFT) [3], Collaborative Topic Regression (CTR) [4] and Ratings Meet Reviews (RMR) [5]. However, these methods cannot extract the deep semantic information from reviews effectively. Moreover, they do not take the consistence of the latent semantic space of user profiles and item representations into account. Consequently, they may be not capable of making recommendations accurately.

© Springer International Publishing AG 2017
D. Liu et al. (Eds.): ICONIP 2017, Part V, LNCS 10638, pp. 134–144, 2017.
https://doi.org/10.1007/978-3-319-70139-4_14

To address problems described above, we propose a method combing ratings and reviews. We extract user profiles and item representations from reviews with Variational Autoencoder (VAE) model [6]. Specifically, the main contributions of this paper can be summarized as the following three aspects:

- To generate better recommendations, we propose a novel method named as Deep Semantic Hybrid Recommendation Method (DSHRM) combing ratings and the deep semantic information of reviews. Moreover, the method can also be used in real-time recommendation task.
- We utilize an effective, unsupervised deep learning model called as VAE to extract deep semantic information from plaintext reviews. Simultaneously, we shape user profiles and item representations by these deep semantic information and make sure both of them are in a consistent latent semantic space.
- Extensive experiments on real-world datasets show that our method significantly outperforms other six state-of-the-art methods.

The remainder of this paper is organized as follows. In Sect. 2, we review relevant prior work on recommendation algorithms. In Sect. 3, we present the framework of our DSHRM. In Sect. 4, we evaluate DSHRM with the state-of-the-art recommendation algorithms. In Sect. 5, we conclude our work.

2   Related Work

Many rating based recommendation methods have achieved great success [1, 2, 7, 8]. Simon Funk proposed a new recommendation algorithm based on Singular Value Decomposition (SVD) [8] in 2006 Netflix Prize and this algorithm was later called as LFM. The main idea of LFM is to associate user interests and items with latent factors. The real world dataset is extremely sparse and degrades the performance of these methods significantly.

Reviews contain rich semantic information about users and items. These semantic information can be a great supplement to the ratings. To improve the performance of recommender system, more and more works focus on combing ratings and semantic information of reviews. Most of them are based on topic model and latent Dirichlet allocation (LDA) [9] is the most prevalent topic model. The LDA-based methods [3–5] model reviews to regularize matrix factorization. However, LDA can be treated as three layers Bayesian model so it has limited ability to extract the deep semantic information from plaintext reviews.

In recent years, deep learning model has been gradually introduced into the recommender system [10–14]. Bag-of-Words Regularized Latent Factor Model (BoWLF) [14] based on a bag-of-words model and Language Model Regularized Latent Factor Model (LMLF) [14] based on a recurrent neural network are proposed to extract the deep semantic information from reviews. However, both of them do not consider the consistence of the latent semantic space of user profiles and item representations. Consequently, precise recommendations cannot be made.

3 Deep Semantic Hybrid Recommendation Method

3.1 Background

Because many methods including our method are based on LFM, we provide a brief description of LFM here. We can approximate the predicted rating as:

$$y_{ij}^{new} = \mu + b_i + b_j + (\mathbf{u}_i)^T \mathbf{v}_j. \tag{1}$$

Notations:

- y_{ij}^{new} is the predicted rating that user i for item j.
- μ is the global average rating of the training set.
- b_i is the bias of user i. b_i is irrelevant to all items. Some users who are more tolerant and pleased with all kinds of good, are used to giving high ratings, while some users who are more demanding and thinking highly of the quality of goods, are used to giving low ratings.
- b_j is the bias of item j. b_j is irrelevant to all users. Some items which are good quality, are used to receiving high ratings, while the items of inferior quality are used to receiving low ratings.
- $\mathbf{u}_i$ is latent factors of user i and $\mathbf{v}_j$ is latent factors of item j. $\mathbf{u}_i$ and $\mathbf{v}_j$ are both k-dimensional vectors. k is the number of latent factor's dimension.

3.2 DSHRM Framework

We propose DSHRM Framework showed in Fig. 1. The framework consists of two main steps:

- **Step 1: Extracting user profiles and item representations.** Basing on reviews from users and items, we utilize a deep neural network named as VAE to extract semantic information from plaintext reviews. Subsequently, we use the deep semantic information to shape user profiles and item representations.
- **Step 2: Rating Prediction.** We introduce user profiles and item representations which are both exhibited by the deep semantic information of reviews. Then we train a rating function for rating prediction.

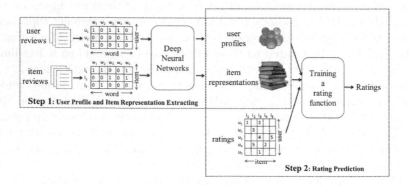

Fig. 1. DSHRM framework

3.3   Extracting User Profiles and Item Representations

Variational Autoencoder (VAE) model is a generative model which inherits the standard autoencoder architecture. VAE uses variational approach for latent representation learning which can be used in unsupervised task. We adopt VAE to extract the latent representation of reviews and use them to shape user profiles and item representations.

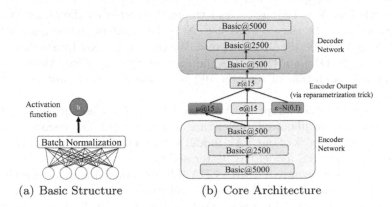

(a) Basic Structure   (b) Core Architecture

Fig. 2. The architecture of our variational autoencoder model for plaintext reviews.

Considering that the representation ability of a single layer neural network is limited, we adopt multi-layer neural network. The core architecture of our multi-layer neural network is shown in Fig. 2(b). z is the latent representation vector which can be approximated by μ, σ and ϵ. μ is the mean; σ is the standard deviation; and ϵ is a sample from $N(0, I)$. μ and σ are linear transformations from the output of the encoder network. Therefore, μ and σ are computed as:

$$\mu = W_\mu h_{lo} + b_\mu, \tag{2}$$

$$\sigma = W_\sigma h_{lo} + b_\sigma, \tag{3}$$

where W_μ and W_σ are weight vectors, h_{lo} is the output of the encoder network, b_μ and b_σ are the biases.

z is the latent representation vector and can be computed as:

$$z = \mu + \sigma \cdot \epsilon. \tag{4}$$

The encoder network, the decoder network, and the hidden vector z connected to the first layer of the decoder network have basic structure shown in Fig. 2(a). For each layer $l \in \{1, \ldots, L\}$ of encoder or decoder, The output of layer l is as follows:

$$\mathbf{h}_l = f(batch\_norm(\mathbf{W}_l \mathbf{h}_l + b_l), \tag{5}$$

where $\mathbf{W}_l$ is the weight vector, b_l is the bias, $batch\_norm(\cdot)$ is the step of Batch Normalization, and $f(\cdot)$ denotes activation function, such as tanh. $\boldsymbol{h}_0$ is the input layer and is shown as the bottom layer in Fig. 2(b). Batch Normalization [15] is added between linear operation and the nonlinear activation function. It accelerates deep network training and helps prevent the vanishing (and exploding) gradient problem.

In practice, after data preprocessing, reviews of users and received reviews for items can be mapped to word vectors. For example, each user or item can be represented as $\boldsymbol{X} = \{w_1, w_2 \ldots w_n\}$ where w_n is whether the review contains the word and n is the size of word list. $\boldsymbol{X}$ is the input of the encoder network and the output of the decoder network. We train the VAE by stochastic gradient descent (SGD) algorithm. $\boldsymbol{\mu}$ is the result output we need. Moreover, $\boldsymbol{\mu}$ is the latent representation of reviews and is regarded as user profiles and item representations.

The reasons why we adopt the VAE model to extract user profiles and item representations from plaintext review are as follows: (1) The encoder network and the decoder network of our VAE model are both multi-layer neural networks. Consequently, our VAE model has better ability to extract deep semantic information from reviews compared with LDA. (2) VAE model is more effective and robust on extracting for global semantic features than standard autoencoders. (3) User word vectors and item word vectors flow through the same VAE model to make sure the latent semantic space is consistent. Therefore, the operation of them will be more reasonable.

3.4   Rating Prediction

To predict the rating of user i for item j, we introduce user profiles and item representations both shaped by the semantic information extracted from reviews. Assuming that user profiles and item representations are both n-dimensional vectors, our rating function is formulated as follows:

$$y_{ij}^{new} = \mu + b_i + b_j + (\boldsymbol{u}_i)^T \boldsymbol{v}_j + \alpha(\boldsymbol{x}_i)^T \boldsymbol{z}_j. \tag{6}$$

Notations:

- $\boldsymbol{x}_i$ is the profile of user i extracted from all reviews posted by user i. $\boldsymbol{x}_i$ is a n-dimensional vector.
- $\boldsymbol{z}_j$ is the representation of item j extracted from all received for item j. $\boldsymbol{z}_j$ is also a n-dimensional vector.
- α is the weight of dot product operation between $\boldsymbol{x}_i$ and $\boldsymbol{z}_j$ in the Eq. 6.

Other parameters is consistent with Eq. 1.

DSHRM is a hybrid method. Equation 6 consists of two key components. One component is a matrix factorization which is the same as Eq. 1, and the other is the semantic operation of the user profile $\boldsymbol{x}_i$ and the item representation $\boldsymbol{z}_j$. As mentioned in Sect. 3.3, $\boldsymbol{x}_i$ and $\boldsymbol{z}_j$ share a consistent latent semantic space while $(\boldsymbol{x}_i)^T \boldsymbol{z}_j$ is approximated as the similarity of them.

The objective function of our method can be formulated as follows,

$$L = min_{b_i,b_j,u_i,v_j} \sum_{i,j \in Train} [(y_{ij} - y_{ij}^{new})^2 + \lambda(b_i^2 + b_j^2 + ||u_i||_2^2 + ||v_j||_2^2)], \quad (7)$$

where Train is the training set, λ is the regularization parameter and $|| \cdot ||_2$ is L^2 norm.

From Eqs. 6 and 7, we know that μ can be computed directly, x_i and z_j can be generated by our VAE model. Therefore, b_i, b_j, u_i and v_j are the only four parameters which should be learned. We use SGD algorithm to learn these parameters. The update rules are:

$$e_{ij} = y_{ij} - y_{ij}^{new}, \quad (8)$$

$$b_i = b_i + \eta(e_{ij} - \lambda \cdot b_i), \quad (9)$$

$$b_j = b_j + \eta(e_{ij} - \lambda \cdot b_j), \quad (10)$$

$$u_i = u_i + \eta(e_{ij}v_j - \lambda u_i), \quad (11)$$

$$v_j = v_j + \eta(e_{ij}u_i - \lambda v_j), \quad (12)$$

where η is learning rate.

Our rating function has the following characteristics: (1) Improving the recommendation accuracy for the long-tail users. Because reviews contain rich semantic information, we can shape the user profiles and the item representations effectively by a few reviews so as to improve the recommendation accuracy for the long-tail users. (2) Alleviating the data sparsity problem. Our method is a hybrid method. When the data is very sparse, the performance of matrix factorization is excessively poor. Nevertheless, the other part of our method is not much affected by the problem of data sparsity because user reviews and item representations can be extracted from only one review. (3) Improving the performance of real-time recommendation. It is difficult for standard LFM to recommend in real time because each recommendation needs to scan all the user behaviors and needs to run multiple iterations for accuracy. In practice, there is only one time to train the model and generate the recommendation results in a single day. Our method can tackle the problem. When the user has a new behavior on the item, user profiles and item representations can be updated in time by our trained VAE model. Through this process, the performance of the real-time recommendation is improved by the estimation of the interest of the user i for the item j by $(x_i)^T z_j$.

4   Experiments

4.1   Datasets

We conduct extensive experiments to evaluate DSHRM's performance. We use 8 real-world amazon datasets[1] to evaluate our method. Table 1 shows statistics

---

[1] https://snap.stanford.edu/data/web-Amazon.html.

Table 1. Statistics of the datasets

Dataset	# users	# items	# ratings (# reviews)	# rating/# users	# rating/# items
Arts	24071	4211	27980	1.162	1.162
Watches	62041	10318	68356	1.101	6.624
Musical instruments	67007	14182	85405	1.274	6.022
Software	68464	11234	95084	1.388	8.463
Office products	110472	14244	138084	1.249	9.694
Patio	166832	19531	206250	1.236	10.560
Pet supplies	160496	17523	217170	1.353	12.393
Beauty	167725	29004	252056	1.502	8.690
Total	827108	120247	1090385	1.318	9.067

of these datasets. Each dataset is one category of product such as arts, watches, musical instruments and so on. Each sample in the dataset includes a user id, an item id, an integer-valued rating (1 to 5 points), a plaintext review and so on. Figure 3 shows statistics of long-tail users of three large-scale datasets.

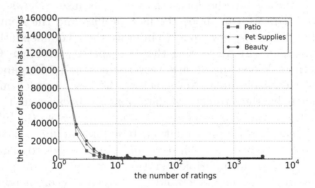

Fig. 3. Statistics of user behaviors

We can find two facts from Table 1. First, these datasets are extremely sparse, which causes the poor performance of the existing models only based on ratings. Second, each user post 1.318 reviews on average and each item receives 9.067 reviews on average. These reviews undoubtedly help us improve the accuracy of our method and the results will be demonstrated later. Figure 3 shows an interesting phenomenon in our datasets that the trade of user behaviors is similar to the power law distribution. Most users have few behaviors and these users can be called as long-tail users.

4.2 Experiment Setting

Preprocess. We randomly divide each dataset into training/validation/test sets in a 80%/10%/10% split consistent with baseline methods. Basing on the reviews, we extract the 5000 most frequent words to build a word list for each dataset. We use a 5000-dimensional word vector to represent a user in the basis of all reviews posted by this user. Likewise, an item is represented by a 5000-dimensional word vector. The word vector is the encoder input and decoder output of VAE.

Baselines and Evaluation Metric. In order to evaluate the performance of our method, we compare it with other six baseline methods: LFM from [1] and SVD++ from [2] which are both only based on ratings; CTR from [4] and RMR from [5] which are both based on LDA; BoWLF and LMLF from [14] which are based on neural networks.

We use LFM and SVD++, implemented by LibRec[2], to predict the rating. The source codes of other baseline methods are unavailable. We report the results from [5] for CTR and RMR, as well as the results from [14] for BoWLF and LMLF. Moreover, our VAE model is implemented by Tensorflow[3].

We adopt mean squared error (MSE) as the evaluation metric,

$$MSE = \frac{1}{|Test|} \sum_{i,j \in Test} (y_{ij} - y_{ij}^{new})^2, \tag{13}$$

where $|Test|$ is the total number of ratings in the test set.

Model Setting and Training. The encoder network and the decoder network of our VAE model are both multi-layer neural networks. The outputs of each layer are 2500 bits, 500 bits and 15 bits respectively. Finally, user profiles and item representations are shaped by 15-dimensional vectors which are the output of the encoder network. Specially, only reviews of the training set are put into VAE model because reviews of the test set can not be observed. We train the VAE by SGD algorithm and tune hyperparameters on the Arts dataset which is the smallest one of all datasets. We keep same hyperparameters of model on other dataset. We set the number of batch, the number of epoch and the learning rate to 100, 3 and 0.001 respectively.

In addition, we use SGD algorithm to learn the parameters of Eq. 7 and use grid search to determine the optimal hyperparameters. Finally, we set hyperparameters α, λ and the learning rate η used in SGD algorithm to 0.5, 0.01 and 0.002. The dimension of user's (item's) latent factors k is set to 5 consistent with baseline methods.

---

[2] https://www.librec.net.
[3] https://www.tensorflow.org.

4.3 Rating Prediction Analysis

Table 2 compares the performance of rating prediction with the approach LFM, SVD++, CTR, RMR, BoWLF, LMLF and DSHRM. First, we can see that our method obtains the lowest MSE on the six datasets and takes third on the other two datasets. Second, the average MSE result of our proposed method is 1.564 which outperforms state-of-the-art methods significantly. In detail, our method improves the prediction accuracy by up to 1.58, 7.55, 4.22, 0.55, 0.77, and 1.77% compared with LFM, SVD++, CTR, RMR, BoWLF and LMLF respectively.

Table 2. MSE results of various methods

Dataset	LFM	SVD++	CTR	RMR	BoWLF	LMLF	DSHRM
Arts	1.395	1.442	1.471	1.371	1.413	1.426	**1.367**
Watches	1.461	1.517	1.491	1.458	1.466	1.473	**1.447**
Musical instruments	1.390	1.465	1.422	1.374	1.375	1.388	**1.370**
Software	2.230	2.525	2.254	**2.173**	2.174	2.203	2.177
Office products	1.625	1.726	1.733	1.638	1.629	1.646	**1.606**
Patio	1.682	1.801	1.720	1.669	1.674	1.680	**1.667**
Pet supplies	1.567	1.700	1.613	1.562	**1.536**	1.544	1.554
Beauty	1.366	1.431	1.361	1.334	1.335	1.370	**1.325**
Average	1.589	1.700	1.633	1.572	1.575	1.591	**1.564**

The reasons why our method performs best are as follows: (1) Reviews contain rich semantic information and indeed help us improve accuracy compared with LFM and SVD++ which are only based on ratings. (2) The VAE model we used can extract the deep semantic information from plaintext review effectively compared with CTR and RMR which are both based on LDA. (3) We ensure the semantic operations of user profiles and item representations are done in a consistent latent semantic space which is more reasonable compared with BoWLF and LMLF which are based on neural networks.

4.4 Long-Tail Users Analysis

To evaluate DSHRM's performance for long-tail users, we conduct experiment on the three large-scale datasets. Figure 4 shows the relationship between MSE (y-axis) and users who have the same number of ratings (x-axis).

We can see that DSHRM performs best for long-tail users. Specifically, the performance of DSHRM is better when user behavior is limited. The main reason is that the deep semantic information extracted from plaintext review is key to shape user profiles and item representations effectively. With the effective user profiles and the item representations, we can predict the ratings accurately.

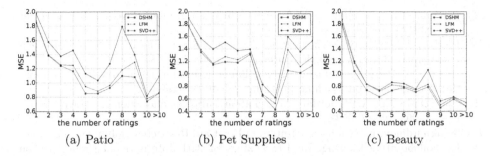

(a) Patio

(a) Patio   (b) Pet Supplies   (c) Beauty

Fig. 4. MSE results of long-tail users

5   Conclusions and Future Work

In this paper, we utilize Variational Autoencoder model to extract deep semantic information from plaintext reviews. We use the deep semantic information to shape user profiles or item representations. Furthermore, we ensure the semantic operations of user profiles and item representations are done in a consistent latent semantic space thus the operations are more reasonable. We propose a deep semantic hybrid method combing user profiles, item representations and ratings. Experiments show that our method performs best compared with other six state-of-the-art methods. In the future, we will investigate other advanced deep learning models or reinforcement learning models to extract user profiles and item representations more effectively to achieve better recommendations.

Acknowledgements. This research is supported by National Natural Science Foundation of China (Grant No. 61375054), Natural Science Foundation of Guangdong Province (Grant No. 2014A030313745), Basic Scientific Research Program of Shenzhen City (Grant No. JCYJ20160331184440545), and Cross fund of Graduate School at Shenzhen, Tsinghua University (Grant No. JC20140001).

References

1. Koren, Y., Bell, R.M., Volinsky, C., et al.: Matrix factorization techniques for recommender systems. J. IEEE Comput. **42**, 30–37 (2009)
2. Koren, Y.: Factor in the neighbors: scalable and accurate collaborative filtering. J ACM Trans. Knowl. Disc. Data **4**, 1–24 (2010)
3. Mcauley, J., Leskovec, J.: Hidden factors and hidden topics: understanding rating dimensions with review text. In: Proceedings of 7th ACM Conference on Recommender Systems, pp. 165–172 (2013)
4. Wang, C., Blei, D.M.: Collaborative topic modeling for recommending scientific articles. In: Proceedings of 17th ACM SIGKDD International Conference on Knowledge Discovery and Data Mining, pp. 448–456. ACM (2011)
5. Ling, G., Lyu, M.R., King, I., et al.: Ratings meet reviews, a combined approach to recommend. In: Proceedings of 8th ACM Conference on Recommender Systems, pp. 105–112 (2014)

6. Kingma, D.P., Welling, M.: Auto-encoding variational Bayes. arXiv preprint arXiv:1312.6114 (2013)
7. Linden, G., Smith, B., York, J., et al.: Amazon.com recommendations: item-to-item collaborative filtering. J. IEEE Internet Comput. **7**, 76–80 (2003)
8. Billsus, D., Pazzani, M.J.: Learning collaborative information filters. In: International Conference on Machine Learning, pp. 46–54 (1998)
9. Blei, D.M., Ng, A.Y., Jordan, M.I.: Latent Dirichlet allocation. J. Mach. Learn. Res. **3**, 993–1022 (2003)
10. Salakhutdinov, R., Mnih, A., Hinton, G.E., et al.: Restricted Boltzmann machines for collaborative filtering. In: Proceedings of 24th International Conference on Machine Learning, pp. 791–798 (2007)
11. Gao, J., Pantel, P., Gamon, M., et al.: Modeling interestingness with deep neural networks. In: Conference on Empirical Methods in Natural Language Processing, pp. 2–13 (2014)
12. Cheng, H.T., Koc, L., Harmsen, J., et al.: Wide & deep learning for recommender systems. In: Proceedings of 1st Workshop on Deep Learning for Recommender Systems, pp. 7–10 (2016)
13. Wang, H., Wang, N., Yeung, D., et al.: Collaborative deep learning for recommender systems. In: Proceedings of 21th ACM SIGKDD International Conference on Knowledge Discovery and Data Mining, pp. 1235–1244 (2015)
14. Almahairi, A., Kastner, K., Cho, K., et al.: Learning distributed representations from reviews for collaborative filtering. In: Proceedings of 9th ACM Conference on Recommender Systems, pp. 147–154 (2015)
15. Ioffe, S., Szegedy, C.: Batch normalization: accelerating deep network training by reducing internal covariate shift. arXiv preprint arXiv:1502.03167 (2015)

A Hybrid Method of Sine Cosine Algorithm and Differential Evolution for Feature Selection

Mohamed E. Abd Elaziz[1,4], Ahmed A. Ewees[2], Diego Oliva[3], Pengfei Duan[1], and Shengwu Xiong[1(✉)]

[1] School of Computer Science and Technology,
Wuhan University of Technology, Wuhan, China
abd_el_aziz_m@yahoo.com, xiongsw@whut.edu.cn
[2] Department of Computer, Damietta University, Damietta, Egypt
ewees@du.edu.eg
[3] Departamento de Ciencias Computacionales, Universidad de Guadalajara,
CUCEI Av. Revolucion 1500, Guadalajara, Jalisco, Mexico
diego.oliva@cucei.udg.mx
[4] Department of Mathematics, Faculty of Science, Zagazig University, Zagazig, Egypt

Abstract. The feature selection is an important step to improve the performance of classifier through reducing the dimension of the dataset, so the time complexity and space complexity are reduced. There are several feature selection methods are used the swarm techniques to determine the suitable subset of features. The sine cosine algorithm (SCA) is one of the recent swarm techniques that used as global optimization method to solve the feature selection, however, it can be getting stuck in local optima. In order to solve this problem, the differential evolution operators are used as local search method which helps the SCA to skip the local point. The proposed method is compared with other three algorithms to select the subset of features used eight UCI datasets. The experiments results showed that the proposed method provided better results than other methods in terms of performance measures and statistical test.

Keywords: Feature selection (FS) · Sine Cosine Algorithm (SCA) · Differential evolution (DE) · Metaheuristic (MH)

1 Introduction

In computational processing, there are some criteria should be considered such as the type of the problem, the purpose of the processing, the size of the problem, and the computational time. The problem's size affects directly on the accuracy of the results and the computational time. Therefore, many researchers work to solve these issues by reducing the dimension of the data to be more homogeneous and coherent. The best method to reduce the problem's dimension is selecting the most important and significant features and deleting the redundant or irrelevant features [1].

© Springer International Publishing AG 2017
D. Liu et al. (Eds.): ICONIP 2017, Part V, LNCS 10638, pp. 145–155, 2017.
https://doi.org/10.1007/978-3-319-70139-4_15

Various techniques are used in dimensionally reduction process; however, the most effective one is using optimization methods such as swarm intelligence (SI) or evolutionary computation approaches to help in escaping from get trapping in local optima. There are several optimization techniques that inspired from the social behavior in natural such as particle swarm optimization (PSO) [2], differential evolution (DE) [3], artificial bee colony (ABC) [4], social spider optimization (SSO) [5], gray wolf optimization (GWO) [7], ant colony optimization (ACO) [6], and sine cosine algorithm (SCA) [8]. Most of these techniques are applied to solve the dimensionality reduction problems. For example, ABC algorithm had been used in [1] to select the most important features and evaluated over ten datasets. Its results had been compared with those results of Genetic Algorithm (GA) as well as PSO and showed high classification accuracy. Bare bones PSO also was used to reduce the features of some datasets and its results proved the best accuracy than using all features. A binary ALO method was introduced by [9] to choose the most important feature. The classification accuracy of this method outperformed those results of bat algorithm, PSO, and GA. Also, SCA was applied by [10] to reduce the problem attributes to get a high classification accuracy. It had been tested over ten datasets against PSO and GA, and showed the best classification accuracy and computational time for most of these datasets.

DE was used as a method of dimensionality reduction in several studies such as DE has been used successful in [11] to improve the classification accuracy and reduce the training time. Yang et al. [12] used DE to select the most important features in the real-time process by using wavelet and statistical features extracted from tool wear. The experimental results of DE outperformed all the compared approaches. Whereas, [13] used DE for feature selection task to increase the face recognition accuracy. The algorithm was evaluated using 10 images and showed the best performance compared with GA. DE also used widely in [14,15]. In addition, there are many research efforts to introduce various methods in this trend such as [16–18]; each of which have its strength and showed good exploration and exploitation in dimensionality reduction tasks; however, they still need further processing and efforts to improve their accuracy and reduce the consuming time.

The hybridization of two techniques is a well-known method to combine the strength of each technique in one approach and benefit from them. Therefore, some successfully studies are listed in the following. [19] introduced a hybrid model that combined DE and artificial bee colony and it performed well against basic algorithms in increasing the accuracy of classification problems and decreasing the classification time. Another hybridization between PSO and DE provided by [20] for feature selection to identify of Diabetic Retinopathy; the results exhibited better accuracy than the compared algorithms. Although there are many techniques used in feature selection, there is no one technique can solve all problems. So, in this paper, we produce a new hybrid algorithm that combined the advantages and strengths of DE and SCA. In which it can be observed from the behavior of the SCA algorithm that it works in exploration better than exploitation. Therefore, its performance can be improved through using an efficient local search algorithm such as DE, which made the SCA avoids getting stuck in optimal local solution. In this way, this modification will lead to improve the convergence of SCA and reduce the time complexity.

This paper is organized as follows, Sect. 2 introduces a brief description of DE and SCA algorithms. The proposed method is explained in Sect. 3. Results and discusses are listed in Sect. 4. In Sect. 5, the conclusion and future work are listed.

2   Preliminaries

2.1   Sine Cosine Algorithm (SCA)

Mirjalili [8] in 2016 proposed a new optimization algorithm based on the forms of sine and cosine in mathematics; which called Sine Cosine Algorithm (SCA).

The SCA begins by generating random solutions. And it searches for the best solution by making several iterations. Meanwhile these iterations, the search space is maintained by adjusting the range of the sine and cosine based on their mathematical forms. These loops are repeated until the stop condition is satisfied. The mathematical form of SCA is determined as follow.

$$x_i^{t+1} = \begin{cases} x_i^t + n_1 \times sin(n_2) \times |n_3 P_i^t - x_i^t|, & n_4 > 0.5 \\ x_i^t + n_1 \times cos(n_2) \times |n_3 P_i^t - x_i^t|, & n_4 \leq 0.5 \end{cases} \tag{1}$$

where x_i^{t+1} is the current solution's position, t indicates the current iteration, n_1, n_2, and n_3 are generated randomly, P_i is a position of the target point in ith dimension, n_4 is generated randomly in $[0, 1]$.

The SCA uses the following equation to balance exploration and exploitation phases.

$$n_1 = c - t \times \frac{c}{t_{max}} \tag{2}$$

where c is a constant and t_{max} is the iterations' length.

2.2   The Basic Differential Evolution

Storn and Price proposed the differential evolution (DE) as an optimization technique [3]. DE has many advantages such as fast convergence, low computational time, and fast and accurate local selection operator. DE starts by generating a population, this population is updated and evaluated in each iteration, and its fitness value is checked if this value is less than the old one the population is updated. As in the following equation, DE generates mutant solution z_i for each solution x_i [3].

$$z_i = x_{r_1} + \beta(x_{r_2} - x_{r_3}), \tag{3}$$

where r_1, r_2, and r_3 indicates random indexes that differ from the running index and $\beta > 0$ indicates the mutation scaling factor $\in [0, 2]$. This equation adds a weighted difference between r_2 and r_3 to r_1.

This step is followed by a crossover phase to generate a new vector v_i that mixed the items of two vectors (*i.e.* z_i and x_i) to increase the diversity of the population vector, as in the following:

$$v_i^t = \begin{cases} z_i^t & if\ \omega \leq C_r\ or\ t = r(i) \\ x_i^t & otherwise \end{cases} \tag{4}$$

where ω is a random value $\in [0, 1]$, C_r is a constant value $\in [0, 1]$ that indicates the crossover probability, t is a current iteration, and $r(i)$ is an integer value that represents a random index.

The selection phase is applied to choose the best solution for using in the next generation; a compassion between x_i and v_i is applied and the best one is used and the other is ignored; as follows:

$$x_i = \begin{cases} v_i & if\ f(v_i) < f(x_i) \\ x_i & otherwise \end{cases} \tag{5}$$

3   The Proposed Method

In this section, the proposed feature selection method that called SCADE is introduced, in which it consists of three stages, initial, updating and the classification, these stages are discussed in the following subsections.

Here, the solution represents as a binary vector with length equal to the number of features of the dataset. The elements of this binary vector determined the feature that must be selected (that corresponding to 1's) and those that ignored (that corresponding to 0's). In which there are several methods can by used to convert the continuous vector into binary while we select the basic one as in the following equation:

$$x_i^{(t+1)} = \begin{cases} 1 & if\ x_i > \sigma \\ 0 & otherwise \end{cases} \tag{6}$$

where $\sigma \in [0, 1]$ is a random value which represents the threshold. Also, to evaluate the performance of the selected features, the fitness function is computed as:

$$f(x_i) = \xi \times Err_{x_i} + (1 - \xi) \times \left(1 - \frac{|S|}{D}\right) \tag{7}$$

where Err_{x_i} represents the classification error for the logistic regression classifier, $|S|$ is the number of selected features and D the total number features. While $\xi \in [0, 1]$ is a random value used to balance between accuracy of classifier (first term Eq. (7)) and the number of selected features (second term in Eq. (7)) [21].

3.1   Initial Stage

In this stage, the population $X \in R^{N \times D}$ is generated randomly as:

$$X = L + (U - L) \odot rand(N, D), \tag{8}$$

where L and U represent the lower and upper bounds of the search domain. The next step is to compute the objective function given in Eq. (7) for each solution and determine the best solution P that corresponding to the best objective function f_P.

3.2 Updating Stage

In this stage, the first step is compute the fitness function for each solution x_i, $(i = 1, 2, \ldots, N)$, as in the following equation:

$$Fit_i = \frac{f_i}{\sum_{i=1}^{N} f_i} \tag{9}$$

Then according to the random value $Prob \in [0, 1]$ the proposed method trends toward the exploration phase or exploitation phase. For example, if the $Prob$ is greater than Fit_i then the current solution x_i is updated according to the DE operators; otherwise, the x_i is updated using the tradition SCA functions. The objective function (Eq. (7)) is computed again and the best solution P is selected, these previous steps are repeated until the stopping condition is reached (Fig. 1).

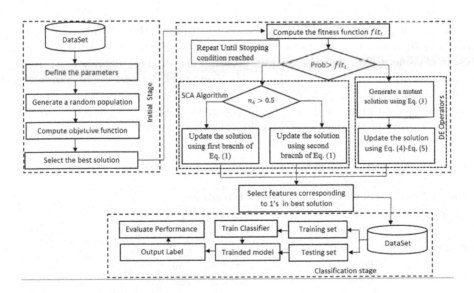

Fig. 1. The Framework of the proposed method

4   Experimental Results and Discussions

4.1   Parameter Setting and Performance Measures

In order to assess the performance of the proposed method, a set of experiments are performed using eight dataset for UCI machine learning site [22], and the description of these datasets is given in Table 1. The results of the proposed approach is compared with other three algorithms namely, SCA [8], SSO [5], and ABC [4].

Table 1. The datasets description

No.	Dataset	Features	Samples
1	Breast	10	699
2	SPECT	40	5000
3	Ionosphere	34	351
4	Wine	13	178
5	Congress	16	435
6	Sensor	255	45
7	Clean 1	166	4767
8	Clean 2	166	6598

In this paper, two classifiers are used to evaluate the performance of the methods to select feature, these classifiers are a logistic regression (Log) and Kstar classifiers [23], and the output performance of these classifiers are computed over 20 runs through calculating the average of the performance measures given in Table 2. In Table 2, N_R, N_{IC} and ND represent the number of runs, true classified samples, and samples in the given dataset, respectively. obj represents a value of the best objective function at i-th run. However, before the

Table 2. The formula of performance measures.

Measure for	Name	Formula
Classification	Error	$Error = \frac{N_{IC}}{ND}$
	F-Measure	$FM = \frac{2(precisian \times Recall)}{(precisian + Recall)}$
Objective function	Average	$Avg_{obj} = \frac{1}{N_R} \sum_{i=1}^{N_R} obj_i$
	Best	$Worst_{obj} = \max_{i=1}^{N_R} obj_i^*$
	Worst	$Best_{obj} = \min_{i=1}^{N_R} obj_i^*$
	Standard deviation	$STD_{obj} = \frac{1}{N_R - 1} \sqrt{\sum_{i=1}^{N_R} N_R(obj_i - Avg_{obj})^2}$
FS	Ratio of selected Feature	$RS = \frac{1}{N_R} \sum_{i=1}^{N_R} \frac{|P|}{D}$
	Time	The time needed to select the best features from the given dataset

classification process, the dataset is split into training and testing set using the 10-fold cross-validation method. By using CV the classifier select nine groups to trained while used one group for testing, this process of selecting the training and testing is repeated ten times and the output is the average of classification error for each run. The implementation of all methods and classifiers are performed using Matlab installed on windows environment with 64-bit support.

4.2 Results and Discussion

Table 3 shows the number of selected feature RS, the time needed to select the features and the classification error using two classifiers for all algorithms. From this table, it can be observed some notes such as, the proposed SCADE method, in general, is better than all other three methods in terms which has an average of $Error$, RS and time equal to 0.064, 0.514 and 333.65 s, respectively. While the SSO algorithm has the second rank in the performance followed by the SCA algorithm and the worst method is the ABC according to our study.

Moreover, according to the ratio of selected feature, the proposed SCADE method gives the better results in Breast, Spect and Clean 2; while SCA algorithm has the smaller RS for Ionosphere, Wine and Clean 1, also, the SSO and ABC have the better RS for Congress and Sensor, respectively. As well as, in terms of time, the proposed method takes the smaller time to achieve the best solution for all datasets except Breast, SPECT and Congress, the SSO is the best. According to classification error $Error$, the SCADE has the smaller error overall datasets using the two classifiers except, when the Log classifier is used for SPECT and Congress datasets, also, when Kstar is used for Clean 1 dataset. In addition, the results of the F-measure FM is given in Fig. 2.

In order to assess the convergence of the methods, the objective function values are evaluated as in Fig. 3; where the $Avg_{obj}, Worst_{obj}, Best_{obj}$ and Std_{obj} are computed. From this figure, it can be seen that the SCADE approach achieves the best value in terms of, $Avg_{obj}, Best_{obj}$ and Std_{obj}; however, the SCA gives the better results in terms of $Worst_{obj}$.

4.3 Statistical Analysis

In this section, the results of the proposed method are further analysis through using a Wilcoxon's rank sum test which used to give a statistical value p-$value$, to accept the null hypothesis (that consider, there is no a significant difference between the median of the pair of compared methods) or to reject this assumption and accept the alterative hypothesis (i.e. there is a significant difference). The results of p-value are 0.0404, 0.1905, 0.0313, and 0.0002 for SCA, SSO, ABC and Full, respectively, at significance level equal to 5%. It can be seen that there exist a significance difference between the proposed SCADE method and SCA, ABC methods and Full, however, there is no significant difference with SSO.

According to all previous results, it can observe that the proposed SCADE method provides better results in terms of all performance measures along the tested datasets, except for SPECT, Clean 2, Wine according to F measure. These

Table 3. The accuracy of the SCADE and all compared algorithms

		SCADE			SCA			SSO			ABC			Full
		Error	FS	Time	Error	FS	Time	Error	FS	Time	Error	FS	Time	Error
Breast	Log	0.026	0.778	119.15	0.064	0.889	123.12	0.058	0.778	100.05	0.059	0.778	221.79	0.112
	Kstar	0.018			0.061			0.075			0.089			0.099
SPECT	Log	0.131	0.591	147.68	0.182	0.636	174.64	0.124	0.682	141.23	0.200	0.727	299.35	0.349
	Kstar	0.218			0.250			0.311			0.270			0.352
Ionosphere	Log	0.036	0.529	110.32	0.077	0.471	130.59	0.068	0.647	141.39	0.076	0.559	196.76	0.129
	Kstar	0.138			0.205			0.162			0.162			0.358
Wine	Log	0.017	0.538	976.40	0.047	0.385	1205.09	0.027	0.462	1185.21	0.050	0.615	1330.84	0.144
	Kstar	0.056			0.172			0.102			0.117			0.190
Congress	Log	0.034	0.375	135.30	0.075	0.563	134.67	0.028	0.313	109.41	0.065	0.438	255.38	0.093
	Kstar	0.022			0.048			0.061			0.065			0.155
Sensor	Log	0.007	0.667	89.64	0.040	0.708	124.00	0.019	0.625	91.66	0.030	0.542	225.23	0.075
	Kstar	0.127			0.106			0.109			0.130			0.248
Clean 2	Log	0.025	0.335	1039.13	0.057	0.371	1217.13	0.030	0.401	1208.35	0.049	0.443	1283.25	0.107
	Kstar	0.065			0.087			0.101			0.141			0.177
Clean 1	Log	0.007	0.299	51.57	0.027	0.222	52.91	0.023	0.240	68.07	0.034	0.234	59.41	0.105
	Kstar	0.096			0.097			0.088			0.098			0.200
Average		0.064	0.514	333.65	0.100	0.531	395.27	0.087	0.518	380.67	0.102	0.542	484.00	0.181

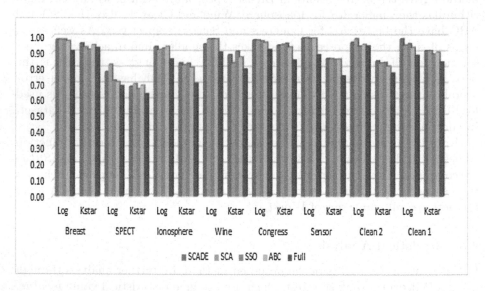

Fig. 2. The average of F-Measure for each classifier overall methods.

results occurred due to the characteristics of DE operators that combined with
the global optimization of the SCA algorithm, therefore, if the SCA algorithm
getting stuck in local point the DE operators help it to skip this point; and this
leads to improve the convergence as seen in the previous results.

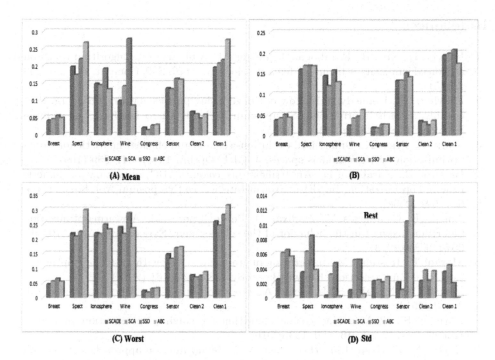

Fig. 3. The results of mean, best, worst and standard deviation of objective function.

5 Conclusion

The feature selection process is an effective phase in decreasing the computational time and helping in increasing the classification accuracy. This paper introduces a new hybrid feature selection method which used the differential evolution (DE) operators as local search method to improve the sine cosine algorithm (SCA). The DE operators make the SCA algorithm to avoid stuck in local optima. The performance of the proposed method had been assessed based on eight UCI datasets and its results were compared with three well-known algorithms. The results showed that the proposed method outperformed the results of the compared algorithms according to different performance measures and Wilcoxon statistical test.

Acknowledgments. This work was in part supported by national key Research & Development Program of China (No. 2016YFD0101903), Nature Science Foundation of Hubei Province (Grant No. 2015CFA059), Science & Technology Pillar Program of Hubei Province (Grant No. 2014BAA146), Science & Technology Cooperation Program of Henan Province (No. 152106000048) and Hubei Collaborative Innovation Center of Basic Education Information technology Services.

References

1. Hancer, E., Xue, B., Karaboga, D., Zhang, M.: A binary ABC algorithm based on advanced similarity scheme for feature selection. Appl. Soft Comput. **36**, 334–348 (2015)
2. Eberhart, R., Kennedy, J.: A new optimizer using particle swarm theory. In: Proceedings of 6th International Symposium on book Micro Machine and Human Science, MHS 1995, pp. 39–43 (1995)
3. Storn, R., Price, K.: Differential evolution-a simple and efficient heuristic for global optimization over continuous spaces. J. Glob. Optim. **11**(4), 341–359 (1997)
4. Basturk, B., Karaboga, D.: An artificial bee colony (ABC) algorithm for numeric function optimization. In: IEEE Swarm Intelligence Symposium, vol. 8, no. 1, pp. 687–697 (2006)
5. Cuevas, E., Cienfuegos, M., Zaldivar, D., Perez-Cisneros, M.: A swarm optimization algorithm inspired in the behavior of the social-spider. Expert Syst. Appl. **40**(16), 6374–6384 (2013)
6. Moradi, P., Rostami, M.: Integration of graph clustering with ant colony optimization for feature selection. Knowl.-Based Syst. **84**, 144–161 (2015)
7. Mirjalili, S., Mirjalili, S.M., Lewis, A.: Grey wolf optimizer. Adv. Eng. Softw. **69**, 46–61 (2014)
8. Mirjalili, S.: SCA: a sine cosine algorithm for solving optimization problems. Knowl.-Based Syst. **96**, 120–133 (2016)
9. Emary, E., Zawbaa, H.M., Hassanien, A.E.: Binary ant lion approaches for feature selection. Neurocomputing **213**, 54–65 (2016)
10. Sindhu, R., Ngadiran, R., Yacob, Y.M., Zahri, N.A.H., Hariharan, M.: Sine-cosine algorithm for feature selection with elitism strategy and new updating mechanism. Neural Comput. Appl. 1–12 (2017)
11. Wang, J., Xue, B., Gao, X., Zhang, M.: A differential evolution approach to feature selection and instance selection. In: Pacific Rim International Conference on Artificial Intelligence, pp. 588–602 (2016)
12. Yang, W.-A., Zhou, Q., Tsui, K.-L.: Differential evolution-based feature selection and parameter optimisation for extreme learning machine in tool wear estimation. Int. J. Prod. Res. **54**(15), 4703–4721 (2016)
13. Maheshwari, R., Kumar, M., Kumar, S.: Optimization of feature selection in face recognition system using differential evolution and genetic algorithm. In: Proceedings of 5th International Conference on Soft Computing for Problem Solving, pp. 363–374 (2016)
14. Chattopadhyay, S., Mishra, S., Goswami, S.: Feature selection using differential evolution with binary mutation scheme. In: 2016 International Conference on Microelectronics, Computing and Communications (MicroCom), pp. 1–6 (2016)
15. Sikdar, U.K., Ekbal, A., Saha, S., Uryupina, O., Poesio, M.: Differential evolution-based feature selection technique for anaphora resolution. Soft. Comput. **19**(8), 2149–2161 (2015)
16. Nakamura, R.Y.M., Pereira, L.A.M., Costa, K.A., Rodrigues, D., Papa, J.P., Yang, X.S.: BBA: a binary bat algorithm for feature selection. In: 2012 25th SIBGRAPI Conference on Graphics, Patterns and Images, pp. 291–297 (2012)
17. Ewees, A.A., El Aziz, M.A., Hassanin, A.E.: Chaotic multi-verse optimizer-based feature selection. Neural Comput. Appl. 1–17 (2017)
18. El Aziz, M.A., Hassanien, A.E.: Modified cuckoo search algorithm with rough sets for feature selection. Neural Comput. Appl. 1–10 (2016)

19. Zorarpaci, E., Özel, S.A.: A hybrid approach of differential evolution and artificial bee colony for feature selection. Expert Syst. Appl. **62**, 91–103 (2016)
20. Balakrishnan, U., Venkatachalapathy, K., Girirajkumar, M.S.: A hybrid PSO-DEFS based feature selection for the identification of diabetic retinopathy. Curr. Diabetes Rev. **11**(3), 182–190 (2015)
21. Wang, X., Yang, J., Teng, X., Xia, W., Jensen, R.: Feature selection based on rough sets and particle swarm optimization. Pattern Recog. Lett. **28**(4), 459–471 (2007)
22. Asuncion, A., Newman, D.H.: UCI machine learning repository. School of Information and Computer Sciences, University of California, Irvine (2010). http://archive.ics.uci.edu/ml/
23. El Aziz, M.A., Hassanien, A.E.: An improved social spider optimization algorithm based on rough sets for solving minimum number attribute reduction problem. Neural Comput. Appl. 1–12 (2017)

Feature Selection Based on Improved Runner-Root Algorithm Using Chaotic Singer Map and Opposition-Based Learning

Rehab Ali Ibrahim[1], Diego Oliva[2], Ahmed A. Ewees[3], and Songfeng Lu[1,4(✉)]

[1] School of Computer Science and Technology, Huazhong University of Science
and Technology, Wuhan 430074, China
rehab100r@yahoo.com, lusongfeng@hust.edu.cn
[2] Departamento de Ciencias Computacionales, Universidad de Guadalajara,
CUCEI Av. Revolucion 1500, Guadalajara, Jalisco, Mexico
diego.oliva@cucei.udg.mx
[3] Department of Computer, Damietta University, Damietta, Egypt
ewees@du.edu.eg
[4] Shenzhen Research Institute, Huazhong University of Science and Technology,
Shenzhen 518063, China

Abstract. The feature selection (FS) is an important step for data analysis. FS is used to reduce the dimension of data by selecting the relevant features; while removing the redundant, noisy and irrelevant features that lead to degradation of the performance. Several swarm techniques are used to solve the FS problem and these methods provide results better than classical approaches. However, most of these techniques have limitations such as slow convergence and time complexity. These limitations occur due that all the agents update their position according to the best one. However, this best agent may be not the optimal global solution for FS, therefore, the swarm getting stuck in a local solution. This paper proposes an improved Runner-Root Algorithm (RRA). The RRA is combined with chaotic Singer map and opposition-based learning to increase its accuracy. The experiments are performed in eight datasets and the performance of the proposed method is compared against swarm algorithms.

Keywords: Feature selection (FS) · Opposition-based learning (OBL) · Metaheuristic algorithms (MH) · Chaotic map · Runner-Root Algorithm (RRA) · Swarm intelligence (SI)

1 Introduction

The data in real-world applications often contains a large number of features with a small number of instances and this high dimensionality results in many problems. For example, increasing the time computation, increasing the memory usage, and degradation of the performance of the classification method. These problems occur because that some of these features are relevant, but, the other

© Springer International Publishing AG 2017
D. Liu et al. (Eds.): ICONIP 2017, Part V, LNCS 10638, pp. 156–166, 2017.
https://doi.org/10.1007/978-3-319-70139-4_16

features are not important (irrelevant). Thus, using the feature selection (FS) [1] approaches are very important to solve these problems through removing them without influence on the representation of the original data. Therefore, FS became recently very important in machine learning [2], for many applications such as bioinformatics [3], medical applications [4], and others [5,6]. However, the FS is hard problem because the search space for a small number of features becomes large, for example, the total number of possible solutions is 2^k for a dataset with k features [7]. In general, there are two categories of FS methods, filter and wrapper methods [8], the main difference between them is in the strategy of selecting the subset of features. The wrapper methods used the learning technique to evaluate the feature subset. However, the wrapper cannot be used to deal with the high-dimensional dataset since it needs a large time to determine the relevant features. However, filters approach does not use the learning techniques in the step of selecting the features, unlike wrapper methods [8]. For these reasons, the wrapper is computationally expensive so it cannot be applied to data with large size, whereas filter algorithms are often less computationally expensive.

In recent years, swarm intelligence (SI) techniques, that inspired by the social behavior of insects, birds, fish, animals and other in the real world, have been used to solve FS problem [5]. Example of such methods including Artificial Bee Colony (ABC) [9], Cuckoo search optimization (CS) [10], Social-spider Optimization [5], and Particle Swarm Optimization (PSO) [11]. The main limitation of SI techniques is that their convergence is slow, especially in the large search domain, because it depends on the value of the best solution. However, it may be not the optimal one but the optimal may be in opposite direction to the current solution. To overcome this drawback, the opposition-based learning (OBL) strategy [12,13] is used to make the SI algorithms improve their convergence to search about the global solution for a specific problem. In which, it make the SI techniques search in the opposite direction to the current solutions, and then determine if the current solutions or their opposite are the best and select them. This strategy makes the solution closer to the optimal solution and the convergence becomes fast. Moreover, there are other methods called chaotic methods are used, also, to improve the convergence of the SI techniques through using the properties of the chaos which make the SI techniques to skip the local point.

This paper proposes an alternative FS approach based on a new SI technique called runner-root algorithm (RRA) [14], which is inspired from the function of the runners and the roots of some plants in nature like strawberry and spider plants. It is one of the popular SI which have been used to solve different optimization problems to determine the optimal global solution and provides acceptable results. By comparing with other SI techniques, The RRA has the same properties, however, at all iterations, in RRA, the number of function evaluations (FE's) is different. Also, it does not need complex operators like those which are applied in GA (mutation and crossover), simple mathematical operators is needed only, as well as, the time and space complexities of RRA are inexpensive [14]. The RRA algorithm has a very good advantage that it does

not apply the same number of the evaluations of the function in the whole iterations. However, with all these characteristics, the initial population of RRA has
the largest effect on its behaviour, and since it is generated randomly from the
uniform distribution, so this may not be optimal choice. Therefore, the main
contribution of this paper is to improve the RRA algorithm through using the
chaotic Singer map with the OBL strategy for generating the best initial population; also, the OBL is used to improve the updated solutions. The proposed
approach is compared with some other FS techniques and the experiment results
illustrate that the proposed approach outperformed other methods in terms of
performance measures such as accuracy, convergence, time and the size of a
subset of the selected features.

The structure of the rest parts of this paper is. The preliminaries about the
Chaotic methods, opposition-based strategy(OBL) and the runner-root algorithm are introduced in Sect. 2. The proposed approach is introduced in Sect. 3.
Section 4 introduces the experimental results and their discussion. In Sect. 5, the
occlusions and future works are presented.

2 Preliminaries

This section briefly provides an introduction for Chaotic methods, the
Opposition-based Learning (OBL) strategy and the runner-root algorithm.

2.1 Chaotic Singer Map

The chaotic methods have an important characteristics that can be used to
improve the convergence optimization methods. These characteristics include,
(1) the stochastic, (2) sensitivity to the initial conditions and (3) ergodicity [15],
and they are converted into chaotic maps. The chaotic Singer map is one of the
most popular chaotic maps, defined as:

$$ch_{i+1} = \mu(7.86ch_i - 23.31ch_i^2 + 28.75ch_i^3 - 13.301875ch_i^4), \quad \mu = 1.07 \tag{1}$$

2.2 Opposition-Based Learning (OBL)

Opposition-based learning (OBL) is a method used to compute the opposite
solution for each solution in the population then the best solution is selected
according to the value of its objective function f. In general, this method is used
to enhance the convergence of the metaheuristic (MH) algorithms to find the
global solution. In which, the MH algorithms update their solution according
to the best agent, however, this agent may be local optimal and this will lead
to make the MH algorithms suffer from slow convergence and increase the time
complexity. The basics of the OBL method are presented in [12], by considering
a real number $x \in [u, l]$ (u and l represent the lower and the upper bound of the
problem), the opposite number of it $\bar{x}$ is defined as:

$$\bar{x} = u + l - x \tag{2}$$

Also, the opposite vector $\overline{\mathbf{x}} \in R^D$ for the real vector $x \in R^D$ can be defined as:

$$\overline{x_j} = u_i + l_j - x_j, j = 1, 2, \ldots, D \tag{3}$$

Finally, in the optimization method, the current solution $\mathbf{x}$ is selected if $f(\mathbf{x})$ is better than $f(\overline{\mathbf{x}})$ otherwise $\overline{\mathbf{x}}$ is selected. Therefore, the population of solutions is updated based on the best values of x and $\overline{\mathbf{x}}$ [13].

2.3 Runner-Root Algorithm RRA

The Runner-Root algorithm [14] is an algorithm for swarm intelligence inspired from the real plant life of plants named running plants which have runners and roots looking for minerals and water resources by developing runners, roots and parts named root hairs. The water resources for the plant can move randomly large steps and the minerals are for the plant can move randomly small steps. Then the plant will search for the best location in a big size area by using the roots and the runners. RRA has a great advantage that if it is trapped in local position, it will start a local search known as re-initialization strategy.

The initial stage of the algorithm is generating random population X (of size N) by uniform distribution where each element in this population is called mother plant $x_{mp_i} = [x_{mp_{i1}}, x_{mp_{i2}}, \ldots, x_{mp_{iD}}]$, $i = 1, 2, \ldots, N$. Then each x_{mp} generates another plant called daughter plant x_{dp}.

$$x_{dp_i}(t) = \begin{cases} x_{mp_1}(t), & i = 1 \\ x_{mp_i}(t) + dis_r \times r_i, & i = 1, 2, \ldots, N \end{cases} \tag{4}$$

where dis_r represents the maximum distance between x_{mp} and x_{dp}; while $r_l \in R^D$ is a random vector. Then the fitness function $(f(x_{dp}))$ of x_{dp} is computed and the best daughter x_{dpb} is determined, and the current x_{dpb} is compared with the previous best to decide whether the global search (in the case of satisfying the inequality (5)) or the local search (in the case of not satisfying the inequality (5)) is started according to the following inequality:

$$\left| \frac{f(x_{dpb}(t)) - f(x_{dpb}(t-1))}{f(x_{dpb}(t-1))} \right| \geq tolerance \tag{5}$$

The local search with random large steps that emulates the function of the roots in the nature begins by calculating a new daughter representation (that called $xper_i$ at the first element $i = 1$ which represents a random change to the i-th element of $x_{dpb}(t)$) as in the following equation:

$$xper_i = diag\{1, \ldots, 1, 1 + dis_r \beta_l, 1, \ldots, 1\} \times x_{dpb}(t) \tag{6}$$

where $\beta_i \in N(0, 1)$ and $diag\{.\}$ represent a random number and a diagonal matrix whose all diagonal elements equal 1 (except the i-th entry equals $1 + dis_r n_i$), respectively. If $f(xper_1) > f(x_{dpb})(t)$ then $x_{dpb}(t)$, otherwise

$x_{dpb}(t) = xper_1$ is not changed. Then the $xper_k$ at $k = 2$ is computed using Eq. (6), as well as, if $f(xper_2) > f(x_{dpb}(t))$ then $x_{dpb}(t)$, otherwise $x_{dpb}(t) = xper_2$. Repeat these steps for the whole entry of $x_{dpb}(t)$. Similarly, the LS with the random small steps is preformed except that the $xper_k$ is updated by the following equation [14]:

$$xper_i = diag\{1, \ldots, 1, 1 + dis_{ro} r_k, 1, \ldots, 1\} \times x_{dpb}(t) \tag{7}$$

where dis_{ro} is a scalar smaller than dis_r, and ru_i is the same for β_l. Here, $x_{dpb}(t)$ is the final output of the LS with random small steps. After performing the LS, for the next iteration, the $x_{mp}(t + 1)$ are chosen based on the $x_{dp}(t)$, by using a hybrid between elite and roulette wheel selection. The elite selection is performed through [14] as:

$$x_{mp}(t + 1) = x_{dpb}(t) \tag{8}$$

Using Eq. (9), the fitness function of x_{dp_k} is computed.

$$fit(x_{dp_i}(t)) = \frac{1}{c + x_{dp_i}(t) - x_{dpb}(t)} \tag{9}$$

where c is a positive real constant which controls the process of selection. In terms of (9), the x_{dpb} has the greatest value of the fitness function, which is equal to $1/c$. The probability of choosing the $x_{dp_i}(t)$ as x_{mp} for the $(t + 1)$-th iteration is computed as:

$$Pro_i = \frac{fit(x_{dp_i}(t))}{\sum_{j=1}^{N} fit(x_{dp_j}(t))} \tag{10}$$

3   The Proposed Method

In this section, we introduce the improvement for the Runner-Root Algorithm (RRA) through using the chaotic Singer map to generate random population, as well as using the OBL strategy to determine the optimal initial population. The steps of the proposed algorithm (which is called CORRA) are illustrated in details as below (also in Fig. 1). The initial population X (with size N) of the proposed method is generated by using the chaotic *Singer* map using Eq. (1) as the following:

$$x_{mp_i} = L_i + (U_i - L_i) * ch_i \tag{11}$$

After that, the opposite mother plant $\bar{x}_{mp}$ for each x_{mp} is generated and before computing the objective function for both $\bar{x}_{mp}$ and x_{mp}, they must be converted to binary vector:

$$x_{mp_i}(t + 1) = \begin{cases} 1 & if \ \frac{1}{1 + e^{-x_{mp_i}(t)}} > \delta \\ 0 & otherwise \end{cases} \tag{12}$$

where $\delta \in [0, 1]$ represents the random threshold value, and the output from Eq. (12) refers to the selected features (i.e. those features that corresponding

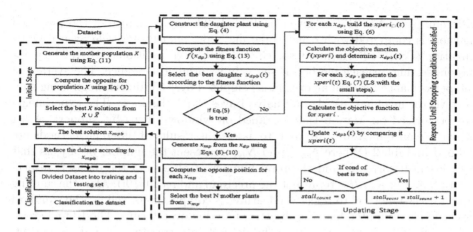

Fig. 1. The Proposed method

to 1's). The objective function of the current solution $x_{mp_i}(t)$ (a subset of features) is computed by using the following equation [5].

$$fit\left(x_{mp_i}(t)\right) = \xi\gamma_{x_{mp_i}(t)} + (1 - \xi)\left(1 - \frac{|x_{mp_i}(t)|}{|C|}\right), \tag{13}$$

where $\gamma_{x_{mp_i}(t)}$ represents the accuracy of classification using naive Bayes (NB) classifier; while $\xi \in [0, 1]$ and $|.|$ represent the parameters which balance between the first term (accuracy of classification) and the second term (number of the selected features) [16]. The best N solutions are selected from the union of both population x_{mp} and its opposite $\bar{x}_{mp}$.

The updating stage started by selecting the best mother plant $x_{mpb}(t)$ based on the objective functions, then the $x_{dp_i}(t)$ is generated using Eq. (4) and its objective function is computed ($fit_{x_{dp_i}}$). The best daughter $x_{dpb}(t)$ is chosen from $x_{dp}(t)$ according to $fit_{x_{dp}}$. The next step is to perform the global search (or local search) if the criteria in Eq. (5) is satisfied (or not satisfied). The LS is performed through computing the $xper_i(t)$ defined in Eq. (6) and this occurred at two levels. In the first level, for random large steps; each $xper_i(t)$ is compared with x_{dpb}, and to update x_{dpb}. The same steps are performed for the second level that is called small steps local search. The mother plants used in the next iteration, are generated from the $x_{dp}(t)$ (using a hybrid between elite and roulette as in Eqs. (8)–(10)). The OBL strategy is used, again to compute the opposite solution for the mother plants. The previous steps are performed again until the stopping criteria are reached (for example, if $stall_{count}$ equal to the maximum stall ($stall_{max}$) and the t_{max} are reached).

4   Experimental Results and Discussions

In this section, the performance of the proposed method is evaluated through using a set of UCI machine learning data sets that contains eight datasets with different properties [17] as given in Table 1. The performance of the proposed method is compared with other three methods, namely, Particle Swarm Optimization (PSO) [18], Harmony search (HS) [19], and traditional RRA. The parameters setting for each method is set as in its original reference. The average of the performance for each algorithm is computed over 35 runs through using support vector machine (SVM) and naive Bayes (NB) classifiers. The experiments are performed on Matlab that installed on windows 10 64-bit support. A set of performance measures are used to assess the results of different methods are defined in Table 2 (Where N_{corr} and N_{DT} represent the number of objects that are classified correctly and the total number of objects in the dataset, respectively. The Pre and Rec represent the precisian and recall [10]). Also, the running time for each method to select the features is computed; as well as, the standard deviation is computed for best objective function overall runs. The 10-fold cross-validation (CV) is used to divide the dataset into training set and testing set. Where CV divides the dataset into 10 groups and at each run selects 9 groups to be a training set and the selected one to be testing set. This is repeated ten times and the average of accuracy overall runs is the output.

The comparison results of the proposed CORRA approach with the other approaches are given in Fig. 2 and Table 3 (Note, the word Full refer to the all number of features). Where, Table 3 shows the results of accuracy for each method along each dataset, from this table it can be observed that the proposed approach method gives better accuracy average (∼0.92) than other methods (as in the last row). In addition, the rest three methods nearly have the same average of accuracy, however, the RRA is better than the other methods with small

Table 1. The datasets description

No.	Dataset	Samples	Features	CORRA		RRA		HS		PSO	
				Time	S_R	Time	S_R	Time	S_R	Time	S_R
1	Ionosphere	351	34	17	119.08	18	182.57	16	915.51	20	348.04
2	Spect	267	22	12	97.69	13	116.81	14	115.89	13	232.83
3	BreastCW	699	10	5	49.87	5	90.06	6	31.73	8	69.13
4	Wine	178	13	6	98.38	7	122.33	6	174.09	8	372.65
5	Sensor	5456	25	14	990.61	17	1056.78	18	982.96	12	1894.97
6	Congress	435	16	5	92.32	6	163.43	10	112.09	8	209.24
7	Clean 2	6598	166	63	1033.68	70	1732.88	54	1879.57	82	1856.15
8	Clean 1	4767	166	32	40.67	42	46.64	61	49.05	40	94.93

Table 2. The formula of performance measures.

Accuracy	$Acc = \frac{N_{corr}}{N_{DT}}$
Average of objective function	$Avg_f = \frac{1}{35} \sum_{i=1}^{35} f_i$
Best objective function	$f_{Best} = \max_{i=1}^{35} f_i^*$
Worst objective function	$f_{Worst} = \min_{i=1}^{35} f_i^*$
Ratio of selected Feature	$S_R = \frac{1}{35} \sum_{i=1}^{35} \frac{|x_{mpb}^i|}{D}$

number of selected features as given in Table 1. Also, Table 1 shows the computational time for each algorithm to select the relevant features according to its strategy. It can be observed that the proposed CORRA method takes the smaller time overall the given datasets, nearly, 315.28 s than other methods; while the RRA needs (~438.93 s) smaller time than PSO (~532.61 s) and HS(~634.74 s) to select the best features. However, along each dataset, the CORRA takes smaller time than other methods in all datasets except two datasets namely, Breast and Sensor where the PSO is the smallest approach. In addition, the NB classifier gives better results than SVM, according to this study, overall datasets. In order to discuss the performance of the proposed method in terms of convergence of the objective function, the average, best, worst and the standard deviation are evaluated overall 30 runs as in Fig. 2. It can be seen from this figure that, in general, the proposed CORRA method has average value of the objective function, nearly, similar to other methods. However, it is better than other methods in most of datasets; if we consider the four measures with eight datasets so we have 32 cases. The proposed method achieves 17 cases, followed by the HS algorithm that gives better values in 8 cases; while the RRA algorithm allocated in third rank with 5 cases and the PSO is the worst algorithm by two cases only. Also, this figure (Fig. 2(D)) gives the evidence that the stability of the proposed method.

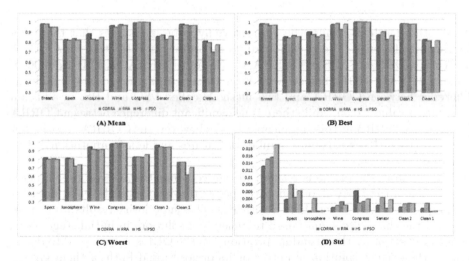

Fig. 2. The results of mean, best, worst and standard deviation of the objective function.

Moreover, a non-parametric statistical test called Wilcoxon's rank sum (WRS) test is used to further statistical analysis the results; to check if there exists a significant difference between the proposed method as the other methods at significance level equal to 5%. Where the null hypothesis assumes that there is no significant difference between median of CORRA and the median

Table 3. The accuracy of the algorithm using NB and SVM classifiers.

		CORRA	RRA	HS	PSO	Full	CORRA	RRA	HS	PSO	Full
Breast	NB	0.9893	0.9649	0.9681	0.9525	0.8882	0.97	0.96	0.98	0.98	0.95
	SVM	0.9598	0.9236	0.9298	0.9239	0.9005	0.92	0.91	0.94	0.95	0.90
SPECT	NB	0.8315	0.7555	0.7522	0.7640	0.6511	0.86	0.87	0.72	0.71	0.66
	SVM	0.7940	0.7500	0.7281	0.7079	0.6478	0.76	0.78	0.80	0.73	0.66
Ionosphere	NB	0.9602	0.9373	0.9116	0.9162	0.8708	0.95	0.95	0.93	0.93	0.89
	SVM	0.8520	0.7949	0.7692	0.8120	0.6419	0.79	0.81	0.76	0.82	0.68
Wine	NB	0.9822	0.9531	0.9467	0.9461	0.8563	1.00	0.98	0.97	0.97	0.94
	SVM	0.8975	0.8305	0.8433	0.8475	0.8098	0.93	0.90	0.90	0.85	0.81
Congress	NB	0.9862	0.9414	0.9517	0.9310	0.9069	0.99	0.97	0.95	0.96	0.93
	SVM	0.9586	0.9248	0.9172	0.9197	0.8448	0.96	0.94	0.91	0.91	0.89
Sensor	NB	0.9813	0.9584	0.9540	0.9545	0.9251	0.90	0.90	0.90	0.90	0.82
	SVM	0.7519	0.7112	0.7064	0.6894	0.5234	0.76	0.76	0.76	0.76	0.65
Clean 2	NB	0.9898	0.9314	0.9382	0.9458	0.8935	0.97	0.94	0.97	0.99	0.93
	SVM	0.9602	0.9182	0.9255	0.9186	0.8226	0.88	0.90	0.87	0.87	0.74
Clean 1	NB	0.9827	0.9780	0.9823	0.9803	0.8950	0.97	0.96	0.99	0.99	0.90
	SVM	0.8963	0.8964	0.8901	0.8905	0.8004	0.94	0.92	0.92	0.92	0.88
Average		0.9296	0.8956	0.8903	0.8940	0.8192	0.91	0.90	0.89	0.89	0.83

of the other methods; while, the alternative hypothesis assumes that there is a significant difference. The p-value results according to WRS test for CORRA vs. RRA, CORRA vs. PSO, CORRA vs. HS and CORRA vs. Full are 0.0675, 0.045, 0.0478 and 0.0014, respectively. From these results, there exist a significant difference between CORRA and two methods called PSO and HS, in addition Full since p-value < 0.05; while there is no significant difference between CORRA and traditional RRA at this significance level. However, the CORRA gives higher accuracy than RRA with small number of features and less computational time over most of datasets.

Finally, from the previous results, we can conclude that the proposed modified version of the RRA algorithm gives better results overall the eight datasets by comparing it with other three algorithms. This superiority results from three facts (1) the chaotic tent map used to generate a solution, in initial stage, gives better position other than random position. (2) the OBL strategy used to determine if the current solution is better or its opposite and both of them gives a good benefit to traditional RRA algorithm which make it fast convergence. (3) the random jump with big steps, in the traditional RRA, makes the solutions to find the best region from the search space and avoids the other region, as well as, if these solutions getting stuck in optimal local point then the re-initialization strategy will be used to redistribute the solutions randomly in the search space.

5   Conclusions and Future Work

Feature selection is very important to deal real-world applications which have a large number of features that can affect the performance of data analysis methods. Datasets contain irrelevant features that must be removed to increase the accuracy of classifiers. The use of metaheuristic approaches has been extended to determine the features that are undesired. However, most of these feature selection methods have drawbacks as slow convergence. In their implementation, they include strategies that update the population using the best current solution. If this current best is not near to the global optimal, the optimization process could be trapped in a local optimum. Therefore, the OBL strategy is applied to explore the search space computing the opposite values of each solution. Moreover, chaotic maps are also used to increase the accuracy of the search process. This paper proposes an improvement of a new swarm technique called RRA by combining it with OBL strategy and chaotic maps. The Experimental results, performed using eight UCI datasets, provide evidence that the improved RRA is able to find better solutions for FS in terms of performance and accuracy. Summarizing the proposed version of RRA is a good alternative for complex optimization problems and it can also be extended for multi-objective optimization problems in future works.

Acknowledgments. This work are supported by the Natural Science Foundation of Hubei Province of China under Grant No. 2016CFB541 and the Applied Basic Research Program of Wuhan Science and Technology Bureau of China under Grant No. 2016010101010003 and the Science and Technology Program of Shenzhen of China under Grant No. JCYJ20170307160458368.

References

1. Guyon, I., Elisseeff, A.: An introduction to variable and feature selection. J. Mach. Learn. Res. **3**, 1157–1182 (2003)
2. Han, J., Kamber, M.: Data Mining: Concepts and Techniques. Morgan Kaufman, San Francisco (2000)
3. Awada, W., Khoshgoftaar, T.M., Dittman, D., Wald, R., Napolitano, A.: A Review of the stability of feature selection techniques for bioinformatics data. In: 2012 IEEE 13th International Conference on Information Reuse Integration (IRI), pp. 356–363 (2012)
4. Chang, P., Lin, J., Liu, C.: An attribute weight assignment and particle swarm optimization algorithm for medical database classifications. Comput. Methods Prog. Biomed. **107**(3), 382–392 (2012)
5. El Aziz, M.A., Hassanien, A.E.: An improved social spider optimization algorithm based on rough sets for solving minimum number attribute reduction problem. Neural Comput. Appl. (2017)
6. Guyon, I., Weston, J., Barnhill, S., Vapnik, V.: Gene selection for cancer classification using support vector machines. J. Mach. Learn. Res. **46**, 389–422 (2002)
7. Sivagaminathan, R.K., Ramakrishnan, S.: A hybrid approach for feature subset selection using neural networks and antcolony optimization. Expert Syst. Appl. **33**, 49–60 (2007)

8. Zhu, Z.X., Ong, Y.S., Dash, M.: Wrapper-filter feature selection algorithm using a memetic framework. IEEE Trans. Cybern. Part B: Cybern. **37**(1), 70–76 (2007)
9. Suguna, N., Thanushkodi, K.: An independent rough set approach hybrid with artificial bee colony algorithm for dimensionality reduction. Am. J. Appl. Sci. **8**(3), 261–266 (2011)
10. El Aziz, M.A., Hassanien, A.E.: Modified cuckoo search algorithm with rough sets for feature selection. Neural Comput. Appl. 1–10 (2016)
11. Inbarani, H.H., Azar, A.T., Jothi, G.: Supervised hybrid feature selection based on PSO and rough sets for medical diagnosis. Comput. Methods Prog. Biomed. **113**, 175–185 (2014)
12. Tizhoosh, H.R.: Opposition-based learning: a new scheme for machine intelligence. In: International Conference on Computational Intelligence for Modelling, Control and Automation and International Conference on Intelligent Agents, Web Technologies and Internet Commerce, vol. 1, pp. 695–701 (2005)
13. Cuevas, E., Oliva, D., Zaldivar, D., Perez-Cisneros, M., Pajares, G., Prez-Cisneros, M., Pajares, G.: Opposition-based electromagnetism-like for global optimization. Int. J. Innov. Comput. Inf. Control **8**, 8181–8198 (2012)
14. Merrikh-Bayat, F.: The runner-root algorithm: a metaheuristic for solving unimodal and multimodal optimization problems inspired by runners and roots of plants in nature. Appl. Soft Comput. **33**, 292–303 (2015)
15. Chuang, L.-Y., Yang, C.-H., Li, J.-C.: Chaotic maps based on binary particle swarm optimization for feature selection. Appl. Soft Comput. **11**(1), 239–248 (2011)
16. Wang, X., Yang, J., Teng, X., Xia, W., Jensen, R.: Feature selection based on rough sets and particle swarm optimization. Pattern Recog. Lett. **28**(4), 459–471 (2007)
17. Asuncion, A., Newman, D.H.: UCI machine learning repository. School of Information and Computer Sciences, University of California, Irvine (2010). http://archive.ics.uci.edu/ml/
18. Kennedy, J., Eberhart, R.C.: Particle swarm optimization. In: Proceedings of IEEE International Conference on Neural Networks 1995, vol. 4, pp. 1942–1948. IEEE Press (1995)
19. Mahdavi, M., Fesanghary, M., Damangir, E.: An improved harmony search algorithm for solving optimization problems. Appl. Math. Comput. **188**(2), 1567–1579 (2007)

LWMC: A Locally Weighted Meta-Clustering Algorithm for Ensemble Clustering

Dong Huang[1], Chang-Dong Wang[2,3,4($\boxtimes$)], and Jian-Huang Lai[2,3,4]

[1] College of Mathematics and Informatics, South China Agricultural University,
Guangzhou, China
huangdonghere@gmail.com
[2] School of Data and Computer Science, Sun Yat-sen University, Guangzhou, China
changdongwang@hotmail.com, stsljh@mail.sysu.edu.cn
[3] Guangdong Key Laboratory of Information Security Technology,
Sun Yat-sen University, Guangzhou, China
[4] Key Laboratory of Machine Intelligence and Advanced Computing,
Ministry of Education, Guangzhou, China

Abstract. The last decade has witnessed a rapid development of the ensemble clustering technique. Despite the great progress that has been made, there are still some challenging problems in the ensemble clustering research. In this paper, we aim to address two of the challenging problems in ensemble clustering, that is, the local weighting problem and the scalability problem. Specifically, a locally weighted meta-clustering (LWMC) algorithm is proposed, which is featured by two main advantages. First, it is highly efficient, due to its ability of working and voting on clusters. Second, it incorporates a locally weighted voting strategy in the meta-clustering process, which can exploit the diversity of clusters by means of local uncertainty estimation and ensemble-driven cluster validity. Experiments on eight real-world datasets demonstrate the superiority of the proposed algorithm in both clustering quality and efficiency.

Keywords: Ensemble clustering · Consensus clustering · Meta-clustering · Local weighting · Scalability

1 Introduction

Ensemble clustering is the process of fusing multiple clusterings, each referred to as a base clustering, into a probably better and more robust consensus clustering [1–12,14–18]. It has proved to be an advantageous clustering technique in dealing with noisy data, finding clusters of arbitrary shapes, handling data from multiple sources, and constructing robust clustering result [16]. Despite its rapid development and significant success, there are still some challenging problems in ensemble clustering that remain to be tackled. In this paper, we pay attention to two of these challenging problems, i.e., the local weighting problem as well as the scalability problem, and propose a novel ensemble clustering algorithm based on locally weighted meta-clustering.

© Springer International Publishing AG 2017
D. Liu et al. (Eds.): ICONIP 2017, Part V, LNCS 10638, pp. 167–176, 2017.
https://doi.org/10.1007/978-3-319-70139-4_17

In the past decade, various ensemble clustering algorithms have been designed by exploiting different techniques [1–12,14–18]. Evidence accumulation clustering (EAC) [2] is one of the most classical ensemble clustering algorithms, which first builds a co-association matrix by considering the frequency that two objects appear in the same cluster among the multiple base clusterings, and then achieves the consensus clustering by means of hierarchical agglomerative clustering. Yi et al. [17] proposed to identify the uncertain pairs in the co-association matrix by a global threshold and recover them by the matrix completion technique. Fern and Brodley [1] formulated the ensemble clustering problem into a bipartite graph partitioning problem, where both clusters and objects are treated as graph nodes. These conventional ensemble clustering algorithms typically treat each base clustering in the ensemble with equal weights, and fail to take into account the different reliability of the ensemble members.

In an clustering ensemble, there may be some low-quality, or even ill, ensemble members, which can significantly degrade the consensus performance. There is a need to evaluate the quality of the base clusterings and weight them accordingly. To this end, Li and Ding [10] proposed a weighted ensemble clustering method based on non-negative matrix factorization (NMF), where the weight of each base clustering is automatically determined in an optimization process. Huang et al. [4] proposed to evaluate the reliability of each base clustering by the normalized crowd agreement index (NCAI), and then presented two weighted ensemble clustering algorithms, termed weighted evidence accumulation clustering (WEAC) and graph partitioning with multi-granularity link analysis (GP-MGLA), respectively. These methods [4,10] typically treat each base clustering as an individual and assign a global weight to each of them, but fail to explore the different reliability of the clusters inside the same base clustering. Different from the methods in [4,10], Huang et al. [8] proposed to estimate the uncertainty of clusters by an entropic criterion, and developed two locally weighted ensemble clustering algorithms, which are capable of evaluating and weighting the clusters in the ensemble without making specific assumption on the data distribution or access to the original data features. However, the algorithms in [8], as well as most of the existing ensemble clustering algorithms [1–4,8–11,14,16–18], work at the object-level, i.e., they use the original objects as the basic operating units, which restricts their scalability for very large datasets. It remains an open problem how to tackle the local weighting issue and the scalability issue at the same time in a unified ensemble clustering framework.

In this paper, we propose a locally weighted meta-clustering (LWMC) algorithm for ensemble clustering. Each cluster is a subset of objects, and can be view as a local region in the dataset. The uncertainty of each cluster is first estimated by considering the distribution of cluster labels in the entire ensemble, and then an ensemble-driven cluster index (ECI) is computed as an indication of the reliability of this cluster. We build a cluster similarity graph (CSG) by treating each cluster as a node and deciding the edge weights with respect to the Jaccard coefficient. With the CSG partitioned into a certain number of meta-clusters, we then propose a locally weighted voting strategy to yield the final clustering result. Different from the conventional meta-clustering algorithm (MCLA) [15], which treats all clusters equally, our LWMC algorithm is able to exploit the different reliability of clusters and weight them accordingly in the meta-clustering process. Experiments are conducted on multiple real-world datasets, which have shown the superiority of our ensemble clustering algorithm in both clustering quality and efficiency.

The rest of the paper is organized as follows. Section 2 describes the proposed ensemble clustering algorithm based on local weighting and meta-clustering. Section 3 reports the experimental results of the proposed algorithm against several baseline algorithms. Section 4 concludes the paper.

2 Proposed Algorithm

In this section, we describe the overall framework of the proposed algorithm. We first present the formulation of the ensemble clustering problem in Sect. 2.1, and then introduce the process of local uncertainty estimation and ensemble-driven cluster validity in Sect. 2.2. Finally, we present a new consensus function based on locally weighted mete-clustering to obtain the consensus result in Sect. 2.3.

2.1 Formulation of the Ensemble Clustering Problem

Ensemble clustering is the process of combining multiple base clusterings into a probably better and more robust consensus clustering. Let $\mathcal{X} = \{x_1, \cdots, x_N\}$ denote a dataset with N objects. The base clusterings for $\mathcal{X}$ can be generated by using different clustering algorithms or using the same algorithm with different parameter settings. Formally, an ensemble of M base clusterings can be denoted as follows:

$$\Pi = \{\pi^1, \cdots, \pi^M\}, \tag{1}$$

where π^m is the m-the base clustering in the ensemble Π. Each base clustering consists of a number of clusters. The m-th base clustering in Π can be denoted as

$$\pi^m = \{C_1^m, \cdots, C_{n^m}^m\}, \tag{2}$$

where C_i^m is the i-th cluster and n^m is the number of clusters in π^m. Further, we can denote the set of all clusters in the ensemble Π as

$$\mathcal{C} = \{C_1, \cdots, C_{N_c}\}, \tag{3}$$

where C_i is the i-th cluster and N_c is the total number of clusters in Π. It is obvious that $N_c = \sum_{m=1}^{M} n^m$. Then, given the ensemble Π, the objective of ensemble clustering is to build a better consensus clustering π^*.

2.2 From Local Uncertainty to Ensemble-Driven Cluster Validity

To deal with the potentially low-quality, or even ill, base clusterings, recently some weighted ensemble clustering approaches [4, 10] have been developed, which is able to (globally) evaluate and weight the base clusterings. However, these methods generally neglect the local diversity inside the same base clustering. A cluster can be viewed as a local region in a base clustering. Even in the same base clusterings, the quality of the clusters may be very different. To locally evaluate and weight the clusters, we follow the practice of [8] and resort to the concept of entropy, which is an important concept in information theory and indicates the uncertainty of a random variable.

Specifically, the uncertainty (entropy) of a cluster is measured by considering the distribution of cluster labels in the entire ensemble. We first consider the case of measuring the entropy of a cluster, say, $C_i \in \mathcal{C}$, w.r.t. one base clustering, say, π^m, which can be computed as follows:

$$H^m(C_i) = - \sum_{C_j^m \in \pi^m} p(C_i, C_j^m) \log_2 p(C_i, C_j^m) \tag{4}$$

with

$$p(C_i, C_j^m) = \frac{|C_i \cap C_j^m|}{|C_i|}, \tag{5}$$

where $\cap$ computes the intersection of two sets (or clusters), and $||$ outputs the number of objects in a set.

Based on the assumption that the base clusterings are independent of each other, we can further compute the entropy of cluster C_i w.r.t. the entire ensemble Π, that is

$$H^\Pi(C_i) = \sum_{m=1}^{M} H^m(C_i). \tag{6}$$

Note that $H^\Pi(C_i)$ indicates the uncertain of cluster C_i w.r.t. the ensemble Π. It is obviously that $H^\Pi(C_i) \in [0, +\infty)$ for any cluster $C_i \in \mathcal{C}$. When the objects in C_i belong to the same cluster in all of the base clusterings, the uncertainty of cluster C_i, i.e., $H^\Pi(C_i)$, reaches its minimum value 0.

With the uncertainty measure of clusters, the ensemble-driven cluster index (ECI) can be computed as follows:

$$ECI(C_i) = e^{-\frac{H^\Pi(C_i)}{\theta \cdot M}}, \tag{7}$$

where $\theta > 0$ is a parameter to adjust the correlation between the cluster uncertainty and the ECI value. It holds that $ECI(C_i) \in (0, 1]$ for any cluster $C_i \in \mathcal{C}$. When the uncertainty of a cluster reaches its minimum 0, its ECI value reaches its maximum 1, which indicates the highest reliability (i.e., the lowest uncertainty) of it with consideration to the ensemble. In our algorithm, the ECI measure acts as a local weighting term to explore the diverse clusters in ensembles.

2.3   Finding Consensus by Locally Weighted Meta-Clustering

In this section, we devise a new consensus function based on locally weighted meta-clustering (LWMC). The cluster similarity graph (CSG) is first constructed, where the clusters in the ensemble are treated as the graph nodes and the edge weights between clusters are computed w.r.t. the Jaccard coefficient. Given two clusters C_i and C_j, the edge weight between them is computed as

$$e_{ij} = \frac{|C_i \bigcap C_j|}{|C_i \bigcup C_j|}. \tag{8}$$

With the CSG graph constructed, the normalized cut (Ncut) algorithm [13] is then adopted to partition the graph into a certain number of meta-clusters, denoted as

$$\mathcal{MC} = \{MC_1, \cdots, MC_K\}, \tag{9}$$

where MC_i denotes the i-th meta-cluster and K is the number of meta-clusters in $\mathcal{MC}$. Each meta-cluster is a set of clusters. The conventional meta-clustering algorithm [15] typically adopts a simple majority voting strategy to assign each object to one of the meta-clusters, which neglects the different reliability of clusters and may be misled by some low-quality clusters. In this paper, we propose a locally weighted voting strategy based on the ECI measure, which is able to exploit the diversity of clusters in ensembles. Given an object o_i and a meta-cluster MC_j, the locally weighted voting score of o_i w.r.t. MC_j is computed as

$$Score(o_i, MC_j) = \frac{1}{|MC_j|} \sum_{C_k \in MC_j} w(C_k) \cdot \mathbf{1}(o_i \in C_k), \tag{10}$$

with

$$w(C_k) = ECI(C_k), \tag{11}$$

$$\mathbf{1}(statement) = \begin{cases} 1, & \text{if statement is true,} \\ 0, & \text{otherwise,} \end{cases} \tag{12}$$

where $w(C_k)$ is the local weighting term, $\mathbf{1}(statement)$ is the voting term, and $|MC_j|$ is the number of clusters in the meta-cluster MC_j. Then, each object will be assigned to the meta-cluster that gives it the highest score, that is

$$MetaCls(o_i) = \arg\max_{MC_j \in \mathcal{MC}} Score(o_i, MC_j). \tag{13}$$

With each object assigned to a meta-cluster by the locally weighted voting strategy, the consensus clustering can be obtained by treating the objects assigned to the same meta-cluster as a final cluster.

3 Experiments

In this section, we conduct experiments on a variety of real-world datasets to evaluate the proposed LWMC algorithm against several other ensemble clustering algorithms.

3.1   Datasets and Evaluation Method

In our experiments, eight real-world datasets are used, namely, *Semeion*, *Steel Plates Faults (SPF)*, *Multiple Features (MF)*, *MNIST*, *Texture*, *ISOLET*, *USPS*, and *Letter Recognition (LR)*. The MNIST and USPS datasets are from Dr. S. Roweis's homepage[1], while the other six datasets are from UCI Machine Learning Repository[2]. The details of these datasets are given in Table 1.

Table 1. The benchmark datasets.

Dataset	Semeion	SPF	MF	MNIST	Texture	ISOLET	USPS	LR
#Object	1,593	1,941	2,000	5,000	5,500	7,797	11,000	20,000
#Class	10	7	10	10	11	26	10	26
#Attribute	256	27	649	784	40	617	256	16

To provide a fair comparison, in each test, we run the proposed algorithm as well as the baseline algorithms a large number of times and report their average performances. At each run, an ensemble of ten base clusterings is generated by k-means with the cluster number k randomly selected in $[2, \sqrt{N}]$ for each base clustering. To quantify the clustering performance, we adopt the normalized mutual information (NMI) [15] as the evaluation measure. Note that a larger NMI value indicates a better clustering result.

3.2   Sensitivity of Parameter θ

In this section, we test the sensitivity of parameter θ. As can be seen in Fig. 1, the proposed algorithm exhibits consistent performance with varying values of parameter θ. Note that the X axis corresponds to $\log_2 \theta$. Empirically, it is suggested that the parameter θ be set to moderate values, e.g., with $\log_2 \theta \in [-2, 0]$, which corresponds to $\theta \in [0.25, 1]$. In the following, we will use $\theta = 0.5$ in the experiments for all datasets.

3.3   Comparison with Base Clusterings

The objective of ensemble clustering is to combine multiple base clusterings to build a better consensus clustering. In this section, we compare the consensus clustering produced by the proposed LWMC algorithm against the base clusterings. As shown in Fig. 2, LWMC is capable of producing significantly better clustering results than the base clusterings. In particular, despite the low quality of the base clusterings for the *SPF* dataset (according to their low NMI scores), the LWMC algorithm can still yield much better consensus clustering results.

---

[1] http://www.cs.nyu.edu/%7eroweis/data.html.
[2] http://archive.ics.uci.edu/ml.

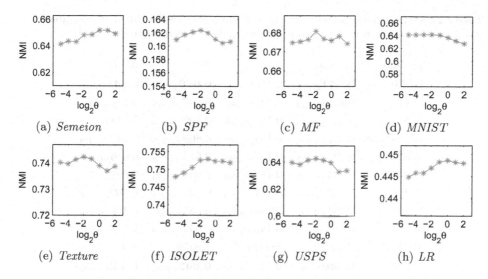

Fig. 1. The average NMI scores over 20 runs by LWMC with varying values of parameter θ. Note that the X axis corresponds to $\log_2 \theta$.

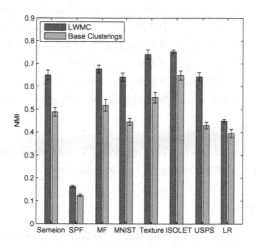

Fig. 2. Average Performances (over 20 runs) of our algorithm and the base clusterings.

3.4 Comparison with Other Ensemble Clustering Algorithms

In this section, we evaluate the performance of the proposed algorithm against eight baseline algorithms, namely, SEC [12], KCC [16], GP-MGLA [4], WEAC (with average-link) [4], EAC (with average-link) [2], CSPA [15], HGPA [15], and MCLA [15]. As shown in Table 2, the proposed LWMC algorithm shows a consistently good performance on the benchmark datasets. Although the WEAC method marginally outperforms our method on the *ISOLET* dataset, yet on all of the other seven datasets our method yield higher, or significantly higher, NMI

Table 2. Average performances (w.r.t. NMI) over 20 runs by different ensemble clustering methods (the best NMI score for each dataset is highlighted in bold).

Method	Semeion	SPF	MF	MNIST
LWMC	**0.649**±0.022	**0.162**±0.006	**0.676**±0.017	**0.641**±0.018
SEC	0.545±0.023	0.132±0.008	0.597±0.018	0.499±0.026
KCC	0.549±0.018	0.131±0.009	0.596±0.016	0.518±0.017
GP-MGLA	0.642±0.026	0.154±0.007	0.669±0.019	0.628±0.030
WEAC	0.644±0.025	0.152±0.009	0.643±0.022	0.616±0.028
EAC	0.641±0.027	0.152±0.009	0.635±0.023	0.601±0.032
CSPA	0.553±0.036	0.117±0.009	0.623±0.019	0.509±0.049
HGPA	0.491±0.027	0.116±0.010	0.535±0.482	0.409±0.028
MCLA	0.583±0.022	0.143±0.011	0.642±0.034	0.563±0.035
Method	Texture	ISOLET	USPS	LR
LWMC	**0.740**±0.020	0.752±0.007	**0.642**±0.019	**0.449**±0.007
SEC	0.644±0.015	0.694±0.016	0.473±0.021	0.412±0.008
KCC	0.644±0.014	0.690±0.009	0.501±0.016	0.409±0.006
GP-MGLA	0.725±0.025	0.749±0.007	0.615±0.038	0.440±0.004
WEAC	0.729±0.028	**0.753**±0.008	0.592±0.038	0.437±0.007
EAC	0.714±0.024	0.742±0.010	0.589±0.041	0.432±0.008
CSPA	0.653±0.022	0.700±0.030	0.509±0.061	0.262±0.170
HGPA	0.492±0.035	0.629±0.023	0.361±0.031	0.361±0.006
MCLA	0.703±0.014	0.723±0.020	0.545±0.030	0.410±0.013

scores than WEAC. To conclude, the proposed LWMC algorithm exhibits overall the best performance when compared to the eight baseline algorithms.

3.5   Execution Time

In this section, we test the execution time of different ensemble clustering algorithms with varying data sizes. In our experiments, different subsets of the *LR* datasets are used, whose sizes range from 0 to 20,000. As shown in Fig. 3, LWMC is the fastest algorithm, whild MCLA is the second fastest one. To process the entire dataset of *LR*, the proposed LWMC method and the MCLA method consume 3.74 and 5.31 s, respectively. Despite the extra time cost of the local weighting component, the proposed LWMC method still runs faster than the conventional MCLA method, probably due to the efficiency of the Ncut algorithm we used for graph partitioning.

The experiments are conducted in MATLAB R2014a 64-bit on a workstation with 8 Intel 2.40 GHz processors and 96 GB of RAM.

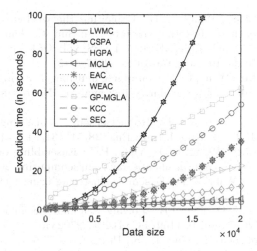

Fig. 3. Execution time of different methods with varying data sizes.

4   Conclusion

This paper proposes a locally weighted meta-clustering (LWMC) algorithm for ensemble clustering. Local uncertainty in ensembles is estimated by exploiting an entropic criterion, based on which the ensemble-driven cluster index can be obtained to evaluate and weight the clusters. Then a meta-cluster based consensus function is proposed, which exhibits two main advantages in the consensus process. First, it works at the cluster-level and is efficient for dealing with large-scale datasets. Second, it is able to exploit the diversity of clusters by incorporating a locally weighted voting strategy in the meta-clustering phase. Experiments on eight real-world datasets have shown the effectiveness and efficiency of the proposed algorithm.

Acknowledgement. This work was supported by NSFC (61602189, 61502543 and 61573387), the Ph.D. Start-up Fund of Natural Science Foundation of Guangdong Province, China (2016A030310457), Guangdong Natural Science Funds for Distinguished Young Scholars (2016A030306014), and the Tip-Top Scientific and Technical Innovative Youth Talents of Guangdong Special Support Program (No. 2016TQ03X542).

References

1. Fern, X.Z., Brodley, C.E.: Solving cluster ensemble problems by bipartite graph partitioning. In: International Conference on Machine Learning (ICML 2004) (2004)
2. Fred, A.L.N., Jain, A.K.: Combining multiple clusterings using evidence accumulation. IEEE Trans. Pattern Anal. Mach. Intell. **27**(6), 835–850 (2005)

3. Huang, D., Lai, J.-H., Wang, C.-D.: Exploiting the wisdom of crowd: a multi-granularity approach to clustering ensemble. In: Sun, C., Fang, F., Zhou, Z.-H., Yang, W., Liu, Z.-Y. (eds.) IScIDE 2013. LNCS, vol. 8261, pp. 112–119. Springer, Heidelberg (2013). doi:10.1007/978-3-642-42057-3_15

4. Huang, D., Lai, J.H., Wang, C.D.: Combining multiple clusterings via crowd agreement estimation and multi-granularity link analysis. Neurocomputing **170**, 240–250 (2015)

5. Huang, D., Lai, J.H., Wang, C.D.: Robust ensemble clustering using probability trajectories. IEEE Trans. Knowl. Data Eng. **28**(5), 1312–1326 (2016)

6. Huang, D., Lai, J.H., Wang, C.D., Yuen, P.C.: Ensembling over-segmentations: from weak evidence to strong segmentation. Neurocomputing **207**, 416–427 (2016)

7. Huang, D., Lai, J., Wang, C.D.: Ensemble clustering using factor graph. Pattern Recogn. **50**, 131–142 (2016)

8. Huang, D., Wang, C.D., Lai, J.H.: Locally weighted ensemble clustering. IEEE Trans. Cybern. (2017). doi:10.1109/TCYB.2017.2702343

9. Iam-On, N., Boongoen, T., Garrett, S., Price, C.: A link-based approach to the cluster ensemble problem. IEEE Trans. Pattern Anal. Mach. Intell. **33**(12), 2396–2409 (2011)

10. Li, T., Ding, C.: Weighted consensus clustering. In: SIAM International Conference on Data Mining (SDM 2008), pp. 798–809 (2008)

11. Li, Y., Yu, J., Hao, P., Li, Z.: Clustering ensembles based on normalized edges. In: Zhou, Z.-H., Li, H., Yang, Q. (eds.) PAKDD 2007. LNCS (LNAI), vol. 4426, pp. 664–671. Springer, Heidelberg (2007). doi:10.1007/978-3-540-71701-0_71

12. Liu, H., Wu, J., Liu, T., Tao, D., Fu, Y.: Spectral ensemble clustering via weighted k-means: theoretical and practical evidence. IEEE Trans. Knowl. Data Eng. **29**(5), 1129–1143 (2017)

13. Shi, J., Malik, J.: Normalized cuts and image segmentation. IEEE Trans. Pattern Anal. Mach. Intell. **22**(8), 888–905 (2000)

14. Singh, V., Mukherjee, L., Peng, J., Xu, J.: Ensemble clustering using semidefinite programming with applications. Mach. Learn. **79**(1–2), 177–200 (2010)

15. Strehl, A., Ghosh, J.: Cluster ensembles: a knowledge reuse framework for combining multiple partitions. J. Mach. Learn. Res. **3**, 583–617 (2003)

16. Wu, J., Liu, H., Xiong, H., Cao, J., Chen, J.: K-means-based consensus clustering: a unified view. IEEE Trans. Knowl. Data Eng. **27**(1), 155–169 (2015)

17. Yi, J., Yang, T., Jin, R., Jain, A.K.: Robust ensemble clustering by matrix completion. In: IEEE International Conference on Data Mining (ICDM 2012), pp. 1176–1181 (2012)

18. Zhong, C., Yue, X., Zhang, Z., Lei, J.: A clustering ensemble: two-level-refined co-association matrix with path-based transformation. Pattern Recogn. **48**(8), 2699–2709 (2015)

PUD: Social Spammer Detection Based on PU Learning

Yuqi Song[1,2], Min Gao[1,2(✉)], Junliang Yu[1,2], Wentao Li[3], Junhao Wen[1,2], and Qingyu Xiong[1,2]

[1] Key Laboratory of Dependable Service Computing in Cyber Physical Society, Chongqing University, Ministry of Education, Chongqing, China
{songyq,gaomin,yu.jl,jhwen,xiong03}@cqu.edu.cn
[2] School of Software Engineering, Chongqing University, Chongqing, China
[3] Faculty of Engineering and Information Technology,
Centre for Artificial Intelligence, School of Software, University of Technology Sydney, Ultimo, Australia
wentao.li@student.uts.edu.au

Abstract. Social networks act as the communication channels for people to share various information online. However, spammers who generate spam information reduce the satisfaction of common users. Numerous notable studies have been done to detect social spammers, and these methods can be categorized into three types: unsupervised, supervised and semi-supervised methods. While the performance of supervised and semi-supervised methods is superior in terms of detection accuracy, these methods usually suffer from the dilemma of imbalanced data since the labeled normal users are far more than spammers in real situations. To address the problem, we propose a novel method only relying on normal users to detect spammers. Firstly, a classifier is built from a part of normal and unlabeled samples to pick out reliable spammers from unlabeled samples. Secondly, our well-trained detector, which is based on the given normal users and predicted spammers, can distinguish between normal users and spammers. Experiments conducted on real-world datasets show that the proposed method is competitive with supervised methods.

Keywords: Spammer detection · Social network · PU Learning

1 Introduction

With the popularity of the social network, users are taking delight in sharing. For example, users can send tweets and comments on Twitter [1]. However, spammers are also planning to benefit from the prosperity by means of advertising, posting nonsenses and spreading fake information. Series of security risks may be caused due to spammers. For instance, users' privacy information can be filched by phishing links and the recommended lists are polluted by spam. Hence, spammer detection has become a significant work in social service.

© Springer International Publishing AG 2017
D. Liu et al. (Eds.): ICONIP 2017, Part V, LNCS 10638, pp. 177–185, 2017.
https://doi.org/10.1007/978-3-319-70139-4_18

By now, social spammer detection has attracted extensive attention from researchers and the industry. Existing efforts are categorized into unsupervised methods, supervised methods, semi-supervised methods, etc. Unsupervised spammer detection methods [2–4] do not need the labeled samples, which can save the cost of labeling. But the absence of labels may lead to the low accuracy. In contrast, supervised methods [1,5–7] and semi-supervised [8,9] methods perform better than unsupervised methods with the supervision of the labels. However, these methods relying on both positive and negative labels fail when there are only one class labels available. In addition, it is time-consuming to label numerous spammers in real situations. In order to resolve this problem, we propose a novel spammer detection method based on Positive and Unlabeled Learning (PU Learning) [10], named PUD. At first, we build a reliable negative (RN) classifier from normal users and unlabeled samples. Then some reliable negative samples are picked out. Secondly, the positive and unlabeled detecting (PUD) classifier is trained on positive and reliable negative samples. The main contributions of this paper are as follows:

- Propose a novel method PUD to detect spammers in social network;
- Evaluate and compare the performance of the proposed PUD method on real-world datasets with supervised methods;
- Discuss the effect of the proportion of positive samples in PUD, which proves PUD can achieve well result merely rely on a few positive samples.

The remainder of this paper is structured as follows. In Sect. 2, we introduce some related work. The problem statement and the illustration of PUD method are shown in Sect. 3. In Sect. 4, we conduct experiments on two real-world datasets. Finally, Sect. 5 concludes this paper and point out the potential future work.

2   Related Work

In this section, we review some related work from current research about social spammer detection and background knowledge about PU Learning.

2.1   Social Spammer Detection Methods

Generally speaking, the notable detection methods can be classified into unsupervised methods, supervised methods and semi-supervised methods according to the amount of needed labeled data.

Unsupervised Detection methods mainly utilize the social network topology to identify the abnormal nodes. The method of combining social relation graphs and user link diagrams was proposed in [3]. Zhang et al. [4] adopted 12 types of topological features in ego network to detect spammers.

Supervised Detection methods usually extract relevant characteristics of users. Benevenuto et al. [1] extracted the user behavior characteristics and tweet content characteristics to detect spammers. A group modeling framework was proposed in [7], which adaptively characterizes social interactions of spammers.

Semi-supervised Detection methods leverage labeled samples and massive unlabeled samples. A hybrid method that aimed to detect multiple spammers from user characteristics and user relationships was proposed in [8]. Li et al. [9] used the Laplace method to extract features, then used the semi-supervised method to train classifier.

Among these methods, Supervised methods outperform the unsupervised methods, but they need abundant labeled data. Semi-supervised methods require labeled and unlabeled data. Either supervised or semi-supervised methods rely on both positive and negative samples. Only a few positive labeled data and plenty of unlabeled data are required in our work.

2.2 Outline of PU Learning

The approach merely adopting positive and unlabeled data is called Positive and Unlabeled Learning or PU Learning. At the beginning, PU Learning mainly aimed to solve the task of text classification [10], then researchers extended this method to other areas. Such as the remote-sensing data classification, the disease gene identification, the Multi-graph learning, etc.

There are massive unlabeled user in real social networks, and the quantity of labeled spammers is much smaller than those of normal users. Furthermore, the cost of marking normal users is cheaper than marking spammers. These characteristics show that PU Learning can be applied in real situations.

PU Learning mainly consists of two steps [10]. Step 1: Identify the reliable negative samples (RN) from the unlabeled samples (U) according to the positive samples (P). Step 2: Construct the binary classifier by positive samples and reliable negative samples.

3   PUD Method

In this section, we will first state the problem of social spammer detection formally. Next, the main steps of the PUD method will be illustrated.

3.1   Problem Statement

Let $\mathbf{X} \in \mathbb{R}^{n \times t}$ be the t features of n users in a social network, and $\mathbf{Y} \in \{0, 1\}^n$ are corresponding labels of users, where $y_i = 0$ indicates the ith account is a spammer and equals to 1 otherwise. U, P, RN represent the unlabeled samples, positive samples and reliable negative samples, respectively. Meanwhile μ, l, r represent the amount of users in the corresponding samples.

The task of the spammer detection can be summarized as follows: Given the features for all n instances and some positive labels, learning a model PUD with well performance to classify an unknown account.

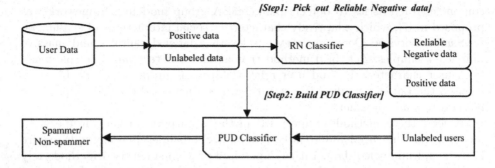

Fig. 1. The framework of PUD

3.2   PUD Framework

The framework of our proposed method consists of two steps, as described in Fig. 1, and each step will be illustrated in detail.

Step 1: Pick out Reliable Negative Samples. Picking out the reliable negative samples is a critical in PU Learning. Theoretically, maximizing the confidence of the negative samples and ensuring the positive samples are correctly classified, we can get a superior classifier [10]. Therefore, it is vital to find as many reliable negative samples as possible in the unlabeled dataset. In the following, we will describe the algorithm more specifically.

In our method, the reliable negative classifier is constructed by Naive Bayes, because it is a mature and popular classified algorithm, while other algorithms are alternative. Naive Bayes learns the joint probability distribution $P_r(X, Y)$ from the training dataset. Before that, it needs to learn priori probability distribution in Eq. (1) and conditional probability distribution in Eq. (2).

$$P_r(Y = c_k), \quad k = 0, 1 \tag{1}$$

$$P_r\left(X = x \mid Y = c_k\right) = P_r\left(X^{(1)} = x^{(1)}, \ldots, X^{(d)} = x^{(d)} \mid Y = c_k\right), \quad k = 0, 1 \tag{2}$$

where c_0 and c_1 denote the labels of positive samples and unlabeled samples, respectively. And then the joint probability distribution $P_r(X, Y)$ is learnt by Eqs. (1) and (2).

Given X, a set of user features, Naive Bayes algorithm calculates the posterior probability distribution in Eq. (3) by the learnt model. The label of x is the one having the highest posterior probability.

$$P_r\left(Y = c_k \mid X = x\right) = \frac{P_r\left(X = x \mid Y = c_k\right) P_r\left(Y = c_k\right)}{\sum_k P_r\left(X = x \mid Y = c_k\right) P_r\left(Y = c_k\right)} \tag{3}$$

The reliable negative classifier can be define as

$$y = f(x) = \arg\max_{c_k} P_r\left(Y = c_k\right) \prod_{j=1}^{d} P_r\left(X^{(j)} = x^{(j)} \mid Y = c_k\right). \tag{4}$$

The process of identifying the reliable negative samples RN from the positive sample P and unlabeled samples U is as follows: first of all, we assign each normal user label 1 to constitute P and some unlabeled users label 0 to form U. Secondly, RN classifier is learnt from αP and βU by Naive Bayes, where l is the amount of αP and r is the amount of RN. We set $\beta = 0.5$, because training and predicting both need plenty of unlabeled samples. Note that, α is an important parameter will be discussed in experiment. Thirdly, we exploit the classifier to identify other unlabeled users, $(1 - \beta)U$. Finally, reliable negative users are pick out from $(1-\beta)U$ until $r = l$, whose predicted labels are spammer. These reliable negative samples will be utilized in PUD classifier.

Step 2: Build PUD Classifier. A binary classifier is build in step 2. from positive and reliable negative samples by Random Forest algorithm to detect spammers. Random Forest is an ensemble algorithm that constructs a multitude of many decision trees at training time and outputs the class that is the mode of the classification of the individual tress. It is efficient for estimating missing data and maintains accuracy when a large proportion of the data are missing. Thus, Random Forest meets the requirements for our methods.

The procedures are as follows: firstly, the PUD classifier is trained by the predicted negative samples RN from RN classifier and given positive samples P. Then the PUD classifier can be utilized to detect spammers: the user is a spammer if the predicted label is negative, otherwise the user is legitimate.

The complete process of PUD method which integrates step 1 and step 2 is shown in Table 1.

Table 1. The complete process of PUD method

Input:
User Feature Matrix $\mathbf{X} = \{\mathbf{x}_1, \mathbf{x}_2, \cdots \mathbf{x}_n\} \in \mathbb{R}^{t \times n}$
User Labels $\mathbf{Y}$
Parameter α, β
Output:
A spammer detection classifier PUD

Step:
1:   $P = \emptyset, \quad U = \emptyset, \quad RN = \emptyset$
2:   for $\mathbf{x}_i$
3:       if $\mathbf{y}_i == 1$
4:           $P = P \cup \mathbf{x}_i$
5:       else
6:           $U = U \cup \mathbf{x}_i$
7:   RN$\leftarrow clf.learn(\alpha P, \beta U)$
8:   RN.predict$((1 - \beta)U)$
9:   while $r < l$
10:      $RN = RN \cup \{ predict == 0 \}$
11: PUD $\leftarrow clf.learn(P, RN)$

---

4   Experiments

In this section, we conduct experiments to evaluate the effectiveness of the proposed PUD method. We first introduce the datasets and metrics. Then we compare the performance of our method with other detection methods. Finally the sensitivity of parameter α will be discussed.

4.1   Datasets and Metrics

Two real datasets provided by Benevenuto [1,5] are used for evaluation. The one is from YouTube [5], includes 188 spammers and 641 legitimate users. Each user has 60 features which are derived from video attributes, individual characteristics of user behavior, and node attributes. The other is from Twitter [1]. This dataset contains 1650 labeled users, 355 of them are spammers. Each user has 62 features which are derived from tweet content and user social behavior.

The experiments are conducted by 5-fold cross validation 10 times, and average value are used to represent the results. We adopt the three frequent used evaluation metrics, i.e., *Precision, Recall* and *F-measure* for performance evaluation.

4.2   Experimental Results

Table 2 reported the performance of PUD method on both datasets. We apply Naive Bayes algorithm to pick out reliable negative samples on YouTube dataset while Logistic Regression is utilized on Twitter. The results show the validity of PUD and prove it is a general and base method.

Table 2. Performance of PUD

	Precision	Recall	F-measure
YouTube	0.786	0.662	0.71
Twitter	0.85	0.69	0.756

In order to further show our proposed method has competitive performance, it is compared with traditional supervised methods which exploit various proportion of labeled spammer in training. Traditional methods include Naive Bayes (NB), Logistic Regression (LR), Decision tree (DT), Random Forest (RF) and Gradient Boosting Decision Tree (GBDT). The results of different methods are displayed in Table 3, and we bold the best values in each dataset.

Based on the results, we make following observations. Firstly, the F-measure of PUD is quite close to the best values in Twitter while Random Forest and Gradient Boosting Decision Tree both need 30% labeled spammers. In YouTube, the F-measure of our method can reach to 71%, it increases over 4.7% than other methods. Secondly, it can be seen that PUD are superior to tradition methods whose labeled spammers are less than 20%. Therefore, the proposed method can relieve the dilemma of imbalanced data.

Table 3. F-measure comparison between PUD and other methods

	Spammer ratio	LR	NB	DT	RF	GBDT	PUD
YouTube	0%	\	\	\	\	\	**0.71**
	1%	0.232	0.269	0.218	0.27	0.25	\
	2%	0.262	0.314	0.246	0.276	0.262	\
	5%	0.39	0.418	0.422	0.53	0.37	\
	10%	0.416	0.432	0.538	0.624	0.478	\
	20%	0.542	0.434	0.618	0.65	0.562	\
	30%	0.644	0.44	0.646	0.678	0.674	\
Twitter	0%	\	\	\	\	\	0.768
	1%	0.214	0.14	0.376	0.24	0.38	\
	2%	0.296	0.21	0.558	0.45	0.548	\
	5%	0.35	0.426	0.644	0.612	0.586	\
	10%	0.36	0.49	0.69	0.706	0.654	\
	20%	0.38	0.51	0.71	0.736	0.72	\
	30%	0.45	0.542	0.716	0.776	**0.78**	\

4.3 Parametric Sensitivity Analysis

Now, we discuss the sensitivity of the parameter α which determines the proportion of positive samples chosen. The experimental results are shown in Fig. 2.

Figure 2(a) shows the fluctuant performance of PUD with the different values of α on the YouTube dataset. It can be observed that the precision increases while the recall reduces as a result of imbalanced data. In order to balance the performance of PUD, we take $\alpha = 0.7$ in experiment, and then the F-measure can reach the optimal state. Figure 2(b) shows the performance on Twitter, and α is set to 0.5 to make the precision and recall balance in experiment.

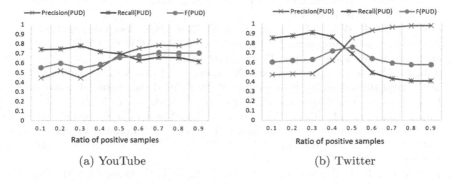

(a) YouTube   (b) Twitter

Fig. 2. Performance of PUD with varying α on datasets

In summary, the proposed method is not always outstanding in F-measure compared with supervised methods, but it can achieve competitive performance without labeled spammers. In addition, the effect of the parameter is analyzed as well. It proves that PUD can get ideal result merely using a few positive samples which reduces the cost of labeling.

5   Conclusion and Future Work

In this paper, we proposed a novel method PUD based on PU Learning, it aims to construct a detection classifier by a few positive samples and plenty of unlabeled data. Our method includes two steps: at first, we pick out reliable negative samples from unlabeled users. After that, the PUD classifier is trained by positive and reliable negative samples. Experimental results on the two real-world datasets show that our approach has competitive performance and prove it is a general and base method. Furthermore, PUD shows its merits in detecting spammers. Thus the proposed method can be applied extensively.

A few possible works remain to be done. We will combine PUD with various state-of-the-art supervised methods to improve the accuracy of spammer detection. Besides, our method can be used to detect fake comments in social networks.

Acknowledgments. The work is supported by the Basic and Advanced Research Projects in Chongqing under Grant No. cstc2015jcyjA40049, the National Key Basic Research Program of China (973) under Grant No. 2013CB328903, the Guangxi Science and Technology Major Project under Grant No. GKAA17129002, and the Graduate Scientific Research and Innovation Foundation of Chongqing, China under Grant No. CYS17035.

References

1. Benevenuto, F., Magno, G., Rodrigues, T., Almeida, V.: Detecting spammers on Twitter. In: Collaboration, Electronic Messaging, Anti-abuse and Spam Conference (CEAS), vol. 6, p. 12 (2010)
2. Gao, H., Hu, J., Wilson, C., Li, Z., Chen, Y., Zhao, B.Y.: Detecting and characterizing social spam campaigns. In: Proceedings of 10th ACM SIGCOMM conference on Internet measurement, pp. 35–47. ACM (2010)
3. Tan, E., Guo, L., Chen, S., Zhang, X., Zhao, Y.: Unik: unsupervised social network spam detection. In: Proceedings of 22nd ACM international conference on Information & Knowledge Management, pp. 479–488. ACM (2013)
4. Zhang, B., Qian, T., Chen, Y., You, Z.: Social spammer detection via structural properties in ego network. In: Li, Y., Xiang, G., Lin, H., Wang, M. (eds.) SMP 2016. CCIS, vol. 669, pp. 245–256. Springer, Singapore (2016). doi:10.1007/978-981-10-2993-6_21
5. Benevenuto, F., Rodrigues, T., Almeida, V., Almeida, J., Gonçalves, M.: Detecting spammers and content promoters in online video social networks. In: Proceedings of 32nd International ACM SIGIR Conference on Research and Development in Information Retrieval, pp. 620–627. ACM (2009)

6. Hu, X., Tang, J., Zhang, Y., Liu, H.: Social spammer detection in microblogging. In: IJCAI, vol. 13, pp. 2633–2639. Citeseer (2013)
7. Wu, L., Hu, X., Morstatter, F., Liu, H.: Adaptive spammer detection with sparse group modeling. In: ICWSM, p. 319–326 (2017)
8. Wu, Z., Wang, Y., Wang, Y., Wu, J., Cao, J., Zhang, L.: Spammers detection from product reviews: a hybrid model. In: 2015 IEEE International Conference on, Data Mining (ICDM), pp. 1039–1044. IEEE (2015)
9. Li, W., Gao, M., Rong, W., Wen, J., Xiong, Q., Ling, B.: LSSL-SSD: social spammer detection with laplacian score and semi-supervised learning. In: Lehner, F., Fteimi, N. (eds.) KSEM 2016. LNCS, vol. 9983, pp. 439–450. Springer, Cham (2016). doi:10.1007/978-3-319-47650-6_35
10. Liu, B., Dai, Y., Li, X., Lee, W.S., Yu, P.S.: Building text classifiers using positive and unlabeled examples. In: 3rd IEEE International Conference on Data Mining, ICDM 2003, pp. 179–186. IEEE (2003)

Discovery of Interconnection Among Knowledge Areas of Standard Computer Science Curricula by a Data Science Approach

Yoshitatsu Matsuda[1]([✉]), Takayuki Sekiya[2], and Kazunori Yamaguchi[1]

[1] Department of General Systems Studies, The University of Tokyo,
3-8-1, Komaba, Meguro-ku, Tokyo 153-8902, Japan
{matsuda,yamaguch}@graco.c.u-tokyo.ac.jp
[2] Information Technology Center, The University of Tokyo,
3-8-1, Komaba, Meguro-ku, Tokyo 153-8902, Japan
sekiya@ecc.u-tokyo.ac.jp

Abstract. Computer Science Curricula 2013 (CS2013) is a widely-used standard curricula of computer science, which has been developed jointly by the ACM and the IEEE Computer Society. CS2013 consists of 18 Knowledge Areas (KAs) such as Programming Languages and Software Engineering. Though it is obvious that there are strong interconnections among the KAs, it was hard to investigate the interconnections objectively and quantitatively. In this paper, the interconnections among the KAs of CS2013 are investigated by a data science approach. For this purpose, a collection of actual syllabi from the world's top-ranked universities was constructed. Then, every actual syllabus is projected to the KA space by a probabilistic model-based method named simplified, supervised Latent Dirichlet Allocation (denoted by ssLDA). Consequently, the following interesting properties of the interconnections among the KAs were discovered: (1) There are the high interconnections among the KAs in each syllabi; (2) A plausible hierarchical structure of the KAs is found by utilizing the interconnections; (3) The structure shows that the KAs are classified into the three principal independent factors (HUMAN, THEORY, and IMPLEMENTATION). The factor of IMPLEMENTATION can be divided into PROGRAMMING and SYSTEM. The factor of SYSTEM can be divided further into DEVICES and NETWORK.

Keywords: Data mining · Computational education · Curriculum analysis · Latent Dirichlet allocation

1   Introduction

The curricula and the syllabi play very important roles in all the educational programs. Fortunately, there is a guideline of curricula in computer science education: Computer Science Curricula 2013 (abbreviated as CS2013) [1]. CS2013 is the latest version of the curricular guideline for undergraduate programs in computer science, which has been developed jointly by the ACM and the IEEE

© Springer International Publishing AG 2017
D. Liu et al. (Eds.): ICONIP 2017, Part V, LNCS 10638, pp. 186–195, 2017.
https://doi.org/10.1007/978-3-319-70139-4_19

Computer Society for over 40 years. CS2013 is widely used as the foundation for designing new curricula, evaluating current ones, and supporting the learning [2–7]. For example, we proposed a web-based visualization system for comparing curricula from different universities by CS2013 [2]. Gluga et al. developed another web-based visualization system named PROGOSS for supporting the students by CS2013 [3]. On the other hand, the appropriateness of CS2013 itself has not been fully discussed. It has been estimated only by massive but subjective survey from the universities in [1].

In this paper, CS2013 is investigated quantitatively by a data science approach, where a collection of actual syllabi from the world's top-ranked universities is utilized. The collection can be constructed manually from the public data at the web sites of the universities. Therefore, the data collection is more objective and more easily available than the survey data from the universities. In our approach, each actual syllabus is regarded as a bag of words and it is projected to a point in the KA space (where each axis corresponds to a KA) by a probabilistic model-based method named simplified, supervised LDA (ssLDA). Then, the properties of the KAs are investigated quantitatively by using the distribution of actual syllabi in the KA space. Our other work investigates the characteristics of the actual curricula of the universities by applying the same method to a similar dataset [8]. On the other hand, this work discovers the interconnections among the KAs of CS2013 from a different viewpoint.

This paper is organized as follows. In Sect. 2, the two backgrounds are described. One is the structure of CS2013 in Sect. 2.1. Another is the method (named ssLDA) projecting the syllabi to the KA space in Sect. 2.2. The data collection and the investigation results are described in Sect. 3. Section 3.1 describes the method for collecting actual syllabi and the properties of the collection. The interconnections among the KAs are investigated in Sect. 3.2. In addition, a hierarchical structure of the KAs is discovered in Sect. 3.3 and the independent principal factors in the KAs are extracted from the structure in Sect. 3.4. The results are discussed in Sect. 3.5. Section 4 concludes this paper.

2 Background

2.1 Structure of CS2013

Here, the structure of CS2013 is described. The main part of CS2013 is called as the Body of Knowledge (BOK). The BOK consists of a set of 18 Knowledge Areas (KAs), where each KA corresponds to a topic area in computer science such as "Algorithm and Complexity." Table 1 shows the names and the abbreviation IDs of KAs. Each KA consists of about 10 Knowledge Units (KUs), where each KU is a short document describing a part of the KA. The total number of KUs in CS2013 is 163. In summary, CS2013 represents each KA (namely, a topic in computer science) as a set of KUs. Each KU is regarded as a bag of words.

Table 1. Knowledge areas of CS2013.

ID	KA
AL	Algorithms and complexity
AR	Architecture and organization
CN	Computational science
DS	Discrete structures
GV	Graphics and visualization
HCI	Human-computer interaction
IAS	Information assurance and security
IM	Information management
IS	Intelligent systems
NC	Networking and communication
OS	Operating systems
PBD	Platform-based development
PD	Parallel and distributed computing
PL	Programming languages
SDF	Software development fundamentals
SE	Software engineering
SF	Systems fundamentals
SP	Social issues and professional practice

2.2   Simplified, Supervised Latent Dirichlet Allocation

In order to investigate the interconnections among the KAs in actual syllabi, each syllabus is projected to "the KA space" by our previously-proposed method named "simplified, supervised latent Dirichlet allocation" (ssLDA). Here, we explain ssLDA and the KA space.

ssLDA is an extension of the well-known LDA [9]. The original LDA is a probabilistic model in natural language processing, which extracts the topics (corresponding to the KAs in this paper) from a given set of bags of words (the KUs). It can project any new bag of words (an actual syllabus) to a point in the extracted topic space (called the KA space in this paper). As the original LDA is an unsupervised method, it can not utilize the KA label on each KU. Therefore, we have developed a supervised version of LDA. Though the supervised LDA was proposed in [10,11], it is too complicated to utilize only the KA labels. ssLDA proposed in [12] is a simplification of the supervised LDA, which is designed for extracting the KA space from the KUs. The generative model of ssLDA is shown in Fig. 1. The model is essentially equivalent to the original LDA except for the given KA label c and the hyper-parameter of the softmax prediction η. ssLDA projects the actual syllabi to the KA space by the following two phases.

1. Inference phase: All the KUs with their corresponding KA labels are given as the training dataset. Then, the optimal β (the degree of relationship between each KA and each word) is estimated by maximizing the likelihood of the generative model of Fig. 1.
2. Prediction phase: The optimal θ (the occurrence probability of each KA) of each actual syllabus is estimated under the optimal β. The estimated θ is regarded as the position $w = (w_k)$ of the syllabus in the KA space.

The first inference phase is given as a simple extension of the original LDA by incorporating the softmax model with the original inference. The second prediction phase is completely equivalent to that of the original LDA. The hyperparameters $\alpha(=1)$ (determining the form of the prior Dirichlet distribution) and $\eta(=50)$ (determining the effect of the softmax model) were set empirically by a cross validation method. See [12] for the details of the algorithm.

Note that a position w in the KA space is given as a probability distribution over the KAs. Therefore, $\sum w_k = 1$ and $w_k \geq 0$ hold. The number of dimensions in the KA space is 18 (the number of the KAs). The degree of freedom is 17 because of $\sum w_k = 1$. w_k represents a weight of the k-th KA in a syllabus. If a syllabus is strongly related to some KAs, the weights of the related KAs are higher than the other weights. Therefore, the interconnections among the KAs are expected to be discovered by investigating the distribution of the actual syllabi in the KA space.

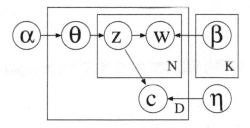

Fig. 1. Generative model of simplified, supervised LDA (cited from [12]).

3   Results

3.1   Data Collection

Here, the method for collecting the actual syllabi is described. First, a well-known university ranking, "Times Higher Education (THE) WORLD UNIVERSITY RANKINGS, Top 100 universities for engineering and technology 2014–2015" [13], was used for selecting the world's 50 top-ranked universities. Then, one department dedicated to computer science was picked up manually from each university. It was found that most of the departments (47/50) published their curricula and syllabi at their web sites. Regarding the other (3/50) departments, they were omitted in this analysis because we could not obtain any information about their syllabi. As there is no common format of curricula unfortunately, we manually downloaded HTML or PDF files from the site of each department. Thus, we collected 3437 actual syllabi in total. One hour work of a person who has no knowledge on the curriculum analysis is sufficient to collect all the syllabi of one department. In other words, the total time for collecting the actual syllabi was about 30–40 person-hours. Six departments did not provide their syllabi in English and they were translated into English by Google Translate (https://translate.google.com/). Note that the translation quality of English sentences is not required because each syllabus is treated as a bag of words. Then, a simple stemming algorithm was applied. Moreover, the words which are not included in any KU of CS2013 were removed because such words have no influence on the KA space. In order to verify this pre-processing, Table 2 shows the simple coverage of CS2013 over all the actual syllabi with the vocabulary size. The coverage rate is about 40%. Considering that the vocabulary size of the actual syllabi was about 8 times as much as that of CS2013, this rate is definitely high. Finally, the syllabi of less then 10 words were removed. Thus, 3077 bags of words were constructed from the actual syllabi and all of them were projected to the KA space by ssLDA in Sect. 2.2.

Table 2. Coverage of CS2013 over all the actual syllabi.

	CS2013	All actual syllabi
Number of words ($=$A)	20,171	478,629
Number of words included in the CS2013 vocabulary ($=$B)	20,171	194,524
Coverage ($=$B/A)	100%	40.6%
Vocabulary size	3,304	25,849

3.2 Interconnection Among KAs

Here, it is verified that there exist the high interconnections among the KAs in the actual syllabi. In order to measure the degree of the interconnection, an index $\lambda(\boldsymbol{w})$ is employed, which is defined as the exponential of the entropy

$$\lambda(\boldsymbol{w}) = e^{H(\boldsymbol{w})}. \tag{1}$$

Here, $H(\boldsymbol{w})$ is Shannon entropy defined as

$$H(\boldsymbol{w}) = -\sum_{k} w_k \log(w_k). \tag{2}$$

The index $\lambda(\boldsymbol{w})$ is 0 (the theoretical minimum) if and only if w_k is 1 for a single k and the others are 0 (where no interconnection among the KAs exists). In contrast, $\lambda(\boldsymbol{w})$ is 18 (the theoretical maximum) if and only if all the 18 w_k's are $\frac{1}{18}$ uniformly. Therefore, the higher $\lambda\,(\boldsymbol{w})$ means the higher interconnections among the KAs in the syllabus. Figure 2 shows the histograms of $\lambda\,(\boldsymbol{w})$ for all the KUs of CS2013 and for all the actual syllabi. In Fig. 2(a), the degree of the interconnection in most of KUs is low. It is not surprising because the learning in ssLDA forces each KU to belong to the single KA given as the label. On the other hand, Fig. 2(b) shows that there exist the high interconnections in most of the actual syllabi. It shows that our actual syllabi-based approach can detect the interconnections among the KAs, which can not be discovered by the KUs.

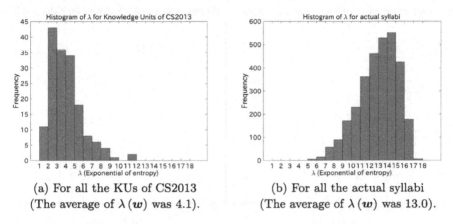

(a) For all the KUs of CS2013   (b) For all the actual syllabi

(The average of $\lambda\,(\boldsymbol{w})$ was 4.1).   (The average of $\lambda\,(\boldsymbol{w})$ was 13.0).

Fig. 2. Histograms of the exponential of the entropy $\lambda\,(\boldsymbol{w})$ for all the KUs of CS2013 (a) and for all the actual syllabi (b).

3.3 Hierarchical Structure of KAs

As shown in Sect. 3.2, there exist the high interconnections among the KAs in actual syllabi. In this section, the interconnections are utilized for discovering a hierarchical structure of the KAs. By letting $\boldsymbol{w}^i = (w_k^i)$ be the position of the i-th actual syllabus in the KA space, each KA k is characterized as a vector $\boldsymbol{v}_k = (w_k^1, w_k^2, \cdots, w_k^L)$ where L is the number of the actual syllabi (namely, $L = 3077$). Then, the distances among the KAs can be estimated as the distances among $\boldsymbol{v}_k$'s. In order to remove the effects of the mean and the variance, each $\boldsymbol{v}_k$ was normalized to $\bar{\boldsymbol{v}}_k$ so that the mean and the variance of $\bar{\boldsymbol{v}}_k$ over all the actual syllabi are 0 and 1, respectively. Then, the distance between the KAs i and j was estimated as the Euclidean distance between $\boldsymbol{v}_i$ and $\boldsymbol{v}_j$, and Ward's method [14,15] was applied to the distances for constructing the hierarchical cluster tree of the KAs. Figure 3(a) shows the constructed structure. It is an intuitively plausible structure. For example, the three KAs related to the factor of HUMAN (HCI, SP, and SE) were grouped together. In addition, the highly-related pairs among the KAs could be found in the structure (AR and SF, PBD and PL, IAS and NC, AL and DS, and so on). On the other hand, it was difficult to extract such a hierarchical structure directly from the KUs. For example, Fig. 3(b) shows a hierarchical structure where the distances among the KAs are estimated by their tf-idf vectors. Here, each KA is given as a long document consisting of its included KUs. We could not find a plausible structure in Fig. 3(b). Though we tried some other methods as well as the tf-idf (the unweighted term frequency, the binary frequency, and so on), no plausible structure could be discovered. It verifies that the actual syllabi are essential for discovering the hierarchical structure.

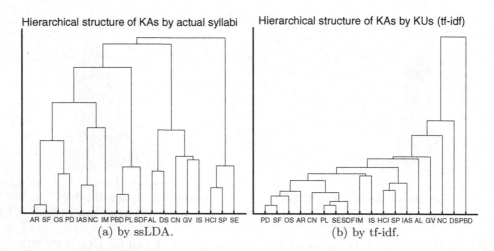

Hierarchical structure of KAs by actual syllabi   Hierarchical structure of KAs by KUs (tf-idf)

(a) by ssLDA.   (b) by tf-idf.

Fig. 3. Hierarchical structures of the KAs by applying ssLDA to the actual syllabi (a) and by using the tf-idf values of the KUs (b).

3.4   Principal Factors in KAs

In this section, the principal factors in the KAs are extracted by focusing on the clusters of the KAs in Fig. 3(a). The KAs can be divided into any number $P \leq 18$ of clusters by the hierarchical structure in Fig. 3(a). By avoiding a cluster including only a single element, the maximum number of P is 5. Table 3 shows the clusters with the labels for $P = 2, 3, 4, 5$. The label naming each common factor of the included KAs was assigned manually. Letting Ω_l ($l = 1, \cdots, P$) be the set of the KAs belonging to the l-th cluster, the weight vector $\boldsymbol{w}$ for each actual syllabus can be reduced to the P-dimensional factors $\boldsymbol{f} = (f_l)$ by

$$f_l = \sum_{k \in \Omega_l} w_k. \tag{3}$$

The space spanned by $\boldsymbol{f}$ is called the factor space. The number of clusters P should be set suitably according to the purpose of analysis. In the following analysis, P is set to 3 in order to visualize easily the distributions of the actual syllabi. Then, there are the three factors of HUMAN, THEORY, and IMPLE-MENTATION. The factor space can be visualized in the two-dimensional equilateral triangle because of $f_1 + f_2 + f_3 = 1$, where each vertex corresponds to a factor. Figure 4 shows the scatter plot of the actual syllabi in the factor space. The distribution of the syllabi is inclined to IMPLEMENTATION because it includes a relatively large number of KAs. Nevertheless, Fig. 4 shows that the distribution is widely dispersed and the weights of the factors are approximately independent of each other over the triangle. It suggests that the three factors are useful for visualizing many actual syllabi and for supporting both the learners and the teachers.

Table 3. Labels of the clusters of KAs extracted from the hierarchical structure in Fig. 3-(a) for the number of clusters $P = 2, 3, 4, 5$.

$P = 2$	COMPUTER				HUMAN
$P = 3$	IMPLEMENTATION			THEORY	
$P = 4$	SYSTEM		PROGRAMMING		
$P = 5$	DEVICES	NETWORK			
KAs	AR SF OS PD	IAS NC IM	PBD PL SDF	AL DS CN GV IS	HCI SP SE

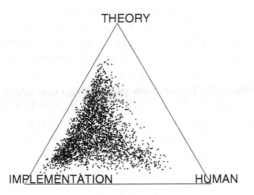

Fig. 4. Distribution of the actual syllabi in the three-dimensional factor space.

3.5 Discussions

Here, the above results are discussed from the two viewpoints of the confirmation of the present premises and the discovery of the surprising facts. First, Fig. 2 shows that each actual syllabus spreads over multiple KAs. In other words, there exist the high interconnections among the KAs of the actual syllabi. Moreover, Fig. 3(a) shows that there is a hierarchical structure among the KAs. CS2013 emphasizes the interconnections among the KAs as follows: "It is important to recognize that Knowledge Areas are interconnected and that concepts in one KA may build upon or complement material from other KAs." ([1], p. 15). These investigation results confirm this assertion. Second, Fig. 4 and Table 3 show that the factor of HUMAN is a principal factor. This factor often has been regarded as a supplement. For example, ETS major field test [16], which is one of the standard tests assessing the skills of students, employs the following three major assessment indicators in computer science: (a) Programming and Software Engineering, (b) Discrete Structures and Algorithms, and (c) Systems.[1] The three indicators seem to correspond to the discovered three factors of PROGRAMMING, THEORY, and SYSTEM in Table 3. However, the factor of HUMAN seems to be underestimated. The investigation results suggest that this underestimation is not consistent with the actual syllabi and the factor of HUMAN should be assessed in more detail.

4   Conclusion

In this paper, a data science approach was applied for the analysis of CS2013, where the actual syllabi were collected and utilized. We projected the actual syllabi to the KA space by ssLDA, investigated the distributions of the projected positions, and discovered some interesting properties about the interconnections among the KAs of CS2013. This approach enables the guidelines for curricula and the actual syllabi to evaluate each other mutually. This mutual feedback is expected to improve both of them effectively. In order to enable any users to project their own syllabi to the same KA space, we have released a new analysis tool with a web-based interface at https://rp3.ecc.u-tokyo.ac.jp/cs2013data/ [17]. We are now investigating CS2013 quantitatively from other viewpoints such as the appropriateness of the Core topics and the difference among the top-ranked universities [8]. We are also planning to construct an automated system for collecting more massive actual syllabi and additional metadata such as required/elective. Moreover, we are planning to develop an objective, data-driven method instead of the current subjective one for naming the principal factors in the KAs. This work was supported by JSPS KAKENHI Grant Number 17H01837.

References

1. ACM/IEEE-CS Joint Task Force on Computing Curricula: Computer science curricula 2013. Technical report, ACM Press and IEEE Computer Society Press (2013)
2. Sekiya, T., Matsuda, Y., Yamaguchi, K.: Analysis of computer science related curriculum on LDA and Isomap. In: Proceedings of 15th Annual SIGCSE Conference on Innovation and Technology in Computer Science Education (ITiCSE 2010), pp. 48–52 (2010)
3. Gluga, R., Kay, J., Lister, R.: PROGOSS: mastering the curriculum. In: Proceedings of Australian Conference on Science and Mathematics Education (formerly UniServe Science Conference) (2012)

---

[1] https://www.ets.org/s/mft/pdf/mft_testdesc_compsci.pdf, accessed: 2017-May-30.

4. Zhao, L., Su, X., Wang, T.: Bring CS2013 recommendations into C programming course. Procedia - Soc. Behav. Sci. **176**, 194–199 (2015)

5. Walker, H.M., Rebelsky, S.A.: Using CS2013 for a department's curriculum review: a case study. J. Comput. Sci. Coll. **29**(5), 138–144 (2014)

6. McGuffee, J.W., Palmer, E.K., Guzman, I.R.: Assessing the tier-1 core learning outcomes of CS2013. In: Proceedings of 47th ACM Technical Symposium on Computing Science Education (SIGCSE 2016), pp. 485–489. ACM (2016)

7. Dai, Y., Asano, Y., Yoshikawa, M.: Course content analysis: an initiative step toward learning object recommendation systems for MOOC learners. In: Proceedings of 9th International Conference on Educational Data Mining (EDM 2016), pp. 347–352 (2016)

8. Matsuda, Y., Sekiya, T., Yamaguchi, K.: Curriculum analysis of CS departments based on computing curricula by simplified, supervised LDA (2017, submitted manuscript)

9. Blei, D.M., Ng, A.Y., Jordan, M.I.: Latent Dirichlet allocation. J. Mach. Learn. Res. **3**, 993–1022 (2003)

10. Wang, C., Blei, D., Li, F.F.: Simultaneous image classification and annotation. In: IEEE Conference on Computer Vision and Pattern Recognition (CPVR 2009), pp. 1903–1910. IEEE (2009)

11. Blei, D.M., McAuliffe, J.D.: Supervised topic models. In: NIPS, vol. 7, pp. 121–128 (2007)

12. Sekiya, T., Matsuda, Y., Yamaguchi, K.: Curriculum analysis of CS departments based on CS2013 by simplified, supervised LDA. In: Proceedings of 5th International Conference on Learning Analytics And Knowledge (LAK 2015), pp. 330–339. ACM (2015)

13. Times Higher Education: The 2014–2015 Times Higher Education World University Rankings, Subject Ranking 2014–15: Engineering & Technology. http://www.timeshighereducation.co.uk/world-university-rankings/2014-15/subject-ranking/subject/engineering-and-IT. Accessed 2 Oct 2015

14. Ward, J.H.: Hierarchical grouping to optimize an objective function. J. Am. Stat. Assoc. **58**(301), 236–244 (1963)

15. Duda, R.O., Hart, P.E., Stork, D.G.: Pattern Classification, 2nd edn. Wiley, Hoboken (2000)

16. Educational Testing Service: ETS®Major Field Tests. https://www.ets.org/mft/. Accessed 21 December 2016

17. Sekiya, T., Matsuda, Y., Yamaguchi, K.: A web-based curriculum engineering tool for investigating syllabi in topic space of standard computer science curricula. In: Proceedings of 2017 IEEE Frontiers in Education Conference (FIE 2017). IEEE (2017, in press)

A Probabilistic Model for the Cold-Start Problem in Rating Prediction Using Click Data

ThaiBinh Nguyen[1]([✉]) and Atsuhiro Takasu[1,2]

[1] Department of Informatics, SOKENDAI
(The Graduate University for Advanced Studies), Tokyo, Japan
{binh,takasu}@nii.ac.jp
[2] National Institute of Informatics, Tokyo, Japan

Abstract. One of the most efficient methods in collaborative filtering is matrix factorization, which finds the latent vector representations of users and items based on the ratings of users to items. However, a matrix factorization based algorithm suffers from the *cold-start* problem: it cannot find latent vectors for items to which previous ratings are not available. This paper utilizes *click data*, which can be collected in abundance, to address the cold-start problem. We propose a *probabilistic item embedding* model that learns item representations from click data, and a model named EMB-MF, that connects it with a *probabilistic matrix factorization* for rating prediction. The experiments on three real-world datasets demonstrate that the proposed model is not only effective in recommending items with no previous ratings, but also outperforms competing methods, especially when the data is very sparse.

Keywords: Recommender system · Collaborative filtering · Item embedding · Matrix factorization

1 Introduction

Rating prediction is one of the key tasks in recommender systems. From research done previously on this problem, matrix factorization (MF) [5,8,12] is found to be one of the most efficient techniques. An MF-based algorithm finds the *vector representations* (latent feature vectors) of users and items, and uses these vectors to predict the unseen ratings. However, an MF-based algorithm suffers from the *cold-start* problem: it cannot find the latent feature vectors for items that do not have any prior ratings; thus cannot recommend them.

To address the *cold-start* problem, many methods have been proposed. Most of them rely on exploiting *side information* (e.g., the contents of items). The collaborative topic model [14] and content-based Poisson factorization [6] use text content information of items as side information for recommending new items. In [15], the author proposed a model for music content using deep neural network and combined it with an MF-based model for music recommendation. However, in many cases, such side information is not available, or is not informative enough (e.g., when an item is described only by some keywords).

© Springer International Publishing AG 2017
D. Liu et al. (Eds.): ICONIP 2017, Part V, LNCS 10638, pp. 196–205, 2017.
https://doi.org/10.1007/978-3-319-70139-4_20

This paper focuses on utilizing *click data*, another kind of feedback from the user. The advantage of utilizing click data is that it can be easily collected with abundance during the interactions of users with the systems. The idea is to identify item representations from click data and use them for rating prediction.

The main contributions of this work can be summarized as follows:

- We propose a *probabilistic item embedding* model for learning item embedding from click data. We will show that this model is equivalent to performing PMF [12] of the positive-PMI (PPMI) matrix, which can be done efficiently.
- We propose EMB-MF, a model that combines the *probabilistic item embedding* and PMF [12] for coupling the item representations of the two models.
- The proposed model (EMB-MF) can automatically control the contributions of prior ratings and clicks in rating predictions. For items that have few or no prior ratings, the predictions mainly rely on click data. In contrast, for items with many prior ratings, the predictions mainly rely on ratings.

2 Proposed Method

2.1 Notations

The notations used in the proposed model are shown in Table 1.

Table 1. Definitions of some notations

Notation	Meaning
N, M	The number of users, number of items
R, S	The rating matrix, PPMI matrix
$\mathcal{R}$	The set of (u, i)-pair that rating is observed (i.e., $\mathcal{R} = \{(u, j)|R_{uj} > 0\}$
$\mathcal{R}_u, \mathcal{R}_i$	The set of items that user u rated, set of users that rated item i
$\mathcal{S}$	The set of (i, j)-pair that $S_{ij} > 0$ (i.e., $\mathcal{S} = \{(i, j)|S_{ij} > 0\}$
$\mathcal{S}_i$	The set of item j that $S_{ij} > 0$ (i.e., $\mathcal{S}_i = \{j|S_{ij} > 0\}$
d	the dimensionality of the feature space
$\mathbf{I}_d$	The d-dimensional identity matrix
ρ_i, α_i	The *item embedding vector* and *item context vector* of item i
θ_u, β_i	The *feature vector* of user u, *feature vector* of item i, in the rating model
μ, b_u, c_i	The *global mean* of ratings, *user bias* of user u, *item bias* of item i
B_{ui}	$b_u + c_i + \mu$
$\rho, \alpha, \theta, \beta, \mathbf{b}, \mathbf{c}$	$\{\rho\}_{i=1}^M, \{\alpha\}_{i=1}^M, \{\theta\}_{u=1}^N, \{\beta\}_{i=1}^M, \{b\}_{u=1}^N, \{c\}_{i=1}^M$
Ω	The set of all model parameters (i.e., $\Omega = \{\rho, \alpha, \theta, \beta, \mathbf{b}, \mathbf{c}\}$)

2.2 Probabilistic Item Embedding Based on Click Data

The motivation behind the use of clicks for learning representations of items is the following: if two items are often clicked in the *context* of each other, they are likely to be similar in their nature. Therefore, analyzing the *co-click* information of items can reveal the relationship between items that are often clicked together.

"*Context*" is a modeling choice and can be defined in different ways. For example, the context can be defined as the set of items that are clicked by the user (*user-based* context); or can be defined as the items that are clicked in a session (*session-based* context). Although we use the user-based context to describe the proposed model in this work, other definitions can also be used.

We represent the association between items i and j via a *link function* $g(.)$, which reflects how strong i and j are related, as follows:

$$p(i|j) = g(\rho_i^\top \alpha_j) p(i) \tag{1}$$

where $p(i)$ is the probability that item i is clicked; $p(i|j)$ is the probability that i is clicked by a user given that j has been clicked by that user. We want the value of $g(.)$ to be large if i and j are frequently clicked by the same users.

There are different choices for the link functions, and an appropriate choice is $g(i,j) = \exp\{\rho_i^\top \alpha_j\}$. Equation 1 can be rewritten as:

$$\log \frac{p(i|j)}{p(i)} = \rho_i^\top \alpha_j \tag{2}$$

Note that $\log \frac{p(i|j)}{p(i)}$ is the point-wise mutual information (PMI) [4] of i and j, and we can rewrite Eq. 2 as:

$$PMI(i,j) = \rho_i^\top \alpha_j \tag{3}$$

Empirically, PMI can be estimated using the actual number of observations:

$$\widehat{PMI}(i,j) = \log \frac{\#(i,j)|\mathcal{D}|}{\#(i)\#(j)} \tag{4}$$

where $\mathcal{D}$ is the set of all item–item pairs that are observed in the click history of all users, $\#(i)$ is the number of users who clicked i, $\#(j)$ is the number of users who clicked j, and $\#(i,j)$ is the number of users who clicked both i and j.

A practical issue arises here: for item pair (i,j) that is not often clicked by the same user, $PMI(i,j)$ is negative, or if they have never been clicked by the same user, $\#(i,j) = 0$ and $PMI(i,j) = -\infty$. However, a negative value of PMI does not necessarily imply that the items are not related. The reason may be because the users who click i may not know about the existence of j. A common resolution to this is to replace negative values by zeros to form the PPMI matrix [3]. Elements of the PPMI matrix S are defined below:

$$S_{ij} = \max\{\widehat{PMI}(i,j), 0\} \tag{5}$$

We can see that item embedding vectors $\boldsymbol{\rho}$ and item context vectors $\boldsymbol{\alpha}$ can be obtained by factorizing PPMI matrix S. The factorization can be performed by PMF [12].

2.3   Joint Model of Ratings and Clicks

In modeling items, we let item feature vector β_i deviate from embedding vector ρ_i. This deviation (i.e., $\beta_i - \rho_i$) accounts for the contribution of rating data in the item representation. In detail, if item i has few prior ratings, this deviation should be small; in contrast, if i has many prior ratings, this deviation should be large to allow more information from ratings to be directed toward the item representation. This deviation is introduced by letting β_i be a Gaussian distribution with mean ρ_i:

$$p(\boldsymbol{\beta}|\boldsymbol{\rho}, \sigma_\beta^2) = \prod_i \mathcal{N}(\beta_i|\rho_i, \sigma_\beta^2) \tag{6}$$

Below is the *generative process* of the model:

1. **Item embedding model**
   (a) For each item i: draw embedding and context vectors

   $$\rho_i \propto \mathcal{N}(0, \sigma_\rho^2 \mathbf{I}), \quad \alpha_i \propto \mathcal{N}(0, \sigma_\alpha^2 \mathbf{I}) \tag{7}$$

   (b) For each pair (i, j), draw S_{ij} of the PPMI matrix:

   $$S_{ij} \propto \mathcal{N}(\rho_i^\top \alpha_j, \sigma_S^2) \tag{8}$$

2. **Rating model**
   (a) For each user u: draw user feature vector and bias term

   $$\theta_u \propto \mathcal{N}(0, \sigma_\theta^2 \mathbf{I}), \quad b_u \propto \mathcal{N}(0, \sigma_b^2) \tag{9}$$

   (b) For each item i: draw item feature vector and bias term

   $$\epsilon_i \propto \mathcal{N}(0, \sigma_\epsilon^2 \mathbf{I}), \quad c_i \propto \mathcal{N}(0, \sigma_c^2) \tag{10}$$

   (c) For each pair (u, i): draw the rating

   $$R_{uj} \propto \mathcal{N}(B_{ui} + \theta_u^\top \beta_j, \sigma_R^2) \tag{11}$$

The *posterior distribution* of the model parameters given the rating matrix R, PPMI matrix S and the hyper-parameters is as follows.

$$
\begin{aligned}
p(\boldsymbol{\Omega}|R, S, \boldsymbol{\Theta}) &\propto P(R|\mathbf{b}, \mathbf{c}, \boldsymbol{\theta}, \boldsymbol{\beta}, \sigma_R^2) P(S|\boldsymbol{\rho}, \boldsymbol{\alpha}, \sigma_S^2) \\
&\quad \times p(\boldsymbol{\theta}|\sigma_\theta^2) p(\boldsymbol{\beta}|\boldsymbol{\rho}) p(\boldsymbol{\rho}|\sigma_\rho^2) p(\boldsymbol{\alpha}|\sigma_\alpha^2) p(\mathbf{b}|\sigma_b^2) p(\mathbf{c}|\sigma_c^2) \\
&= \prod_{(u,i)\in\mathcal{R}} \mathcal{N}(R_{ui}|\theta_u^\top \beta_i, \sigma_R^2) \prod_{(i,j)\in\mathcal{S}} \mathcal{N}(S_{ij}|\rho_i^\top \alpha_j, \sigma_S^2) \\
&\quad \times \prod_u \mathcal{N}(\theta_u|\mathbf{0}, \sigma_\theta^2) \prod_i \mathcal{N}(\beta_i|\rho_i, \sigma_\beta^2) \prod_i \mathcal{N}(\rho_i|\mathbf{0}, \sigma_\rho^2) \\
&\quad \times \prod_j \mathcal{N}(\alpha_j|\mathbf{0}, \sigma_\alpha^2) \prod_u \mathcal{N}(b_u|\mathbf{0}, \sigma_b^2) \prod_i \mathcal{N}(c_i|\mathbf{0}, \sigma_c^2)
\end{aligned} \tag{12}
$$

where $\boldsymbol{\Theta} = \{\sigma_\theta^2, \sigma_\beta^2, \sigma_\rho^2, \sigma_\alpha^2, \sigma_b^2, \sigma_c^2\}$.

2.4 Parameter Learning

Since learning the full posterior of all model parameters is intractable, we will learn the maximum a posterior (MAP) estimates. This is equivalent to minimizing the following error function.

$$
\mathcal{L}(\Omega) = \frac{1}{2} \sum_{(u,i)\in\mathcal{R}} [R_{ui} - (B_{ui} + \theta_u^\top \beta_i)]^2 + \frac{\lambda}{2} \sum_{(i,j)\in\mathcal{S}} (S_{ij} - \rho_i^\top \alpha_j)^2
$$

$$
+ \frac{\lambda_\theta}{2} \sum_{u=1}^N ||\theta_u||_F^2 + \frac{\lambda_\beta}{2} \sum_{i=1}^M ||\beta_i - \rho_i||_F^2 + \frac{\lambda_\rho}{2} \sum_{i=1}^M ||\rho_i||_F^2 \tag{13}
$$

$$
+ \frac{\lambda_\alpha}{2} \sum_{j=1}^M ||\alpha_j||_F^2 + \frac{\lambda_b}{2} \sum_{u=1}^N b_u^2 + \frac{\lambda_c}{2} \sum_{i=1}^M c_i^2
$$

where $\lambda = \sigma_R^2/\sigma_S^2, \lambda_\theta = \sigma_R^2/\sigma_\theta^2, \lambda_\beta = \sigma_R^2/\sigma_\beta^2, \lambda_\rho = \sigma_R^2/\sigma_\rho^2$, and $\lambda_\alpha = \sigma_R^2/\sigma_\alpha^2$.

We optimize this function by coordinate descent; which alternatively, updates each of the variables $\{\theta_u, \beta_i, \rho_i, \alpha_j, b_u, c_i\}$ while the remaining are fixed.

For θ_u: given the current estimates of the remaining parameters, taking the partial deviation of $\mathcal{L}(\Omega)$ (Eq. 13) with respect to θ_u and setting it to zero, we obtain the update formula:

$$
\theta_u = \Big(\sum_{i\in\mathcal{R}_u} \beta_i \beta_i^\top + \lambda_\theta \mathbf{I}_d \Big)^{-1} \sum_{i\in\mathcal{R}_u} (R_{ui} - B_{ui})\beta_i \tag{14}
$$

Similarly, we can obtain the update equations for the remaining parameters:

$$
\beta_i = \Big(\sum_{u\in\mathcal{R}_i} \theta_u \theta_u^\top + \lambda_\beta \mathbf{I}_d \Big)^{-1} \Big[\lambda_\beta \rho_i + \sum_{u\in\mathcal{R}_i} (R_{ui} - B_{ui})\theta_u \Big] \tag{15}
$$

$$
b_u = \frac{\sum_{i\in\mathcal{R}_u} R_{ui} - \Big[\mu|\mathcal{R}_u| + \sum_{i\in\mathcal{R}_u} (c_i + \theta_u^\top \beta_i)\Big]}{|\mathcal{R}_u| + \lambda_b} \tag{16}
$$

$$
c_i = \frac{\sum_{u\in\mathcal{R}_i} R_{ui} - \Big[\mu|\mathcal{R}_i| + \sum_{u\in\mathcal{R}_i} (b_u + \theta_u^\top \beta_i)\Big]}{|\mathcal{R}_i| + \lambda_c} \tag{17}
$$

$$
\rho_i = \Big[\lambda \sum_{j\in\mathcal{S}_i} \alpha_j \alpha_i^\top + (\lambda_\beta + \lambda_\rho)\mathbf{I}_d\Big]^{-1} \Big(\lambda_\beta \beta_i + \lambda \sum_{j\in\mathcal{S}_i} S_{ij}\alpha_j\Big) \tag{18}
$$

$$
\alpha_j = \Big(\lambda \sum_{i\in\mathcal{S}_j} \rho_i \rho_i^\top + \lambda_\alpha \mathbf{I}_d\Big)^{-1} \Big(\lambda \sum_{i\in\mathcal{S}_j} S_{ij}\rho_i\Big) \tag{19}
$$

Computational Complexity. For user vectors, as analyzed in [7], the complexity for updating N users in an iteration is $\mathcal{O}(d^2|\mathcal{R}^+| + d^3 N)$. For item vector updating, we can also easily show that the running time for updating M items in an iteration is $\mathcal{O}(d^2(|\mathcal{R}^+| + |\mathcal{S}^+|) + d^3 M)$. We can see that the computational complexity linearly scales with the number of users and the number of items. Furthermore, this algorithm can easily be parallelized to adapt to large scale data. For example, in updating user vectors θ, the update rule of user u is independent of other users' vectors, therefore, we can update θ_u in parallel.

2.5 Rating Prediction

We consider two cases of rating predictions: **in-matrix** prediction and **out-matrix** prediction. In-matrix prediction refers to the case where we predict the rating of user u to item i, where i has not been rated by u but has been rated by at least one of the other users; while out-matrix prediction refers to the case where we predict the rating of user u to item i, where i has not been rated by any users (i.e., only click data is available for i). The missing rating r_{ui} can be predicted using the following formula:

$$\hat{r}_{ui} \approx \mu + b_u + c_i + \theta_u^\top \beta_i \tag{20}$$

3 Empirical Study

3.1 Datasets, Competing Methods, Metric and Parameter Settings

Datasets. We use three public datasets of different domains with varying sizes. The datasets are: (1) *MovieLens 1M*: a dataset of user-movie ratings, which consists of 1 million ratings in the range 1–5 to 4000 movies by 6000 users, (2) *MovieLens 20M*: another dataset of user-movie ratings, which consists of 20 million ratings in range 1–5 to 27,000 movies by 138,000 users, and (3) *Bookcrossing*: a dataset for user-book ratings, which consists of 1,149,780 ratings and clicks to 271,379 books by 278,858 users.

Since Movielens datasets contain only rating data, we artificially create the click data and rating data following [2]. Click data is obtained by binarizing the original rating data; while the rating data is obtained by randomly picking with different percentages (10%, 20%, 50%) from the original rating data. Details of datasets obtained are given in Tables 2 and 3.

From each dataset, we randomly pick 80% of rating data for training the model, while the remaining 20% is for testing. From the training set, we randomly pick 10% as the validation set.

As discussed in Sect. 2, we consider two rating prediction tasks: *in-matrix* prediction and *out-matrix* prediction. In evaluating the in-matrix prediction, we ensure that all the items in the test set appear in the training set. In evaluating the out-matrix prediction, we ensure that none of the items in the test set appear in the training set.

Table 2. Datasets obtained by picking ratings from the *Movielens 1M*

Dataset	% rating picked	Density of rating matrix (%)
ML1-10	10%	0.3561
ML1-20	20%	0.6675
ML1-50	50%	1.6022

Table 3. Datasets obtained by picking ratings from *Movielens 20M*

Dataset	% rating picked	Density of rating matrix (%)
ML20-10	10%	0.0836
ML20-20	20%	0.1001
ML20-50	50%	0.2108

Competing Methods. We compare EMB-MF with the following methods.

– *State-of-the-art methods* in rating predictions: PMF [12], SVD++ [8][1].
– *Item2Vec+MF*: The model was obtained by training item embedding and MF separately. First we trained an Item2Vec model [1] on click data to obtain the item embedding vectors ρ_i. We then fixed these item embedding vectors and used them as the item feature vectors β_i for rating prediction.

Metric. We used Root Mean Square Error (RMSE) to evaluate the accuracy of the models. RMSE measures the deviation between the rating predicted by the model and the true ratings (given by the test set), and is defined as follows.

$$RMSE = \sqrt{\frac{1}{|Test|} \sum_{(u,i) \in Test} (r_{ui} - \hat{r}_{ui})^2} \tag{21}$$

where $|Test|$ is the size of the test set.

Parameter Settings. In all settings, we set the dimension of the latent space to $d = 20$. For PMF and SVD++, Item2Vec+MF, we used a grid search to find the optimal values of the regularization terms that produced the best performance on the validation set. For our proposed method, we explored different settings of hyper-parameters to study the effectiveness of the model.

3.2 Experimental Results

The test RMSE results for in-matrix and out-matrix prediction tasks are reported in Tables 4 and 5.

From the experimental results, we can observe that:

– The proposed method (EMB-MF) outperforms all competing methods for all datasets on both in-matrix and out-matrix predictions.

---

[1] The results are obtain by using the LibRec library: http://librec.net/.

Table 4. Test RMSE of in-matrix prediction. For EMB-MF, we fixed $\lambda = 1$ and used the validation set to find optimal values for the remaining hyper-parameters

Methods	ML-1m			ML-20m			Bookcrossing
	ML1-10	ML1-20	ML1-50	ML20-10	ML20-20	ML20-50	
PMF	1.1026	0.9424	0.8983	1.0071	0.8663	0.8441	2.1663
SVD++	0.9825	0.9066	0.8871	0.8947	0.8348	0.8191	1.6916
Item2Vec+MF	0.9948	0.9135	0.8984	0.9098	0.8527	0.8355	1.9014
EMB-MF (our)	**0.9371**	**0.8719**	**0.8498**	**0.8767**	**0.8299**	**0.8024**	**1.6558**

Table 5. Test RMSE of out-matrix prediction. For EMB-MF, we fixed $\lambda = 1$; the remaining hyper-parameters are determined using the validation set. Only *Item2Vec* is compared, because PMF and SVD++ cannot be used for out-matrix prediction

Methods	ML-1m			ML-20m			Bookcrossing
	ML1-10	ML1-20	ML1-50	ML20-10	ML20-20	ML20-50	
Item2Vec+MF	1.0986	1.0365	1.039	1.0128	0.9582	0.9784	1.7027
EMB-MF (our)	**1.0312**	**1.0059**	**1.0132**	**0.9729**	**0.9422**	**0.9494**	**1.6828**

- EMB-MF, SVD++ and Item2Vec+MF are much better than PMF, which use rating data only. This indicates that exploiting click data is a key factor to increase the prediction accuracy.
- For all methods, the accuracies increase with the density of rating data. This is expected because the rating data is reliable for inferring users' preferences.
- In all cases, the differences between EMB-MF with the competing methods are most pronounced in the most sparse subsets (ML1-10 or ML20-10). This demonstrates the effectiveness of EMB-MF on extremely sparse data.
- EMB-MF outperforms Item2Vec+MF although these two models are based on similar assumptions. This indicates the advantage of training these models jointly, rather than training them independently.

Impact of Parameter λ_β. λ_β is the parameter that controls the deviation of β_i from the item embedding vector ρ_i (see Eqs. 6 and 13). When λ_β is small, the value of β_i is allowed to diverge from ρ_i; in this case, β_i mainly comes from rating data. On the other hand, when λ_β increases, β_i becomes closer to ρ_i; in this case β_i mainly comes from click data. The test RMSE is given in Table 6.

We can observe that, for small values of λ_β, the model produces low prediction accuracy (high test RMSE). The reason is that when λ_β is small, the model mostly relies on the rating data which is very sparse and cannot model items well. When λ_β increases, the model starts using click data for prediction, and the accuracy will increase. However, when λ_β reaches a certain threshold, the accuracy starts decreasing. This is because when λ_β is too large, the representations of items mainly come from click data. Therefore, the model becomes less reliable for modeling the ratings.

Table 6. Test RMSE of in-matrix prediction task by the proposed method over different values of λ_β while the remaining hyper-parameters are fixed.

λ_β	0.1	1.0	10.0	20.0	50.0	100.0	1000.0
ML1-10	1.1301	0.9971	0.9318	0.9381	0.9527	0.9651	0.9924
ML1-20	1.1107	0.9963	0.8911	0.8756	0.8723	0.8769	0.8885
ML1-50	0.9798	0.9193	0.8634	0.8545	0.8512	0.8539	0.8626

4   Related Work

Exploiting click data for addressing the cold-start problem has also been investigated in the literature. Co-rating [11] combines explicit (rating) and implicit (click) feedback by treating explicit feedback as a special kind of implicit feedback. The explicit feedback is normalized into the range $[0, 1]$ and is summed with the implicit feedback matrix with a fixed proportion to form a single matrix. This matrix is then factorized to obtain the latent vectors of users and items.

Wang et al. [13] proposed Expectation-Maximization Collaborative Filtering (EMCF) which exploits both implicit and explicit feedback for recommendation. For predicting ratings for an item, which does not have any previous ratings, the ratings are inferred from the ratings of its neighbors according to click data.

The main difference between these methods with ours is that they do not have a mechanism for balancing the amounts of click data and rating data when making predictions. In our model, these amounts are controlled depending on the number of previous ratings that the target items have.

Item2Vec [1] is a neural network based model for learning item embedding vectors using *co-click* information. In [10], the authors applied a word embedding technique by factorizing the shifted PPMI matrix [9], to learn item embedding vectors from click data. However, using these vectors directly for rating prediction is not appropriate because click data does not exactly reflect preferences of users. Instead, we combine item embedding with MF in a way that allows rating data to contribute to item representations.

5   Conclusion

In this paper, we proposed a probabilistic model that exploits click data for addressing the cold-start problem in rating prediction. The model is a combination of two models: (i) an item embedding model for click data, and (ii) MF for rating prediction. The experimental results showed that our proposed method is effective in rating prediction for items with no previous ratings and also boosts the accuracy of rating prediction for extremely sparse data.

We plan to explore several ways of extending or improving this work. The first direction is to develop a full Bayesian model for inferring the full posterior distribution of model parameters, instead of point estimation which is prone to overfitting. The second direction we are planning to pursue is to develop an online learning algorithm, which updates user and item vectors when new data are collected without retraining the model from the beginning.

Acknowledgments. This work was supported by a JSPS Grant-in-Aid for Scientific Research (B) (15H02789, 15H02703).

References

1. Barkan, O., Koenigstein, N.: Item2Vec: neural item embedding for collaborative filtering. In: 26th IEEE International Workshop on Machine Learning for Signal Processing, pp. 1–6 (2016)
2. Bell, R.M., Koren, Y.: Scalable collaborative filtering with jointly derived neighborhood interpolation weights. In: Proceedings of the 7th IEEE International Conference on Data Mining, pp. 43–52 (2007)
3. Bullinaria, J.A., Levy, J.P.: Extracting semantic representations from word co-occurrence statistics: a computational study. Behav. Res. Methods, 510–526 (2007)
4. Church, K.W., Hanks, P.: Word association norms, mutual information, and lexicography. Comput. Linguist. 22–29 (1990)
5. Gopalan, P., Hofman, J.M., Blei, D.M.: Scalable recommendation with hierarchical Poisson factorization. In: Proceedings of the 31st Conference on Uncertainty in Artificial Intelligence, pp. 326–335 (2015)
6. Gopalan, P.K., Charlin, L., Blei, D.: Content-based recommendations with Poisson factorization. In: Proceedings of the 27th Advances in Neural Information Processing Systems, pp. 3176–3184 (2014)
7. Hu, Y., Koren, Y., Volinsky, C.: Collaborative filtering for implicit feedback datasets. In: Proceedings of the 8th IEEE International Conference on Data Mining, pp. 263–272 (2008)
8. Koren, Y.: Factorization meets the neighborhood: a multifaceted collaborative filtering model. In: Proceedings of the 14th ACM SIGKDD International Conference on Knowledge Discovery and Data Mining, pp. 426–434 (2008)
9. Levy, O., Goldberg, Y.: Neural word embedding as implicit matrix factorization. In: Proceedings of the 27th International Conference on Neural Information Processing Systems, pp. 2177–2185 (2014)
10. Liang, D., Altosaar, J., Charlin, L., Blei, D.M.: Factorization meets the item embedding: regularizing matrix factorization with item co-occurrence. In: Proceedings of the 10th ACM Conference on Recommender Systems, pp. 59–66 (2016)
11. Liu, N.N., Xiang, E.W., Zhao, M., Yang, Q.: Unifying explicit and implicit feedback for collaborative filtering. In: Proceedings of the 19th ACM International Conference on Information and Knowledge Management, pp. 1445–1448 (2010)
12. Mnih, A., Salakhutdinov, R.R.: Probabilistic matrix factorization. In: 20th Advances in Neural Information Processing Systems, pp. 1257–1264 (2008)
13. Wang, B., Rahimi, M., Zhou, D., Wang, X.: Expectation-maximization collaborative filtering with explicit and implicit feedback. In: Tan, P.-N., Chawla, S., Ho, C.K., Bailey, J. (eds.) PAKDD 2012. LNCS, vol. 7301, pp. 604–616. Springer, Heidelberg (2012). doi:10.1007/978-3-642-30217-6_50
14. Wang, C., Blei, D.M.: Collaborative topic modeling for recommending scientific articles. In: Proceedings of the 17th ACM SIGKDD International Conference on Knowledge Discovery and Data Mining, pp. 448–456 (2011)
15. van den Oord, A., Dieleman, S., Schrauwen, B.: Deep content-based music recommendation. In: Proceedings of the Advances in Neural Information Processing Systems, vol. 26, pp. 2643–2651 (2013)

Dynamic Forest Model for Sentiment Classification

Mingming Li[1,2], Jiao Dai[1(✉)], Wei Liu[1], and Jizhong Han[1]

[1] Institute of Information Engineering, Chinese Academy of Sciences, Beijing, China
{limingming,daijiao,liuwei1119,hanjizhong}@iie.ac.cn
[2] School of Cyber Security, University of Chinese Academy of Sciences,
Beijing, China

Abstract. Sentiment classification is a useful approach to analyse the emotional polarity of user reviews, and method based on machine learning has achieved a great success. In the era of Web2.0, the emotional intensity of terms will change with time and events, while a large number of Out-Of-Vocabulary (OOV) terms are appearing. But the method of machine learning pays little attention to them because they focus to reduce the computational complexity. To address the problem, we proposed a dynamic forest model, which can describe the emotional intensity of the term in character granularity, and can append OOV dynamically and adjust their emotional intensity value. Experiments show that in the Chinese environment, our model greatly boosts the performance compared with the method based machine learning, while the time is saved by halves.

Keywords: Sentiment classification · Machine learning · Out-Of-Vocabulary · Sentiment lexicon · Dynamic forest

1 Introduction

With the flourish of Web2.0 based applications, it becomes more convenient for people to take part in the online reviews. At the same time, analyzing emotional tendency from subjective reviews becomes a research hotspot. Based on the text of the sentiment classification [1–4] is a useful approach to find the user sentiment polarity (i.e., positive or negative).

At present, the research works about the sentiment classification can be divided into three categories, sentiment lexicons method [5,6], method based on machine learning [7,9–13] and mixed method [15–17]. Lexicons will assign a sentiment polarity for each word, so we can count up the number of positive words and negative words of the given document respectively, and then according to some rules to determine the document polarity. Based on machine learning method is to obtain a classification model by supervised learning. Feature extraction is essential to machine learning, and it decides the upper limit of the model. There are some common features such as POS (part-of-speech tagging),

© Springer International Publishing AG 2017
D. Liu et al. (Eds.): ICONIP 2017, Part V, LNCS 10638, pp. 206–215, 2017.
https://doi.org/10.1007/978-3-319-70139-4_21

TF-IDF, word2Vec, etc. Mixed method combined above two methods to improve the effect. The above methods has achieved great success in the static situation. However, emotional tendency of the term is not immutable, and it will change and even reversal with the occurrence of major events, time, etc. In addition, there are large Out-Of-Vocabulary (OOV) terms appearing, and their emotional trends are unknown. The processing speed of the classification should be also quick in order to meet the requirement of real-time. Methods mentioned above are based on the assumption that emotional intensity of the term is invariably, so it is difficult to meet above environmental requirements.

To address above problems, we present a sentiment classification model based on dynamic forest. This model can describe emotional intensity in character granularity, and it can reduce some dependence on tokenism. So we can predict the polarity of word-group only according to those prefixes. At the same time, we can append and adjust emotional intensity of term dynamically. Experiments show that the classification effect is improved obviously. F1 of our model is 26% higher than the method based on sentiment lexicons [5], and 3% higher than the method based machine learning (i.e., SVM, GBDT, Neural Network, Logistic Regression), but total time is only a half. Besides that, the time will become one-fourth of the original after parallel processing.

The rest of this paper is organized as follows. Section 2 introduces the current methods of sentiment classification. Section 3 introduces the proposed model with details. Section 4 gives the experimental results and analysis. Section 5 concludes the paper.

2 Related Work

The method of common sentiment lexicons is easy to implement, but the effect of prediction is bad. Further research work is to enrich sentiment lexicons, such as Yong-Mei et al. [6] to give an emotional intensity for each term, and improve the classification accuracy. But the emotional intensity is static, and it cannot be changed with the environment dynamically. In machine learning, feature extraction is significant. The feature of POS is dependent on the artificial structure; Kouloumpis et al. [7] designed some comparison experiments used a variety of features for emotional analysis of Twitter, and results show that the POS features have little effect on micro-blogging data. TF-IDF reflects the importance of a word in the text set, Paltoglou and Thelwall [8] enhanced the performance by combing TF-IDF feature. The better way to get a vector for a document is to use word2Vec model which considers the context of terms. In a specific scenario, we should consider some specific features, for instance, Pak and Paroubek [9] consider text information, user information, subject information, etc. for the short text classification; Tang et al. [13] consider some specific emotional words, and cascade other features for text classification. In addition, we can obtain a better classification model by adjustment parameter, stacking [14] or bending. Yang and Yang [15] uses the sentiment lexicons to build a Bayesian classifier for a large number of Chinese text classification; Jie et al. [16] uses some Chinese semantic rules and method based on machine learning; Zhilin et al. [17] uses lexical theme features, text features and emotional tendencies of characteristics to improve the classification effect.

These methods make a good effect in most of text classification. But there has been little attention to the OOV and the change of emotional intensity. Besides that, it takes a long time by re-training to meet the need for real-time webservices.

3   Proposed Model

We proposed a sentiment classification model based on dynamic forest, which can describe emotional intensity in character granularity. Firstly, we init a forest using some sentiment lexicons. Then, we learn a classification model using the training set, and compute the error of prediction. At the same time, the polarity of OOV is computed according to the way in Sect. 3.4. Finally, the error and OOV will be feedback to the forest. After several iterations, a forest will be generated. When new corpus is added, it can be trained directly on existing model parameters. This method of dynamic incremental updating can respond to the needs of real-time online classification. The following will be a detailed description of the forest structure, some methods of initing forest, error updating, and OOV appending.

We define a term set, denote as $W = \{w_1, w_2, \ldots, w_L\}$, $|w_i|$ is the length of i-th term, and $|W| = L$. All Chinese characters make up a set C. Each term contains some characters, such as $w_i = [c_1^1, c_k^2, \ldots, c_m^{|w_i|}] = [c^1, c^2, \ldots, c^j, c^{|w_i|}]$, $c_m^j, c^j \in C$, $w_i \in W$, $j = index(c)$. Now we use $\theta_{w_i,j}$ to represent the emotional intensity of c which is the j-th character of w_i, so the emotional intensity of w_i is $\sum_{j=1}^{|w_i|} \theta_{w_i,j}$. Given a document $d = \{w_1, w_2, ..., w_n\}$, $w_i \in W$, the sentiment score of document is the sum of all w_i, and we use a function to map value into $[-1, 1]$, as:

$$score(d) = \tanh(\sum_{i=1}^{n} \sum_{j=1}^{|w_i|} \theta_{w_i,j}). \tag{1}$$

Our task is to find the optimization θ which can make the error of prediction minimum.

$$\arg\min_{\theta} \sum_{d \in D} |tag(d) - score(d)| = \arg\min_{\theta} \sum_{d \in D} |E(d)| \tag{2}$$

where D is the document set, $tag(d)$ is the label of the document d, $tag(d) \in \{-1, 1\}$. From the Fig. 1, we know that it is difficult to calculate the θ, because many terms share same θ, which have the same prefix. If we update θ in some terms, the other which have had a better score before may become more worse. In addition, the number of θ will increase during the training process due to the OOV, while we cannot know those in advance. In order to simplify this problem, we define a threshold value $\xi \in [0, 0.05]$, and we only require that for each d, $-\xi < E(d) < \xi$ is satisfied.

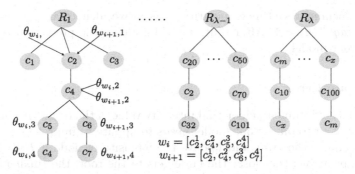

Fig. 1. The struct of forest; for instance, w_i, w_{i+1} have the same prefix, and they share common parameter of θ on c_1 and c_2, respectively; the real number of θ is $|w_i| + |w_{i+1}| - samePrefix(w_i, w_{i+1}) = 4 + 4 - 2 = 6$.

3.1 Init Forest

In this paper, we use the structure of forest which is consisted of many trees to storage data. The number of trees is λ, The λ has little impact on the accuracy, but much influence on the training, the following we will give the reason. Each node in the tree consists of two pieces of information: character and emotional intensity. We use the existing sentiment lexicons to build a forest and init the emotional intensity for each character.

Given a term from lexicon, denote as $w_i = [c^1, c^2, \ldots, c^j, c^{|w_i|}]$, and its label is $tag(w_i)$. Firstly, we find the root node position of w_i by $mod(c^1, \lambda)$. And then, the other character $c^2, \ldots, c^{|w_i|}$ will be appended to the tree in order. The initial emotional intensity can be calculated as:

$$\theta'_{w_i,j} = \frac{tag(w_i)}{|w_i|} + \theta_{w_i,j} \tag{3}$$

where $\theta_{w_i,j}$ is the old value, and default value is zero; $\theta'_{w_i,j}$ is the new value, it is affected by some terms together which have the same prefix.

3.2 Compute Document Polarity

We have completed the initialization work in the previous section, and now we consider to compute the document polarity using the forest. If all characters of the term are in the same branch of the tree, the score can be calculated by formula (1) directly. But if there are in the cross branch of the tree, we need to split the term to several sub-terms, and each sub-term can be computed by the above method. In addition, if the term are not in the forest, we can ignore it, and then add it to the candidate list of the OOV; if its prefix is in the forest, we can use its prefix to replace it, or we find the most similar term from the forest to replace it using the method of dynamic programming. If $socre(d) < 0$, we mark the document sentiment polarity as negative, denote as $tag^{pre}(d) = -1$; otherwise $tag^{pre}(d) = 1$. After that, we will evaluate the prediction result, and to update our model.

3.3   Update Error

In the actual environment, emotional intensity value of the term will change with time, events and so on. So our model needs to update to improve the accuracy. We will compute the error and update the intensity of node if $E(d) \notin [-\xi, \xi]$. There is a principle: the closer the node gets to the root, the smaller the error to update is; and vice versa. The node closer to the root may be the common prefix for more words, and its value will be more profound, so we have to update discreetly; the node closer to the leaf can be more representative a term, so its value can be updated more with little impact on others.

Given a document d, we can compute its error $E(d)$, and then we compute the error for each w_i according to the weight of the term in the document, denote as $error_{w_i}$,

$$error_{w_i} = E(d)\frac{|f(w_i)|}{\sum_{k=1}^{n}|f(w_k)|}. \tag{4}$$

Then we will update the character error of w_i as:

$$\theta'_{w_i,j} = \frac{level_{c^j} * error_{w_i} * rate}{|w_i|} + \theta_{w_i,j} \tag{5}$$

where $level_{c^j}$ is the depth of the c^j in the forest, $rate$ is a hyper-parameter, $rate \in [0.03 \sim 0.05]$. We can find that $level_{c^j} = j$ if all c^j of the w_i are in the same branch; but at many times, they are in the different branch, so we have to split the w_i to several sub-terms, and all c^j of each sub-term are in the same branch. For example, w_i is splited to two sub-terms, $w_i^1 = [c^1, c^2, \ldots, c^s]$, $w_i^2 = [c^{s+1}, c^{s+2}, \ldots, c^{|w_i|}]$, $w_i = [w_i^1, w_i^2]$; the error of $c^1, c^2, \ldots, c^s$ can be updated by formula (5), and echo character of the w_i^2 will be updated as follows:

$$\theta'_{w_i^2,x} = \left(\frac{|w_i^1| + 1}{|w_i|} * error_{w_i^1}\right) * \frac{level_{c^x}}{|w_i| - |w_i^1|} * rate + \theta_{w_i^2,x} \tag{6}$$

where $|w_i^1| = s$, $x \in \{s+1, \ldots, |w_i|\}$. We can know that the formula (6) is a recursive form of (5) in nature, so we only need to treat w_i^2 as a new term, and make its error as the first item of the formula (6).

In addition, our model can be updated by an incremental approach, which can be completed in a short time. We need to collect many new training data that has the sentiment polarity at first; then we will iterate our model using the existing parameters according to the method that previous sections had introduced. If the super-parameter is not good for now, we can also adjust it. In addition, this method is also quite suitable for small corpus environment.

3.4 Dynamic Append OOV

We add the OOV to the candidate list during the process of training and prediction, and then append it to the forest according to the initialization way if the frequency of new terms are more than the threshold. For the training data, $tag(d)$ is known in advance, so we can assignment for the emotional intensity indirectly by formula (3); but during the prediction process, we do not know that. There are two ways to deal with this problem. One way is using the $tag^{pre}(d)$ as the real $tag(d)$, because the new term does not happen alone, we can analyze the whole polarity from its context. Second way is using the N-Gram method to find the most similar term, and to replace itself.

4 Experiments

In order to verify the availability and high efficiency of our model, we design some comparison experiments, and we will analyze and evaluate the experimental results.

4.1 Datasets

We use three sentiment lexicons in our model, NTUSD[1], HowNet[2] and Chinese Praised Dictionary[3]. Table 1 summarizes the statistics on the three lexicons. Our datasets are from Datatang[4]: the dataset of takeaway comments is consisted of 4000 positive and 4000 negative, and the dataset of hotel comments contains of 3000 positive data and 3000 negative data. In addition, we use 3.1G data of Sohu news[5] and 1.1G Chinese data in wiki[6] to train the Word2Vec model.

Table 1. Statistics of three sentiment lexicons

Sentiment lexicons	Positive	Negative	Totals
NUTSUD	2810	8276	11086
HowNet	4566	4370	8936
Praised dictionary	5567	4468	10035

4.2 Experimental Setup

We design a comparison test with the method of sentiment lexicons and machine learning respectively. We implement some classical methods of machine learning,

---

[1] http://www.datatang.com/data/11837.

[2] http://www.keenage.com.

[3] http://yynl.jsnu.edu.cn/_t307/0c/b4/c541a3252/page.htm.

[4] http://www.datatang.com.

[5] http://www.sogou.com/labs/resource/cs.php.

[6] http://licstar.net/archives/262.

such as KNN, Linear SVM, Non-Linear SVM, Logistic Regression, Decision Tree, GBDT, AdaBoost, Neural Network, and so on, and then we select the optimal comparison with our model. The input vector of machine learning is pre-trained by Word2Vec with dimension of 400, and all parameters of machine learning are default by the package of sklearn. The performance is measured with precision (P), recall (R), F1, and total time (the sum of pre-processing, training, and forecasting, excluding word2vec). In order to explain the contribution of OOV, we design a group comparative experiments: appending OOV or not appending OOV in hotel comments. In addition, we design some experiments to show the time of training and explain the concurrency of our model by multi-thread.

4.3 Experimental Results

We implement a sentiment lexicons model, our proposed model and some machine learning model on takeaway dataset and hotel dataset, respectively. In our model, $\lambda = 200$, $rate = 0.04$, $\xi = 0.04$; We also adopt an early stopping strategy, which stops the training process if the iterations are meeting the set value (default value is 20). Experimental results are shown in Table 2.

Table 2. Results of different methods

Method	Hotel dataset				Takeaway dataset			
	P	R	F1	Time (s)	P	R	F1	Time (s)
Sentiment lexicons	0.65	0.58	0.47	2.210	0.68	0.66	0.64	2.568
KNN	0.81	0.79	0.79	16.655	0.79	0.78	0.78	15.845
Linear SVM	0.84	0.83	0.83	20.587	0.87	0.88	0.87	17.062
RBF SVM	0.68	0.50	0.34	32.655	0.81	0.69	0.65	23.675
Decision tree	0.71	0.70	0.71	13.474	0.79	0.79	0.79	14.378
Neural network	0.81	0.81	0.81	14.153	0.88	0.87	0.87	13.407
Random forest	0.80	0.79	0.79	11.029	0.83	0.82	0.82	12.866
Logistic regression	0.82	0.83	0.82	11.233	0.87	0.86	0.86	14.058
AdaBoost	0.81	0.80	0.80	23.996	0.83	0.83	0.83	22.289
GBDT	0.83	0.83	0.83	29.178	0.86	0.86	0.86	26.506
Proposed model	**0.86**	**0.86**	**0.86**	**4.519**	**0.90**	**0.90**	**0.90**	**7.968**

As we can see that our proposed model outperforms the other methods. We can find more observations from the results. The total time of sentiment lexicons method is shortest, because it is a hash-based way. Linear SVM and GBDT has a better performance on F1 in takeaway dataset, but they need long time for training and prediction. Our proposed model have a better effect on F1 than others, which are 3% more than the GBDT and 39% more than sentiment lexicons, while total time is saved at least half. In the hotel dataset, our proposed model has also a best performance, the F1 is 3% higher than Neural Network, and time is saved by halves.

From the Fig. 2, we can know that our model is as stable as the method of sentiment lexicons on the different size of dataset, and the F1 is higher than others obviously. OOV plays an important rule on the effect of F1 in our model; the F1 is improved 2% in hotel dataset, and it will be more if there are more OVV words.

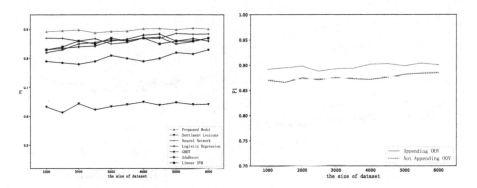

Fig. 2. The result on different size of dataset (left); the effect of OOV (right)

In order to explain the effect of iteration and the number of trees λ, we design some comparison as the Fig. 3. We can find that the F1 of our model will increase rapidly during the training process, and then tend to a stable value. For instance, the F1 closes to 98% in the hotel dataset when the iteration is above 20, and 95% in the takeaway dataset when the iteration is above 40. In addition, it is easy to seen that the λ pay little rule on the F1 from the right of Fig. 3. Our model is based on a forest, if λ is large, each branch will have fewer trees, and if λ is small conversely. Changing λ is only effect the location of the terms, but not change of the θ.

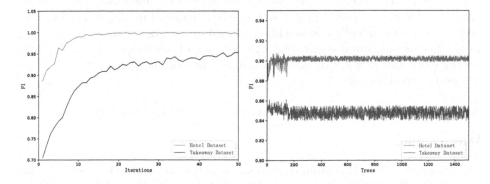

Fig. 3. The effect of iterations (left); the effect of trees (right)

On the other hand, querying and updating time will be changed with the location of terms. So we use multi-thread to train and predict data renewedly. As shown in Fig. 4, the time becomes a quarter while we use 16 threads instead of 1 thread in hotel dataset. With the increasing of λ, terms will be more possible spread along more branches, so the speed is quick. We can find a better λ for each dataset, in our model, $\lambda_{hotel} = 1000$, $\lambda_{takeaway} = 1200$.

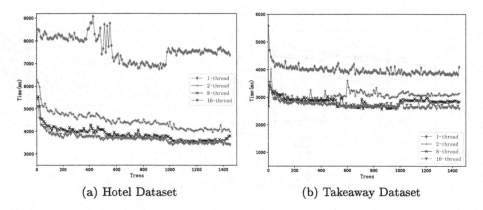

(a) Hotel Dataset   (b) Takeaway Dataset

Fig. 4. The effect of multi-thread

5   Conclusion

In this paper, we focus on an environment that emotional intensity will change with some events or time, and more OOV will appear. We proposed a model based on dynamic forest, which can describe the emotional intensity in character granularity. And it is convenient to append OOV and update the intensity. Besides, our model is easy to be parallelized, so it can also save a lots time. There are some advances of our model. Firstly, error feedback is imposed to well-directed word rather than all related neurons, and this leads to faster converge. The second, there are less parameters in the dynamic model for the share of term-pairs with the same prefix, so it can be calculated more quickly. The last, OOV can be dynamically added while it should be retrained with the neural network. The experimental results show the efficacy of our proposed model. There is still much future work to do, for example, we can cut some branches which are useless to accelerate.

Acknowledgments. We would like to thank Sougou for its news data.

References

1. Abbasi, A., Chen, H., Salem, A.: Sentiment analysis in multiple languages: feature selection for opinion classification in web forums. ACM Trans. Inf. Syst. (TOIS) **26**(3), 12 (2008)

2. Pang, B., Lee, L., et al.: Opinion mining and sentiment analysis. Found. Trends® Inf. Retrieval **2**(1–2), 1–135 (2008)
3. Liu, B., Zhang, L.: A survey of opinion mining and sentiment analysis. In: Aggarwal, C., Zhai, C. (eds.) Mining Text Data, pp. 41–463. Springer, Heidelberg (2012). doi:10.1007/978-1-4614-3223-4_13
4. Vinodhini, G., Chandrasekaran, R.: Sentiment analysis and opinion mining: a survey. Int. J. **2**(6), 282–292 (2012)
5. Taboada, M., Brooke, J., Tofiloski, M., Voll, K., Stede, M.: Lexicon-based methods for sentiment analysis. Comput. Linguist. **37**(2), 267–307 (2011)
6. Yong-Mei, Z., Yang Jia-Neng, Y.A.M.: A method on building Chinese sentiment Lexicon for text sentiment analysis. J. ShanDong Univ. (Eng. Sci.) **43**(6), 27–33 (2013)
7. Kouloumpis, E., Wilson, T., Moore, J.D.: Twitter sentiment analysis: the good the bad and the OMG! Icwsm **11**(538–541), 164 (2011)
8. Paltoglou, G., Thelwall, M.: A study of information retrieval weighting schemes for sentiment analysis. In: Proceedings of the 48th Annual Meeting of the Association for Computational Linguistics, pp. 1386–1395. Association for Computational Linguistics (2010)
9. Pak, A., Paroubek, P.: Twitter as a corpus for sentiment analysis and opinion mining. In: LREc, vol. 10 (2010)
10. Vo, D.T., Zhang, Y.: Target-dependent Twitter sentiment classification with rich automatic features. In: IJCAI, pp. 1347–1353 (2015)
11. Sriram, B., Fuhry, D., Demir, E., Ferhatosmanoglu, H., Demirbas, M.: Short text classification in Twitter to improve information filtering. In: Proceedings of the 33rd International ACM SIGIR Conference on Research and Development in Information Retrieval, pp. 841–842. ACM (2010)
12. Banerjee, S., Ramanathan, K., Gupta, A.: Clustering short texts using Wikipedia. In: Proceedings of the 30th Annual International ACM SIGIR Conference on Research and Development in Information Retrieval, pp. 787–788. ACM (2007)
13. Tang, D., Wei, F., Yang, N., Zhou, M., Liu, T., Qin, B.: Learning sentiment-specific word embedding for Twitter sentiment classification. In: ACL, vol. 1, pp. 1555–1565 (2014)
14. Shoushan, L., ChuRan, H.: Chinese sentiment classification based on stacking combination method. J. Chin. Inf. Process. **24**(5), 56–61 (2010)
15. Yang, D., Yang, A.M.: Classification approach of Chinese texts sentiment based on semantic Lexicon and naive Bayesian. Jisuanji Yingyong Yanjiu **27**(10) (2010)
16. Jie, J., Rui, X., et al.: Microblog sentiment classification via combining rule-based and machine learning methods. Acta Scientiarum Naturalium Universitatis Pekinensis **53**(2), 247–254 (2017)
17. Li, L., Cao, D., Li, S., Ji, R.: Sentiment analysis of Chinese micro-blog based on multi-modal correlation model. In: 2015 IEEE International Conference on Image Processing (ICIP), pp. 4798–4802. IEEE (2015)

A Multi-attention-Based Bidirectional Long Short-Term Memory Network for Relation Extraction

Lingfeng Li[✉], Yuanping Nie, Weihong Han, and Jiuming Huang

College of Computer, National University of Defense Technology, Changsha, China
{lilingfeng11,yuanpingnie}@nudt.edu.cn, hanweihong@gmail.com,
jiuming.huang@qq.com

Abstract. Compared to conventional methods, recurrent neural networks and corresponding variants have been proved to be more effective in relation extraction tasks. In this paper, we propose a model that combines a bidirectional long short-term memory network with a multi-attention mechanism for relation extraction. We designed a bidirectional attention mechanism to extract word-level features from a single sentence and chose a sentence-level attention mechanism to focus on features of a sentence set. Our experiments were conducted on a public dataset to evaluate the performance of the model. The experimental results demonstrate that the multi-attention mechanism can make full use of all informative features of a single sentence and a sentence set and our model achieves state-of-the-art performance.

Keywords: Relation extraction · Bidirectional long short-term memory · Multi-attention mechanism

1 Introduction

Information extraction aims to discover structured information from unstructured or semi-structured documents, and it can be broken down into two main subtasks: name entity recognition and relation extraction [1]. A relation extraction task detects and classifies semantic relations within a set of artifacts, typically from documents. Existing relation extraction methods that employ hand-crafted features from lexical resources are usually based on pattern matching and rely on natural language process (NLP) tools. Although they have achieved good performance [2], there are still two critical problems that need to be solved when using conventional relation extraction methods: (i) it is very difficult to decide which features to extract and costly to extract them manually; and (ii) it is inevitable that extra errors will be introduced when utilizing NLP tools to extract high-level features, such as dependency relations and hypernyms etc.

Deep neural networks with attention mechanisms have obtained remarkable effect in computer vision and NLP tasks, such as machine translation [3] and question answering [4] etc. For relation extraction, state-of-the-art models [5,6]

© Springer International Publishing AG 2017
D. Liu et al. (Eds.): ICONIP 2017, Part V, LNCS 10638, pp. 216–227, 2017.
https://doi.org/10.1007/978-3-319-70139-4_22

employ different attention mechanisms that focus on features of different levels. However, their attention models both apply single attention mechanism on deep neural networks aimed at different levels.

In this paper, a bidirectional long short-term memory (BLSTM) network is combined with a multi-attention mechanism that includes a bidirectional (two-way) attention mechanism and a sentence-level attention mechanism. The bidirectional attention mechanism can extract features directly from the outputs of the forward and backward LSTM network, which means that the bidirectional attention mechanism will not be affected by extra errors from previous computations and will preserve more informative features. The sentence-level attention mechanism can gather information from a sentence set and reduce the effect of incorrect labelling.

The architecture of the model proposed in this paper is illustrated in Fig. 1. The "BLSTM+BSATT" is used to represent the model, where the "BLSTM" represents the bidirectional LSTM network and the "BSATT" is an abbreviation of "bidirectional and sentence-level attention". More details of the model are provided in Sect. 3.

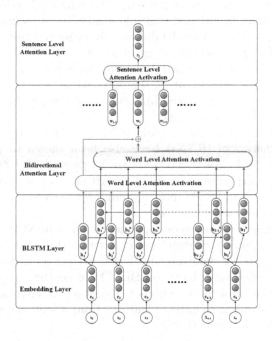

Fig. 1. Architecture of the BLSTM+BSATT model

Experiments were performed with a publicly available benchmark dataset[1] to evaluate the performance of the model with the multi-attention mechanism.

---

[1] http://iesl.cs.umass.edu/riedel/ecml.

The performance of our BLSTM network with the multi-attention mechanism was compared against those of naive versions of BLSTM networks and existing methods. The experimental results show that the multi-attention mechanism plays an important role in our model and our model achieves state-of-the-art performance in relation extraction.

The rest of the paper is organized as follows: Sect. 2 covers the related work; Sect. 3 describes the details of the model. The experimental evaluations and results are presented in Sect. 4. Finally, Sect. 5 provides the conclusion to this paper.

2   Related Work

In most cases, a relation extraction task is treated as a classification problem. Compared to traditional features-based methods and kernel-based methods, which have dominated the supervised methods, deep neural networks are gaining ground.

Zeng et al. [7] used a convolutional deep neural network to extract lexical and sentence-level features and predict the relation between entity pairs. Nguyen and Grishman [8] utilized multiple window sizes to improve the model proposed by Zeng et al. [7]. A model dubbed piecewise convolutional neural networks (PCNNs) was proposed by Zeng et al. [9] to handle the problems that arise when using distant supervision for relation extraction. Socher et al. [10] used a recursive neural network along sentences parse trees to classify relations. Hashimoto et al. [11] weighted the importance of phrases in recursive neural networks to improve the performance.

Besides convolutional neural networks (CNNs) and recursive neural networks, recurrent neural networks (RNNs) have also been shown to perform well in relation extraction. A bidirectional RNN model proposed by Zhang and Wang [12] was simple but outperformed state-of-the-art methods in relation extraction. Xu et al. [13] proposed an LSTM model leveraging the shortest dependency path to identify the relation of two entities.

Attention mechanisms are becoming a new hotspot in NLP research, such as machine translation and intelligent answer systems [15], and many researchers have applied attention mechanisms in relation extraction tasks. Zhou et al. [5] used an attention mechanism based on BLSTM to capture the most important semantic information in a single sentence. Lin et al. [6] introduced a sentence-level attention mechanism to extract features from a sentence set and improve the performance of relation extraction.

3   Model

In this section, the multi-attention-based BLSTM network is described in detail. The architecture of our model consists of four components: (i) the embedding layer to encode words into real-valued vectors; (ii) the BLSTM layer to obtain high-level features from the embedding layer; (iii) the bidirectional attention layer to merge the features from each time step; and (iv) the sentence-level attention layer to gather information from a sentence set and predict the relation between entity pairs.

3.1 Embedding Layer

Distributed word representations can capture fine-grained semantic and syntactic regularities. Therefore, each word is encoded into a real-valued vector by looking up the embedding matrix $W^{emb} \in \Re^{d^w} \times \mid V \mid$ in the embedding layer, where d^w is the dimension of the vectors and $\mid V \mid$ denotes the size of the vocabulary of the whole data set. There are several methods that can be used to train the embedding model and obtain the matrix. However, in order to make our experiments more convincing, we adopted one of the most commonly used word embeddings techniques pretrained by Pennington et al. [16].

3.2 BLSTM Layer

RNNs perform well when dealing with sequential data because hidden units inside a RNN can make previous computations available to the present computation. However, with longer sequences, RNNs are unable to learn information from previous steps owing to the vanishing gradient problem. To make each recurrent unit adaptively capture features of different time steps, Hochreiter and Schmidhuber [17] introduced the LSTM unit. The architecture of a LSTM unit is illustrated as Fig. 2, where i, f and o are the input, forget, and output gates, respectively, and c and $\tilde{c}$ denote the memory cell and the new memory cell content.

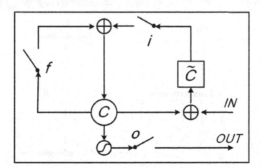

Fig. 2. Schematic of the architecture of a LSTM unit

The unit at $t-$th word receives an $n-$dimensional input vector x_t; the previous hidden state h_{t-1} and the present hidden state h_t are calculated as follows:

$$f_t = \sigma(W_f x_t + U_f h_{t-1} + V_f c_{t-1}) \tag{1}$$

$$i_t = \sigma(W_i x_t + U_i h_{t-1} + V_i c_{t-1}) \tag{2}$$

$$\tilde{c}_t = \tanh(W_c x_t + U_c h_{t-1}) \tag{3}$$

$$c_t = f_t c_{t-1} + i_t \tilde{c}_t \tag{4}$$

$$o_t = \sigma(W_o x_t + U_o h_{t-1} + V_o c_t) \tag{5}$$

$$h_t = o_t \tanh(c_t) \tag{6}$$

where σ denotes the logistic function, W and U are weight matrices and V is a diagonal matrix. The extent to which the existing memory c_{t-1} is forgotten is modulated by the forget gate f_t, and the degree to which the new memory content $\tilde{c}_t$ is added to the memory cell c_t is decided by the input gate i_t. The output gate o_t determines the amount of memory content exposure.

Besides a simple forward output $\overrightarrow{h_t}$, a backward output $\overleftarrow{h_t}$ also needs to be calculated. The backward computation calculates the outputs from the tail of the sequence data to the head of it. Equations (1)–(6) are used to calculate the forward and backward outputs, $\overrightarrow{h_t}$ and $\overleftarrow{h_t}$, which are restored as the inputs of the next layer.

3.3   Bidirectional Attention Layer

In our model, an attention mechanism is used to extract features from a single sentence, which has been proved to be effective in relation classification by Zhou et al. [5]. Figure 3 shows how the attention mechanism works.

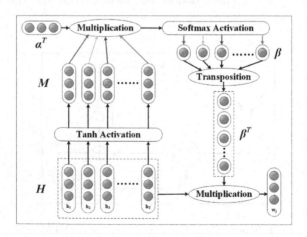

Fig. 3. Process of the attention mechanism

The output w_j is computed by:

$$M_w = \tanh(H) \tag{7}$$

$$\beta = \text{softmax}(\alpha^T M) \tag{8}$$

$$w_j = H\beta^T \tag{9}$$

where $H \in \Re^{d^w \times T}$ is the input matrix, which consists of vectors $[h_1, h_2, ..., h_T]$ produced by the BLSTM layer, where d^w is the dimension of the vectors and T is the length of the sentence. α is a trained parameter vector and α^T is the transpose. The dimensions of vector α and β are d^w and T, respectively. The output of the layer is w_j representing the features of the single sentence j.

We entered the outputs $H^f = [\overrightarrow{h_1}, \overrightarrow{h_2}, \overrightarrow{h_3}, ..., \overrightarrow{h_T}]$ and $H^b = [\overleftarrow{h_1}, \overleftarrow{h_2}, \overleftarrow{h_3}, ..., \overleftarrow{h_T}]$ of the BLSTM network into the bidirectional attention layer, and obtained the w_j^f and w_j^b respectively. w_j^f and w_j^b are combined by element-wise sum to form the output matrix w_j (see Eq. (10)).

$$w_j = w_j^f \bigoplus w_j^b \tag{10}$$

3.4   Sentence-Level Attention Layer

Since sentences containing same entity pair (entity1, entity2) and expressing same relation r may also carry many noise or useless information, we used a sentence-level attention layer to extract the features from a sentence set, which contains the information about the relation r. Similar to the previous layer, the representation s_j of a sentence set is calculated by the following equations:

$$M_s = \tanh(W) \tag{11}$$

$$\gamma = \text{softmax}(\delta^T M_s) \tag{12}$$

$$s_j^* = RM_s\gamma^T + b \tag{13}$$

$$s_j = \text{softmax}(s_j^*) \tag{14}$$

where $W \in \Re^{d^w \times m}$ (m is the number of sentences in the set) is formed by the outputs w_1, w_2, w_3,..., w_m from the previous layer, δ and R are trained parameters and b is a bias. Let n_r be the number of predefined relation types and the dimensions of vector δ, b and matrix R are d^w, n_r and $n_r \times d^w$ respectively.

The last softmax activation is used to output the prediction s_j and the cost function of the model is the cross-entropy of the true class labels y defined as:

$$C = -\frac{1}{n_r}\sum[y\ln s_j + (1-y)\ln(1-s_j)] + \frac{\lambda}{2n_r}\sum_\omega \omega^2 \tag{15}$$

The $L2$ regularization term is added to the cost function, where $\lambda > 0$ is known as the regularization parameter and $\sum_\omega \omega^2$ is the sum of the squares of the weights ω in the softmax activation. In order to optimize the cost, we chose Adam [18] to minimize the objective function.

4   Experiments

Our experiments were designed to evaluate the performance of the multi-attention mechanism in our model as well as the overall performance of the model in relation extraction. To this end, the dataset and the experimental setting details are first introduced. Then, the effect of the multi-attention mechanism is evaluated. Finally, the performance of our model is analyzed and compared with other methods.

4.1   Dataset and Setting Details

Experimental Dataset. Our experiments were conducted on the dataset[2] developed by Riedel et al. [19], the relations in which are extracted from the New York Times corpus, and Freebase is used as the knowledge base (Freebase was shut down in 2015, but the dataset still has great research value). There are 53 predefined relation types including a special relation "NA" indicating that the relation between the two entities does not belong to any predefined relations.

The training data contains 570088 sentences, 291010 entity pairs, and 156664 relational facts, and the testing data contains 172448 sentences, 96678 entity pairs, and 6444 relational facts. There are 156664 and 6444 relational facts in the training and testing data, respectively, which means that 72.5% and 96.3% of sentences in the two data sets have no relational facts. This will not affect the performance of our model [2]. Although the incomplete nature of Freebase means that some relations are incorrectly marked, the huge amount of data ensures the reliability of the experiments.

Setting Details. The dropout mechanism [20] and the $L2$ regularization are combined to avoid overfitting, and the dropout is employed on the output of the sentence-level attention layer.

In the experiments, we adopted position indicators[3] to mark the positions of entities, which use "< e1 >", "< /e1 >", "< e2 >" and "< /e2 >" to indicate entity pairs. Assume that we have the following sentence: "Last season under Carthon, the Browns scored the fewest points in the League.", and the entities are "Carthon" and "Browns". After simple pretreatments, the sentence above will be expressed as: "last season under < e1 > carthon < /e1 >, the < e2 > browns < /e2 > scored the fewest points in the league.". The four position indicators are regarded as single words and will be transformed into word vectors.

An overview of the important parameters is shown in Table 1, and other parameters of the model are initialized randomly.

4.2   Effect of the Multi-attention Mechanism

To demonstrate the effect of the multi-attention mechanism, experiments were designed to compare the performance of different parts of the model through

---

[2] http://iesl.cs.umass.edu/riedel/ecml.
[3] http://www.kozareva.com/downloads.html.

Table 1. Parameter setting details

Parameter	Value
Dimension of embeddings	100
Batch size	50
Epoch	5
Adam learning rate	0.001
Dropout	0.5
$L2$ regularization parameter λ	0.0001

held-out evaluation. We retained the result of a naive version of the BLSTM network as a baseline, and presented the performance of the BLSTM network with max-pooling (BLSTM+MP). Meanwhile, the performance of the two different types of BLSTM network with attention mechanisms were also compared: (i) BLSTM + bidirectional attention (BLSTM+BATT); (ii) BLSTM + bidirectional and sentence-level attentions (BLSTM+BSATT). The precision-recall curves of above models are shown in Fig. 4.

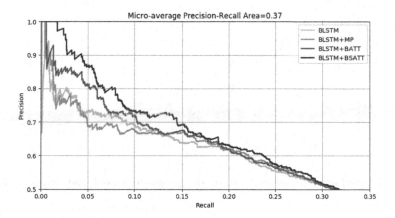

Fig. 4. The precision-recall curves of BLSTM, BLSTM+MP, BLSTM+BATT, and BLSTM+BSATT networks

From Fig. 4, these can be observed:

(1) Compared to the naive BLSTM network or the BLSTM+MP network, the BLSTM networks with attention mechanisms perform better, as expected.
(2) Bidirectional attention improves the relation extraction performance to a certain extent, but the effectiveness is limited. It is speculated that two factors contribute to this result: (i) the effectiveness of the attention mechanism is weak, so the improvement in the general performance is limited;

(ii) in the BLSTM+BSATT network, we apply attention first, and combine the two results through element-wise sum, which may weaken the ability of the attention mechanism to focus on the most important features.

(3) Sentence-level attention results in better performance on the basis of bidirectional attention, which means that sentence-level attention captures the important features of all sentences in the sentence set.

4.3 Performance of the Model

We evaluated the performance of our model through an automatic held-out evaluation and a manual evaluation proposed by Mintz et al. [2]. In the held-out evaluation, relation instances extracted from the testing data were compared with those in the whole data set.

The precision-recall curves of the different methods are illustrated in Fig. 5. The CNN+SATT and the PCNN+SATT networks were designed by Lin et al. [6].

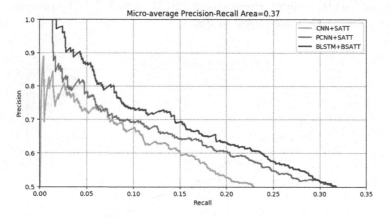

Fig. 5. The performance comparison of the proposed model and other traditional methods

The experimental results show that:

(1) The BLSTM+BSATT network is superior to the CNN+SATT network over the entire range of recall, and it outperforms the PCNN+SATT network in most cases. This provides convincing evidence that our model is superior to existing methods.

(2) When the recall is less than 0.15, the precision of the BLSTM+BSATT network is higher than those of the CNN+SATT network and the PCNN+SATT network by at least 0.1 and 0.05, respectively, but when the recall is greater than 0.15, the precision advantage of our model declines, which indicates that the effectiveness of bidirectional attention is limited when dealing with a large amount of data.

For manual evaluation, we only considered the top 300 extraction results returned by the model. The Precision@100, Precision@200, Precision@300 and the mean of them for each model are shown in Table 2.

Table 2. Precision@N for relation extraction in different subdatasets

Test datasets	One-sentence dataset				Two-sentence dataset				All-sentence dataset			
Precision@N(%)	100	200	300	Mean	100	200	300	Mean	100	200	300	Mean
CNN+SATT	76.2	65.2	60.8	67.4	76.2	65.7	61.1	68.0	76.2	68.6	59.8	68.2
PCNN+SATT	73.3	69.2	60.8	67.8	77.2	71.6	66.1	71.6	76.2	73.1	67.4	72.2
BLSTM+BSATT	**81.0**	**72.2**	**69.3**	**74.2**	**82.2**	**75.1**	**72.0**	**76.4**	**83.2**	**76.8**	**72.1**	**77.4**

According to Lin et al. [6], sentence-level attention lies in the entity pairs containing multiple instances, so we compared the performance of CNN+SATT, PCNN+SATT, and BLSTM+BSATT on the dataset where entity pairs have more than one instance. The "One-sentence dataset" means that for each entity pair, one sentence was chosen randomly to form the subdataset and used to predict relation. Meanwhile, we selected two sentences randomly and all sentences to form the "Two-sentence dataset" and the "All-sentence dataset", respectively.

Table 2 shows the Precision@N for the three compared models and we can clearly observe that: (i) compared to CNN+SATT and PCNN+SATT, the model we have proposed showed a 3% to 7.7% improvements in all cases; and (ii) the performance of the BLSTM+BSATT network drops gradually from Precision@100 to Precision@300, which also indicates that when the scale of the extracted data becomes larger, the performance of the bidirectional attention mechanism degenerates.

The functionality of the multi-attention mechanism showed above is not very satisfactory, and one main reason may lay on that the attention vector we adopted is universe for all sentences, entities and relations. In this case, probably the attention mechanism only learns to ignore the words generally useless for all kinds of relations and the effectiveness of attention is limited.

From the above, we can draw the following conclusions: (i) the BLSTM network with the multi-attention mechanism achieves the state-of-the-art performance in relation extraction; (ii) the bidirectional attention mechanism has an excellent ability to extract features from a single sentence and exhibits better performance in relation extraction; and (iii) when the number of sentences in a sub-dataset increases, the sentence-level attention mechanism plays a more important role and the contribution of the bidirectional attention mechanism is limited.

5   Conclusion

In this paper, we propose a novel BLSTM model with a multi-attention mechanism for relation extraction. Compared to CNNs, RNNs are more suitable for dealing with sequential data and a BLSTM model performs well when learning relations within a long sentence unaffected by the vanishing gradient problem. Meanwhile, with the help of the multi-attention mechanism, our model is able to capture the most significant features of a single sentence and a sentence set. Although our model does not rely on any NLP tools or extra resources, it still outperforms the state-of-the-art methods in relation extraction.

In the future, we will explore our work in the following directions:

- How to improve the effectiveness of attention mechanisms to capture the features from target sentences or sentence sets will be our research priority.
- With the increase of the model complexity, the enhancement in the computation speed is an important research direction.
- We do not use any NLP tools in this paper in order to avoid introducing extra errors, but we will attempt to integrate the model with other mechanisms, such as the shortest dependency path [21].
- Relation extraction shares similarities with other related tasks, therefore, the model we have proposed in this paper could be applied to other NLP tasks.

Acknowlegement. This research was supported by the National Natural Science Foundation of China (NSFC) under the project Nos. 61502517, 61672020 and 61662069.

References

1. Aggarwal, C.C., Zhai, C. (eds.): Mining Text Data. Springer Science & Business Media, Berlin (2012)
2. Mintz, M., Bills, S., Snow, R., Jurafsky, D.: Distant supervision for relation extraction without labeled data. In: the Joint Conference of the 47th Annual Meeting of the Association for Computational Linguistics and the 4th International Joint Conference on Natural Language Processing of the Asian Federation of Natural Language Processing Associations, vol. 2, pp. 1003–1011 (2009)
3. Bahdanau, D., Cho, K., Bengio, Y.: Neural machine translation by jointly learning to align and translate. arXiv preprint arXiv:1409.0473 (2014)
4. Shang, L., Lu, Z., Li, H.: Neural responding machine for short-text conversation. arXiv preprint arXiv:1503.02364 (2015)
5. Zhou, P., Shi, W., Tian, J., Qi, Z., Li, B., Hao, H., Xu, B.: Attention-based bidirectional long short-term memory networks for relation classification. In: The 54th Annual Meeting of the Association for Computational Linguistics, p. 207 (2016)
6. Lin, Y., Shen, S., Liu, Z., Luan, H., Sun, M.: Neural relation extraction with selective attention over instances. In: The 54th Annual Meeting of the Association for Computational Linguistics, vol. 1, pp. 2124–2133 (2016)
7. Zeng, D., Liu, K., Lai, S., Zhou, G., Zhao, J.: Relation classification via convolutional deep neural network. In: International Conference on Computational Linguistics, pp. 2335–2344 (2014)
8. Nguyen, T.H., Grishman, R.: Relation extraction: perspective from convolutional neural networks. In: Conference of the North American Chapter of the Association for Computational Linguistics–Human Language Technologies, pp. 39–48 (2015)
9. Zeng, D., Liu, K., Chen, Y., Zhao, J.: Distant supervision for relation extraction via piecewise convolutional neural networks. In: Conference on Empirical Methods in Natural Language Processing, pp. 1753–1762 (2015)

10. Socher, R., Huval, B., Manning, C.D., Ng, A.Y.: Semantic compositionality through recursive matrix-vector spaces. In: The 2012 Joint Conference on Empirical Methods in Natural Language Processing and Computational Natural Language Learning, pp. 1201–1211. Association for Computational Linguistics (2012)
11. Hashimoto, K., Miwa, M., Tsuruoka, Y., Chikayama, T.: Simple customization of recursive neural networks for semantic relation classification. In: Conference on Empirical Methods in Natural Language Processing, pp. 1372–1376 (2013)
12. Zhang, D., Wang, D.: Relation classification via recurrent neural network. arXiv preprint arXiv:1508.01006 (2015)
13. Xu, Y., Mou, L., Li, G., Chen, Y., Peng, H., Jin, Z.: Classifying relations via long short term memory networks along shortest dependency paths. In: Conference on Empirical Methods in Natural Language Processing, pp. 1785–1794 (2015)
14. Nie, Y., An, C., Huang, J., Yan, Z., Han, Y.: A bidirectional LSTM model for question title and body analysis in question answering. In: Data Science in Cyberspace, pp. 307–311. IEEE Press (2016)
15. Nie, Y.P., Han, Y., Huang, J.M., Jiao, B., Li, A.P.: Attention-based encoder-decoder model for answer selection in question answering. Frontiers Inf. Technol. Electron. Eng. 18(4), 535–544 (2017)
16. Pennington, J., Socher, R., Manning, C.D.: Glove: global vectors for word representation. In: Conference on Empirical Methods in Natural Language Processing, vol. 14, pp. 1532–1543 (2014)
17. Hochreiter, S., Schmidhuber, J.: Long short-term memory. Neural Comput. 9(8), 1735–1780 (1997)
18. Kingma, D., Ba, J.: Adam: a method for stochastic optimization. arXiv preprint arXiv:1412.6980 (2014)
19. Riedel, S., Yao, L., McCallum, A.: Modeling relations and their mentions without labeled text. In: Balcázar, J.L., Bonchi, F., Gionis, A., Sebag, M. (eds.) ECML PKDD 2010. LNCS, vol. 6323, pp. 148–163. Springer, Heidelberg (2010). doi:10.1007/978-3-642-15939-8_10
20. Srivastava, N., Hinton, G.E., Krizhevsky, A., Sutskever, I., Salakhutdinov, R.: Dropout: a simple way to prevent neural networks from overfitting. J. Mach. Learn. Res. 15(1), 1929–1958 (2014)
21. Fundel, K., Kner, R., Zimmer, R.: RelEx−Relation extraction using dependency parse trees. Bioinformatics 23(3), 365 371 (2007)

Question Recommendation in Medical Community-Based Question Answering

Hong Cai, Cuiting Yan$^{(\boxtimes)}$, Airu Yin, and Xuesong Zhao

College of Computer and Control Engineering, Nankai University, Tianjin, China
1006601845@qq.com, yct9301@163.com, yinar@nankai.edu.cn, 846506115@qq.com

Abstract. The medical community question answering system (MCQA) which is a new kind of medical information exchange platform is becoming more and more popular. Due to the number of patients is much more than the doctors, resulting in many patients can not get timely answers to their questions. Similar question recommendation is a common approach to solve this problem. The contributions of this paper are two-fold: (1) we propose a Siamese CNN model which measure correlation between questions and answers. (2) We first apply word2vec to learn the semantic relations between words and then construct a similar question retrieval model with answers. The study above can achieve a good performance in the real MCQA data set. It shows that our method can effectively extract similar questions recommendation list, shorten user's time to wait for an answer and improve user experience as well.

Keywords: Medical community question answering system · Similar question retrieval · Correlation between questions and answers · Convolutional neural network

1 Introduction

With the great popularity of computer network and the improvement of human health consciousness, human are increasingly inclined to seek medical advice on the Internet. A large number of redundant and similar questions have been accumulated in MCQA. Users can obtain the answers of similar questions in advance as a reference by retrieving similar questions, which greatly shorten the waiting time for answers of users and improve user experience.

Retrieval of similar questions is the key point for similar questions recommendation. The traditional retrieval algorithm, such as LMIR [1], Okapi BM25 [2] can not solve the problem of mismatched characters. By means of machine learning [3–5] and natural language processing [6], textual semantic information can be obtained by large scale data training. Zhang et al. [7] presents that there is a topic distribution between questions the answers. Xue et al. [8] adopt the maximum likelihood estimation of the characteristic words in the answers. The more characteristic words of question overlap in the answer, the more relevant of question and answer are. Jeon et al. [9] conducts an experiment to find that put questions, descriptions and answers three parts together to retrieve outperform the single part.

© Springer International Publishing AG 2017
D. Liu et al. (Eds.): ICONIP 2017, Part V, LNCS 10638, pp. 228–236, 2017.
https://doi.org/10.1007/978-3-319-70139-4_23

In medical community question answering system, the retrieved objects consist of two parts: question and answer. The definition of similar question is: (1) Two questions are semantic similarity. (2) One question and answer of another question are highly correlation. In this paper, the Siamese CNN model is used to measure the correlation between the questions and answers. Constructing a similar question retrieval model which adding answers can effectively improve the retrieval result.

The remainder of the paper is structured as follows. Section 2 introduces the correlation model between questions and answers based on the Siamese CNN. In Sect. 3, we propose a similar question retrieval model which adding answers. Section 4 describes experimental settings and results. Section 5 is the conclusion of the paper.

2 The Correlation Model Between Questions and Answers

For a given question Q and its corresponding answer A, the model is performed by learning Q and A distributed vector representations, and then the similarity algorithm is used to measure the correlation between Q and A.

In this paper, we propose a Siamese convolutional neural network (CNN) which consists of two CNN with the same structure and parameter. In order to ensure the consistency of questions and corresponding answers in semantic space. The objective function of the model is to minimize the distance between questions and correlation answers and maximize the distance between questions and uncorrelation answers in the semantic space (Fig. 1).

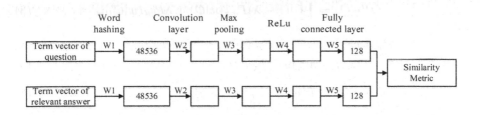

Fig. 1. Siamese CNN model

We consider $C = (q_i, a_i, y_i)$, $i = 1, 2, \ldots, n$, describing question q_i, answer a_i and label of q_i, a_i pair y_i, $y_i \in \{-1, 1\}$. We define the loss function of one sample

$$\text{loss}(q_i, a_i) = \begin{cases} 1 - \cos(q_i, a_i) & y_i = 1 \\ \max(0, \cos(q_i, a_i) - m) & y_i = -1 \end{cases}, \tag{1}$$

where $\cos(q_i, a_i)$ is correlation between q_i and a_i, m means similarity gap. The smaller value the m is, the more negative samples are involved in training.

The loss function of this model is the sum of whole data loss functions:

$$L(\Lambda) = \sum_{(q_i, a_i, y_i) \in C} \text{loss}(q_i, a_i). \tag{2}$$

3 Similar Question Retrieval Model with Answers

We input a query question q, the model retrieve the top n similar questions Q and measure the correlation between q and answers in Q. The final score consists of the similarity of q and questions as well as the correlation of q and answers (Fig. 2).

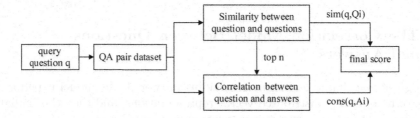

Fig. 2. Similar question retrieval model with answers

The correlation model between questions and answers has been proposed in Sect. 2. Here, we introduce a similar question retrieval model. In this paper, we improve translation language model (TRLM) [9] by maximum transition probability instead of average transition probability.

$$P(q|Q) = \prod_{w \in q} [(1 - \lambda) P_{mx}(w|Q) + \lambda P_{ml}(w|C)], \tag{3}$$

$$P_{mx}(w|Q) = (1 - \beta) P_{ml}(w|Q) + \beta P_{tr}(w|Q), \tag{4}$$

$$t_{max} = \arg \max_{t \in D} T(w|t), \tag{5}$$

$$P_{tr}(w|Q) = T(w|t_{max}) P(t_{max}|Q), \tag{6}$$

where λ, β are priori parameters, t_{max} denotes the word closest to w in question set Q, $P(t_{max}|Q)$ is the maximum likelihood estimation of t_{max} in Q.

$\forall w_i \in V$, $V = (w_1, w_2, \ldots, w_n)$ represents the vocabulary of all words in corpus C and map V into vector spaces by word2vec, $\boldsymbol{V} = (v(w_1), v(w_2), \ldots, v(w_n))$, where $v(w_i)$ is represented the Word2vec vector of word i. The similarity [10] of word i and word j is

$$P_{sim}(w_i|w_j) = \cos(v(w_i), v(w_j)) = \frac{v(w_i) \cdot (w_j)}{|v(w_i)||v(w_j)|}. \tag{7}$$

The definition for similar words of w_i is

$$sim(w_i) = \{w_j | P_{sim}(w_i|w_j) > 0, w_j \in V\}. \tag{8}$$

Hence we have the transition probability which is expressed as

$$P_{tr}(w_i|w_j) = \begin{cases} \dfrac{\cos(v(w_i), v(w_j))}{\sum\limits_{w' \in sim(w_j)} \cos(v(w_i), v(w_j))} & w_i \in sim(w_j) \\ 0 & otherwise \end{cases} . \qquad (9)$$

The main body of MCQA is question answering. The definition of similarity between question Q_1 and Q_2 needs to be satisfied with the high similarity between Q_1 and Q_2 as well as the high correlation between Q_1's and Q_2's answers.

For a query question q, first, we retrieve top n similar questions of q by similar question retrieval model. Second, we use the Siamese CNN model to measure the correlation between q and top n similar questions' answers. Finally, the final score is obtained by combining the similarity of questions with the correlation of questions and answers.

$$score(q, (Q_i, A_i)) = \alpha sim(q, Q_i) + (1 - \alpha)cons(q, A_i), \qquad (10)$$

where $sim(q, Q_i)$ is the similarity of query question q and Q_i, $cons(q, A_i)$ is the correlation of query question q and answer A_i, α is a priori parameter that is used to balance $sim(q, Q_i)$ and $cons(q, A_i)$.

4 Experimental Results

In order to prove the qualities of our proposed methods, we conduct several experiments to compare our approaches with other state-of-the-art methods.

4.1 Datasets

Data Preprocessing. Because of no existing public large-scale question and answer data in MCQA, the experimental dataset in this paper is crawled from Chinese MCQA websites, include http://www.xywy.com/, http://www.haodf.com/ and http://www.120ask.com/. Crawler framework is Java+HttpClient+Jsoup. Due to large-scale data and question always belongs to one category, the classification of questions can effectively reduce the size of candidate retrieval questions [11].

This paper use the shallow CNN classification model to classify questions based on character level. The character vector [12,13] input can not only reduce the dimension of the feature vector, but also correct the typos and wrong spelling problems. It is proved that the shallow CNN classification model can automatically extract multiple text features and avoid the ambiguity of Chinese word segmentation.

Experimental Dataset. The final experimental data set consists of 460,000 questions and answers about "dermatologica" category. Each data include title, question detail and answer. We combine title with detail as a question data and connect all the answers of this question as a answer data. We randomly divide the QA pairs into two disjoint sets: test set consist 1000 questions and answers, the rest of dataset is training set.

4.2   Compared Algorithms

We compare similar questions retrieval model with answers to traditional methods, Vector Space Model (VSM) [14], Query Likelihood Model (LMIR) [1], Probability Retrieval Model (Okapi BM25) [2], Translation Based Model (TRLM) [9], so as to show the effectiveness of our proposed approaches.

4.3   Metrics

We employ three popular metrics, Mean Average Precision (MAP), Mean Reciprocal Rank (MRR) and Precision@k to evaluate the retrieval quality.

$$Precision@K = \frac{|R_k|}{k}, \tag{11}$$

$$MAP = \frac{1}{|T|} \sum_{q \in T} \frac{1}{n_q} \sum_{k=1}^{n_q} Precision(R_k), \tag{12}$$

$$MRR = \frac{1}{|T|} \sum_{i=1}^{|T|} \frac{1}{r_i}. \tag{13}$$

where T denotes the number of whole data set, n_q is the number of relevant questions from $q.r_i$ shows the location of the first occurrence of the relevant problem in the retrieval result. $Precision@K$ represents the proportion of top k related questions in the retrieval result.

4.4   Experimental Settings

The similarity model between questions. We use Word2vec and skip-gram algorithm [15] with negative sampling to training word vector. Chinese Wikipedia and our dataset are included in corpora. The word vector dimension of 300, window size of 5 and minimum count of 5.

The correlation model between questions and answers. Siamese CNN Model has 3 layer, the size of convolution kernel is 3, 128 kernel, 3 stride of MAX pooling. The vector dimension between question and answer is 50. In order to prevent overfitting problems, we set dropout to 0.5 in the training process and 0 in the test process. Training dataset is divided into several sub sets so as to improve the training efficiency of the model where batch_size = 100 and num_epoch = 300. We choose to use a learning rate of 0.01 and "adam" optimizer.

Similar questions retrieval model with answers. First, we retrieve top 20 similar questions. Second, measuring score between query question and top 20 questions' answers. At last, adding two part scores by balancing parameter α.

4.5 Experimental Results

Table 1 shows results of similarity model between questions. VSM only measures word frequency, perform worse than other methods. Our method is based on semantic extension of query question as well as using the maximum transition probability of word to get latent semantic and perform better.

Table 2 shows the correlation results between questions and answers. Our method outperform in MAP and P@5. Siamese CNN is supervised learning, word features are mapped to multiple feature spaces by convolution kernels. Word2vecTrans Model perform better on MRR that means the proportion of related questions in several space is higher.

Table 1. Results of similarity model between questions.

q&Q Model	MAP	MRR	P@5
VSM	0.409	0.604	0.413
LMIR	0.593	0.721	0.613
Okapi BM25	0.662	0.816	0.693
Our method Word2vecTrans	**0.727**	**0.863**	**0.746**

Table 2. Results of correlation model between question and answers.

q&A Model	MAP	MRR	P@5
LMIR	0.554	0.678	0.568
Okapi BM25	0.624	0.741	0.649
Word2vecTrans	0.648	**0.783**	0.663
Our method Siamese CNN	**0.655**	0.736	**0.668**

Compared Table 1 with Table 2, under the single model, the question part contributes more to the similarity than the answer part. Next, we put two single models together.The results of combined model are showed in Table 3.

In all evaluation measures, our approaches (Word2vecTrans + Siamese CNN) outperforms the other combine models and single models. Under the same similarity model between questions, Siamese CNN model works better than other q&A models in our experiments. It shows that Siamese CNN can extract the latent semantic relations between the query question and corresponding answers.

4.6 Impact of Parameter α

In our methods proposed in this paper, the parameter α plays an important role. It balances similarity model of questions and correlation model of questions and answers. In this part, we select well performance methods Word2vecTrans with Siamese CNN and Okapi BM25 with Siamese CNN. With different value of α, Fig. 3 shows the impact of α on metrics.

Table 3. Results of similar questions retrieval model with answers.

q&Q Model	q&A Model	MAP	MRR	P@1	P@5	P@10
LMIR	LMIR	0.582	0.719	0.591	0.595	0.588
	Okapi BM25	0.587	0.702	0.568	0.613	0.591
	Word2vecTrans	0.612	0.754	0.636	0.636	**0.604**
	Siamese CNN	**0.618**	**0.817**	**0.750**	**0.645**	0.593
Okapi BM25	LMIR	0.634	0.783	0.681	0.636	0.622
	Okapi BM25	0.658	0.772	0.659	0.686	0.670
	Word2vecTrans	0.666	0.776	0.659	0.681	**0.665**
	Siamese CNN	**0.661**	**0.811**	**0.727**	**0.691**	0.645
Word2vecTrans	LMIR	0.681	0.803	0.704	0.718	0.668
	Okapi BM25	0.687	0.821	0.727	0.709	0.691
	Word2vecTrans	0.695	0.828	0.750	0.722	0.686
	Siamese CNN	**0.733**	**0.861**	**0.795**	**0.754**	**0.727**

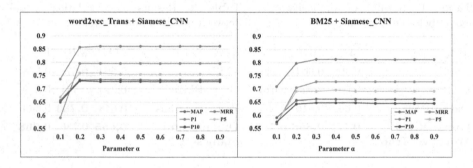

Fig. 3. Impact of α

We observe that the value of α impacts the results significantly. $\alpha = 5$ make the model performance well. The results illustrate the similar questions retrieval model with answers can effectively improve the accuracy by extracting similar answer sets of related questions.

Extracting answers of similar questions from the trained similarity model can reduce the size of candidate set and improve the retrieval efficiency, so that users who ask question in Medical Question Answering System can effectively reduce the time to obtain answers and enhance the user experience. Considering the similar questions and corresponding answers information, we recommend similar questions for users. Our approaches improve precision of recommendation questions and optimize information acquired by users, so that users can get more accurate similar questions list from MCQA as well as optimize the quality of users requirements information.

5   Summary

In this paper, a similar question retrieval model with answers is proposed to solve similar question recommendation in medical community question answering system. Because of no existing public large-scale question and answer data in MCQA, the experimental data in this paper is crawled from Chinese MCQA websites. Similar questions recommendation focuses on how to calculate the similarity between questions and questions. In this paper, we proposed a similar question retrieval model with answers which combines the improved translation language model and the Siamese CNN model. Our approaches can effectively improve the accuracy of retrieval, help users get relevant information in time and optimize the performance of Medical Question Answering System as well.

Acknowledgments. This work is supported by the National Science Foundation of China(No. U1633103), the Open Project Foundation of Information Technology Research Base of Civil Aviation Administration of China (No. CAAC-ITRB-201502).

References

1. Zhai, C., Lafferty, J.: A study of smoothing methods for language models applied to information retrieval. ACM Trans. Inf. Syst. (TOIS) **22**(2), 179–214 (2004)
2. Surhone, L.M., Tennoe, M.T., Henssonow, S.F.: Okapi BM25. Betascript Publishing (2010)
3. Zhang, D., Lee, W.S.: Question classification using support vector machines. In: Proceedings of the 26th Annual International ACM SIGIR Conference on Research and Development in Informaion Retrieval, pp. 26–32. ACM (2003)
4. Nguyen, T.T., Nguyen, L.M., Shimazu, A.: Using semi-supervised learning for question classification. Inf. Media Technol. **3**(1), 112–130 (2008)
5. Moschitti, A., Quarteroni, S., Basili, R., et al.: Exploiting syntactic and shallow semantic kernels for question answer classification. In: Annual Meeting-Association for Computational Linguistics, vol. 45, p. 776 (2007)
6. Goldberg, Y., Levy, O.: word2vec explained: deriving Mikolov et al'.s negative-sampling word-embedding method. arXiv preprint arXiv:1402.3722 (2014)
7. Zhang, K., Wu, W., Wu, H., et al.: Question retrieval with high quality answers in community question answering. In: ACM International Conference on Information and Knowledge Management, pp. 371–380. ACM (2014)
8. Xue, X., Jeon, J., Croft, W.B.: Retrieval models for question and answer archives. In: Proceedings of the 31st Annual International ACM SIGIR Conference on Research and Development in Information Retrieval, pp. 475–482. ACM (2008)
9. Jeon, J., Croft, W.B., Lee, J.H.: Finding similar questions in large question and answer archives. In: ACM International Conference on Information and Knowledge Management, pp. 84–90 (2005)
10. Hinton, G.E.: Learning distributed representations of concepts. In: Proceedings of the Eighth Annual Conference of the Cognitive Science Society, Amherst, MA, vol. 1, p. 12 (1986)
11. Cai, L., Zhou, G., Liu, K., et al.: Large-scale question classification in cQA by leveraging Wikipedia semantic knowledge. In: Proceedings of the 20th ACM International Conference on Information and Knowledge Management, pp. 1321–1330. ACM (2011)

12. Bian, J., Gao, B., Liu, T.-Y.: Knowledge-powered deep learning for word embedding. In: Calders, T., Esposito, F., Hüllermeier, E., Meo, R. (eds.) ECML PKDD 2014. LNCS, vol. 8724, pp. 132–148. Springer, Heidelberg (2014). doi:10.1007/978-3-662-44848-9_9
13. Hu, B., Tang, B., Chen, Q., et al.: A novel word embedding learning model using the dissociation between nouns and verbs. Neurocomputing **171**, 1108–1117 (2016)
14. Salton, G., Wong, A., Yang, C.S.: A vector space model for automatic indexing. Commun. ACM **18**(11), 613–620 (1975)
15. Mikolov, T., Sutskever, I., Chen, K., et al.: Distributed representations of words and phrases and their compositionality. In: Advances in Neural Information Processing Systems, pp. 3111–3119 (2013)

A Visual Analysis of Changes to Weighted Self-Organizing Map Patterns

Younjin Chung$^{(\boxtimes)}$, Joachim Gudmundsson, and Masahiro Takatsuka

School of IT, Faculty of Engineering and IT,
The University of Sydney, Camperdown, NSW 2006, Australia
ychu2895@uni.sydney.edu.au

Abstract. Estimating output changes by input changes is the main task in causal analysis. In previous work, input and output Self-Organizing Maps (SOMs) were associated when conducting causal analysis of multivariate and nonlinear data. Based on the SOM association, a weight distribution of the output conditional on a given input was obtained over the output map space. Such a weighted SOM pattern of the output changes when the input changes. In order to analyze the pattern change, it is important to measure the difference of the patterns. Many methods have been proposed for measuring the dissimilarity of patterns; however, it is still a major challenge to identify how patterns are different. In this paper, we propose a visual approach for analyzing changes to weighted SOM patterns. This approach extracts features that represent the difference of patterns by change and facilitates overall and detailed comparisons of pattern changes. Ecological data are used to demonstrate the usefulness of our approach and the experimental results show that it visualizes the change information effectively.

Keywords: Self-Organizing map · Weighted SOM pattern · Pattern dissimilarity · Information visualization · Pattern change analysis

1   Introduction

Analyzing causality is one of the central tasks of science since it influences the decision making in such diverse domains as ecological, social and health sciences. Causality is the relationship between two events, if changes of one (cause) trigger changes of the other (effect) [1]. In our previous work [2], a causal analysis model was developed for analyzing causality of multivariate and nonlinear data (unlabeled in nature). In that model, different SOMs [3] for input and output data sets were networked using a weight association based on the connection prototype feature vector similarity. Given the SOMs, the similarity weights, conditional on a given input, could be assigned to the neurons of the output SOM. Such a weighted SOM pattern of the output is described by two information types: (1) the weight distribution and (2) the property (prototype feature vector) distribution. To assess the output pattern changes by the input changes, it is crucial to measure the pattern difference based on the two information types.

© Springer International Publishing AG 2017
D. Liu et al. (Eds.): ICONIP 2017, Part V, LNCS 10638, pp. 237–246, 2017.
https://doi.org/10.1007/978-3-319-70139-4_24

There have been many attempts to quantitatively measure the dissimilarity between two distributions (patterns) for a variety of applications, for example image retrieval [4]. The most used families of functions are the Minkowski and the Shanon's entropy families [4,5]. However, these functions do not match the perceptual pattern dissimilarity well since they only compare the weights of the corresponding fixed bins [6]. The families do not use the similarity information across neighboring bins such as adaptive neurons of weighted SOM patterns. When considering the spatial information across bins, the Quadratic Form Distance (QFD) [7] and the Earth Mover's Distance (EMD) [6] are the most used functions. The QFD tends to overestimate the dissimilarity of patterns as the weight of each bin is simultaneously compared with the weights across all bins [6]. On the other hand, the EMD uses the ground distance of feature vectors across bins for the minimum weight flow and provides better perceptual matches. The EMD, as the others, provides a numerical value to define only a notion of the total resemblance of patterns. However, the total dissimilarity alone cannot differentiate between patterns having the same feature vector distance and weight distribution with different properties. It is difficult to compare the dissimilarity in terms of the pattern property when analyzing pattern changes in causal analysis by relying on such a quantitative measure.

Hence, in this paper, we propose a visual approach for analyzing changes to weighted SOM patterns by representing the dissimilarity of patterns in comparable forms. In the visual approach, change features are extracted to define the perceptual dissimilarity of weighted SOM patterns with the property change information in a measurable form. A *global* SOM visualization that represents all possible pattern changes for a given data space is used for comparing the overall difference of pattern changes. For analyzing the detailed pattern change, a star glyph shape and colors are integrated to represent the *local* information of the pattern change. Using the visual approach, the relative difference and the tendency of pattern changes can be analyzed by capturing the overall and the detailed dissimilarity information of pattern changes. When there is no property difference over a pattern change, possible feature values that gain the pattern change can also be observed by exploring specific regions in the weighted SOM patterns. Ecological domain data are used to demonstrate that our approach is useful for the comparison of weighted SOM pattern changes in causal analysis. The experimental results show that our approach visualizes the change information effectively for analyzing causal effects.

2   Background

A Self-Organizing Map (SOM) [3] projects high dimensional data onto a low dimensional (typically 2-dimensional) grid map space. A set of neurons in the map, which are prototype feature vectors adaptively projected for original feature vectors, reflect the data properties. Using the causal analysis model in our previous work [2], a similarity weight distribution is estimated on the property distribution of the output SOM for a given input. Figure 1 shows an example illustrating the two information types in a weighted SOM pattern: (1) the weight distribution and (2) the property distribution. A 3×3 SOM in Fig. 1(a) is used as it is easy for visualizing the 3-dimensional RGB color property and position. Based on the SOM, several weighted SOM patterns ((b)–(g) in Fig. 1) are created by different color opacity values (weights). Such a weighted SOM pattern (S) can be depicted as $S = \{(v_i, w_i)\}$, $i = 1,..,n$, where n is the number of neurons ($n = 9$); v_i is ith prototype feature vector ($v_i = \{(c_k)_i\}$, $k = 1,..,m$, where m is the number of features ($m = 3$); c_k is the kth component of the prototype feature vector) and w_i is ith weight, representing the two information types.

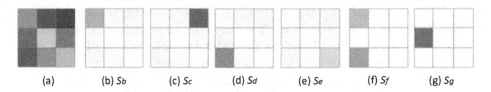

(a)   (b) S_b (c) S_c (d) S_d (e) S_e (f) S_f (g) S_g

Fig. 1. (a) A 3×3 SOM of a RGB color data space. (b)–(g): Six weighted SOM patterns of the SOM in (a) by the changes made from S_b to S_c–S_e and from S_f to S_g.

The perceptual dissimilarity between the weighted SOM patterns in Fig. 1 can be measured by observing the color properties in the highlighted neurons. The patterns S_c and S_d show that they are different patterns by their different color properties although they have the same weight and distance relation (total dissimilarity) to their changes from S_b. The pattern S_e has the furthest map distance from S_b; however, it shows a smaller difference (total dissimilarity) with the similar color property to the change by the border effect of a 2-dimensional SOM [3]. The pattern S_g shows the change in two ways by the two different color properties highlighted in S_f. Such pattern differences can be explained using the visualized 3-dimensional color property in the overall map space. Nonetheless, it is still difficult to compare the size and the direction of the pattern changes and it becomes harder to measure such differences in higher dimensions. In an attempt to handle these issues, we propose a visual approach for analyzing changes to weighted SOM patterns of high dimensional data.

3   Our Approach

In this section, we propose a visual analysis approach that extracts features to define the difference of patterns by change and uses two visualization methods for the overall and the detailed comparisons of weighted SOM pattern changes.

3.1   Pattern Change Features

The difference of weighted SOM patterns is featured by the total dissimilarity and the property dissimilarity in our study. Thus, we introduce a metric function, called Property EMD (PEMD) to measure the property dissimilarity based on the capability of the EMD for the total dissimilarity measure of weighted SOM patterns. The PEMD measures the individual feature difference of data to represent the property dissimilarity in a pattern change.

According to [6], the EMD between two weighted SOM patterns P and Q is defined as follows:

$$EMD(P,Q) = \frac{\sum_{i=1}^{n} \sum_{j=1}^{n} f_{ij} d_{ij}}{\sum_{i=1}^{n} \sum_{j=1}^{n} f_{ij}}, \tag{1}$$

where d_{ij} is the ground distance function and f_{ij} is the minimum cost flow under constraints: $\forall i, j: f_{ij} \geq 0$, $\forall i: \sum_{j=1}^{n} f_{ij} \leq w_{p_i}$, $\forall j: \sum_{i=1}^{n} f_{ij} \leq w_{q_j}$, and $\sum_{i=1}^{n} \sum_{j=1}^{n} f_{ij} = \min\{\sum_{i=1}^{n} w_{p_i}, \sum_{j=1}^{n} w_{q_j}\}$. The weighted SOM patterns P and Q are based on the same SOM; thus, they have the same number (n) of neurons and their weights sum to 1. Based on the EMD, the difference of a feature c_k in Q for given P can be measured by a function as follows:

$$PEMD_{c_k}(Q|P) = \frac{\sum_{i=1}^{n} \sum_{j=1}^{n} f_{ij} d_{ij} (c_{kj} - c_{ki})}{\sum_{i=1}^{n} \sum_{j=1}^{n} f_{ij} d_{ij}}. \tag{2}$$

The change direction of the feature c_k is accounted for by its difference from P to Q. The PEMD is then defined as the resulting feature difference of the data to the change normalized by the total work flow of the EMD avoiding larger differences between the pattern changes. The pattern difference to the pattern change is then defined by the EMD and the PEMD features.

The individual PEMDs between the weighted SOM patterns in Fig. 1 are measured for the property comparison. The EMD is measured and scaled by the maximum EMD of the data space ($\sqrt{m}$ for the m-dimensional data space with the value range $[0, 1]$), denoted by EMD^N for the overall pattern comparison. As seen in Table 1, the pattern changes can be explained by coupling the change features. The property difference between the patterns S_c and S_d changed from S_b can be explained by the different PEMD values while they show the same EMD^N. The EMD^N for the pattens S_f and S_g shows that they are not the same pattern while their PEMD values show no property difference between them. However, it is not easy to compare the relative difference of the pattern changes by the measures. It is also difficult to identify the two ways of the pattern change from S_f to S_g as the property information is lost in the measures. This shows

Table 1. The scaled EMD (EMD^N) and the PEMDs of the pattern changes in Fig. 1.

Pattern change	EMD^N	$PEMD_R$	$PEMD_G$	$PEMD_B$
From S_b to S_c	0.4491	0	-1	$+1$
From S_b to S_d	0.4491	$+1$	-1	0
From S_b to S_e	0.2245	$+0.5$	0	$+0.5$
From S_f to S_g	0.4082	0	0	0

that the pattern difference can be further explained by exploring the possible feature values that gain the pattern change. In the next section, we introduce two visualization methods for comparing and analyzing the overall and the detailed pattern changes based on the change features.

3.2   Pattern Change Visualizations

The pattern changes in causal analysis can be efficiently compared in a global space of all possible pattern changes. Thus, we first develop a *global* visualization method using a SOM that approximates the global space. A *global* SOM can be trained by generating artificial data which can cover the entire range of all possible pattern changes of a data space. The $m + 1$ change features are used to represent the global space of the pattern changes for the m-dimensional data space. The EMD^N values fall into the range $[0, 1]$ and the PEMD value range is varied depending on the feature value range. For the weighted SOM patterns in Fig. 1, the feature values are normalized between 0 and 1. Thus, the PEMD values are ranged between -1 and 1, and scaled to the range $[0, 1]$.

A *global* SOM for the pattern changes in Fig. 1 is trained with the size of 12×14 by generating 1000 artificial data for the change features (EMD^N, $PEMD_R$, $PEMD_G$ and $PEMD_B$). The size of the SOM is determined using a heuristic formula as proposed in [8]. Figure 2(a) shows the *global* SOM for the pattern changes with their Best Matching Units (BMUs). As there is no change for the reference pattern itself, its change always shows the same BMU ($S0$). Using the *global* SOM, we can capture that the patterns S_c and S_d change from S_b by relatively similar size but with different directions. The pattern change

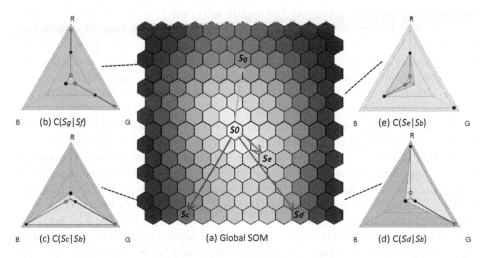

Fig. 2. Our visual approach for the pattern changes in Fig. 1. (a): The *global* SOM with the change BMUs and paths from the reference pattern, $S0$. The darker the gray shade the higher the EMD^N value. (b)–(e): The *local* views of the pattern changes. (Color figure online)

$C(S_g|S_f)$ can also be compared with the others in the different change relation. It shows the overall difference of $C(S_g|S_f)$ is relatively smaller than $C(S_c|S_b)$ and $C(S_d|S_b)$. This *global* visualization facilitates the simultaneous comparison of the relative difference between the weighted SOM pattern changes by their BMU locations from the reference BMU. The component planes [8] of the *global* SOM can be used to understand how a pattern is changed in detail. However, the multiple component visualizations can make the comparison harder when the dimensionality grows as many different views have to be switched [9].

Due to the drawbacks with the current solution, we develop a *local* visualization method to effectively analyze the detailed pattern changes. It is also designed to present the possible feature changes such in $C(S_g|S_f)$ by selecting regions of interest in the patterns. The *local* visualization uses a star glyph object with colors to represent the difference between a pair of weighted SOM patterns by change. A star glyph [10] has m evenly angled branches emanating from a central point in the same ordering of m dimensions (the fixed orientation). The length of each branch marks the value changed along the dimension, and the value points are connected creating a bounded polygon shape. A star glyph shape created by the individual PEMDs is imposed on the m branch frame reflecting the overall dissimilarity of the property change. The absolute PEMD value for each feature is scaled in the range $[0.1, 0.9]$ to improve the visualization. The average of the PEMD values is used to indicate the direction of the overall property change by applying it to the color saturation. The property change shape is filled with the direction color; red for increase, blue for decrease and white for no change. The property changes that have the same shape with any opposite feature change directions can be differentiated by the direction color. The EMD^N is visualized by filling its gray scale in the frame. The possible value changes in each feature can be visualized depending on the region selection (e.g. highlighted regions). The possible changes are indicated by red and blue for increase and decrease, respectively, between the minimums and the maximums in the selected regions. The minimum and the maximum feature values are indicated by empty dots for the reference pattern and full dots for the changed pattern. The line colored with red or blue from the center to 0.1 indicates increase or decrease of the overall individual feature change, respectively.

As the pattern changes are approximated on the *global* SOM, the detailed change information can be viewed by navigating the *global* SOM and selecting the BMUs. Figure 2(c) and (d) show the detailed changes of S_c and S_d obtained from S_b and their difference can simply be compared by the different property change shapes in the fixed orientation. The pattern change $C(S_e|S_b)$ in Fig. 2(e) shows a relatively small increase by the shape size and the direction color with the value increasing in the features R and B. For $C(S_g|S_f)$ in Fig. 2(b), the frame color shows that they are not the same pattern while the shape shows no property difference. This can then be explained by observing the possible changes in the features R and G by the same size increase and decrease which make no difference in the property change. Using the *local* visualization, it is easy to recognize the detailed difference between the pattern changes by the property change shape variation together with the colored change information.

4 Experimental Results

In this section, we test our approach by applying it to the ecological domain data[1] [11] for analyzing changes of the output pattern by changes in the input. The physical and biological SOMs were trained using 10×12 hexagonal grids by the minimum values of quantization and topological errors. The physical input SOM was associated with the biological output SOM. Among the physical features, it is known that *Embeddedness* ($P4$) has a strong impact on the biotic integrity [12]. Thus, for our experiments, we varied the physical input values to examine the impact of $P4$ on the biological output. Each physical feature value was increased by 1 and 2 standard deviations (SD) for the first and the second changes, respectively, while the others were fixed at the given value (-0.5SD). The standardized Z-score values of the data used in the data analysis were converted to the T-score values for visualization.

Figure 3 shows the weighted biological output SOM patterns obtained by the physical input changes. For better viewing of a weighted SOM pattern, we integrated hue colors and star glyph shape objects, which are perceptually orthogonal [13], for the weight and the property distributions in the output SOM. The perceptual dissimilarity of the normalized hue colors (0 to 240 degree for high to low weight) indicates a clear boundary of the weights and provides a better region selection. The graphical object mapping of the prototype feature vector components into a star glyph shape makes it easier to recognize a neuron property by perceiving the multiple dimensions of the data. The pattern $S0$ in Fig. 3(a) was obtained by the given physical input. The reddish regions

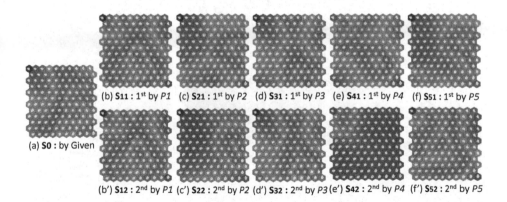

(a) S0 : by Given

(b) S11 : 1st by P1   (c) S21 : 1st by P2   (d) S31 : 1st by P3   (e) S41 : 1st by P4   (f) S51 : 1st by P5

(b') S12 : 2nd by P1 (c') S22 : 2nd by P2 (d') S32 : 2nd by P3 (e') S42 : 2nd by P4 (f') S52 : 2nd by P5

Fig. 3. The weighted biological output SOM patterns by the given input in (a), by the first changed inputs in (b)–(f) and by the second changed inputs in (b')–(f'). (Color figure online)

---

[1] The ecological features: $B1$ (Shredders), $B2$ (Filtering-Collectors), $B3$ (Collector-Gathers), $B4$ (Scrapers) and $B5$ (Predators) for Biological data set; $P1$ (Elevation), $P2$ (Slope), $P3$ (Stream Order), $P4$ (Embeddedness) and $P5$ (Water Temperature) for Physical data set. The feature values are all standardized for the total 130 data.

were selected in each pattern and more than one region were observed in $S0$ showing the high possibility of having different outputs for the given input. The pattern changes were measured and visualized in Fig. 4 using our approach. For the relative pattern change analysis, a *global* SOM was trained with the heuristic size of 14×16 by generating 2000 artificial data [8]. The change features used for training the *global* SOM are EMD^N, $PEMD_{B1}$, $PEMD_{B2}$ $PEMD_{B3}$, $PEMD_{B4}$ and $PEMD_{B5}$. More information can be added in the *local* view and we added the significance information, measured over the difference of every weighted biological feature distribution using the Kolmogorov-Smirnov test [14]. The insignificant changes are indicated by yellow and cyan while the significant changes are indicated by red and blue for increase and decrease, respectively.

As seen in Fig. 4(a), the BMUs of the biological output pattern changes are located in the different neurons for the different physical input changes. The patterns changed by $P1$ and $P3$ are shown to be similar in the first change (S_{11} and S_{31}) and the second change (S_{12} and S_{32}) without any significant

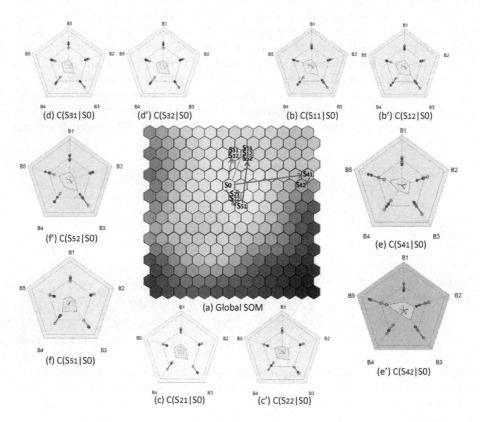

Fig. 4. Our approach. (a): The global SOM. The same indices for the 1st changes ((b)–(f)) and the 2nd changes ((b')-(f')) in Fig. 3 are used to indicate their local views. (Color figure online)

difference between both the changes. The patterns changed by $P2$ and $P5$ are also shown to be similar in the first change (S_{21} and S_{51}). However, compared with the first change, S_{52} shows a very different change while S_{22} remains the same. The pattern changes by $P5$ (*Water Temperature*) could be interpreted as different compositions of biological entities for different water temperatures rather than having an impact on the biological quality. The patterns S_{41} and S_{42} changed by $P4$ show the largest changes compared with the others. The change paths are shown in the *global* SOM with the *local* views of the pattern changes listed in each BMU. The similarity of the pattern changes was also clearly captured using the property change shape and the change direction color variations between the *local* views. Based on the selection of the reddish regions, the possible feature changes are shown in each frame. For the output pattern changes by $P4$, $C(S_{42}|S0)$ shows a larger change than $C(S_{41}|S0)$ with a similar tendency of the property change. It also shows the biological features $B1$ and $B4$ become significant and the changes in $B1$, $B2$, $B3$ and $B5$ become larger when the value of $P4$ is increased. In S_{41} and S_{42}, both increase and decrease are seen in $B4$, that could be misinterpreted as not being effected by $P4$.

The analytical results show that the impact of $P4$ is significant and much stronger than the other physical factors. Throughout the experiments, we could derive the impact of $P4$ (*Embeddedness*) as its increase lowers the balance of the biological composition of the ecosystem while the others change the biological quality much less than $P4$. The change of $B4$ could also be explained in analyzing the causal effects of $P4$ by its possible changes shown in the *local* views. Therefore, the impacts of physical factors on the biological quality could effectively be analyzed using our approach. The relative difference of the overall output pattern changes could be simultaneously compared using the *global* SOM visualization. The detailed property difference and tendency of the output pattern changes could be further explained using the *local* visualization. Such change information can effectively support well-informed decision making.

5 Conclusion

In this paper, we have presented a visual approach for analyzing weighted output SOM pattern changes by input changes in causal analysis. We elucidated the idea of analyzing changes to weighted SOM patterns by extracting and comparing the change features, the total and the property dissimilarities. Our proposed approach measures and visualizes the difference of pattern changes by the *global* and the *local* representations. Using the *global* visualization, the overall and simultaneous comparison of relative pattern changes can be facilitated in the process of causal analysis. The *local* visualization can facilitate the detailed comparison of pattern changes by exploring regions of interest and capturing possible changes to support a better interpretation of causality. Throughout the experiments, we have shown that our approach is useful for measuring and comparing weighted output SOM pattern changes by input changes. The experimental results showed that our approach provides the change information in an effective visual way when analyzing causal effects of multivariate and nonlinear data. We believe the proposed approach can be used in diverse application domains and currently applying it to a large case study in the health domain.

Acknowledgments. This research was partially supported by HMR+ SPARC Implementation Funding (BMRI 2015) under the Project G181478 and ARC's Discovery Project funding scheme (DP150101134).

References

1. May, W.E.: Knowledge of causality in Hume and Aquinas. Thomist **34**, 254–288 (1970)
2. Chung, Y., Takatsuka, M.: A causal model using self-organizing maps. In: Arik, S., Huang, T., Lai, W.K., Liu, Q. (eds.) ICONIP 2015. LNCS, vol. 9490, pp. 591–600. Springer, Cham (2015). doi:10.1007/978-3-319-26535-3_67
3. Kohonen, T.: Self-Organizing Maps. Information Sciences, 3rd edn. Springer, Heidelberg (2001). doi:10.1007/978-3-642-56927-2
4. Cha, S.H.: Comprehensive survey on distance/similarity measures between probability density functions. Int. J. Math. Models Methods Appl. Sci. **1**, 300–307 (2007)
5. Kullback, S.: Information Theory and Statistics. Courier Corporation, North Chelmsford (1997)
6. Rubner, Y., Tomasi, C., Guibas, L.J.: The earth mover's distance as a metric for image retrieval. Int. J. Comput. Vis. **40**(2), 99–121 (2000)
7. Niblack, C.W., Barber, R., Equitz, W., Flickner, M.D., Glasman, E.H., Petkovic, D., Yanker, P., Faloutsos, C., Taubin, G.: Qbic project: querying images by content, using color, texture, and shape. Proc. SPIE **1908**, 173–187 (1993)
8. Vesanto, J., Himberg, J., Alhoniemi, E., Parhankangas, J.: SOM Toolbox for Matlab 5. Technical report A57, Neural Networks Research Centre, Helsinki University of Technology (2000)
9. Vesanto, J.: SOM-based data visualization methods. Intell. Data Anal. **3**(2), 111–126 (1999)
10. Ward, M.O.: Multivariate data glyphs: principles and practice. In: Handbook of Data Visualization. Springer Handbooks Computational Statistics, pp. 179–198. Springer, Heidelberg (2008). doi:10.1007/978-3-540-33037-0_8
11. Giddings, E.M.P., Bell, A.H., Beaulieu, K.M., Cuffney, T.F., Coles, J.F., Brown, L.R., Fitzpatrick, F.A., Falcone, J., Sprague, L.A., Bryant, W.L., Peppler, M.C., Stephens, C., McMahon, G.: Selected physical, chemical, and biological data used to study urbanizing streams in nine metropolitan areas of the United States, 1999–2004. Technical report Data Series 423, National Water-Quality Assessment Program, U.S. Geological Survey (2009)
12. Novotny, V., Virani, H., Manolakos, E.: Self organizing feature maps combined with ecological ordination techniques for effective watershed management. Technical Report 4, Center for Urban Environmental Studies, Northeastern University, Boston (2005)
13. Wong, P.C., Bergeron, R.D.: 30 years of multidimensional multivariate visualization. In: Scientific Visualization, Overviews, Methodologies, and Techniques, pp. 3–33. IEEE Computer Society (1997)
14. Press, W.H., Teukolsky, S.A., Vetterling, W.T., Flannery, B.P.: Numerical Recipes in C: The Art of Scientific Computing, 2 edn. Cambridge University Press, Cambridge (1992)

Periodic Associated Sensor Patterns Mining from Wireless Sensor Networks

Md. Mamunur Rashid[1]([✉]), Joarder Kamruzzaman[1,2], Iqbal Gondal[1,2], and Rafiul Hassan[3]

[1] Faculty of Information Technology, Monash University, Clayton, Australia
{md.rashid,joarder.kamruzzaman,iqbal.gondal}@monash.edu
[2] ICSL, Federation University, Ballarat, Australia
[3] King Fahd University of Petroleum and Minerals, Dhahran, Saudi Arabia
mrhassan@kfupm.edu.sa

Abstract. Mining interesting knowledge from the massive amount of data gathered in wireless sensor networks is a challenging task. Works reported in literature all-confidence measure based associated sensor patterns can captures association-like co-occurrences and the strong temporal correlations implied by such co-occurrences in the sensor data. However, when the user given all-confidence threshold is low, a huge amount of patterns are generated and mining these patterns may not be space and time efficient. Temporal periodicity of pattern appearance can be regarded as an important criterion for measuring the interestingness of associated patterns in WSNs. Associated sensor patterns that occur after regular intervals is called periodic associated sensor patterns. Even though mining periodic associated sensor patterns from sensor data stream is extremely important in many real-time applications, no such algorithm has been proposed yet. In this paper, we propose a compact tree structure called Periodic Associated Sensor Pattern-tree (PASP-tree) and an efficient mining approach for finding periodic associated sensor patterns (PASPs) from WSNs. Extensive performance analyses show that our technique is time and memory efficient in finding periodic associated sensor patterns.

Keywords: Wireless sensor networks · Data mining · Periodicity · Knowledge discovery · Associated sensor pattern

1 Introduction

Recently data mining techniques have received a great deal of attention to extract interesting knowledge from WSN [1]. These techniques have shown to be a promising tool to improve WSN performance and quality of services (QoS) [2]. Loo et al. [3] and Romer [4] have focused on extracting pattern regarding the phenomenon monitored by the sensor nodes, in which the mining techniques are applied to the sensed data received from the sensor nodes and stored in a central database. Sensor-association rules was proposed in [5] where patterns are

© Springer International Publishing AG 2017
D. Liu et al. (Eds.): ICONIP 2017, Part V, LNCS 10638, pp. 247–255, 2017.
https://doi.org/10.1007/978-3-319-70139-4_25

extract regarding the sensor nodes rather than the area monitored by the WSN. An example of sensor association rules could be $(s_1, s_2 \rightarrow s_3, 85\%, \lambda)$ which means that if sensor s_1 and s_2 detect events within λ time interval, then there is 85% of chance that s_3 detects events within same time interval. However, these rules often generate a huge number of rules, most of which are non-informative or fail to reflect the true correlation among data objects. To resolve this issue Rashid et al. [6] proposed a new type of sensor behavioral pattern called associated sensor patterns that captures association-like co-occurrences and the strong temporal correlations implied by such co-occurrences in the sensor data.

Another important criterion for identifying the interestingness of associated patterns might be the shape of occurrence, i.e., whether they occur periodically, non-periodically, or mostly in specific time interval in the sensor database. An associated pattern that occurs after periodic intervals in WSNs called as periodic associated sensor patterns. Periodic associated sensor patterns can be used for predicting the source of future events. By knowing the source of future event, we can detect the faulty nodes easily from the network. For example, we are expecting to get event from a particular node, and it does not occur. It also may be used to identify the source of the next event in the case of emergency preparedness class of applications. Periodic associated sensor patterns also can identify a set of temporally correlated sensors. This knowledge can be helpful to overcome the undesirable effects (e.g., missed reading) of the unreliable wireless communications.

Traditional associated pattern mining methods fail to discover such periodic associated sensor patterns because they only focus on the high frequency pattern. In [7], Tanbeer et al. proposed an algorithm to mine periodic frequent pattern from transactional database. For periodicity calculation they used maximum period (*maxPrd*) of interval of the patterns. They mined only those patterns which are appearing regularly throughout the database. But in many real-world applications, it is difficult for the patterns to appear regularly without any interruptions. Therefore in erroneous or noisy environment, *maxPrd* measure for regularity calculation is not effective. Recently, Rashid et al. [8] have introduced a problem of discovering regularly frequent patterns that follow a temporal regularity in their occurrence characteristics from sensor data. They used variance of interval time between pattern occurrences in database instead of *maxPrd*. For regularly frequent patterns mining, they use a tree structure, called a RSP-tree (Regularly Sensor Pattern tree), which capture the database contents in a highly compact manner with one database scans. In this model, at first they mine the frequent patterns and then discover the regular patterns. Therefore, huge amount of candidate patterns are generated and mining these huge number of patterns may not efficient in real time. Even though mining periodic frequent pattern [7] and regularly frequent sensor patterns [8] are closely related to our work, but we cannot directly applied these for finding periodic associated sensor patterns from sensor data because these patterns do not consider the all-confidence threshold.

Motivated from the above demand, in this paper, we develop a single-pass tree structure, called the PASP-tree (periodic associated sensor patterns tree), that can capture important knowledge from the stream contents of sensor data in a very compact manner. Using FP-growth [9] like pattern-growth approach, PASP-tree can efficiently mine the associated sensor patterns in sensor stream data for user given min_all_conf, min_sup and max_var thresholds. To the best of our knowledge, PASP-tree is the first effort to mine periodic associated sensor patterns over sensor stream data. Extensive performance study shows that our proposed technique is very efficient in discovering periodic associated sensor patterns over sensor data stream.

2   PASPs Mining Problem in Wireless Sensor Networks

Let $S = \{s_1, s_2, ..., s_n\}$ be the set of sensors in a specific WSN. We assume that the time is divided into equal-sized slots $t = \{t_1, t_2, ..., t_q\}$ such that $t_{j+1} - t_j = \lambda, j \in [1, q-1]$ where λ is the size of each time slot. A set $P = \{s_1, s_2, ..., s_p\} \subseteq S$ is called a pattern of a sensors.

An epoch is a tuple $e(e_{ts}, Y)$ such that Y is a pattern of the event detecting sensors that report events within the same time slot and e_{ts} is the epoch's time slot. A sensor database SD shown in Table 1, is a set of epochs $E = \{e_1, e_2, ..., e_m\}$ with $m = |SD|$, i.e., total number of epochs in SD. If $X \subseteq Y$, it is said that X occurs in e and denoted as $e_j^X, j \in [1, m]$. Let $E^X = \{e_j^X, ..., e_k^X\}$, where $j \leq k$ and $j, k \in [1, m]$ be the ordered set of epochs in which pattern X has occurred in SD. Let e_s^X and e_t^X, where $j \leq s < t \leq k$ be the two consecutive epochs in E^X. The number of epochs or time difference between e_t^X and e_s^X, can be defined as a period of X, say p^X, i.e., $p^X = \{e_t^X - e_s^X\}$. Let $P^X = \{p_1^X, ..., p_s^X\}$ be the set of periods for patterns X. For simplicity in period computation, assume the first and last epochs in SD as *null* with $e_f = 0$ and $(e_l = e_m)$ respectively.

Table 1. A sensor database (SD)

TS	Epoch	TS	Epoch
1	$s_1 s_2 s_3 s_4 s_7 s_8$	4	$s_1 s_2 s_4 s_7$
2	$s_1 s_5 s_6$	5	$s_1 s_2 s_4 s_5$
3	$s_2 s_5 s_6 s_7 s_8$	6	$s_1 s_2 s_3 s_4 s_7$

Definition 1 *(Periodicity of Patten X):* For a given E^X, let P^X be the set of all periods of X, i.e., $P^X = \{p_1^X, p_2^X ..., p_n^X\}$, where n is the total number of periods in P^X. Then the average period value of pattern X is represented as, $\bar{p}^X = \sum_{k=1}^{N} \frac{p_k^X}{n}$ and its variance as $\sigma^X = \sum_{k=1}^{N} \frac{(p_k^X - \bar{p}^X)^2}{n}$. The periodicity of X can be denoted as $Per(X) = \sigma^X$ (variance of periods for pattern X).

The interestingness measure *all-confidence* of a pattern X is defined as follows:

$$all - confidence = \frac{Sup(X)}{Max\_sensor\_Sup(X)} \tag{1}$$

Definition 2 *(Associated Pattern):* A pattern is called an associated pattern, if its *all-confidence* is greater than or equal to the given minimum *all-confidence* threshold.

Definition 3 *(Periodic Associated sensor Pattern):* A pattern is called a periodic associated sensor pattern if it satisfies both of the following three conditions: (i) its all-confidence value is no less than a user-given min_all_conf (ii) its support is no less than a user-given min_sup and (iii) its periodicity is no greater than a user-given max_var.

Given a SD, min_sup, min_all_conf and max_var constraints, the objective is to discover the complete set of interesting patterns in SD having than support no less min_sup and min_all_conf and periodicity no more than max_var.

3   Proposed Tree Structure and Algorithm

In this section, we discuss the construction and mining process of the periodic associated sensor pattern tree (PASP-tree) for finding periodic associated sensor patterns. The PASP-tree construction has two phases: insertion phase and restructuring-compression phase. The step-by-step construction process of the PASP-tree based on the sensor database of Table 1 shown in Fig. 1(a–e). For the figure simplicity, we do not show the node traversal pointers in the tree. Similar to an FP-tree [9], each node in a PASP-tree represents a sensor set in the path from the root up to that node. An important feature of a PASP-tree is that, in the tree structure it maintains the appearance information for each epoch. To explicitly track such information, it keeps a list of TS (time-slot) information only at the last sensor-node for an epoch. Such a node is denoted as tail-node. Hence, a PASP-tree maintains two types of nodes; say ordinary node and tail node. The former are types of nodes used in FP-tree that do not maintain TS information. On the other hand, the latter type used in RSP-tree [8], can be defined as follows:

Definition 4 *(tail node):* Let $e = y_1, y_2, ..., y_n$ be an epoch that is sorted according to the SL-list order. If e is inserted into PASP-tree in this order, then the node of the tree that represents item y_n is defined as the tail-node for e and it explicitly maintains e's TS. Irrespective of the node type, no node in PASP-tree needs to maintain a support count value like FP-tree. Each node in the PASP-tree maintains parents, children, and node traversal pointers. So, the structures of an ordinary node and a tail node are given as follows: For ordinary node: M, where M is the item name of the node. For tail node: $M[e_1, e_2, ..., e_n]$, where M is the sensor name of the node and $e_i, i\epsilon[1, n]$, is an epoch TS in the TS-list, indicating that M is the tail-node for epoch e_i.

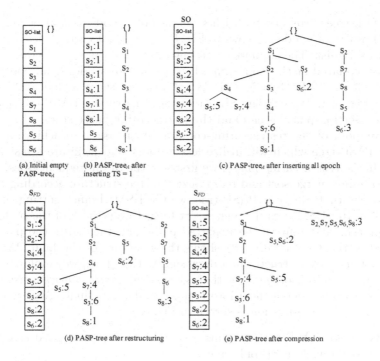

Fig. 1. PASP-tree construction

From above definitions and the PASP-tree node structure we can deduce the following lemma.

Lemma 1: A tail-node in an PASP-tree inherits an ordinary node; but not vice versa.

Proof: The structure of an ordinary node states that it exactly maintains three types of pointers: a parent pointer, a list of child pointers, and a node traversal pointer. A tail-node maintains all such information like an ordinary node. It also maintains the TS-list, which is additional information. Since the TS-list is not maintained in an ordinary node, so we can say, there is an ordinary node in every tail-node and in contrast, no tail-node in an ordinary node.

For the insertion phase, PASP-tee arranges the sensors according to sensors' appearance order in the database and is built by inserting every epoch in database one after another. At this stage we call it PASP-tree$_A$, which simply maintains a sensor order-list (SO-list). The SO-list includes each distinct sensor found in all epochs in database according to their appearance and contains support value of each item in the database. Initially the PASP-tree is empty and starts construction with *null* root node shown in Fig. 1(a). Using SD in Table 1 as an example, the first epoch (i.e., TS = 1) $\{s_1 s_2 s_3 s_4 s_7 s_8\}$ is inserted into the tree $< \{\} \rightarrow s_1 \rightarrow s_2 \rightarrow s_3 \rightarrow s_4 \rightarrow s_7 \rightarrow s_8 : 1 >$ as-it-is manner. Thus the first branch of the tree is constructed with s_1 as the initial node (just after root node) and $s_8 : 1$ as the tail node is shown in Fig. 1(b). Hence, it carries the TS (i.e., 1) epoch in its TS-list. The support count entries for sensors s_1, s_2, s_3, s_4, s_7 and s_8 are also updated at the same time. In this way, after adding all epochs (TS $= 2$, TS $= 3$, TS $= 4$, TS $= 5$, and TS $= 6$), we get the complete PASP-$tree_A$ shown in Fig. 1(c). We call the SL-list of the constructed PASP-$tree_A$ as SL. Here, the insertion phase is end and the restructuring phase starts.

The purpose of the restructuring-compression phase is to achieve a highly compact PASP-tree which will utilize less memory and facilitate a fast mining process. In the restructuring phase, we first sort the SL in frequency-descending order S_{FD} using merge sort and reorganize the tree structure according to S_{FD} order. For restructuring our PASP-tree, we use BSM (branch sorting method) proposed in [10]. BSM uses the merge sort to sort every path of the prefix tree. This approach, first remove the unsorted paths and then sorts the paths and reinserted to the tree. Figure 1(d) shows the structure of the final PASP-tree that we obtained by restructuring operation. At this stage, we employ a simple but effective compression process that selects the *same support sensor nodes* in each branch and merge them into a single node. The final PASP-tree, after restructuring and compression is shown in Fig. 1(e).

Property 1: An PASP-tree contains a complete set of associated sensor projection for each epoch in SD only once.

Now we describe the mining process of our proposed algorithm. Similar to the FP-growth [10] mining approach, we recursively mine the PASP-tree of decreasing size to generated regularly frequent patterns by creating conditional pattern-bases (PB) and corresponding conditional trees (CT) without additional database scan. Then generate the frequent pattern from the conditional tree. At last we check the regularity of generated frequent pattern to find regularly frequent sensor pattern. Before discussing these operations in details, we explore the following important property and lemma of an PASP-tree like RSP-tree [8].

Property 2: The TS-list in a PASP-tree maintains the occurrence information for all the nodes in the path (from that tail-node to the root) at least in the epochs of the list.

Lemma 2: Let $X = b_1, b_2, ..., b_n$ be a path in a PASP-tree where node bn is the tail node that carries the TS-list of the path. If the TS-list is pushed-up to node b_{n-1}, then the node b_{n-1} maintain the occurrence information of the path $X' = b_1, b_2, ..., b_{n-1}$ for the same set of epochs in TS-list without any loss.

Proof: Based on Property 2, the TS-list at node bn maintains the occurrence information of the path Z' at least in epochs it contains. So, the same TS-list at node b_{n-1} exactly maintains the same epoch information for Z' without any lose.

While constructing PB and CT for a sensor s the TS-lists of all tail-nodes of s are pushed-up to respective parent nodes in PASP-tree and PB respectively.

Since the mining operation is performed in bottom-up manner in S_{FD}, the TS push-up operation ensures the complete mining for the next item. For example database shown in Table 1, suppose the $min\_sup = 3$, $min\_all\_conf = 0.55$ and $max\_var = 1.0$. We start building the conditional pattern-base and conditional trees from the sensor at the bottom of the S_{FD} list (Fig. 1(e)). The three bottom sensors s_6, s_8 and s_3 do not satisfy the min_sup threshold. Therefore, at first the conditional pattern-base tree of s_5 is created by taking all the branches prefixing the sensor s_5 as shown in Fig. 2(a). Sensor s_5 creates branches $(s_1 s_2 s_4:5)$, $(s_2 s_7:3)$ and $(s_1:2)$ where number after ":" indicates each sub-pattern occurring TS. The conditional-tree of s_5 is empty, because s_1, s_2, s_4 and s_7 do not satisfy the given min_sup and min_all_conf thresholds. Now prefix and conditional-tree for s_7 is created in Fig. 2(b). Its conditional-tree contains two path $(s_1 s_2 s_4 : 1, 4, 6)$ and $(s_2 : 3)$ and the generated associated sensor patterns for s_7 are $s_1 s_7 : 1, 4, 6$, $s_2 s_7 : 1, 3, 4, 6$, $s_4 s_7 : 1, 4, 6$ and $s_1 s_4 s_7 : 1, 4, 6$. Then, we calculate the periodicity of $s_1 s_7$, $s_2 s_7$, $s_4 s_7$ and $s_1 s_4 s_7$ by Definition 1 and get their respective regularity values are 1.25, 0.56, 1.25 and 1.25. Since $Per(s_2 s_7) < 1.0$, then $s_2 s_7$ is the only periodic associated sensor pattern. Similar process is repeated for other sensors in the PASP-tree to find the complete set of periodic associated sensor pattern which are shown in Fig. 2(c–e). Table 2 shows the overall mining process of periodic associated pattern for our example sensor database in Table 1.

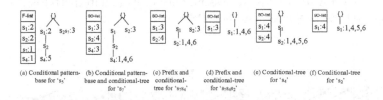

(a) Conditional pattern-base for 's_5' (b) Conditional pattern-base and conditional-tree for 's_7' (c) Prefix and conditional-tree for '$s_7 s_4$' (d) Prefix and conditional-tree for '$s_7 s_4 s_2$' (e) Conditional-tree for 's_4' (f) Conditional-tree for 's_2'

Fig. 2. Conditional pattern-base and conditional tree construction with the PASP-tree

Table 2. Mining the PASP-tree by creating conditional (sub-) pattern base

Sensor	Conditional pattern-base	Conditional-tree	ASP	PASP
s_5	$(s_1 s_2 s_4)$, (s_2, s_7), (s_1)	-	-	-
s_7	$(s_1 s_2 s_4)$, (s_2)	$<s_1, s_2>$, $<s_4>$	$s_1 s_7$, $s_2 s_7$, $s_4 s_7$, $s_1 s_4 s_7$, $s_2 s_4 s_7$, $s_1 s_2 s_4 s_7$	$s_2 s_7$
s_4	$(s_1 s_2)$	$<s_1 s_2>$	$s_1 s_4$, $s_2 s_4$, $s_1 s_2 s_4$	$s_1 s_4$, $s_2 s_4$ $s_1 s_2 s_4$
s_2	s_1	$<s_1>$	$s_1 s_2$	$s_1 s_2$

4   Experimental Results

In this section, we present the experimental results on periodic associated sensor patterns on PASP-tree. We compare its performance with the existing method [8]. Our programs are written in Microsoft Visual C++ and run with Windows 7 on a 2.66 GHz machine with 4 GB of main memory. To evaluate the performance of our proposed approach, we have performed experiments on IBM synthetic dataset (T10I4D100K) and real life dataset musroom from frequent itemset mining dataset repository [11]. Context and objects in these datasets are similar to the epochs and sensors in the terminology of this paper. We also used another dataset containing real WSN data from Intel Berkely Research Lab [12] which is widely used by many research community [5,6].

Here, we show the effectiveness of PASP-tree in mining periodic associated sensor pattern mining in terms of execution time. To analysis the execution time performance, experiments were conducted with a mining request for the given datasets by varying the min_sup and max_var values. In our first experiment, for PASP-tree we fixed $min\_all\_conf = 10\%$ and $max\_var = 0.4\%$ and varied min_sup values. For RSP-tree, we set $max\_var = 0.4\%$ and varied min_sup values. The x-axis in each graph shows the change of min_sup value in the form of percentage of database size and the y-axis indicates the overall execution time. The overall execution time is the total of the construction time, tree restructuring time (for PASP-tree only) and the mining time. Figure 3 shows that, when the min_sup value is increased the mining time for both trees decreased, however PASP-tree outperforms the RSP-tree in terms of overall execution time all of the case. Similar observation is found when we varied max_var which is shown in Fig. 4. Since, PASP-tree tested less number of candidate patterns; therefore, its overall runtime is always less than RSP-tree.

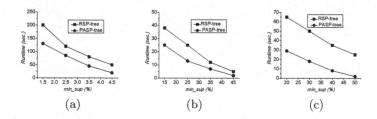

Fig. 3. Runtime comparison: PASP-tree v/s RSP-tree by varying min_sup.

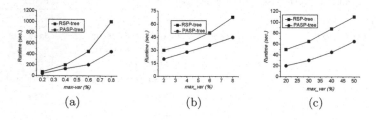

Fig. 4. Runtime comparison: PASP-tree v/s RSP-tree by varying max_var.

5   Conclusion

The key contribution of this paper is to provide an efficient method for mining periodic associated sensor patterns from WSNs data using a prefix tree called PASP-tree. We evaluated the performance of PASP-tree over diverse datasets. Extensive performance analyses show that PASP-tee is very efficient for periodic associated sensor pattern mining and significantly better than the existing method. The proposed technique is suitable for application in many other domains, such as mobile communications [13] and financial applications [14].

References

1. Mahmood, A., Shi, K., Khatoon, S., Xiao, M.: Data mining techniques for wireless sensor networks: a survey. Int. J. Distrib. Sens. Netw. **2013** (2013). doi:10.1155/2013/406316
2. Tan, P.-N.: Knowledge discovery from sensor data. Sensors **23**, 14–19 (2006)
3. Loo, K.K., Tong, I., Kao, B.: Online algorithms for mining inter-stream associations from large sensor networks. In: Ho, T.B., Cheung, D., Liu, H. (eds.) PAKDD 2005. LNCS, vol. 3518, pp. 143–149. Springer, Heidelberg (2005). doi:10.1007/11430919_18
4. Romer, K.: Distributed mining of spatio-temporal event patterns in sensor networks. In: EAWMS/DCOSS, pp. 103–116 (2006)
5. Boukerche, A., Samarah, S.A.: Novel algorithm for mining association rules in wireless ad-hoc sensor networks. IEEE Trans. P & D. Syst. 865–877 (2008)
6. Rashid, M.M., Gondal, I.: Mining associated patterns from wireless sensor networks. IEEE Trans. Comput. **45**, 638–651 (2016)
7. Tanbeer, S.K., Ahmed, C.F., Jeong, B.-S., Lee, Y.-K.: Discovering periodic-frequent patterns in transactional databases. In: Theeramunkong, T., Kijsirikul, B., Cercone, N., Ho, T.-B. (eds.) PAKDD 2009. LNCS (LNAI), vol. 5476, pp. 242–253. Springer, Heidelberg (2009). doi:10.1007/978-3-642-01307-2_24
8. Rashid, M.M., Gondal, I., Kamruzzaman, J.: Regularly frequent patterns mining from sensor data stream. In: Lee, M., Hirose, A., Hou, Z.-G., Kil, R.M. (eds.) ICONIP 2013. LNCS, vol. 8227, pp. 417–424. Springer, Heidelberg (2013). doi:10.1007/978-3-642-42042-9_52
9. Han, J., Pei, J., Yin, Y.: Mining frequent pattern without candidate generation. In: ACM SIGMOD International Conference on Management of Data, pp. 1–12 (2000)
10. Tanbeer, S.K., Ahmed, C.F., Jeong, B.-S.: Efficient single-pass frequent pattern mining using a prefix-tree. Inf. Sci. **179**(5), 559–583 (2009)
11. Frequent itemset mining repository. http://fimi.cs.helsinki.fi/data/. Accessed 20 Mar 2017
12. Intel Lab data. http://db.csail.mit.edu/labdata/labdata.html. Accessed 20 Mar 2017
13. Haider, A., Gondal, I., Kamruzzaman, J.: Dynamic dwell timer for hybrid vertical handover in 4G coupled networks. In: VTC Spring, pp. 1–5 (2011)
14. Hassan, R., Ramamohanarao, K., Kamruzzaman, J., Rahman, M., Hossain, M.: A HMM-based adaptive fuzzy inference system for stock market forecasting. Neurocomputing **104**, 10–25 (2013)

Online Multi-label Passive Aggressive Active Learning Algorithm Based on Binary Relevance

Xizhi Guo, Yongwei Zhang, and Jianhua Xu[✉]

School of Computer Science and Technology, Nanjing Normal University,
Nanjing 210023, Jiangsu, China
guoxizhi520@sina.com, zywei1030@gmail.com, xujianhua@njnu.edu.cn

Abstract. Online multi-label learning is an efficient classification paradigm in machine learning. However, traditional online multi-label methods often need requesting all class labels of each incoming sample, which is often human cost and time-consuming in labeling classification problem. In order to tackle these problems, in this paper, we present online multi-label passive aggressive active (MLPAA) learning algorithm by combining binary relevance (BR) decomposition strategy with online passive aggressive active (PAA) method. The proposed MLPAA algorithm not only uses the misclassified labels to update the classifier, but also exploits correctly classified examples with low prediction confidence. We perform extensive experimental comparison for our algorithm and the other methods using nine benchmark data sets. The encouraging results of our experiments validate the effectiveness of our proposed method.

Keywords: Online active learning · Multi-label classification · Passive aggressive · Binary relevance

1 Introduction

Online learning represents an important family of efficient and scalable machine learning algorithms for large-scale applications [1]. In the past ten years, a variety of online learning algorithms have been proposed, including first-order and second-order online learning algorithms, e.g., perceptron algorithm [2], passive-aggressive (PA) algorithm [3] and adaptive regularization of weights (AROW) [4], confidence-weighted (CW) [5], and so on. In many real-world applications especially for mining real-life data streams (e.g., spam email filtering), acquiring the true class labels from an oracle is often time-consuming due to the unavoidable interaction between the classifier and the environment. This has motivated the recent study of online active learning, which reduces time-consuming of acquiring the true class labels by avoiding requiring to class labels of every incoming samples in an online learning setting. Recently, online active learning is applied well in the binary classification, such as an efficient online active learning algorithm [6], perceptron-based active learning [7] and passive-aggressive active (PAA) learning [8].

© Springer International Publishing AG 2017
D. Liu et al. (Eds.): ICONIP 2017, Part V, LNCS 10638, pp. 256–266, 2017.
https://doi.org/10.1007/978-3-319-70139-4_26

All aforementioned algorithms could tackle traditional online single-label (binary [9] or multi-class [10, 11]) classification problem. Multi-label classification, an emerging topic in machine learning, has received increasing attention in recent years. However, a few online multi-label classification methods have been proposed [12–15]. In order to further improve multi-label classification effectiveness, active learning is also applied to online multi-label learning due to its efficiency in saving labeling cost by exploiting the redundancy in samples. For example, Hua et al. [16] and Sheng [17] have studied online multi-label active learning and have demonstrated the advantages of online active learning in their empirical evaluation.

In this paper, we present a new online multi-label passive aggressive active (MLPAA) learning algorithm based on binary relevance (BR) decomposition strategy [18]. The BR approach has a low computational cost. Additionally this strategy does not affect other predicted labels for some given sample, which makes it very easy to implement online multi-label classification. Moreover, Due to exploring the principle of passive-aggressive learning, MLPAA algorithm not only decides when the classifier should make a query appropriately, but also attempts to fully exploit the potential of every queried label in updating the classification model.

The rest of this paper is organized as follows: Sect. 2 reviews related work and Sect. 3 provides the proposed MLPAA algorithm. Sections 4 and 5 are associated with experimental results and conclusions, respectively.

2 Related Work

We review some existing methods on online multi-label learning in this section. Accelerated non-smooth stochastic gradient descent (ANSGD) [14] is presented to minimize the ranking loss function in the primal form using the accelerated non-smooth stochastic gradient descent. Moreover, a Bayesian online multi-label classification framework (BOMC) [15] is proposed to learn a probabilistic linear classifier. The likelihood is modeled by a graphical model similarly to $TrueSkill^{TM}$ [19], and inference is based on Gaussian density filtering with expectation propagation.

In online multi-label classification problem, active learning is also more and more concerned because of its advantages. For example, online multi-label active learning for large scale multimedia annotation [17] proposes a scalable framework for annotation-based video search as well as a novel approach to enable large-scale semantic concept annotation. Two dimensional active learning (2DAL) [16] proposes to select sample-label pairs, rather than only samples, to minimize a multi-label Bayesian classification error bound. This new active learning strategy considers not only the sample dimension but also the label dimension.

Traditional online multi-label methods often need to request all class labels of each incoming sample, which were often human cost and time-consuming in labeling classification problem. Fortunately, active learning is efficiency in saving labeling cost by exploiting the redundancy in samples. Although existing works mentioned above are used to solve online multi-label learning, only a few online multi-label active learning algorithms are studied. In order to tackle these problems, we propose a novel online MLPAA algorithm by combining BR strategy with online PAA method.

3   Online Multi-label Passive Aggressive Active Learning Algorithm

In this section, we present a new method for online multi-label active learning to reduce human and time costs, and to further improve the efficiency of the classification.

3.1   Multi-label Classification Setting

Without loss of generality, in online multi-label learning, let $\mathbb{R}^d$ be a d-dimensional input real space and $Q = \{1, 2, ..., q\}$ denote a finite set of class q labels and 2^Q all possible subsets of Q.

Assume a training data set to be $D = \{(\mathbf{x}_1, L_1), ..., (\mathbf{x}_t, L_t), ...\}$, where $\mathbf{x}_t \in \mathbb{R}^d$ and $L_t \in 2^Q$ represent the t-th sample and its relevant labels respectively. Moreover, the complement of L, i.e., $\bar{L}_t = Q/L_t$, is referred to as a set of irrelevant labels of $\mathbf{x}_t$. Additionally, the t-th sample is also annotated by a binary label vector $\mathbf{y}_t = [y_{t,1}, ..., y_{t,i}, ..., y_{t,q}]^T \in \{-1, +1\}^q$, where the i-th relevant label is $y_{t,i} = +1$ and the irrelevant label is $y_{t,i} = -1$.

3.2   Online Passive-Aggressive Active Learning (PAA) Method

In this sub-section, we mainly review the online PAA algorithm for binary case [8], where $\mathbf{x}_t$ is labeled by $y_t \in \{+1, -1\}$. The online PA algorithm is to minimize the cumulative loss for a certain prediction task from the sequentially arriving training samples and it updates the model $\mathbf{w}_{t+1}$ by solving three variants of the optimization task:

$$\min_{\mathbf{w}} F(\mathbf{w}) = \frac{1}{2}\|\mathbf{w} - \mathbf{w}_t\|^2 + C\ell_t(\mathbf{w}_t; (\mathbf{x}_t, y_t)), \tag{1}$$

where $C > 0$ is a penalty cost parameter, and the loss function becomes,

$$\ell_t(\mathbf{w}_t; (\mathbf{x}_t, y_t)) = \max(0, 1 - y_t(\mathbf{w}_t \cdot \mathbf{x}_t)). \tag{2}$$

We define the Lagrangian of the optimization problem in Eq. (1). The solutions to the optimization problem in Eq. (1) has a simple closed-form solution, $\mathbf{w}_{t+1} = \mathbf{w}_t + \tau_t y_t \mathbf{x}_t$, and the stepsize τ_t is calculated as,

$$\tau_t = \min(C, \ell_t(\mathbf{w}_t; (\mathbf{x}_t, y_t))/\|\mathbf{x}_t\|^2). \tag{3}$$

Based on PA, the PAA algorithm adopts a simple yet effective randomized rule to decide whether the label of an incoming instance should be queried or not, and then employs PA algorithm to exploit the labeled sample for updating the online classifier. In particular, for an incoming sample $\mathbf{x}_t$ at the t-th round, the PAA algorithm first computes its prediction margin, i.e., $p_t = \mathbf{w}_t \cdot \mathbf{x}_t$, by the current classifier, and then decides if the class label should be queried according to a Bernoulli random variable $Z_t \in \{0,1\}$ with probability being equal to $\rho/(\rho + |p_t|)$, where ρ indicates a smooth parameter. If the outcome is $Z_t = 0$, the class label will not be queried and the classifier is not updated; otherwise, the class label is queried and the outcome y_t is disclosed. Whenever the class label of an incoming sample is queried, the PAA algorithm will try the best effort to exploit the potential of this example for updating the classifier.

3.3   Online Multi-label Passive Aggressive Active Learning (MLPAA) Algorithm

In this sub-section, by combining BR approach with the former online PAA method, we proposed a new multi-label algorithm for online active learning.

In online multi-label classification, the current classifier predicts a label sub-set of sample $\mathbf{x}_t$, the loss function is calculated according to its true label subset, and finally the classifier is updated or not. However, there are two key challenges when handle multi-label classification by the method of online active learning algorithm: (i) for multi-label samples, how should a classifier deal with its weights of an incoming sample? and (ii) when the sample is queried and disclosed, how does the classifier exploit multiple labels of a sample? We proposed online MLPAA algorithm to tackle the above problems as follows. In particular, MLPAA algorithm first define a weighted matrix $\mathbf{W}_t = [\mathbf{w}_{t,1}, ..., \mathbf{w}_{t,i}, ..., \mathbf{w}_{t,q}]^{d \times q}$, where $\mathbf{W}_0 = \mathbf{0}$ and the weighted vector $\mathbf{w}_{t,i}$ corresponds to the i-th label. For an incoming sample $\mathbf{x}_t$ at the t-th round, like PAA learning algorithm, MLPAA first computes its prediction margin for each label, i.e., $P_{t,i} = \mathbf{w}_{t,i} \cdot \mathbf{x}_t (i = 1, ..., 1)$, by the current classifier $\mathbf{W}_t$.

In multi-label active learning setting, if an sample is selected, we provide all the labels for this sample once, which may not be better than interaction with human multiple times by querying different label for the same sample. Because there is no specific information to query the sample label, this method is not conducive to the classification of samples. Thus, in order to take full advantage of the characteristic of each sample, the MLPAA algorithm decides if each class label of sample should be queried according to a Bernoulli random variable $Z_t \in \{0, 1\}$ and its probability can be defined as,

$$\rho/(\rho + |P_{t,i}|). \tag{4}$$

And then the classifier can regard $\mathbf{x}_t$ as q independently single-label samples. Each single-label sample may have a class label, i.e., for i-th class label, its weighted vector can be updated by,

$$\mathbf{w}_{t+1,i} = \mathbf{w}_{t,i} + \tau_t y_t \mathbf{x}_t, \tag{5}$$

using decomposition strategy. Thus q linear binary classifiers for predicting $\mathbf{x}_t$ are defined as,

$$\hat{y}_{t,i} = \text{sgn}(f(\mathbf{x}_t, \mathbf{w}_{t,i})), \tag{6}$$

where sgn$(\cdot)$ is an indicator function. Thus, q linear classifiers are regarded as q PAA classifiers to decide whether weighted $\mathbf{w}_{t,i}$ should be updated or not. Finally, we summarize the detailed steps of the proposed online MLPAA algorithm in Algorithm 1.

Algorithm 1. Online multi-label passive-aggressive active learning algorithm

---

Require: Set two key parameters: C and ρ.
 1: Initialize the weighted matrix $\mathbf{W} = \mathbf{0}$;
 2: **for** $1, ..., T (T$ is the number of samples) **do**
 3:     **for** $1, ..., q$ **do**
 4:         sample $\mathbf{x}_t$, set $P_{t,i} = \mathbf{w}_{t,i} \cdot \mathbf{x}_t$;
 5:         Calculate a Bernoulli random variable $\mathbf{Z}_t \in \{0, 1\}$, according to $\rho/(\rho + |P_{t,i}|)$;
 6:         **if** $\mathbf{Z}_t = 1$ **then**
 7:             Calculate the loss function according to Eq.(2);
 8:             Update weights according to Eq.(3) and Eq.(5);
 9:         **end if**
10:     **end for**
11: **end for**

---

4   Experiments

We divide this section into three subsections, including an introduction of multi-label datasets, experimental setting and experimental analysis.

4.1   Datasets

To compare our algorithm with the aforementioned online multi-label classification methods, we conduct extensive experiments on a variety of benchmark datasets from web multi-label classification datasets. Table 1 shows the details of multi-label datasets used in our experiments such as the number of samples in training and test sets, dimensions of features, the number of labels, and average labels. All these datasets can be downloaded from Mulan datasets website[1]. These datasets are chosen fairly in order to cover various sizes of datasets, which include four different domains: image, text, video and biology, to validate the generalization ability of the proposed algorithm.

4.2   Experimental Setting

To show the effectiveness of MLPAA, we compare our algorithm with MLPEA [20], MLRPE [21], MLRPA [8], in which the performance of different algorithms is evaluated by five evaluation metrics [22], including Hamming loss, average

---

[1]   http://mulan.sourceforge.net/datasets-mlc.html.

Table 1. Statistics for nine benchmark multi-label datasets.

Dataset	Domain	Train	Test	Features	Classes	Average labels
Enron	Text	1123	579	1001	53	3.38
Scene	Image	1211	1196	294	6	1.07
Yeast	Biology	1500	917	103	14	4.24
Rcv1v2	Text	3000	3000	47236	101	2.88
Bibtex	Text	4480	2515	1836	159	2.40
Corel5k	Image	4500	500	499	174	3.52
Tmc2007	Text	21519	7077	49060	22	2.16
Tmc2007-500	Text	21519	7077	500	22	2.16
Mediamill	Video	30993	19914	120	101	4.38

precision, coverage, subset accuracy and ranking loss. Moreover, for the ranking loss, coverage and Hamming loss, the smaller value indicates the better performance, and for the other two measures, the bigger value the better performance. All the compared algorithms learn q linear classifiers for the multi-label classification tasks. In our experiments, we choose the penalty parameter $C = 1$ [3] for all algorithms except random algorithms. The smoothing parameter is set as $2^{[-10,0,10]}$ in order to examine varied sampling situations. The way of online is used on the training, and offline is used on the testing. All the experiments were executed about 10 runs of different random permutations for each dataset. The results were reported by averaging over these 10 runs.

4.3 Experimental Analysis

In this sub-section, we compare our online MLPAA with MLRPA, MLRPE, and MLPEA in testing data sets. Due the space limitation, we only describe the comparison of results of two metrics (e.g., Hamming loss and ranking loss) on the nine datasets. Figure 1 shows the Hamming loss of the four different algorithms for online multi-label active learning. We can observe some important results in Fig. 1. First of all, we find out that the online multi-label active learning algorithms outperform their corresponding random multi-label learning algorithms, which demonstrates the good performance and efficiency of the multi-label active learning strategies. MLPAA can exploit those requested labeled data, especially those samples that are correctly classified but with low confidence in multi-label classification problem. Finally, with the continuous increase of query proportion, we also observe that two Hamming losses of our MLPAA and MLPEA algorithms have not much change. Especially, our MLPAA has been in a straight line on large datasets almost (e.g., Bibtex, Tmc2007, Tmc2007-500, and Mediamill).

Additionally, to further examine the online predictive performance, Fig. 2 summarizes the ranking loss performance of the four diverse algorithms for online multi-label learning. With the increase of query proportion, we find that our

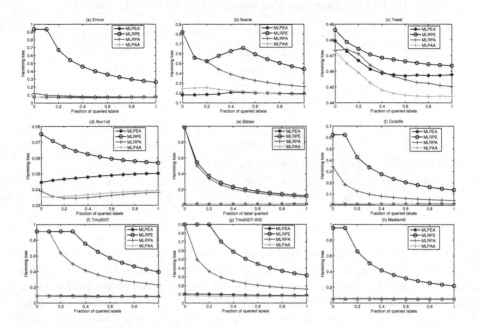

Fig. 1. The Hamming loss of four different algorithms on nine datasets.

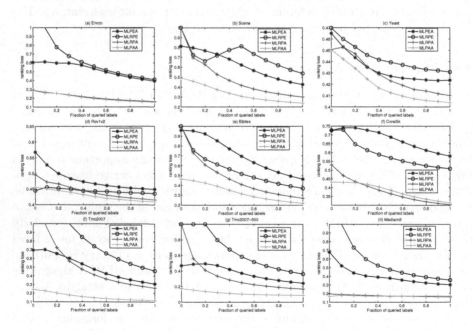

Fig. 2. The ranking loss of four different algorithms on nine datasets.

MLPAA algorithm is better than other algorithms except the Corel5k dataset, especially, in text datasets (e.g., Tmc2007, Bibtex, and Enron). Besides, we can observe that our method outperforms the MLPEA method in all nine datasets. Finally, we also find that the MLPEA algorithm is not better than random version algorithm in some datasets (e.g., Bibtex and Rcv1v2).

To further evaluate the online MLPAA algorithm, Tables 2, 3, 4, 5 and 6 show detailed metrics of four different algorithms for online multi-label active learning on nine datasets. We adjust ρ to make the percent of queried samples about near 20% and compare all the algorithms on a fair platform. The bold elements indicate the best performance in Tables 2, 3, 4, 5 and 6.

Table 2. Hamming loss (%) of MLPAA and other algorithms

DataSet	MLPAA	MLRPA	MLPEA	MLRPE
Enron	**5.90 ± 0.20**	6.60 ± 0.20	7.30 ± 0.30	10.10 ± 0.10
Scene	**15.63 ± 1.90**	19.45 ± 4.50	19.78 ± 4.50	19.71 ± 4.80
Yeast	**42.89 ± 0.50**	44.67 ± 0.90	44.72 ± 0.70	46.08 ± 0.80
Rcv1v2	**4.00 ± 0.00**	4.70 ± 0.00	4.90 ± 0.10	5.70 ± 0.20
Bibtex	**1.40 ± 0.00**	1.60 ± 0.10	1.70 ± 0.10	2.70 ± 0.20
Corel5k	**1.00 ± 0.00**	1.10 ± 0.00	1.20 ± 0.20	1.30 ± 0.00
Tmc2007	**7.30 ± 0.20**	7.90 ± 0.40	7.70 ± 0.40	9.10 ± 0.70
Tmc2007-500	**7.10 ± 0.30**	8.40 ± 0.40	7.30 ± 0.20	9.20 ± 0.50
Mediamill	**4.60 ± 0.90**	5.40 ± 1.20	4.60 ± 0.80	5.60 ± 1.30

Table 3. Subset accuracy (%) of MLPAA and other algorithms

DataSet	MLPAA	MLRPA	MLPEA	MLRPE
Enron	**9.30 ± 1.10**	6.80 ± 1.00	3.90 ± 1.70	1.10 ± 0.90
Scene	**30.03 ± 5.20**	23.89 ± 7.00	22.82 ± 6.60	19.12 ± 7.50
Yeast	**0.60 ± 0.80**	0.00 ± 0.00	0.10 ± 0.10	0.00 ± 0.00
Rcv1v2	**5.07 ± 0.00**	3.70 ± 0.00	4.90 ± 0.10	5.70 ± 0.20
Bibtex	**12.34 ± 0.40**	10.12 ± 0.70	9.50 ± 1.40	3.60 ± 1.50
Corel5k	**0.40 ± 0.20**	0.20 ± 0.10	0.10 ± 0.10	0.10 ± 0.10
Tmc2007	**20.89 ± 1.00**	17.84 ± 1.10	18.62 ± 1.50	13.41 ± 1.80
Tmc2007-500	**22.56 ± 1.20**	17.24 ± 1.50	21.81 ± 1.10	14.78 ± 1.50
Mediamill	**1.40 ± 2.00**	0.50 ± 1.00	1.40 ± 1.80	0.40 ± 0.60

In Table 2 on Hamming loss, our MLPAA works the best on all datasets, comparing with other three algorithms, and the two online multi-label active

Table 4. Ranking loss (%) of MLPAA and other algorithms

DataSet	MLPAA	MLRPA	MLPEA	MLRPE
Enron	**11.82 ± 0.50**	13.23 ± 0.40	13.24 ± 0.40	26.89 ± 3.60
Scene	**17.14 ± 3.20**	22.03 ± 4.50	23.89 ± 6.20	23.93 ± 6.20
Yeast	**38.56 ± 0.70**	40.62 ± 1.60	41.15 ± 1.10	42.68 ± 1.10
Rcv1v2	**40.04 ± 1.70**	41.43 ± 1.80	44.40 ± 2.20	43.73 ± 2.40
Bibtex	**12.47 ± 0.40**	15.52 ± 0.70	21.93 ± 0.70	21.89 ± 0.70
Corel5k	31.71 ± 1.70	**28.23 ± 0.90**	57.89 ± 2.90	47.33 ± 1.50
Tmc2007	**8.40 ± 0.40**	9.20 ± 0.50	10.53 ± 0.90	12.04 ± 1.20
Tmc2007-500	**6.30 ± 0.50**	7.90 ± 0.90	9.40 ± 1.50	10.91 ± 1.40
Mediamill	17.13 ± 3.80	**15.81 ± 3.90**	24.83 ± 4.90	24.82 ± 4.90

Table 5. Average precision (%) of MLPAA and other algorithms

DataSet	MLPAA	MLRPA	MLPEA	MLRPE
Enron	**63.67 ± 1.10**	60.04 ± 1.40	50.41 ± 2.80	46.07 ± 4.70
Scene	**74.45 ± 5.20**	51.46 ± 1.80	50.72 ± 1.30	50.72 ± 1.30
Yeast	**51.42 ± 1.20**	40.57 ± 1.60	41.16 ± 1.10	42.81 ± 1.10
Rcv1v2	**24.22 ± 0.00**	15.32 ± 0.30	15.34 ± 0.30	13.26 ± 0.90
Bibtex	**42.44 ± 0.40**	40.63 ± 1.20	37.54 ± 1.20	33.02 ± 1.00
Corel5k	**20.45 ± 1.20**	19.75 ± 1.00	11.63 ± 1.20	11.86 ± 1.10
Tmc2007	**72.28 ± 0.70**	70.14 ± 0.90	69.72 ± 1.20	66.13 ± 1.70
Tmc2007-500	**82.58 ± 1.20**	74.43 ± 1.80	76.28 ± 1.70	71.52 ± 2.10
Mediamill	**51.69 ± 7.10**	46.59 ± 8.70	46.46 ± 6.90	43.65 ± 7.40

Table 6. Coverage (%) of MLPAA and other algorithms

DataSet	MLPAA	MLRPA	MLPEA	MLRPE
Enron	**30.11 ± 1.00**	31.67 ± 0.80	57.43 ± 3.90	53.90 ± 4.70
Scene	**16.12 ± 2.50**	16.09 ± 3.70	21.74 ± 5.20	20.87 ± 2.20
Yeast	**68.07 ± 1.20**	70.53 ± 1.80	70.52 ± 1.30	72.27 ± 1.20
Rcv1v2	**61.03 ± 2.00**	62.34 ± 1.90	66.07 ± 2.40	66.12 ± 2.50
Bibtex	**26.12 ± 0.80**	26.36 ± 1.10	35.37 ± 1.10	36.07 ± 1.20
Corel5k	13.73 ± 4.70	**12.56 ± 2.30**	18.33 ± 4.10	16.50 ± 3.30
Tmc2007	**16.04 ± 0.60**	16.84 ± 0.60	19.22 ± 1.40	20.92 ± 1.80
Tmc2007-500	**15.54 ± 0.70**	17.74 ± 1.20	20.73 ± 2.30	22.16 ± 2.00
Mediamill	44.23 ± 6.60	**41.44 ± 6.40**	59.10 ± 6.80	51.92 ± 3.90

learning algorithms outperform their random version in terms of Hamming loss results, respectively. From Table 3 for subset accuracy, our MLPAA performs the best on four datasets. In Table 4 for average precision, all the best ranking loss values are from our MLPAA in nine datasets. According to Table 5, our MLPAA works the best on nine datasets but Mediamill and Corel5k. Finally, in Table 6 for coverage, except two datasets such as Corel5k and Mediamill, our MLPAA algorithm is also the best on other seven datasets.

5   Conclusions

In this paper, we investigate online multi-label active learning techniques for machine learning and mining data stream. We present a new online MLPAA learning algorithm for multi-label classification. The proposed algorithm is based on the decomposition strategy, which means that the MLPAA algorithm can be decomposed into PAA algorithm via using BR. The proposed algorithm could also overcome the drawback of the perceptron-based active learning algorithm that could waste queried labeled samples which are correctly classified but with low prediction confidence. The experimental results on nine datasets illustrate that our proposed method works better than other algorithms. For future work, we will take the label correlations and unbalance into account in online multi-label classification problem.

Acknowledgement. This work was supported by the Natural Science Foundation of China (NSFC) under Grant 61273246.

References

1. Hoi, S.C.H., Wang, J.L., Zhao, P.: LIBOL: a library for online learning algorithms. J. Mach. Learn. Res. **15**, 495–499 (2014)
2. Rosenblatt, F.: The perceptron: a probabilistic model for information storage and organization in the brain. Psychol. Rev. **65**(1), 386–408 (1958)
3. Crammer, K., Dekel, O., Keshet, J., et al.: Online passive-aggressive algorithms. J. Mach. Learn. Res. **7**, 551–585 (2006)
4. Crammer, K., Kulesza, A., Dredze, M.: Adaptive regularization of weight vectors. Mach. Learn. **91**(2), 155–187 (2013)
5. Crammer, K., Dredze, M., Pereira, F.: Exact convex confidence-weighted learning. In: Proceedings of the NIPS, pp. 345–352 (2008)
6. Liu, D., Zhang, P., Zheng, Q.: An efficient online active learning algorithm for binary classification. Pattern Recogn. Lett. **68**(1), 22–26 (2015)
7. Dasgupta, S., Kalai, A., Monteleoni, C.: Analysis of perceptron-based active learning. J. Mach. Learn. Res. **10**, 281–299 (2009)
8. Lu, J., Zhao, P., Hoi, S.C.H.: Online passive aggressive active learning and its applications. In: Proceedings of the ACML, pp. 266–282 (2014)
9. Zhao, P., Hoi, S.C.H., Jin, R.: Double updating online learning. J. Mach. Learn. Res. **12**, 1587–1615 (2011)
10. Crammer, K., Dredze, M., Kulesza, A.: Multi-class confidence weighted algorithms. In: Proceedings of the EMNLP, pp. 496–504 (2009)

11. Fink, M., Shwartz, S.S.; Singer, Y., et al.: Online multiclass learning by interclass hypothesis sharing. In: Proceedings of the ICML, pp. 313–320 (2006)
12. Crammer, K., Singer, Y.: A family of additive online algortihtms for category ranking. J. Mach. Learn. Res. **3**, 1025–1058 (2003)
13. Higuchi, D., Ozawa, S.: A neural network model for online multi-task multi-label pattern recognition. In: Mladenov, V., Koprinkova-Hristova, P., Palm, G., Villa, A.E.P., Appollini, B., Kasabov, N. (eds.) ICANN 2013. LNCS, vol. 8131, pp. 162–169. Springer, Heidelberg (2013). doi:10.1007/978-3-642-40728-4_21
14. Park, S., Choi, S.: Online multi-label learning with accelerated nonsmooth stochastic gradient descent. In: ICASSP, pp. 3322–3326 (2013)
15. Zhang, X., Graepel, T., Herbrich, R.: Bayesian online learning for multi-label and multi-variate performance measures. In: Proceedings of the AISTATS, pp. 956–963 (2013)
16. Hua, X.S., et al.: Two-dimensional multi-label active learning with an efficient online adaptation model for image classification. IEEE Trans. Pattern Anal. Mach. Intell. **31**(10), 1880–1897 (2009)
17. Hua, K., Sheng, X., Qi, G.: Online multi-label active learning for large-scale multimedia annotation. Technical report from Microsoft Research (2008). https://www.microsoft.com/en-us/research/wp-content/uploads/2008/06/tr-2008-103.pdf
18. Gibaja, E., Ventura, S.: A tutorial on multi-label learning. ACM Comput. Surv. **47**(3), 1–38 (2015). Article No. 51
19. Herbrich, R., Minka, T., Graepel, T.: TrueskillTM: a Bayesian skill ranking system. In: Proceedings of the NIPS, pp. 569–576 (2006)
20. Cesa-Bianchi, N., Gentile, C., Zaniboni, L.: Worst-case analysis of selective sampling for linear classification. J. Mach. Learn. Res. **7**, 1205–1230 (2006)
21. Cesa-Bianchi, N., Lugosi, G.: Prediction, Learning, and Games. Cambridge University Press, Cambridge (2006)
22. Zhang, M.L., Zhou, Z.H.: A review on multi-label learning algorithms. IEEE Trans. Knowl. Data Eng. **26**(8), 1819–1837 (2014)

Predicting Taxi Passenger Demands Based on the Temporal and Spatial Information

Sang Ho Kang, Han Bin Bae, Rhee Man Kil$^{(\boxtimes)}$, and Hee Yong Youn

College of Software, Sungkyunkwan Univesity,
2066, Seobu-ro, Jangan-gu, Suwo-si, Gyeonggi-do 16419, Korea
{sh21kang,rmkil,youn7147}@skku.edu, bhb0722@nate.com

Abstract. This paper presents a new method of predicting taxi passenger demands in the central city areas of Seoul and New York based on the temporal and spatial information on predicted values. For the efficiency of the city's taxi system, investigating the taxi passenger demands is required mainly in the large scaled cities. From this context, this paper proposes a prediction model of combining the conditional transition distribution and the neighboring information on taxi passenger demands. As a result, the proposed method provides higher prediction performances than other methods of homogeneous prediction models.

Keywords: Taxi passenger demands · Poisson process · Co-occurrence matrix · Temporal and spatial information

1 Introduction

Taxis play an important role for mass transportation which provides comfortable and direct services for customers. From this context, the prediction of taxi passenger demands provides the valuable insight of balancing the supply and demand of mass transportation to city planners. According to the Seoul Institute of Technology [1], Seoul Taxi's load factor is 39.3%, far below the 50% offered by the Ministry of Land, Transport and Tourism. Load factor is the ratio of time spent driving the passenger during the total operation time. It means that the taxi driver who started the business is alone without a guest for more than 6 h in 10 h. This implies that the taxi in Seoul is overpowered. On the other hand, there are many customers whose taxi requests are not satisfied during the rush hour or in the midnight. To resolve this problem, a taxi should be placed to balance supply and demand. From this point of view, predicting taxi passenger demands becomes an important issue.

For this issue, the vehicular traffic flow was analyzed using an adaptive linear model referred as the ARIMA process [2]. The taxi pickup patterns was analyzed by the k means clustering method [3] and the Poisson rates of taxi passenger demands were also investigated and used for the prediction model [4]. In these

© Springer International Publishing AG 2017
D. Liu et al. (Eds.): ICONIP 2017, Part V, LNCS 10638, pp. 267–274, 2017.
https://doi.org/10.1007/978-3-319-70139-4_27

prediction methods, the taxi flow for various features including the temporal information was investigated.

In this work, the goal of the prediction model is to find the demand patten of passengers over time. Once an appropriate predictive model is established, there are many benefits. First, if the taxi is properly placed according to the predicted value, passengers no longer have to wait a long time. Secondly, it is possible to reduce the amount of time a taxi runs without passengers. Thirdly, taxi drivers no longer have to rely on experience to pick up more passengers. Appropriate predictive models will lead to better trust and benefit. From this context, this paper proposes a prediction model of combining not only the temporal but also the spatial information. In this model, the conditional transition distribution is used to predict the taxi passenger demands in the next period of time as the temporal information and these predicted values in the neighborhood is combined to improve the prediction accuracy of taxi passenger demands. As a result, the proposed method provides higher prediction performances than other methods of homogeneous prediction models.

This paper is organized as follows: in Sect. 2, a method of locally predicted taxi passenger demands using the co-occurrence matrix is described, Sect. 3 presents a method of combing locally predicted taxi passenger demands, Sect. 4 shows simulation results for predicting taxi passenger demands in the central city areas of Seoul and New York, and finally, Sect. 5 presents the conclusion.

2   Predicting Taxi Passenger Demands with Co-occurrence Matrices

In the taxi traffic flow, taxi arrivals can be considered as independent random variables. Then, the taxi arrivals during a certain period of time can be interpreted as a Poisson process with a rate λ [5]; that is, the number of taxi arrivals $N(T)$ during a certain period of time T follows a Poisson distribution with a parameter λT. In this case, the probability mass function (PMF) of $N(t)$ is given by

$$P\{N(T) = k\} = e^{-\lambda T} \frac{(\lambda T)^k}{k!}, \tag{1}$$

where k represents the number of taxi arrivals during the time T.

However, the taxi arrivals dependent on external environments such as the day of the week, time of the day, weather condition, etc. In this respect, the transition of $N(T)$ is investigated for the prediction of taxi passenger demands. In general, the number of taxi arrivals $N(T)$ is greatly influenced by the previous time of $N(T)$ such as the hidden Markov model. From this point of view, a random variable $X(n)$ at the time index n is defined by the number of taxi arrivals during $(n-1)T$ and nT. Then, the transition of these random variables can be represented by the joint PMF of $X(n)$ and $X(n+1)$ which can be

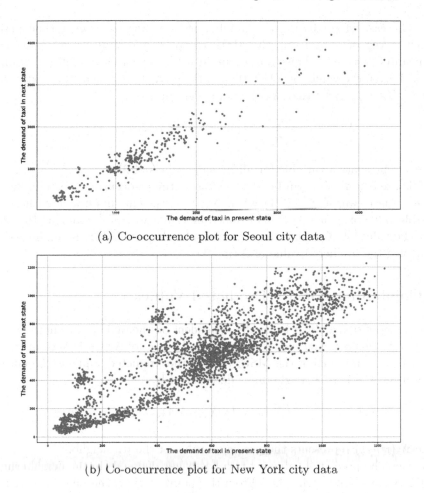

(a) Co-occurrence plot for Seoul city data

(b) Co-occurrence plot for New York city data

Fig. 1. Co-occurrence plot between the present and next states

illustrated as the co-occurrence matrix of the values of these two random variables. As an example of co-occurrence matrix, the co-occurrence plots for taxi passenger demands in Seoul and New York are illustrated in Fig. 1. From these co-occurrence plots, the corresponding co-occurrence matrices are made using the properly determined bin size. In our approach, the bin size is determined in such a way of maximizing the prediction performance.

For the prediction of a random variable X, the predicted value c in such a way of minimizing the mean square error (MSE) is determined as the mean of a random variable; that is,

$$\mathcal{E} = E[(X - c)^2] \tag{2}$$

is minimized when

$$\frac{d\mathcal{E}}{dc}\big|_{c=\hat{c}} = -2E[X] + 2\hat{c} = 0. \tag{3}$$

In the case of predicting the number of taxi arrivals of next time period in such a way of minimizing the MSE, the predicted value is determined by the mean of $X(n+1)$. This can be obtained by the conditional distribution of $X(n+1)$ given the values of $X(n)$; that is, the predicted value at the next period of time $\hat{X}(n+1)$ given that $X(n) = x_0$ is determined by

$$\hat{X}(n+1) = \sum_x x P\{X(n+1) = x | X(n) = x_0\}, \tag{4}$$

where x represents a variable for all possible values of $X(n+1)$.

This predicted value can be obtained using the co-occurrence matrix by calculating the mean of $X(n+1)$ when $X(n) = x_0$. This implies that the most probable value of $X(n+1)$ is determined in such a way of minimizing the MSE when the value of $X(n)$ is given. As a result, the total MSE is minimized by calculating the predicted value of $\hat{X}(n+1)$.

3   Combining Locally Predicted Values

The number of taxi arrivals is usually dependent on the taxi arrivals in the neighborhood. In this context, the locally predicted values of $X(n+1)$ is combined in such a way of minimizing the MSE; that is, the predicted value at the current position $\tilde{X}_0(n+1)$ of k neighbors is determined by

$$\tilde{X}_0(n+1) = f\left(\hat{X}_0(n+1), \hat{X}_1(n+1), \cdots, \hat{X}_k(n+1)\right), \tag{5}$$

where $\hat{X}_i(n+1)$ represents the predicted value of the ith neighbor.

In fact, the predicted values of taxi passenger demands in the neighborhood are highly correlated each other. From this point of view, the functional form f is determined as a linear function; that is, $\tilde{X}_0(n+1)$ is determined by

$$\tilde{X}_0(n+1) = \sum_{i=0}^{k} w_i \hat{X}_i(n+1) + w_b, \tag{6}$$

where w_i and w_b represent the ith weight associated $\hat{X}(n+1)$ and the bias term, respectively.

Then, the weight vector $\mathbf{w} = [w_b, w_0, \cdots, w_k]^t$ is determined in such a way of minimizing the MSE for l data described by

$$\mathcal{E} = \frac{1}{l} \sum_{i=1}^{l} \left(X_0(i) - \hat{X}(i)\right)^2, \tag{7}$$

where $X_0(i)$ and $\hat{X}_0(i)$ represent the actual and predicted values at the current position, respectively.

Let the ith input vector be

$$\hat{\mathbf{x}}(i) = [1, \hat{X}_0(i), \cdots, \hat{X}_k(i)]^t.$$

Then, the optimal weight vector $\mathbf{w}^*$ is determined by

$$\mathbf{w}^* = R^{-1}P, \tag{8}$$

where

$$R = \frac{1}{l}\sum_{i=1}^{l}\hat{\mathbf{x}}(i)\hat{\mathbf{x}}(i)^t \quad \text{and} \tag{9}$$

$$P = \frac{1}{l}\sum_{i=1}^{l}\hat{\mathbf{x}}(i)X_0(i). \tag{10}$$

The whole procedure of predicting taxi passenger demands using the temporal and spatial information is described as follows:

Predicting the taxi passenger demands by combining the predicted values using co-occurrence matrices

Step 1. From the taxi passenger demands data, construct the co-occurrence matrix of $X(n)$ versus $X(n+1)$.

Step 2. For the given data of $X(n)$, determine the predicted values of $\hat{X}(n+1)$ using (4).

Step 3. Determine the weight vector of (8) for the predicted values.

Step 4. After the construction of the prediction model, the prediction of the taxi passenger demand in the next period of time is determined as follows:

4.1. For the current taxi passenger demand $X(n) = x_0$, determine $\hat{X}(n+1)$ using (4).

4.2. Combine the predicted values in the neighborhood using (6).

4.3. Then, the predicted value is determined by $\tilde{X}_0(n+1)$.

This step is repeated for the given data of prediction problem.

The proposed algorithm improves the prediction performance using the predicted values in the neighborhood. This is possible because the taxi traffic flow tends to have a high correlation (linear correlation) with the taxi traffic flow in the neighborhood. From this context, the linear model is used to combine the predicted values in the neighborhood. As a result, the accuracy of predicted values is improved.

4   Simulation

For the prediction problems of taxi passenger demands, two data sets from two cities of Seoul and New York were collected. First, Seoul taxi data set was collected from Seoul Information Communication Plaza [1]. The data set was divided into 150 m intervals for the roads in Seoul and Gyeonggi province, and the number of ridings was calculated for every 30 min. These data included the date, time, and location information. For our problem of predicting taxi passenger demands, the portion of Seoul between latitudes 37.49° to 37.63° and longitudes 126.87° to 127.07° was considered and split into a square of 0.01° × 0.01° and assigned a zone number for each. In this data set, the data from December 1, 2014 to November 30, 2015 were used as the training data whereas the data from December 1, 2015 to November 30, 2016 were used as the test data. In the case of New York city, the data set was collected from the NYC site [6]. The data set covered the latitudes 40.70° through 40.84° and longitudes −74.01° through −73.94° of New York city and this area was divided into several zones of 0.01° × 0.01° squares. The number of taxi ridings were recorded every hour. In this data set, the first 70% of data (January 1 through March 28) and the remaining 30% of data (March 29 through April 30) were used as the training and testing data, respectively.

For the prediction of these data sets, the co-occurrence matrices (COC) with the bin sizes of 21 and 29 for the cities of Seoul and New York were made, respectively. These bins were determined using the grid search for the optimal performance of prediction. Then, the predicted values were determined by (4) and these predicted values were combined by (6) for 8 neighbors.

For the comparison of the proposed method of combining the temporal and spatial (CTS) information of predicted values, simulation for predicting taxi passenger demands using the mean of a Poisson process (PP) and the support vector regression (SVR) with Gaussian kernel functions [7] were also made. In the case of the mean of a Poisson process, the mean value for each week day and time was calculated. In the case of SVR, the number of taxi passenger demands were trained for the feature of week day and time. In this training, the proper learning parameters of kernel width γ and control parameter C were determined using the grid search. As a result, $C = 10^{-1}$, $\gamma = 2^{-1}$ were selected. To evaluate the prediction performances of these methods, the following root mean square error (RMSE) and the coefficient of determination R^2 were used:

$$RMSE = \left(\frac{1}{l} \sum_{l=1}^{l} (y_i - \hat{y})^2 \right)^{1/2}, \tag{11}$$

where y_i and $\hat{y}_i$ represent the true and predicted values of taxi passenger demands, respectively.

$$R^2 = 1 - \frac{\sum_{l=1}^{l} (y_i - \hat{y}_i)^2}{\sum_{l=1}^{l} (y_i - \bar{y})^2}, \tag{12}$$

where $\bar{y}$ represents the sample mean of true values y_i, $i = 1, \cdots, l$.

The simulation results using the PP, SVR, COC, and CTS methods in the cases of Seoul and New York cities were illustrated in Tables 1 and 2, respectively.

These simulation results have shown that (1) the combination of the temporally and spatially predicted values provides the best prediction performances in both Seoul city and New York city data, and (2) other than the CTS method, the COC method provides the better performance than other methods in Seoul city data whereas the PP method provides the better performances than other methods. These results imply that the PP and COC methods provides the relatively good performances for this problem of predicting taxi passenger demands compared with the complexity of learning in the SVR method. Furthermore, the combination of the temporal and spatial information; that is, combining the predicted values in the neighborhood is quite effective in this problem. This is mainly due to the fact that the taxi traffic flow is very much correlated in the neighborhood areas.

Table 1. Simulation results for predicting taxi passenger demands in Seoul.

Evaluation measure	PP	SVR	COC	CTS
R^2	0.9098	0.8731	0.9515	**0.9572**
RMSE	532.57	631.91	390.64	**366.69**

Table 2. Simulation results for predicting taxi passenger demands in New York.

Evaluation measure	PP	SVR	COC	CTS
R^2	0.9581	0.9571	0.9469	**0.9662**
RMSE	64.38	65.13	72.47	**57.75**

5   Conclusion

A new method of predicting taxi passenger demands is proposed based on the temporal and spatial information on predicted values. For the efficiency of the city's taxi system, investigating the taxi passenger demands is required mainly in the large scaled cities. From this context, this paper proposes a prediction model based on the conditional distribution of transition represented by the co-occurrence matrix. Furthermore, for the improvement of prediction accuracy, the predicted values in the neighborhood is combined in such a way of minimizing the MSE. As a result, the proposed method showed higher prediction performances than other methods of homogeneous prediction models. Through the simulation for predicting taxi passenger demands in the central city areas of Seoul and New York, the effectiveness of the proposed method has been established. These results have shown that the combining strategy of the temporal and spatial information on predicted values is quite effective in the problems of predicting taxi passenger demands.

Acknowledgments. This work was supported by Institute for Information & communications Technology Promotion (IITP) grant funded by the Korea government (MSIT) (No. B0717-17-0070).

References

1. Seoul Information Planning Team: Information on Seoul Metropolitan Taxi Operational Analysis Data, 25 January 2017. Seoul Metropolitan Government, 14 June 2017. http://data.seoul.go.kr/openinf/fileview.jsp?infId=OA-12066
2. Williams, L., Hoel, L.: Modeling and forecasting vehicular traffic flow as a seasonal ARIMA process: theoretical basis and empirical results. J. Transp. Eng. **50**, 159–175 (2003)
3. Lee, J., Shin, I., Park, G.: Analysis of the passenger pick-up pattern for taxi location recommendation. In: International Conference on Networked Computing and Advanced Information Management, pp. 199–204 (2008)
4. Luis, M., João, G., Michel, F., Luís, D.: A predictive model for passenger demand on a taxi network. In: International IEEE Conference on Intelligent Transportation Systems, pp. 1014–1019 (2012)
5. Leon-Garcia, A.: Probability, Statistics, and Random Process for Electrical Engineering, 3rd edn. Pearson Prentice Hall, Upper Saddle River (2009)
6. Taxicab and Livery Passenger Enhancement Programs. TLC Trip Record Data, 14 February 2017. http://www.nyc.gov/html/tlc/html/about/trip-record-data.shtml
7. SVMlight ver 6.02. Cornell University. http://svmlight.joachims.org

Combining the Global and Local Estimation Models for Predicting PM_{10} Concentrations

Han Bin Bae, Tae Hyun Kim, Rhee Man Kil$^{(\boxtimes)}$, and Hee Yong Youn

College of Software, Sungkyunkwan Univesity,
2066, Seobu-ro, Jangan-gu, Suwon-si 16419, Gyeonggi-do, Korea
bhb0722@nate.com, fanpa9@gmail.com, {rmkil,youn7147}@skku.edu

Abstract. This paper presents a new way of predicting timely air pollution measure such as the PM_{10} concentration in Seoul based on a new method of combining the global and local estimation models. In the proposed method, the structure of nonlinear dynamics of generating air pollution data series is analyzed by investigating the attractors in the phase space and this structure is used to build the prediction model. Then, the global estimation model such as the network with Gaussian kernel functions is trained for the air pollution series data. Furthermore, the local estimation model which will recover the errors of the global estimation model using the on-line adaptation method, is also adopted. As a result, the proposed prediction model combining the global and local estimation models provides robust performances of predicting PM_{10} concentrations.

Keyword: PM_{10} concentration, Phase space analysis, Time series prediction, Gaussian kernel functions, Global and local estimation models

1 Introduction

Particle matters (PMs) suspended in the atmosphere can be greatly influenced to human health. Specifically, smaller particle matters with an aerodynamic diameter less than $10\,\mu\mathrm{m}$ are major concerns for a measure of air pollution. For this air pollution measure, the PM_{10} and $PM_{2.5}$ concentrations are widely used. This paper is investigating the prediction model for predicting timely averaged PM_{10} concentrations in the central city areas of Seoul.

For the prediction of these air pollution measures such as the PM_{10} and $PM_{2.5}$ concentrations, the prediction models using the artificial neural networks [1-6] for the environmental features such as the wind speed, temperature, humidity, rain fall, cloud cover, etc., have been usually adopted. In these models, the prediction performances are very much dependent upon the settings of environmental features. In this context, this paper presents a new way of constructing the prediction model by analyzing the nonlinear dynamics of air pollution data series. For this purpose, the nonlinear dynamics of PM_{10} concentration series is

© Springer International Publishing AG 2017
D. Liu et al. (Eds.): ICONIP 2017, Part V, LNCS 10638, pp. 275–284, 2017.
https://doi.org/10.1007/978-3-319-70139-4_28

investigated using the phase space analysis to determine the structure of prediction models. As a nonlinear prediction model, a network with Gaussian kernel functions is considered since this model is a nonparametric estimation model and good for incremental learning due to the locality of kernel functions. Furthermore, the dynamics of generating air pollution data is usually changing slowly. From this point of view, a local estimation model which will compensate the global estimation model which are trained for the given data series, are adopted. The purpose of this local estimation model is to recover the errors of global estimation model for a temporal window of local data using the on-line adaptation method. In this local estimation model, the linear regression model is adopted since the model of error compensation is almost linear once the global estimation model identifies the nonlinear dynamics of air pollution data series. As a result, the proposed prediction model of combining the global and local estimation models provides robust performances of predicting PM_{10} concentrations.

This paper is organized as follows: in Sect. 2, the phase space analysis of PM_{10} concentration series dynamics to determine the input structure of prediction models, is described, Sect. 3 presents an algorithm for predicting PM_{10} concentration series, Sect. 4 shows simulation results for predicting timely averaged PM_{10} concentrations, and finally, Sect. 5 presents the conclusion.

2 Phase Space Analysis of PM_{10} Concentration Series

The goal of time series prediction is to predict future values from a series of past values observed at regular time intervals. With no information on the state space (or phase space), we can only observe $x(t)$. Moreover, we usually consider the sampled discrete time series data $x(t_k)$; that is,

$$x(t_k) = x(t_0 + k\Delta t) \equiv x(k), \quad k = 0, 1, 2, \cdots \tag{1}$$

where t_0, k, and Δt represent the initial time, sampling step, and sampling interval, respectively. Here, let us construct E-dimensional vectors

$$\boldsymbol{x}_{\tau,E}(k) = [x(k), x(k - \tau), \ldots, x(k - (E - 1)\tau)]^t, \tag{2}$$

where τ and E represent the delay time measured as the unit of Δt and embedding dimension, respectively.

For the prediction of future values, we assume that the state vectors in the reconstructed state space are governed by a nonlinear function f; that is,

$$x(t_k + P) = f(\boldsymbol{x}_{\tau,E}(k)) \tag{3}$$

where P represents the prediction step. Then, our goal is to make a prediction model $\hat{f}$ estimating the unknown target function f. However, since the geometric structure of f is determined by the embedding parameters τ and E, the performance of the prediction model is strongly affected by the choice of the embedding parameters. In this context, we consider to determine the embedding parameters using the measure related to the smoothness of the target function.

For the analysis of time series data, let us denote the rth nearest neighbor of the vector $\boldsymbol{x}_{\tau,E}(k)$ by $\boldsymbol{x}_{\tau,E}^r(k)$. Since the difficulty of estimation depends on the smoothness of the target function f, we define the gradient of f at each point $\boldsymbol{x}_{\tau,E}(k)$ by

$$\Delta f(\boldsymbol{x}_{\tau,E}^r) = \frac{\left| f(\boldsymbol{x}_{\tau,E}(k)) - f(\boldsymbol{x}_{\tau,E}^r(k)) \right|}{\left\| \boldsymbol{x}_{\tau,E}(k) - \boldsymbol{x}_{\tau,E}^r(k) \right\|}, \tag{4}$$

where the norm in the denominator represents the Euclidean distance in $\mathbb{R}^E$. In this paper, $r = 1$ is used; that is, the nearest neighbor of the state vector. Here, the smoothness measure $S(\tau, E)$ of a target function f [7] is defined by using an average of the gradient values for time series $x(t_k)$, $k = 0, 1, \cdots, n-1$; that is,

$$S(\tau, E) = 1 - \frac{1}{n - (E-1)\tau} \sum_{k=(E-1)\tau}^{n-1} \Delta f(\boldsymbol{x}_{\tau,E}^1(k)). \tag{5}$$

The smoothness measure $S(\tau, E)$ is closely related to the unfolding of the orbits of the vectors in the reconstructed state space. If the embedding dimension E is too small, the state space is yet restricted to the low dimensional space. On the other hand, if the embedding dimension E is large enough, the nearest neighbor vectors are obtained from the same orbit, which results in the small gradient values. From these observations, the optimal embedding dimension and delay time can be determined by identifying the points where the smoothness measure $S(\tau, E)$ changes rapidly from the smaller values to the larger values.

For the analysis of next 4 h prediction for PM$_{10}$ concentration series, the smoothness measure of (5) is calculated for every delay time τ between 1 and 24, and every embedding dimension E between 4 and 20. The calculated smoothness measure of PM$_{10}$ concentration series is illustrated in Fig. 1. From the plot of smoothness measure, the smallest embedding dimension is selected first. Here, the proper embedding dimension is selected as $E = 7$. Then, for the given embedding dimension $E = 7$, the proper delay time is selected as $\tau = 4$. As a result, the model of four hour delay step prediction is determined by

$$x(t+4) = f\left(x(t), x(t-4), \cdots, x(t-24) \right). \tag{6}$$

From this phase space analysis, it is evident that the information of 4 data within a window of 24 h provides an important clue to determine the next 4 h prediction of PM$_{10}$ concentration series.

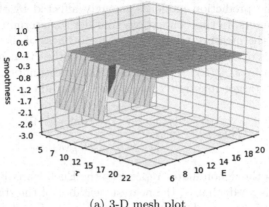

(a) 3-D mesh plot

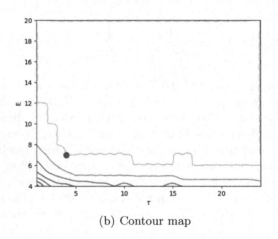

(b) Contour map

Fig. 1. The smoothness measure for PM_{10} concentration series when the prediction step is four hours.

3   Prediction Model for PM_{10} Concentration Series

As a nonlinear prediction model for PM_{10} concentration series, a network with Gaussian kernel function is selected since this network is able to perform non-linear and nonparametric estimation and good for incremental learning due to the locality of kernel functions. The suggested prediction model $\hat{f}$ with m kernel functions is described by

$$\hat{f}(\boldsymbol{x}) = \sum_{i=1}^{m} w_i \psi_i(\boldsymbol{x}), \quad \psi_i(\boldsymbol{x}) = e^{-||\boldsymbol{x}-\boldsymbol{\mu}_i||^2/2\sigma_i^2}, \tag{7}$$

where w_i represents the connection weight between the output and the ith kernel function ψ_i in which $\boldsymbol{\mu}_i$ and σ_i represent the mean and standard deviation, respectively.

In (7), we need to determine the parameters of m and also kernel related parameters w_i, μ_i, σ_i. For kernel related parameters, [8] suggested an efficient estimation method in such a way of minimizing the mean square error (MSE). The optimal structure of Gaussian kernel function networks (GKFNs) [9]; that is, the number of kernel fuctions m was also determined by estimating the noise variance of the time series data which was used as a stopping criteria for the incremental learning of Gaussian kernel functions. Here, in the case of time series $x(t_k + P)$, it can be described by

$$x(t_k + P) = f(\boldsymbol{x}_{\tau,E}(k)) + \epsilon, \tag{8}$$

where f represent an embedded function for the generation of time series data and ϵ represents a noise term expressed as a random variable with mean 0 and variance σ^2.

Then, for $x(t_k + P)$, the predicted value $\hat{x}(t_k + P)$ is described by

$$\hat{x}(t_k + P) = \hat{f}(\boldsymbol{x}_{\tau,E}(k)) \tag{9}$$

and the error term e is given by

$$e(t_k + P) = x(t_k + P) - \hat{x}(t_k + P). \tag{10}$$

In the case of air pollution data, the distribution of above errors is slowly changing because of the dynamics of generating air pollution data. From this point of view, the local estimation model is adopted to recover the errors of global estimation model for a temporal window of local data using the on-line adaptation method. In this local estimation model, the linear regression model is adopted since the model of error compensation is almost linear once the global estimation model identifies the nonlinear dynamics of air pollution data series. Here, set

$$\tilde{t}_k = t_k + P. \tag{11}$$

Then, the local estimation model is given by

$$\hat{e}(\tilde{t}_k) = \boldsymbol{w}^t \cdot \boldsymbol{e}(k), \tag{12}$$

where

$$\boldsymbol{w} = [w_b, w_0, w_1, \cdots, w_{E-1}]^t \quad \text{and} \tag{13}$$

$$\boldsymbol{e}(k) = [1, e(t_k), e(t_k - \tau_l), \cdots, e(t_k - (E_l - 1)\tau_l)]^t. \tag{14}$$

In this model, the delay time τ_l and embedding dimension E_l are searched in the region of (τ_l, E_l) where the smoothness measures is less than or equal to -1. Then, for the given (τ_l, E_l), the optimal weight vector $\boldsymbol{w}^*$ is determined in such a way of minimizing the mean square error (MSE); that is, minimizing

$$\mathcal{E}(l) = \frac{1}{l} \sum_{k=1}^{l} \left(e(\tilde{t}_k) - \hat{e}(\tilde{t}_k) \right)^2, \tag{15}$$

where l represents the window size for error data.

The final estimates of (τ_l, E_l) and $\boldsymbol{w}^*$ are determined when the MSE of (15) is minimized. For the adaptation of weight vectors, the $(k+1)$th recursive update of $\boldsymbol{w}$ is determined by

$$\boldsymbol{w}_{k+1} = \boldsymbol{w}_k + \frac{B_k \boldsymbol{e}(k)}{1 + \boldsymbol{e}(k)^t B_k \boldsymbol{e}(k)} e(\tilde{t}_k) \quad \text{and} \tag{16}$$

$$B_{k+1} = B_k - \frac{B_k \boldsymbol{e}(k)\boldsymbol{e}(k)^t B_k}{1 + \boldsymbol{e}(k)^t B_k \boldsymbol{e}(k)}. \tag{17}$$

After the adaption of weight vector $\boldsymbol{w}$ for the data in a moving window, the compensation error $\hat{e}$ is determined by (12) and the predicted value is determined by

$$\tilde{x}(\tilde{t}_k) = \hat{x}(\tilde{t}_k) + \hat{e}(\tilde{t}_k) \tag{18}$$

The whole procedure of predicting air pollution data series using the combined model of the global and local estimation networks is described as follows:

Construction of the Prediction Model for Air Pollution Data Series

Step 1. Determine the structure of time series prediction model:
   1.1 From the given PM_{10} concentration series $x(t_k)$, $k = 0, 1, \cdots, n-1$ and prediction time P, determine the values of smoothness measure of (5).
   1.2 Determine the embedding dimension E and delay time τ when the value of smoothness measure is large (usually, greater than -1) at the smaller embedding dimension.

Step 2. Determine the nonlinear prediction model using the values of E, τ, and P; that is,

$$\hat{x}(t_k + P) = \hat{f}(\boldsymbol{x}_{\tau,E}(k)),$$

where $\hat{f}$ is given by (7).

Step 3. Training of the global estimation model:
   3.1 Train the nonlinear prediction model using an incremental learning algorithm such as [8].
   3.2 The optimal number of Gaussian kernel functions is determined by investigating the confidence interval for the noise variance [9].

Step 4. Determine the local estimation model: for the region of (τ_l, E_l) where the smoothness measures is less than or equal to -1, apply the following procedure:
   4.1 After the construction of the global estimation model, the errors of (10) in the global estimation models for the latest $l + E_l \tau_l$ data and the corresponding l error vectors of (14) are collected.
   4.2 The weights of the local estimation model are updated for the collected error data using (16) and (17).
   4.3 The estimated error $\hat{e}(t_k + P)$ is updated as (12).
   4.4 Finally, the predicted value $\tilde{x}(\tilde{t}_k)$ is determined by

$$\tilde{x}(t_k + P) = \hat{x}(t_k + P) + \hat{e}(t_k + P).$$

   4.5 Determine the MSE of (15).

Step 5. The estimates of (τ_l, E_l) and w^* are determined when the MSE of (15) is minimized.

The whole diagram of the proposed algorithm is illustrated in Fig. 2. For the given time series and prediction step P, the input structure (τ, E) of the prediction model is determined by the phase space analysis of time series. Then, the global estimation model such as the GKFN is trained for the given time series. Once the global estimation model is constructed, the proposed algorithm improves the prediction performance using the error compensation implemented by the local estimation model. In the proposed algorithm, the linear estimation model is adopted in the local estimation model because of the efficiency of computation. Furthermore, the model of error compensation tends to have an almost linear form when the global estimation model is well fitted to the nonlinear dynamics of time series.

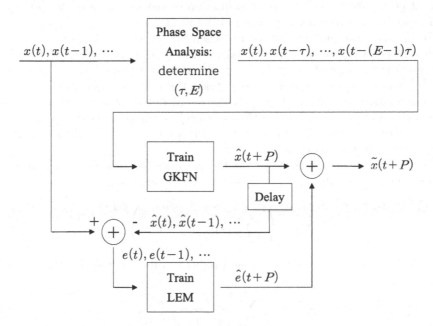

Fig. 2. The schematic diagram of time series prediction using the global and local estimation models.

4   Simulation

For the simulation of predicting PM$_{10}$ concentration series, the data set was collected from Air Korea [10] which provides measurement data of atmospheric environment reference materials measured in 322 air monitoring networks installed in 97 cities across the country in various forms and provides them to the public in real time. Among various atmospheric data, the PM$_{10}$ concentration series data in the central city area of Seoul from the 1st Jan 2015 to the 31st May 2016 were collected. Then, the first 1 year data were used as the training data and the remaining 5 months data were used as the test data.

For the construction of the prediction model, the and delay time τ and embedding dimension E for various values of prediction steps $P = 2, 4, \cdots, 12$ were determined by using (5) in the first place. Then, for the optimal structure of GKFN model, the optimal number of kernels was selected by investigating the noise variance [9]. In our approach, the optimal number of kernels was selected when the train error reached the upper bound of the confidence interval of noise variance. Then, for the local estimation model of (10), the proper embedding dimension and delay time were selected using the smoothness measure of (5) and the local estimation model was trained initially using the last portion of the training data. The estimates of the delay time τ_l and embedding dimension E_l of the local estimation model were determined in such a way of minimizing the MSE of (15). Afterwards, the local estimation model was updated adaptively at each time for the given data by using (16) and (17).

To compare the performance of proposed model, the epsilon support vector regression (ε-SVR) model was trained by using the same train and test data sets. For the training of the ε-SVR, the Scikit-learn toolkit [11] was used. The learning parameters were selected as it takes the best performances. To handle the non-linearity of data patterns, the kernel of radial basis function was used with the kernel parameter $\gamma = 1/E$ and to select the parameter of sum of the slack variables C, the grid search as $C = 10^i$ for $i = -7, -6, ..., 7$ was used. As a result, $C = 10$ was selected.

For the evaluation of these methods, the following root mean square error (RMSE) and the coefficient of determination R^2 were used:

$$RMSE = \left(\frac{1}{n} \sum_{k=1}^{n} \left(x(\tilde{t}_k) - \hat{x}(\tilde{t}_k) \right)^2 \right)^{1/2}, \tag{19}$$

where $x(\tilde{t}_k)$ and $\hat{x}(\tilde{t}_k)$ are the true and predicted values at the time $\tilde{t}_k = t_k + P$, respectively.

$$R^2 = 1 - \sum_{k=1}^{n} \frac{\left(x(\tilde{t}_k) - \hat{x}(\tilde{t}_k) \right)^2}{\left(x(\tilde{t}_k) - \bar{x}(\tilde{t}_k) \right)^2}, \tag{20}$$

where $\bar{x}(\tilde{t}_k)$ is the sample average of $x(\tilde{t}_k)$.

The simulation results for the prediction of PM_{10} concentration series were summarized in Table 1. These simulation results have shown that (1) the combination of global and local estimation (CGLE) methods provides the best prediction performances, (2) the performance improvement becomes evident when the prediction step P is increasing, and (3) the prediction performances of the SVR and GKFN models becomes lower when the prediction step P is increasing. This implies that the proposed method of combining the global and local estimation models is quite effective by providing the robust prediction performances in these types of time series prediction problems.

Table 1. Simulation results for the prediction of PM$_{10}$ concentration series

E	τ	P	R^2			$RMSE$		
			ε-SVR	GKFN	CGLE	ε-SVR	GKFN	CGLE
7	3	2	0.5587	0.7083	**0.7345**	28.24	22.96	**21.90**
7	4	4	0.1438	0.4244	**0.5586**	39.33	32.25	**28.23**
7	6	6	−0.0120	0.2036	**0.4802**	42.30	37.50	**30.31**
7	7	8	−0.1084	0.1686	**0.5321**	44.22	38.34	**28.72**
8	5	10	−0.2988	0.1792	**0.5215**	48.32	43.24	**29.12**
8	4	12	−0.4126	0.1578	**0.6072**	50.13	38.71	**26.43**

5 Conclusion

The problems of predicting timely averaged PM$_{10}$ concentrations involve the time series analysis and also optimization of regression models. In this work, the series of timely averaged PM$_{10}$ concentrations is analyzed by the phase space analysis method [7]. As the prediction model, a network with Gaussian kernel functions [8] is selected for this model is good for incremental learning due to the locality of kernel functions. For the further improvement of prediction accuracy, a local estimation model which will compensate the global estimation model which are trained for the given data series, are adopted. The purpose of this local estimation model is to recover the errors of global estimation model for a temporal window of local data using the on-line adaptation method. As a result, the proposed prediction model of combining the global and local estimation models provides robust performances of predicting the time series data. Through the simulation for predicting timely averaged PM$_{10}$ concentrations, the effectiveness of the proposed prediction model has been demonstrated. The proposed model can also be applied to various problems of time series prediction.

Acknowledgments. This work was supported by Institute for Information & communications Technology Promotion (IITP) grant funded by the Korea government (MSIT) (No. B0717-17-0070).

References

1. Gardner, M., Dorling, S.: Artificial neural networks (the multilayer perceptron) - a review of applications in the atmosphere sciences. Atmos. Environ. **32**, 2627–2636 (1998)
2. Patricio, P., Alex, T., Jorge, R.: Prediction of PM$_{2.5}$ concentrations several hours in advance using neural networks in Santiago, Chile. Atmos. Environ. **34**, 1189–1196 (2000)
3. Jiang, D., Zhang, Y., Hu, X., Zeng, Y., Tan, J., Shao, D.: Progress in developing an ANN model for air pollution index forecast. Atmos. Environ. **38**, 7055–7064 (2004)

4. Hooyberghs, J., Mensink, C., Dumont, G., Fierens, F., Brasseur, O.: A neural network forecast for daily average PM_{10} concentrations in Belgium. Atmos. Environ. **39**, 3279–3289 (2005)
5. Grivas, G., Chaloulakou, A.: Artificial neural network models for prediction of PM_{10} hourly concentrations in the greater area of Athens, Greece. Atmos. Environ. **40**, 1216–1229 (2006)
6. Cao, Z., Yu, S., Xu G., Chen B., Principe J.: Multiple adaptive kernel size KLMS for Beijing PM2.5 prediction. In: International Joint Conference on Neural Networks, pp. 1403–1407 (2016)
7. Kil, R., Park, S., Kim, S.: Time series analysis based on the smoothness measure of mapping in the phase space of attractors. In: International Joint Conference on Neural Networks, vol. 4, pp. 2584–2589 (1999)
8. Kil, R.: Function approximation based on a network with kernel functions of bounds and locality: an approach of non-parametric estimation. ETRI J. **15**, 35–51 (1993)
9. Kim, D.K., Kil, R.M.: Stock price prediction based on a network with Gaussian kernel functions. In: Lee, M., Hirose, A., Hou, Z.-G., Kil, R.M. (eds.) ICONIP 2013. LNCS, vol. 8227, pp. 705–712. Springer, Heidelberg (2013). doi:10.1007/978-3-642-42042-9_87
10. Air Korea, PM_{10} and $PM_{2.5}$ Concentrations Information on South Korea. http://www.airkorea.or.kr/
11. Pedregosa, F., et al.: Scikit-learn: machine learning in python. J. Mach. Learn. Res. **12**, 2825–2830 (2011)

Anomaly Detection for Categorical Observations Using Latent Gaussian Process

Fengmao Lv[1], Guowu Yang[1], Jinzhao Wu[2(✉)], Chuan Liu[1], and Yuhong Yang[3]

[1] The Bid Data Center, The School of Computer Science and Engineering,
University of Electronic Science and Technology of China,
Chengdu 611731, Sichuan, People's Republic of China
fengmaolv@126.com, {guowu,liuchuan}@uestc.edu.cn
[2] Guangxi Key Laboratory of Hybrid Computation and IC Design Analysis,
Guangxi University for Nationalities,
Nanning 530006, Guangxi, People's Republic of China
gxmdwjzh@aliyun.com
[3] The School of Statistics, University of Minnesota, Minneapolis, MN 55455, USA
yangx374@umn.edu

Abstract. Anomaly detection is an important problem in many applications, ranging from medical informatics to network security. Various distribution-based techniques have been proposed to tackle this issue, which try to learn the probabilistic distribution of conventional behaviors and consider the observations with low densities as anomalies. For categorical observations, multinomial or dirichlet compound multinomial distributions were adopted as effective statistical models for conventional samples. However, when faced with small-scale data set containing multivariate categorical samples, these models will suffer from the curse of dimensionality and fail to capture the statistical properties of conventional behavior, since only a small proportion of possible categorical configurations will exist in the training data. As an effective bayesian nonparametric technique, categorical latent Gaussian process is able to model small-scale categorical data through learning a continuous latent space for multivariate categorical samples with Gaussian process. Therefore, on the basis of categorical latent Gaussian process, we propose an anomaly detection technique for multivariate categorical observations. In our method, categorical latent Gaussian process is adopted to capture the probabilistic distributions of conventional categorical samples. Experimental results on categorical data set show that our method can effectively detect anomalous categorical observations and achieve better detection performance compared with other anomaly detection techniques.

Keywords: Anomaly detection · Categorical data · Bayesian nonparametric model · Gaussian process · Data-efficient learning

1 Introduction

Anomaly detection aims at detecting patterns in data which do not accord with defined pattern of conventional behavior [7]. In tasks such as intrusion detection

© Springer International Publishing AG 2017
D. Liu et al. (Eds.): ICONIP 2017, Part V, LNCS 10638, pp. 285–296, 2017.
https://doi.org/10.1007/978-3-319-70139-4_29

[6], fraud detection [21] or scientific experiments, anomaly detection is an important work, since these samples may be dangerous to systems or critical for scientific research. According to the availability of labeled data, anomaly detection technique can operate in 3 modes: supervised, semi-supervised and unsupervised anomaly detection [7]. In this paper, we mainly focus on semi-supervised anomaly detection, which assume that the training data has labeled instances for only the conventional class. In semi-supervised anomaly detection, the typical method is to build a model for the class corresponding to conventional behavior, and use the model to identify anomalies in the test data.

Distribution-based methods are the common techniques for semi-supervised anomaly detection, which aim at capturing the probabilistic distribution of conventional behavior, and consider the observations with low densities as anomalies [8,22]. So far, various statistical techniques, ranging from parametric methods [2,18] to non-parametric methods [1,15] have been adopted to learn distributions of conventional behaviors. For continuous data, Gaussian distributions [18], mixtures of Gaussians [2,14] or kernel function based methods [15] are commonly used. The samples with low probability densities are considered as anomalous samples and detected out. For categorical data, multivariate bernoulli or dirichlet compound multinomial (DCM) distributions are adopted as alternative methods to model the distributions [13,17,19]. However, when faced with small-scale data set containing multivariate categorical samples, these models suffer from the curse of dimensionality and detectors based on these models fail to work, since the number of possible categorical values grows exponentially with the number of attributes and only a small proportion of possible categorical values are covered by the data set [10].

Motivation: as an effective embedding technique, categorical latent Gaussian process (CLGP), was proposed to model categorical data in the machine learning community [10]. Similar to deep generative models like variational autoencoders (VAEs) [12] or generative adversarial networks (GANs) [11], CLGP is a latent variable model, which assumes that the original categorical observations are generated from a continuous manifold. However, VAEs or GANs are data-consuming since they are modeled with deep neural networks. CLGP can overcome this problem by relating the latent space to the observed data space through Gaussian process. Due to the bayesian non-parametric characteristics of Gaussian processes [16], CLGP can handle small-scale data well. Therefore, CLGP can be an effective tool to capture distributions of conventional behaviors when faced with small-scale training set.

Contribution: In this paper, we propose an anomaly detection method based on CLGP. In our method, the conventional observations are assumed to be generated from the probabilistic generative model - CLGP. For a newly obtained observation, we use CLGP model built on conventional observations to estimate its probabilistic density, and the inverse of density is assigned as the anomaly score. The observation with high anomaly score is considered as the anomaly. Experimental results on categorical data set show that our method can effectively detect anomalous categorical observations and achieve better detection performance compared with other anomaly detection techniques.

This paper is organized as follows. Section 2 reviews the CLGP model. Section 3 presents our anomaly detection method. Section 4 shows the experimental results on anomaly detection tasks. Finally, Sect. 5 summarizes this paper.

2 CLGP Revisited

In this paper, we use the notation $\mathbf{Y} = [\mathbf{y}_1, ..., \mathbf{y}_N]^T$ to represent the categorical data, where $\mathbf{y}_n$ indicates the n-th sample containing D attributes. As a categorical feature, the d-th variable of $\mathbf{y}_n$, y_{nd}, takes an integer value from 0 to K_d. For ease of notation, we assume that the cardinalities of each categorical variable are identical and denoted as K. Similar to deep generative models, CLGP is a latent variable model, which assumes that the categorical data are generated from a continuous latent space. To describe the corresponding latent representations for the data in $\mathbf{Y}$, the notation $\mathbf{X} = [\mathbf{x}_1, ..., \mathbf{x}_N]^T$ is used, where $\mathbf{x}_n \in R^Q$.

Specifically, the categorical variable y_{nd} is assumed to be generated through a categorical distribution with probability given by a Softmax with weights $\mathbf{f}_{nd} = (f_{nd1}, ..., f_{ndK})$. Each weight f_{ndk} is the output of a nonlinear function $\mathcal{F}_{dk}$ at latent point $\mathbf{x}_n$: $\mathcal{F}_{dk}(\mathbf{x}_n)$. The nonlinear function $\mathcal{F}_{dk}$ does not follow a parametric formula such as linear functions or neural networks, but a Gaussian process prior $\mathrm{GP}(0, \mathbf{K}_d)$. Each point, $\mathbf{x}_n$, is assumed to be generated from isotropic Gaussian distribution prior $\mathrm{N}(0, \sigma^2\mathbf{I})$. As done in bayesian Gaussian process latent variable model [9], to make the inference of CLGP tractable, a set of M inducing inputs $\mathbf{Z} \in R^{M \times Q}$ lying in the latent space is considered, with their corresponding outputs $\mathbf{U} \in R^{M \times D \times K}$ lying in the weight space (together with $\mathbf{f}_{nd}$). $\mathbf{Z}$ is considered as a sparse approximation to $\mathbf{X}$. Please refer to [9,20] for more details of inducing points. The following equations describe the generative model of CLGP:

$$\mathbf{x}_n \overset{\text{iid}}{\sim} \mathrm{N}(0, \sigma^2\mathbf{I})$$

$$\mathcal{F}_{dk} \overset{\text{iid}}{\sim} \mathrm{GP}(0, \mathbf{K}_d)$$

$$f_{ndk} = \mathcal{F}_{dk}(\mathbf{x}_n), \quad u_{mdk} = \mathcal{F}_{dk}(\mathbf{z}_m)$$

$$y_{nd} \sim \mathrm{Softmax}(\mathbf{f}_{nd}),$$

(1)

where $n = 1, 2, ..., N$, $d = 1, 2, ..., D$, $k = 1, 2, ..., K$ and $m = 1, 2, ..., M$. Variational inference [5] is adopted to train a CLGP model. Following [9], the posterior over the unobserved variables, $P(\mathbf{X}, \mathbf{F}, \mathbf{U}|\mathbf{Y})$, is approximated with a variational distribution factorized as

$$q(\mathbf{X}, \mathbf{F}, \mathbf{U}) = q(\mathbf{X})q(\mathbf{U})p(\mathbf{F}|\mathbf{X}, \mathbf{U}),$$

(2)

with

$$q(\mathbf{U}) = \prod_{d=1}^{D} \prod_{k=1}^{K} \mathrm{N}(\mathbf{u}_{dk}|\boldsymbol{\mu}_{dk}, \boldsymbol{\Sigma}_d)$$

$$q(\mathbf{X}) = \prod_{n=1}^{N} \prod_{i=1}^{Q} \mathrm{N}(x_{ni}|m_{ni}, s_{ni}^2),$$

(3)

where $\boldsymbol{\mu}_{dk}$, $\boldsymbol{\Sigma}_d$, m_{ni} and s_{ni}^2 are the variational parameters. Through the Jensen's inequality, $\log p(\mathbf{Y})$ can be approximated by its lower bound:

$$\begin{aligned}
\log p(\mathbf{Y}) \geq & \int q(\mathbf{X})q(\mathbf{U})p(\mathbf{F}|\mathbf{X},\mathbf{U}) \\
& \cdot \log \frac{p(\mathbf{X})p(\mathbf{U})p(\mathbf{F}|\mathbf{X},\mathbf{U})p(\mathbf{Y}|\mathbf{F})}{q(\mathbf{X})q(\mathbf{U})p(\mathbf{F}|\mathbf{X},\mathbf{U})} d\mathbf{X}d\mathbf{F}d\mathbf{U} \\
= & -\mathrm{KL}(q(\mathbf{X})||p(\mathbf{X})) - \mathrm{KL}(q(\mathbf{U})||p(\mathbf{U})) \\
& + \int q(\mathbf{X})q(\mathbf{U})p(\mathbf{F}|\mathbf{X},\mathbf{U}) \log p(\mathbf{Y}|\mathbf{F})d\mathbf{X}d\mathbf{F}d\mathbf{U} \\
:= & \mathcal{L}_{\mathbf{Y}}.
\end{aligned}$$

(4)

To maximize $\mathcal{L}_{\mathbf{Y}}$, parameters including hyper-parameters ($\boldsymbol{\theta}$ for $\mathbf{K}_d$) and variational parameters ($\mathbf{Z}$, $\boldsymbol{\mu}_{dk}$, $\boldsymbol{\Sigma}_d$, m_{ni} s_{ni}^2) need to be optimized. The details of CLGP can be found in [10].

3   Anomaly Detection Technique Based on CLGP

To simplify the notations, we use $\mathbf{Y} = [\mathbf{y}_1, ..., \mathbf{y}_N]^T$ to represent the training data, in which all the samples are conventional samples. The notation $\mathbf{y}_*$ is adopted to represent a newly obtained observation, and our purpose is to determine whether $\mathbf{y}_*$ is anomalous or not.

In general, 2 phases are needed to design a semi-supervised anomaly detection method: (1) Firstly, we need to build a model on the conventional behaviors, which reflects the characteristic of conventional observations; (2) Then, strategies on estimating anomaly scores for newly obtained observations are under consideration and observations with high anomaly scores are suspect of being anomalies. In our method, we assume that the conventional samples in $\mathbf{Y}$ are generated according to the probabilistic generative model of CLGP. Therefore, to model the behaviors of conventional observations, we need to build a CLGP model on the training data $\mathbf{Y}$, which is equivalent to maximizing the lower bound of $\log p(\mathbf{Y})$, expressed in Eq. (4). In [10], the maximization of $\log p(\mathbf{Y})$ has been discussed at full length. Therefore, in this part, we will mainly focus on how to estimate the anomaly scores of $\mathbf{y}_*$.

As CLGP is a probabilistic generative model, it is natural to adopt the inverse of the probability densities as the anomaly scores. As indicated in [4], for a statistical anomaly detection technique, the observations occur in low probability regions of the stochastic generative model are suspect of being anomalies and assigned with high anomaly scores. Suppose $\mathbf{y}_*$ is generated from the CLGP model constructed on $\mathbf{Y}$, and our interest is to evaluate the conditional density $p(\mathbf{y}_*|\mathbf{Y})$. It is easy to see that $p(\mathbf{y}_*|\mathbf{Y})$ can be represented as the ratio of two marginal likelihoods:

$$p(\mathbf{y}_*|\mathbf{Y}) = \frac{p(\mathbf{y}_*, \mathbf{Y})}{p(\mathbf{Y})}. \tag{5}$$

By the use of Jensen's inequality, both $\log p(\mathbf{Y})$ and $\log p(\mathbf{y}_*, \mathbf{Y})$ can be approximated by their lower bounds:

$$\begin{aligned} \log p(\mathbf{Y}) &\approx \mathcal{L}_{\mathbf{Y}}, \\ \log p(\mathbf{y}_*, \mathbf{Y}) &\approx \mathcal{L}_{\mathbf{y}_*, \mathbf{Y}}. \end{aligned} \tag{6}$$

Similar to Eq. (4), $\mathcal{L}_{\mathbf{y}_*, \mathbf{Y}}$ has the following representation:

$$\begin{aligned} \mathcal{L}_{\mathbf{y}_*, \mathbf{Y}} = &- \text{KL}(q(\mathbf{x}_*, \mathbf{X}) \| p(\mathbf{x}_*, \mathbf{X})) - \text{KL}(q(\mathbf{U}) \| P(\mathbf{U})) \\ &+ \int q(\mathbf{x}_*, \mathbf{X}) q(\mathbf{U}) p(\mathbf{f}_*, \mathbf{F} | \mathbf{x}_*, \mathbf{X}, \mathbf{U}) \\ &\cdot \log p(\mathbf{y}_*, \mathbf{Y} | \mathbf{f}_*, \mathbf{F}) \, d\mathbf{x}_* d\mathbf{X} d\mathbf{f}_* d\mathbf{F} d\mathbf{U}. \end{aligned} \tag{7}$$

Since both $p(\mathbf{x}_*, \mathbf{X})$ and $q(\mathbf{x}_*, \mathbf{X})$ decompose across data, $\mathcal{L}_{\mathbf{y}_*, \mathbf{Y}}$ can be further represented as

$$\begin{aligned} \mathcal{L}_{\mathbf{y}_*, \mathbf{Y}} = &- \text{KL}(q(\mathbf{X}) \| p(\mathbf{X})) - \text{KL}(q(\mathbf{U}) \| p(\mathbf{U})) \\ &+ \int q(\mathbf{X}) q(\mathbf{U}) p(\mathbf{F} | \mathbf{X}, \mathbf{U}) \log p(\mathbf{Y} | \mathbf{F}) d\mathbf{X} d\mathbf{F} d\mathbf{U} \\ &+ \int q(\mathbf{x}_*) q(\mathbf{U}) p(\mathbf{f}_* | \mathbf{x}_*, \mathbf{U}) \log p(\mathbf{y}_* | \mathbf{f}_*) d\mathbf{x}_* d\mathbf{f}_* d\mathbf{U} \\ &- \text{KL}(q(\mathbf{x}_*) \| p(\mathbf{x}_*)) \\ = & \mathcal{L}_{\mathbf{Y}} + H_*, \end{aligned} \tag{8}$$

$$\begin{aligned} H_* = \sum_{d=1}^{D} \int &q(\mathbf{x}_*) q(\mathbf{U}_d) p(\mathbf{f}_{*d} | \mathbf{x}_*, \mathbf{U}_d) \log p(y_{*d} | \mathbf{f}_{*d}) \\ &d\mathbf{x}_* d\mathbf{f}_{*d} \mathbf{U}_d - \text{KL}(q(\mathbf{x}_*) \| p(\mathbf{x}_*)), \end{aligned} \tag{9}$$

where

$$q(\mathbf{x}_*) = \prod_{i=1}^{Q} \text{N}(x_{*i} | m_{*i}, s_{*i}^2), \tag{10}$$

$$p(\mathbf{f}_{*d} | \mathbf{x}_*, \mathbf{U}_d) = \prod_{k=1}^{K} \text{N}(f_{*dk} | \mathbf{a}_{*d} \mathbf{u}_{dk}, b_{*d}), \tag{11}$$

with

$$\begin{aligned} \mathbf{a}_{*d} &= \mathbf{K}_{d,*M} \mathbf{K}_{d,MM}^{-1}, \\ b_{*d} &= \mathbf{K}_{d,**} - \mathbf{K}_{d,*M} \mathbf{K}_{d,MM}^{-1} \mathbf{K}_{d,M*}. \end{aligned} \tag{12}$$

In the optimization of $\mathcal{L}_{\mathbf{y}_*,\mathbf{Y}}$, the term of $\mathcal{L}_{\mathbf{Y}}$ are kept fixed with its value optimized in the phase of building CLGP model on $\mathbf{Y}$, in order to reduce calculation amount. To evaluate the quantity of $\mathcal{L}_{\mathbf{y}_*,\mathbf{Y}}$, we only maximize H_* over m_{*i} and s_{*i}^2 as an approximation. This simplification allows us to process data in batch when faced with multiple observations being detected. With the Gaussian distributions defined in Eqs. (1) and (3), the KL divergence terms in H_* can be calculated analytically. However, due to the Softmax function $p(y_{*d}|\mathbf{f}_{*d})$ involved in the first term of H_*, the integration can not provide a closed-form solution. To handle this intractability, reparametrization and sampling tricks can be utilized, which is particularly introduced in [10,12].

As a consequence, $p(\mathbf{y}_*|\mathbf{Y})$ has the following approximation:

$$p(\mathbf{y}_*|\mathbf{Y}) \approx \frac{e^{\mathcal{L}_{\mathbf{y}_*,\mathbf{Y}}}}{e^{\mathcal{L}_{\mathbf{Y}}}}$$
$$= e^{\mathcal{L}_{\mathbf{y}_*,\mathbf{Y}} - \mathcal{L}_{\mathbf{Y}}} \qquad (13)$$
$$= e^{H_*}.$$

The inverse of $p(\mathbf{y}_*|\mathbf{Y})$ is used to evaluate the anomaly score of $\mathbf{y}_*$:

$$S = e^{-H_*}. \qquad (14)$$

An observation with a high anomaly score, S, will be suspect of being an anomaly and detected out. To maximize H_*, we choose RMSPROP as the optimization technique and adopt Theano to automatically obtain the gradients.

4   Experimental Results

In this section, experiments on categorical data are conducted to evaluate the performance of our proposal. The data are from the UCI Machine Learning Repository.

In our experiments, we compare our method with 2 distribution based methods - naive bayes over multinomial (NB-M) and naive bayes over DCM (NB-DCM), both of which are effective for modeling the probabilistic distributions of small-scale categorical data. Firstly, the conventional observations are modeled by CLGP, NB-M and NB-DCM respectively. Then, the test data containing anomalous observations is used to evaluate the detection ability of each model. To demonstrate the performance of our method on small-scale data, the training datasets in our experiments contain very few (less than 300) conventional observations. Since the observations are usually unbalanced in detection tasks, we make 10% of the testing samples be anomalies. We adopt radial basis kernel with automatic relevance determination (ARD) for the Gaussian process mapping. The hyper-parameters are determined through an independent random split of datasets. σ^2 in Eq. (1) represents the variance of prior distribution of latent space, and it is fixed at 1. In the optimization process, the initial values of the hyper-parameters $\boldsymbol{\theta}$ for $\mathbf{K}_d$ and the standard deviation (SD) s_{ni} for $q(\mathbf{X})$

are set to 0.1. The initializations of mean values m_{ni} and μ_{dk} in Eq. (4) are produced by Gaussian distributions with mean 0, SD 1 and mean 0, SD 0.01, respectively. Due to the randomness in the initialisation of the parameters, for each newly obtained observation $\mathbf{y}_*$, $\mathcal{L}_{\mathbf{y}_*, \mathbf{Y}}$ is optimized 3 times in the testing phase and the largest lower bound is adopted to evaluate the anomaly score of $\mathbf{y}_*$. For all the problems, the maximum iteration of RMSPROP in the training phase and testing phase are set to 100 and 30, respectively. All the results are averaged over 20 different random splits of datasets. The experiments are run on Mac OS 10.9.5 with 2.6 GHz Intel core i5 and 8 GB 1600 MHz DDR3.

Referring to [3], to compare the performance of anomaly detection approaches, we can evaluate the number of anomalies included among the testing observations with top-k anomaly scores. Additionally, we are also interested in the receiver operating characteristic (ROC) curves for each detection model, and take the area under curve (AUC) into consideration when comparing different techniques. The ROC curves are drawn according to the anomaly score assigned to each testing observation.

4.1   Detecting Poisonous Mushrooms

This data set has 22 categorical attributes which describe the various characteristics of a mushroom and a class label which denotes whether a mushroom is edible or poisonous. The 22 categorical attributes include the characteristic of the cap, gill, stalk, veil *et al.* The poisonous mushrooms are considered as the anomalies. Our aim is to learn the probabilistic distribution of conventional observation and detect out the poisonous mushrooms. To demonstrate the performance of our method on small data, 300 samples of edible mushroom are randomly selected to form the training set. The remaining 3908 edible mushrooms are used as the testing samples, and 434 poisonous mushrooms are randomly selected as the anomalies. We average the experimental results over 20 random split of the dataset, with a 15 dimensional latent space and 50 inducing points. As can be seen in Table 1, our proposal obtains larger AUC than the compared methods (the ROC curves averaged over 20 random splits of the mushroom dataset are shown in Fig. 1). Table 2 shows the detection results obtained by each model on the mushroom data set. It is shown that our method almost always obtains the best detection performances for different settings of top ratios. Our method can detect all the anomalies when top ratio is 24%. In contrast, NB-M and NB-DCM fail to achieve this goal within top ratio of 26%. In conclusion, for the

Table 1. Comparison of each detection method in terms of AUC (averaged over 20 different random splits of datasets)

Dataset	Averaged AUC (10^{-1})		
	NB-M	NB-DCM	Ours
Mushroom	9.35 ± 0.10	9.69 ± 0.12	9.76 ± 0.24
SPECT heart	9.24 ± 0.72	9.19 ± 0.69	9.45 ± 0.77

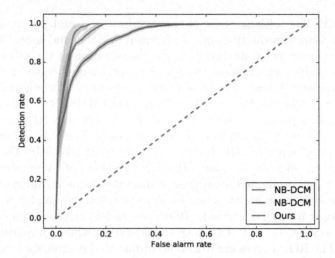

Fig. 1. ROC curves for different detection techniques averaged over 20 random splits of the mushroom dataset. The shaded areas represent the variance of each curve: ±0.5 × standard deviation.

Table 2. Detected poisonous samples in mushroom dataset (The top ratio is percentage of testing samples with top-k anomaly scores. The coverage is the proportion of number of anomalies included among the top-k samples to total number of anomalies. The results are averaged over 20 random splits of the mushroom dataset.)

Top ratio (number of observations)	Number of anomalies included (coverage)		
	Ours	NB-M	NB-DCM
3% (130)	82(18.89%)	114(26.35%)	109(25.12%)
6% (261)	195(44.93%)	199(45.85%)	207(47.70%)
9% (261)	304(70.05%)	199(45.85%)	207(47.70%)
12% (521)	376(86.64%)	296(68.26%)	346(79.72%)
15% (651)	407(93.78%)	326(75.12%)	378(87.10%)
18% (782)	425(97.93%)	349(80.44%)	395(91.01%)
21% (912)	431(99.31%)	361(83.21%)	411(94.70%)
24% (1042)	434(100.00%)	377(86.81%)	424(97.70%)
26% (1129)	434(100.00%)	386(88.97%)	430(99.08%)

mushroom data set, our method can achieve desirable detection results when faced with small-scale conventional data set.

4.2   Detecting Heart Patients

This dataset describes the cardiac Single Proton Emission Computed Tomography (SPECT) images of the heart patients and healthy controls. To obtain this dataset, the SPECT image of each subject was processed and summarized with 22 binary variables. In our experiments, the heart patients are considered as the anomalies. Our aim is to learn the probabilistic distribution of the healthy controls and detect out the heart patients. To further evaluate the detection ability of our method on extremely small data, only 20 samples among the healthy controls are randomly selected as the training set. The remaining 35 healthy controls are used as the test samples, and 4 heart patients are randomly selected as the anomalous samples to be detected out. The number of inducing points and the latent dimension are set to 15 and 3, respectively. From Table 1, it can be seen that our method achieves the largest AUC value on the SPECT heart data (the ROC curves averaged over 20 random splits of the SPECT heart dataset are shown in Fig. 2). Table 3 shows the percentage of the detected anomalies. Since the testing data contains only 4 anomalies, the average numbers of anomalies included in testing samples with top-k anomaly scores are documented as decimals. As can be seen, the detection results obtained by NB-DCM and NB-M are dissatisfactory, with coverage of less than 91% within top ration of 31%. For different top ratios, our method always achieves the best detection performances. To sum up, though the training data contains an extremely small amount of conventional samples, our method still works well as a anomaly detection method.

4.3   Limitations and Future Works

The computational efforts of each method are shown in Table 4. As can be seen, the computational costs of our method are much larger compared with the others since the matrix operations contained in Gaussian process are very time-consuming. One of our future works will focus on reducing the computational cost of our method in distributed frameworks.

Table 3. Detected heart patients in SPECT heart dataset (The results are averaged over 20 random splits of the SPECT heart dataset. Since the testing data contains only 4 anomalies, the average numbers of anomalies included in testing samples with top-k anomaly scores are documented as decimals.)

Top ratio (number of observations)	Number of anomalies included (coverage)		
	Ours	NB-M	NB-DCM
10% (4)	3.00(75.00%)	2.75(68.75%)	2.75(68.75%)
13% (5)	3.25(81.25%)	2.75(68.75%)	2.75(68.75%)
16% (6)	3.50(87.50%)	2.88(71.88%)	2.75(68.75%)
19% (7)	3.63(90.63%)	3.00(75.00%)	2.88(71.88%)
22% (9)	3.75(93.75%)	3.50(87.50%)	3.38(84.38%)
25% (10)	3.75(93.75%)	3.50(87.50%)	3.50(87.50%)
28% (11)	3.75(93.75%)	3.50(87.50%)	3.50(87.50%)
31% (12)	3.88(96.88%)	3.63(90.63%)	3.63(90.63%)

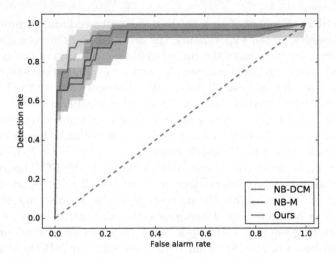

Fig. 2. ROC curves for different detection techniques averaged over 20 random splits of the SPECT heart dataset. The shaded areas represent the variance of each curve: $\pm 0.5 \times$ standard deviation.

Table 4. The computational cost of each detection method (averaged over 20 runs)

Dataset	Method	Computational cost (s)	
		Training	Testing
Mushroom	NB-M	1.06×10^{-2}	7.49×10^{-1}
	NB-DCM	45.72	5.64×10^{-1}
	Ours	1178.18	4282.47
SPECT heart	NB-M	7.69×10^{-3}	6.01×10^{-3}
	NB-DCM	8.56×10^{-1}	6.43×10^{-3}
	Ours	380.81	153.63

Additionally, the detection phase takes a large computational cost since an optimization process is needed to assign the anomaly scores. We will further try to decrease the time overhead of detection phase by modeling the latent space with a recognition model as done in variational auto-encoders.

5   Conclusion

In this paper, we propose a semi-supervised anomaly detection method for categorical observations. Our proposal is a distribution-based technique, and we use CLGP to estimate the distribution of conventional behaviors. CLGP is a probabilistic generative model, and the conventional categorical observations are considered to be generated from a continuous manifold through Gaussian process. Due to the bayesian non-parametric properties of Gaussian process, CLGP can well capture the probabilistic distributions of conventional behaviors, even when faced with small-scale conventional sample set. The experimental results on categorical data show that our method can effectively detect anomalous categorical observations and achieve better detection performance compared with other distribution-based techniques.

Acknowledgments. This paper is supported by the National Natural Science Foundation of China under grant No. 61572109, No. 11461006 and No. 61402080. The authors would like to thank the anonymous reviewers for their helpful and constructive comments.

References

1. Abolhasanzadeh, B.: Gaussian process latent variable model for dimensionality reduction in intrusion detection. In: Electrical Engineering (2015)
2. Agarwal, D.: Detecting anomalies in cross-classified streams: a Bayesian approach. Knowl. Inf. Syst. **11**(1), 29–44 (2007)
3. Aggarwal, C.C., Yu, P.S.: Outlier detection for high dimensional data. In: ACM Sigmod Record, vol. 30, pp. 37–46. ACM (2001)
4. Anscombe, F.J.: Rejection of outliers. Technometrics **2**(2), 123–146 (1960)
5. Beal, M.J.: Variational algorithms for approximate Bayesian inference. University of London United Kingdom (2003)
6. Butun, I., Morgera, S.D., Sankar, R.: A survey of intrusion detection systems in wireless sensor networks. IEEE Commun. Surv. Tutor. **16**(1), 266–282 (2014)
7. Chandola, V., Banerjee, A., Kumar, V.: Anomaly detection: a survey. ACM Comput. Surv. (CSUR) **41**(3), 15 (2009)
8. D'Alconzo, A., Coluccia, A., Ricciato, F., Romirer-Maierhofer, P.: A distribution-based approach to anomaly detection and application to 3G mobile traffic. In: GLOBECOM, pp. 1–8 (2009)
9. Damianou, A.C., Titsias, M.K., Lawrence, N.D.: Variational inference for latent variables and uncertain inputs in Gaussian processes. J. Mach. Learn. Res. **17**(1), 1425–1486 (2016)
10. Gal, Y., Chen, Y., Ghahramani, Z.: Latent Gaussian processes for distribution estimation of multivariate categorical data. In: Proceedings of the 32nd International Conference on Machine Learning, ICML2015, pp. 645–654 (2015)
11. Goodfellow, I., Pouget-Abadie, J., Mirza, M., Xu, B., Warde-Farley, D., Ozair, S., Courville, A., Bengio, Y.: Generative adversarial nets. In: Advances in Neural Information Processing Systems, pp. 2672–2680 (2014)
12. Kingma, D.P., Welling, M.: Auto-encoding variational bayes. arXiv preprint arXiv:1312.6114 (2013)
13. Kudo, D., Waizumi, Y., Nemoto, Y.: Network traffic anomaly detection using multinomial distribution model according to service. Gastroenterology **148**(4), S-500–S-501 (2015)
14. Laxhammar, R., Falkman, G., Sviestins, E.: Anomaly detection in sea traffic - a comparison of the Gaussian mixture model and the kernel density estimator. In: International Conference on Information Fusion, pp. 756–763. IEEE Computer Society (2009)
15. Oliveira, H., Caeiro, J.J., Correia, P.L.: Improved road crack detection based on one-class parzen density estimation and entropy reduction. In: 2010 17th IEEE International Conference on Image Processing (ICIP), pp. 2201–2204 (2010)

16. Orbanz, P., Teh, Y.W.: Bayesian nonparametric models. In: Sammut, C., Webb, G.I. (eds.) Encyclopedia of Machine Learning, pp. 81–89. Springer, Boston (2011)
17. Ranganathan, A.: PLISS: detecting and labeling places using online change-point detection. Auton. Robots **32**(4), 351–368 (2010)
18. Shewhart, W.A.: Economic Control of Quality of Manufactured Product. Van Nostrand, New York City (1931)
19. Swarnkar, M., Hubballi, N.: OCPAD: one class Naive Bayes classifier for payload based anomaly detection. Expert Syst. Appl. **64**, 330–339 (2016)
20. Titsias, M.K.: Variational learning of inducing variables in sparse Gaussian processes. In: AISTATS, vol. 5, pp. 567–574 (2009)
21. Van Vlasselaer, V., Bravo, C., Caelen, O., Eliassi-Rad, T., Akoglu, L., Snoeck, M., Baesens, B.: APATE: a novel approach for automated credit card transaction fraud detection using network-based extensions. Decis. Support Syst. **75**, 38–48 (2015)
22. Wang, W., Zhang, B., Wang, D., Jiang, Y., Qin, S., Xue, L.: Anomaly detection based on probability density function with Kullback-Leibler divergence. Sig. Process. **126**, 12–17 (2016)

Make Users and Preferred Items Closer: Recommendation via Distance Metric Learning

Junliang Yu[1,2], Min Gao[1,2(✉)], Wenge Rong[3], Yuqi Song[1,2], Qianqi Fang[1,2], and Qingyu Xiong[1,2]

[1] School of Software Engineering, Chongqing University, Chongqing 400044, China
{yu.jl,gaomin,songyq,fqq0429,xiong03}@cqu.edu.cn
[2] Key Laboratory of Dependable Service Computing in Cyber Physical Society (Chongqing University), Ministry of Education, Chongqing 400044, China
[3] School of Computer Science and Engineering, Beihang University, Beijing 100191, China
w.rong@buaa.edu.cn

Abstract. Recommender systems can help to relieve the dilemma called information overload. Collaborative filtering is a primary approach based on collective historical ratings to recommend items to users. One of the most competitive collaborative filtering algorithm is matrix factorization. In this paper, we proposed an alternative method. It aims to make users be spatially close to items they like and be far away from items they dislike, by connecting matrix factorization and distance metric learning. The metric and latent factors are trained simultaneously and then used to generate reliable recommendations. The experiments conducted on the real-world datasets have shown that, compared with methods only based on factorization, our method has advantage in terms of accuracy.

Keywords: Recommendation · Distance metric learning · Collaborative filtering · Matrix factorization

1 Introduction

The rapid development of the Internet makes information exchange conveniently, but it also leads to a dilemma called information overload. As an automatic system that can recommend an appropriate item, the recommender system now plays an increasingly remarkable role to facilitate valuable information acquisition. Typically, recommender systems are predicated on collaborative filtering (CF), a technique that relies on collective historical ratings to predict items which will be positively rated by the active user [9].

Matrix factorization [7] is a type of CF-based methods that has been extensively applied due to its scalability and efficiency. Most existing recommender systems are based on matrix factorization [7], a basic model that has been extensively used owing to its scalability and efficiency. Generally, matrix factorization decomposes the user-item rating matrix into a user-preference matrix and an

© Springer International Publishing AG 2017
D. Liu et al. (Eds.): ICONIP 2017, Part V, LNCS 10638, pp. 297–305, 2017.
https://doi.org/10.1007/978-3-319-70139-4_30

item-characteristics matrix based on the observed ratings. Each user or item is denoted with a low-dimensional vector called the latent factors. Then, the obtained latent factors are used to complement the rating matrix with their dot products, which can be considered as a type of interaction between users and items. However, in most cases matrix factorization fails to capture the finer grained preference information as it has some problems in explaining the dot product of the latent factors derived from the decomposition.

In this paper, we proposed a novel method having good interpretability, which is based on matrix factorization as well. It allows us to get a spatial understanding of the latent factor space and how users and items are positioned inside it. The principle of the proposed method is that distance reflects likability; namely, in our model, the closer two points are (one is the user, the other is an item), the more likely is consumption or a click. However, in our model we use Mahalanobis distance to replace Euclidean distance to measure the gap between locations because of the previous research [10,13,14] which have shown that a good distance metric can remarkably improve the performance of learners that rely on spatial data. Therefore, distance metric learning is incorporated in our method [8]. In contrast to general distance metric learning, our model needs to learn the samples (latent factors) and the distance metric simultaneously, which makes the model scalable. At the end of training, users are spatially close to their preferred items and far away from their disliked items. The experiments conducted on real-world dataset show that our method are superior to the state-of-the-art methods in terms of accuracy.

The remainder of this paper is organised as follows. In Sect. 2, the background is briefly introduced. Section 3 will focus on the proposed method. Section 4 reports the experimental results. Finally, in Sect. 5 we conclude this paper and point out some potential future work.

2 Background

2.1 Related Work

Generally, most of the model-based CF methods are on the basis of matrix factorization [7], in which the user-item rating matrix $R \in \mathbb{R}^{m \times n}$ is decomposed into user latent preference matrix $U \in \mathbb{R}^{m \times k}$ and item latent characteristics matrix $V \in \mathbb{R}^{n \times k}$ by using observed ratings. And the unknown ratings in the original rating matrix can be complemented by the dot products of the learned matrices. Since matrix factorization is prone to incorporate prior knowledge, lots of derivative methods [1,6,12] were brought up, which have been proven effective.

Due to the lack of interpretability for matrix factorization, there have some models take distance into consideration [3–5]. However, they use Euclidean distance to measure the gap and do not learn a distance metric to capture the interactions between different dimensions.

Among them, the Euclidean embedding model [5] is the most relevant work to ours. The idea of the model is that all items and users can be embedded in a unified Euclidean space, and the learned coordinates of users and items should reflect the negative correlation between the distance and the given rating. It means, the higher a given rating is, spatially the closer the user and the item are. However, whether the Euclidean distance can measure the gap between two locations precisely is questionable. Meanwhile, no constraints were imposed to guarantee that the learned distance is desired.

In [3], an item-ranking-based recommender was proposed. This model learns a joint metric space to encode not only users' preferences but also the user-user and item-item similarity. The authors focus on the collaborative filtering problem for implicit feedback and use collective metric learning to capture the interactions among all points in the metric space. They also demonstrate their model's ability to integrate various types of item features and prove that their model can uncover the fine-grained relationships among users' preferences. However, this model does not learn a metric matrix to reveal the relations among different dimensionalities.

2.2   Distance Metric Learning

Distance metric learning is crucial in real-world application. Generally, the training samples of distance metric learning will be cast into pairwise constraints. The target of distance metric learning is to learn a distance metric subjected to a given set of constraints in a global sense. Let the distance metric denoted by matrix $A \in \mathbb{R}^{k \times k}$, and the distance between any two samples x and y expressed by

$$d_A^2(x, y) = \|x - y\|_A^2 = (x - y)A(x - y)^T. \tag{1}$$

In Eq. 1, A has to be a positive semi-definite matrix to keep the distance non-negative and symmetric. The global optimization problem with constraints can be stated as

$$\min_{A \in \mathbb{R}^{k \times k}} \sum_{(x,y) \in S} \|x - y\|_A^2,$$
$$s.t. \quad A \succeq 0, \quad \sum_{(x,y) \in D} \|x - y\|_A^2 \geq \beta, \tag{2}$$

where S denotes the set of equivalent constraints in which x and y belong to the same class, and D denotes the set of inequivalent constraints in which x and y belong to different classes, and β is a constant to restrict the minimum distance between data points in different classes.

3   Recommendation via Distance Metric Learning

In this section we put forward a **Recommendation** model connecting **F**actorization and **D**istance metric learning called **RFD**.

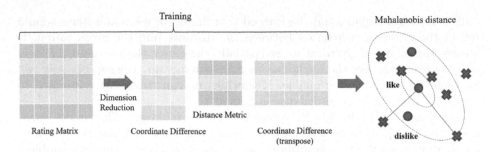

Fig. 1. Users (circles) and items (crosses) are embedded in a low dimensional space. The more closer the two symbols are, the more likely that the user prefers the item.

3.1 Preliminaries

In recommender systems, we have a user set $\mathbf{U} = \{u_1, \ldots u_m\}$, and an item set $\mathbf{I} = \{i_1, \ldots i_n\}$. Ratings given by users on items are marked with the matrix $\mathbf{R} = [r_{u,i}]_{m \times n}$, and $r_{u,i}$ denotes the rating from user u on item i. The task of a recommender can be summarized as follows: given a user u and an item i, using the observed ratings in $\mathbf{R}$ to predict the missing $r_{u,i}$.

3.2 Model Definition

As the latent factors in traditional matrix factorization are difficult to explain, we think something comprehensible should be integrated. The inspiration for RFD is that distance reflects likability. The latent factors in RFD can be explained as the coordinates of users and items rather than the characteristics. Since we consider that all dimensions in the latent factors are correlative, we introduce a distance metric matrix to capture the interactions between different dimensions. Different from the general distance metric learning, our model regards the latent factors as the training samples which are not prepared at first, and classifies labels into two types (like and dislike) according to the ratings expressed by users. Note that, we consider that if a user rates an item with a higher score, it shows the item is positively rated, otherwise the item is negatively rated. And the sets of pairwise constraints are constructed as follows: given a user and an item, if the user positively rated the item, the pair will be distributed to the set of equivalent constraints, otherwise they will be distributed to the set of inequivalent constraints.

To reach the goal of our model, we need to learn the latent factors and the distance metric simultaneously. Firstly, we initialize two k-rank matrices filled with random values denoting the user latent matrix and the item latent matrix respectively, and define a matrix $\in \mathbb{R}^{k \times k}$ to be the distance metric. The Mahalanobis distance between users and items can be calculated by the dot products of the differences of latent factors and the distance metric. During the training stage, constraints are imposed to guarantee that users should be spatially close to their liked items, and be far away from their disliked items. Finally, the obtained latent factors can be interpreted as points in the low dimensional space, and the distance calculated can be used to generate understandable recommendations. The overall process is shown in Fig. 1.

According to the conception above, in our model, the predicted rating is defined as

$$\hat{r}_{ui} = \mu + b_u + b_i - (x_u - y_i)\mathbf{A}(x_u - y_i)^T \tag{3}$$

where x_u and $y_i \in \mathbb{R}^k$ are the coordinates of user u and item i, the subscript u and i specify the active user and the active item, μ represents the overall average rating, and b_u and b_i indicate the deviations of user u and item i. Instead of learning a positive semi-definite matrix A directly, we can learn $W \in \mathbb{R}^{k \times k}$ under the condition that $WW^T = A$. Furthermore, W does not need to be positive semi-definite, which makes the problem can be solved with generic approaches. Hence, we can learn latent factors and the distance metric by solving the following optimization problem:

$$\mathcal{L} = \sum_{u,i}(r_{ui} - \mu - b_u - b_i + (x_u - y_i)WW^T(x_u - y_i)^T)^2$$

$$+ \lambda(b_u^2 + b_i^2) + \alpha\{ \sum_{(u,i) \in N} [\beta - \|x_u - y_i\|_A^2]_+ \tag{4}$$

$$+ \sum_{(u,i) \in P} \|x_u - y_i\|_A^2 \}$$

where P is the set of pairs contains user u and his positively rated items, N is the set of pairs contains user u and his negatively rated items, the last two terms are constraints used to adjust the distance into an appropriate range, $[z]_+ = max(z; 0)$ is the standard hinge loss, λ controls the magnitudes of the biases, and α restricts the influence of the constraints.

There have been many optimizers available to find good solutions for this problem [2,11]. In our work, we use stochastic gradient descent, because it works very efficiently in case of redundant data. A local minimum of the objective function given by Eq. 4 can be found by performing gradient descent in b_u, b_i, x_u, y_i, and W,

$$\frac{\partial \mathcal{L}}{\partial b_u} = \lambda b_u - e_{ui} \qquad \frac{\partial \mathcal{L}}{\partial b_i} = \lambda b_i - e_{ui}$$

$$\frac{\partial \mathcal{L}}{\partial y_i} = -(e_{ui} \pm \alpha)(x_u - y_i)(WW^T + W^TW)$$

$$\frac{\partial \mathcal{L}}{\partial x_u} = (e_{ui} \pm \alpha)(x_u - y_i)(WW^T + W^TW) \tag{5}$$

$$\frac{\partial \mathcal{L}}{\partial W} = (e_{ui} \pm \alpha)W(x_u - y_i)^T(x_u - y_i)$$

where e_{ui} is the gap between the real rating and the predicted result, γ is the step size, and the exact operator of the plus-minus sign is determined by the sign of the term related to distance in Eq. 4.

Compared with matrix factorization, RFD is is more understandable. As for the training cost, apart from the latent factors, RFD merely needs to learn an additional distance metric matrix with $k \times k$ elements. Since k is a small number which is consistent with the dimension of the latent factor of users and items, the cost will not be computationally expensive.

Table 1. The ranking quality of all methods

	FilmTrust				MovieLens			
Metric	PMF	EE	SVD	**RFD**	PMF	EE	SVD	**RFD**
Recall@20	11.53%	13.74%	12.92%	**15.06%**	3.83%	4.60%	3.44%	**5.21%**
Recall@50	16.14%	21.78%	19.29%	**24.17%**	4.33%	6.97%	4.67%	**7.27%**

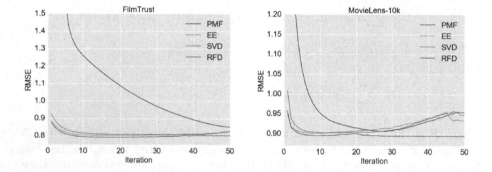

Fig. 2. Performance of all methods

4   Experiments and Analysis

4.1   Experimental Setup

Two real-world datasets are used in our experiments. The smaller one is FilmTrust crawled from the entire FilmTrust website, and the larger one is MovieLens-10k. The FilmTrust dateset consists of 35,497 ratings (05–4.0 scale) by 1,508 users on 2,071 movies, and the MovieLens-10k dataset consists of 100,004 ratings (1–5 scale) by 671 users on 8,417 movies.

In our experiments, the Root Mean Square Error (RMSE) is chosen to measure the prediction error of all methods. Besides, we also use Recall@k to measure the quality of the recommendation list, which is more important to users.

In order to show the performance improvement of our RFD method, we compare our method with PMF [12], SVD [7], and the Euclidean embedding (EE) model. In all experiments, we set the parameters of these methods according to their best performance on the validation set.

To ensure the creditability of our experiments, we employed a 5-fold cross validation strategy, all of the values presented in next section are the average of the 5 folds.

4.2   Performance for Predicting Missing Ratings

In this part, we compare the performance of RFD with that of the baselines. We set $\alpha = 0.1$, $\beta = 0.25$ for RFD, step size $\gamma = 0.005$, dimension $d = 20$ and the regularization parameter $\lambda = 0.01$ for all the methods.

Figure 2 shows the change of RMSE on the test set as the stochastic gradient descent algorithm proceeds. We can clearly observe that RFD are superior to the baselines on both two datasets. It should be noted that, the predicted rating equations of RFD, SVD, and EE include the part of the global average rating, therefore they can obtain better initial predictions and converge fast. In addition, we can see that all the baselines overfit at the last stage of the training process. However, we may consider that, as RFD has utilized the distance metric to capture the interactions between different dimensions, it tends to keep stable rather than overfit.

In reality, the recommendations for users are usually presented as a recommendation list. Thus, we prefer the ranking metrics rather than the RMSE. In the preprocessing stage, we binarize the explicit rating data by keeping the top 25% ratings. In the prediction stage, we use Eq. 3 to generate ranking scores for users on all items. Sorting the ranking scores for all users, we can get the recommendation lists. In Table 1, we list the Recall@20 and Recall@50 of all methods. As users rate few of the items and the candidate items are numerous, the figures in Table 1 may look small. However, it is inevitable and can also reflect the capacity of different methods. From Table 1, we can see RFD significantly outperforms the baselines on both two datasets, despite the fact that it is not fine tuned.

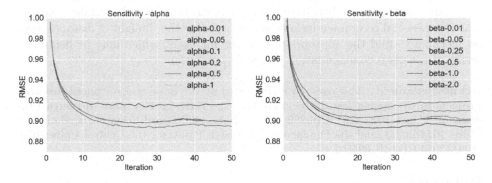

Fig. 3. Change of the distance

4.3 Investigation for the Sensitivity of Parameters

In our proposed method, two new hyper-parameters are introduced. They are α and β, which are used to control the influence of the constraints and to restrict the minimum distance between a user and his negatively rated items, respectively. In this subsection, we will investigate the sensitivity of these two parameters. First, we fix $\beta = 0.25$, and change the value of α to see the change of the performance. Then, we fix $\alpha = 0.1$, to observe the influence of the change of β. We conduct the experiment on MovieLens-10k, and the other settings are the same as those in Sect. 4.2.

Figure 3 shows the variation of the learning curves with different values of α and β. From the left part of Fig. 3, we can see our RFD model achieves the best performance when $\alpha = 0.05$, while smaller value like $\alpha = 0.01$ can potentially degrade the model performance. This indicates that that fusing the distance greatly improves the recommendation accuracy. From the right part of Fig. 3, we observe that RFD achieves the best performance when $\beta = 0.5$, while large values will lead to obvious performance degradation. In Eq. 3 we can see the predicted rating is determined by the global average rating, the biases, and the distance together. As most of users tend to express positive opinions, choosing a smaller β can obtain a good performance. In fact, the value of this parameter, to a great extent, depends on the distribution of the dataset.

5 Conclusion

In this paper, we proposed a novel recommendation method based on distance metric learning, it aims to minimize the distance between a user and his positively rated items and to maximize the distance between a user and his negatively rated items. The experiments show that, RFD outperforms the-state-of-art methods in terms of the recommendation accuracy.

Similarly, the distance between two users can be defined as well. With the growing popularity of the online social platform, social network based approaches to recommendation have become a hot topic in the field of recommender systems. The future social recommendations can incorporate the thought of distance metric learning to find the appropriate distance for users, which may be helpful in terms of friends recommendation and links prediction.

Acknowledgment. This research is supported by the Basic and Advanced Research Projects in Chongqing (cstc2015jcyjA40049), the National Key Basic Research Program of China (973) (2013CB328903), the National Natural Science Foundation of China (61472021), and the Fundamental Research Funds for the Central Universities (106112014 CDJZR 095502).

References

1. Hernando, A., Bobadilla, J., Ortega, F.: A non negative matrix factorization for collaborative filtering recommender systems based on a Bayesian probabilistic model. Knowl. Based Syst. **97**, 188–202 (2016)

2. Hocke, J., Martinetz, T.: Global metric learning by gradient descent. In: Wermter, S., Weber, C., Duch, W., Honkela, T., Koprinkova-Hristova, P., Magg, S., Palm, G., Villa, A.E.P. (eds.) ICANN 2014. LNCS, vol. 8681, pp. 129–135. Springer, Cham (2014). doi:10.1007/978-3-319-11179-7_17

3. Hsieh, C.K., Yang, L., Cui, Y., Lin, T.Y.: Collaborative metric learning. In: Proceedings of the 26th International Conference on World Wide Web, pp. 193–201. ACM (2017)

4. Jamali, M., Ester, M.: A matrix factorization technique with trust propagation for recommendation in social networks, pp. 135–142 (2010)

5. Khoshneshin, M., Street, W.N.: Collaborative filtering via Euclidean embedding. In: ACM Conference on Recommender Systems, pp. 87–94 (2010)

6. Koren, Y.: Collaborative filtering with temporal dynamics. Commun. ACM **53**(4), 89–97 (2010)

7. Koren, Y., Bell, R.M., Volinsky, C.: Matrix factorization techniques for recommender systems. IEEE Comput. **42**(8), 30–37 (2009)

8. Kulis, B., et al.: Metric learning: a survey. Found. Trends ®. Mach. Learn. **5**(4), 287–364 (2013). Now Publishers, Inc.

9. Lu, J., Wu, D., Mao, M., Wang, W., Zhang, G.: Recommender system application developments: a survey. Decis. Support Syst. **74**, 12–32 (2015)

10. Nguyen, B., Morell, C., De Baets, B.: Supervised distance metric learning through maximization of the Jeffrey divergence. Pattern Recog. **64**, 215–225 (2017). Elsevier

11. Qian, Q., Jin, R., Yi, J., Zhang, L., Zhu, S.: Efficient distance metric learning by adaptive sampling and mini-batch stochastic gradient descent (SGD). Mach. Learn. **99**(3), 353–372 (2015)

12. Salakhutdinov, R., Mnih, A.: Probabilistic matrix factorization. In: International Conference on Neural Information Processing Systems, pp. 1257–1264 (2007)

13. Wang, J., Deng, Z., Choi, K., Jiang, Y., Luo, X., Chung, F., Wang, S.: Distance metric learning for soft subspace clustering in composite kernel space. Pattern Recogn. **52**, 113–134 (2016)

14. Wu, P., Hoi, S.C.H., Zhao, P., Miao, C., Liu, Z.: Online multi-modal distance metric learning with application to image retrieval. IEEE Trans. Knowl. Data Eng. **28**(2), 454–467 (2016)

Deep Bi-directional Long Short-Term Memory Model for Short-Term Traffic Flow Prediction

Jingyuan Wang[1], Fei Hu[1,2], and Li Li[1(✉)]

[1] School of Computer and Information Science, Southwest University,
Chongqing, China
wjykim@email.swu.edu.cn, lily@swu.edu.cn
[2] Network Centre, Chongqing University of Education, Chongqing, China

Abstract. Short-term traffic flow prediction plays an important role in intelligent transportation system. Numerous researchers have paid much attention to it in the past decades. However, the performance of traditional traffic flow prediction methods is not satisfactory, for those methods cannot describe the complicated nonlinearity and uncertainty of the traffic flow precisely. Neural networks were used to deal with the issues, but most of them failed to capture the deep features of traffic flow and be sensitive enough to the time-aware traffic flow data. In this paper, we propose a deep bi-directional long short-term memory (DBL) model by introducing long short-term memory (LSTM) recurrent neural network, residual connections, deeply hierarchical networks and bi-directional traffic flow. The proposed model is able to capture the deep features of traffic flow and take full advantage of time-aware traffic flow data. Additionally, we introduce the DBL model, regression layer and dropout training method into a traffic flow prediction architecture. We evaluate the prediction architecture on the dataset from Caltrans Performance Measurement System (PeMS). The experiment results demonstrate that the proposed model for short-term traffic flow prediction obtains high accuracy and generalizes well compared with other models.

Keywords: Traffic flow prediction · Long short-term memory · Deep learning · PeMS

1 Introduction

Intelligent transportation system (ITS) is an important part of smart city [1]. Short-term traffic flow prediction is a core technology in ITS. The aim of short-term traffic flow prediction is to predict the number of vehicles within a given time interval on the basis of the historical traffic information. The time intervals are usually in the range of 5 to 30 min. The study of short-term traffic flow prediction has an important significance in real-time route guidance and reliable traffic control strategies [2,3]. Great efforts have been made to cope with the short-term traffic flow prediction problem in the past decades.

© Springer International Publishing AG 2017
D. Liu et al. (Eds.): ICONIP 2017, Part V, LNCS 10638, pp. 306–316, 2017.
https://doi.org/10.1007/978-3-319-70139-4_31

The traditional methods for short-term traffic flow prediction are parametric approaches. In [4,5], time series methods, such as the autoregressive integrated moving average (ARIMA), were employed to forecast short-term traffic flow. Sun et al. [6] proposed a Bayesian network approach to predict short-term traffic flow. Yu et al. [7] used Markov chain model for short-term traffic flow prediction. Due to the stochasticity and nonlinearity of the traffic flow, parametric approaches cannot describe traffic flow precisely.

More and more researchers tried to apply nonparametric approaches to short-term traffic flow prediction, for nonparametric approaches can capture the complicated nonlinearity of the traffic flow and take the uncertainty into consideration. Castro Neto et al. [8] employed support vector regression (SVR) to predict short-term traffic flow under typical and atypical traffic conditions. In [9], a locally weighted learning (LWL) method was proposed. Neural network (NN) models were reported in [10]. Owing to the ability of dealing with high-dimensional data, flexible model structure, strong generalization and learning ability of deep learning methods, many deep learning models and structures were applied for traffic flow prediction. Huang et al. [2] incorporated multitask learning (MTL) into deep belief networks (DBN) for traffic flow prediction. Lv et al. [11] proposed a stacked auto-encoder (SAE) model. Tian and Pan [12] used long short-term memory (LSTM) recurrent neural network to forecast short-term traffic flow prediction, which could automatically determine the optimal time lags. The limitations of the current deep learning models are the shallow structure unable to mine deep features and the weak sensitivity to time-aware traffic flow data.

Unlike the existing approaches, we propose a deep architecture for traffic flow prediction in this paper. For unsupervised feature learning, we propose a deep bi-directional long short-term memory (DBL) by introducing long short-term memory (LSTM) recurrent neural network, residual connections, deeply hierarchical networks and bi-directional traffic flow. A regression layer is used above the DBL for supervised prediction. Additionally, we adopt dropout training method to avoid overfitting problem. In other words, our model is able to mine the deep features of traffic flow and take full advantage of time-aware traffic flow data.

The rest of this paper is organized as follows. Section 2 formalizes the problem of traffic flow prediction and introduces long short-term memory (LSTM) recurrent neural network model. In Sect. 3, we present the traffic flow prediction architecture. Section 4 shows the experimental settings and the experimental results. In Sect. 5, we discuss the key components of our model. Finally, Sect. 6 is the conclusion.

2   Related Work

2.1   Short-Term Traffic Flow Prediction

Traffic flow prediction is a typical temporal and spatial process. The traffic flow prediction problem can be stated as follows. The traffic flow of the i_{th} observation point (road, segment or station) at the t_{th} time interval is denoted as $f_{i,t}$. At time t', the prediction task is to forecast the traffic flow $f_{i,t'+1}$ at time $t' + 1$, which is based on the traffic flow sequence $F = \{f_{i,t} | i \in O, t = 1, 2, \ldots, t'\}$ in the past. O is the full set of observation points. The prediction time interval is the interval between time t and $t + 1$, which is denoted as Δt. According to the length of the prediction time interval, the traffic flow prediction can be divided into three types: long-term, mid-term and short-term traffic flow prediction. Short-term traffic flow prediction has a significant meaning in real-time route guidance and reliable traffic control strategies. 15-min is recommended as short-term prediction interval by the Highway Capacity Manual [13].

2.2 Long Short-Term Memory Network

Long Short-Term Memory (LSTM) [14] is an effective approach to predict short-term traffic flow, which takes advantage of the three multiplicative units in the memory block to determine the optimal time lags dynamically. The LSTM prediction structure is composed of one input layer, one recurrent hidden layer whose basic unit is memory block, and one output layer. Memory blocks are a set of recurrently connected subnets. The memory block consists of one or more self-connected memory cells and three multiplicative units: the input gate, output gate and forget gate. The multiplicative gates allow LSTM memory cells to keep the information for long periods of time. The following equations mathematically abstract the process and the notations are illustrated in Table 1 [12].

$$h_t = H(W_{xh}x_t + W_{hh}h_{t-1} + b_h) \tag{1}$$

$$y_t = W_{hy}h_t + b_y \tag{2}$$

$$i_t = \sigma(W_{xi}x_t + W_{hi}h_{t-1} + W_{ci}c_{t-1} + b_i) \tag{3}$$

Table 1. Notations for LSTM prediction model

Notation	Definition
x	The input historical traffic flow sequence
i, f, o, c	The input gate, the forget gate, the output gate and the memory cell
h	The hidden vector sequence
y	The output predicted traffic flow sequence
W	The weight matrices (e.g. W_{xh} is the input-hidden weight matrix)
b	The bias vectors (e.g. b_h is hidden bias vector)
H	The hidden layer function
$\sigma(\cdot)$	The standard logistic sigmoid function
$g(\cdot), h(\cdot)$	The transformations of function $\sigma(\cdot)$

$$f_t = \sigma(W_{xf}x_t + W_{hf}h_{t-1} + W_{cf}c_{t-1} + b_f) \tag{4}$$

$$c_t = f_tc_{t-1} + i_tg(W_{xc}x_t + W_{hc}h_{t-1} + b_c) \tag{5}$$

$$o_t = \sigma(W_{xo}x_t + W_{ho}h_{t-1} + W_{co}c_t + b_o) \tag{6}$$

$$h_t = o_th(c_t) \tag{7}$$

$$\sigma(x) = \frac{1}{1 + e^{-x}} \tag{8}$$

$$g(x) = \frac{4}{1 + e^{-x}} - 2 \tag{9}$$

$$h(x) = \frac{2}{1 + e^{-x}} - 1 \tag{10}$$

3 Our Prediction Architecture

To capture the deep features of traffic flow and take full advantage of time-aware traffic flow data, we propose a deep bi-directional long short-term memory (DBL) model on the basis of Sect. 2.2. Additionally, we introduce the DBL model, regression layer and dropout training method into a traffic flow prediction architecture. In Sect. 3.1, we will show the details of the DBL model. And the traffic flow prediction architecture will be explained in Sect. 3.2.

3.1 Deep Bi-directional Long Short-Term Memory Model

The structure of deep bi-directional long short-term memory model is shown in Fig. 1(a). The input n-length historical traffic flow sequence is denoted as $x^0 = \{x_0^0, x_1^0, x_2^0, \ldots, x_{n-1}^0\}$, where $x_i^0 (i = 0, 1, 2, \ldots, n - 1)$ is the traffic flow at the t_{th} time interval. $x^i (i - 1, 2, \ldots, m)$ is the output of the i_{th} layer. BiLSTM is the bi-directional long short-term memory network, where $\overrightarrow{LSTM}$ encodes the input sequence from the start to the end and $\overleftarrow{LSTM}$ encodes the input sequence from the end to the start. Due to the strong ability to handle the sequential data, biLSTM has been successfully applied in natural language processing and image processing [15,16]. By using biLSTM, the traffic flow information of both directions can be taken into consideration.

In this paper, we use 6 biLSTM layers. The deep hierarchy structure always results in the gradient vanishing problem. To achieve the idea of modeling differences between an intermediate layers output and the targets, we introduce residual connections among the DBL layers in a stack (the red line shown in Fig. 1(a)). Residual connections performed well in the past [17,18]. With residual connections in the DBL model, the equations are as follows.

$$c_t^i, h_t^i = biLSTM^i(c_{t-1}^i, h_{t-1}^i, x_t^{i-1}; \theta^i) \tag{11}$$

$$x_t^i = x_t^{i-1} + h_t^i \tag{12}$$

$$c_t^{i+1}, h_t^{i+1} = biLSTM^{i+1}(c_{t-1}^{i+1}, h_{t-1}^{i+1}, x_t^i; \theta^{i+1}) \tag{13}$$

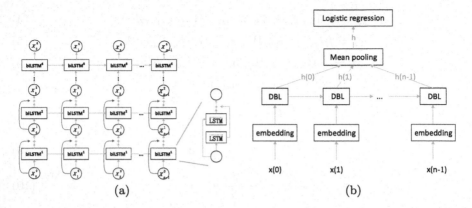

Fig. 1. (a) The structure of deep bi-directional long short-term memory model. (b) The structure of the prediction architecture.

where c_t^i and h_t^i are the memory states and hidden states of $biLSTM^i$ at the t_{th} time interval for the i_{th} layer, respectively; x_t^i is the input at the t_{th} time interval for the i_{th} layer; θ^i is the set of parameters of $biLSTM^i$ for the i_{th} layer.

3.2 The Traffic Flow Prediction Architecture

As shown in Fig. 1(b), the prediction architecture mainly consists of four parts: the embedding layer, the DBL, the mean pooling layer and the logistic regression layer. $\{x(0), x(1), \ldots, x(n-1)\}$ is the input which represents traffic flow data, and each $x(i)(i = 0, 1, \ldots, n-1)$ is a piece of traffic flow data at a time interval encoded by one-hot representation. The traffic flow data is mapped into a space of same dimension, which is a 64-dimensional vector space. After the DBL encodes the time-aware traffic flow information, a sequence $\{h(0), h(1), \ldots, h(n-1)\}$ is produced. Then, the mean pooling layer extracts mean values of the sequence over time intervals. Besides, the mean pooling layer makes the features encoded into a vector h. The vector h is fed into the logistic regression layer at the top of the prediction architecture.

To avoid overfitting problem [19] and improve the generalization capability of the model, we adopt the dropout method [20,21] in the embedding layer. The key idea of the dropout method is to randomly drop units (along with their connections) from the neural network during training, which can prevent units from co-adapting too much. During training, dropout samples from numerous different thinned networks. When testing, it becomes easy to approximate the effect of averaging the predictions of all these thinned networks and it can be achieved by a single unthinned network with smaller weights.

4 Experiments and Results

4.1 Experimental Settings

There are mainly two types of traffic flow data in the real world [2]. The first type is the loop detector data, which is collected by sensors on each road, such as inductive loops. The second type is the entrance-exit station data, which is collected at the entrance and exit of a road segment. The prediction task for the first type of data is to forecast the traffic flow on each road or segment, while the prediction task for another type is to forecast the traffic flow in each station, particularly the exit station.

In this paper, we evaluate our model and other comparison models on the dataset from Caltrans Performance Measurement System (PeMS) [22]. PeMS is the most widely used dataset in traffic flow prediction. PeMS constantly collects loop detector data in real time for more than 8100 freeway locations throughout the State of California. Thus, the PeMS dataset is a typical dataset of the loop detector data and the prediction task for PeMS is to forecast the traffic flow on each road or segment. We use the data of five months (from July to November) in 2016 as the training set and the later one month (December) as the testing set.

The models used in comparison experiments are explained as follows.

- **ARIMA:** autoregressive integrated moving average;
- **SVM:** support vector machine [8];
- **DBN:** deep belief network [2];
- **SAE:** stacked auto-encoder [11];
- **LSTM:** long short-term memory recurrent neural network [12];
- **BiLSTM:** we remove the deep hierarchy from the proposed architecture;
- **DBL:** the proposed prediction architecture.

4.2 Experimental Results

To evaluate the effectiveness of the traffic flow prediction models, we use two performance indexes, which are the Mean Absolute Percentage Error (MAPE) and the Root Mean Square Error (RMSE). According to them, we can evaluate the relative error and the absolute error [11]. They are defined as follows.

$$MAPE(f, \hat{f}) = \frac{1}{n} \sum_{i=1}^{n} \frac{|f_i - \hat{f}_i|}{f_i} \tag{14}$$

$$RMSE(f, \hat{f}) = \left[\frac{1}{n} \sum_{i=1}^{n} (|f_i - \hat{f}_i|)^2 \right]^{\frac{1}{2}} \tag{15}$$

where f is the observation (real) value of traffic flow, and $\hat{f}$ is the prediction value of traffic flow.

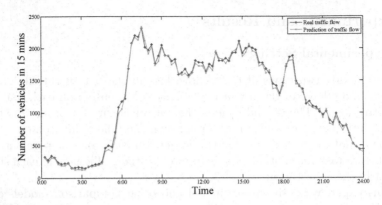

Fig. 2. The comparison between real traffic flow and prediction of traffic flow (Color figure online)

4.2.1   The Prediction Accuracy Comparison

To evaluate the prediction accuracy of the DBL model, we use DBL to predict 15-min interval traffic flow of a whole day (December 1, 2016) using the data collected from No. 311974 observation road on D03-5 freeway in California. As shown in Fig. 2, the red line represents the real traffic flow, and the blue line shows the prediction of traffic flow. From Fig. 2, we can notice that the performance of the DBL model is quiet good during most of the day. Besides, there are mainly three fluctuating periods, which are around 7:00, 12:00 and 18:00 respectively. Those fluctuating periods are all peak traffic periods during which the performance of the DBL model is not as stable as other periods. The MAPE of DBL is 4.83% and the RMSE of DBL is 46.01, which manifests that DBL obtains a high prediction accuracy.

The performance of seven models is tested and the results of them are listed in Table 2. As shown in Table 2, both MAPE and RMSE of DBL are lowest among the prediction models. We can notice that bi-directional traffic flow improves the performance of LSTM, while the deep hierarchy enhances the prediction ability of biLSTM.

Table 2. The results of traffic flow prediction with 15-min time interval

Models	MAPE(%)	RMSE
ARIMA	9.13	61.86
SVM	6.93	53.89
DBN	8.04	59.34
SAE	7.79	56.43
LSTM	6.21	50.32
BiLSTM	5.52	48.21
DBL	**4.83**	**46.01**

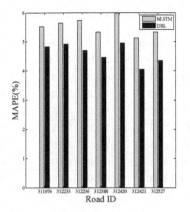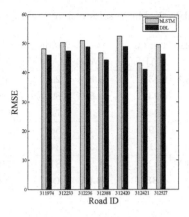

Fig. 3. The MAPE and RMSE comparison between biLSTM and DBL

4.2.2   The Effect of Deep Hierarchy

To verify the effectiveness of the deep hierarchy, we use biLSTM and DBL to predict 15-min interval traffic flow of a whole day on different roads. Due to the incompleteness of the data, only 876 observation roads of 1278 observation roads on D03-5 freeway in California have complete data. We evaluate the proposed method with the baseline model on these 876 observation roads and achieve the similar results. Next, we will illustrate the details for seven randomly selected roads and omit the details of others owing to the space limit. The MAPE and RMSE comparison between biLSTM and DBL is shown in Fig. 3. In our experiment, we notice that for the prediction model to obtain a high accuracy, the prediction network structure should be deep enough to capture deep features from the historical traffic information, which is similar to previous observations that for neural machine translation systems, the deep encoder and decoder RNNs significantly outperform shallow encoder and decoder RNNs [17]. According to the results, we can find that the MAPE and the RMSE of DBL are both lower than those of biLSTM, which means that the deep hierarchy is an effective part to improve the prediction accuracy.

4.2.3   The Generalization Capability of DBL

The selection of time interval has a great impact on the prediction accuracy of short-term traffic flow. Thus, the generalization capability of traffic flow prediction models can be evaluated by predicting traffic flow with different time intervals. The time intervals of short-term traffic flow prediction are usually no more than 30 min. We use 15-min, 20-min, 25-min and 30-min time intervals to verify the generalization capability of those prediction models and the results are shown in Table 3. As we can see from Table 3, the MAPE and RMSE of DBL are all lowest among the models with different time intervals, which demonstrates that DBL generalizes well.

Table 3. The results of traffic flow prediction with different time intervals

Models	15-min		20-min		25-min		30-min	
	MAPE(%)	RMSE	MAPE(%)	RMSE	MAPE(%)	RMSE	MAPE(%)	RMSE
ARIMA	9.13	61.86	9.22	68.32	9.24	85.59	9.63	124.21
SVM	6.93	53.89	6.89	65.13	6.92	79.23	7.23	106.41
DBN	8.04	59.34	7.72	67.35	8.10	82.32	8.45	115.20
SAE	7.79	56.43	7.56	62.27	7.43	75.30	7.63	104.32
LSTM	6.21	50.32	6.34	60.28	6.48	71.47	6.76	93.78
BiLSTM	5.52	48.21	5.81	59.89	5.71	69.49	5.60	87.21
DBL	**4.83**	**46.01**	**5.04**	**57.86**	**4.90**	**67.09**	**4.98**	**76.85**

5   Discussion

Compared with parametric models, neural network methods can capture the complicated nonlinearity of the traffic flow and take the uncertainty into consideration.

Through memory cells and three multiplicative units, long short-term memory (**LSTM**) recurrent neural network can automatically determine the optimal time lags when predicting traffic flow. Besides, the multiplicative gates allow LSTM memory cells to keep the traffic flow information for long periods of time.

Deep features are hidden in the historical traffic information that are hard to extract, but our prediction architecture can mine them with the deeply hierarchical neuron networks; **bi-directional traffic flow** helps mine time-aware traffic information from forward and backward directions; **residual connections** can improve the gradient flows and help avoid gradient vanishing with a highly deep neuron networks.

The **dropout** probabilities are used in the embedding layer as a learned function of the input, which helps to avoid the overfitting problem and make the model generalize well.

Those key components greatly improve the traffic flow prediction model to obtain a higher accuracy and the experiments have proved their effectiveness and efficiency. In addition to the traffic flow prediction, our proposed deep prediction architecture has potential applications in other fields such as traffic prediction in wireless networks, the study of shifting meaning of words, the analysis of changes in people's spending habits and etc.

6   Conclusion and Future Work

In this paper, we propose a deep bi-directional long short-term memory (DBL) model which is able to capture the deep features of traffic flow and take full advantage of time-aware traffic flow data. Additionally, we introduce the DBL, regression layer and dropout training method into a traffic flow prediction architecture. To verify the performance of the DBL model, the dataset from PeMS is used in the proposed model and other six comparison models (ARIMA, SVM, DBN, SAE, LSTM, BiLSTM). In the exprimental results, DBL obtains high accuracy. Besides, both the MAPE and RMSE of DBL are lowest among the comparison models with different prediction intervals, which demonstrates that DBL is effective and generalizes well.

We are planning to take more factors, such as the situation of rainfall and the size of vehicle, into consideration for a much higher accuracy. And the whole model and its optimization strategy will be evaluated on more data.

References

1. Vlahogianni, E.I., Karlaftis, M.G., Golias, J.C.: Optimized and meta-optimized neural networks for short-term traffic flow prediction: a genetic approach. Transp. Res. Part C: Emerg. Technol. **13**(3), 211–234 (2005)
2. Huang, W., Song, G., Hong, H., Xie, K.: Deep architecture for traffic flow prediction: deep belief networks with multitask learning. IEEE Trans. Intell. Transp. Syst. **15**(5), 2191–2201 (2014)
3. Abadi, A., Rajabioun, T., Ioannou, P.A.: Traffic flow prediction for road transportation networks with limited traffic data. IEEE Trans. Intell. Transp. Syst. **16**(2), 653–662 (2015)
4. Moorthy, C.K., Ratcliffe, B.G.: Short term traffic forecasting using time series methods. Transp. Plan. Technol. **12**(1), 45–56 (1988)
5. Thomas, T., Weijermars, W., Van Berkum, E.: Predictions of urban volumes in single time series. IEEE Trans. Intell. Transp. Syst. **11**(1), 71–80 (2010)
6. Sun, S., Zhang, C., Yu, G.: A Bayesian network approach to traffic flow forecasting. IEEE Trans. Intell. Transp. Syst. **7**(1), 124–132 (2006)
7. Yu, G., Hu, J., Zhang, C., Zhuang, L., Song, J.: Short-term traffic flow forecasting based on Markov chain model. In: Proceedings of IEEE Intelligent Vehicles Symposium, pp. 208–212 (2003)
8. Castro-Neto, M., Jeong, Y.S., Jeong, M.K., Han, L.D.: Online-SVR for short-term traffic flow prediction under typical and atypical traffic conditions. Expert Syst. Appl. **36**(3), 6164–6173 (2009)
9. Shuai, M., Xie, K., Pu, W., Song, G., Ma, X.: An online approach based on locally weighted learning for short-term traffic flow prediction. In: Proceedings of the 16th ACM SIGSPATIAL International Conference on Advances in Geographic Information Systems, p. 45 (2008)
10. Chan, K.Y., Dillon, T.S., Singh, J., Chang, E.: Neural-network-based models for short-term traffic flow forecasting using a hybrid exponential smoothing and Levenberg-Marquardt algorithm. IEEE Trans. Intell. Transp. Syst. **13**(2), 644–654 (2012)
11. Lv, Y., Duan, Y., Kang, W., Li, Z., Wang, F.Y.: Traffic flow prediction with big data: a deep learning approach. IEEE Trans. Intell. Transp. Syst. **16**(2), 865–873 (2015)
12. Tian, Y., Pan, L.: Predicting short-term traffic flow by long short-term memory recurrent neural network. In: IEEE International Conference on Smart City/SocialCom/SustainCom (SmartCity), pp. 153–158 (2015)
13. Highway Capacity Manual, Special report, vol. 1, pp. 5–7, Washington D.C. (2000)
14. Hochreiter, S., Schmidhuber, J.: Long short-term memory. Neural Comput. **9**(8), 1735–1780 (1997)

15. Cornegruta, S., Bakewell, R., Withey, S., Montana, G.: Modelling radiological language with bidirectional long short-term memory networks (2016)
16. Bin, Y., Yang, Y., Shen, F., Xu, X., Shen, H.T.: Bidirectional long-short term memory for video description. In: ACM on Multimedia Conference, pp. 436–440 (2016)
17. Wu, Y., Schuster, M., Chen, Z., Le, Q.V., Norouzi, M., Macherey, W., Klingner, J.: Google's neural machine translation system: bridging the gap between human and machine translation. arXiv preprint arXiv:1609.08144 (2016)
18. He, K., Zhang, X., Ren, S., Sun, J.: Deep residual learning for image recognition. In: Proceedings of the IEEE Conference on Computer Vision and Pattern Recognition, pp. 770–778 (2016)
19. Hawkins, D.M.: The problem of overfitting. J. Chem. Inf. Comput. Sci. **44**(1), 1 (2004)
20. Hinton, G.E., Srivastava, N., Krizhevsky, A., Sutskever, I., Salakhutdinov, R.R.: Improving neural networks by preventing co-adaptation of feature detectors. Comput. Sci. **3**(4), 212–223 (2012)
21. Srivastava, N., Hinton, G., Krizhevsky, A., Sutskever, I., Salakhutdinov, R.: Dropout: a simple way to prevent neural networks from overfitting. J. Mach. Learn. Res. **15**(1), 1929–1958 (2014)
22. Caltrans Performance Measurement System (PeMS). http://pems.dot.ca.gov

Odor Change of Citrus Juice During Storage Based on Electronic Nose Technology

Xue Jiang, Pengfei Jia$^{(\boxtimes)}$, Siqi Qiao, and Shukai Duan

Southwest University, Chongqing 400715, China
jiapengfei200609@126.com

Abstract. In order to master the law of citrus juice odor components changes during the storing process, electronic nose composed of metal-oxide semiconductor (MOS) sensors array is used to monitor the odor during valencia oranges juice storing process. A self-made electronic nose system and experiment are described in detail, after data preprocessing, extreme learning machine (ELM) is used for analysis on samples. Analysis result indicates that the odor synthesized curve derived from the electronic nose technology can reflect overall trend of odor during valencia oranges juice storing process truly and effectively, and the experimental results prove that the E-nose can correctly distinguish the current stage of the stored valencia oranges juice and the classification accuracy of test data set is 96.29% when ELM is used as the classifier, which shows that the E-nose can be successfully applied to the qualitative analysis of citrus.

Keywords: Electronic nose · Citrus juice · Odor · Extreme learning machine

1 Introduction

Citrus industry is the largest fruit industry in the world, and orange juice is the most popular fruit beverage. Orange juice contains a lot of aroma components, which is one of the reasons why it is widely welcomed by consumers. Therefore, citrus flavor is an important characteristic of its quality [1]. Traditionally, the flavor of citrus is the comprehensive feeling of the human senses to the aroma and taste. The grade and value of citrus are largely determined by its flavor.

The traditional methods for citrus quality detection include sensory analysis and precision instrument analysis. Sensory analysis, after a long period of development and improvement, the analysis and evaluation, comparison and quality control of different products have been realized, which has made an indelible contribution to the development of society. However, the differences, fatigue and subjective problems of assessors will affect the scientific and objectivity of the sensory analysis results in a certain extent. Methods for the precision instruments analysis include gas chromatography (GC), gas chromatography with mass spectrometry (GC-MS), gas chromatography-olfactometry (GC-O) and so on [2]. These methods are objective, have good repeatability, and have become the main means of analysis and detection. However, these relevant instruments are expensive with low penetration rate, can only confined to well funded research platform, and need tedious pre-treatment steps and have strict requirements on the operator's skill and instrumental analysis skill. The real

© Springer International Publishing AG 2017
D. Liu et al. (Eds.): ICONIP 2017, Part V, LNCS 10638, pp. 317–326, 2017.
https://doi.org/10.1007/978-3-319-70139-4_32

production in urgent need of an analytical instrument for detection and identification of citrus juice flavor conveniently and quickly, in addition, giving an objective quality attribute evaluation.

Electronic nose (E-nose) is an expert system which is composed of an array of gas sensors and a corresponding artificial intelligence technique. It can be used for detection and analysis of many volatile components. Compared with traditional chemical methods, the detection is fast and convenient, can save a lot off manpower, material resources and financial resources. It is effective in dealing with problems analysis, and has been introduced to many fields such as food engineering [3], disease diagnosis [4], environmental control [5], explosive detection [6], spaceflight applications [7] and so on. As an intelligent sensory technology, E-nose has the advantages of strong independence, no fatigue, good repeatability and fast detection speed. It don't need to sample pretreatment, has little or no need to use any organic solvents, which is conducive to environmental protection, and in line with the needs of citrus flavor analysis testing requirements. The electronic nose is a kind of instrument which imitates the human olfactory organ. The object of the test is the gas/odor, which can realize the identification of the citrus or citrus juice, and realize the overall evaluation of the citrus or citrus juice aroma. Oshita et al. conducted the E-nose test when the "La France" pears were not mature, and analyzed the pears in three different stages using chemical analysis methods (GC, GC-MS) [8]. The results showed that the E-nose was able to distinguish three kinds of pears in different storage time and had a strong correlation with other analysis results. Corrado Di Natale et al. used E-nose to study the quality of citrus and apple, and showed that using E-nose could easily distinguish citrus species together with apple species and predict the extent of damage, and it could predict the storage time as well [9].

In this paper, E-nose is used to detect and identify the valencia orange juice, and extreme learning machine (ELM) is used as the classifier of E-nose to give quality attribute evaluation of the test data. The rest of this paper is organized as follows, the customized electronic nose system is introduced in Sect. 2. PS, experimental part is described clearly in Sect. 3. ELM is used to classify the test data, and the classification results are analyzed in Sect. 4. Finally, we will give the conclusion of this paper in Sect. 5.

2 Electronic Nose System

In this experiment, the E-nose system is designed independently and can be used to collect and monitor citrus flavor during storage. As shown in Fig. 1, the experimental platform mainly consists of the E-nose system, a PC, a temperature-humidity controlling chamber, a flow meter and an air pump. There are two ports on the sidewall of the chamber, and the target gas and the clean air are put into the chamber through ports 1 and 2, respectively. Data collected from the sensor array can be saved in a PC through a joint test action group port with its related software. An image of the experimental setup is shown in Fig. 2.

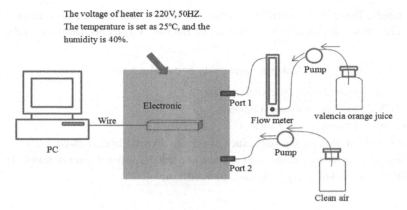

The voltage of heater is 220V, 50HZ.
The temperature is set as 25°C, and the
humidity is 40%.

Fig. 1. Schematic diagram of the experimental system.

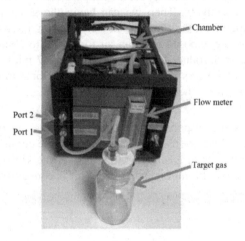

Fig. 2. Image of the experimental setup.

2.1 Sensor Selection

According to research reports, on the basis of the chemical composition of the aroma compounds in the valencia orange juice, it can be divided into hydrocarbons, alcohols, aldehydes, ketones, esters and so on [10]. Taking into account the gas sensor array of E-nose system can respond to all the chemical compositions of the target gas or smell and can output the measurable physical changes, the sensor array is usually composed with multiple gas sensors which have cross sensitivity, broad spectrum response and high sensitivity. Semiconductor gas sensor has many advantages, such as good physical and chemical stability, long life and low cost. Therefore, we choose semiconductor gas sensor to build the sensor array. The number of sensors in the array must be able to ensure the full coverage of the volatile odor, but also make the array to meet the high cost performance, size and power consumption and other indicators in line with the actual needs. Therefore, 15 semiconductor gas sensors are selected to form the sensor array. The sensitive characteristics of part of the gas sensors are shown in Table 1.

Table 1. Sensitivity of partial sensors

Sensors	Example response characteristics
TGS813	Methane, propane, isobutane
TGS822	Ethanol, organic solvent
MS1100	Aromatic compounds, such as toluene, formaldehyde, benzene
MQ135	Ammonia, sulfide, benzene, acetone, toluene, ethanol, carbon monoxide
MP503	Alcohol, smog, isobutane, formaldehyde

2.2   Selection of E-Nose Experimental Condition

Before sampling experiments, we firstly set the temperature and humidity of the chamber as 25°C and 40%, respectively. Then we can begin the gas sampling experiments. A single sampling experiment will implement the following three phases:

Phase 1:   All sensors are exposed to clean air for 5 min to obtain the baseline;

Phase 2:   Target gas is introduced into the chamber for 7 min;

Phase 3:   The array of sensors is exposed to clean air for 5 min again to wash the sensors and make them recover their baseline.

Figure 3 illustrates the response of sensors when valencia orange juice odor is introduced into the chamber. We can see that each response curve rises obviously from the fifth minute when the target gas begins to pass over the sensor array, and recovers to the baseline after the twelfth minute when clean air is conveyed to wash the sensors.

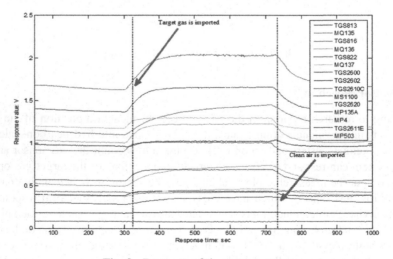

Fig. 3. Response of the sensors array.

As can be seen from Fig. 3, the maximum value of the response of each sensor is able to be observed when the sampling time is 7 min. The average value of the data from 5 to 12 min in each cycle is extracted as the stable value.

3 Experimental Part

3.1 Experimental Pretreatment

In this project, pick the valencia orange with the same maturity, the dynamic headspace method was used to collect the smell of citrus. Firstly, the cold-pressed technology was used to obtain the valencia orange juice. Then filtration, sterilization and canning processes are adopted in turn. There is a tap under the pot, a mixer in the pot and the nitrogen is filled in the top. Stir the juice every 15 days, and then turn on the tap to get the samples at different stages.

Before each sampling experiment, the E-nose should be self-test, and the power of the sensor array preheating for 2–3 h. After that, the normal odor detection can be carried out. In the process of detection, the response curve of the sensor array to the orange juice odor was real-time displayed on the odor collection software.

3.2 Data Analysis

In each of the 12 min odor collection cycle, collect data once every 30 s. Firstly, select the average data between each period of 5–12 min as stable value. Then, a 15 dimensional original data matrix is established by using the stable value of the response the electronic nose system. After 15 days, the juice odor is sampled again. We had a total of four times sampling experiment, that is, we obtained four storage stages of valencia orange odor data. Then we randomly select 70% of the samples of each gas to establish the training data set of classifier, and the rest are used as the test data set. Detailed information is shown in Table 2. In Fig. 4, we can see that the sensor responses of four samples extracted from the four stages. Obviously, the response of each phase is different, and this is the premise that we used to classify.

Table 2. Amount of samples in training set and test set.

Sample stage	Training set	Test set
First stage	17	7
Second stage	17	7
Third stage	17	7
Fourth stage	17	7
ALL-4	68	28

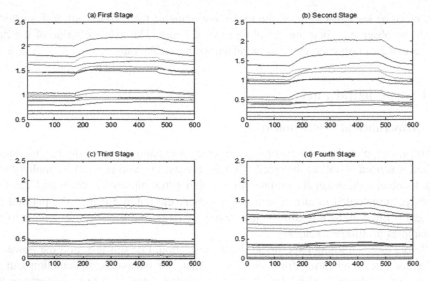

Fig. 4. Response of the four stage.

4 Data Analysis and Results

4.1 Extreme Learning Machine

ELM, a fast learning algorithm for single hidden layer feed-forward neural networks (SLFNs) which randomly chooses the input weights and analytically determines the output weights of SLFNs, was first proposed by Huang et al. [11]. So far, ELM has been widely used in many applications, such as sales forecasting [12], mental task [13] and face recognition [14]. Qiu et al. used an E-nose which consists of a sampling apparatus, a detector unit containing the array of sensors and pattern-recognition software to characterize five types of strawberry juices based on different processing approaches [15]. So, we choose the ELM to carry on the stage analysis of citrus odor.

Figure 5 shows the extreme learning machine, which is divided into the input layer, the hidden layer and the output layer, each layer contains many "neurons", there exists a connection between layers of neurons, and the data to be processed is first multiplied by the value of the input layer neuron, the output layer output discriminant results at last.

In our experiment, firstly, four stages of the valencia orange odor samples (collected at each stage) are collected to obtain the database of different stages. And then the samples are normalized and standardized to prepare the training set and the test set, the training set is the input of ELM. After training and learning, using the test set to test the ELM, if the test set recognition rate is more than 90%, it is considered that the overrun ELM has reached the correct rate of target detection.

4.2 Results and Discussion

In ELM, the only parameter is the number of hidden nodes in the hidden layer of SLFN, which is normally obtained by a trial and error method. Thus, 100 experiments

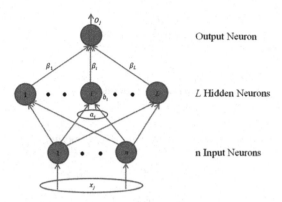

Fig. 5. Extreme leaning machine

were carried out according to the number of hidden nodes in the hidden layer from 1 to 100. The best performance of all results will be regarded as the final classification results of ELM. Finally, the program is repeated for 10 times among which the best result will be the final result. The results are shown in Table 3.

Table 3. The odor phase classification accuracy of valencia orange juice (%)

Sample stage	Training set	Test set
First stage	99.08	97.23
Second stage	99.25	96.57
Third stage	98.99	95.89
Fourth stage	99.05	96.23
ALL-4	99.06	96.29

We can see the maximum of classification accuracy in test data set is up to 96.29. It is within the range of error allowed. The feasibility of the E-nose for the detection of citrus juice in different storage periods is proved. Compared with the traditional chemical detection methods, such as GC-MS technology which needs the gas chromatography-mass spectrometry for analysis and identification, and retrieves chemical composition of volatile substances from computer spectrum library, then the peak area normalization method was used for quantitative analysis and the relative percentage content of each component was obtained. This method can only obtain qualitative and quantitative results of some or several components in the sample under test. In addition, the GC-O technique uses a well trained sensory evaluation operator's nose as a detector, followed by olfactory analysis and multiple experimental records and then perform statistical analysis. This method relies on artificial sensory evaluation, but the human nose has fatigue, the results will differ from time, environment, and individual. The electronic nose technology hardly needs the use of organic solvent, and can obtain the whole information of the volatile components in the sample. Moreover, the analysis speed is fast, and the determination is not affected by subjective factors, so it is more objective. The flow chart of data analysis process of ELM is shown in Fig. 6.

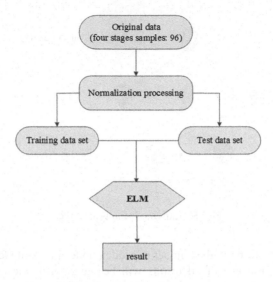

Fig. 6. Flow diagram of experiment.

From Fig. 7, it can be seen that the classification rate gradually improves with the number of hidden nodes from 1 to 35 and from 81 to 96, while the classification rate gradually declines with the number of hidden nodes from 52 to 81. Moreover, ELM can achieve the best classification accuracy of 96% when is the number of hidden nodes are 48 and 52.

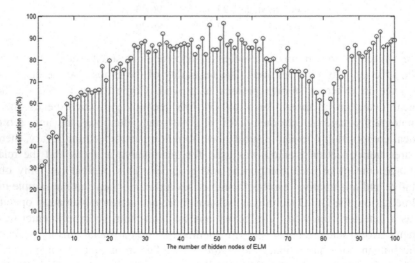

Fig. 7. Performance of ELM according to the number of hidden nodes from 1 to 100

5   Conclusion

It has been proved that E-nose is an effective way to detect the citrus juice. The E-nose can make a rapid analysis of the odor of citrus juice with different storage phase, reflect the overall information about the odor of the samples, which is different from the GC-MS which has complex analysis process and massive data processing, but also is more accurate and more objective than GC-O flavor profile. ELM is employed in the identification of the storage stage of citrus juice and we obtain good classification accuracy, which further shows the effectiveness of the E-nose in the identification of citrus juice and provides an important basis for us to better use the E-nose to determine the freshness of citrus juice. Of course, over a longer period of time and under realistic challenging conditions, the application of the E nose in the citrus juice still have a lot of research to be done, such as the optimization of sensor and pattern recognition technology. On the basis of the research in this paper, we can further analyze the odor of citrus and citrus juice with different varieties and maturity and accurately identify the citrus juice. It can also be used to distinguish the different varieties of citrus juice, different processing types of orange juice and also different degrees of enrichment of reducing juice, which can provide reference for the future application of online detection of citrus juice processing, fruit juice anti-counterfeiting, quality grading.

References

1. Hofsommer, H.J.: New technological aspects, pt. 5: quality aspects of concentrated orange juice. Honeybee Neurobiol. Behav. pp. 117–122 (1992)
2. Goodner, K.L., Jella, P., Rouseff, R.L.: Determination of vanillin in orange, grapefruit, tangerine, lemon, and lime juices using GC-olfactometry and GC-MS/MS. J. Agric. Food Chem. **48**(7), 2882–2886 (2000)
3. Loutfi, A., Coradeschi, S., Mani, G.K., et al.: Electronic noses for food quality: a review. J. Food Eng. **144**, 103–111 (2015)
4. Chapman, E.A., Thomas, P.S., Stone, E., Lewis, C., Yates, D.H.: A breath test for malignant mesothelioma using an electronic nose. Eur. Respir. J. **40**(2), 448 (2012)
5. Romain, A.C., Nicolas, J.: Long term stability of metal oxide-based gas sensors for E-nose environmental applications: an overview. Sens. Actuators B: Chem. **146**(2), 502–506 (2010)
6. Norman, A., Stam, F., Morrissey, A., et al.: Packaging effects of a novel explosion-proof gas sensor. Sens. Actuators B: Chem. **95**(1), 287–290 (2003)
7. Young, R.C., Buttner, W.J., Linnell, B.R., Ramesham, R.: Electronic nose for space program applications. Sens. Actuators B: Chem. **93**, 7–16 (2003)
8. Oshita, S., Shima, S., Haruta, T., et al.: Discrimination of odors emanating from 'La France' pear by semi-conducting polymer sensors. Comput. Electron. Agric. **26**(2), 209–216 (2000)
9. Natale, D., Macagnano, C.A., Martinelli, E., et al.: The evaluation of quality of post harvest orange and apples by means of an electronic nose. Sens. Actuators B: Chem. **78**(1), 26–31 (2001)
10. Moshonas, M.G., Shaw, P.E.: Quantitative determination of 46 volatile constituents in fresh unpasteurized orange juice using dynamic headspace gas chromatography. J. Agric. Food Chem. **42**(7), 1525–1528 (1994)

11. Huang, G.B., Zhu, Q.Y., Siew, C.K.: Extreme learning machine: a new learning scheme of feedforward neural networks. In: Proceedings of 2004 IEEE International Joint Conference Neural Networks, vol. 2, pp. 985–990. Budapest, Hungary (2004)
12. Sun, Z.L., Choi, T.M., Au, K.F., et al.: Sales forecasting using extreme learning machine with applications in fashion retailing. Decis. Support Syst. **46**(1), 411–419 (2009)
13. Liang, N.Y., Saratchandran, P., Huang, G.B.: Classification of mental tasks from EEG signals using extreme learning machine. Int. J. Neural Syst. **16**(1), 29–38 (2006)
14. Mohammed, A.A., Minhas, R., Wu, Q.M.J., et al.: Human face recognition based on multidimensional PCA and extreme learning machine. Pattern Recogn. **44**(10–11), 2588–2597 (2011)
15. Qiu, S., Gao, L., Wang, J.: Classification and regression of ELM, LVQ and SVM for E-nose data of strawberry juice. J. Food Eng. **144**, 77–85 (2015)

A Tag-Based Integrated Diffusion Model
for Personalized Location Recommendation

Yaolin Zheng[1,2], Yulong Wang[1,2], Lei Zhang[1,2], Jingyu Wang[1,2],
and Qi Qi[1,2(✉)]

[1] State Key Laboratory of Networking and Switching Technology,
Beijing University of Posts and Telecommunications,
Beijing 100876, People's Republic of China
zhengyaolinbupt@163.com
[2] EBUPT Information Technology Co., Ltd.,
Beijing 100191, People's Republic of China

Abstract. The location based services have attracted millions of users to share their locations via check-ins. It is highly important to recommend personalized POIs (Points-Of-Interest) to users in terms of their preference learned from historical data. In current research work, users' check-in behavior is wildly used to model user's preference. However, the sparsity of the check-in data makes it difficult to capture users' preferences accurately. This paper proposes a tag-based integrated diffusion recommender system for location recommendation, considering not only social influence but also venue features. Firstly, we model user location preference by combining the preference extracted from check-ins data and short text tips, where sentiment analysis techniques are used. Furthermore, we collect venue information by merging descriptions and tips and then generate tags of each venue, which are processed using keyword extraction approaches. Then we apply the recommendation algorithm with user's initial preference and obtain the final integrate diffusion results for each user, recommending top-N venues by descending order. We conduct experiments on Foursquare datasets of two cities, the results on both datasets show that our recommender system can produce better performance, providing more personalized and higher novel recommendations.

Keywords: Location recommendation · Sentiment analysis · Diffusion · Keyword extraction · Tag

1 Introduction

With the booming of location sharing services such as Foursquare, location recommendation is becoming emerging research topic. In location sharing services, users are allowed to check in at a venue and leave tips to comment on a venue, which make it possible to learn users' preference. It is crucial to recommend POIs(Points-Of-Interest) to users. On the one hand, it helps users know new POIs they may like, guiding them to places they may want. On the other hand, it also helps content provider delivery advertisements to targeted users and improve business profits. Apparently, users' historical behavior can reflect their preference. Nevertheless, there are many shortcomings

D. Liu et al. (Eds.): ICONIP 2017, Part V, LNCS 10638, pp. 327–337, 2017.
https://doi.org/10.1007/978-3-319-70139-4_33

for merely using check-in data especially negative feedback is given. The state-of-the-art location recommendation approaches consider more about inter-user relationships, getting recommendation results depends on social influence [1–3]. But in fact, location recommendation needs to consider more factors such as geographical constraint, temporal influence, venue category and reviews, etc.

In this paper, we propose a tag-based integrated diffusion (TBID) recommendation algorithm, providing personalized location recommendations with the help of user's check-in data, venue information and tips on venues. An extra information besides check-in data is used to initialize users' personalized preference, and a diffusion-based recommendation algorithm on user-venue-tag tripartite graph is applied to capture both social and inter-venue influence. Firstly, we consider both check-ins and comments of venues in user preference model. The text-based sentiment analysis techniques are used to analyze one's sentiment in tips and a unified preference model from both check-ins and tips is obtained. Furthermore, venues can construct a similarity according to their features. Some venue information can be included as tags, namely, categories and tags. On top of that, tips of a venue can also be seen as venue's description, which can be a supplement for venue information. We use text feature extraction techniques to expand tags for venues. We believe that venue similarity can also influence recommendation performance, therefore, a graph based diffusion model for location recommendation is proposed to capture the influence from both inter-users and inter-venues.

The rest of this article is organized as follows. In Sect. 2, we briefly survey the related works in recommender systems, especially in location recommendations. Section 3 detail the proposed TBID recommender system. In Sect. 4, we conduct a series of experiments using data collected from Foursquare for evaluation. Conclusion and future work is presented in Sect. 5.

2   Related Work

The work in this paper is closed related to general recommender system and location recommendation, and some natural language processing work is done as well, including sentiment analysis and feature extraction from short text. In the following, we briefly review the related work.

A wide range of research work has been done in building recommender systems. The essence purpose of recommendation technologies is establishing the relationship between user and item. Therefore, an intermediary entity is needed since the direct relationship is always unknown. The type of intermediary entity can fall into three categories: user, item, and feature [4]. Features-based approaches use features of recommended item as intermediary entities, such as, tag. Tags are generally chosen informally and personally by the items' creator or by its viewer. Many systems, such as Delicious, music website Last.fm, video website Hulu use social tagging to categorize their information items and help users to share them. There are variety of approaches to build a recommender system with this extra information. First and foremost, tag information can be used in traditional recommend algorithm as an extension, such as CF [5], content-based recommendation. Moreover, these historical users' activities are usually represented by the connections in a user-item bipartite graph [6], likewise, tag relevance can describe an item-tag bipartite graph. Therefore, some graph-based model can be used to make a recommendation, such as FolkRank algorithm [7], diffusion based recommendation algorithm on graphs [8]. Besides, some other algorithms are well used, such as probabilistic latent semantic analysis (PLSA) approach [9], tensor factorization [10]. Experimental results indicate that diversification and novelty of recommendations are significantly enhanced by exploiting the tag information. Since all these works have proved that extra accessorial information can be very useful in recommender system, we believe that considering influence between items through tags can improve recommendation performance.

With the emerging location based services, there has been great research passion in POIs or generally location recommendations. Personalized location recommendation aims to providing users with the most pertinent venues by considering individual's preference. Various recommendation algorithms are used for personalized location recommendation such as collaborative filtering [1, 11], matrix factorization [12, 13] and recommendation with random walk [2, 3] and so on. Besides, Ye et al. [14] proposed Geo-Measured friend-based collaborative filtering (GM-FCF) approach for location recommendation based on collaborative ratings of commonly visited places made by social friends and geospatial characteristics. In [15], social, categorical, geographical, sequential, and temporal characteristics are considered to exploit their influences for recommending personalized POIs. To consider users' preferences, geographical influence and personalized ranking, Cheng et al. [16] proposed a unified POI recommendation framework, which unifies all of them together, resulting in a better performance. Furthermore, Fang et al. [17] proposed a spatial-temporal context-aware personalized location recommendation system, offering a particular user a set of location items by considering personal interest, local preference, and spatial-temporal context influence.

To recommend locations that user maybe interested in, users' historical activities (e.g. check-in) is important to reveal their preference. However, merely based on user check-ins may introduce bias on measuring users' preference. In Foursquare, users are allowed leave tips on venues, such information can be well used. Hence, we conduct sentiment analysis of users' tips to extract their positive and negative feedback about venues, which can enhance user's sentiment and make a more accurate preference. Additionally, keyword extraction method is applied to extract tags from venue tips, description. In this paper, we extend a diffusion-based recommendation algorithm on user-item-tag tripartite graphs for location recommendation, considering both user social influence and venue feature information, and initialize user's preference by fusing check-in preference and sentiment preference.

3   Tag-Based Integrated Diffusion Model

In this section, we present the proposed tag-based integrated diffusion model. Firstly, a hybrid preference approach is proposed to initialize user's preference by unifying preference in both check-ins and tips. Secondly, some kind of natural language processing method is done to generate venue tags from venue information and descriptions. Finally, we explain an extended diffusion algorithm on tripartite graph for personalized location recommendation.

3.1 Hybrid User Preference

There is no denying that users' preference about a place can be extracted from their historical behavior. For Foursquare users, they interact with the venues through check-ins and tips. In this paper, we propose a fusion approach to combines user's preference, considering both check-ins preference and tip sentiment.

In Foursquare, users can post tips at specific venues, commenting on their previous experiences as a feedback. Besides, users can also leave tips at some interested venues. Dictionary based sentiment analysis method are used in our work to process tips. Firstly, the tips are split into sentences and add POS (part of speech) tags to the words in each sentence. Then we obtain a sentiment score for each word by looking it up in SentiWordNet with the corresponding part of speech tag. The overall sentiment score is the sum of all the sentiment score of each word in a tip and is normalized into $[-1, 1]$, where -1 represent the most negative sentiment, while 1 represent the most positive sentiment. The Implementation is based on SentiWordNet3.0 and NLTK toolkit.

We believe that the more times someone checked in a venue, the more he/she like this place. What's more, as the sentiment score contains more specific information about a user's preference about a venue, it should be considered together with the check-ins to characterize a user's preference. Therefore, the final preference score can be calculated based on the number of check-ins and sentiment scores obtained from tips. The fusion process as follows: (1) The original preference score equals to the number of check-ins plus 1 point, which means the minimum of the score is 2 point. (2) Sentiment score is used to amend the original score. Original score will be increased or decreased by sentiment score expect that one time check-in circumstance, in this case, sentiment influence is more important as feedback, since one time activity cannot reveal precise information about user's real feeling about the venue.

$$P_c = Count(check\_ins) + 1 \tag{1}$$

$$P_{final} = \begin{cases} P_c + 2 \times P_s, & if \ P_c = 2 \\ P_c + P_s, & otherwise \end{cases} \tag{2}$$

Based on the above fusion criteria, a user-venue preference matrix combining preference extracted from both check-ins and tips is constructed.

3.2 Venue Tags Generation

In foursquare, venues have their detailed information, e.g. description (provided by venue owner), categories, tags. Categories and tags are some key words about the venue, can be used directly as feature tags in our algorithm. Hence, we get original venue tags from these two kind of information after words splitting and invalid words (e.g. stop words) filtering. Besides, the venue description that is regarded as a simple introduction reveals important features of the venue. Moreover, the venue tips can also be seen as description provided by users. Feature extraction technologies are used to extract tags from these descriptions.

The procedure of tag extraction is shown in as follows. Firstly, the descriptions are split into sentence and the POS is identified. When the number of sentences about a venue more than 3, considering words collocation in sentences, bigram words are found first by using likelihood ratio and POS tag, e.g., "ping pong", "free pizza", "French toast". Then we use a numerical statistic term frequency (TF) to extract important features of the venue, which means the more times the words occur, the more important they are to this venue. Otherwise, simple filtering work to be done based on POS of words, in this case, we only keep noun as tag in the sentences. Last but not the least, some noise and invalid tags (e.g., stop words, punctuations, letters) are filtered. After extending the original venue tags, some synonyms are merged by stemming the words. So far, we get the final venue tags, including the venue characteristics and features extracted from descriptions.

3.3 Diffusion on Tripartite Graph

The tripartite graph representation can be described by two adjacent matrices, A and B, for user-venue and venue-tag relations. Considering a user-venue bipartite graph first, suppose that a kind of interest resource is initially located on venues for a target user, representing user's preference on various venues. Since the bipartite network itself is without weight, each venue will averagely distribute its resource to all neighboring users, and then each user will equally redistribute the received resource to all his/her related venues. The diffusion process consists of two steps: first from venue to user, then back to venue. Denoting $\vec{f}$ the initial resource vector on items, the final resource vector, f_j', can be obtained as follows after diffusion:

$$f_j' = \sum_{l=1}^{n} \frac{a_{lj}}{N_v(U_l)} \sum_{s=1}^{m} \frac{f_s}{N_u(V_s)}, \qquad j = 1, 2, \ldots, m \qquad (3)$$

where $N_u(V_s) = \sum_{i=1}^{n} a_{is}$ is the number of neighboring users for venue V_s, and $N_v(U_l) = \sum_{j=1}^{m} a_{lj}$ is the number of check-in venues for user U_l. f_s is the amount of resource located on V_s, m is the number of all venues, and n is the number of users.

Likewise, on the venue-tag bipartite graph, each venue will equally distribute its resource to all related tags, and then each tag will redistribute the resource received to all its neighboring venues. Thus, given the initial resource, the final resource vector, f_j'', is:

$$f_j'' = \sum_{l=1}^{r} \frac{b_{jl}}{N_v(T_l)} \sum_{s=1}^{m} \frac{f_s}{N_T(V_s)}, \qquad j = 1, 2, \ldots, m \qquad (4)$$

where $N_T(V_s) = \sum_{l=1}^{r} b_{sl}$ is the number of neighboring tags for venue V_s, and $N_v(T_l) = \sum_{j=1}^{m} b_{jl}$ is the number of neighboring venues for tag T_l. r is the number of all tags.

A simplest way is adopted to integrate the diffusion results on user-venue and venue-tag bipartite graphs. The final resource score is defined as a linear superposition f' of and $\vec{f}''$.

$$\vec{f^*} = \lambda \vec{f'} + (1 - \lambda)\vec{f''} \tag{5}$$

where $\lambda \in [0, 1]$ is a tunable parameter. $\vec{f'}, \vec{f''}$ is the vector derived from Eqs. (3) and (4), respectively. In the extreme cases $\lambda = 0$ and $\lambda = 1$, the integrated algorithm degenerates to the pure diffusions on user-venue and venue-tag bipartite graphs, considering merely inter-users or inter-venue influence.

Given a target user U_i, the initial resource vector $\vec{f}$ is the ith row of preference matrix. Since the different initial resource vectors for different users have captured the personalized preferences, the final resource located on venues can reflect the preference of the user as well, connections between venues are built through users and tags, and those venues with highest scores are more possibly checked in one day. These venues with top-N highest scores are recommended to U_1.

4   Experiments

4.1   Data Description

In this work, we use a collection of Foursquare check-ins lasting for 8 months (from 1 October 2012 to 31 May 2013). Since personal check-in information is not available publicly, in other words, they can only be accessed from one's own account. Hence, we collected users' ID from public dataset published by others, and then captured check-ins by crawling Foursquare user API (But for now, the user check-ins API is only supported for the developer self). The whole dataset includes 50,813 users and 3,564,144 check-ins globally, among which there are 21,890 users and 3,176,507 check-ins in the United States. Consequently, we select two big cities in the US, namely New York and California, and then extract tips of the venues in these two cities, as well as venue information. The procedure of data processing is as follows in detail.

Firstly, noise and invalid check-in data was filtered, at the same time, some check-ins at the same venue happened in an hour were merged. Then we select users as test users who have performed at least one check-in per week (these users are regarded as active users), statistically, check-ins of active users account for 2/3 of the total. In addition, the private venues information was filtered, check-ins which were performed over these venues were also excluded, since private places are not to be recommended. Note that tips maybe observed even user didn't check in at that venue, since user maybe interested in venue like that and leave a tip depend on others or check-in date is out of our period. The data statistics is shown in Table 1.

4.2   Metrics

Four different metrics are used for recommendation evaluation: Precision, Recall, F1 Score and Novelty. The precision is the proportion of recommendations that are good recommendations. The recall is the number of good recommendations divided by the number of positive results that should have been returned (testing set). The F1 score (also balanced F score) can be interpreted as a weighted average of the precision and

Table 1. Dataset statistics

Dataset	New York	California
Users	2084	1681
Venues	60386	59345
Check-ins	171132	139672
Tips	22711	19005

recall, where an F1 score reaches its best value at 1 and worst at 0. It considers both the precision and the recall of the test to compute the final score. Novelty quantifies the capacity of an algorithm to generate novel and unexpected results, that is to say, to recommend less popular venues.

To test the algorithm performance, data set for each city is divided into two parts: The training set contains 80% of check-ins of every user and the remaining 20% of check-ins constitutes the testing set. The training set is treated as known information used for learn users' preference and generating recommendations, while the testing set is used for testing recommendation results, metrics are employed here to evaluate if the algorithm performance is good.

$$precision = \frac{\sum_{u \in U} |R(u) \cap T(u)|}{\sum_{u \in U} |R(u)|} \tag{6}$$

$$recall = \frac{\sum_{u \in U} |R(u) \cap T(u)|}{\sum_{u \in U} |T(u)|} \tag{7}$$

$$F1 = \frac{2 \times precision \times recall}{precision + recall} \tag{8}$$

where U is the test users set, R(u) and T(u) represent recommendation venue set and testing venue set for user u, respectively.

$$Novelty = \frac{1}{mL} \sum_u \sum_{I_r \in I_R^u} k(I_r) \tag{9}$$

where $k(I_r)$ is the degree of the item I_r, m is the number of users, L is the length of the recommended list. For a specific user, the smaller the average degree of venues in the recommended list is, the higher the novelty is.

4.3   Evaluation

The performance results are calculated with different length of recommendation list. Figure 1 shows the top-N (N = 20) recommendation performance results (i.e., precision, recall, F1 score) of two datasets for $\lambda \in [0, 1]$. Obviously, when the inter-user influence parameter is set $\lambda = 0.8$, resulting in best performance, other cases with different N as well. The F1 score between precision and recall is the harmonic mean of precision and recall, sharing the same tendencies of them.

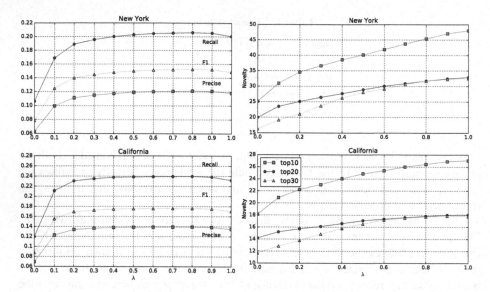

Fig. 1. Metrics (precise, recall and F1) for different λ of two cities and novelty versus λ for typical top-N: N = 10, N = 20, N = 30

In order to evaluate the proposed algorithm for location recommendation, we compare the performance with classic item-based collaborative filtering and pure bipartite diffusions. The results of New York dataset and California dataset are shown in Table 2, from which we can see (1) diffusion algorithm performs better than item-based CF for location recommendation, (2) pure diffusion on user-venue graph ($\lambda = 1$) performs better than venue-tag bipartite graph ($\lambda = 0$) and (3) integrated diffusion on tripartite graph ($\lambda = 0.8$) provide more accurate recommendations than the

Table 2. Performance comparison between different approaches

Algorithm		Collaborative filtering			Diffusion on user-venue graph			Diffusion on venue-tag graph			Diffusion on tripartite graph		
City	Metric	Precision	Recall	F1	Precision	Recall	F1	Precision	Recall	F1	Precision	Recall	F1
New York	Top10	5.48 %	4.66 %	0.0504	16.14 %	13.71 %	0.1483	9.60 %	8.15 %	0.0882	16.53 %	14.13 %	0.1524
	Top20	4.75 %	8.07 %	0.0598	11.81 %	20.07 %	0.1487	6.24 %	10.60 %	0.0785	12.13%	20.60 %	0.1527
	Top30	4.28 %	10.91 %	0.0615	9.80 %	24.98 %	0.1487	4.7 %	11.99 %	0.0676	10.01 %	25.52 %	0.1438
	Top50	3.43 %	14.60 %	0.0556	7.26 %	30.83 %	0.1175	3.27 %	13.90 %	0.0530	7.33 %	31.13 %	0.1187
	Top100	1.72 %	14.60 %	0.0307	4.28 %	36.38 %	0.0766	1.93 %	16.42 %	0.0346	4.29 %	36.46 %	0.0768
California	Top10	7.61 %	6.54 %	0.0704	17.92 %	15.41 %	0.1657	10.68 %	9.18 %	0.0988	18.28 %	15.72 %	0.1691
	Top20	6.36 %	10.94 %	0.0804	13.47 %	23.18 %	0.1704	6.94 %	11.94 %	0.0878	13.93 %	23.96 %	0.1761
	Top30	5.55 %	14.29 %	0.0799	11.28	29.10 %	0.1625	5.29 %	13.64 %	0.0762	11.53 %	29.76 %	0.1662
	Top50	4.28 %	18.41 %	0.0694	8.35 %	35.89 %	0.1354	3.64 %	15.66 %	0.0591	8.50 %	36.58 %	0.1380
	Top100	2.14 %	18.41 %	0.0383	4.86 %	41.76 %	0.0870	2.14 %	18.43 %	0.0384	4.89 %	42.06 %	0.0876

pure cases. (4) As the length of recommendation list decreases, the precision increases while the call decreases, but the F1 score reaches maximum when it's set 20. Note that CF algorithm will meet a situation lacking of recommendation results when the length of recommendation list increases, resulting in a negative impact on the performance.

As is illustrated in the right part of Fig. 1, all curves are monotone, which means the algorithm depending more on tags (the smaller λ is) can provide higher novel recommendations. Consequently, usage of tag information can enhance the chance of discovering individuals' interest. It's easy to increase novelty by giving up on accuracy, the goal is making the balance in this trade-off that enhance these aspects while still achieving a fair match of the user's interests.

5 Conclusion and Future Work

In this paper, we propose a sentiment enhanced integrated diffusion algorithm with the help of tag information for personalized location recommendation. Both check-in data and user tips are considered to model user's preference, simultaneously, both user similarity and inter venue similarity through tags are combined to make recommendations. Experimental results demonstrate that diffusion based algorithm has better performance than classic CF in location recommendation domain. Moreover, the usage of tag information can improve accuracy of recommendations, as well as diversification and novelty, indicating the advantage of integration. The parameter λ is adjustable depends on the quality of users' historical activity data and venue tags.

We have considered the social influence and inter-venue influence, while there are some other characteristics that affect the visiting behaviors of users not being taken into consideration, such as geographical influence, temporal influence and so on. The geographical information (i.e., latitude and longitude coordinates) of POIs significantly affects users' check-in behaviors in that physical interactions are required for users to visit some place, and no one can deny that time is also a very important factor influencing human activities at different times on weekends and weekdays. These factors remain research for future study.

Acknowledgments. This work was jointly funded by: (1) National Natural Science Foundation of China (Nos. 61421061, 61372120, 61671079, 61471063); (2) Beijing Municipal Natural Science Foundation (No. 4152039).

References

1. Ye, M., Yin, P., Lee, W.C., Lee, D.L.: Exploiting geographical influence for collaborative point-of-interest recommendation. In: Proceedings of the 34th International ACM SIGIR Conference on Research and Development in Information Retrieval, pp. 325–334. ACM (2011)
2. Noulas, A., Scellato, S., Lathia, N., Mascolo, C.: A random walk around the city: new venue recommendation in location-based social networks. In: ASE/IEEE International Conference on Social Computing & ASE/IEEE International Conference on Privacy, Security, Risk and Trust, pp. 144–153 (2012)

3. Wang, H., Terrovitis, M., Mamoulis, N.: Location recommendation in location-based social networks using user check-in data. In: Proceedings of the 21st ACM SIGSPATIAL International Conference on Advances in Geographic Information Systems, pp. 374–383. ACM (2013)

4. Vig, J., Sen, S., Riedl, J.: Tagsplanations: explaining recommendations using tags. In: Proceedings of the 14th International Conference on Intelligent User Interfaces. ACM (2009)

5. Parra, D., Brusilovsky, P.: Collaborative filtering for social tagging systems: an experiment with CiteULike. In: Proceedings of the third ACM Conference on Recommender Systems, pp. 237–240. ACM (2009)

6. Zhou, T., Ren, J., Medo, M., Zhang, Y.C.: Bipartite network projection and personal recommendation. Phys. Rev. E **76** (2007)

7. Hotho, A., Jäschkz, R., Schmitz, C., Stumme, G.: FolkRank: a ranking algorithm for folksonomies. In: LWA, vol. 1, pp. 111–114 (2006)

8. Zhang, Z.K., Zhou, T., Zhang, Y.C.: Personalized recommendation via integrated diffusion on user–item–tag tripartite graphs. Phys. A **389**, 179–186 (2010)

9. Wetzker, R., Umbrath, W., Said, A.: A hybrid approach to item recommendation in folksonomies. In: Proceedings of the WSDM 2009 Workshop on Exploiting Semantic Annotations in Information Retrieval, pp. 25–29. ACM (2009)

10. Rendle, S., Marinho, L.B., Nanopoulos, A., et al.: Learning optimal ranking with tensor factorization for tag recommendation. In: Proceedings of the 15th ACM SIGKDD International Conference on Knowledge Discovery and Data Mining, pp. 727–736. ACM (2009)

11. Yuan, Q., Cong, G., Ma, Z., et al.: Time-aware point-of-interest recommendation. In: Proceedings of the 36th international ACM SIGIR Conference on Research and Development in Information Retrieval, pp. 363–372. ACM (2013)

12. Cheng, C., Yang, H., King, I., Lyu, M.R.: Fused matrix factorization with geographical and social influence in location-based social networks. In: AAAI Conference on Artificial Intelligence (2012)

13. Yang, D., Zhang, D., Yu, Z., et al.: A sentiment-enhanced personalized location recommendation system. In: ACM Conference on Hypertext & Social Media, pp. 119–128. ACM (2013)

14. Ye, M., Yin, P., Lee, W.C.: Location recommendation for location-based social networks. In: Proceedings of the 18th SIGSPATIAL International Conference on Advances in Geographic Information Systems, pp. 458–461. ACM (2010)

15. Zhang, J.D., Chow, C.Y.: Point-of-interest recommendations in location-based social networks. SIGSPATIAL Spec. **7**(3), 26–33 (2016). ACM

16. Cheng, C., Yang, H., King, I., Lyu, M.R.: A unified point-of-interest recommendation framework in location-based social networks. ACM Trans. Intell. Syst. Technol. **8**(1) (2016). ACM

17. Fang, Q., Xu, C., Hossain, M.S., Muhammad, G.: STCAPLRS: a spatial-temporal context-aware personalized location recommendation system. ACM Trans. Intell. Syst. Technol. 7(4) (2016). ACM

18. Liu, B.: Opinion mining and sentiment analysis. In: Liu, B. (ed.) Web Data Mining. Data-Centric Systems and Applications, pp. 459–526. Springer, Heidelberg (2012). doi:10.1007/978-3-642-19460-3_11

19. Bifet, A., Frank, E.: Sentiment knowledge discovery in Twitter streaming data. In: Pfahringer, B., Holmes, G., Hoffmann, A. (eds.) DS 2010. LNCS, vol. 6332, pp. 1–15. Springer, Heidelberg (2010). doi:10.1007/978-3-642-16184-1_1

20. Thelwall, M., Buckley, K., Paltoglou, G.: Sentiment in Twitter events. J. Am. Soc. Inf. Sci. Technol. (2011). DBLP

21. You, Q.: Sentiment and emotion analysis for social multimedia: methodologies and applications. In: Proceedings of the 2016 ACM on Multimedia Conference, pp. 1445–1449. ACM (2016)
22. Pontes, T., Vasconcelos, M., Almeida, J., Kumaraguru, P., Almeida, V.: We know where you live: privacy characterization of foursquare behavior. In: Proceedings of the 2012 ACM Conference on Ubiquitous Computing, pp. 898–905. ACM (2012)
23. Jannach, D., Zanker, M., Felfernig, A., Friedrich, G.: Recommender Systems: An Introduction. Cambridge University Press, New York (2010)
24. Ricci, F., Rokach, L., Shapira, B., et al.: Recommender Systems Handbook. Springer, Boston (2011). doi:10.1007/978-0-387-85820-3

Relationship Measurement Using Multiple Factors Extracted from Merged Meeting Events

Zeng Chen[1(✉)], Keren Wang[1], and Zheng Yang[2]

[1] National Key Laboratory of Science and Technology on Blind Signal Processing, Chengdu, China
chenzengll@tsinghua.org.cn
[2] School of Software, Tsinghua University, Beijing, China

Abstract. With the popularity of mobile phones and mobile applications, it becomes possible to collect large-scale mobility data and do research on human mobility. Among these research, relationship mining from location information is a hot topic which has plenty of applications including marketing applications, social studies and even terrorist discovery. This paper focuses on measuring the relationship strength of user pairs according to their meeting events. A novel method using multiple factors extracted from merged meeting events is proposed for measuring relationship. Firstly, meeting events are merged and each merged meeting event is represented by several features, from which multiple factors can be drawn. Specifically, the duration factor and the diameter factor are proposed for measuring relationship on the basis of merged meeting events. Finally, a model synthesizing multiple factors (including location entropy factor, location personal factor, temporal factor, duration factor and diameter factor) is proposed to quantify the relationship between users in an unsupervised way. Experimental results on three different real datasets demonstrate that our method performs significantly more favorable than existing methods on the effectiveness.

Keywords: Relationship measurement · Merged meeting event · Multiple factors · Spatiotemporal

1 Introduction

In recent years, the popularity of mobile phones has grown rapidly, resulting in gaining large amounts of spatiotemporal data to study human behavior and social networks being possible. Basically, a mobile phone can record locations by confirming its base station [1]. Besides, lots of social network services (e.g. Facebook) allow users to share locations with posting geo-tagged microblogs or check-ins [2].

Mining human relationship is one of the most attracting topics among those research on human mobility [3]. The study could result in some appealing applications, such as advertisement targeting [4], friendship recommendation [5] and even anomaly detection [6], et al.. In recent years, extensive research studying the interplay between social networks and human mobility has been published. For example, Lambiotte et al.

© Springer International Publishing AG 2017
D. Liu et al. (Eds.): ICONIP 2017, Part V, LNCS 10638, pp. 338–347, 2017.
https://doi.org/10.1007/978-3-319-70139-4_34

[10] and Descioli et al. [11] revealed that users' social ties are highly related with their distance, i.e., users living close to each other are more likely to have strong relationship tie. Amounts of previous works aimed at understanding human mobility using social relationship [12, 13]. And Cho et al. [14] discovered that short-ranged periodical travel is irrelevant with users' social network while long distance movement is more influenced by social ties.

Early studies mainly took the meeting frequency as the measurement. Eagle et al. [15] found that visiting the same place nearly at the same time (i.e. meeting) is a strong indicator for friends. Crandall et al. [16] found that the meeting frequency is positively correlated with the social tie, i.e., a larger meeting frequency between two users indicate a higher probability that they are friends.

Further studies suggested that not all meeting events are equally important in measuring relationship using mobility data. Pham et al. [8] proposed an entropy based model which took the diversity of meeting locations into consideration to penalize random meeting events at popular locations. Additionally, Wang et al. [9] introduced personal mobility background and temporal factor. They proposed PGT method to weight each meeting event according to personal, global and temporal factor. Zhang and Pang [17] utilized the physical distance to predict friendship and improve the performance of differentiating friends and non-friends with no meeting event. However, the method is limited when locations are recorded as discrete numbers in which case it's impossible to calculate distance between users. Cheng et al. [18] proposed two supervised models, which extracted features from spatial and temporal dimensions using weighting method stated in [8]. Valverderebaza et al. [19] combined mobility data and social network information to predict links, and achieved a favorable experimental results.

Our idea is inspired by the fact that social relationship have great effects on human mobility. For example, friends tends to visit same places together more frequently [7]. Thus, we focus on the formulation and weighting of meeting events according to several heuristic factors, and synthesizes all the meeting events between two users for measuring their relationship. The major contributions of this paper are as follows. 1. A method for merging meeting events is proposed, which results in merged meeting events more close to real-life meeting events. 2. Calculations of location entropy weight, location personal weight and temporal weight from literature [8, 9] are modified to be applicable to merged meeting events. Additionally, novel duration factor and diameter factor are introduced. 3. A novel EPTDD model is proposed to synthesize multiple factors for measuring relationship between users.

The next section demonstrates preliminaries corresponding to our method, as well as the formulation of the merged meeting event. Section 3 presents our novel EPTDD model, which uses multiple factors for measuring relationship between users. Section 4 demonstrates experiments on three datasets, which evaluate the effectiveness of different factors in Sect. 3 and compare the performance of our method with existing methods. The conclusions are drawn in the last section.

2 Preliminaries

Denote n users by $U = \{U_i, i = 1, 2, \ldots, n\}$. As for mobility mining, a user U_i can be represented by a sequence of history location records $S_i = \{(t_r^i, loc_r^i), r = 1, 2, \ldots, R_i\}$, where (t_r^i, loc_r^i) is the timestamp and location respectively, and R_i is the total record number of user U_i. With different capturing tools and methods, the location loc_r^i may be a discrete number, or a continuous geographical position denoted by (lat, lon).

A meeting event occurs when two users U_i and U_j visit nearly the same location within a short time interval, which is formalized by

$$\begin{cases} loc_p^i \sim loc_q^j \\ |t_p^i - t_q^j| < \tau \end{cases} \tag{1}$$

where $loc_p^i \sim loc_q^j$ indicates that two locations are quite near each other. Specifically, we get $loc_p^i = loc_q^j$ when locations are discrete, and $dist\left(loc_p^i, loc_q^j\right) < Dist_{th}$ when locations are continuous, where $Dist_{th}$ is a manual threshold.

A *detected meeting event* formed by $(t_p^i, loc_p^i) \in S_i$ and $(t_q^j, loc_q^j) \in S_j$ can be represented as $oe_{i,j;k} = (t_{i,j;k}^1, t_{i,j;k}^2, loc_{i,j;k}^1, loc_{i,j;k}^2)$ where $(t_{i,j;k}^1, loc_{i,j;k}^1)$ is a record among these two with $t_{i,j;k}^1 = \min(t_p^i, t_q^j)$ while $(t_{i,j;k}^2, loc_{i,j;k}^2)$ is the other one.

Given two users U_i and U_j, all detected meeting events between them can be denoted by $OE_{ij} = \{oe_{i,j;k}, k = 1, 2, \ldots, R_{ij}^o\}$, where R_{ij}^o is number of detected meeting event records between U_i and U_j. Most existing works directly mine relationship from OE_{ij} [8, 9, 16]. Nevertheless, since a long-term meeting event may be logged multiple times when location records are dense in the time domain, we suggest that detected meeting events occurring in neighboring times should be merged for reducing the perplexity. Specifically, a sub-sequence of detected meeting events $\{oe_{i,j;k_1}, oe_{i,j;k_1+1}, \ldots, oe_{i,j;k_2}\}$ should be merged when it satisfy $\left|t_{i,j;k_1}^2 - t_{i,j;k_2}^2\right| < \tau_s$ and we get the merged meeting event $e_{i,j;h} = \left(t_{i,j;h}, loc_{i,j;h}^1, loc_{i,j;h}^2, dur_{i,j;h}, dia_{i,j;h}\right)$, where $t_{i,j;h}$, $loc_{i,j;h}^1$, $loc_{i,j;h}^2$, $dur_{i,j;h}$ and $dia_{i,j;h}$ denote the starting time, the location of U_i, the location of U_j, the duration, and the diameter representing the maximum moving distance of the merged meeting event. For simplicity, we use sub-script h to denote $i, j; h$. Then the merged meeting event feature can be calculated as:

$$\begin{cases} t_h = t_{k_1}^1 \\ (loc_h^1, loc_h^2) = rand((loc_{k_1}^1, loc_{k_1}^2), \ldots, (loc_{k_2}^1, loc_{k_2}^2)) \\ dur_h = t_{k_2}^2 - t_{k_1}^1 \\ dia_h = \begin{cases} \max\limits_{\substack{k_1 \le r,s \le k_2 \\ \alpha=1,2}} (dist(loc_r^\alpha, loc_s^\alpha)) & \text{for continuous locations} \\ \sum\limits_{k_1 \le m,n \le k_2} \delta(loc_m^1 - loc_n^1) & \text{for discrete locations} \end{cases} \end{cases} \tag{2}$$

where $rand(i, j, k, \ldots)$ is a function that selects an element from the inputs $i, j, k, \ldots$ randomly. It should be noted that dia_h denotes the number of different locations of a meeting event when locations are discrete.

3   Proposed EPTDD Model for Relationship Measurement

Let $OE_{ij} = \{oe_h, h = 1, 2, \ldots R_{ij}^o\}$ and $E_{ij} = \{e_h, h = 1, 2, \ldots R_{ij}^e\}$ $(R_{ij}^e \leq R_{ij}^o)$ denote the sequence of detected meeting events and merged meeting events between user U_i and user U_j respectively. One of the most simple relationship measuring metrics is the detected meeting frequency [16]:

$$G_{o,MF}(OE_{ij}) = |OE_{ij}| \qquad (3)$$

In order to weight each merged meeting event from more diverse aspects, this section presents our EPTDD (Entropy, Personal, Temporal, Duration and Diameter) model, which synthesizes five factors of merged meeting events for relationship measurement.

3.1   Location Entropy Factor and Location Personal Factor

Let $S_i(loc) = \{(t_q^i, loc_q^i) \in S_i : loc_q^i \sim loc\}$ be the set of location records of user U_i visiting location loc. Considering that there is a big difference on the total number of location records of different users, we calculate the probability for user U_i visiting location loc by following other than using the records number directly:

$$P(i, loc) - \frac{|S_i(loc)|}{|S_i|} \Big/ \sum_i \frac{|S_i(loc)|}{|S_i|} \qquad (4)$$

Similar to Pham's method [8], we use the exponential function of the Shannon entropy to model the effect of each location's entropy factor:

$$\omega_{ij}^{e'}(loc) = \exp(-g(loc)) = \exp\left(\sum_{i:P(i,loc)\neq 0} P(i, loc) \cdot \log P(i, loc)\right) \qquad (5)$$

For a merged meeting event $e_h = (t_h, loc_h^1, loc_h^2, dur_h, dia_h) \in E_{ij}$, we use the following equation to measure the location entropy weight of each real meeting event:

$$\omega_{ij}^e(e_h) = \sqrt{\omega_{ij}^{e'}(loc_h^1) \cdot \omega_{ij}^{e'}(loc_h^2)} \qquad (6)$$

With the meeting frequency replaced by the sum of location entropy weight, the relationship between users U_i and U_j can be calculated by

$$G_{e,1}(E_{ij}) = \sum \omega_{ij}^e(e_h) = \bar{\omega}_{ij}^e \times |E_{ij}| \tag{7}$$

The location personal weight is primitively calculated in [9]:

$$\omega_{ij}^p(e_h) = -\log(\rho(i, loc_h^1) \cdot \rho(j, loc_h^2)) \tag{8}$$

where $\rho(i, loc)$ represents the probability for user U_i visiting a location loc.

Thus, the location-personal-factor-based relationship measurement is calculated by

$$G_{e,2}(E_{ij}) = \sigma_{e_h}\{\omega_{ij}^p(e_h)\} \times \sum_{e_h} \omega_{ij}^e(e_h) \tag{9}$$

Since the maximum operator in [9] excessively emphasizes the meeting event which happens at rarely visited locations, the standard deviation operator $\sigma\{\bullet\}$ employed here on all location personal weights.

3.2 Temporal Factor

Since meeting events temporally close to each other are merged into a single one merged meeting event and in result, primitive time-interval-based temporal factor in [9] becomes inapplicable. Thus, a new method for modelling the impact of temporal factor is proposed as following.

Similar as location personal factor, we define the temporal weight as:

$$\omega_{ij}^t(e_h) = -\log(\rho(i, t_h) \cdot \rho(j, t_h)) \tag{10}$$

where $\rho(i, t)$ represent the density function for U_i to have a meeting event at time t:

$$\rho(i, t) = \sum_{e_h \in M_i} \exp(-c_t \cdot |t_h - t|)/|M_i| \tag{11}$$

where c_t is a parameter and $M_i = \{e_h : e_h \in E_{ij}, \forall j\}$ denotes all meeting events of users U_i with any other user. With (10) and (11), the temporal weight penalized meeting events at a regular time when the pairs have meetings with other users frequently.

Accordingly, we obtain the relationship measurement as follows:

$$G_{e,3}(E_{ij}) = \sigma_{e_h}\{\omega_{ij}^t(e_h)\} \times \sigma_{e_h}\{\omega_{ij}^p(e_h)\} \times \sum_{e_h} \omega_{ij}^e(e_h) \tag{12}$$

3.3 Duration Factor

A short-term meeting event is likely to occur occasionally, while a long-term meeting event seems relatively more significant. Thus we directly correlate the weight of a merged meeting event with its duration. To restrict the boundary of the weight, we set the weight of a merged meeting event to be zero when its duration is zero, and to capture the improvement more precisely when the duration is small, we use the logarithmic function to calculate the duration weight of a meeting event $e_h \in E_{ij}$:

$$\omega_{ij}^d(e_h) = \log(c_d \cdot dur_h + 1) \tag{13}$$

To combine the weights from all the above factors, a new relationship measure for user U_i and U_j can be obtained as:

$$G_{e,4}(E_{ij}) = \sigma_{e_h}\{\omega_{ij}^t(e_h)\} \times \sigma_{e_h}\{\omega_{ij}^p(e_h)\} \times \sum_{e_h}(\omega_{ij}^e(e_h) \cdot \omega_{ij}^d(e_h)) \tag{14}$$

3.4 Diameter Factor

In real life, a person may travel with other people that have a strong relationship with him, and we call the phenomenon *accompanying movement* which probably doesn't exist if pairs have a meeting event by chance. As mentioned in Sect. 3, the feature dia_h is used to model the accompanying movement. Similar as duration weight, we use the logarithmic function of dia_h to calculate the diameter weight of a merged meeting event $e_h \in E_{ij}$:

$$\omega_{ij}^m(e_h) = \log(c_m \cdot dia_h + 1) \tag{15}$$

Accordingly, our novel relationship measurement called *EPTDD* combining the above 5 factors can be calculated as:

$$G_{e,5}(E_{ij}) = \sigma_{e_h}\{\omega_{ij}^m(e_h)\} \times \sigma_{e_h}\{\omega_{ij}^t(e_h)\} \times \sigma_{e_h}\{\omega_{ij}^p(e_h)\} \times \sum_{e_h}(\omega_{ij}^e(e_h) \cdot \omega_{ij}^d(e_h)) \tag{16}$$

4 Experiments

4.1 Datasets and Metrics

To evaluate the effectiveness of our method, three datasets are employed in our experiments, i.e., MIT Reality Mining dataset, Gowalla dataset and Brightkite dataset. Note that locations are recorded as discrete numbers in MIT dataset, and geo-graphical positions in Gowalla dataset and Brightkite dataset.

The MIT dataset contains records of 106 users, and only 87 users having record numbers larger than 200 are remained. Gowalla dataset contains 107,902 users, and Brightkite dataset contains 58,228 users. Top 5000 users (ranked by the number of location history records) are extracted from both datasets for experiments. Each dataset contains a social network of friendships, which can serve as the ground truth for our evaluation. Other statistics of three datasets are shown in Table 1.

Table 1. Statistics of extracted datasets

	MIT dataset	Gowalla	Brightkite
# of users	87	5,000	5,000
# of friend pairs	47	27,678	68,228
# of location records	291,176	2,563,771	3,508,326
Average # of location	3346	512	701

In order to evaluate the performance of relationship measuring methods, four metrics are employed here, including Accuracy, F1 score, AUC (area under the ROC curve), and AP (area under the P-R curve). Those metrics can be calculated as followings:

$$
Precision = \frac{|TP|}{|TP| + |FP|}, \ Recall = \frac{|TP|}{|TP| + |FN|}
$$
$$
Acc = \frac{|TP| + |TN|}{|TP| + |FP| + |TN| + |FN|}, F1 = \frac{2 \cdot Precision \cdot Recall}{Precision + Recall}
$$
(17)

where TP, FP, TN and FN denote true positive, false positive, true negative, and false negative respectively. Besides, P-R (Precision-Recall) curves are drawn to display the performance of our method and compare our method with others in a visual way.

4.2 Effectiveness Analysis

In this subsection, we evaluate the effectiveness of our weighting methods for five factors, among which the weighting methods for location entropy factor, location personal factor and temporal factor are compared with that in [9], while duration factor and diameter factor are newly proposed in the paper. Figure 1 draws cumulative distribution functions (CDFs) of different factors for friend and non-friend pairs on MIT dataset.

Figure 1(a) and (b) compare CDFs of location entropy weight $\bar{\omega}_{ij}^g$ in [9] and $\bar{\omega}_{ij}^e$ in this paper. Compared with $\bar{\omega}_{ij}^g$, the largest gap between the two curves in Fig. 1(b) is much larger illustrating that $\bar{\omega}_{ij}^e$ is likely to be more effective for differentiating friends and non-friends.

Figure 1(c) and (d) show CDFs of $\max(\omega_{ij}^p)$ and $\sigma\{\omega_{ij}^p\}$. We can see that the largest gap of the two curves in Fig. 1(d) is little larger than that in Fig. 1(c). Compared with $\bar{\omega}_{ij}^e$ in Fig. 1(b), the value of $\max(\omega_{ij}^p)$ is relatively large which excessively emphasize location personal weight of each meeting events. Thus, it's reasonable to use standard deviation instead of maximum value to calculate location personal weight.

Figure 1(e) and (f) plot CDFs of the average temporal weight used in [9] and the standard deviation of temporal weights calculated by (10). In Fig. 1(e), the two curves are nearly coincidence. However, Fig. 1(f) shows a much larger gap between the two curves and illustrates the effectiveness of our temporal weight.

Figure 1(g) and (h) plot CDFs of $\bar{\omega}_{ij}^d$ and $\sigma_{e_h}\{\omega_{ij}^m(e_h)\}$ respectively. In Fig. 1(g), the largest gap between the two curves is 27%, while in Fig. 1(h), the largest gap

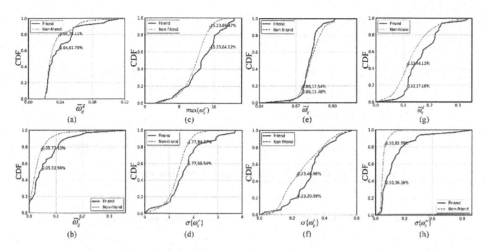

Fig. 1. Comparison of the CDF for friend and non-friend pairs. (a): $\bar{\omega}_{ij}^g$ in [9]; (b): $\bar{\omega}_{ij}^e$ in this paper; (c): $\max(\omega_{ij}^p)$ in [9]; (d): $\sigma\{\omega_{ij}^p\}$ in this paper; (e): $\bar{\omega}_{ij}^t$ in [9]; (f): $\bar{\omega}_{ij}^t$ in this paper; (g): $\bar{\omega}_{ij}^d$ that newly proposed in this paper; (h): $\sigma\{\omega_{ij}^m\}$ that newly proposed in this paper.

reaches up to 46.19%, stating that the duration and diameter factor are valid factors for measuring relationship in this dataset.

4.3 Experiment Results

Three relationship measuring methods are considered in this section, i.e., MF (meeting frequency) method, PGT method [9], and our EPTDD method. Those methods are evaluated on MIT dataset, Gowalla dataset, and Brightkite dataset, respectively. Parameters are set as follows: for MIT dataset, we set $\tau = 3$ min because the average time interval is about 1.5 min and $\tau_s = 0.5$ h to ensure the detected meeting events belonging to a real meeting event can be merged into a single one; for Gowalla and Brightkite dataset, we set $\tau = 1$ h, $\tau_s = 2$ h because time intervals are much larger than that in MIT dataset, and we set $Dist_{th} = 50$ m which seems to be the proximate distance that a person can walk in a minute.

Figure 2 demonstrates P-R curves of all methods on three datasets respectively, and Table 2 shows corresponding performance metrics, where best results among three methods is bolded.

As we can see, all methods perform much better on Gowalla dataset and Brightkite dataset than MIT dataset. Reasons for the phenomenon may be as follows:

(1) Friends tend to check-in together. Locations in Gowalla and Brightkite are check-ins shared by users whereas locations in MIT dataset are collected actively which means pairs visiting the same location in Gowalla and Brightkite are more likely to be friends than that in MIT dataset.

(2) Locations are recorded as the ids of base stations in MIT dataset, which are less precise than GPS geo-locations recorded in other two datasets, leading to an unfavorable performance on measuring relationship on MIT dataset.

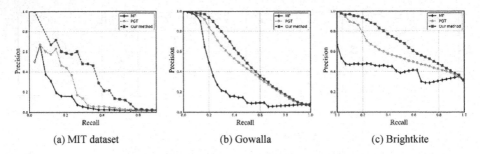

(a) MIT dataset   (b) Gowalla   (c) Brightkite

Fig. 2. The P-R curves from three datasets

Table 2. Performance comparison (all results are expressed as percentages)

Method		Acc	F1	AUC	AP		Acc	F1	AUC	AP		Acc	F1	AUC	AP
MF	MIT	98.60	18.82	61.10	8.39	Gowalla	94.57	28.70	67.26	27.37	Brightkite	70.05	51.58	63.91	40.17
PGT	dataset	98.64	28.57	64.10	14.46		94.88	46.43	83.93	47.85		74.27	55.76	74.20	58.19
EPTDD		**98.69**	**40.00**	**71.78**	**25.79**		**95.23**	**49.74**	**84.93**	**52.50**		**78.49**	**61.01**	**80.13**	**68.58**

Obviously, our EPTDD method outperforms MF method and PGT method on all tested datasets and all considered performance metrics. Specifically, as for F1-score metric, our method introduces an improvement of 40.01% on MIT dataset, 7.13% on Gowalla dataset, and 9.42% on Brightkite dataset.

Additionally, our EPTDD method achieves larger improvement on MIT dataset than that on Gowalla dataset and Brightkite dataset. This may be because that locations recorded in MIT dataset are denser, from which accompanying movements are more likely to be detected. Thus the merged meeting events become more credible and contribute more to the relationship measurement.

5 Conclusion

Aiming at measuring relationship strength between mobile users based on their spatio-temporal mobility data, the novel EPTDD method is proposed in this paper, which merges detected meeting events according to their occurring times, and combines multiple factors together for measuring relationship in an unsupervised way. Experimental results on three real-life datasets demonstrate that our method performs significantly better than existing methods, and achieves favorable results when locations of users are collected densely. In future, we will present a supervised method combining factors considered in this paper, and research on predicting friendship relationship on larger social networks.

Acknowledgements. This work is supported by National Natural Science Foundation of China (No. 61361166009).

References

1. Asgari, F., Gauthier, V., Becker, M.: A survey on human mobility and its applications. arXiv: Social and Information Networks (2013)
2. Bao, J., Zheng, Y., Wilkie, D., Mokbel, M.: Recommendations in location-based social networks: a survey. GeoInformatica **19**(3), 525–565 (2015)
3. Wang, D., Pedreschi, D., Song, C.: Human mobility, social ties, and link prediction. In: 17th Knowledge Discovery and Data Mining, San Diego, CA, USA, pp. 1100–1108 (2011)
4. Dhar, S., Varshney, U.: Challenges and business models for mobile location-based services and advertising. ACM (2011)
5. Zheng, V.W., Zheng, Y., Xie, X., Yang, Q.: Collaborative location and activity recommendations with GPS history data. In: WWW 2010, Raleigh, USA, pp. 1029–1038 (2010)
6. Ge, Y., Xiong, H., Liu, C., Zhou, Z.H.: A taxi driving fraud detection system. In: International Conference on Data Mining, pp. 181–190. IEEE Computer Society (2011)
7. Tang, J., Chang, Y., Liu, H.: Mining social media with social theories: a survey. SIGKDD Explor. **15**(2), 20–29 (2014)
8. Pham, H., Shahabi, C., Liu, Y.: EBM: an entropy-based model to infer social strength from spatiotemporal data. In: International Conference on Management of Data (2013)
9. Wang, H., Li, Z., Lee, W.C.: PGT: measuring mobility relationship using personal, global and temporal factors. In: IEEE ICDM, Shenzhen, China, pp. 570–579 (2014)
10. Lambiotte, R., Blondel, V.D., de Kerchove, C., Huens, E., Prieur, C., Smoreda, Z., Van Dooren, P.: Geographical dispersal of mobile communication networks. Phys. A-Stat. Mech. Appl. **387**(21), 5317–5325 (2008)
11. Descioli, P., Kurzban, R., Koch, E.N.: Best friends alliances, friend ranking, and the MySpace social network. Perspect. Psychol. Sci. **6**(1), 6–8 (2011)
12. Zhang, D., Vasilakos, A.V., Xiong, H.: Predicting location using mobile phone calls. ACM Spec. Interest Group Data Commun. **42**(4), 295–296 (2012)
13. Pang, J., Zhang, Y.: Exploring communities for effective location prediction. In: International World Wide Web Conferences (2015)
14. Cho, E., Myers, S.A., Leskovec, J.: Friendship and mobility: user movement in location-based social networks. In: 17th ACM SIGKDD International Conference on Knowledge Discovery and Data Mining, pp. 1082–1090 (2011)
15. Eagle, N., Pentland, A., Lazer, D.: Inferring friendship network structure by using mobile phone data. PNAS **106**(36), 15274–15278 (2009)
16. Crandall, D., Backstrom, L., Cosley, D., Suri, S., Huttenlocher, D.P., Kleinberg, J.: Inferring social ties from geographic coincidences. PNAS **107**(52), 22436–22441 (2010)
17. Zhang, Y., Pang, J.: Distance and friendship: a distance-based model for link prediction in social networks. In: Cheng, R., Cui, B., Zhang, Z., Cai, R., Xu, J. (eds.) APWeb 2015. LNCS, vol. 9313, pp. 55–66. Springer, Cham (2015). doi:10.1007/978-3-319-25255-1_5
18. Cheng, R., Pang, J., Zhang, Y.: Inferring friendship from check-in data of location-based social networks. In: International Conference on Advances in Social Networks Analysis and Mining, pp. 1284–1291. IEEE (2015)
19. Valverdebaza, J., Roche, M., Poncelet, P., de Andrade Lopes, A: Exploiting social and mobility patterns for friendship prediction in location-based social networks. In: International Conference on Pattern Recognition (2016)

Reinforcement Label Propagation Algorithm Based on History Record

Kai Liu[(⊠)], Yi Zhang, Kai Lu, Xiaoping Wang, and Xin Wang

Science and Technology on Parallel and Distributed Processing Laboratory,
College of Computer, National University of Defense Technology,
Changsha 410073, People's Republic of China
kailiu@nudt.edu.cn

Abstract. With the continuous development of Internet, social networks are becoming more and more complex, and the research on these complex networks has attracted many researchers' attention. A large number of community discovery algorithms have emerged, among which the label propagation algorithm is widely used because of its simplicity and efficiency. However, this algorithm has poor stability due to the randomness in the label propagation process. To solve the problem, we propose a reinforcement label propagation algorithm (RLPA) in this paper. In RLPA, a similarity matrix is generated from the historical records of classification, which can be adopted to obtain the final result of community detection. The experimental results show that our algorithm can not only get better performance in accuracy, but also has higher stability.

Keywords: Data mining · Community discovery · Label propagation algorithm

1 Introduction

With the continuous development of the Internet and social software, the relationship between people is getting closer, and the social network is also expanding. In recent years, more and more researchers have begun to carry out related researches on these complex networks, and community discovery has become a hot spot. In general, the process of community discovery is to place the nodes into several communities in order to make the closely connected nodes in the same community.

At present, several community discovery algorithms have been proposed such as fast unfolding algorithm [1] and label propagation algorithm [2] (LPA). Since the computational complexity of LPA is near linear time in the number of edges and it works well in community discovery, this algorithm is widely used in complex network analysis. However, it also has some shortcomings because of the randomness. In propagation process, this randomness may lead to the accident that small communities will be swallowed up, thus affecting the accuracy of the community detection and making the result of the label propagation algorithm unstable.

To solve this problem, LPA could be applied several times in practical application. Then these community classification results are judged by using a metric named modularity. The final result is often the one whose modularity reached the maximum.

D. Liu et al. (Eds.): ICONIP 2017, Part V, LNCS 10638, pp. 348–356, 2017.
https://doi.org/10.1007/978-3-319-70139-4_35

However, this metric can not objectively measure the accuracy of community discovery according to the discussion in [3]. Therefore, we could not get the optimal result in this way to overcome the problem caused by the randomness in LPA algorithm.

In this paper, we propose a reinforcement label propagation algorithm, which use several historical records from some label propagation processes to generate a similarity matrix. In the final label transmission process, when there are some labels to randomly select, we could use this similarity matrix as a basis of judgment, thus avoiding the problem caused by randomness. The experimental results show that our algorithm could not only enhance the accuracy of community detection but also has higher stability.

2 Related Work

LPA is a simple community detection algorithm that uses network structure alone as its guide and does not require any prior information about the communities. For an undirected graph with n nodes $G = \{V, E\}$, where V represents the set of vertex and E represents the edge set. In the process of conventional label propagation, every node is given a different label which represents a unique community firstly. Then the node labels are updated in a random order at every iteration. When updating the label of node x, we need to calculate the frequency of label appearing in all neighbor nodes of x. Next, the node x will be set to the label with the highest frequency. If there are multiple candidate labels, randomly selection will be executed. The termination condition of LPA is that the labels are no longer changed after several iterations or the upper bound of iteration is reached.

The problem of this algorithm is that it treats each node equally, which makes the label propagation too arbitrary. As shown in Fig. 1, the randomness of this algorithm makes it easy for small communities to be annexed, which will reduce the accuracy of the results. Although label propagation algorithm is simple and efficient, the stability of the results is not good enough.

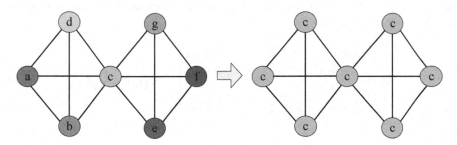

Fig. 1. An example of community annexation causes by LPA. The left is the original graph in which every node has a unique label and the right is an unreasonable result of this algorithm.

In order to solve this problem, some improved algorithms are proposed. For example, the method BMLPA [4] is designed by Wu et al. which uses a new concept named belonging coefficient to complete label selection. Another method SLPA [5] spreads labels according to dynamic interaction rules mimicking human pairwise communication behavior. A more intuitive algorithm [3] is designed to facilitate real-time community detection by incorporating the priority about the vertex, edge, and community size. Recently, a novel method named UWLPA [6] is proposed to adopt the local topological structure in label propagation.

Although these algorithms optimize some details of label propagation, most of them set the weight of the label according to the degree of node. However, it's not possible to get a good classification result only by considering the weight of label in this way. At the same time, these optimization methods make the original simple and efficient label propagation algorithm become complicated.

3   Reinforcement Label Propagation Algorithm (RLPA)

This section will discuss the details of algorithm implementation. Firstly, the generation process of similarity matrix is introduced. Then we will describe how to apply this similarity matrix to the RLPA algorithm. Finally, time complexity of our method is analyzed.

3.1   Similarity Matrix

In order to represent the relationship between nodes, we use the similarity matrix as a reference. This similarity matrix is generated by several traditional label propagation processes. The number of column and row in this matrix is equal to the total amount of nodes in the graph. We use the conventional label propagation algorithm to classify the nodes several times, and record the results of each classification.

In these history records, if node i and node j are assigned to the same community $k(i, j)$ times, then the similarity degree between i and j is set $k(i, j)/T$, where T is the number of history records. Therefore, each value in the matrix represents the impact factor between the two nodes and the impact factor of node i on node j is represented by $a(i, j)$. After similarity matrix A generated, we will use this matrix to optimize the label propagation algorithm in next subsection.

$$A = \begin{bmatrix} a(1,1) & a(1,2) & \dots & a(1,n) \\ a(2,1) & a(2,2) & \dots & a(2,n) \\ \vdots & \vdots & \ddots & \vdots \\ a(n,1) & a(n,2) & \dots & a(n,n) \end{bmatrix} \qquad a(i,j) = \frac{k(i,j)}{T} \qquad (1)$$

3.2   Algorithm Design

In our proposed algorithm, the label update for node x is also determined by the frequency of label occurrence. Different from traditional method, our selecting process is based on the similarity matrix with the help of a concept named label influence factor.

For the label t appearing on a neighbor of node x, the influence factor of label t to node x represented by $F(t, x)$ is the sum of several impact factors in similarity matrix, which used to indicate the impact of all neighbor nodes with same label t on node x.

$$F(t,x) = \sum_{i=1}^{n} a(i,x) \times sign(i,t)$$

$$sign(i,t) = \begin{cases} 1 & label(i) = t \\ 0 & label(i) \neq t \end{cases}$$

(2)

Algorithm1. The Reinforcement Label Propagation Algorithm

Input: $G=\{V, E\}$, T, *maxturn*
Output: Array *label*, the label for every node
1: $t = 1$
2: **while** $t \leq T$ **do**
3:      Execute traditional label propagation algorithm.
4:      Record the result of classification.
5:      $t = t + 1$
6: **end while**
7: Generate the similarity matrix using Equation (1)
8: Initialize the *label(v)* = v for each node in vertex set V
9: $i = 1$
10: **while** $i \leq$ *maxturn* **do**
11:      *termination* $= 1$
12:      Arrange a random order X for label updating
13:      **for each** $x \in X$ **do**
14:          Find all neighbor nodes of x and keep them in $S(x)$
15:          Find the most frequency label c on the node in $S(x)$, keep them in $C(x)$ if many
16:          **if** $|C(x)| > 1$ **then**
17:              **for** each label $p \in C(x)$
18:                  Calculate the influence factor according to the Equation (2)
19:              **end for**
20:              $c =$ the label with the maximum influence factor in $C(x)$
                 Choose one randomly if multiple candidate
21:          **end if**
22:          **if** *label(x)* $\neq c$ **then**
23:              *label(x)* $= c$
24:              *termination* $= 0$
25:          **end if**
26:      **end for**
27:      **if** *termination* $= 1$ **then;**
28:          **break**
29:      **end if**
30:      $i = i + 1$
31: **end while**

After improvement, the label propagation process in our method becomes: when updating the label of node x, the frequency of each label appearing in the neighbor of x is calculated. Then the most frequently occurring label will be selected. If multiple candidates exist, the label with the maximum influence factor finally wins. The details of our method are shown in Algorithm 1.

3.3   Time Complexity

In the generation of similarity matrix, traditional label propagation algorithm is executed T times. Because the time complexity of LPA is $O(m)$ as discussed in [2], where m is the number of edges, this initialization process requires $O(Tm)$.

As for the final process of label propagation, the outer loop is controlled by the user defined parameter *maxturn*, which is usually a small constant. And the inner loop is dominated by the number of nodes n. In each operation of inner loop, we first group neighbors according to the labels $[O(d_x)]$ for every node x, where d_x is the degree of node x. Then the groups with maximum size are picked and the label with highest influence factor is assigned to x, requiring a worst-case time of $O(d_x)$. Therefore the time complexity of final label propagation is $O(m)$. Through the above analysis, we can conclude that the time complexity of entire algorithm is $O(Tm)$.

4   Experiment

In this section, some experiments will be conducted in several data sets to evaluate the reinforcement label propagation algorithm proposed in this paper. All code implemented in these experiments are programmed in Matlab R2013a

Dataset. In our experiments, some social networks are used whose details are shown in the Table 1. These data sets can be split into two categories, one of which is with correct classification results including Zarachy karate network [7], Dolphins network [8], the dataset of Polbooks [9], American College football network [10], and the network of Political blogs (Polblogs) [11]. Another category without ground truth includes Arxiv GR-QC collaboration network (Ca-GrQc) and Arxiv HEP-TH collaboration network (Ca-HepTh) [12].

Table 1. The datasets used in experiments.

Dataset	Vertices	Edges	Class	With Ground-truth
Karate	34	78	2	Yes
Dolphins	62	159	2	Yes
Polbooks	105	441	2	Yes
Football	115	616	12	Yes
Polblogs	1490	19090	3	Yes
Ca-GrQc	5242	28980	Unknown	No
Ca-HepTh	9877	51971	Unknown	No

4.1   Accuracy and Quality Analysis

For the social networks with the ground-truth, normalized mutual information (NMI) will be used to measure the clustering accuracy. We fix $T = 10$ in Algorithm 1 and repeat 50 times for average NMI value.

The Fig. 2 presents the accuracy performance of our algorithm. The maximum performance improvement occurs in the Karate for an average NMI of 0.80 and a promotion of 17%. As for the other data sets, RLPA performs better all the time and it

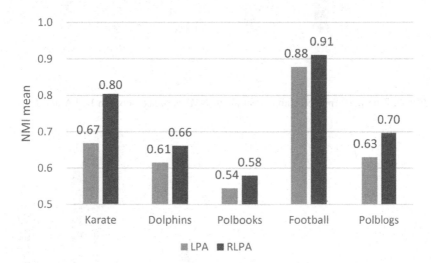

Fig. 2. The performance comparison about NMI mean between RLPA and LPA.

has an average promotion of 7%, 6%, 4% and 10%.

As shown in Fig. 3, the stability of our algorithm is also better for all the data sets. Because NMI variance decreased by 24%, 16%, 49%, 67%, and 97% respectively, the fluctuation of our results is greatly reduced.

For all the networks, we also use modularity as a metric to evaluate the quality of community detection and the T in Algorithm 1 is still fixed to 10. With similarity to the last experiments, we repeat 50 times for every performance. In Fig. 4, the obvious performance improvements occur in the Karate and Polblogs with the promotion of 9% and 11%. As for the other data sets, all implementations also achieve superior quality according to this metric.

Considering both experiments about NMI and modularity, we could conclude that RLPA can improve the overall performance in terms of accuracy and quality.

4.2   The Impact of Parameter

In previous experiments, the parameter T is fixed to 10 all the time. In fact, it still has an impact on the results of community detection. Next, the value of T in Algorithm 1 is set

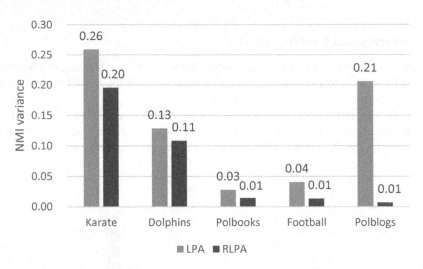

Fig. 3. The performance comparison about NMI variance between RLPA and LPA.

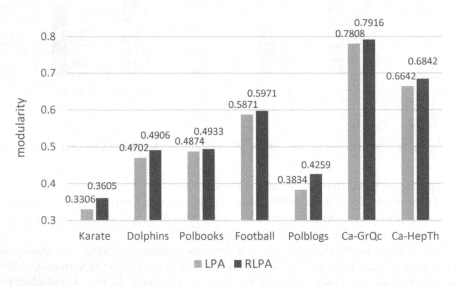

Fig. 4. The performance comparison about modularity between RLPA and LPA.

to range from 20 to 200 with each step of 20. In this subsection, the data set Karate and Dolphins will be chosen to implement RLPA algorithm and we also repeat 50 times for every NMI value.

As can be seen from Fig. 5, the line of Karate increase on the whole in spite of a little fluctuation. When the parameter T reaches 200, the NMI value runs up to 93.48% which is a ideal result. The similar trend has also occurred on the curve of Dolphins.

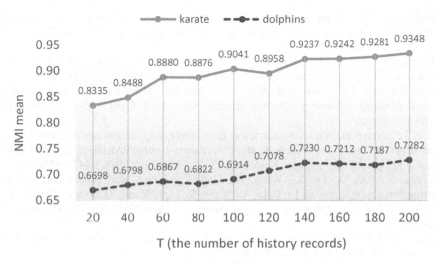

Fig. 5. The effect on the accuracy due to the change of parameter.

That is to say, when more history records obtained, we could get more accurate result using RLPA.

5   Conclusion

In this paper, we propose a reinforcement label propagation algorithm (RLPA) using several classification histories. Based on these history records, a similarity matrix is generated. With the help of this matrix, we could overcome the problems caused by the randomness in the traditional LPA algorithm. Experimental results show that our method could not only increase the accuracy of classification, but also greatly reduce the fluctuation of results.

Actually, this method can also be applied to other LPA modification algorithms in order to eliminate randomness. In the future, we will explore the application of our algorithm and try to achieve acceleration by introducing parallel calculation scheme.

Acknowledgement. This work is supported by National High-tech R&D Program of China (863 Program) under Grants 2015AA01A301, 2015AA010901, and 2015AA01A301, by program for New Century Excellent Talents in University by National Science Foundation (NSF) China 61272142, 61402492, 61402486, 61379146, 61272483, by the open project of State Key Laboratory of High-end Server & Storage Technology (2014HSSA01).

References

1. Blondel, V.D., Guillaume, J.L., Lambiotte, R., Lefebvre, E.: Fast unfolding of communities in large networks. J. Stat. Mech.: Theory Exp. **2008**(10), 155–168 (2008)

2. Raghavan, U.N., Albert, R., Kumara, S.: Near linear time algorithm to detect community structures in large-scale networks. Phys. Rev. E: Stat., Nonlinear Soft Matter Phys. **76**(3 Pt 2), 036106 (2007)
3. Leung, I.X., Hui, P., Li, P., Crowcroft, J.: Towards real-time community detection in large networks. Phys. Rev. E: Stat. Nonlinear Soft Matter Phys. **79**(6 Pt 2), 066107 (2008)
4. Wu, Z.H., Lin, Y.F., Gregory, S., Wan, H.Y., Tian, S.F.: Balanced multi-label propagation for overlapping community detection in social networks. J. Comput. Sci. Technol. **27**(3), 468–479 (2012)
5. Xie, J., Szymanski, B.K.: Towards linear time overlapping community detection in social networks. In: Tan, P.-N., Chawla, S., Ho, C.K., Bailey, J. (eds.) PAKDD 2012 Part II. LNCS, vol. 7302, pp. 25–36. Springer, Heidelberg (2012). doi:10.1007/978-3-642-30220-6_3
6. Wang, X., Jian, S., Lu, K., Wang, X.: Unified weighted label propagation algorithm using connection factor. In: Li, J., Li, X., Wang, S., Li, J., Sheng, Q.Z. (eds.) ADMA 2016. LNCS (LNAI), vol. 10086, pp. 434–444. Springer, Cham (2016). doi:10.1007/978-3-319-49586-6_29
7. Zachary, W.W.: An information flow model for conflict and fission in small groups. J. Anthropol. Res. **33**(4), 452–473 (1977)
8. Lusseau, D., Schneider, K., Boisseau, O.J., Haase, P., Slooten, E., Dawson, S.M.: The bottlenose dolphin community of doubtful sound features a large proportion of long-lasting associations. Behav. Ecol. Sociobiol. **54**(4), 396–405 (2003)
9. Newman, M.E.: Modularity and community structure in networks. Proc. Natl. Acad. Sci. **103**(23), 8577–8582 (2006)
10. Girvan, M., Newman, M.E.: Community structure in social and biological networks. Proc. Natl. Acad. Sci. **99**(12), 7821–7826 (2002)
11. Adamic, L.A., Glance, N.: The political blogosphere and the 2004 US election: divided they blog. In: Proceedings of the 3rd international workshop on Link discovery, pp. 36–43. ACM (2005)
12. Leskovec, J., Kleinberg, J., Faloutsos, C.: Graph evolution: densification and shrinking diameters. ACM Trans. Knowl. Discov. Data (TKDD) **1**(1), 2 (2007)

A Hybrid Approach for Recovering Information Propagational Direction

Xiang-Rui Peng[1,2], Ling Huang[1,2], and Chang-Dong Wang[1,2(✉)]

[1] School of Data and Computer Science, Sun Yat-sen University, Guangzhou, China
pengxr6@mail2.sysu.edu.cn, huanglinghl@hotmail.com,
changdongwang@hotmail.com
[2] Guangdong Key Laboratory of Information Security Technology,
Guangzhou, China

Abstract. With the rapid development of network technology, people are communicating with each other through a variety of network access, such as computer, mobile phone, tablet, etc., for the sharing of information and interactive behavior. The flow of information is directional, but this directionality is usually hidden. In recent years, link prediction technology has been developed very rapidly in social network analysis. The active and passive of the relationship, in social network, could be identified via undirected relationship network structure. However, this approach only focuses on the topological structure while ignoring the information shared between individuals, which is not suitable for study in terms of information propagation. To solve this problem, we propose a hybrid approach termed DRHM to recover the information sharing direction in networks. It combines not only topology structure but also node content. Since the algorithm is based on edge structure, it is equally applicable to large-scale data set. The experiment has demonstrated that our algorithm performs well in information propagational network.

Keywords: Information propagation · Direction prediction · Hybrid

1 Introduction

With the recent development of modern digital and internet technology, the information sharing is explosive growth. Everyone receive information from others as well as sharing their own knowledge to the outside world. This kind of information exchange behavior between individuals gradually forms many relationship network structure, such as social network, citation network and collaboration network [3,4]. These relationship network has received a large amount of research interests based on different studies of relationship property between individuals, including strong and weak ties [2,12,13], positive and negative signed network [1,11]. The focus of this paper is on the directionality of the network structure, i.e. recovering or determining the direction from undirected network. The directionality of edge plays a key role in many data mining applications,

© Springer International Publishing AG 2017
D. Liu et al. (Eds.): ICONIP 2017, Part V, LNCS 10638, pp. 357–367, 2017.
https://doi.org/10.1007/978-3-319-70139-4_36

such as Recommendation System [9], Community Detection [8,10] and so on. Although almost all of these applications are using the direction property of edge, not all kinds of edge direction can be easily obtained. For example, as the number of people sharing information with each other increases, the direction of information propagation between individuals gradually becomes blurred. In citation network, while one author cites one paper of another author, it is assumed that there is a directed information passing from one to another. But when they cite more than one paper of each other, it is difficult to determine the direction of information propagation correctly. The knowledge transmit relationship between them has been hidden. Therefore, it is not a simple thing to correctly dig out the propagational direction from the information sharing precess, even it is believed that a better understanding of these propagational directions might be very significant in many network analysis tasks. Very recently, Zhang et al. [7] has developed one effective method for recovering the direction of undirected social ties via topological structure. However, it uses only topological structure, which is not able to recover the direction of information propagational process.

In this paper, we propose a novel approach to estimate the direction tendency in information propagational process, termed DRHM. Firstly, one observation on the popular DBLP dataset is conducted, which is a typical knowledge transmit network storing millions of computer science literatures as well as their references. Some directional-related patterns have been obtained from both node content and topological structure. Then these patterns are integrated together, based on which an algorithm is designed to estimate the knowledge sharing direction tendency of each undirected edge in this information propagation network. Extensive experiments have demonstrated that our approach has achieved satisfactory performance on learning the knowledge sharing direction.

2   The Proposed Model

In this section, we first introduce some patterns which are observed from real data set. Then, from the node content perspective, we propose a similarity model to estimate the direction probability as well as ternary relation transmit model and degree model in topological structure. For clarity, we define that, in author citation network, $i \leftarrow j$ denotes author i has been cited by author j, i.e. a directed edge. The node set that node i points to and is pointed by are denoted as N_i^+ and N_i^-. When one author i has been cited by author j frequently, we say author i has shared information to author j and denote it by $i \Rightarrow j$.

2.1   Data Set Observation

Since we aim to study some patterns that can be used to estimate direction between two individuals, we will observe the ground-truth data set which is extracted from the preprocessed DBLP data. The nodes of this data set represent authors and directed edges represent the reference relationship between authors. For one edge, the head node of one edge is called citing person, and tail node is called cited person. Each node contains a vector which characterizes the features of the corresponding author that we concluded from the papers they published. We observe a series of opposite events and calculate the proportion between them to conclude some properties of directed network structure.

Similarity Observation. First of all, from the content perspective, we will introduce the function $Sim(*)$ measuring the similarity between two nodes according to the corresponding node content:

$$Sim(x,y) = \frac{1}{m} \sum_{i=1}^{m} \frac{\min\{F_i(x), F_i(y)\}}{\max\{F_i(x), F_i(y)\}} \tag{1}$$

where $F \in \mathbb{R}^{n \times m}$ is the feature matrix, where each row represents the feature vector of the corresponding node.

Table 1. The frequency of each event in the ground-truth data set.

Event	$[S_{N_v^+}^u]$	$[S_{N_v^-}^u]$	$[S_{N_u^+}^v]$	$[S_{N_u^-}^v]$	$[T^+]$	$[T^-]$
Frequency	71.16%	28.84%	24.93%	75.07%	99.2%	0.8%
Event	$[D_{u>v}^+]$	$[D_{u>v}^-]$	$[D_{u>v}^{in+}]$	$[D_{u>v}^{in-}]$	$[D_{u>v}^{out+}]$	$[D_{u>v}^{out-}]$
Frequency	59.36%	40.64%	78.31%	21.69%	33.90%	66.10%

After obtaining the similarity, for one reference between cited person u and citing person v, denoted by $u \leftarrow v$, two pairs of opposite events can be observed.

- $[S_{N_v^+}^u]$ or $[S_{N_v^-}^u]$: $S_{u \frown \{N_v^+ - u\}}$ is larger than $S_{u \frown N_v^-}$ or the opposite.
- $[S_{N_u^+}^v]$ or $[S_{N_u^-}^v]$: $S_{v \frown N_u^+}$ is larger than $S_{v \frown \{N_u^- - v\}}$ or the opposite;

where $S_{u \frown N_v^+}$ denotes the average similarity between author u and N_v^+. The equal situation is not considered due to the reason that the proportion of this situation is very small which can be ignored. The distribution of these events is shown in Table 1. Accordingly, we have the following proposition.

Proposition 1 (Similarity observation). *For one directed edge $u \leftarrow v$, the head node u is more similar to N_v^+ than N_v^-, and the tail node v is more similar to N_u^- than N_u^+.*

This proposition can be explained as follows. For one reference $u \leftarrow v$, the cited person u will be more similar to the others who v is citing, than who v is cited by, while the citing person v is more similar to the authors who cite u than the authors who are cited by u.

Ternary Transmit Relationship Observation. Secondly, we study the ternary transmit relationship of the ground-truth data set. A ternary structure is defined as three interconnected authors, say u, k and v. While there are two references, $u \leftarrow k$ and $k \leftarrow v$, we note $u \leftarrow v$ as positive transmission, and $v \leftarrow u$, negative transmission. Similarly, a pair of opposite events of each ternary structure can be observed as follows.

- $[T^+]$ or $[T^-]$: while $u \leftarrow k$ and $k \leftarrow v$, $e_{u,v}$ is a positive transmission or negative one;

The frequency statistic is shown in Table 1. It is obvious that if we have two references passing through three authors, the reference relationship between the first author and rear author is almost certain, i.e. it is always a positive transmission. Accordingly, we have the following proposition.

Proposition 2 (Ternary transmit relationship observation). *For a ternary structure, the directions of the connections between the nodes will not intend to be a loop.*

General speaking, the cited person has a very small possibility to cite the paper of the author who has cited the citing person.

Node Degree Observation. Since the node degree is a simple but significant perspective of topological structure, we will explore the relation between node degree and author references. Three pairs of opposite events will be observed as

- $[D^+_{u>v}]$ or $[D^-_{u>v}]$: the degree of u is larger than that of v or the opposite;
- $[D^{in+}_{u>v}]$ or $[D^{in-}_{u>v}]$: the in-degree of u is larger than that of v or the opposite;
- $[D^{out-}_{u>v}]$ or $[D^{out-}_{u>v}]$: the out-degree of u is larger than that of v or the opposite;

Similarly, the proportion of equal situation is small and can be ignored. The frequency statistic of the ground-truth data set is shown in Table 1. We simply compare these observations and obtain the following proposition.

Proposition 3 (Node degree observation). *For one directed edge $u \leftarrow v$, the in-degree of node u usually outnumbers that of v, while in terms of out-degree, v will outnumbers u.*

It could be explained that, in most cases, the author who has a high preference to cite the papers of others will be more likely to be a citing person, i.e. the head node of one directed edge, as well as the author who has been cited by others frequently will be the cited person, i.e. the tail node of one directed edge.

2.2   The Patterns

Similarity Pattern. According to Proposition 1, it is found that the direction of one undirected edge is somewhat related with its surrounding nodes' content, which is significant for learning the edge direction from a content-based perspective. For an edge $i \leftarrow j$, since the tail node j, i.e. citing person j, is more similar to N_i^-, who have cited author i, compared with N_i^+, the direction tendency of the edge is positively correlated with the similarity between the tail node j and N_i^-. Thus, we define $SP1(i, j)$ to estimate the positive direction tendency of $i \Rightarrow j$, i.e. the possibility that i will share information to j, as follows.

$$SP1(i, j) = \frac{1}{|N_i^-|} \sum_{k \in N_i^-} Sim(j, k) \tag{2}$$

Similarly, the direction tendency of $i \Rightarrow j$ is positively correlated with the similarity between the head node i and N_j^+. Thus, the direction tendency of $i \Rightarrow j$ could be estimated by

$$SP2(i, j) = \frac{1}{|N_j^+|} \sum_{k \in N_j^+} Sim(i, k) \tag{3}$$

Using Eqs. (2) and (3), we can estimate the direction tendency of one edge from a content-based perspective. The estimate value of this perspective will be helpful for inferring direction of undirected network.

Ternary Transmit Relationship Pattern. As observed in Sect. 2.1, it is almost impossible for the connections between ternary nodes to be a loop. In other words, for every three authors who have reference relations with each other, the reference relations between them will not form a loop. Thus, while we estimate the direction tendency between node i and node j, the remaining nodes which form a ternary structure will probably provide some help. Here, we define $TP(i, j)$ to characterize the direction tendency of $i \Rightarrow j$ as follows.

$$TP(i, j) = \frac{1}{|N_i \cap N_j|} \sum_{k1 \in N_i^-} \sum_{k2 \in N_j^+} \sigma(k1, k2) \tag{4}$$

where the value of $\sigma(x, y)$ is equal to 1 if x and y are the same, and 0 otherwise. All neighbors of node i are denoted by N_i.

Node Degree Pattern. As we have demonstrated in Sect. 2.1, for a directed edge, the node that has higher in-degree will be more possible to be a head node rather than the tail node. This observation can help us to recognize the relationship between two authors. That is, the one who has cited many others has a high preference to be a citing person, while the one who has been cited by many others will tend to have a high possibility to be the cited person. Accordingly, we can estimate the direction tendency of $i \Rightarrow j$ as follows.

$$DP(i, j) = \frac{D_{in}(i) - D_{in}(j)}{D_{in}(i) + D_{in}(j)} + \frac{D_{out}(i) - D_{out}(j)}{D_{out}(i) + D_{out}(j)} \tag{5}$$

where $D_{out}(i)$ and $D_{in}(i)$ are the out-degree and in-degree of node i respectively.

Algorithm 1. Direction Recovering using Hybrid Model

1: **Input:** Feature Matrix F, Undirected Edge Set E, initialize pseudo directed edge percentage m, maximum Iteration max_iter, average error weight coefficient θ, truncated error ϕ
2: **Initialization:**
3: Sort undirected edge set by **degree rate**, assign preceding m percent of these sorted edges to a confirm direction, update N_i^+ and N_i^-.
4: **repeat**
5:    **for** $\forall e_{ij} \in E$ **do**
6:        Calculate $P(i,j)$ and $P(j,i)$ via Eqs. (6) and (7).
7:    **end for**
8:    Compute average error ae via Eq. (8).
9:    Assign direction to the undirected edges, update N_i^+ and N_i^-.
10: **until** The average error is smaller than truncated error ϕ or max_iter is reached.
11: Update directed Edge set E_{dir} via N_i^+ and N_i^-.
12: **Output:** Directed edge Set E_{dir}

2.3 Hybrid Model

We have proposed four patterns from two perspectives to estimate the positive direction tendency between node i and node j, namely the two similarity patterns for the node content, as well as the ternary transmit relation pattern and the node degree pattern for topological structure. To obtain one unified model, these four patterns are linearly combined to form a hybrid model $HM(i,j)$, which represents the total positive direction tendency of $i \Rightarrow j$ as follows.

$$HM(i,j) = \lambda_1 SP1(i,j) + \lambda_2 SP1(i,j) + \lambda_3 TP(i,j) + \lambda_4 DP(i,j) \qquad (6)$$

where $\lambda_1, \lambda_2, \lambda_3$ and λ_4 are used to tune the contribution of four patterns, satisfying $\lambda_1 + \lambda_2 + \lambda_3 + \lambda_4 = 1$. In our experiments, we set $\lambda_1 = \lambda_2 = \lambda_3 = \lambda_4 = 0.25$, i.e. equal contribution. However, the direction tendency is determined not only by positive direction tendency but also reversed direction tendency. While we have obtained the positive direction tendency function $HM(i,j)$, the reversed direction tendency can be obtained as $HM(j,i)$. Therefore, by normalizing the direction tendency into a probability, we have

$$P(i,j) = \frac{HM(i,j)}{HM(i,j) + HM(j,i)}, P(j,i) = \frac{HM(j,i)}{HM(i,j) + HM(j,i)} \qquad (7)$$

where $P(i,j)$ denotes the positive direction tendency $i \Rightarrow j$ and $P(j,i)$ is the reversed direction tendency, satisfying $P(i,j) + P(j,i) = 1$.

2.4 Algorithm

In the previous section, we have proposed the hybrid model for estimating the direction tendency of one undirected edge. In this section, we will apply it to figure out most possible direction of edges from undirected network. The whole process of our algorithm is shown in Algorithm 1. Some detailed description of the steps will be given below.

- **Initialization.** In Sect. 2.1, we have observed that the direction of one edge is strongly associated with the in-degree and out-degree of its two endpoints, while it is weakly associated with the total degree. However, the degree ratio, i.e. the proportion of two endpoints in an undirected edge, is still helpful for initializing the direction in an undirected network. Therefore, we use the degree ratio to initialize the direction of m percentage undirected edges as $DR(i, j) = \frac{deg(i)}{deg(j)}$ where $deg(i)$ is the degree of node i [7].
- **Direction assignment.** We will determine that author u is a cited person by whom author v is cited when $P(u, v) - P(v, u) - \delta > 0$.
- **Calculation of Average Error** ae **and Dynamic threshold** δ. The deviation of directional tendency of different undirected edges could be very different, which leads to hard assignment of the edge direction. Since we assign direction via the threshold δ, a fixing value may not lead to a better performance. Therefore, we generate the dynamic threshold δ by weighting the average error of the positive and reversed directional tendency of the remaining undirected edges. The average error ae is calculated as:

$$ae = \frac{1}{|E|} \sum_{e_{ij} \in E} |P(i, j) - P(j, i)| \tag{8}$$

Then the dynamic threshold δ of the t-th iteration is defined as $\delta^t = \theta \times ae^t$, where θ is the average error weight coefficient to weight average error.

3 Experiment

3.1 Data Preparation

Since the citation network is a typical representative for the transmission of knowledge (information), we use the DBLP data set in our experiments. The goal is to figure out information propagation direction between authors in citation network via not only considering topological structure but also the node content. In particular, we regard the title and abstract of the paper as the node content and crawl them from internet. The crawled papers are published in 14 conferences/journals from 2001 to 2016, including MOBICOM, SIGCOMM, CoNext, Sensys, ICML, IJCAI, TOG, SIGGRAPH, KDD, CIKM, VLDB, TODS, TOIS and SIGGMOD.

After constructing the paper citation network from subset of DBLP, we exclude the papers which have no citation relationship with others, i.e. outlier. The well-known **TF-IDF** [5,6] is used to represent each paper and construct the paper feature matrix $F_p \in \mathbb{R}^{n^p \times m}$, where n^p is the number of papers and m is the number of features. Then, the author citation network (ACN) will be constructed according to the paper citation network. In particular, we obtain the author citation relationship via the paper citation relationship, e.g. there is a tie between author u and v as long as one of them has cited no less than one paper of the other. The ACN contains 24273 vertex and 571789 edges. The feature vector of each author can be constructed by the papers they have published. The i-th feature of author u who has published k papers is defined as $F_{a_u}^i = \max(F_{p_1}^i, F_{p_2}^i, \ldots, F_{p_k}^i)$, where $F_{p_j}^i$ is the i-th feature of the j-th paper that author u has published.

Since we explore the information propagation direction between individuals in citation network, the number of papers cited by authors usually is no less than one and it is not reasonable to determine one information propagation direction only according to one single reference. Therefore, we extract a part of references from the complete author reference data set to be the ground-truth to evaluate the performance of the algorithms. The extraction process contains the following two steps as follows.

- **step1:** Start with the complete author reference data set E_a, if one paper of author i has been cited by author j, there is a directed node pair of $\{i, j\} \in E_a$. We extract only the node pairs which have appeared more than twice. The reason is that author i has surely shared information to author j if author j has cited author i more than twice.
- **step2:** Next, we remove the duplicated node pairs, i.e. remove $e_1 = \{i1, i2\}$ and $e_2 = \{i2, i1\}$ if $e_1, e_2 \in E_a$, since the duplicated information may lead to weak tendency which can not be identified clearly.

After the above extraction process, we can obtain some node pairs, i.e. ground-truth, which contain the credible information sharing direction. The edge number of ground-truth is 57860, nearly 10% of the total edges of ACN, covering 12623 vertexes.

3.2   Parameter Analysis

In this section, parameter analysis is conducted to show the impact of the three parameters on the performance of the proposed method, namely initialization pseudo directed edge percentage m, average error weight coefficient θ and maximum iteration number max_iter. When analyzing one parameter, the remaining two parameters are fixed. In order to better observe the convergence situation, we set the truncate error $\phi = 0$.

Parameter Analysis on m. The initialization pseudo directed edge percentage m determines how much should we assign the pseudo direction before iteration via degree ratio. As shown in Fig. 1(a), by setting the other two parameters $\theta = 0.5$ and $max\_iter = 5$, it is found that the algorithm has achieved the best performance when $m = 0.3$, i.e. 30% of pseudo directed edge is the most suitable, while overall relative stable results have been achieved.

Parameter Analysis on θ. Parameter θ represents the coefficient to weight average error. While we use a dynamic threshold δ to relax the edge direction

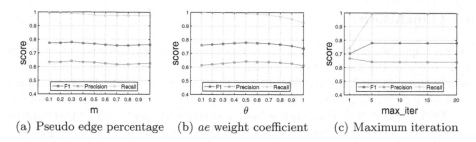

(a) Pseudo edge percentage   (b) ae weight coefficient   (c) Maximum iteration

Fig. 1. Parameter analysis: the values of evaluation measures as a function of the three different parameters.

assignment, the value of θ determines how much this dynamic threshold is close to average error. As shown in Fig. 1(b), by fixing $m = 0.3$ and $max\_iter = 5$, it is found that the algorithm achieves the best trade-off between *Precision* and *Recall* when $\theta = 0.5$.

Parameter Analysis on max_iter. Parameter max_iter determines how many iterations the algorithm will run. The larger iteration number is, the more undirected edge will be assigned direction. We try different iteration number and find that the algorithm will be very stable when the number of iterations is more than 5, as shown in Fig. 1(c). That is why we set $max\_iter = 5$ when analyzing m and θ.

3.3 Comparison Results

In this section, we compare the proposed DRHM approach with one of the algorithms proposed by Zhang et al. [7], ReDirect-T/SF, in running one iteration. One of the reasons why we choose ReDirect-T/SF is that the variance of degree of ACN is more than 2000. The ReDirect-T/SF algorithm will achieve a better performance as mentioned [7]. Another reason is that, both of Re-Direct-T/SF and our approach are based on edge structure. The comparison results on the ACN data set are shown in Table 2.

Table 2. Comparison results on the ACN dataset.

Approach	$F1$	*Precision*	*Recall*	TP
ReDirect-T/SF	0.3605	0.4069	0.3027	8073
DRHM	**0.7016**	**0.6670**	**0.7401**	**28561**

According to the values of TP (true positive), it shows that, our approach outperforms the Re-Direct-T/SF in one iteration in recovering more undirected edges in citation network. It confirms that the node content plays an indispensable role to shape the information propagational network.

4   Conclusion

In this paper, we study how to restore the flow direction of information between individuals in the information sharing network, according to the interrelationship between individuals and the information content shared by the individuals. By observing some special patterns from the popular DBLP dataset, we propose a hybrid method that combines topological structure and node content information. The benefits of our approach can be concluded as: Firstly, we propose an unsupervised learning method to infer information direction which does not require a large number of data sets to be trained in advance. Besides, our approach outperforms the topological structure based method, and is more suitable in knowledge transmit network. Experiments conducted on the DBLP dataset have confirmed the effectiveness of our method.

Acknowledgment. This work was supported by NSFC (No. 61502543) and Tip-top Scientific and Technical Innovative Youth Talents of Guangdong special support program (No. 2016TQ03X542).

References

1. Li, J.H., Li, P.Z., Wang, C.D., Lai, J.H.: Community detection in complicated network based on the multi-view weighted signed permanence. In: 14th IEEE International Symposium on Parallel and Distributed Processing with Applications, pp. 1589–1596. IEEE Press, New York (2016)
2. Ding, Y., Huang, L., Wang, C.-D., Huang, D.: Community detection in graph streams by pruning zombie nodes. In: Kim, J., Shim, K., Cao, L., Lee, J.-G., Lin, X., Moon, Y.-S. (eds.) PAKDD 2017 Part I. LNCS (LNAI), vol. 10234, pp. 574–585. Springer, Cham (2017). doi:10.1007/978-3-319-57454-7_45
3. Wang, C.D., Lai, J.H., Yu, P.S.: Dynamic community detection in weighted graph streams. In: 13th SIAM International Conference on Data Mining, pp. 151–161. SIAM, Philadelphia (2013)
4. Wang, C.D., Lai, J.H., Yu, P.S.: NEIWalk: community discovery in dynamic content-based networks. IEEE Trans. Knowl. Data Eng. **26**, 1734–1748 (2013)
5. Luhn, H.P.: The automatic creation of literature abstracts. IBM J. Res. Dev. **2**, 159–165 (1958)
6. Robertson, S.E., Jones, K.S.: Relevance weighting of search terms. J. Am. Soc. Inf. Sci. **27**, 129–146 (1976)
7. Zhang, J., Wang, C.K., Wang, J.M., Yu, J.X., Chen, J., Wang, C.P.: Inferring directions of undirected social ties. IEEE Trans. Knowl. Data Eng. **28**, 3276–3292 (2016)
8. Liu, L.Y., Xu, L.L., Wang, Z., Chen, E.H.: Community detection based on structure and content: a content propagation perspective. In: 15th IEEE International Conference on Data Mining, pp. 271–280. IEEE Press, New York (2015)
9. Deng, Z.H., Wang, Z.H., Zhang, J.: ROBIN: a novel personal recommendation model based on information propagation. Expert Syst. Appl. **40**, 5306–5313 (2013)
10. Velden, T., Yan, S.Y., Lagoze, C.: Mapping the cognitive structure of astrophysics by infomap clustering of the citation network and topic affinity analysis. Scientometrics **111**, 1033–1051 (2017)

11. Leskovec, J., Huttenlocher, D.P., Kleinberg, J.M.: Predicting positive and negative links in online social networks. In: 19th International Conference on World Wide Web, pp. 641–650. ACM, New York (2010)
12. Kahanda, I., Neville, J.: Using transactional information to predict link strength in online social networks. In: 3th International Conference on Weblogs and Social Media, pp. 74–81. AAAI, Palo Alto (2009)
13. Granovetter, M.S.: The strength of weak ties. Am. J. Sociol. **78**, 1360–1380 (1973)

Geo-Pairwise Ranking Matrix Factorization Model for Point-of-Interest Recommendation

Shenglin Zhao[1,2]([✉]), Irwin King[1,2], and Michael R. Lyu[1,2]

[1] Shenzhen Key Laboratory of Rich Media Big Data Analytics and Application,
Shenzhen Research Institute, The Chinese University of Hong Kong, Shenzhen, China
{slzhao,king,lyu}@cse.cuhk.edu.hk
[2] Department of Computer Science and Engineering,
The Chinese University of Hong Kong, Shatin, N.T., Hong Kong

Abstract. Point-of-interest (POI) recommendation that suggests new locations for people to visit is an important application in location-based social networks (LBSNs). Compared with traditional recommendation problems, e.g., movie recommendation, geographical influence is a special feature that plays an important role in recommending POIs. Various methods that incorporate geographical influence into collaborative filtering techniques have recently been proposed for POI recommendation. However, previous geographical models have struggled with a problem of *geographically noisy POIs*, defined as POIs that follow the geographical influence but do not satisfy users' preferences. We observe that users in the same geographical region share many POIs, and thus we propose the *co-geographical influence* to filter *geographically noisy POIs*. Furthermore, we propose the *Geo-Pairwise Ranking Matrix Factorization (Geo-PRMF)* model for POI recommendation, which incorporates *co-geographical influence* into a personalized pairwise preference ranking matrix factorization model. We conduct experiments on two real-life datasets, i.e., Foursquare and Gowalla, and the experimental results reveal that the proposed approach outperforms state-of-the-art models.

Keywords: POI Recommendation · Matrix factorization · Geographical influence · Pairwise ranking

1 Introduction

Point-of-interest (POI) recommendation has been being driven by soaring development of location-based social network (LBSN) services such as Foursquare and Facebook Places. A typical LBSN allows users to check-in at their locations, make friends, and share information. POI recommendation in LBSNs aims to help users explore new and interesting places in a city through an LBSN service. When you go shopping, for instance, you can easily find detailed downtown shopping mall information and nearby food shops using POI recommendation; and doing so not only improves users' experiences, but also provides merchants with new chances to target customers.

© Springer International Publishing AG 2017
D. Liu et al. (Eds.): ICONIP 2017, Part V, LNCS 10638, pp. 368–377, 2017.
https://doi.org/10.1007/978-3-319-70139-4_37

Due to the importance of POI recommendation, various methods have been proposed to tackle this task [1,8,11–13,18,20]. Inspired by the conventional recommendation systems, e.g., Netflix's movie recommendation system, a user-POI matrix is constructed that treats POIs as items and users' check-in frequencies as rating values. Then collaborative filtering techniques are used to recommend POIs. In addition, geographical influence has been incorporated as an important factor into the proposed POI recommendation systems to improve performance [1,2,11,16,19]. However, previous models designed to capture geographical influence have struggled with the problem of *geographically noisy POIs*.

Existing geographical influence models suffer from the problem of *geographically noisy POIs*, as they recommend new POIs that are close to those where the user has checked-in, depending solely on the user-POI geographical relationship. Here, we give an example of a *geographically noisy POI*. Suppose a user likes to visit shops and restaurants near his/her home, and as such generates many check-ins at these places. Meanwhile, a hotel is also located near the user's home. According to previous geographical influence models, the hotel should be recommended as it is near the POIs where the user has checked-in. However, people live in their own houses, and do not typically want to visit a hotel nearby. Hence, the hotel is defined as a *geographically noisy POI*, which follows the geographical influence but does not satisfy the user's preference.

In this paper, we propose the *co-geographical influence* to address the problem of *geographically noisy POIs*. We observe that users acting in the same region share many POIs. Two students attending the same university, for example, may not know each other, but may check into many of the same POIs, such as popular restaurants and night clubs around the university. Each user's check-in behavior enhances each shop's popularity, attracting more people. Inspired by this observation, we propose the *co-geographical influence*, which assumes that users follow similar visiting patterns in close areas.

Furthermore, we propose the *Geo-Pairwise Ranking Matrix Factorization (Geo-PRMF)* model to tackle the POI recommendation problem. Inspired by [19,20], we treat users' check-ins as implicit feedback and learn the system via personalized pairwise preference ranking. The preference is implicitly embedded in pairs (checked-in, unchecked-in), with users assumed to have stronger interest in the checked-in POIs than in the unchecked-in POIs. We exploit the *co-geographical influence* to refine the preference pair set, which reduces the complexity cost. Specifically, our model filters the *geographically noisy POIs*, which are unresolved in existing geographical influence models [1,11,16].

The contributions of this paper are summarized as follows. First, we propose the *co-geographical influence* to overcome the problem of *geographically noisy POIs* hindering previous geographical influence models. Moreover, we propose the *Geo-PRMF* model, which incorporates *co-geographical influence* into a personalized pairwise preference ranking model to learn user preference and performs better than state-of-the-art models.

2 Related Work

In this section, we first demonstrate the recent progress of POI recommendation. Then, we show how previous studies have modeled geographical influence. Finally, we explore how our proposed model relates to the prior work.

POI Recommendation. POI recommendation has attracted intensive academic attention recently. Most of the proposed methods have used collaborative filtering (CF) techniques, including the memory and model-based methods, to recommend POIs. The researchers in [11,14,15] employ the user-based CF to recommend POIs, whereas, other studies leverage the model-based CF, including the Matrix Factorization (MF) technique [1,8,9,18]. Specifically, the researchers in [8,9] model the check-ins as implicit feedback and use the weighted regularized MF for POI recommendation. Unlike the researchers in [8,9], those in [3,7,19,20] model implicit feedback via a pairwise ranking method, which exhibits better performance.

Geographical Influence. Geographical influence plays an important role in POI recommendation, as users' activity in LBSNs is limited by geographical constraints. To capture geographical influence, the researchers assume that the co-occurrence of POIs follows a specific distribution. On the one hand, studies in [1,4,16] suppose the checked-in POIs follow a Gaussian distribution and propose Gaussian distribution based models; those in [11,14] employ the power law distribution model; and studies in [15] leverage the kernel density estimation model to learn the distribution. On the other hand, the researchers in [8,9] incorporate geographical influence into a weighted regularized MF model. The work in [19] incorporate the geographical influence into a ranking model and propose a hierarchical geographical pairwise ranking for POI recommendation. The core idea of the proposed geographical influence models has based on the intuition that a user prefers the visit new POIs nearby where the user has checked-in.

Connection to Prior Work. Prior studies have captured the geographical influence to recommend new POIs, prioritizing proximity to the user's activity center or previous checked-in POIs. This creates the problem of *geographically noisy POIs*. We propose the *co-geographical influence* to overcome this problem. Moreover, due to the success of using pairwise preference ranking to model the check-in activity, we propose the *Geo-PRMF* model, which incorporates *co-geographical influence* into a pairwise preference ranking model to learn users' POI preferences.

3 Model

In this section, we first propose *co-geographical influence* to address the problem of *geographically noisy POIs*. Then, we propose the *Geo-PRMF* model, which incorporates *co-geographical influence* into a pairwise preference ranking model for recommending POIs.

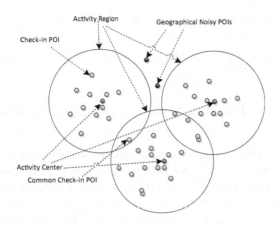

Fig. 1. Demonstration of user check-in pattern

3.1 Co-geographical Influence

For illustration purposes, we define several terms as follows.

Definition 1 (Geographical activity center) *A geographical activity center is the POI with the highest check-in probability based on geographical influence.*

Definition 2 (Geographical neighbors). *Geographical neighbors are users who have close geographical activity centers.*

Definition 3 (Geographically noisy POI). *A geographically noisy POI is the POI near a user's geographical activity center but not preferred by the user.*

Figure 1 demonstrates the user check-in pattern and the problem of *geographically noisy POIs*. Previous studies [1,16] have shown that most people live and have fun in constrained activity regions. According to this kind of geographical characteristic, previous work constructs the user-POI geographical relation: a POI that is near a user's *geographical activity center* is geographically preferred [1,16]. However, this assumption is easily affected by *geographically noisy POIs*, as shown in Fig. 1. Some POIs are geographically near a user's *geographical activity center* but they do not match the user's check-in pattern, such as the hotel example mentioned in Sect. 1.

Co-geographical influence depicts the user-user geographical relation instead of the user-POI relation. We observe that *geographical neighbors* share many POIs. Specifically, the Jaccard similarity between *geographical neighbors* is about 10 times higher than between random users. The model not only considers a user's geographical feature but also extracts geographical relation between two users. We follow the discovery that a user's checked-in POIs distribute around some activity center(s) [1,16]. Hence, we expect the POIs in which a user is interested to be located in the range where the user's *geographical neighbors* have checked-in. This helps to filter out the *geographically noisy POIs*. As a result, the candidate POI set for a user consists of POIs where the user's *geographical neighbors* have checked-in but he/she has not yet. *Co-geographical influence* exploits the common check-in pattern among *geographical neighbors* to filter out *geographically noisy POIs*.

3.2   Geo-Pairwise Ranking Matrix Factorization (Geo-PRMF) Model

We propose the *Geo-PRMF* model, which incorporates *co-geographical influence* into a pairwise ranking model. Due to the success of pairwise preference ranking in modeling the check-in activity as implicit feedback in prior work [7,18,20], we utilize the Bayesian personalized ranking criteria [10] to learn user preference on POIs. Moreover, we exploit *co-geographical influence* to classify the unrated POIs as comparable POIs and unrelated POIs. We assume that the POIs where a user's *geographical neighbors* have checked-in are comparable and others are unrelated. Therefore, we only make use of the comparable POIs to generate the pairwise preference set, and discard the unrelated ones, recommending POIs from the comparable POI candidate set. Based on this assumption, we extract the refined pairwise preference set and candidate POI set as follows:

1. We map a *geographical activity center* for a user and identify the top k *geographical neighbors* by nearby centers.
2. We consider only the POIs checked-in to by the user's *geographical neighbors* but not yet by the user to be comparable and any others to be unrelated.
3. We generate triplets (user u, checked-in POI l_i, comparable POI l_j) as refined preference set $\mathcal{P}$, and comparable POIs as candidate set $\mathcal{L}_u^c$.

Then, we can learn the user's preferences from the refined pairwise preference set and recommend POIs from the candidate POI set.

We formulate the POI recommendation problem as follows. Let $\mathcal{U}$ be the set of users and $\mathcal{L}$ be the set of POIs. The pairwise preference of user u prefers POI l_i over l_j, is defined as $l_i \succ_u l_j$. Then, we define the pairwise preference set $\mathcal{P} := \{l_i \succ_u l_j | l_i \in \mathcal{L}_u^+ \wedge l_j \in \mathcal{L}_u^c\}$, where L_u^+ denotes the POIs where user u has checked-in, and $\mathcal{L}_u^c$ denotes the POIs where *geographical neighbors* of user u have checked-in but u has not. Now training the POI recommendation system is to learn the pairwise preference relationships in $\mathcal{P}$,

$$\arg\max_{\Theta} \prod_{(u,l_i,l_j) \in \mathcal{P}} p(l_i \succ_u l_j | \Theta), \tag{1}$$

where $p(l_i \succ_u l_j)$ is the probability of a user preferring POI l_i over l_j, and Θ denotes the model's learning parameters.

We employ the biased MF to model the user preference on POI. Then, the preference score function of user u on POI l_i is formulated as,

$$f(u, l_i) = U_u^T L_{l_i} + b_{l_i}, \tag{2}$$

where $U_u, L_{l_i} \in R^d$ are latent feature vectors for user u and POI l_i respectively, and b_{l_i} is the estimation bias. Furthermore, we estimate the probability function of $p(l_i \succ_u l_j)$ via a sigmoid function, $p(l_i \succ_u l_j) = \sigma(f(u, l_i) - f(u, l_j))$, where σ is the sigmoid function $\sigma(x) = 1/(1 + \exp(-x))$. Thus, it is not hard to gain the objective function by minimizing the negative log likelihood

$$
\begin{aligned}
&\mathcal{O}(b_{l_i}, b_{l_j}, L_{l_i}, L_{l_j}, U_u) \\
&= -\sum_{u=1}^{|\mathcal{U}|} \sum_{l_i \in \mathcal{L}_u^+} \sum_{l_j \in \mathcal{L}_u^c} \ln \sigma(U_u^T(L_{l_i} - L_{l_j}) + b_{l_i} - b_{l_j}) + \frac{\lambda_1}{2} \|U_u\|^2 \\
&\quad + \frac{\lambda_1}{2} \|L_{l_i}\|^2 + \frac{\lambda_1}{2} \|L_{l_j}\|^2 + \frac{\lambda_2}{2} \|b_{l_i}\|^2 + \frac{\lambda_2}{2} \|b_{l_j}\|^2,
\end{aligned}
\tag{3}
$$

where λ_1 and λ_2 are the regularization parameters.

We adopt the stochastic gradient decent (SGD) method to learn the parameters in Eq. (3). We define a common expression as $z = \frac{1}{1+\exp(U_u^T(L_{l_i}-L_{l_j})+b_{l_i}-b_{l_j})}$. Then, the parameters are updated as follows,

$$
\begin{aligned}
b_{l_i} &\leftarrow b_{l_i} + \gamma \cdot (z - \lambda_1 \cdot b_{l_i}), \\
b_{l_j} &\leftarrow b_{l_j} + \gamma \cdot (-z - \lambda_2 \cdot b_{l_j}), \\
L_{l_i} &\leftarrow L_{l_i} + \gamma \cdot (z \cdot U_u - \beta_1 \cdot L_{l_i}), \\
L_{l_j} &\leftarrow L_{l_j} + \gamma \cdot (-z \cdot U_u - \beta_2 \cdot L_{l_j}), \\
U_u &\leftarrow U_u + \gamma \cdot (z \cdot L_{l_i} - z \cdot L_{l_j} - \alpha \cdot U_u).
\end{aligned}
\tag{4}
$$

After learning the parameters, the *Geo-PRMF* model predicts a user's check-in preference at a given POI according to the score computed by Eq. (2). We first rank the POIs in candidate set in terms of check-in preference, then recommend the top N POIs for a specific user. Algorithm 1 demonstrates how to recommend POIs through *Geo-PRMF* model.

Complexity Analysis. There are two steps to recommend POIs: model training and item recommendation. The complexity of training *Geo-PRMF* model is $O(d \cdot |S|)$, the same order as the BPR-MF model [10], where d denotes the latent factor vector dimensionality and $|S|$ denotes the number of samples. *Geo-PRMF* has an advantage over other models at the item recommendation step. For general MF-based recommendation models, the time complexity of the item recommendation step is $O(|\mathcal{U}| \cdot |\mathcal{L}| \cdot d)$. The item recommendation time complexity of the *Geo-PRMF* model is $O(|\mathcal{U}| \cdot |\mathcal{L}^c| \cdot d)$ with $|\mathcal{L}^c|$ denoting the average number of candidate POIs for a user. As $|\mathcal{L}^c|$ is much less than $|\mathcal{L}|$, the *Geo-PRMF* consumes less calculation than other models at the item recommendation step.

Algorithm 1. POI Recommendation via Geo-PRMF Model
***
Input: Preference set $\mathcal{P}$, learning parameters $b_{l_i}, b_{l_j}, L_{l_i}, L_{l_j}$,
      U_u, regularization parameters λ_1, λ_2
Output: recommended POI set S_N for each user
1: Randomly initialize parameters $b_{l_i}, b_{l_j}, L_{l_i}, L_{l_j}, U_u$
2: **repeat**
3:    Draw $(u, l_i, l_j) \in \mathcal{P}$ uniformly
4:    Update parameters with Eq. (4)
5: **until** convergence
6: **for** u in $\mathcal{U}$ **do**
7:    **for** l in $\mathcal{L}_u^c$ **do**
8:        Predict user preference score with Eq. (2)
9:    Recommend N POIs S_N with higher preference score
***

4 Experiment

4.1 Data Description and Experimental Setting

Two real-world datasets are used in the experiment: Foursquare data in [5] and Gowalla data in [4]. We extract the data from March to October in 2010 from both datasets, filter the POIs checked-in to by less than 5 users, and then choose users who have checked-in more than 10 times as our samples. Table 1 shows the data statistics. We randomly choose 80% of each user's check-ins as training data, and use the remaining 20% for test data. Following [8,20], we use **precision** and **recall** to measure the model performance.

Table 1. Data statistics

Source	#users	#POIs	#check-ins	Density
Foursquare	18,106	17,145	1,330,247	0.001
Gowalla	107,091	144,034	2,581,093	0.0001

4.2 Baseline Methods

Given that the proposed method aims to construct an effective MF-based model for POI recommendation, we select BiasedMF [6] and BPR-MF [10] as the basic comparable models. Moreover, to show the advantage of our proposed model in capturing geographical influence, we compare it with fused MF with multi-center Gaussian model (MGMMF) [1] and joint model with geographical influence and MF (GeoMF) [8], which are state-of-the-art POI recommendation methods capturing geographical influence.

4.3 Experimental Results

In the following, we demonstrate the performance comparison on precision@N and recall@N between the baseline models and our proposed *Geo-PRMF* model. We set the latent factor vector dimension as 20 for all compared models.

We evaluate different models for both datasets on top-5 and top-10 POI recommendation tasks. Figure 2 shows the obtained results, from which we make the following observations. (1) The proposed *Geo-PRMF* model achieves the best performance, with advantages over the MGMMF and the GeoMF at capturing geographical influence by filtering out *geographically noisy POIs*. Compared with the best baseline competitor, the *Geo-PRMF* model achieves at least 5% improvements on precision@5 and recall@5, and at least 7% improvements on precision@10 and recall@10 for both datasets. (2) *Geo-PRMF*, MGMMF, and GeoMF perform better than BiasedMF and BPR-MF, which demonstrates the effectiveness of capturing geographical influence.

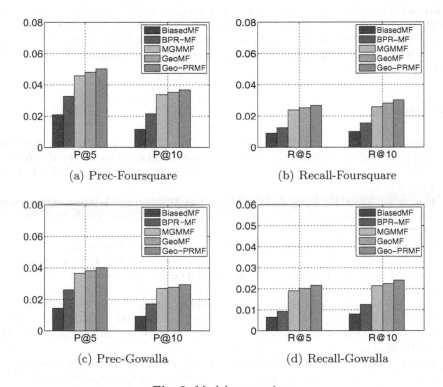

(a) Prec-Foursquare   (b) Recall-Foursquare

(c) Prec-Gowalla   (d) Recall-Gowalla

Fig. 2. Model comparison

5 Conclusion and Future Work

In this paper, we propose the *Geo-PRMF* model to tackle the POI recommendation problem. We first present *co-geographical influence*, which reduces *geographically noisy POIs* and significantly shrinks the candidate set for a specific user. Moreover, we propose the *Geo-PRMF* model, which incorporates *co-geographical influence* into a pairwise ranking model. Finally, we conduct elaborated experiments on two real-life LBSN datasets to verify our proposed model. The experimental results show that our proposed *Geo-PRMF* model outperforms state-of-the-art models.

In the future, we will improve the *Geo-PRMF* model in the following aspects. We may design an adaptive way to select the number of activity centers to improve the performance. Furthermore, we may consider users' comments or location category features to further improve the overall recommendation performance. In addition, a new application in LBSNs [17] has appeared recently, which uses the check-in data to mine business opportunities. We will consider to exploit the check-in characteristics to enhance the business mining application.

Acknowledgments. The work described in this paper was partially supported by the Research Grants Council of the Hong Kong Special Administrative Region, China (Nos. CUHK 14203314 and CUHK 14234416 of the General Research Fund), and 2015 Microsoft Research Asia Collaborative Research Program (Project No. FY16-RES-THEME-005).

References

1. Cheng, C., Yang, H., King, I., Lyu, M.R.: Fused matrix factorization with geographical and social influence in location-based social networks. In: Proceedings of the Twenty-Sixth AAAI Conference on Artificial Intelligence, pp. 17–23. AAAI Press (2012)
2. Cheng, C., Yang, H., King, I., Lyu, M.R.: A unified point-of-interest recommendation framework in location-based social networks. ACM Trans. Intell. Syst. Technol. (TIST) **8**(1), 10 (2016)
3. Cheng, C., Yang, H., Lyu, M.R., King, I.: Where you like to go next: successive point-of-interest recommendation. In: Proceedings of the Twenty-Third International Joint Conference on Artificial Intelligence, pp. 2605–2611. AAAI Press (2013)
4. Cho, E., Myers, S.A., Leskovec, J.: Friendship and mobility: user movement in location-based social networks. In: Proceedings of the 17th ACM SIGKDD International Conference on Knowledge Discovery and Data Mining, pp. 1082–1090. ACM (2011)
5. Gao, H., Tang, J., Liu, H.: Exploring social-historical ties on location-based social networks. In: Sixth International AAAI Conference on Weblogs and Social Media. AAAI (2012)
6. Koren, Y., Bell, R., Volinsky, C.: Matrix factorization techniques for recommender systems. Computer **42**(8), 30–37 (2009)
7. Li, X., Cong, G., Li, X.L., Pham, T.A.N., Krishnaswamy, S.: Rank-GeoFM: a ranking based geographical factorization method for point of interest recommendation. In: Proceedings of the 38th International ACM SIGIR Conference on Research and Development in Information Retrieval, pp. 433–442. ACM (2015)

8. Lian, D., Zhao, C., Xie, X., Sun, G., Chen, E., Rui, Y.: GeoMF: joint geographical modeling and matrix factorization for point-of-interest recommendation. In: ACM SIGKDD International Conference on Knowledge Discovery and Data Mining, pp. 831–840. ACM (2014)

9. Liu, Y., Wei, W., Sun, A., Miao, C.: Exploiting geographical neighborhood characteristics for location recommendation. In: ACM International Conference on Conference on Information and Knowledge Management, pp. 739–748 (2014)

10. Rendle, S., Freudenthaler, C., Gantner, Z., Schmidt-Thieme, L.: BPR: Bayesian personalized ranking from implicit feedback. In: Proceedings of the Twenty-Fifth Conference on Uncertainty in Artificial Intelligence, pp. 452–461. AUAI Press (2009)

11. Ye, M., Yin, P., Lee, W.C., Lee, D.L.: Exploiting geographical influence for collaborative point-of-interest recommendation. In: Proceedings of the 34th International ACM SIGIR Conference on Research and Development in Information Retrieval, pp. 325–334. ACM (2011)

12. Yin, H., Cui, B., Sun, Y., Hu, Z., Chen, L.: LCARS: a spatial item recommender system. ACM Trans. Inf. Syst. (TOIS) **32**(3), 11 (2014)

13. Yin, H., Cui, B., Zhou, X., Wang, W., Huang, Z., Sadiq, S.: Joint modeling of user check-in behaviors for real-time point-of-interest recommendation. ACM Trans. Inf. Syst. (TOIS) **35**(2), 11 (2016)

14. Yuan, Q., Cong, G., Ma, Z., Sun, A., Thalmann, N.M.: Time-aware point-of-interest recommendation. In: Proceedings of the 36th International ACM SIGIR Conference on Research and Development in Information Retrieval, pp. 363–372. ACM (2013)

15. Zhang, J.D., Chow, C.Y.: GeoSoCa: exploiting geographical, social and categorical correlations for point-of-interest recommendations. In: Proceedings of the 38th International ACM SIGIR Conference on Research and Development in Information Retrieval, pp. 443–452. ACM (2015)

16. Zhao, S., King, I., Lyu, M.R.: Capturing geographical influence in POI recommendations. In: Lee, M., Hirose, A., Hou, Z.-G., Kil, R.M. (eds.) ICONIP 2013 Part II. LNCS, vol. 8227, pp. 530–537. Springer, Heidelberg (2013). doi:10.1007/978-3-642-42042-9_66

17. Zhao, S., King, I., Lyu, M.R., Zeng, J., Yuan, M.: Mining business opportunities from location-based social networks. In: Proceedings of the 40th International ACM SIGIR Conference on Research and Development in Information Retrieval, pp. 1037–1040. ACM (2017)

18. Zhao, S., Lyu, M.R., King, I.: Aggregated temporal tensor factorization model for point-of-interest recommendation. In: Hirose, A., Ozawa, S., Doya, K., Ikeda, K., Lee, M., Liu, D. (eds.) ICONIP 2016 Part III. LNCS, vol. 9949, pp. 450–458. Springer, Cham (2016). doi:10.1007/978-3-319-46675-0_49

19. Zhao, S., Zhao, T., King, I., Lyu, M.R.: Geo-Teaser: geo-temporal sequential embedding rank for point-of-interest recommendation. In: Proceedings of the 26th International Conference on World Wide Web Companion, pp. 153–162. International World Wide Web Conferences Steering Committee (2017)

20. Zhao, S., Zhao, T., Yang, H., Lyu, M.R., King, I.: STELLAR: spatial-temporal latent ranking for successive point-of-interest recommendation. In: Proceedings of the Thirtieth AAAI Conference on Artificial Intelligence, pp. 315–321. AAAI Press (2016)

A Method to Improve Accuracy of Velocity Prediction Using Markov Model

Ya-dan Liu[1], Liang Chu[1], Nan Xu[1(✉)], Yi-fan Jia[1], and Zhe Xu[2]

[1] State Key Laboratory of Automotive Simulation and Control, Jilin University,
Changchun 130022, China
nanxu@jlu.edu.cn
[2] R&D Center, China FAW Group Corporation, Changchun 130011, China

Abstract. In order to predict the velocity in driving cycle, first-stage Markov chain (MC) predictor method is adopted. In the traditional Markov prediction model, only one state transition matrix was used to predict the speed. However it will produce a larger error to use the same matrix for predicting speed in different categories of driving cycles. Random Markov-Chain (RMC) model is adopted to improve the accuracy, but the accuracy is still not enough. In this paper, we propose that the state transition matrices in RMC model are divided into two categories: city and highway. Before the prediction, we use the neural network to choose state transition matrix by judging the kinematic parameters of velocity in driving cycles. The simulation results show that the effect of prediction using the state transition matrix after neural network classification is more accurate than no classification. Therefore, the improved RMC model can increase the accuracy of velocity prediction effectively.

Keywords: Neural network · Random Markov-Chain model · Velocity prediction · Classification · State transition matrix

1 Introduction

In the development of the vehicle, fuel economy has been one of the most important issues that have been widely concerned. Sophisticated energy management strategies have been well developed to deal with this problem. In hybrid vehicles, the energy management strategies determine whether the engine and every motor work or not. So the energy distribution path is diverse, scholars need to predict the consumption of energy in the next step, and then to use the dynamic planning method to find the energy distribution path that has the minimum fuel consumption. Prediction for velocity is one of the most critical parts of energy prediction [1, 2] under such condition.

For the velocity prediction, a lot of methods have been proposed. The exponentially varying velocity predictor method [3] has the advantage of simple structure and fast calculation, but the deviation is too large so that the forecast is not good. Using neural networks to predict [4, 5] can improve the prediction accuracy. However this measure to ensure the accuracy of prediction will make the structure complicated, leading to the slow calculation. It is not very good in the vehicle. The traditional Markov-chain model [6–8] is the most commonly used predictor model. This model constructs the state

© Springer International Publishing AG 2017
D. Liu et al. (Eds.): ICONIP 2017, Part V, LNCS 10638, pp. 378–386, 2017.
https://doi.org/10.1007/978-3-319-70139-4_38

transition matrix before prediction, so the calculation speed is faster and the effect of prediction is better in the actual prediction process.

In recent years, it has been tried to improve the prediction accuracy of the Markov-chain model again. For the traditional Markov model, the more historical data is used, the higher the accuracy of the prediction is. The random Markov-chain model is based on the traditional Markov model, whose state transition matrix is updated by continuously collecting the data during the driving process. Compared to the traditional Markov model, it will improve the accuracy of 1.56% for UDDS after updating. However, this approach is not obvious to improve the prediction accuracy. The increase of the Markov order can further improve the prediction accuracy, but the corresponding structure is complicated. The first order and third-order Markov models are used to predict the velocity, and the results indicate that the prediction using third-order Markov model is better than first order, the accuracy is improved significantly. But the calculation time is longer, the third-order model uses 163.787 s compare to the 0.310 s using in the first order. Do not use the higher order Markov model if there are other methods generally. Both the traditional Markov and random Markov models use all kinds of cycles to train one state transition matrix without classification, when the differences are large among all of the cycles, the state transition matrix will be not accurate and produce a larger deviation for the prediction. Therefore, this paper proposes to classify the state transition matrix. Due that the differences between city and highway cycles are large, the state transition matrix is divided into two categories: city and highway.

The remainder of this brief is organized as follows. In Sect. 2, the prediction model of velocity is introduced, including the traditional MC model, RMC model and improved RMC model. Section 3 describes to use the neural network for classification and determines the structure of the neural network. Comparison results for model between with and without classification are illustrated in Sect. 4. Finally, the conclusion is drawn in Sect. 5.

2 Velocity Prediction Using Markov-Chain

The Markov process is a kind of stochastic problem. Since the speed of vehicle is random at any time that meets the Markov nature, the MC model is often used for velocity prediction in the traditional method. In this section, the traditional MC model, RMC model and improved RMC model will be introduced, respectively.

2.1 Traditional Velocity Prediction Using Markov-Chain

It is worth attention that the traditional Markov state of transmission is to consider all the data, including various categories of driving cycles, without classification. The data uses six kinds of driving cycles, including both city cycles and highway cycles. Suppose the vehicle velocity and acceleration are discretized into p and q intervals and indexed by i and j, respectively. V_{k+m-1} is the velocity at time step k, and a_{k+m} is the next step acceleration. The MC process is defined by an emission probability matrix $T \in R^{p \times q}$ with

$$[T]_{ij} = \Pr\left[a_{k+m} = \bar{a}_j \mid V_{k+m-1} = \bar{V}_i\right] \tag{1}$$

Wherein $i \in \{1, \ldots, p\}, j \in \{1, \ldots, q\}, m \in \{1, \ldots, L_p\}$. L_p is the length of prediction time.

For the traditional way, RMC model [9] is used to improve the accuracy of the MC model. This model can update constantly the state transition matrix to improve the accuracy of the prediction through real-time detection of speed. However, the accuracy of prediction for RMC model is still not enough.

2.2   An Improved Markov Model

In the traditional MC model, all the data are used to train one state transition matrix to predict velocity. But there exist some drawbacks in some driving cycles. For example, when the vehicle drives in the city all the time, the prediction will be affected due that state transition matrix includes the part of highway. The accuracy of prediction will be produced a large deviation. Therefore, this paper proposes to classify the state transition matrix in the MC model. During the training, the driving cycles are divided into two categories including city cycles and highway cycles, corresponding to the state transition matrix respectively. Using three kinds of city cycles and three kinds of highway cycles to train, the state transition matrices are respectively in Fig. 1.

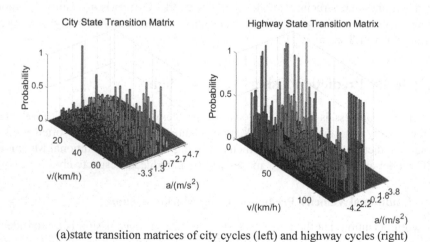

(a)state transition matrices of city cycles (left) and highway cycles (right)

Fig. 1.   State transition matrices after classification

From the two state transition matrices, it can be seen that the differences are large between city and highway. So the state transition matrices should be used separately.

3 Classification Using Back Propagation Neural Network

Back Propagation neural network (BP-NN) is one of the most widely used neural network models, it has a strong ability to nonlinear mapping, compared with Radial Basis Function neural network and Perception. When the neural network is used for classification, the output result is only 0 or 1, which corresponds to the city cycles or highway cycles. BP-NN is suitable for dealing with such problems.

The BP-NN is generally composed of an input layer, the intermediate layer, and an output layer, wherein the intermediate layer may be one or more layers. In the past, the input layer was generally selected according to the subjective will when BP-NN was used for classification. Since there was no theoretical basis, scholars need to adjust the test constantly until the results met the requirement. However this way will usually cause a large error rate. In order to improve the classification effect, the number of input nodes will be determined by feature selection [10].

For traditional neural networks, the number of neurons for middle-layer has been studied [11, 12]. The upper limit of the number of neurons for middle-layer is [13]:

$$N_{hid} \leq N_{train}/R \times (N_{in} + N_{out}) \tag{2}$$

wherein, N_{in} is the number of neurons for input; N_{out} is the number of neurons for output; N_{train} is the number of sample; $5 \leq R \leq 10$;

And the simplified formula which is obtained by least squares method [14] is:

$$S = \sqrt{m(n+3)} + 1 \tag{3}$$

wherein, S is the number of neurons for hidden layer; m is the number of neurons for input; n is the number of neurons for output.

4 Simulation and Results

In this section, we provide a comparative analysis of velocity prediction between traditional RMC model and improved model. Our discourse begins to classify the driving cycles and decide which state transition matrix to be used. Simulating the velocity predicted in final.

4.1 Neural Network Classification Results

The number of input nodes in the BP-NN is seven after filtering [15–20] using feature selection, and the number of neurons in the middle layer is selected as five according to the formula. The output layer is the result of classification, so it is only one neuron. Neural network simulation samples are the three kinds of city standard cycles and three highway standard cycles. The more the number of neural network samples, the better the training results. In order to improve the number of sample, each standard cycle is evenly divided into six equal parts. And we need ensure that the mean values of the new samples were the almost same as the original samples, respectively. Then 18 samples are randomly selected from the 36 samples, including 9 city cycles and 9 highway cycles. The test results are shown in Fig. 2:

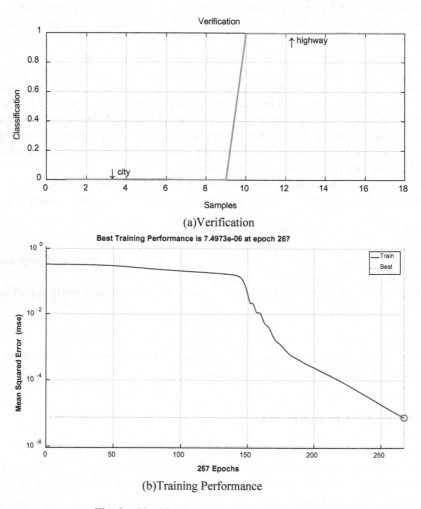

(a)Verification

(b)Training Performance

Fig. 2. Classification using BP neural network

From the result of verification in Fig. 2, it can be seen that the first nine samples are judged as city cycles, and the others are judged as highways cycles. So the BP-NN has been built successfully at the first step. Then training performance shows that the neural network training is convergent, the result error is less than 1×10^{-4}. Therefore, the training effect is good, and the driving cycle categories can be classified well.

4.2   Results of Speed Prediction

After classification by neural network, we can determine to use which state transition matrix when prediction. This paper adopts five working cycles to train, including three kinds of city cycles and two kinds of highway driving cycles. The prediction time is four seconds. The simulation results are shown in Fig. 3.

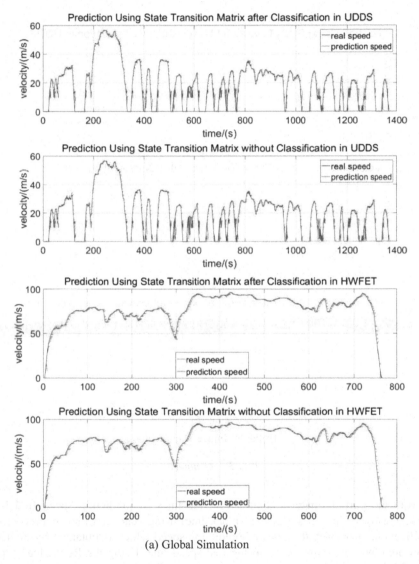

(a) Global Simulation

Fig. 3. Comparison of UDDS and HWFET cycles using state transition matrix with and without classification (Color figure online)

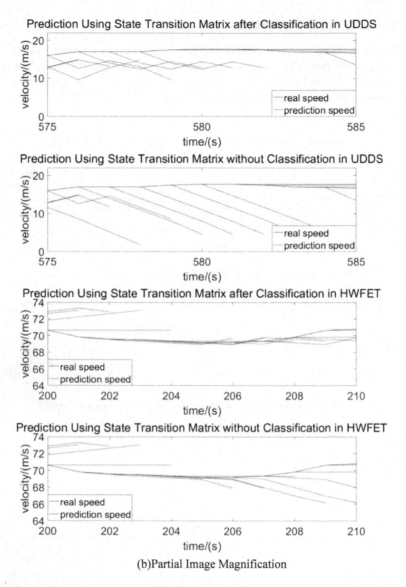

(b)Partial Image Magnification

Fig. 3. (*continued*)

From the partial image magnification, the lines of prediction speed (red lines) deviate greatly from real speed line (black line) using state transition matrix without classification. However, the lines of prediction speed which is obtained by improved model are obviously convergent to the real speed line. Using the Root Mean Square Error (RMSE) to calculate the error between the predicted results and the real results, it can be seen that the RMSE of all driving cycles with classification is lower than without classification (Table 1).

Table 1. Simulation results

Driving cycle	Running time (s)	Whether classification for matrix or not	RMSE	Increase the percentage (%)
UDDS	1370	Yes	63.36	9.79
		No	70.23	
HWFET	766	Yes	26.44	18.16
		No	32.31	

It can be seen that the effect of using the classified state transition matrix is better than that of the traditional RMC model. Therefore, the method proposed in this paper to classify the state transition matrix in the model can improve the prediction effect.

5 Conclusions

In this paper, we propose an improved RMC model to predict the velocity of vehicle. The driving cycles are divided into city cycles and highway cycles. Before the prediction, we use the neural network that based on feature selection to determine the category of driving cycle. Then we use the corresponding state transition matrix to predict. The simulation results show that the predictive effect of matrix classification is better than that of traditional prediction. Therefore, the classification of prediction matrix proposed can improve the prediction accuracy well.

Acknowledgement. Supported by Jilin Province Science and Technology Development Fund (20150520115JH); Energy Administration of Jilin Province [2016]35.

References

1. Sun, C., Hu, X., Moura, S.J., Sun, F.: Velocity predictors for predictive energy management in hybrid electric vehicles. IEEE Trans. Control Syst. Technol. **23**(3), 1197–1204 (2015). IEEE Press, New York
2. Zhao, D., Dai, Y., Zhang, Z.: Computational intelligence in urban traffic signal control: a survey. IEEE Trans. Syst. Man Cybern. **42**(4), 485–494 (2012). IEEE Press, New York
3. Borhan, H.A., Vahidi, A.: Model predictive control of a power-split hybrid electric vehicle with combined battery and ultracapacitor energy storage. In: American Control Conference, vol. 1, no. 4, pp. 5031–5036. IEEE Press, New York (2010)
4. Hagan, M.T., Demuth, H.B., Beale, M.H.: Neural Network Design. PWS-Kent, Boston (1996)
5. Shi, G., Liu, D., Wei, Q.: Energy consumption prediction of office buildings based on echo state networks. Neurocomputing **216**, 478–488 (2016). Elsevier Science Publishers B.V., Amsterdam
6. Bichi, M., Ripaccioli, G., Di Cairano, S., Bernardini, D., Bemporad A., Kolmanovsky, I.V.: Stochastic model predictive control with driver behavior learning for improved powertrain control. In: 49th IEEE Conference on Decision and Control, pp. 6077–6082. IEEE Press, New York (2010)

7. Moura, S.J., Fathy, H.K., Callaway, D.S., Stein, J.L.: A stochastic optimal control approach for power management in plug-in hybrid electric vehicles. IEEE Trans. Control Syst. Technol. **19**(3), 545–555 (2008). IEEE Press, New York

8. Pan, D.: Multi-scale Prediction of Urban Recycling Cycle and Driving Cycle of Hybrid Electric Vehicle. Beijing Institute of Technology (2015)

9. Ripaccioli, G., Bernardini, D., Cairano, S.D., Benporad, A., Kolmanovsky, I.: A stochastic model predictive control approach for series hybrid electric vehicle power management. In: American Control Conference, pp. 5844–5849. IEEE Press, New York (2010)

10. Tang, J., Alelyani, S., Liu, H.: Feature selection for classification: a review. In: Documentacion Administrativa, pp. 313–334 (2014)

11. Jiao, B., Ye, M.: The method of determining the number of hidden layer in BP neural network. J. Shanghai Motor Univ. **16**(3), 113–116 (2013)

12. Shen, H.: Determination of the number of implied layer units in BP neural network. J. Tianjin Univ. Technol. **24**(5), 13–15 (2008)

13. Liu, X.: Applied adaptive control. Northwestern Polytechnical University Press (2003)

14. Gao, D.: Hidden node pruning algorithm for forward multilayer neural networks. J. Electron., 114–115 (1997)

15. Lv, R.: HEV Control Strategy Based on Cycle and Driving Intention Recognition. Dalian University of Technology (2013)

16. Kohavi, R., John, G.H.: Wrappers for feature subset selection. Artif. Intell. **97**(1–2), 273–324 (1997)

17. Belkin, M., Niyogi, P.: Laplacian eigenmaps and spectral techniques for embedding and clustering. In: International Conference on Neural Information Processing Systems, vol. 14, pp. 585–591. MIT Press Cambridge (2001)

18. He, X., Cai, D., Niyogi, P.: Laplacian score for feature selection. In: International Conference on Neural Information Processing Systems, vol. 18, pp. 507–514. MIT Press Cambridge (2005)

19. Wan, J., Yang, M., Chen, Y.: Discriminative cost sensitive Laplacian score for face recognition. Neurocomputing **152**, 333–344 (2015)

20. Li, L., Guo, Y., Yi, P.: Analyzing the principles for choosing dimensionless methods. J. Syst. Manag., 1040–1045 (2016)

Strength Analysis on Safety-Belt ISOFIX Anchorage for Vehicles Based on HyperWorks and Ls-Dyna

Peicheng Shi[✉], Suo Wang, and Ping Xiao

Anhui Engineering Technology Research Center of Automotive New Technique,
Anhui Polytechnic University, Wuhu 241000, China
shipeicheng@126.com, 1159461528@qq.com, tlxp95@163.com

Abstract. We, per the national standard GB14167-2013 of the People's Republic of China about strength test for ISOFIX anchorage on vehicle seats and taking a new vehicle seat product as research object with the finite element analysis theory, established the finite element model for ISOFIX anchorage on vehicle seats; obtained the stress and strain nephogram of vehicle seats based on HyperWorks software for forward force test and oblique force test; thus provided reference for structural optimization design by analyzing and forecasting the weak parts of vehicle seats.

Keywords: Vehicle · Safety belt · ISOFIX analysis · HyperWorks · Ls-Dyna

1 Introduction

As the prevalence of vehicles increases, the number of casualties caused by vehicles is soaring, which is largely caused by vehicles' insufficient safety performance. As far as the vehicle safety is concerned, apart from strong body frame, the vehicle inner structure, especially the seats, is very important to protect passenger in accidents [1]. The ultimate goal of this paper is to improve the safety and comfort performance and reach equilibrium between cost reduction and corporate competitiveness improvement. We established an ISOFIX anchorage system by taking a new ISOFIX anchorage for vehicles as the research object [2]. ISOFIX (Innernational Standards Organization FIX) refers to a system connecting the child restraint system and the vehicle. As a finite element model, it consists of two rigid connection points on the vehicle and two rigid connection devices on the child restraint system. We obtained the stress and strain nephogram of vehicle seats based on HyperWorks software for forward force test and oblique force test. Improvement plan was also proposed after analyzing and forecasting the weak parts of vehicle seats, enabling their structural design and safety factors to meet requirements in GB14167-2013 [3].

2 Static Strength Test of ISOFIX Anchorage

National Standard GB14167-2013 has the following requirements for the strength of ISOFIX anchorage system [3].

© Springer International Publishing AG 2017
D. Liu et al. (Eds.): ICONIP 2017, Part V, LNCS 10638, pp. 387–396, 2017.
https://doi.org/10.1007/978-3-319-70139-4_39

a. Devices only equipped with lower ISOFIX anchorage received forward force test and oblique force test as SFAD (static force application device) as per 5.5.2.2. During the application period, no SFAD displacement in forward and an oblique direction was allowed to exceed 125 mm. Permanent deformation and partial cracking were allowed, but lower ISOFIX anchorage and surrounding areas cannot be invalidated within the specified time.

b. Devices equipped with ISOFIX top tether anchorage received 50 ± 5 KN restrain between SFAD and upper anchorage as per 5.5.2.3 and 8 ± 0.25 KN horizontal forward force as per 5.5.2.1.2. During the application period, no SFAD displacement in forward and an oblique direction was allowed to exceed 125 mm. Permanent deformation and partial cracking were allowed, but lower ISOFIX anchorage and surrounding areas cannot be invalidated within the specified time.

The test method connected SFAD (as shown in Fig. 1) and ISOFIX anchorage system and all ISOFIX positions on the same row were tested as well [3].

3 Establishment of Finite Element Model for Vehicle Seats

3.1 Geometric Model for Vehicle Seats

To facilitate the calculation by HyperWorks software, necessary simplification treatment should be made on this model in combination with structural characteristics of vehicle seat to improve the calculation accuracy. Before introducing the vehicle seat model, we repaired the boundary and holes and removed chamfer of such model in CATIA V5 6R2015 [4, 5]. This can also be achieved through geometry clean in Hypermesh.

3.2 Introduction of Vehicle Seat Model

We established the three-dimensional solid modeling in CATIA V5 6R2015 software, saved such model as *.igs format and used a special *.igs interface read-in model provided by HyperWorks to judge there's no geometric distortion in such model [6].

3.3 Mesh Generation of Vehicle Seat Model

We introduced the vehicle sear in special preprocessor and postprocessor models in HyperWorks and used the middle surface of Hypermesh to extract Midsurface function for vehicle seats. However, the geometry quality of middle surface extraction may be poor. After that, we used Hypermesh geometry cleaning tool to eliminate redundant surfaces, repair defects, compress boundaries of adjacent surfaces and seam free edges [4, 6]. Only by this way can we get qualified mesh after mesh generation of this model in a larger and more reasonable area.

Here the mesh generation was automatically done using Automesh model in Hypermesh, most of which were quadrilateral shell meshes along with triangle meshes as transition and the latter should be avoided for it may cause huge local rigidity and inaccurate calculation accuracy. For greater calculation accuracy of the finite element model for vehicle seats, we set the element size as 10 mm and checked the quality of QI unit during the mesh generation to obtain a reasonable finite element analysis model. We finally divided vehicle seats into 128225 units and 139541 nodes [7, 8].

3.4   Material Model

In this paper, the seat frame defined when establishing the CAE model is mostly nonlinear materials [8].

To put it simply, nonlinear material occurs when its stress δ and strain ε are not subject to a linear relationship. It is caused by nonlinear elasticity, plasticity, material loss and failure mechanism to a great extent. There is a linear relationship between stress and strain in linear elastic material:

$$\delta = E \cdot \varepsilon \tag{1}$$

In the formula, stress is denoted as δ with unit of Pa; strain is denoted as ε with dimension of one, an elementary quantity measuring the deformation level at a point; elasticity modulus denoted as E with unit of Pa. However, E is no longer a constant matrix in nonlinear materials, but a variable quantity.

Steel is one of the most common seat materials and this paper used CR980, SAPH 440 and Q345 in seat frame. Its stress and strain follow the Hooke's law at the elastic stage and their increment relationship follows the incremental theory (flow theory) at the plastic stage [9]. Figure 2 presents the stress-strain diagram (δ-ε curve) of low-carbon steel stretching. However, due to the setting of software program, using this curve for finite element analysis may often cause failure or errors in software calculation, so material mechanical model, namely material constitutive relation, may be required for simplification. It is a vital parameter in finite element simulation calculation. Only by inputting accurate material parameters can we obtain accurate calculation result. Under normal circumstances, the stress-strain relationship measured by ideal one-dimensional test shall be used to replace or change that under complex conditions.

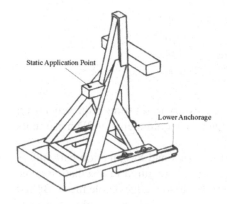

Fig. 1. SFAD

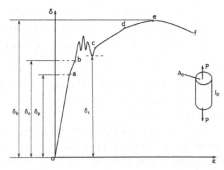

Fig. 2. Stress-strain curve when low-carbon steel stretches

In Fig. 2, ob: elastic stage; bc: yield stage; ce: hardening stage; ef: local deformation stage; δ_p: proportional limit (Pa); δ_e: elastic limit (Pa); δ_s: yield limit (Pa); δ_b: tensile strength (Pa); A_0: cross section area of test sample (m^2); P: applied load (N); l_0: original length of test sample (m);

ε: stress with dimension of one; δ: strain (Pa).

The finite element model for safety-belt anchorage strength test mostly applies the elastic-plastic material MAT_24. There're three methods to control the model: bilinear control method, inputting 8 groups of effective stress strain points and inputting the effective stress-strain curve [8, 10]. In this paper, metal materials in seat frame are basically set by bilinear control method. Figure 3 sets the window for parameters of tube material CR980, wherein Rho-material density of 7.80 * 10–9 kg mm^{-3}, E-elasticity modulus of 210000 Mpa, NU-Poisson's ratio of 0.30, SIGY-yield strength of 850 Mpa and other parameters use default values.

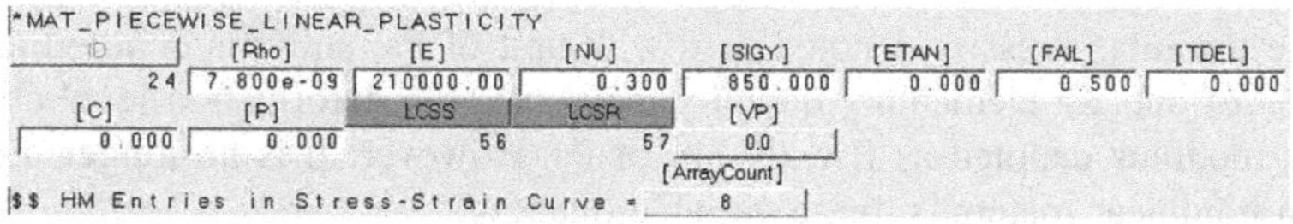

Fig. 3. Setting of tube material CR980 parameters

Table 1 shows the materials of seat parts.

Table 1. Thickness and materials of parts

Name	Thickness	Material	Elongation at break
Seat tube	1.5 mm	CR980	0.08
Innerior panel	1.5 mm	HC600/980QP	0.20
Junction plate	1.5 mm	HC600/980QP	0.20
Bidet holder	1.5 mm	HC600/980QP	0.20
Inner slide-way	1.8 mm	S500MC	0.12
Outer slide-way	1.8 mm	S500MC	0.12
Dropper	φ6.0 mm	Q345	0.20
Dropper holder	6 mm	Q345	0.20
Foundation	2.5 mm	SAPH440	0.26
Tube	1.2 mm	CR980	0.08

To ensure the accuracy of CAE analysis, it is necessary to set some requirements on mesh quality and model establishment method. Quality requirements of 2D mesh units are shown in Table 2.

The seat and the vehicle floor are connected by bolts and the junction part suffers great stress, so in the CAE analysis, washer processing is required in installation holes to automatically create the polygon (greater than or equal to 6 edges) hole mesh. Bolted connection applies the rigid unit for simulation, while the frame assembly and connections between seat foot and vehicle body are all connected by solder joint units in

Table 2. Quality requirements of 2D mesh units

Warping degree	Length-width ratio	Torsion resistance	Jacobian	Triangle	Quadrangle
>5.00	>5.00	>60.00	<0.70	Minimum angle < 20.00° Maximum angle > 120.00°	Minimum angle < 45.00° Maximum angle > 35.00°

the model. The calculation takes no consideration for failure in solder joints and hinged shafts. Empirical analysis tells us ISOFIX strength analysis does not have great impact due to the weak seat foam. To save the calculation time, we merely modeled the seat frame, as shown in Fig. 4. The final CAE analysis model is shown in Fig. 5.

Fig. 4. Seat frame assembly   **Fig. 5.** Seat mesh model

3.5   Application Condition

As per the requirements of GB14167-2013, safety-belt anchorage application test conditions are shown in Table 3.

Table 3. Application conditions

Condition	Test direction	Test anchorage	Lord force/KN
1	Forward 0° ± 5°	Two lower anchorages	8KN ± 0.25KN
2	Oblique (right) 75° ± 5°	Two lower anchorages	5KN ± 0.25KN
3	Oblique (left) 75° ± 5°	Two lower anchorages	5KN ± 0.25KN

The application curve is shown in Fig. 6.

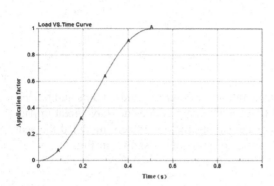

Fig. 6. Application curve (Y-axis is for application factor, namely the percentage of application)

Fig. 7. Forward seat stress distribution diagram

3.6 CAE Evaluation Standard

In accordance with the requirements in National Standard GB14167-2013, during the application period, no SFAD displacement in forward and an oblique direction was allowed to exceed 125 mm. Permanent deformation and partial cracking were allowed, but lower ISOFIX anchorage and surrounding areas cannot be invalidated within the specified time. In CAE analysis, displacement of SFAD application direction is set 80% safety margin according to the law [3], i.e. maximum displacement is 100 mm. Judging based on past experience, the structure of sheet metal parts may not be invalidated if the plastic deformation is controlled less than 20%. For this reason, plastic deformation in ISOFIX region is controlled within 20% in CAE analysis.

4 CAE Analysis Result

4.1 Test Analysis Under Working Condition 1

According to the test conditions set by the working condition 1, we got the seat stress distribution shown in Fig. 7 by submitting the analysis model for Ls-Dyna calculation. From the Fig. 7, we can know that when testing under working condition 1, the seat deformation is slight on the whole; the main load suffered by seat is mainly on the connection of foundation holder and outer & inner slide-way bolts. The maximum stress is at the connection of foundation holder and inner slide-way bolt, reaching 0.13, which may not cause failure or destroy. Figure 8 shows the stress distribution of other key seat parts, according to which we know that these parts also remain in a safe scope.

The test analysis under working condition 1 not only observes the stress of key parts and check if there's any failure in parts, but measure whether the displacement

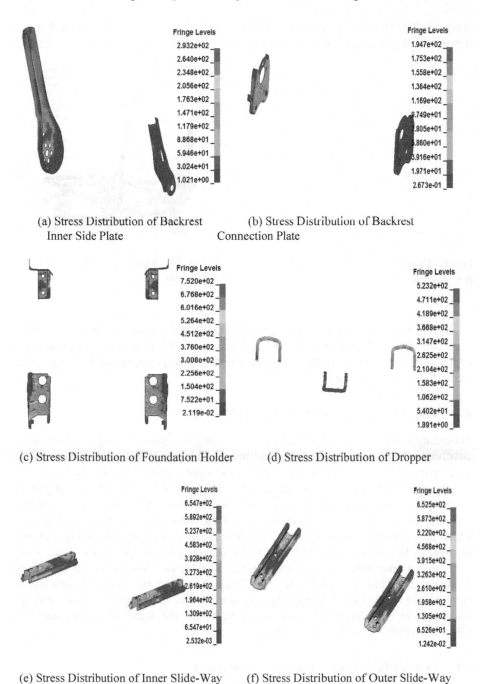

(a) Stress Distribution of Backrest
Inner Side Plate

(b) Stress Distribution of Backrest
Connection Plate

(c) Stress Distribution of Foundation Holder

(d) Stress Distribution of Dropper

(e) Stress Distribution of Inner Slide-Way

(f) Stress Distribution of Outer Slide-Way

Fig. 8. Stress distribution diagram of key parts

curve of SFAD (i.e. test points) to exceed the safety margin of 100 mm or not. Figure 9 shows the displacement curve of SFAD points in the test under working condition 1.

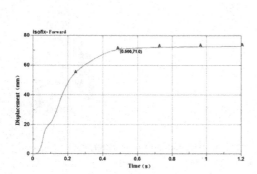

Fig. 9. Displacement curve of SFAD (i.e. test points) under 100% stress

Fig. 10. Right seat stress distribution diagram

In Fig. 9, the working condition 1 loads 8000 N and reaches full for merely 0.50 s, at which the displacement achieves its maximum 71.0 mm. The displacement in application direction is smaller than the standard safety margin of 100 mm, which meets the legal regulations.

4.2   Test Analysis Under Working Condition 2

According to the test conditions set by the working condition 2, we got the seat stress distribution shown in Fig. 10 by submitting the analysis model for Ls-Dyna calculation. From the Fig. 10, we can know that when testing under working condition 1, the seat deformation is slight on the whole; the main load suffered by seat is mainly on the connection of foundation holder and outer & inner slide-way bolts. The maximum stress is at the connection of foundation holder and inner slide-way bolt, reaching 0.11, which may not cause failure or destroy. Figure 11 shows the stress distribution of other key seat parts, according to which we know that these parts also remain in a safe scope.

Figure 12 presents the distribution curve of SFAD (i.e. test points) under working condition 2. In Fig. 12, the working condition 2 loads 5000 N and reaches full for merely 0.50 s, at which the displacement achieves its maximum 48.8 mm. The displacement in application direction is smaller than the standard safety margin of 100 mm, which meets the legal regulations.

4.3   Test Analysis Under Working Condition 3

According to the test conditions set by the working condition 3 and after submitting the analysis model for Ls-Dyna calculation, we know that the stress and strain results may not cause failure or destroy to the whole seat and parts, either. Figure 13 presents the distribution curve of SFAD (i.e. test points) under working condition 3. In Fig. 13, the

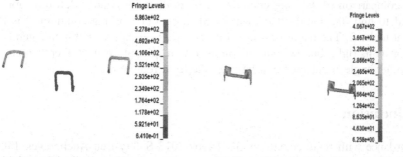

(a) Stress Distribution of Dropper   (b) Stress Distribution of Dropper Holder

Fig. 11. Stress distribution diagram of key parts

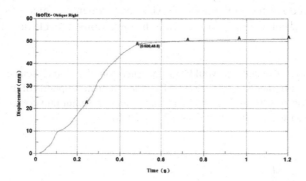

Fig. 12. Displacement curve of SFAD (i.e. test points) under 100% stress

working condition 3 loads 5000 N and reaches full for merely 0.60 s, at which the displacement achieves its maximum 73.8 mm. The displacement in application direction is smaller than the standard safety margin of 100 mm, which meets the legal regulations.

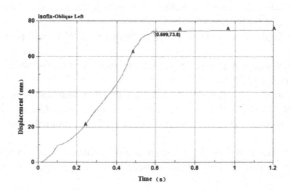

Fig. 13. Displacement curve of SFAD (i.e. test points) under 100% stress

By comparison of working condition 2 (right direction) and 3 (left direction), it's not hard to find that the displacement of SFAD under working condition 3 is larger than that under 2. The major reason is that the seat is equipped with a unilateral angle regulator on its right, but without that on its left. In spite of this fact, there's no need to add another angle regulator for it has met the legal requirement.

5   Conclusion

In accordance with requirements in GB 14167-2013 Safety-Belt Anchorages, ISOFIX Anchorages Systems and ISOFIX Top Tether Anchorages for Vehicles, this paper used CATIA and Hypermesh software to establish the finite element model for vehicle ISOFIX forward and oblique force tests and Ls-Dyna for solution. The safety evaluation of new vehicle seats based on CAE analysis can greatly improve the design efficiency and reduce the number of physical experiments, which is of great significance to reduce development costs, shorten the development cycle and improve the test pass rate.

Acknowledgments. The authors would like to thank Anhui province science and technology research key project (Grant No. 1604a0902158), of the financial supports of the National Natural Science Foundation of China (Grant No. 51575001) and of Anhui university scientific research platform innovation team building projects (2016–2018).

References

1. Kim, H., Lee, Y., Yang, S., et al.: Structural analysis on variable characteristics of automotive seat frame by FEA. Int. J. Precis. Eng. Manuf.-Green Technol. **11**, 75–79 (2016)
2. Hou, Y., Zhou, L., Cui, D., Xie, S., et al.: Research on child seat ISOFIX based on LS-DYNA. Automob. Technol. **01**, 42–46 (2016)
3. Safety-belt Anchorages, ISOFIX Anchorages Systems and ISOFIX Top Tether Anchorages for Vehicles. GB 14167-2013 of the People's Republic of China (2013)
4. Chen, H.N., Chen, H., Wang, L.J.: Analysis of vehicle seat and research on structure optimization in front and rear impact. World J. Eng. Technol. **2**, 92–99 (2014)
5. Wei, G.: Simulation Research on Safety of Car Rear Seat Frame. Beijing University of Chemical Technology, Beijing (2013)
6. Dang, X.: The Study of Vehicle Seat Safety With LS-DYNA. Ningbo University, Ningbo (2014)
7. Lu, J.: System Strength Analysis of Vehicle Seatbelt Anchorage. Shenyang University of Technolgy, Shenyang (2016)
8. Liang, R.: Research on Strength of an Automobile Front Seat. Jilin University, Jilin (2014)
9. Himmetoglu, S., Acar, M., Bouazza-Marouf, K., Taylor, A.J.: Car seat design to improve rear-impact protection. Proc. Inst. Mech. Eng. Part D J. Automob. Eng. **225**(4), 441–459 (2011)
10. Ramanathan, B., Hu, J., Reed, M.P.: A computational study of seat and seatbelt performance for protecting 6–12 year-old children in frontal crashes. Int. J. Veh. Des. **70**(1), 29–44 (2016)

Robust Adaptive Beamforming in Uniform Circular Array

Xin Song[1(✉)], Ying Guan[1], Jinkuan Wang[2], and Jing Gao[1]

[1] Engineering Optimization and Smart Antenna Institute,
Northeastern University, Qinhuangdao 066004, China
sxin78916@mail.neuq.edu.cn
[2] School of Computer Science and Engineering, Northeastern University,
Shenyang 110819, China

Abstract. Phase-mode transformation (PMT) is a commonly used technique to convert a uniform circular array (UCA) into a virtual uniform linear array (ULA). This method restores the Vandemonde structure of the steering vector and makes it easy to apply many existing beamforming algorithms to UCA. One such method is the famous Minimum Variance Distortionless Response (MVDR) algorithm, in which the array gain is equal to unity in the direction of arrival of the desired signal. However, due to the approximation errors of the PMT and signal steering vector mismatches, the performance of these algorithms degrades. To address these two issues, in this paper we develop a robust recursive updating algorithm based on worst-case performance optimization. We show that the proposed algorithm belongs to the class of the diagonal loading technique and the transformation matrix belongs to a certain ellipsoid set. Using the Lagrange multiplier method, we have also derived closed-form solution to the weight vector. Our robust algorithm has low implementation complexity and makes the mean output array SINR consistently close to the optimal one. Numerical experiments have shown that our method outperforms the MVDR algorithm.

Keywords: Worst-case performance optimization · Steering vector mismatches · Robust adaptive beamforming · Phase-mode transformation

1 Introduction

Uniform circular array (UCA) is a structure frequently used in practical situations such as radar, sonar, navigation, radio astronomy, and wireless conmmunications, due to its isotropic coverage of the azimuth angle. It is well known that circular arrays possess important advantages such as broadband processing and electronic steering, including direction of arrival (DOA) estimation and adaptive beamforming, etc. Despite the benefits, circular arrays suffer from a major shortcoming and the array steering vector does not have the Vandermonde structure [1]. Hence, many adaptive beamforming and DOA estimation methods have to be modified in order to be applied in circular arrays [2]. The phase-mode transformation technique is used to overcome this problem. Using this phase-mode transformation, DOA estimation and adaptive beamforming methods can be applied to the virtual uniform linear array (ULA).

© Springer International Publishing AG 2017
D. Liu et al. (Eds.): ICONIP 2017, Part V, LNCS 10638, pp. 397–406, 2017.
https://doi.org/10.1007/978-3-319-70139-4_40

There are several efficient approaches to design robust adaptive beamformers, such as the linearly constrained minimum variance (MV) beamformer [3], the eigenspace-based beamformer [4], and the projection beamforming techniques [5]. The popular class of robust beamforming techniques called diagonal loading (DL) [6]. In these methods the array covariance matrix is loaded with an appropriate multiple, called the loading level, of the identity matrix in order to satisfy the imposed quadratic constraint. However, it is somewhat difficult to relate the loading level with the uncertainty bounds of the array steering vector. But the above methods can not be expected to provide sufficient robustness improvements.

Among more recent, some new robust adaptive beamforming algorithms are proposed [7, 8]. In [7], a new MVDR robust adaptive beamforming technique was developed, which uses as little as possible and easy to obtain imprecise prior information. In [8], the essence of the proposed approach is to estimate the difference between the actual and presumed steering vectors and to use this difference to correct the erroneous presumed steering vector. It is emphasized that the correction of errors in estimating the steering vector and in the phase-mode transformation matrix are not considered in the algorithms mentioned above.

In this paper, we propose a robust beaforming algorithm. Our approach is based on worst-case performance optimization and phase-mode transformation. It is assumed that signal steering vector pertains to a ball set. The transformation matrix is covered by the ellipsoid sets. Due to considering the error correction, the proposed algorithm has some advantages to the traditional methods. The excellent performance of the proposed algorithm is demonstrated as compared with conventional algorithms via several example.

2 Background

Consider a uniform circular array with M sensors. We assume that arrived signals are narrowband with known frequency. The array observation vector is given by at time k

$$
\begin{aligned}
x(k) &= s(k) + i(k) + n(k) \\
&= s_0(k)a + i(k) + n(k)
\end{aligned}
\tag{1}
$$

where $s(k)$, $i(k)$ and $n(k)$ are the desired signal, interference, and noise components, respectively. Here, $s_0(k)$ is the desired signal waveform, and a is the signal steering vector. It is assumed that arrived signals and the array are coplanar. So, we can show steering vector a by $a(\theta_0)$, where θ_0 is the azimuth angle of arrival of the desired signal. We assume that there are D interferences in the practical environment, which are narrowband incoherent plane waves, impinging from directions $\{\theta_1, \theta_1, \cdots, \theta_D\}$. So, the interference vector $i(k)$ is written in the following form

$$
i(k) = A_i s_i(k)
\tag{2}
$$

where A_i is the matrix having columns that are steering vectors of interference signals, and $s_i(k)$ is a vector of the interference signal, that is

$$A_i = [a(\theta_1), \ldots, a(\theta_D)] \tag{3}$$

$$s_i(k) = [i_1(k), \ldots, i_D(k)]^{\mathrm{T}} \tag{4}$$

We assume that q is the wavenumber of the received signal and r is the radius of the array, so the steering vector $a(\theta_i)$ is written as

$$a(\theta_i) = [e^{jqr\cos(\theta_i)}, \ldots, e^{jqr\cos(\theta_i - (2\pi(M-1)/m))}]^T \tag{5}$$

where.

The output of a narrowband beamformer in a UCA is given by

$$y(k) = w^H x(k) \tag{6}$$

where $w = [w_1, \cdots, w_M]^T$ is the complex vector of beamformer weights, and $(\cdot)^T$ and $(\cdot)^H$ stand for the transpose and Hermitian transpose, respectively.

The cost function of MVDR algorithm is described by the following optimization problem [3]:

$$\min_{w} w^H R_{i+n} w \text{ subject to } w^H a = 1 \tag{7}$$

where

$$R_{i+n} = E\{(i(k) + n(k))(i(k) + n(k))^H\} \tag{8}$$

is the $M \times M$ interference-plus-noise covariance matrix.

The optimum weight vector can be found

$$w_{\mathrm{opt}} = \frac{R_{i+n}^{-1} a}{a^H R_{i+n}^{-1} a} \tag{9}$$

From the Eq. (9), it is known that the traditional algorithm is quite sensitive to signal steering vector mismatches. In addition, the exact interference-plus-noise covariance matrix R_{i+n} is unavailable. Usually, the sample covariance matrix

$$R_N = \frac{1}{N} \sum_{k=1}^{N} x(k) x^H(k) \tag{10}$$

is used to replace R_{i+n} in (10), where N is the training sample size. Therefore, the performance degradation of MVDR algorithm can arise owing to the small training sample size and signal steering vector mismatches.

3   Robust Recursive Algorithm in Uniform Circular Array

3.1   Covering Ellipsoid Set

In a UCA, the signal steering vector does not have the Vandermonde structure, which causes some difficulties in applying a number of DOA estimation and adaptive beamforming methods [1]. We can use phase-mode transformation to convert UCA to a virtual uniform linear array. The phase-mode transformation is done by premultiplying the array observation vector by matrix $\boldsymbol{\Psi}_0 = \boldsymbol{PT}$, where

$$
T = \begin{bmatrix}
1 & \varphi^{-h} & \cdots & \varphi^{-(M-1)h} \\
\vdots & \vdots & \vdots & \vdots \\
1 & \varphi^{-1} & \cdots & \varphi^{-(M-1)} \\
1 & 1 & \cdots & 1 \\
1 & \varphi^{1} & \cdots & \varphi^{(M-1)} \\
\vdots & \vdots & \vdots & \vdots \\
1 & \varphi^{h} & \cdots & \varphi^{(M-1)h}
\end{bmatrix}
\tag{11}
$$

and

$$
P = \mathrm{diag}\left[\frac{1}{\sqrt{M}j^{-h}P_{-h}(qr)}, \cdots, \frac{1}{\sqrt{M}j^{h}P_{h}(qr)}\right]
\tag{12}
$$

where $\varphi = e^{j(2\pi/M)}$, $l = 2h+1$ is size of the virtual array, and $P_i(\cdot)$ is the Bessel function of the first kind of order i. The parameter h must be satisfied as following criterion [9]

$$
h = \max\left\{v \middle| v \le \frac{M-1}{2}, \frac{|P_{v-M}(qr)|}{||P_v(qr)||} \le \rho_t\right\}
\tag{13}
$$

where ρ_t is a small value.

We multiply the steering vector $\boldsymbol{a}(\theta_i)$ by transformation matrix $\boldsymbol{\Psi}_0$ and convert it to the steering vector of the virtual ULA with the Vandemonde structure

$$
\begin{aligned}
\hat{\boldsymbol{a}}(\theta_i) &= \boldsymbol{\Psi}_0 \boldsymbol{a}(\theta_i) \\
&= [e^{-jh\theta_i}, e^{-j(h-1)\theta_i}, \cdots, e^{j(h-1)\theta_i}, e^{jh\theta_i}]^T
\end{aligned}
\tag{14}
$$

where $\hat{\boldsymbol{a}}(\theta_i)$ is the virtual steering vector. Form (14), we note that $\hat{\boldsymbol{a}}(\theta_i)$ is independent of the frequency f.

Considering the errors in transformation matrix $\boldsymbol{\Psi}_0$, this matrix is instead of the matrix $\boldsymbol{\Psi}$ defined as

$$\boldsymbol{\Psi} = \boldsymbol{\Psi}_0 + \Delta\boldsymbol{\Psi} \tag{15}$$

where $\Delta\boldsymbol{\Psi} = \Delta\boldsymbol{PT}$ is the error matrix due to frequency mismatch, here the matrix $\boldsymbol{T}$ is defined in (11).

We assume that in practical applications, the unknown mismatch vector $\Delta\boldsymbol{a}$ is norm-bounded by some known constant $\xi > 0$, that is [10],

$$\|\Delta\boldsymbol{a}\| \le \xi \tag{16}$$

Furthermore, we assume that the norm of the error matrix Δ can be bounded by some known constant r, $\|\Delta\| \le r$. Then, the actual covariance matrix is [11]

$$\hat{\boldsymbol{R}}_{xx} = \boldsymbol{R}_{xx} + r\boldsymbol{I} \tag{17}$$

where $\boldsymbol{R}_{xx} = E[\boldsymbol{x}(k)\boldsymbol{x}^H(k)]$ is the theoretical array correlation matrix of the array output vector.

In the case that the phase-mode transformation is applied to the UCA outputs, the array correlation matrix $\boldsymbol{G}$ is rewritten as [1]

$$\begin{aligned}
\boldsymbol{G} &= (\boldsymbol{\Psi}_0 + \Delta\boldsymbol{\Psi})\hat{\boldsymbol{R}}_{xx}(\boldsymbol{\Psi}_0 + \Delta\boldsymbol{\Psi})^H \\
&= \Delta\hat{\boldsymbol{R}}_{xx}\Delta\boldsymbol{\Psi}^H + \Delta\hat{\boldsymbol{R}}_{xx}\boldsymbol{\Psi}_0^H + \boldsymbol{\Psi}_0\hat{\boldsymbol{R}}_{xx}\Delta\boldsymbol{\Psi}^H + \boldsymbol{\Psi}_0\hat{\boldsymbol{R}}_{xx}\boldsymbol{\Psi}_0^H
\end{aligned} \tag{18}$$

where the matrix $\boldsymbol{G}$ is composed of four parts, the last part $\boldsymbol{\Psi}_0\hat{\boldsymbol{R}}_{xx}\boldsymbol{\Psi}_0^H$ is known, the other three matrices are not known and they belong to certain ellipsoid sets.

The first part in the right-hand side of (18) is $\boldsymbol{G}_1 = \Delta\hat{\boldsymbol{R}}_{xx}\Delta\boldsymbol{\Psi}^H$. First, this ellipsoid is obtained, which covers all choices for the column in matrix $\boldsymbol{G}_1$, which belongs to the following ellipsoid set

$$\begin{aligned}
\boldsymbol{E}_{1,i}(\boldsymbol{0}, \boldsymbol{Q}_{1,i}) &= \left\{ \boldsymbol{Q}_{1,i}^{1/2}\boldsymbol{v} \mid \|\boldsymbol{v}\| \le 1 \right\} \\
\boldsymbol{Q}_{1,i}^{1/2} &= \delta_{-h+i-1}\,\varepsilon\hat{\boldsymbol{R}}_{xx}
\end{aligned} \tag{19}$$

We consider the second matrix $\boldsymbol{G}_2 = \Delta\hat{\boldsymbol{R}}_{xx}\boldsymbol{\Psi}_0^H$, which belongs to the following ellipsoid set

$$\begin{aligned}
\boldsymbol{E}_{2,i}(\boldsymbol{0}, \boldsymbol{Q}_{2,i}) &= \left\{ \boldsymbol{Q}_{2,i}^{1/2}\boldsymbol{v} \mid \|\boldsymbol{v}\| \le 1 \right\} \\
\boldsymbol{Q}_{2,i}^{1/2} &= \varepsilon|\boldsymbol{P}_{-h+i-1}|\hat{\boldsymbol{R}}_{xx}
\end{aligned} \tag{20}$$

The third matrix in the right-hand side of (18) is $\boldsymbol{G}_3 = \boldsymbol{\Psi}_0\hat{\boldsymbol{R}}_{xx}\Delta\boldsymbol{\Psi}^H$, which belongs to a certain ellipsoid set

$$E_{3,i}(0, Q_{3,i}) = \left\{ Q_{3,i}^{1/2} v \mid \|v\| \leq 1 \right\}$$

$$Q_{3,i}^{1/2} = \delta_{-h+i-1} \Psi_0 \hat{R}_{xx} \tag{21}$$

Considering the errors in transformation matrix, the array steering vector b in circular arrays is defined as

$$b = \Delta\Psi a + \Psi_0 a \tag{22}$$

Next, we will solve the ellipsoid covering the vector $b_1 = \Delta\Psi a$. The matrix $b_1 b_1^H$ is written

$$b_1 b_1^H = \Delta\Psi a a^H \Delta\Psi^H \tag{23}$$

We can compute the matrix $\Delta\Psi\Delta\Psi^H$

$$\Delta\Psi\Delta\Psi^H = \Delta P \Delta P^H = \begin{bmatrix} |\tau_{-h}|^2 & \cdots & 0 \\ \vdots & \ddots & \vdots \\ 0 & \cdots & |\tau_h|^2 \end{bmatrix} \leq \begin{bmatrix} \delta_{-h}^2 & \cdots & 0 \\ \vdots & \ddots & \vdots \\ 0 & \cdots & \delta_h^2 \end{bmatrix} \tag{24}$$

Applying (24), we can obtain the following inequation

$$\beta^2 \begin{bmatrix} \delta_{-h}^2 & \cdots & 0 \\ \vdots & \ddots & \vdots \\ 0 & \cdots & \delta_h^2 \end{bmatrix} - b_1 b_1^H \geq 0 \tag{25}$$

where the parameter $\beta^2 = \|a\|_2^2$.

Consequently, we conclude that the vector $b_1 = \Delta\Psi a$ belongs to the ellipsoid B_1

$$B_1(0, P_1) = \left\{ P_1^{1/2} v \mid \|v\| \leq 1 \right\}$$

$$P_1^{1/2} = \beta \begin{bmatrix} \delta_{-h} & \cdots & 0 \\ \vdots & \ddots & \vdots \\ 0 & \cdots & \delta_h \end{bmatrix} \tag{26}$$

3.2   Set-Based Worst-Case Approach

The optimization problem is described as [10]–[11]

$$\min_{w} w^H \hat{R}_{xx} w \quad \text{subject to} \quad |w^H a - 1|^2 = \xi^2 w^H w \tag{27}$$

In UCA, the problem (27) of finding optimum weight vector is converted to the following problem by using (18) and (22)

$$\min_{w} \; \hat{w}^H G \hat{w} \text{ subject to } |\hat{w}^H b - 1|^2 = \hat{\xi}^2 \hat{w}^H \hat{w} \tag{28}$$

The solution to (28) can be derived by minimizing the Lagrange function

$$H(\hat{w}, \phi) = \hat{w}^H G \hat{w} + \phi \left(\hat{\xi}^2 \hat{w}^H \hat{w} - \hat{w}^H b b^H \hat{w} + \hat{w}^H b + b^H \hat{w} - 1 \right) \tag{29}$$

where ϕ is a Lagrange multiplier. The gradient vector of $H(\hat{w}, \phi)$ is given by

$$\mathbf{V} = \left(G + \phi \hat{\xi}^2 I - \phi b b^H \right) \hat{w} + \phi b \tag{30}$$

The updated weight vector becomes

$$\hat{w}(k+1) = \hat{w}(k) - \mu [B \hat{w}(k) + \phi b] \tag{31}$$

where $B = G + \phi \hat{\xi}^2 I - \phi b b^H$, the step size μ is given by [12, 13]

$$\mu = \frac{\sigma \mathbf{V}^H(k) \mathbf{V}(k)}{\mathbf{V}^H(k) G \mathbf{V}(k)} \tag{32}$$

The parameter σ is added to improve the numerical stability of the algorithm. For a practical system, σ should be adjusted during the initial of the system and it should satisfy $0 < \sigma < 1$ [14].

The gradient vector of $H(\hat{w}, \phi)$ is equal to zero and we can obtain the optimum weight vector

$$\hat{w}_{\text{opt}} = \upsilon \phi \left(G + \phi \hat{\xi}^2 I \right)^{-1} b \tag{33}$$

where $\upsilon(\phi) = \dfrac{\phi}{\phi b^H \left(G + \phi \hat{\xi}^2 I \right)^{-1} b - 1}$.

From (33), we note that the proposed algorithm belongs to the class of diagonal loading technique, and the loading factor is complicated.

4 The Experiment Result Analysis

In this section, we present some simulation results to demonstrate the performance of the improved robust recursive algorithm by matlab software. We assume a uniform circular array with $M = 13$ omni-directional sensors. For each scenario, 100 simulation runs are used to obtain each simulation point. We assume that signals are in the frequency band [12 kHz, 13 kHz]. The centre frequency $f_c = 12.5$ kHz.

In all examples, two interfering sources are assumed with plane wavefronts and the directions of arrival impinging from $-50°$ and $50°$, respectively.

Example 1: Exactly known signal steering vector
In this example, the plane wave signal is assumed to impinge on the array from $\theta = 0°$. Figure 1 displays the performance of the methods tested versus the number of

snapshots for the fixed SNR = 10 dB and $\hat{\xi} = 3.7$. Figure 2 shows the performance of these algorithms versus the SNR for the fixed training data size $N = 100$. In this scenario, we note that the proposed robust algorithm outperforms MVDR algorithm and makes all the values of SINR close to the optimal one. Moreover, the robust recursive beamforming algorithm offers faster convergence rate.

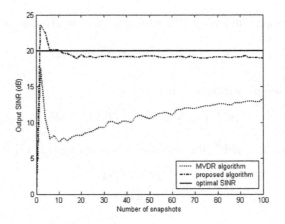

Fig. 1. Output SINR versus N in no mismatch.

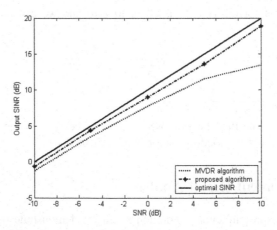

Fig. 2. Output SINR versus SNR in no mismatch.

Example 2: Signal look direction mismatch

We assume that both the presumed and actual signal spatial signatures are plane waves impinging from the DOAs 0° and 3°, respectively. This corresponds to a 3° mismatch in the signal look direction. Figure 3 displays the performance of the methods tested versus the number of snapshots for SNR = 10 dB and $\hat{\xi} = 3.7$. The performance of these algorithms versus the SNR for the fixed training data size $N = 100$ is shown in

Fig. 4. When there is a mismatch in the signal steering vector, MVDR algorithm treats the desired signal as an interference signal, and tries to place null in the direction of the desired signal. Therefore, the output SINR degrades severely. Also, the degradation in the performance of MVDR algorithm is more significant in high SNR scenarios. However, our proposed robust algorithm provides excellent robustness against signal steering vector mismatches and the small training sample size. The example shows that the proposed algorithm has better performance than MVDR algorithm.

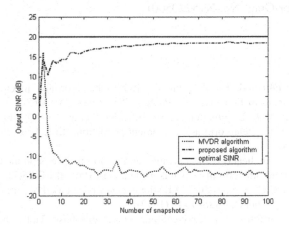

Fig. 3. Output SINR versus N in 3° mismatch.

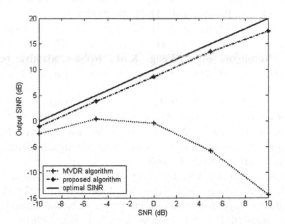

Fig. 4. Output SINR versus SNR in 3° mismatch.

5 Conclusions

In this paper, a novel robust adaptive beamforming has been developed based on worst-case performance optimization and phase-mode transformation in uniform circular array, which is robust not only against uncertainty in signal steering vector and small training sample size, but also against errors in phase-mode transformation method. We can prove that ellipsoid sets cover the transformation matrix columns. When the performance of algorithms in the frequency band is investigated, it is seen that the output SINR of the proposed algorithm is better than conventional algorithms. The robust algorithm provides improved output performance and offers faster convergence rate. Simulation results demonstrate that the proposed algorithm enjoys a significantly improved performance as compared with the conventional algorithms.

This work is supported by the National Nature Science Foundation of China under Grant nos. 61473066 and 61403069, the Fundamental Research Funds for the Central Universities under Grant No. N152305001.

References

1. Mohsen, A., Mahmood, K., Zakiyeh, A.: Robust beamforming in circular arrays using phase-mode transformation. IET Sig. Process. **7**(8), 693–703 (2013)
2. Jiang, X., Zeng, W.J., Yasotharan, A., So, H.C., Kirubarajan, T.: Quadratically constrained minimum dispersion beamforming via gradient projection. IEEE Trans. Sig. Process. **63**(1), 192–205 (2015)
3. Godara, L.C.: Application of antenna arrays to mobile communication, Part II: Beam-forming and direction-of-arrival considerations. Proc. IEEE **85**(7), 1213–1216 (1997)
4. Chang, L., Yeh, C.C.: Performance of DMI and eigenspace-based beamformers. IEEE Trans. Antennas Propag. **40**(11), 1336–1347 (1992)
5. Feldman, D.D., Griffiths, L.J.: A projection approach to robust adaptive beamforming. IEEE Trans. Sig. Process. **42**(4), 867–876 (1994)
6. Zou, Q., Yu, Z.L., Lin, Z.A.: Robust algorithm for linearly constrained adaptive beamforming. IEEE Sig. Process. Lett. **11**(1), 26–29 (2004)
7. Kim, S.J., Magnani, A., Mutapcic, A., Boyd, S.P., Luo, Z.Q.: Robust beamforming via worst-case SINR maximization. IEEE Trans. Sig. Process. **56**(4), 1539–1547 (2008)
8. Hassanien, A., Vorobyov, S.A., Wong, K.M.: Robust adaptive beamforming using sequential programming: an iterative solution to the mismatch problem. IEEE Sig. Process. Lett. **15**, 733–736 (2008)
9. Wax, M., Sheinvald, J.: Direction finding of coherent signals via spatial smoothing for uniform circular arrays. IEEE Trans. Antennas Propag. **42**(5), 613–620 (1994)
10. Vorobyov, S.A., Gershman, A.B., Luo, Z.Q.: Robust adaptive beamforming using worst-case performance optimization: a solution to the signal mismatch problem. IEEE Trans. Sig. Process. **51**(2), 313–324 (2003)
11. Shahbazpanahi, S., Gershman, A.B., Luo, Z.Q., Wong, K.M.: Robust adaptive beamforming for general-rank signal models. IEEE Trans. Sig. Process. **51**(9), 2257–2269 (2003)
12. Elnashar, A.: Efficient implementation of robust adaptive beamforming based on worst-case performance optimization. IET. Sig. Process. **2**(4), 381–393 (2008)
13. Elnashar, A., Elnoubi, S., Elmikati, H.: Further study on robust adaptive beamforming with optimum diagonal loading. IEEE Trans. Antennas Propag. **54**(12), 3647–3658 (2006)
14. Attallah, S., Abed-Meraim, K.: Fast algorithms for subspace tracking. IEEE Trans. Signal Process. Lett. **8**(7), 203–206 (2006)

Evaluating Accuracy in Prudence Analysis for Cyber Security

Omaru Maruatona[1], Peter Vamplew[2], Richard Dazeley[2],
and Paul A. Watters[3(✉)]

[1] PwC, Melbourne, Australia
omaru.maruatona@pwc.com
[2] Federation University, Ballarat, Australia
[3] La Trobe University, Melbourne, Australia
P.Watters@latrobe.edu.au

Abstract. Conventional Knowledge-Based Systems (KBS) have no way of detecting or signalling when their knowledge is insufficient to handle a case. Consequently, these systems may produce an uninformed conclusion when presented with a case beyond their current knowledge (brittleness) which results in the KBS giving incorrect conclusions due to insufficient knowledge or ignorance on a specific case. Prudence Analysis (PA) has been shown to be a viable alternative to brittleness in Ripple Down Rules (RDR) knowledge bases. To date, there have been two approaches to Prudence; attribute-based and structural-based prudence. This paper introduces Integrated Prudence Analysis (IPA), a novel Prudence method formed by combining these methods.

Keywords: IPA · Expert systems · Prudence analysis

1 Introduction

Most conventional Knowledge-Based Systems (KBS) are incapable of indicating when a case they are processing is beyond their expertise [1, 2]. Instead, these systems arbitrarily produce one of the pre-existent conclusions. It would be ideal in such situations, if the system could somehow signal insufficient knowledge to resolve the given case. Such an indication by the KBS would allow the administrator/expert to inspect the case further and add specific knowledge if necessary.

One of the earliest attempts to directly address brittleness was a technique named Prudence [3]. Since then, a number of prudence methods have been tested to enable a KBS to produce an additional signal indicating that the case being processed may be beyond the system's knowledge, prompting the administrator or knowledge engineer to further examine the case.

The rest of the paper is arranged as follows: Sect. 2 presents an introduction of Ripple Down Rules (RDR) and a summarised review of Prudence Analysis in RDR methods including the earliest Prudence methods and the subsequent introductions of two of the three prudence methods evaluated in this paper (Rated MCRDR and Ripple Down Models). In Sect. 3, IPA Analysis (IPA) is introduced.

D. Liu et al. (Eds.): ICONIP 2017, Part V, LNCS 10638, pp. 407–417, 2017.
https://doi.org/10.1007/978-3-319-70139-4_41

2 MCRDR Prudence

Three prudence methods (structural, attribute-based and IPA) use a Multiple Classification RDR (MCRDR) KB. MCRDR is an advancement of RDR that handles multiple classification domains and has been shown to produce a more compact KB with fewer redundancies than single class RDR [4].

2.1 RDR: A Brief Review

RDR was introduced in 1988 as an alternative to KBS' slow approach to knowledge acquisition and maintenance [5]. An extension of the original single class RDR, known as MCRDR was introduced to handle multiple domains (Kang et al. 1995). RDR and MCRDR's advantage over other KBS is their integrated knowledge acquisition and maintenance [4]. In RDR and MCRDR, the domain expert directly interacts with the system to add knowledge and maintain the system [6]. In conventional KB's, maintenance and knowledge acquisition require the services of a domain expert who interacts with the system through a costly knowledge engineer [5].

This paper introduces a third group of prudence methods, Integrated Prudence Analysis (IPA), formed by combining the other prudence methods. IPA was formed on the premise that combining structural and attribute methods in some strategic way could improve the Prudence Accuracy of either method. This proposition was initially suggested during evaluations of a structural prudence method Rated MCRDR, and an attribute-based prudence method, Ripple Down Models [7]. The IPA system combines key components of Rated MCRDR and Ripple Down Models (RDMs).

2.2 Ripple Down Models (RDM)

An extension of the profile based prudence system [8] was re-introduced by Prayote and Compton [9]. Ripple Down Models is an attribute-based prudence method, built on the principle that through inferencing, RDR also effectively partitions a search space into smaller sub-regions. Consequently, as new rules are added, new homogeneous partitions are created, and the data in each is uniformly distributed [1].

In RDR, a corresponding class is returned after a case is processed. Ripple Down Models extends and modifies this concept such that after RDR inferencing, a model is returned instead of a class, known as a Situated Profile (SP), containing profiles describing each of the current cases' attributes. Each SP comprises the same number of profiles as the attributes. If a dataset has n attributes, each SP will contain n profiles. The use of profiles is adopted from general ID systems' profile modeling classification where a system either defines a normal behaviour profile (Anomaly Detection) or maintains a list of known unacceptable profiles (Signature Detection).

A profile in the RDM context is a behaviour pattern, representing a homogenous subspace in the domain data. In Prayote's [1] application, a profile consisted of a homogenous pattern of network traffic where each profile represented a particular situation in the network. A new SP is added whenever a new situation is identified. If the current case corresponds to a pre-existing situation, a matching SP is retrieved using through inference. The retrieved SP is then passed to the Outlier Estimation (OE) module - each profile is screened for anomalies.

RDMs have three main components; RDR modeling, OE and Final RDR classification. A case is introduced to the system through the RDR modeling component - the RDR engine retrieves an appropriate SP for the case. An RDR KBS always has a default conclusion so the RDR modeling component has a default SP in case there is no match. The SP is then passed to the OE components where each attribute value is searched for possible outliers. The OD results are passed to the RDR decision base where an RDR inferencing process classifies the case as either anomalous or not. The RDR classification is first confirmed with an expert and is stored in the RDR decision base if correct. Otherwise, a new SP and a new rule are created, as shown in Fig. 1.

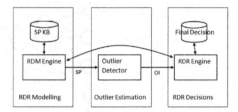

Fig. 1. The main components of Ripple Down Models.

The SP passed from the RDM Modeling component is analysed by an OD algorithm to detect outliers within Profiles. Each profile is screened by one of two algorithms and a binary indication of whether the case's attribute value for that profile is an outlier is returned. If a profile models a categorical attribute, the Outlier Estimation for Categorical Attributes (OECA) algorithm is used to detect if the case's value for the attribute is an outlier. For numerical profiles, the Outlier Estimation with Backward Adaptability (OEBA) method is applied and a binary outlier index is returned. For each SP of n profiles, the Outlier Detector (OEBA and OECA) returns a set of n binary flags indicating the status of each of the case's attributes. Each algorithm estimates whether the current case is an outlier by comparing its attributes against the previously observed values for each attribute for this SP. A possible limitation of both algorithms is that they assume that all attributes are independent.

After the OD methods process the SP, a set of binary outlier indexes representing each profile is passed to the RDR KB for a final classification of the case. The re-classification of a case is done because OEBA and OECA only assess each attribute in isolation. Some individual attributes may be more important than others and this is why an additional RDR inferencing will give an expert a chance to justify a conclusion based on particular attributes. The second RDR KB also allows the expert to confirm or correct OEBA/OECA results, hence reducing false positives.

2.3   Rated Multiple Classification Ripple Down Rules (Rated MCRDR)

The idea behind structural-based Rated MCRDR is that additional context could be deduced from a pattern of firing terminating rules in an MCRDR structure. Dazeley and Kang [2] adds that finding this context between the structure's individual paths can enhance understanding of the KB's domain. A hybrid system known as Rated MCRDR can determine these correlations and their usefulness in learning. Rated MCRDR's hybrid architecture combines MCRDR outputs with an Artificial Neural Network (ANN) to maximize the learning ability of MCRDR with the generalization ability of the ANN. A schematic of Rated MCRDR is illustrated in Fig. 2.

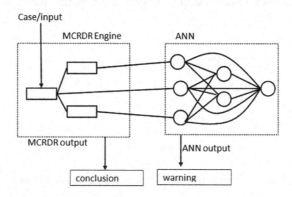

Fig. 2. Illustration of Rated MCRDR's main components.

After processing by the MCRDR engine, an indexing mechanism converts the MCRDR outputs into a set of binary inputs for the ANN. The ANN is used as a secondary classifier where MCRDR outputs are indexed and passed to the ANN and the output is used to substantiate the MCRDR conclusion. This is where the prudence capability of Rated MCRDR is derived, using a secondary classifier (ANN) to provide additional context and support the validation of each MCRDR conclusion.

Each rule path is associated to an ANN input neuron. The input neurons are switched on for every firing rule path. The ANN's size is determined by the indexed rule paths. For example, for a dataset with k rule paths, the input size will be up to $k + 1$, where k is the number of indexed paths [2]. Since the ANN inputs are directly connected to the indexed MCRDR conclusions, when a new conclusion fires in the MCRDR structure, a matching input will have to be introduced to the ANN. When a new input node is added, new hidden neurons may also be added. To conserve currently learned content, the ANN employs shortcut connections from a new input neuron directly to each output node. The new connections adjust weights needed to produce a particular output. The network is capable of adjusting when a new input is added to the network while still retaining previous information.

3 IPA

Combining Rated MCRDR and RDMs leverages strengths of the two systems and eliminate some individual vulnerabilities, and take advantage of the supplementary rule path context extraction of Rated MCRDR and partition based outlier detection methods of RDMs. IPA is a novel prudence method and consists of an MCRDR engine, an ANN (from Rated MCRDR) and an OD unit (from RDMs). The ANN is meant to extract context from paths of fired MCRDR rules while OEBA and OECA screens attribute inconsistencies to detect outliers. Figure 3 illustrates the IPA system.

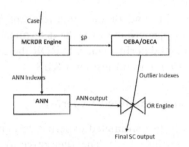

Fig. 3. The IPA system.

The OR Engine combines Rated MCRDR's ANN output with the RDMs' aggregated outlier index through a logical OR operation. The method is a generic connection of Rated MCRDR and RDMs to an MCRDR engine with the MCRDR engine serving both the OEBA/OECA outlier detectors and the ANN. The indexing of rule paths (for the ANN) and the creation of SPs (for the outlier detectors) is generated from a single MCRDR engine. Each time a new MCRDR rule is added, a new SP is created and a new ANN index is generated. The output of the complementary classifier is the OR result of the ANN index and the OEBA/OECA aggregated outlier index.

One of the contributions of this research is the development of a multiple classifications implementation of RDMs. The original version of RDMs [1] used a Single Classification RDR engine. Redeveloping a Multiple Classification version of RDMs was done for a number of reasons. First, it was developed primarily to enable comparisons and evaluations against Rated MCRDR, which is a Multiple Classifications system. The conversion to MC-RDMs then allowed for comparisons between two conceptually different prudence methods using the same MCRDR KB. Another reason for having MC-RDMs was informed by Richards' [4] observation that even for single classification domains, MCRDR produced a more compact KB with fewer redundancies than single class RDR. Finally, the RDM system to handle both single classification and multiple classification problems.

The only difference between the single classification and multiple classification implementations of RDMs is the use of OEBA/OECA as a complementary classifier to MCRDR in a similar manner to the ANN in the Rated MCRDR system. The Outlier Detection component (OEBA/OECA) in the original RDMs are not used as complementary classifiers but serve a more cautionary task, which the secondary RDR engine can choose to consider or ignore. In this project, a set of binary indexes for each SP were summed and if the aggregated outlier index was above some threshold, the case was classified as an anomaly. The outlier detection methods (OECA/OEBA) were used as the ANN is used in Rated MCRDR, where the systems effectively produce two classifications; the MCRDR classification and the complementary classifier (ANN or OEBA/OECA) classification.

4   Evaluating IPA

There exists a range of specific and organisation-relevant purposes for evaluating a KBS but usually the greater objective involves the determination of the system's actual performance against the specified performance. In RDR the common approach is to use a simulated expert in both training and testing a KBS.

4.1   RDR Evaluation Using Simulated Expertise

RDR evaluations consistently use simulated expertise to enable faster, easily controllable repeatable and cheap testing [1, 8]. The alternative to simulated expertise is the use of actual human experts, who are slower and costlier and less consistent. A simulated expert is a secondary KBS used as a source of expertise in assessing a knowledge acquisition tool. The knowledge acquisition method being trained/assessed acquires knowledge from the simulated expert.

When building a KBS, a simulated expert can be used as a source of expertise for the new KBS. With RDR, the new KBS will typically only have a default rule at the start and will incrementally add new rules (from the simulated expert) to match incoming data. The default rule will be returned initially when the KBS has no other rules and when none of the available rules match the current case.

When a case is first introduced to the new KBS, the default rule fires and returns a default conclusion. Meanwhile, the same case is fed to the simulated expert, which will return the correct conclusion. The new KBS's conclusion is then compared to the simulated expert's conclusion. If they match, then the new KBS is assumed to have the right rule for the case. If they do not match, the rule(s) fired by the simulated expert are added to the new KBS for the current case. The process is repeated until the new KBS matches the simulated expert or until the new KBS has learnt all the rules from the simulated expert. This will obviously depend on whether the training data covers all the rules in the simulated expert [10]. Note that the same training data-set is used to build both the simulated expert and the KBS, to ensure that the simulated expert has adequate knowledge of all of the data encountered by the KBS.

A common approach of the main prudence systems has been online evaluation where the prudent KBS is evaluated during its development. Algorithm 1 summarizes the process used for online evaluation. This process was used to evaluate the Rated MCRDR, RDMs and IPA systems, and has been used previously [1].

Algorithm 1.

    1.Accept a new case.

    2. Evaluate case against assessed KB.

    3. Evaluate case against simulated expert.

    4. If the case is correctly classified (If simulated expert conclusion matches KBS):

    5. Increment True Classifications (TC)

    6. If the case is not correctly classified (If simulated expert conclusion does not match KBS):

    7. Increment False Classifications (FC)

    8. Add rule (or rules) to new KB to correspond to new case

    9.Go to step 1 for new case

A simulated expert represents a complete version of the KB being built or a version faultless enough to be used as a benchmark against another KBS. In RDR research, most simulated experts were built by induction from a range of datasets using a learning algorithm including C4.5, InductRDR and were built from decision trees. An approach to specifying levels of expertise in a simulated expert is detailed in Compton and Cao [8]. Here, the highest level of expertise involves selecting the top four conditions from the intersection between the simulated expert rule traces and the difference list for the current case. The next level of expertise involves choosing a single condition from the intersection of the simulated expert rules and the case's difference list and the lowest level of expertise (the 'dumb' expert) selects all conditions from the case's difference list without reference to the simulated expert's rules. The use of varying levels of expertise in Compton, Preston and Kang's case was to determine the effect of smart and dumb experts on the size and accuracy of the KBS being built. Using varying levels of accuracy for simulated experts also models the fact that in real life, experts are humans who themselves are not always accurate.

4.2 Prudence Evaluation Metrics

Prudence systems can be evaluated across three main metrics: classification accuracy, also known as Simple Accuracy, Relative Accuracy and Prudence Accuracy. The classifier accuracy (or simple accuracy) determines the system's ability to correctly classify a case. Dazeley and Kang [2] refer to this metric as classification. The Simple Accuracy metric is based on two measures: True Classification (TC) and False Classification (FC), evaluated as follows:

- TC: Assigned to a case if the system correctly classified the case
- FC: Assigned to a case where the system failed to pick the right class

After the measures have been recorded, a system's Simple Accuracy is calculated as a proportion of the system's correct classifications (TC) on the whole dataset:

$$Acc = TC/(TC + FC)$$

The ultimate objective of evaluating most systems is to determine how precise they are in prediction. For learning systems such as Rated MCRDR, RDMs and IPA, it may also be worthwhile to assess the system not just in terms of correct predictions but also relative to its simulated expert.

The metric of Relative Accuracy determines a learning system's accuracy as a proportion of the expert's accuracy. Given the use of simulated experts of varying expertise, it is logical that the training system's ultimate accuracy will be influenced by its expert's competence level. The formula for Relative Accuracy is defined like so:

$$Relative\ Accuracy = S_{Acc}/SE_{Acc}$$

where S_{Acc} is the system's Simple Accuracy and SE_{Acc} is the simulated expert's accuracy.

Simple Accuracy (SA) has often been criticized for excluding class proportions and therefore not capturing the whole essence of the classifier's performance [11]. This is potentially problematic in domains where the data is well skewed with outlier cases comprising an average of less than 30% of the data. Consequently, it was suggested that accuracy should incorporate the classifier's rate of positive and negative detections to avoid unbalanced performance ratings on skewed datasets [11].

Balanced Accuracy or Prudence Accuracy incorporates the proportion of negative and positive classes in a dataset and avoids the problem of assigning classifiers exaggerated performance ratings on skewed data [8]. Balanced Accuracy is therefore essentially a sum of the average accuracies of the positive (Sensitivity) and negative (Specificity) classes. The equation below defines the formula for Balanced Accuracy. The issue with KBS evaluation methods is that there exists a range of specialised methods and each evaluation approach suits the KBS it was designed for (or tested on; [7]). This makes it quite a challenge to adapt a uniform approach for KBS, let alone customise one method for a different KBS. It would be desirable and convenient to combine some of the evaluation methods into a single evaluation approach that compares and ranks KBS based on a number of multiple relevant aspects. Evaluation techniques also differ according to domain. For example, false negatives may matter more in a domain where missing errors could be catastrophic. In other systems, some other metric may be worth more than others depending on the system's area of application. For this paper, Balanced Accuracy adopts a default weighting for the metrics but can obviously be varied according to the domain of the given prudence system.

$$Balanced\ Accuracy = 0.5(Se + Sp)$$

where $Se = \frac{TP}{TP + FN}$ and $Sp = \frac{TN}{TN + FP}$

Prudence is an example of a situation where an imbalance between classes is likely to occur, because we would expect cases requiring warnings to be far less common than cases which do not require warnings, particularly later in the incremental development of the RDR tree. Therefore we have used Balanced Accuracy in evaluating the effectiveness of the prudence systems, and in this context will refer to it as Prudence Accuracy. Here the confusion matrix incorporates a prudence system's warnings and whether they were issued at appropriate times. The relatively common True Positive (TP), True Negative (TN), False Positive (FP) and False Negative (FN) measures were used to evaluate individual Rated MCRDR and RDM predictions and are consistent with the measures which the original inventors of Rated MCRDR and Ripple Down Models used in their evaluations of the systems [1]:

- FP: assigned to a case if the system produced a warning incorrectly.
- FN: assigned if a warning was required but the system failed to do so.
- TP: assigned to a case if a warning was produced correctly.
- TN: assigned to a case if the system did not produce a warning when it was not supposed to.

The evaluation process used for each system in this paper is as follows for each randomised dataset:

An MCRDR classifier is built and evaluated using the simulated expert for each dataset. The classification accuracy and balanced accuracy are then measured using the process described in the algorithm for online evaluation. The classification accuracy serves as the base MCRDR accuracy to be compared against balanced accuracy when the system's prudence feature is enabled. In this way, it can then be established whether incorporating prudence has any negative impact on the systems' classification accuracy. The prudence accuracy and classifier accuracy of the system are measured as the system using the process defined in the algorithm given below:

Algorithm 2.
1. Accept a new case from randomised data.
2. Evaluate case against assessed KB.
3. Apply prudence
4. Evaluate case1 against simulated expert.
5. If the case is correctly classified (If simulated expert conclusion matches KBS)
6. If warning is issued, increment FP
7. Else increment TN
8. If the case is not correctly classified (If simulated expert conclusion does not match KBS):
9. If warning is issued, increment TP and Add rule (or rules) to new KB to correspond to new case
10. Else increment FN
11. Go to step 1 for new case.

After a system's performance measures have been collected, the prudence accuracy can then be calculated according to the Balanced Accuracy formulae defined earlier in this section. Alternatively, this can be also calculated from the pseudo-code above by adding the TNs and FPs. The two measures effectively represent a correct classification (TC).

For online evaluation, the system's accuracy would not be expected to be equal to the simulated expert's accuracy since the system initially starts with no knowledge at all and is incrementally built and evaluated. The system will therefore miss at least one example of each class type before a rule covering the class type is added to the MCRDR rule-base. For example, it was observed with the Iris dataset that given a perfect simulated expert, the system will misclassify one of each type of iris plant until the rule covering this particular class of cases is added to the KB. Consequently, even if the simulated expert is 100% accurate, the system will incorrectly classify at least one of each class just before the relevant knowledge is added.

5 Conclusion

The objective of this paper was twofold. First, the paper sought to present results from comprehensive evaluations of two known prudence methods: Rated MCRDR and RDMs. Secondly, the paper introduced a novel prudence type known as IPA, which is formed by combining the structural and attribute-based prudence methods. Consequently, a new prudence system IPA was developed from a merger of Rated MCRDR and RDMs. New metrics for evaluation of prudence analysis were also proposed. What remains now is to undertake a real-world evaluation of these different techniques, to evaluate whether IPA can improve accuracy, especially in areas of data mining that rely heavily on highly accurate, rule-based systems, including phishing [12], malware analysis [13], and fraud [7]. Contemporary prudence analysis techniques have been applied to fruitfully to identify compromised user accounts [14], and validation approaches require further examination [15], especially at runtime [16].

References

1. Prayote, F.: Knowledge based anomaly detection. University of New South Wales, Ph.D. thesis (2007)
2. Dazeley, R., Kang, B.: Rated MCRDR: finding non-linear relationships between classifications in MCRDR. In: 3rd International Conference on Hybrid Intelligent Systems, pp. 499–508. IOS Press, Melbourne (2003)
3. Edwards, G., Kang, B., Preston, P., Compton, P.: Prudent expert systems with credentials: managing the expertise of decision support systems. Int. J. Bio-Med. Comput. **40**, 125–132 (1995)
4. Richards, D.: Two decades of ripple down rules research. Knowl. Eng. Rev. **24**(2), 159–184 (2009)
5. Compton, P., Jansen, R.: Knowledge in context: a strategy for expert system maintenance. In: Barter, C.J., Brooks, M.J. (eds.) AI 1988. LNCS, vol. 406, pp. 292–306. Springer, Heidelberg (1990). doi:10.1007/3-540-52062-7_86
6. Kang, B., Compton, P., Preston, P.: Multiple classification ripple down rules: evaluation and possibilities. In: 9th Banff Knowledge Acquisition for Knowledge Based Systems Workshop, Banff, pp. 17–26 (1995)
7. Maruatona, O., Vamplew, P., Dazeley, R.: RM and RDM, a preliminary evaluation of two prudent RDR techniques. In: Richards, D., Kang, B.H. (eds.) PKAW 2012. LNCS, vol. 7457, pp. 188–194. Springer, Heidelberg (2012). doi:10.1007/978-3-642-32541-0_16

8. Compton, P., Cao, T.M.: Evaluation of incremental knowledge acquisition with simulated experts. In: Sattar, A., Kang, B.-h. (eds.) AI 2006. LNCS, vol. 4304, pp. 39–48. Springer, Heidelberg (2006). doi:10.1007/11941439_8

9. Prayote, A., Compton, P.: Detecting anomalies and intruders. In: Sattar, A., Kang, B.-h. (eds.) AI 2006. LNCS, vol. 4304, pp. 1084–1088. Springer, Heidelberg (2006). doi:10.1007/11941439_127

10. Metz, C.E.: Basic principles of ROC analysis. In: Seminars in Nuclear Medicine, pp. 283–298 (1978)

11. Phua, C., Lee, V., Smith, K., Gayler, R.: A comprehensive survey of data mining-based fraud detection research. Artif. Intell. Rev. (2005)

12. Layton, R., Watters, P.A., Dazeley, R.: Automatically determining phishing campaigns using the USCAP methodology. In: Proceedings of the 5th APWG E-crime Research Summit (2010)

13. Alazab, M., Venkatraman, S., Watters, P.A., Alazab, M.: Zero-day malware detection based on supervised learning algorithms of API call signatures. In: Proceedings of the 9th Australian Data Mining Conference (2011)

14. Amin, A., Anwar, S., Shah, B., Khattak, A.M.: Compromised user credentials detection using temporal features: a prudent based approach. In: Proceedings of the 9th International Conference on Computer and Automation Engineering, pp. 104–110 (2017)

15. Haq, I.U., Gondal, I., Vamplew, P., Layton, R.: Generating Synthetic Datasets for Experimental Validation of Fraud Detection (2016)

16. Finlayson, A., Compton, P.: Run-time validation of knowledge-based systems. In: Proceedings of the Seventh International Conference on Knowledge Capture, pp. 25–32 (2013)

A Bayesian Posterior Updating Algorithm in Reinforcement Learning

Fangzhou Xiong[1,2], Zhiyong Liu[1,2,3,5(✉)], Xu Yang[1], Biao Sun[4], Charles Chiu[6], and Hong Qiao[1,2,3,4,5]

[1] The State Key Lab of Management and Control for Complex Systems, Institute of Automation, Chinese Academy of Science, Beijing 100190, China
zhiyong.liu@ia.ac.cn
[2] School of Computer and Control, University of Chinese Academy of Sciences (UCAS), Beijing 100049, China
[3] CAS Centre for Excellence in Brain Science and Intelligence Technology (CEBSIT), Shanghai 200031, China
[4] University of Science and Technology Beijing, Beijing 100083, China
[5] Cloud Computing Center, Chinese Academy of Sciences, DongGuan 523808, Guangdong, China
[6] School for Higher and Professional Education, Chai Wan, Hong Kong, China

Abstract. Bayesian reinforcement learning (BRL) is an important approach to reinforcement learning (RL) that takes full advantage of methods from Bayesian inference to incorporate prior information into the learning process when the agent interacts directly with environment without depending on exemplary supervision or complete models of the environment. BRL tackles the problem by expressing prior information in a probabilistic distribution to quantify the uncertainty, and updates these distributions when the evidences are collected. However, the expected total discounted rewards cannot be obtained instantly to maintain these distributions after each transition the agent executes. In this paper, we propose a novel idea to adjust immediate rewards slightly in the process of Bayesian Q-learning updating by introducing a state pool technique which could improve total rewards that accrue over a period of time when this pool resets appropriately. We show experimentally on several fundamental BRL problems that the proposed method can perform substantial improvements over other traditional strategies.

Keywords: Bayesian reinforcement learning · Bayesian Q-learning · State pool technique

1 Introduction

As a rapidly growing branch in artificial intelligence, RL is a learning problem where an agent tries to behave optimally when interacting with the environment so as to finish a task through achieving its goal step by step [1,2]. One of the major challenges in RL is the trade-off between exploration of untested actions

© Springer International Publishing AG 2017
D. Liu et al. (Eds.): ICONIP 2017, Part V, LNCS 10638, pp. 418–426, 2017.
https://doi.org/10.1007/978-3-319-70139-4_42

and exploitation of actions that are known to be good. Fortunately, BRL offers a solution to address this exploration-exploitation problem by maintaining an explicit distribution over unknown parameters to quantify the uncertainty [3].

Instead of focusing on learning point estimation of the parameters in traditional RL, BRL tries to transfer prior information encoded relevant domain knowledge into a form of probabilistic distribution to represent unknown parameters. When the agent interacts with the environment, rewards are obtained to update these distributions. After that, according to the latest distribution the agent makes a decision by selecting an action to land up in next state [4].

There are several techniques to select an action. Undirected approaches (e.g. *epsilon-greedy exploration* and *Boltzmann exploration* [2]) usually choose random actions occasionally with no exploration-specific knowledge. Dearden et al. [5] propose Q-value sampling technique by extending to solve bandit problems to multi-state RL problems, and use a myopic value of perfect information (*VPI*) criterion to offer another policy for action selection. Wang et al. [6] present an efficient "sparse sampling" technique for Bayes optimal decision. Brafman et al. [7] propose a *R-MAX* algorithm to explore under the assumption that unknown states provide maximal rewards.

In this paper, we mainly concentrate on the problem of how to update the estimation of distributions over Q-values. Roughly speaking, the agent maintains these distributions over random variables with a tuple of hyperparameters. Each action is selected based on prior distributions which are updated by the received rewards. However, these rewards usually require to be expected and total, which are totally different from the practical situation where available rewards are only local and instantaneous after each action execution [5]. In order to tackle this problem, some sampling techniques are introduced, such as the Thompson Sampling algorithm [8] which suggests a natural Bayesian approach to sample a parameter from the posterior.

From the aspect of the immediate reward, the paper proposes a new idea to refine its effect with almost no change in numerical values, so that the agent could explore state-action space more deep under the same condition, which leads to more total discounted rewards. More specifically, we introduce a state pool which records the "known" and "unknown" states based on whether they have been visited. In addition, in order to incorporate more state information between different sequential episodes, we maintain the state pool by resetting it to a empty collection every few episodes.

2   Background

We consider a basic concept of Markov Decision Processes (MDPs) with infinite horizon represented by a 5-tuple $M = (\mathcal{S}, \mathcal{A}, \mathcal{P}, \mathcal{R}, \gamma)$, where $\mathcal{S}$ is a set of possible states, $\mathcal{A}$ is a set of possible actions, $\mathcal{P}$ is a state transition function that captures the probability of reaching the next state s' after we select action a at state s according to the policy π which denotes a mapping from state s to action a, $\mathcal{R}$ is a reward function that maps state-action pairs to a bounded subset of $\mathcal{R}$, and

$\gamma \in (0,1)$ is the discount factor specifying the effect of the current decision on the future rewards. The agent's goal in RL is to maximize the total discounted reward R by

$$R = \sum_{k=0}^{\infty} \gamma^k r_{t+k} \tag{1}$$

where r_t is the reward received at time t, and $\gamma \in (0,1)$ is the discount factor. The state-action value function $Q^\pi(s,a) = \mathbb{E}[R_t|s_t = s, a_t = a]$ is the expected reward for selecting action a in state s and following policy π.

In this work, we mainly focus on the Bayesian Q-learning (BQL). As for traditional Q-learning, it works by keeping running estimates that are updated at each step, i.e., when action a is executed in state s and transferred to next state s' with the immediate reward r, the Q-value would be updated following Q-learning updating rule:

$$\hat{Q}(s,a) \leftarrow (1-\alpha)\hat{Q}(s,a) + \alpha(r + \gamma \max_{a'} \hat{Q}(s',a')). \tag{2}$$

BQL utilizes probability distributions to represent the uncertainty over the estimated Q-value at each state. The agent executes actions based on these distributions, and receives rewards to update these priors, which could be considered as a process of keeping and propagating distributions over Q-values [9]. In order to denote the total discounted reward R_t more explicit, we formally let $R_{s,a}$ be a random variable when action a is selected in state s and an optimal policy π is adopted thereafter. Obviously, we want to learn the value $\mathbb{E}[R_{s,a}]$ to achieve expected rewards, i.e. optimal state-action function $Q^*(s,a) = \mathbb{E}[R_{s,a}]$. According to [5], $R_{s,a}$ can be assumed to have a normal distribution with mean $\mu_{s,a}$ and precision $\tau_{s,a}$.

For comparisons of subsequent experiments, we consider the same normal-gamma distribution as prior distribution $p(\mu_{s,a}, \tau_{s,a})$:

$$p(\mu_{s,a}, \tau_{s,a}) \propto \tau^{\frac{1}{2}} e^{-\frac{1}{2}\lambda\tau(\mu-\mu_0)^2} \tau^{\alpha-1} e^{\beta\tau} \tag{3}$$

which could be represented by $p(\mu, \tau) \sim NG(\mu_0, \lambda, \alpha, \beta)$ with a tuple of hyperparameters $\rho = \langle \mu_0, \lambda, \alpha, \beta \rangle$. Naturally, we only need to maintain these hyperparameters to represent and update the prior distributions of $R_{s,a}$.

The policy based on the normal-gamma distribution for action selection we will adopt is called myopic value of perfect information algorithm (*VPI*). The idea of this policy is to balance the expected rewards from exploration in the form of improved policies against the expected cost of performing a potential suboptimal action [5].

Once the policy is executed according to the estimated distributions, the immediate rewards will be collected to calculate the posterior probability density. Now we review a updating rule for calculating posterior distribution over $R_{s,a}$ called moment updating method (*Mom*) [5] which will be used in our algorithm. The *Mom* method randomly samples values $R_{t+1}^1, R_{t+1}^2, \ldots, R_{t+1}^n$ from the prior distribution, and assums these n samples contribute equally to solve two moments as:

$$M_1 = \mathbb{E}[r + \gamma R_{t+1}] = r + \gamma \mathbb{E}[R_{t+1}] \tag{4}$$

$$M_2 = \mathbb{E}[(r + \gamma R_{t+1})^2] \\ = r^2 + 2\gamma r \mathbb{E}[R_{t+1}] + \gamma^2 \mathbb{E}[R_{t+1}^2]. \tag{5}$$

According to the fact that the posterior distribution still is a normal-gamma distribution. We have $p(\mu, \tau | R^1, R^2, \ldots, R^n) \sim NG(\mu_0', \lambda', \alpha', \beta')$ where $\mu_0' = \frac{\lambda \mu_0 + n M_1}{\lambda + n}$, $\lambda' = \lambda + n$, $\alpha' = \alpha + \frac{1}{2}n$, and $\beta' = \beta + \frac{1}{2}n(M_2 - M_1^2) + \frac{n\lambda(M_1 - \mu_0)^2}{2(\lambda + n)}$.

3   The Proposed Method

Given the prior density $P(x)$ and the evidence D collected from the observations in the process of state transition, the posterior probability density $P(x|D)$ for model hyperparameters will be updated. Theoretically, we could apply the Bayes Theorem to solve the posterior distribution:

$$P(x|D) = \frac{p(D|x)P(x)}{p(D)}. \tag{6}$$

However, this updating is complicated by the fact that the available observations are local and immediate rewards, whereas the distribution over Q-value is a distribution over total discounted rewards. Hence we cannot use the Bayes Theorem directly.

In order to analyze the update for Q-value more conveniently, now we can rewrite the formula (1) as:

$$R_t = r + \gamma R_{t+1} \tag{7}$$

where R_t is a random variable denoting the total discounted reward from time t at state s. If the agent follows optimal policy with the best action a thereafter, then R_t is distributed as $R_{s,a}$. Since $Q^*(s, a) = \mathbb{E}[R_{s,a}] = \mu_{s,a}$, then the random variable $R_{s,a}$ can be utilized to update the posterior probability density. Now we employ $p(\mu, \tau)$ to generate samples randomly, which results in a moment updating for updating the estimate of the Q-value. Nevertheless, the best action a is not always performed, and there still enjoys a huge potential for improving effectiveness of updating. The existing updating rule only considers the rewards corresponded to the second term in the right part of formula (7), thus it is natural to take the first term into account, i.e., the immediate reward r.

We argue that the immediate reward should be relevant to the state that the agent resides. Therefore, there are two cases to be dealt with: (a) if the agent stays at next state s' that has not been visited before, the immediate reward r is encouraged to increase so that the posterior probability for executing an action a will increase, and (b) if s' has been stepped into previously, r should be reduced a little contrasted to the normal immediate reward so that the corresponding probability density for action a will be decreased. Both of cases will lead to more exploration in the learning process, which is crucial for episode problems in RL. In principle, the immediate reward should be determined by the observation after every state transition. Now we actually modify this value only to assist in updating the posterior probability which benefits for action selection. Nonetheless, the calculation method for the total discounted reward still adopts the formula (1).

Specifically, the paper introduces a state pool P to record the "known" and "unknown" states based on whether they have been visited before so as to modify the immediate reward. After each distribution updating has finished, the "unknown" state will be added into the state pool which leads to the updating for the state pool, and this state becomes the "known" thereafter. Thus we can rewrite the immediate reward:

$$ r_p = \begin{cases} r + \varepsilon & \text{if } s' \text{ not in } P \\ r - \varepsilon & \text{if } s' \text{ in } P \end{cases} $$

where r and ε denote the primitive reward and a small positive number, respectively.

Algorithm 1. Bayesian Q-learning with the state pool technique

---

Require: initial state $s \leftarrow s_0$, final state S_F,
   episode $e \leftarrow 0$, discounted factor γ,
   hyperparameters $\rho \leftarrow \langle \mu_0, \lambda_0, \alpha_0, \beta_0 \rangle$, policy π,
   state pool $P \leftarrow \varnothing$, interval K, small positive number ε.
Ensure: total reward R
1: **for** step $i = 0$ to N **do**
2: Execute action a at state s based on policy π.
3: Generate next state s' and immediate reward r.
4: Calculate new reward r_p:

$$ r_p = \begin{cases} r + \varepsilon \; \text{if } s' \text{ not in } P \\ r - \varepsilon \; \text{if } s' \text{ in } P \end{cases} $$

5: Update state pool P: $P = P \cup s'$
6: Maintain policy π by updating ρ with Mom
7: **if** $s' = S_F$ **then**
8:   Calculate $R_e = \sum \gamma r$, $s = s_0$, $e = e + 1$
9:   **if** $e \bmod K = 0$ **then**
10:    $P \leftarrow \varnothing$
11:   **end if**
12: **else**
13:   $s \leftarrow s'$
14: **end if**
15: **end for**
16: Calculate total reward $R = \sum_{j=0}^{e} R_j$.

Nonetheless, there is a problem that how long we should keep updating for the state pool. On the one hand, if we reset the state pool at one or more steps in one episode, there is no need for resetting since the executed action during one episode would not always behave optimal and it is necessary to bring in more episodes to support the posterior updating. On the other hand, if the state pool is reset at the end of total step, repetitive states will be visited continuously which possibly causes that the state pool can be increased to a full status only after several episodes, then subsequent updating for state pool has no significance to conduct. As a consequence of these two aspects, the paper proposes to maintain the state pool on every K episode by resetting it to a empty collection, which integrates state information between different sequential episodes and considers them as a unit with state pool technique to improve total discounted rewards. Thus, the proposed algorithm is summarized in Algorithm 1.

4 Experimental Illustration

4.1 Experiment Setting

There are 3 different episode experiments [10] conducted to evaluate the proposed algorithm.

Chain. Figure 1 presents the 5-state "Chain" problem that aims at achieving as many rewards as possible over fixed steps. Two available actions are labeled on the arcs followed by the immediate rewards. With probability 0.2, the agent slips and executes the opposite action. Once the final state 5 is visited, the agent starts at state 1 again.

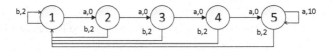

Fig. 1. The "Chain" problem.

Loop. Figure 2 illustrates the 9-state "Loop" problem with the purpose of maximizing total rewards. Similarly, the arcs are labeled with the actions and associated rewards. Performing action a will lead to the traversal of the right loop, and choosing action b repeatedly causes traversal of left loop. State 0 is set both for start state and final state. Hence once it is visited, the next state resets to itself again.

Maze. Figure 3 shows the 264-state "Maze" problem. The reward is measured by the number of flags collected before the agent reaches the goal. In this figure, S stands for the start state, G marks the goal, and F illustrates the location of flag. The agent can move up, down, left and right in the maze except staying at current state when it hits the wall or moves out of the maze. With probability 0.1, the agent slips and conducts an action that goes in a perpendicular direction. Once the agent arrives at the goal, it will immediately return to the start.

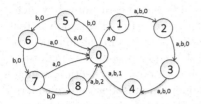

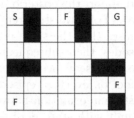

Fig. 2. The "Loop" problem.   **Fig. 3.** The "Maze" problem.

The algorithms we have utilized are as follows:

ϵ-**greedy.** Q-learning with epsilon-greedy exploration policy.
VPI+Mom. Bayesian Q-learning with VPI policy and moment updating.
VPI+SP. Bayesian Q-learning with VPI policy and state pool technique (SP).

4.2   Experimental Results and Discussion

The experiments are evaluated by accumulated total rewards obtained during the learning process which comprises 1000 steps for Chain and Loop, and 20000 steps for Maze. Table 1 represents correlative results by running 10 times for Chain, Loop, and Maze respectively.

Table 1. Results of accumulated rewards with average and standard deviation.

Chain	Avg	Dev	Loop	Avg	Dev	Maze	Avg	Dev.
ϵ-greedy	1288.0	46.3	ϵ-greedy	180.1	8.7	ϵ-greedy	540.4	105.3
VPI+Mom	1530.4	153.6	VPI+Mom	198.7	0.9	VPI+Mom	153.9	7.3
VPI+SP0[1]	1527.2	96.5	VPI+SP0[1]	272.2	70.0	VPI+SP0[1]	455.9	23.5
VPI+SP	1560.6	52.1	VPI+SP	359.2	18.5	VPI+SP	845.6	45.9

[1]This mark stands for using state pool technique without reset operation.

The first two experiments are designed to state the significance of existence for state pool technique so that the subsequent experiment can be performed reasonably to evaluate the improvements for total rewards received. Table 1 presents the relevant results.

To be specific, in the first experiment of Chain, the agent has doubtlessly traversed all states when it lands up in final state. There is no need to consider the exploration about whether the next state is an unknown state since all states will be labeled as "known" when it arrives at the goal. In addition, to reset the problem every few episodes is also in vain as the state pool actually is maintained to make full use of the information about unknown states between different sequential episodes. Therefore, it turns out that the agent achieves similar accumulated rewards, i.e., 1530.4 for MVPI+ Mom policy and 1560.6 with proposed technique.

In the second experiment of Loop, even if the start and the goal share the same state, there are commonly some unexplored states during one episode. Naturally, it leaves the space for improvement with state pool technique. As the Table 1 shows, the agent obtains the average reward of 359.2 in 10 runs when we set interval $K = 10$, which outperforms than ϵ-greedy policy and moment updating rule.

In the last experiment, a maze problem is designed to show that Bayesian Q-learning with proposed technique outperforms the traditional methods. In the maze, the agent tries to conduct sufficient exploration to carry 3 flags to the goal, which pushes it to make more attempts to visit unknown states. Once the agent has been stuck in a corner of the maze, more steps will be executed. Furthermore, the prior for each action shares the same distribution at the initialization phase, thus the agent treats them equally so that it will cost more steps when struggled with a corner, which tends to achieve less accumulated rewards when given the same total steps. Therefore, apart from the action, the state information has to be considered, i.e. the proposed state pool technique. Obviously, the results in Table 1 testify the effectiveness of the proposed algorithm.

Moreover, in the case of Loop and Maze, if we adopt state pool method without reset operation, the total discounted reward will be largely affected. These results manifest that it is crucial to integrate the state information between sequential episodes by state pool technique with a specific form, such as reset operation, which can result in more total discounted rewards.

5   Conclusions

In this paper, the proposed state pool technique enables the agent to obtain more rewards in Bayesian Q-learning domain. It has been evaluated over 3 experiments that the agent with the proposed technique has a potential ability to update the posterior distribution by model hyperparameters towards high rewards so that it will achieve substantial improvements in accumulated rewards. In the future work, we will pay more attention to bridge the relationship between the action selections and state information, and hope to make some improvements for the action initializations by borrowing ideas from previous episodes, thus the agent will not easily visit useless states when performs some explorations, which ultimately leads to more accumulated rewards.

Acknowledgments. This work is partly supported by NSFC grants 61375005, U1613213, 61210009, MOST grants 2015BAK35B00, 2015BAK35B01, Guangdong Science and Technology Department grant 2016B090910001.

References

1. Kaelbling, L.P., Littman, M.L., Moore, A.W.: Reinforcement learning: a survey. J. Artif. Intell. Res. **4**, 237–285 (1996)
2. Sutton, R.S., Barto, A.G.: Reinforcement Learning: An Introduction. MIT press, Cambridge (1998)
3. Ghavamzadeh, M., Mannor, S., Pineau, J., Tamar, A.: Bayesian reinforcement learning: a survey. Found. Trends? Mach. Learn. **8**(5–6), 359–483 (2015)
4. Vlassis, N., Ghavamzadeh, M., Mannor, S., Poupart, P.: Bayesian reinforcement learning. Reinforcement Learning **12**, 359–386 (2012)
5. Dearden, R., Friedman, N., Russell, S.: Bayesian Q-learning. In: The Association for the Advancement of Artificial Intelligence, pp. 761–768 (1998)
6. Wang, T., Lizotte, D., Bowling, M., Schuurmans, D.: Bayesian sparse sampling for on-line reward optimization. In: Proceedings of the 22nd international conference on Machine learning, pp. 956–963 (2005)
7. Brafman, R.I., Tennenholtz, M.: R-max-a general polynomial time algorithm for near-optimal reinforcement learning. J. Mach. Learn. Res. **3**(Oct), 213–231 (2002)
8. Chapelle, O., Li, L.: An empirical evaluation of Thompson sampling. In: Advances in neural information processing systems, pp. 2249–2257 (2011)
9. Strens, M.: A Bayesian framework for reinforcement learning. In: International Conference on Machine Learning, pp. 943–950 (2000)
10. Castronovo, M., Ernst, D., Couëtoux, A., Fonteneau, R.: Benchmarking for Bayesian reinforcement learning. PloS One **11**(6), e0157088 (2016)

Detecting Black IP Using for Classification and Analysis Through Source IP of Daily Darknet Traffic

Jinhak Park[1], Jangwon Choi[1], and Jungsuk Song[1,2($\boxtimes$)]

[1] Korea Institute of Science and Technology Information, Daejeon, Korea
[2] Korea University of Science and Technology, Daejeon, Korea
{painstars,jwchoi,song}@kisti.re.kr

Abstract. Recently, the community is recognizing to an importance of network vulnerability. Also, through the using this vulnerability, attackers can acquire the information of vulnerable users. Therefore, many researchers have been studying about a countermeasure of network vulnerabillty. In recent, the darknet is a received attention to research for detecting action of attackers. The means of darknet are formed a set of unused IP addresses and no real systems of connect to the darknet. In this paper, we proposed an using darknet for the detecting black IPs. So, it was choosen to classification and analysis through source IP of daily darknet traffic. The proposed method prepared 8,192 destination IP addresses in darknet space and collected the darknet traffic during 1 months. It collected total 277,002,257 in 2016, August. An applied results of the proposed process were seen for an effectiveness of pre-detection for real attacks.

Keywords: Darknet · Network vulnerabillty · Detection of black IP

1 Introduction

The Internet is an important infra resource that it controls the economy and society of our country. Also, it is a providing convenience and an efficiency of the everyday life. But, the Internet for developed part of everyday life is threatened through an intelligent and an advanced various attacks. Therefore, a detection of black IP is important. The black IP is known for a result of malicious action of attacker. Because, it have an effectiveness of pre-detection for real attacks. So, many researchers have been studying about the countermeasure [1–7]. But, these research take disadvantage of high cost, detection time, difficulty of management, etc. Therefore, in this paper focus on the detection of black IP. Because, it take advantage of low cost, easily and early detection. So, the early detection time is important compared with a detection time of antivirus solutions. In this paper, a proposed method is an using source IPs in the collected darknet packets.

We propose for how to detect black IP using for classification and analysis through source IPs of daily darknet traffic. The main contribution of proposed

© Springer International Publishing AG 2017
D. Liu et al. (Eds.): ICONIP 2017, Part V, LNCS 10638, pp. 427–433, 2017.
https://doi.org/10.1007/978-3-319-70139-4_43

process is an early detection for sign of attacks. The proposed process prepared 8,192 destination IP addresses in darknet space and collected the darknet traffic during 2016, August. These collected the number of darknet packets 277,002,257 and unique source IPs 8,392,962. Case of an using to monthly duplicate IPs, black IPs were predicted 18 of total 34. It is percentage of 52.94%.

The rest of paper is organized as follows. In Sect. 2, we give a brief description of existing approaches related to the darknet system. In Sect. 3, we present the proposed process and the experimental results are given in Sect. 4. Finally, we explain conclusions for given advantages in Sect. 5.

2   Related Work

Recently, many researchers have been studying about the darknet. An analysis of darknet traffic consist of two methods to classification of traffic and pattern search through analyzed traffic. In this paper targets to classification of darknet traffic. The darknet has been used for studying and developing the countermeasures against malicious activities on the Internet. Among them, three popular systems were proposed in Refs. [1–3].

Many studies have been devoted to characterizing common anomalous events in the darknet. [8] is a combined method of IDS and darknet, [9] is a how to collect darknet traffic. Also, darknet have massive data. Therefore, recently statistical analysis is important [10], method of [11] classify the darknet traffic either normal or abnormal.

3   Proposed Process

3.1   All Kinds of Collected Darknet Packets

The common kinds of collected darknet packets are composed of scanning packets, real attack packets, misconfiguration packets, etc. Firstly, the scanning packets are equal to a pre-investigation for real attacks. Also, the network scanning is constituted for a majority of darknet data. The scanning activities are the result of reconnaissance and DRDoS activities. Attackers scan the Internet to identify vulnerabilities and running services with an intent to compromise them. Secondly, we could detect packets of real attack through known patterns. Therefore, trends and patterns for a some attack were identified. Finally, the misconfiguration packets are occurred by errors and data management in network communications. Users often can enter an composing wrong IP address and port information in a network environment. In practice, these errors were observed that deployed networks suffer from well known errors and faulty configuration. As a result, the scanning packets and the real attack packets are taken for a large percentage of darknet packets.

3.2   Proposed Method

We need to choose one of various darknet information due to need for choice and concentration. Actually, the darknet information is composed to such as source IP, source port, destination IP, destination port, payload, event time, etc. Consequently, this paper have chosen focusing on source IPs in the darknet packets. Therefore, we chose top 10 of source IPs of daily detected packets. Because, the more inflow packets by the detected source IPs is, the more high probability of malicious actions is. Also, this paper is initial phase to verify efectiveness of the proposed method. So, if we change a setup like a period and a number of source IPs, then it can be generated for another results. Figure 1 shows an overall proposed process.

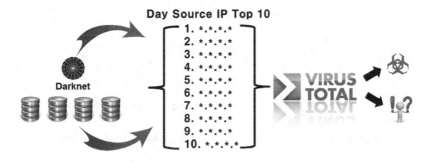

Fig. 1. An overall proposed process

The VirusTotal could provides verification results of various antivirus solutions. So, the chosen top 10 of source IPs are analyzed through VirusTotal site either malicious or normal IPs. And then, we make a comparison between a

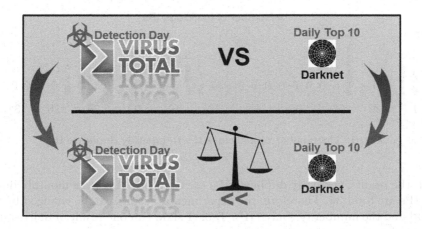

Fig. 2. A main idea of detecting black IPs

detection day by VirusTotal and a collected day of darknet packets. Finally, if a day of detected source IP was before a detection day by VirusTotal, then we can early know a malicious IP through darknet. Figure 2 shows a main idea of detecting black IPs.

4   Experimental Results

We show real experimental results of the proposed process. The proposed process prepared the 8,192 destination IP addresses in darknet space and the collected the darknet traffic in 2016, August. These collected the number of darknet packets 277,002,257 and unique source IPs 8,392,962.

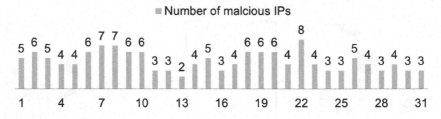

Fig. 3. The number of malicious IPs to daily duplicate IPs

An analyzed result by VirusTotal, the experimental results focus on a daily source IPs in August. Figure 3 shows the number of malicious IPs to daily duplicate IPs. Figure 4 shows the number of malicious IPs to monthly duplicate IPs. Previously mentioned, daily duplicate IPs are an independent IP of each day. Also, monthly duplicate IPs are an independent IP of each month.

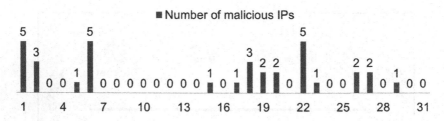

Fig. 4. The number of malicious IPs to monthly duplicate IPs

In the results, the daily duplicate IPs are total 142. Also, the monthly duplicate IPs are total 34. Therefore, the experimental results can be changed by the period and the number of source IPs. Table 1 shows an one result of daily source IPs in August.

Table 1. An one result of daily source IPs in August

2016. 08. 20				
Source IP	The first detection day	Malicious detection of antivirus solutions	Malicious URL	Malicious downloaded from this IP address
207.244.*.169	2016-06-10	1/67	1	0
61.216.*.14	2016-08-24	1/68	1	0
144.76.*.110	2013-11-06	22/47	over of 10	over of 10
61.240.*.65	2017-03-31	2/65	1	0
61.240.*.66	2017-03-28	2/65	1	0
61.183.8.138	2016-08-22	3/50	1	0

The VirusTotal results are detected malicious IPs through Latest detected URLs, Latest detected files that were downloaded from this IP address, Latest Detected Files that communicate with this IP address, etc. So, if many antivirus solutions could decide malicious IPs then we can get confidence results. In the result, daily duplicate IPs are predicted 111 of total 142. It is percentage of 72.17%. Monthly duplicate IPs are predicted 18 of total 34. It is percentage of 52.94%. Therefore, experimental results can be changed depend on the selection of duplicate IPs.

Table 2. A top 10 source IPs of 2016, August

2016, August				
rank	Source IP	Count	Detection to Virustotal	The first detection day
1	**61.240.*.65**	**1,231,980**	**O**	**2017-03-31**
2	113.108.*.31	1,066,500	X	-
3	**61.240.*.66**	**1,004,578**	**O**	**2017-03-31**
4	60.191.*.135	832,138	X	-
5	89.163.*.121	804,880	X	-
6	**125.64.*.200**	**640,734**	**O**	**2017-04-19**
7	209.58.*.109	609,060	X	-
8	207.244.*.169	608,743	O	2016-06-10
9	158.69.*.42	581,486	X	-
10	5.255.*.133	537,926	X	-

Additionally, if we have been analyze a lengthy period of time then we can get another results. For example, case of focus on monthly source IPs, Table 2 shows a top 10 source IPs in Aug, 2016. In the result, we could know in advance 3(rank 1, 3, 6) of 4 malicious IPs(rank 1, 3, 6, 8).

5 Conclusions

We proposed practical classification and analysis through using darknet information. Also, an using the real darknet data, it find for a detecting black IPs. But, an analysis of darknet packets is a difficulty handling about overfull data. So, it can get confidence results by top 10 source IPs thereby choice and concentration. On the basis of statistical analysis, we focus on the source IPs in the darknet packets. Also, through experimental results, if we have been analyze a lengthy period of time then we can get improved results. Applying to security control, malicious and suspicious IPs must have managed and duplicate IPs must have managed. Because, suspicious IPs will be detected malicious IPs. If condition of proposed process is changed such as period of time, day, week, month, etc. and top 10, top 50, top 100, etc. then results can be changed. Therefore, optimal conditions can be found through a various experiment. It need to a various combination of conditions. As a result, if applying to security control system, then it will be improve efficiency and accuracy.

References

1. Moore, D., Shannon, C., Voelker, G., Savage, S.: Network telescopes. Technical report, CAIDA (2004)
2. Yegneswaran, V., Barford, P., Plonka, D.: On the design and use of internet sinks for network abuse monitoring. In: Jonsson, E., Valdes, A., Almgren, M. (eds.) RAID 2004. LNCS, vol. 3224, pp. 146–165. Springer, Heidelberg (2004). doi:10. 1007/978-3-540-30143-1_8
3. Cooke, E., Bailey, M., Watson, D., Jahanian, F., Nazario, J.: The internet motion sensor-a distributed blackhole monitoring system. In: NDSS 2005, pp. 167–179 (2005)
4. Spitzner, L.: The Honeynet project: trapping the hackers. Mag. Secur. Priv. **99**, 15–23 (2003)
5. Abbasi, F.H., Harris, R.J.: Experiences with a generation III virtual Honeynet. In: Telecommunication Networks and Applications Conference 2009, pp. 1–6. IEEE Press (2009)
6. Kim, H.S., Choi, S.-S., Song, J.: A methodology for multipurpose DNS Sinkhole analyzing double bounce emails. In: Lee, M., Hirose, A., Hou, Z.-G., Kil, R.M. (eds.) ICONIP 2013. LNCS, vol. 8226, pp. 609–616. Springer, Heidelberg (2013). doi:10.1007/978-3-642-42054-2_76
7. Lee, H.-G., Choi, S.-S., Lee, Y.-S., Park, H.-S.: Enhanced Sinkhole system by improving post-processing mechanism. In: Kim, T., Lee, Y., Kang, B.-H., Ślęzak, D. (eds.) FGIT 2010. LNCS, vol. 6485, pp. 469–480. Springer, Heidelberg (2010). doi:10.1007/978-3-642-17569-5_46
8. Choi, S., Kim, S., Park, H.: A fusion framework of IDS alerts and darknet traffic for effective incident monitoring and response. Appl. Math. Inf. Sci. **11**, 417–422 (2017)
9. Song, J., Choi, J.-W., Choi, S.-S.: A malware collection and analysis framework based on darknet traffic. In: Huang, T., Zeng, Z., Li, C., Leung, C.S. (eds.) ICONIP 2012. LNCS, vol. 7664, pp. 624–631. Springer, Heidelberg (2012). doi:10.1007/ 978-3-642-34481-7_76

10. Choi, S., Song, J., Kim, S., Kim, S.: A model of analyzing cyber threats trend and tracing potential attackers based on darknet traffic. Secur. Commun. Netw. **7**, 1612–1621 (2013)
11. Ko, S., Kim, K., Lee, Y., Song, J.: A classification method of darknet traffic for advanced security monitoring and response. In: Loo, C.K., Yap, K.S., Wong, K.W., Beng Jin, A.T., Huang, K. (eds.) ICONIP 2014. LNCS, vol. 8836, pp. 357–364. Springer, Cham (2014). doi:10.1007/978-3-319-12643-2_44

A Linear Online Guided Policy Search Algorithm

Biao Sun[1,2], Fangzhou Xiong[2,3], Zhiyong Liu[2,3,4,5(✉)],
Xu Yang[2], and Hong Qiao[1,2,3,4,5]

[1] University of Science and Technology Beijing, Beijing 100083, China
[2] The State Key Lab of Management and Control for Complex Systems,
Institute of Automation, Chinese Academy of Science, Beijing 100190, China
zhiyong.liu@ia.ac.cn
[3] School of Computer and Control, University of Chinese Academy of Sciences
(UCAS), Beijing 100049, China
[4] CAS Centre for Excellence in Brain Science and Intelligence Technology (CEBSIT),
Shanghai 200031, China
[5] Cloud Computing Center, Chinese Academy of Sciences,
DongGuan 523808, Guandong, China

Abstract. In reinforcement learning (RL), the guided policy search
(GPS), a variant of policy search method, can encode the policy directly
as well as search for optimal solutions in the policy space. Even though
this algorithm is provided with asymptotic local convergence guarantees,
it can not work in a online way for conducting tasks in complex environments since it is trained with a batch manner which requires that
all of the training samples should be given at the same time. In this
paper, we propose an online version for GPS algorithm, which can learn
policies incrementally without complete knowledge of initial positions for
training. The experiments witness its efficacy on handling sequentially
arriving training samples in a peg insertion task.

Keywords: Reinforcement learning · Policy search · Online learning

1 Introduction

Reinforcement learning (RL) provides robotics a framework with a set of tools
for the design of sophisticated and hard-to-engineer behaviors to interact with
realistic world. It enables a robot to autonomously discover an optimal behavior
through trial-and-error interactions with its environment. As an important field
in reinforcement learning, policy search methods have been used in robotics for
a wide range of tasks, such as manipulation [1], grasping [2], and locomotion [3],
which scale RL into high dimensional continuous action spaces by using parame-
terized policies to avoid bootstrapping introduced by traditional value-function
approximation. However, direct policy search usually requiring numerous samples to find optimal policy which is impractical for robot learning [4].

Guided policy search (GPS) tackles the issue of sample efficiency by introducing trajectory optimization to guide the policy search away from poor local

© Springer International Publishing AG 2017
D. Liu et al. (Eds.): ICONIP 2017, Part V, LNCS 10638, pp. 434–441, 2017.
https://doi.org/10.1007/978-3-319-70139-4_44

optima [5]. This approach commonly uses trajectory-centric method to generate suitable samples at all training conditions to guide the learning process and train complex, high-dimensional policies [6–8]. Nevertheless, GPS method cannot cope with incremental data processing due to its framework.

The GPS algorithm would cause a problem that it cannot learn continuously based on previous policies when the environment changes. In order to learn a new condition, for example, a task of peg insertion with a new initial position in our experiment, the procedure for optimizing new condition should be added to all steps. As discussed in the next section, new condition's learning process will directly affect global policy optimization in the outer loop, which could obviously have an influence on other local policies indirectly through linear global policy π'_θ. These mutual impacts could be considered as the intrinsic characteristics of GPS from the view of initial conditions for training policies. Therefore, traditional GPS needs to learn all conditions from scratch, which is however hard-to-satisfied in some real applications. To alleviate this drawback, there is a great need to learn policies in an incremental way instead of the strict requirements with acquiring all initial conditions together.

The issue of online learning in GPS with multiple initial conditions can be taken as a part of lifelong learning since its learning never ends as new condition appears continuously. Lifelong learning has been explored for reinforcement learning [9]. Recently, an efficient policy gradient method for lifelong learning has been proposed [10]. In this paper, we aim to optimize GPS with an online learning form.

The main contributions of this paper are twofold, with the first to propose a novel framework for online GPS, and the second to give an effective algorithm to implement the idea. The proposed algorithm can utilize incremental information to search in policy parameter space in the process of interacting with the environment, which makes robot able to adjust to the changed conditions especially in industrial environments. Section 2 gives a brief review on related works, and Sect. 3 proposes the detailed algorithm. Following some preliminary experimental illustrations in Sects. 4, and 5 concludes this paper.

2   Background and Related Works

Reinforcement learning seeks to find a policy π to control an agent in a stochastic environment to finish some specific tasks. Instead to maximize the total rewards, RL often solves a optimal policy by minimizing the total costs under the policy trajectory distribution, given by

$$J(\theta) = \sum_{t=1}^{T} \mathbb{E}_{\pi_\theta}[\ell(x_t, u_t)]. \tag{1}$$

where π_θ stands for the policy distribution and $\ell(x_t, u_t)$ is the cost in step t.

The principle of the GPS is to use a series of special controllers to optimize π_θ. Since the special controllers generate guiding samples that guide policy search to the regions of high rewards, the GPS can efficiently train a deep neural network with fewer samples than direct policy search [2]. The expected cost minimizations can be rewritten as the following constrained problem,

$$\min_{p,\pi_\theta} \mathbb{E}_p[\ell(\tau)] \; s.t. \; p(u_t|x_t) = \pi_\theta(u_t|x_t) \; \forall x_t, u_t, t. \tag{2}$$

This optimization can be decomposed into two loops: the inner loop to optimize local policies and the outer loop to minimize the distance between global policy and each local policy. The algorithm is summarized in Algorithm 1. The inner loop optimizes the local policies respectively as follows:

$$p_i \leftarrow \operatorname*{argmin}_{p_i} \mathbb{E}_{p_i(\tau)}[\sum_{t=1}^{T} \ell(x_t, u_t)] \; s.t. \; D_{KL}(p_i(\tau)||\pi_\theta'(\tau)) \le \epsilon. \tag{3}$$

This subproblem can be solve by using local RL methods such as iterative linear-Gaussian regulator (iLQG) or path integral method [11].

The outer loop optimizes global policy to mimic each local policy by minimizing the KL divergence between them, which is given by

$$\min_\theta \sum_{m=1}^{M} \sum_{t=1}^{T} \mathbb{E}_{p_m(x_t,m)}[\mathbb{D}_{KL}(\pi_\theta(u_t|x_{t,m})||p_m(u_t|x_{t,m}))]. \tag{4}$$

where the number of local policy is M and each trajectory has T steps.

Algorithm 1. GPS contains two loops
___
1: **for** optimizing iteration to make pegging successfully **do**
2:    **for** position $i \in \{0, ..., M\}$ **do**
3:        c-step:$p_i \leftarrow \operatorname{argmin}_{p_i} \mathbb{E}_{p_i(\tau)}[\sum_{t=1}^{T} \ell(x_t, u_t)]$
4:            $s.t. \; D_{KL}(p_i(\tau)||\pi_\theta'(\tau)) \le \epsilon$
5:    **end for**
6:    s-step:$\pi_\theta \leftarrow \operatorname{argmin}_\theta \sum_{t,i,j} \mathbb{D}_{KL}(\pi_{\theta(u_t|x_{t,i,j})}||p_i(u_t|x_{t,i,j}))$
7:            (via supervised learning)
8: **end for**

In general, the optimization of each local policy p_i depends on the last optimized global policy π_θ' as shown in the inner loop, which is a time-varying linear Gaussian function generated by the global policy π_θ. In addition, the optimization for global policy π_θ depends on the samples generated by all local polices as shown in the outer loop. The global policy will be guided to find optimum along with the local policies optimized to optimal values.

3   The Proposed Method

3.1   Linear Online Framework for Guided Policy Search

In this section, we propose an online framework for GPS. The basic idea is that learning at a new condition should separate the interaction effects of the global policy and local policies optimization, because that it would influence both the inner and outer loops. Thus the global policy should be asynchronously learned from each single local policy. Specifically, in the inner loop we optimize a single local policy directly as follows:

$$p \leftarrow \operatorname*{argmin}_{p} \mathbb{E}_{p_\tau} \sum_{t=1}^{T} \ell(x_t, u_t). \tag{5}$$

And in the outer loop, we optimize the global policy with samples generated by each local polices. In order to learn continuously, the global policy should keep remembering the previously learned policy. Therefore, the previously learned policy should be considered to join into optimization.

The framework works like Fig. 1. The global policy learns by combining the policies learned at current condition, we denote this policy as p_{cur} and the previously learned policy as p_{pre}. The previous policies would have an influence in the learning process of the global policy all the time. Hence, the global policy could always remember previous policies regardless of how many new conditions it learns continuously.

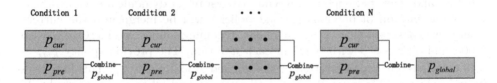

Fig. 1. The framework of online GPS.

However, the main problem is that it is hard to directly combine the global policy and the local policies, because the form of global policy would be enough complicated to represent numerous different local policies, such as neural network with multiple layers. In general, the local policy would be designed like a linear form as simple as possible to complete tasks easily and quickly.

3.2 Online Linear Guided Policy Search

In this subsection, a linear online guided policy search (LOLGPS) is proposed based on the online GPS framework shown in the Fig. 1.

Algorithm 2 summarizes our method. The inner loop, which represents learning to complete task at one condition, is the optimization for the local policies. The outer loop, which represents learning at the conditions one by one, is the optimization of the global policy. In the inner loop, we construct sample dataset D_i by running acting policy, notes as p_{act}. It is a joint controller combining the previous learnt policy into current condition, and the form is given as follows,

$$p_{act} = \alpha p_{pre} + \beta p_{cur}, \tag{6}$$

Algorithm 2. Linear online guided policy search (LOLGPS)

1: Initialize: $p_{pre} \leftarrow 0$
2: (outer loop)
3: **for** position $m = 0$ to M **do**
4:     $p_{cur} \leftarrow$ *arbitrary policy*
5:     (inner loop)
6:     **repeat**
7:         **if** first iteration **then**
8:             Generate samples D_i by running controller $p_{act} = p_{cur}$
9:         **else**
10:             Generate samples D_i by running jointly controller $p_{act} = \alpha p_{pre} + \beta p_{cur}$
11:         **end if**
12:         Fit linear-Gaussian dynamics $p_i(x_{t+1}|x_t, u_t)$
13:         $p_{cur} \leftarrow \text{argmin}_p \, \mathbb{E}_{p(\tau)} \sum_{i=i}^{\tau} l(x_t, u_t) \quad s.t. \ D_{kl}(p_{act}||p_{pre})$
14:         $p_{pre} \leftarrow$ use GMM to fit previous π_θ
15:     **until** get enough successful samples for specific task as D_g
16:     $\pi_\theta \leftarrow \text{argmin}_\theta \, D_{kl}(\pi_\theta||p_{act})$ by using D_g
17: **end for**

---

where α and β are parameters that can be constant or dynamic and control how much information of previous learned policies will be merge into the current policy. we constrain that $\alpha + \beta = 1$. For dynamic manner, we increase α with 0.005 and decrease β with 0.005 in each iteration. The previous policy p_{pre} is a linear Gaussian representation of the learned global policy π'_θ. This loop uses iterative Linear-Quadratic Regulator (iLQR) method [8] to optimize gradually to complete task at the current condition. When using the samples affected by p_{pre}, the optimization of p_{cur} is constrained to be close to the previous policies. In this context, the equation 5 can be rewritten in step 10 as follows

$$p_{cur} \leftarrow \underset{p}{\text{argmin}} \, \mathbb{E}_{p(\tau)} \sum_{i=i}^{\tau} l(x_t, u_t) \quad s.t. \ D_{kl}(p_{cur}||p_{pre}) \leq \epsilon. \tag{7}$$

In the outer loop, we optimize global policy to mimic the local policies in current condition. To train the global policy, the samples would be collected only from the successful trajectories (line 12), as those samples that are not in successful trajectories tend to direct the global policy to a bad local optimum. Since samples are generated from the successful trajectories, the global policy would extract the representation of current policy whose form may be quite different from optimized local current policy. Using samples generated jointly by local policies and previous policy, the global policy optimization is constrained to be close to p_{act}. It is already known that $p_{act} = \alpha p_{pre} + \beta p_{cur}$ and the optimization of p_{cur} is constrained close to p_{pre}. Therefore, the global optimization also depends on p_{pre}. In this case, the step 13 can be rewritten as

$$\pi_\theta \leftarrow \underset{\theta}{\text{argmin}} \, D_{KL}(\pi_\theta||p_{cur}) + D_{KL}(\pi_\theta||p_{pre}). \tag{8}$$

Thus, the proposed method could remember historic policies while learning at new conditions continuously.

4   Experimental Illustration

4.1   Experiment Setting

A series of experiments are conducted to evaluate the proposed algorithm. The task involves a peg insertion manipulation which requires controlling a 7 DoF 3D arm to insert a tight-fitting peg into a hole. The environment simulator uses the same one in [8].

In order to generate sufficient trajectories with 100 steps, we execute the linear Gaussian controller 5 times at each initial position. For the proposed method, samples are generated from 9 optimized phases once the local policy is capable to finish pegging insertion (Fig. 2).

Fig. 2. Peg insertion.

Moreover, the policies are represented by a fully connected neural network with 5 layers and 42 units in each hidden layer.

The cost function is given by [2]

$$\ell(x_t, u_t) = \frac{1}{2} w_u \left\| u_t \right\|^2 + w_p \ell_{12}(p_{x_t} - p^*), \tag{9}$$

where u_t is the action of robot, p_{x_t} is the position of end effector for state x_t, and the norm $\ell_{12}(z)$ is calculated by $\frac{1}{2} \left\| z \right\|^2 + \sqrt{\gamma + z^2}$. This cost function consists of two parts, with the first one weighted by w_u to encourage the less acting energy, and the other weighted by w_p to encourage the peg to reach the hole precisely.

In addition to the cost-function values, the performances of the algorithms are also evaluated by the resulted distance between end-effector positions and target positions. In the following experiments, if the distance is smaller than a base line which is around 0.06, the task is considered to be successful.

4.2   Experimental Results and Discussion

We have carried out a set of experiments to evaluate the efficacy of the proposed method. The experiment is carried out to evaluate the performance of LOLGPS.

Experiment on LOLGPS. The proposed LOLGPS method is evaluated with different number of initial information in an incremental way. In the training phase, we first set 4 initial positions as illustrated by the blue points in Fig. 3, and randomly select two positions from the first and second training areas, respectively. In the testing phase, we randomly select 50 positions in each testing area to construct out testing sets.

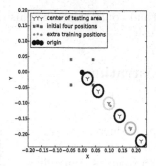

Fig. 3. Illustration of the training and testing positions, where all circles represent testing area. The green and red circles indicate the first and second extra training area, respectively.

Figure 4 illustrates the experimental results. Both the cost values and resulted distances are presented to evaluate the algorithm. The horizontal axis represents the distance between the origin and the center of each testing area. The vertical axis on the left shows the average distance between the bottom of the peg and the hole. The vertical axis on the right represents the total costs corresponding to the cost-function we used. Notes that distance_train0, distance_train1 and distance_train2 stand for the distance to target with different training areas, i.e., initial area, the first and second training area, respectively, while cost_train0, cost_train1 and cost_train2 denote corresponding cost-function values. First, it is observed that both of the resulted distance to target and the cost values increase along with the increment of testing distance. It implies that the performance becomes worse when the testing position deviates from the originally learned one. Second, it shows that once the agent has incrementally learned at new training positions, it can not only finish the tasks around the new positions, but also can still remember learned policies around the previously positions (see for instance distance_train2 and cost_train2 in Fig. 4). In other words, the proposed LOLGPS algorithm shows an ability of learning incrementally in its working environment.

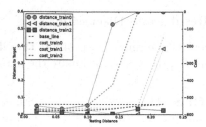

Fig. 4. Results for LOLGPS algorithm with different amounts training positions.

5 Conclusions and Future Works

In this paper, we have proposed a novel framework of GPS from the view of online learning, which enables the agent to learn policies continuously. Particularly, a LOLGPS method with linear representation has been evaluated to finish peg insertion task on simulated environment. It has been shown that the agent with proposed LOLGPS method has a potential ability of resistance to forget as given conditions appear sequentially along with interacting with environment.

Since we have separated the learning process of different condition, it is natural to think that whether the policy learning in current conditions can benefit from the learning results in other conditions. In the future work, we will pay more attention to analyze the relationship between conditions and hope to transfer the previously learned policies into current learning process so that can reduce the number of roll-out.

Acknowledgments. This work is partly supported by NSFC grants 61375005, U1613213, 61702516, 61210009, MOST grants 2015BAK35B00, 2015BAK35B01, Guangdong Science and Technology Department grant 2016B090910001.

References

1. Kalakrishnan, M., Righetti, L., Pastor, P., Schaal, S.: Learning force control policies for compliant robotic manipulation. In: Proceedings of the 29th International Conference on Machine Learning (2012)
2. Levine, S., Finn, C., Darrell, T., Abbeel, P.: End-to-end training of deep visuomotor policies. J. Mach. Learn. Res. **17**(39), 1–40 (2016)
3. Endo, G., Morimoto, J., Matsubara, T., Nakanishi, J., Cheng, G.: Learning CPG-based biped locomotion with a policy gradient method: application to a humanoid robot. Int. J. Robot. Res. **27**(2), 213–228 (2008)
4. Deisenroth, M.P., Neumann, G., Peters, J., et al.: A survey on policy search for robotics. Found. Trends Robot. **2**(1–2), 1–142 (2013)
5. Levine, S., Koltun, V.: Guided policy search. In: Proceedings of the 30th International Conference on Machine Learning, pp. 1–9 (2013)
6. Levine, S., Abbeel, P.: Learning neural network policies with guided policy search under unknown dynamics. In: Advances in Neural Information Processing Systems, pp. 1071–1079 (2014)
7. Levine, S., Koltun, V.: Variational policy search via trajectory optimization. In: Advances in Neural Information Processing Systems, pp. 207–215 (2013)
8. Montgomery, W.H., Levine, S.: Guided policy search via approximate mirror descent. In: Advances in Neural Information Processing Systems, pp. 4008–4016 (2016)
9. Sutton, R.S., Koop, A., Silver, D.: On the role of tracking in stationary environments. In: Proceedings of the 24th international conference on Machine learning, pp. 871–878 (2007)
10. Ruvolo, P., Eaton, E.: ELLA: An efficient lifelong learning algorithm. In: Proceedings of the 30th International Conference on Machine Learning, pp. 507–515 (2013)
11. Chebotar, Y., Kalakrishnan, M., Yahya, A., Li, A., Schaal, S., Levine, S.: Path integral guided policy search. In: International Conference on Robotics and Automation, pp. 3381–3388 (2017)

Detection of Botnet Activities Through the Lens of a Large-Scale Darknet

Tao Ban[1]([⊠]), Lei Zhu[2], Jumpei Shimamura[3], Shaoning Pang[2], Daisuke Inoue[1],
and Koji Nakao[1]

[1] National Institute of Information and Communications Technology,
Tokyo 184-8795, Japan
bantao@nict.go.jp
[2] Unitec Institute of Technology, Auckland 1025, New Zealand
[3] Clwit Inc., Tokyo 140-0001, Japan

Abstract. The growing cyber-threats from botnets compel us to devise proper countermeasures to detect infected hosts in an efficient and timely manner. In this paper, botnet-host identification is approached from a new perspective: by exploring the temporal coincidence in botnet activities visible in the darknet, botnet probing campaigns and botnet hosts can be detected with high accuracy and efficiency. The insights to botnet behavioral characteristics and automated detection results obtained from this study suggest a promising expedient for botnet take-down and host reputation management on the Internet.

Keywords: Botnet detection · Darknet analysis · Abrupt change detection · Pattern classification

1 Introduction

The vicious activities of a great variety of malware programs have led to serious incidents that damage to the Internet infrastructure and user's digital assets. This is particularly true for *botnets,* a network of compromised hosts controlled remotely by botmasters to carry out nefarious activities. As the most prominent cyber-threats to date, botnets are known to be the primary means through which distributed denial of service (DDoS) attacks, theft of personal data, click fraud, and spamming are committed, causing billions of dollars of annual damage [5,10]. Defending and response against botnets is a pressing need to protect the endangered communication systems and digital assets, and had attracted much research and operational attention in the past decade.

A botnet distinguishes itself from the other intrusion forms in two aspects [11,22]. First, it is goal-directed, e.g., gaining financial profits by providing spamming and DDoS-attack services to *botnet consumers.* Second, it works under a Command and Control (C&C) scheme: a *botmaster* (originator of the botnet) interacts with *bots* (infected and controlled machines) via C&C channels to command them to join the intrusion campaigns. A botnet is constructed and

© Springer International Publishing AG 2017
D. Liu et al. (Eds.): ICONIP 2017, Part V, LNCS 10638, pp. 442–451, 2017.
https://doi.org/10.1007/978-3-319-70139-4_45

managed in several stages such as probe and exploitation, C&C server rallying, self-update and synchronization, and a wide range of malicious activities [9].

The increased difficulty in detecting botnet-related traffic has motivated assorted studies aiming for efficient detection of botnet traffic and thereby locating bots and C&C servers in operational networks. Previous approaches can be categorized into four groups. The first group employs fine-grained analysis such as deep packet inspection (DPI) to detect bot-related communications [12,24]. Based on the fact that a majority of botnets utilize Internet Relay Chat (IRC) channels and show discriminating characteristics from normal IRC traffic, the second group consists of IRC-based detection methods [13,19,20]. The third group refers to DNS-related detection approaches that exploit the regularities of DNS queries to identify bots [4,7,8], where botnet detection is implemented by anomaly detection on the temporal similarity, intensity, and periodicity of hosts that queries the same domain. The last group includes extended methods which combine multiple approaches in the above for a more comprehensive study [1,24].

In this paper, we present a novel approach to detect botnet activities from a novel perspective. The analysis is done on the data collected from unused subspaces of the Internet, namely, a *darknet*, which provides a global overview of cyber-threats in the Internet with affordable monitoring costs. The propose approach focuses on strong temporal coincidence of probing activities as a result of the coordinated behavioral characteristics of bots. Implemented in a two-stage framework, it can not only detect the grouped activities of bots in a timely manner, but also identify individual hosts that play in the campaigns with high accuracy.

The remainder of the paper is organized as follows: Sect. 2 provides background information on darknet-related research. Section 3 presents the proposed approach for botnet-activity detection and botnet-host classification. Section 4 describes the experiment results. Section 5 draws the conclusion.

2   Related Work on Darknet Monitoring

A darknet, *a.k.a.* network telescope, blackhole monitors, sinkholes, or background radiation monitors, is a portion of routed, allocated IP space that contains no advertised services [2,17]. Because of the absence of legitimate hosts connected to a darknet, any network traffic observed on it is aberrant: it is the result of either a malicious probing activity or a mis-configuration. Assorted works have employed darknets to help identify the types and sources of malicious traffics on the larger network of which they form a part. In related works presented in [3,14], darknets are typically used to host flow collectors, backscatter detectors, packet sniffers, with worthwhile improvement in detection rates and cut-down in false positive rates reported.

To realize early security warning and mitigation of cyber-threats, we have been developing and operating the NICTER (Network Incident analysis Center for Tactical Emergency Response) project [16,17,21] for more than a decade.

By means of monitoring a distributed global-scale darknet and static-dynamic analysis of a large corpus of malware programs, NICTER binds the results of both macroscopic and microscopic analysis to obtain richer knowledge about cyber-threats in the Internet and deploy it to protect user networks. The macroscopic component (MacS) focuses on capturing the trend of malicious activities observed at darknet sensors. The microscopic component (MicS) makes use of honeypots and email-traps to capture malware programs in the wild.

In [17], Inoue et al. present the primary analysis engines including Change Point Detector (CPD), Self-Organizing Map (SOM) analyzer, and Incident Forecast (IF) engine. To detect a rapid change in monitored traffic in a timely manner, CPD implements a time series analysis engine that uses two-stage on-line discounting learning based on the Auto-Regression (AR) model. The SOM analyzer is a clustering and visualization engine designed for classifying as well as detecting unknown malware variants by means of characterization of their network behaviors. The IF is a forecasting engine for predicting the amount of traffic for future incidents several hours ahead so that prompt reactions can be enabled for the coming incidents. Refer to [17] for more information on analysis engines involved in NICTER.

3   Detection of Temporally Coordinated Botnet Probes

To C&C a network of compromised hosts to perform intrusive activities, a botmaster usually controls the botnet through a group of C&C servers [15]. As a botmaster generally controls a large amount of bots at the same time, the bots often tend to response to the command simultaneously or with a very similar latency [6]. If the darknet happens to capture the probe, then multiple hosts with similar behaviors will show up in a coordinated fashion. This could be a reasonable explanation for the increased observation of coordinated probing activities on the darknet. Investigation to these coordinated activities may reveal the root cause of the probes and enable population estimation of botnets.

3.1   Active Epoch Detection Using Improved CUSUM

The number of distinct hosts observed on the darknet provides a probabilistic estimate of the number of probing hosts in the Internet. In the time series of number of unique probing hosts observed within a unit time period, a coordinated probing event appears as a abrupt change that rises and drops quickly, e.g., the spike in Fig. 1a. We make use of an improved Cumulated Sum (CUSUM) algorithm [18] to detect the abrupt change points on a number-of-Source-IP time series (hereafter referred as a SIP time series) for detecting the abrupt changes associated with coordinated probes.

Given a SIP time series, $X_t = \{x_1, x_2, ...\}$, the Active Epoch (AE) detection algorithm scans X_t from the beginning to the end to detect the abrupt change points associated with the starting and ending points of probe campaigns. During the process, a status variable y_t is assigned as 1 if t is within an AE or 0 if at a

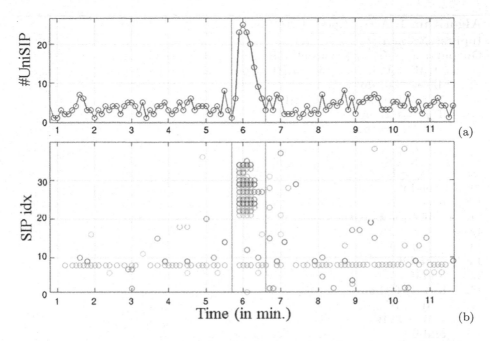

Fig. 1. Active epoch detection results. (a) A time series of number of unique hosts. Period between two red lines indicate the detected active epoch. (b) Host activities during the observation. Each circle stands for a packet observed from a host. (Color figure online)

"normal" level. A sliding window with length L is employed to track the normal level of X_t. To incorporate it with the most recent information, we define a sliding window

$$\boldsymbol{W}_t = \{x_i|i = t_1, t_2, ...t_L, i < t, i = \arg\max_i\}, \tag{1}$$

to be the rightmost L points before time t that are labeled as "normal". Let the mean of all $x_i \in \boldsymbol{W}_t$ be W_t, which provides an estimation of normal level at time t. To apply CUSUM to detect the AEs, we define *alert level* Σ_t as an accumulated sum,

$$\Sigma_t = max(0, \Sigma_{t-1} + (X_t - \alpha W_t)), \tag{2}$$

where α is a parameter controlling the sensitivity of detection. Then the accumulated sum in (2) is compared with a threshold θ, to trigger the y_t status.

As we can see from (2), for a "normal" period when X_t fluctuates around the mean value, if $\alpha > 1$, Σ_i is less likely to accumulate, since $X_t - \alpha W_t < 0$ happens frequently. On the other hands, when the detector reaches an abrupt change point, where X_t grows much higher than αW_t, Σ_i will accumulate quickly. Once Σ_i exceeds θ, the detector triggers a start point of AE by setting $y_t = 1$ and stops updating W_t. This window-update mechanism guarantees that W_t always represents an expected level for normal status. For an AE window that

---

Algorithm 1. Active Epoch Detection

Inputs: $\boldsymbol{X}_t$, L, α, θ
Outputs: $Z = \{(s_j, e_i)\}$: pairs of starting and ending points of AEs
 1: $i = 1$, $\boldsymbol{W} = [0, 0, \ldots, 0]$, $W = E(\boldsymbol{W}_0)$, $y = 0$, $\boldsymbol{\Sigma}_0 = 0$, $j = 0$;
 2: **while** $x_i \in \boldsymbol{X}_t$ **do**
 3:     $\boldsymbol{\Sigma}_i = \max(0, \boldsymbol{\Sigma}_{i-1} + (t_i - \alpha W))$;
 4:     **if** $y == 0$ **then**
 5:         **if** $\boldsymbol{\Sigma}_i > \theta$ **then**
 6:             $j = j + 1$;
 7:             $s_j = i$;
 8:             $y = 1$;
 9:         **end if**
10:     **else**
11:         **if** $\boldsymbol{\Sigma}_i < \boldsymbol{\Sigma}_{i-1}$ **then**
12:             $e_j = i$;
13:             $y = 0$;
14:         **end if**
15:     **end if**
16:     **if** $y == 0$ **then**
17:         $\boldsymbol{W} = \boldsymbol{W} \setminus x_{t_1} \cup x_i$;
18:         $W = E(\boldsymbol{W})$;
19:     **end if**
20:     $i = i + 1$;
21: **end while**

---

satisfies $y_t = 1$, the detector uses information of normal status to search for the ending point of the AE. In conventional CUSUM, the termination of an AE is reached when Σ_t falls below θ, which generally leads to delayed detection of the ending point. To avoid such unnecessary latency, we determine ending point by checking the condition $\Sigma_t < \Sigma_{t-1}$, i.e., when Σ_t starts to fall below αW_t. Once the ending point of an AE is reached, the detector will set $y_t = 0$ so that W_t will be updated in the iteration.

The pseudo-code of Active Epoch detection is given in Algorithm 1.

3.2   Botnet Classification Using Temporal-Coincidence Features

A further look at the host activities in the AE is shown in Fig. 1b. In the graph, probing hosts are randomly ordered along the vertical axis with all packets sent by each host printed as small circles along the horizontal time axis. It is easy to observe the coincidence in the host activities within the AE. Nevertheless, hosts that are apparently not involved in the coordinated probe are also observed in the AE, e.g., the host marked as green at the bottom in the graph that is continuously sending out probing packets to the darknet.

We make use of a simple classifier to differentiate coordinated hosts from the rest. Two features that reflect the temporal coincidence of the host activities are defined as follows. The first feature is the rate of probing packets observed in

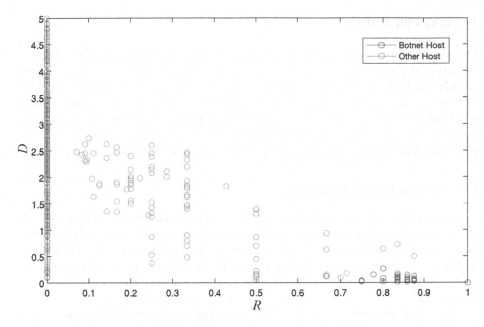

Fig. 2. Scatter plot of the hosts in the 2-D feature space. A red circle indicates a botnet host and a green one indicates host of the other type. (Color figure online)

the AE, namely,

$$R = N_{AE}/N, \tag{3}$$

where N is the number of probing packets observed in a longer time (11 times of AE in the experiment) embracing the AE. The second feature is the average deviation of all packets from the AE, normalized by the length of the AE, namely,

$$D = \mathrm{mean}(d_i)/|AE|, \tag{4}$$

where $|AE|$ denotes the length of the AE. The deviation of the ith packet from the AE, d_i, is computed as

$$d_i = \min(\mathrm{abs}(t_i - s_{AE}), \mathrm{abs}(t_i - e_{AE})). \tag{5}$$

Here, s_{AE} and e_{AE} are the starting and ending times of the AE.

An exemplary result of the feature extraction is shown in Fig. 2. In the figure, red circles denote coordinated botnet hosts and green ones denote the rest. As can be seen, all the bot-like hosts are located at the bottom-right corner, and most of remaining hosts are located close to the D axis. The separability between samples of different classes indicates discriminant information between bots and normal hosts are well captured by the two features. Using these two features, a binary-class classifier can be implemented to automatically pick out probing hosts that are most likely involved in a botnet campaign. In the experiment, we make use of a two-class Support Vector Machine [23] for classification.

4   Experiments

The AE detection and classifier are evaluated on manually labelled datasets collected at different destination ports of two class-B darknet sensors hosted by NICTER. As a preprocessing step, all the probing hosts observed on the darknet are grouped based on the network services (identified by the destination port) probed by the hosts. Then data are manually labelled by domain experts regarding the AEs and hosts involved in the botnet campaigns. The AE detection and host-classification algorithms are evaluated upon the labelled data.

4.1   Active-Epoch Detection

Filtering and justification of the detection results is done by removing insignificant events caused by noises and justifying the starting and ending of the events. Figure 1a shows an example of the detection results. The spike detected at the center of the time series is marked as an AE in the graph. The hosts that are performing probing activities within this span have a large chance to be involved in a botnet.

Table 1 shows the "optimal" event detection results obtained on each port of interest – ports without botnet probing activities observed are omitted from the table. The "optimality" is defined upon highest-recall-then-highest-precision criterion: In the trade off between false positives and false negatives, we prefer a lower false negative rate first as we do not want to miss any botnet events.

Table 1. Bot-event detection results

Port	Precision	Recall	F-measure	Lag
0	100.00%	100.00%	100.00%	1.00
22	100.00%	100.00%	100.00%	2.00
23	100.00%	81.82%	90.00%	10.42
25	87.50%	100.00%	93.33%	1.29
80	82.72%	100.00%	90.54%	2.54
139	93.33%	100.00%	96.55%	1.11
445	90.00%	100.00%	94.74%	1.89
1433	100.00%	100.00%	100.00%	1.00
3389	83.78%	96.88%	89.86%	2.84
5900	91.67%	100.00%	95.65%	1.27
8506	100.00%	100.00%	100.00%	1.50

As shown in the table, on most all destination ports (except 23 and 3389), the AE-detection algorithm obtained 100% recall rate. Meanwhile, the precision is kept at an reasonable level whereas false alarm rate is kept pretty low.

4.2 Botnet-Host Classification

To evaluate the performance of botnet-host classification, we conduct two groups of 10 times 10-fold cross-validation (CV) test. The first group is carried out on the truth events (manually labeled AEs) and the second group is done based on the results of the AE-detection algorithm. Tables 2 and 3 show the means and standard deviations of accuracy and geometric-mean (G-mean) obtained in two tests, respectively. Note that the datasets tend to be imbalanced on some destination ports, therefore G-mean is adopted and the classifiers are tuned for the best performance in terms of G-mean value. In the table, the results are compared with another SVM classifier which uses flow-based features defined in [2].

Table 2. Botnet-host classification results based on manual AE detection

Port	Accuracy		G-mean	
	Coordination	Behavior	Coordination	Behavior
0	100.00% ± 0.00	100.00% ± 0.00	100.00% ± 0.00	100.00% ± 0.00
22	99.91% ± 0.14	76.32% ± 1.92	99.91% ± 0.14	75.01% ± 2.03
23	98.52% ± 0.52	89.80% ± 1.23	94.49% ± 1.94	58.92% ± 3.98
25	98.90% ± 1.29	91.17% ± 3.22	99.05% ± 1.26	83.83% ± 5.87
80	98.51% ± 0.36	61.52% ± 1.32	98.36% ± 0.40	49.20% ± 1.61
139	99.93% ± 0.20	94.80% ± 1.73	99.64% ± 1.04	77.76% ± 7.15
445	98.67% ± 0.37	79.06% ± 1.45	98.91% ± 0.31	72.39% ± 1.79
1433	100.00% ± 0.00	97.05% ± 2.39	100.00% ± 0.00	96.19% ± 3.61
3389	90.65% ± 0.95	59.53% ± 1.54	90.29% ± 0.96	57.64% ± 1.58
5900	98.67% ± 0.49	74.00% ± 2.19	98.15% ± 0.36	48.44% ± 1.08
8506	99.46% ± 0.31	88.81% ± 1.34	99.69% ± 0.18	0.00% ± 0.00

Table 3. Botnet-host classification results based on automated AE detection

Port	Accuracy		G-mean	
	Coincidence	Behavior	Coincidence	Behavior
0	100.00% ± 0.00	100.00% ± 0.00	100.00% ± 0.00	100.00% ± 0.00
22	98.19% ± 0.61	76.38% ± 1.80	98.26% ± 0.59	74.70% ± 2.04
23	95.23% ± 0.83	90.35% ± 1.22	86.83% ± 2.57	59.00% ± 4.37
25	94.30% ± 2.99	88.21% ± 4.09	95.76% ± 2.21	82.83% ± 6.62
80	94.71% ± 0.62	60.12% ± 1.54	94.38% ± 0.67	49.54% ± 1.97
139	93.61% ± 1.75	93.63% ± 1.63	89.35% ± 2.85	91.40% ± 2.22
445	98.11% ± 0.49	78.75% ± 1.35	98.25% ± 0.46	72.16% ± 1.67
1433	98.90% ± 1.72	96.59% ± 2.88	99.16% ± 1.33	95.67% ± 4.27
3389	89.23% ± 0.95	59.18% ± 1.62	88.82% ± 1.01	57.58% ± 1.66
5900	91.18% ± 0.76	71.30% ± 2.26	89.03% ± 0.31	58.09% ± 0.70
8506	98.50% ± 0.42	94.02% ± 0.86	92.19% ± 3.03	0.00% ± 0.00

As can be read from the tables, the proposed coincidence-based approach shows much better result on the evaluation dataset. Accuracies and G-mean values higher or close to 90% in all cases indicate that the bots have fairly good separability from the rest in the 2D-feature space. For the behavior-based approach, comparable results are only obtained in several ports, namely, ports 0, 139, and 1433. This may be caused by the fact that the bots and other hosts share similar behavioral patterns and cannot be easily differentiated without coincidence-related information. Despite of a slightly degenerated performance based on the automated AE-detection results, the proposed approach shows the capability for a fully automated botnet detection scheme.

5 Conclusion

In summary, botnet activities result in high temporal coincidence observable in the darknet. We propose an abrupt-change-detection algorithm which can detect botnet-probe campaigns with a high detection rate. Based on the botnet-campaign detection result, temporal features which measure the coincidence between hosts prove to carry essential discriminant information to differentiate hosts that play in the coordinated botnet probes from the rest. In our future work, we plan to use the valuable information obtained from this study to facilitate botnet take-down and host reputation management in the Internet.

References

1. Abu Rajab, M., Zarfoss, J., Monrose, F., Terzis, A.: A multifaceted approach to understanding the botnet phenomenon. In: Proceedings of the 6th ACM SIG-COMM conference on Internet measurement (IMC 2006), pp. 41–52. ACM (2006). http://doi.acm.org/10.1145/1177080.1177086
2. Ban, T., Zhu, L., Shimamura, J., Pang, S., Inoue, D., Nakao, K.: Behavior analysis of long-term cyber attacks in the darknet. In: Huang, T., Zeng, Z., Li, C., Leung, C.S. (eds.) ICONIP 2012. LNCS, vol. 7667, pp. 620–628. Springer, Heidelberg (2012). doi:10.1007/978-3-642-34500-5_73
3. Benson, K., Dainotti, A., Claffy, K., Aben, E.: Gaining insight into as-level outages through analysis of internet background radiation. In: Proceedings of the 2012 ACM Conference on CoNEXT Student Workshop, pp. 63–64 (2012)
4. Bilge, L., Kirda, E., Kruegel, C., Balduzzi, M.: EXPOSURE: finding malicious domains using passive DNS analysis. In: 18th Annual Network and Distributed System Security Symposium, NDSS 2011, San Diego, CA, USA, 6–9 February 2011. http://www.eurecom.fr/publication/3281
5. Cho, C.Y., Domagoj, B., Shin, E.C.R., Song, D.: Inference and analysis of formal models of botnet command and control protocols. In: Computer and Communications Security (CCS 2010), pp. 426–439. ACM (2010)
6. Choi, H., Lee, H., Lee, H., Kim, H.: Botnet detection by monitoring group activities in DNS traffic. In: Proceedings of the 7th IEEE International Conference on Computer and Information Technology, pp. 715–720 (2007)
7. Choi, H., Lee, H.: Identifying botnets by capturing group activities in DNS traffic. Comput. Netw. 56(1), 20–33 (2012). http://dx.doi.org/10.1016/j.comnet.2011.07.018

8. Choi, H., Lee, H., Kim, H.: Botgad: detecting botnets by capturing group activities in network traffic. In: Proceedings of the Fourth International ICST Conference on COMmunication System softWAre and middlewaRE, COMSWARE 2009, pp. 2:1–2:8. ACM (2009). http://doi.acm.org/10.1145/1621890.1621893

9. Dagon, D., Gu, G., Lee, C.P.: A taxonomy of botnet structures. In: Lee, W., Wang, C., Dagon, D. (eds.) Botnet Detection. Advances in Information Security, vol. 36, pp. 143–164. Springer, Boston (2008). doi:10.1007/978-0-387-68768-1_8

10. Dainotti, A., King, A., Claffy, K., Papale, F., Pescapè, A.: Analysis of a "/0" stealth scan from a botnet. In: Internet Measurement Conference, IMC 2012, pp. 1–14. ACM (2012)

11. Friess, N., Aycock, J., Vogt, R.: Black market botnets. In: Proceedings of the MIT Spam Conference, pp. 1–8 (2010)

12. Gu, G., Porras, P., Yegneswaran, V., Fong, M., Lee, W.: Bothunter: detecting malware infection through ids-driven dialog correlation. In: USENIX Security Symposium, SS 2007, pp. 1–16. USENIX Association (2007)

13. Gu, G., Yegneswaran, V., Porras, P., Stoll, J., Lee, W.: Active botnet probing to identify obscure command and control channels. In: 2009 Annual Computer Security Applications Conference (ACSAC 2009), pp. 241–253 (2009)

14. Harder, U., Johnson, M.W., Bradley, J.T., Knottenbelt, W.J.: Observing internet worm and virus attacks with a small network telescope. Electr. Notes Theor. Comput. Sci. **151**(3), 47–59 (2006)

15. Hyslip, T., Pittman, J.: A survey of botnet detection techniques by command and control infrastructure. JDFSL **10**(1), 7–26 (2015)

16. Inoue, D., Eto, M., Yoshioka, K., Baba, S., Suzuki, K., Nakazato, J., Ohtaka, K., Nakao, K.: Nicter: an incident analysis system toward binding network monitoring with malware analysis. In: Proceedings of the 2008 WOMBAT Workshop on Information Security Threats Data Collection and Sharing, pp. 58–66 (2008)

17. Inoue, D., Yoshioka, K., Eto, M., Yamagata, M., Nishino, E., Takeuchi, J., Ohkouchi, K., Nakao, K.: An incident analysis system NICTER and its analysis engines based on data mining techniques. In: Köppen, M., Kasabov, N., Coghill, G. (eds.) ICONIP 2008. LNCS, vol. 5506, pp. 579–586. Springer, Heidelberg (2009). doi:10.1007/978-3-642-02490-0_71

18. Lai, T.L.: Sequential change-point detection in quality control and dynamical systems. J. R. Stat. Soc. Ser. B **57**(4), 613–658 (1995)

19. Mazzariello, C.: IRC traffic analysis for botnet detection. In: 2008 Fourth International Conference on Information Assurance and Security (ISIAS 2008), pp. 318–323 (2008)

20. Mizoguchi, S., Kugisaki, Y., Kasahara, Y., Hori, Y., Sakurai, K.: Implementation and evaluation of bot detection scheme based on data transmission intervals. In: 2010 6th IEEE Workshop on Secure Network Protocols (NPSec), pp. 73–78 (2010)

21. Nakao, K., Yoshioka, K., Inoue, D., Eto, M.: A novel concept of network incident analysis based on multi-layer ovservation of malware activities. In: Proceedings of The 2nd Joint Workshop on Information Security (JWIS07), pp. 267–279 (2007)

22. Puri, R.: Bots & botnet: an overview. http://www.sans.org/readingroom/whitepapers/malicious/1299.php

23. Vapnik, V.N.: The Nature of Statistical Learning Theory. Springer, New York (1995). doi:10.1007/978-1-4757-2440-0

24. Yen, T.-F., Reiter, M.K.: Traffic aggregation for malware detection. In: Zamboni, D. (ed.) DIMVA 2008. LNCS, vol. 5137, pp. 207–227. Springer, Heidelberg (2008). doi:10.1007/978-3-540-70542-0_11

Time Series Analysis

An Altered Kernel Transformation for Time Series Classification

Yangtao Xue, Li Zhang$^{(\boxtimes)}$, Zhiwei Tao, Bangjun Wang, and Fanzhang Li

School of Computer Science and Technology & Joint International Research
Laboratory of Machine Learning and Neuromorphic Computing, Soochow University,
Suzhou 215006, China
zhangliml@suda.edu.cn

Abstract. Motivated by the great efficiency of dynamic time warping
(DTW) for time series similarity measure, a Gaussian DTW (GDTW)
kernel has been developed for time series classification. This paper proposes an altered Gaussian DTW (AGDTW) kernel function, which takes
into consideration each of warping path between time series. Time series
can be mapped into a special kernel space where the homogeneous data
gather together and the heterogeneous data separate from each other.
Classification results on transformed time series combined with different
classifiers demonstrate that the AGDTW kernel is more powerful to represent and classify time series than the Gaussian radius basis function
(RBF) and GDTW kernels.

Keywords: Dynamic time warping · Gaussian dynamic time warping
kernel · Time series classification · Gaussian radius basis function kernel ·
Warping path

1 Introduction

A time series is a successive sequence of data points, each of which represents
a value in a uniform time interval [1]. During the last decade, time series are
being generated at an unparalleled speed from almost all application domains,
e.g., medical electrocardiogram [2], sign language recognition [3], stock time
series analysis and prediction [4], information study, atmospheric monitoring,
etc. Thus, time series classification and clustering have attracted extensive attention in data mining. Our work focuses on the classification of time series.

There are two key issues for time series classification: time series representation and classification [1]. Many techniques have been proposed for representing
time series, such as discrete Fourier transformation (DFT) [5], piecewise aggregate approximation (PAA) [6], piecewise linear approximation (PLA) [7], singular value decomposition (SVD) [5], piecewise vector quantized approximation
(PVQA) [8] and kernel transformation [1].

As for classifier, the most simple classifier is the nearest neighbor classification
(NNC), which has been widely used in almost all classification tasks, including

© Springer International Publishing AG 2017
D. Liu et al. (Eds.): ICONIP 2017, Part V, LNCS 10638, pp. 455–465, 2017.
https://doi.org/10.1007/978-3-319-70139-4_46

time series data mining [8]. Support vector machine (SVM) was pioneered by Cortes and Vapnik in 1995 [9]. Since SVM has an excellent performance for data with small scale, nonlinear and high dimension, SVM has been applied to different studies on classification including time series [10]. SVM can find an optimal balance between the accuracy and the complexity of model applying the structural risk minimization principle. Sparse representation based classifier (SRC) [11] has been developed in recent years, and can obtain promising consequence in many applications, such as face recognition [1] and image classification [10].

Different from other data mining areas, time series classification requires to consider time warping [1] when calculating similarity between time series. Although the Euclidean distance is widely used as similarity measure in many fields, it is sensitive to time warping and may not measure the similarity of time series well [13]. Berndt and Clifford [13] introduced dynamic time warping (DTW) to overcome the limitation of the Euclidean distance for time series. In addition, many improved measures have been proposed, such as longest common subsequence (LCSS) [14], edit distance with real penalty (ERP) [15], and time warp edit distance (TWED) [16]. However, DTW is still admitted as one of the best similarity measure of time series during two decades.

Motived by the success of DTW in time series, a Gaussian dynamic time warping (GDTW) kernel was constructed for SVM and kernel SRC [10], which can be taken as an invariant of the Gaussian radius basis function (GRBF). The GDTW kernel takes into consideration time warping, but it treats DTW only as a distance measure and ignores the path information. Moreover, the GDTW kernel is not a positive semi-definite kernel, which would undermine the performance of SVM when applying it.

To remedy it, this paper proposes an altered Gaussian DTW (AGDTW) kernel function based on the warping step information, which is the most important thing in DTW. Furthermore, we prove that the proposed kernel is a positive semi-definite kernel. In experiments, the performance of different kernel functions (GRBF, GDTW, and AGDTW) with different classifiers (NNC, SVM and SRC) on time series classification tasks is investigated.

2   Related Work

The Gaussian dynamic time warping kernel was proposed based on the Gaussian RBF kernel and motivated by the success of DTW measurement precision in time series. Before we give the definition of GDTW kernel, we first introduce the GRBF kernel so that the relationship between them can be seen more clearly. Given two time series $\mathbf{x}$ and $\mathbf{y}$, the Gaussian RBF kernel is defined as:

$$K_{rbf}(\mathbf{x}, \mathbf{y}) = \exp\left(-\frac{\|\mathbf{x} - \mathbf{y}\|^2}{\sigma^2}\right) \tag{1}$$

where σ is the kernel parameter, and $\|\mathbf{x} - \mathbf{y}\|$ denotes the Euclidean distance between $\mathbf{x}$ and $\mathbf{y}$. Note that $\|\mathbf{x} - \mathbf{y}\|$ only works when both $\mathbf{x}$ and $\mathbf{y}$ have the same length.

If the length of $\mathbf{x}$ is not equal to that of $\mathbf{y}$, then we need dynamic time warping (DTW). The GDTW kernel is constructed by introducing DTW into kernel mapping to replace the Euclidean distance in (1). Namely, the GDTW kernel has the form

$$K_{dtw}(\mathbf{x}, \mathbf{y}) = \exp\left(-\frac{d_{dtw}(\mathbf{x}, \mathbf{y})^2}{\sigma^2}\right) \tag{2}$$

where $d_{dtw}(\mathbf{x}, \mathbf{y})$ represents the DTW distance between $\mathbf{x}$ and $\mathbf{y}$. Here, time series $\mathbf{x}$ and $\mathbf{y}$ can have different length that benefits from DTW. The detail of DTW is given in Sect. 3.

The GDTW kernel has been imported in SVM but along with unstable classification performance. The main reason is that the GDTW kernel is not a positive semi-definite (PSD) kernel [17].

3   Altered Gaussian Dynamic Time Warping Kernel Time Series Classification

Although the GDTW kernel takes into consideration time warping by replacing the Euclidean distance with the DTW distance, it treats DTW only as a distance measure and ignores the path information which is significant for DTW. In order to improve the kernel transformation for time series, this paper proposes an altered Gaussian dynamic time warping (AGDTW) kernel transformation which uses the information of whole time warping path.

3.1   Dynamic Time Warping

Suppose there are two time series $\mathbf{x}$ and $\mathbf{y}$ with length m and n, respectively, where $\mathbf{x} = [x_1, x_2, \ldots, x_m]^T$ and $\mathbf{y} = [y_1, y_2, \ldots, y_n]^T$. In order to align two time series, DTW constructs a $m \times n$ cost matrix where the i-th row and the j-th column entry in the cost matrix represents the alignment between x_i to y_j, or the Euclidean distance between x_i and y_j. Let the two-tuple (i, j) be the index of the cost matrix.

Figure 1 illustrates the corresponding warping relationship. The warping path W is defined as the optimal mapping between $\mathbf{x}$ and $\mathbf{y}$, namely

$$W = \{w_1, w_2, \ldots, w_S\}, \quad max(m, n) \le S \le m + n \tag{3}$$

where the two-tuple $w_s = (i_s, j_s)$ denotes the i_s-th row and the j_s-th column entry of the cost matrix. Furthermore, the path W must satisfy two conditions, boundary condition and continuity condition.

Boundary condition: The warping path must start from $w_1 = (1, 1)$ and end at $w_S = (m, n)$.

Continuity condition: The adjacent path element must have continuity:

$$w_s = (i_s, j_s), \quad w_{s+1} = (i'_{s+1}, j'_{s+1}), \quad i \le i' \le i+1, \quad j \le j' \le j+1 \tag{4}$$

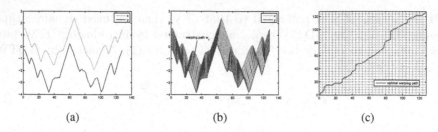

Fig. 1. The illustration of DTW. (a) Two original time series **x** and **y**; (b) the schematic diagram of DTW warping of **x** and **y**; (c) the cost matrix and the optimal warping path.

Therefore, the DTW distance is the optimal path between x and y. Namely,

$$Dist_{dtw}(\mathbf{x}, \mathbf{y}) = Dist_{dtw}(W) = \sum_{s=1}^{S} Dist_{eu}(x_{i_s}, y_{j_s}) \tag{5}$$

where $Dist_{eu}(x_{i_s}, y_{j_s})$ denotes the Euclidean distance between x_{i_s} and y_{j_s}.

The optimal path can be found by dynamic programming to minimize the cumulative distance $D(i, j)$, which is defined as

$$D(i,j) = Dist_{eu}(x_i, y_j) + min\{D(i-1,j), D(i,j-1), D(i-1,j-1)\} \tag{6}$$

where $D(0,0) = 0$, $D(0,j) = \infty$, $D(i,0) = \infty$, $0 \le i \le m$, $0 \le j \le n$.

Finally, we have the value of the DTW distance:

$$Dist_{dtw}(\mathbf{x}, \mathbf{y}) = D(m, n) \tag{7}$$

3.2 Altered Gaussian Dynamic Time Warping Kernel Transformation

From the GDTW kernel (2), we can see that the GDTW kernel directly uses the DTW distance instead of the Euclidean distance. Since the DTW distance is not a metric, which does not satisfy the triangle inequality, the GDTW kernel is not positive semi-definite. Here, we propose an altered GDTW (AGDTW) kernel, which adapts the warping path generated in the process of computing the DTW distance.

Given the warping path $W = \{w_1, w_2, \ldots, w_S\}$ for two time series **x** and **y**, where $w_s = (i_s, j_s)$, $\max(m, n) \le S \le m + n$, $0 \le i_s \le m$, $0 \le j_s \le n$, the AGDTW kernel is defined as:

$$K_{agdtw}(\mathbf{x}, \mathbf{y}) = \sum_{s=1}^{S} \exp\left(-\frac{Dist_{eu}(x_{i_s}, y_{j_s})^2}{\sigma^2}\right) \tag{8}$$

where $\sigma \neq 0$ is the kernel parameter.

We give a theorem to show the property of the AGDTW kernel in the following.

Theorem 1: The altered Gaussian dynamic time warping kernel (8) is a positive semi-definite kernel.

Proof: One of closure properties of kernels given in [18] is that if K_1 and K_2 are two positive semi-definite kernels, then $K_1 + K_2$ is positive semi-definite. Let $K_{i_s} = \exp\left(-\frac{Dist_{eu}(x_{i_s}, y_{j_s})^2}{\sigma^2}\right)$. Thus, the AGDTW kernel (8) can rewritten as

$$K_{agdtw}(\mathbf{x}, \mathbf{y}) = \sum_{s=1}^{S} K_{i_s} \tag{9}$$

Since K_{i_s} is a GRBF kernel with one-dimensional variable, the AGDTW kernel can be taken as the sum of many GRBF kernels. It is well known that the GRBF kernel is a positive semi-definite kernel. In other words, K_{i_s} is positive semi-definite. Thus, the AGDTW kernel is also positive semi-definite.

This completes the proof.

3.3   Classification

Given a test time series $\mathbf{y} \in \mathbb{R}^m$ and training time series dataset $X = \{(\mathbf{x}_i, v_i)\}_{i=1}^{N}$, where $\mathbf{x}_i \in \mathbb{R}^m$, $v_i \in \{1, 2, \ldots, C\}$ denotes the label of $\mathbf{x}_i$, m and N are the length and the number of the time series, respectively. We take the kernel function as the nonlinearly mapping function. Then the time series dataset can be transformed in a specified space. The transformed time series training dataset is defined as $\phi(X) = \{(\phi(\mathbf{x}_i), v_i)\}_{i=1}^{N}$, and the test time series $\mathbf{y}$ is defined as:

$$\phi(\mathbf{y}) = [K(\mathbf{x}_1, \mathbf{y}), K(\mathbf{x}_2, \mathbf{y}), \ldots, K(\mathbf{x}_N, \mathbf{y})]^T \tag{10}$$

where the kernel function K is the AGDWT kernel K_{agdtw}.

The existing classifiers can be used for the classification of transformed time series training and test dataset, such as NNC, SVM, and SRC. In the following, we briefly introduce NNC and SRC.

Let $\mathbf{X} = [\mathbf{X}_1, \mathbf{X}_2, \ldots, \mathbf{X}_C]$ be the transformed training sample matrix, where $\mathbf{X}_c = [\phi(\mathbf{x}_{c1}), \phi(\mathbf{x}_{c2}), \ldots, \phi(\mathbf{x}_{cN_c})]$ is the sample sub-matrix belonging to class c, and $\mathbf{x}_{cj}$ denotes the jth training sample of class c, N_c is the class number and C is the class number.

NNC defines the minimum distance of $\{d(\phi(\mathbf{y}), \phi(\mathbf{x}_{cj}))|j = 1, 2, \ldots, N_k\}$ as the distance of $\phi(\mathbf{y})$ to class c, namely $d(\phi(\mathbf{y}), \mathbf{X}_c)$. As a result, $\mathbf{y}$ can be classified as the category with the minimal $d(\phi(\mathbf{y}), \mathbf{X}_c)$. SRC first calculates the sparse coefficient α_c for each training sample subset $\mathbf{X}_c$, and the distance of $\phi(\mathbf{y})$ to the class c can be defined as $d(\phi(\mathbf{y}), \mathbf{X}_c) = \|\phi(\mathbf{y}) - \mathbf{X}_c \alpha_c\|_2$, where $\|.\|_2$ is the 2-norm.

4   Simulation Experiments

This section validates the transformation performance of the AGDTW kernel on time series. We compare it with the GRBF and GDTW kernels on the classification tasks for time series. All the time series datasets used in this experiment can be found in http://www.cs.ucr.edu/~eamonn/time_series_data/.

4.1   Kernel Transformation

In order to verify the performance of different kernel transformation, we make use of NNC which is one of the most common classification tools. In this experiment, we compare the kernel transform performance by NNC combined with different kernels, or GRBF-1NN, GDTW-1NN and AGDTW-1NN. The classification performance is measured by the classification error rate (CER) defined as

$$CER = \frac{\text{The number of wrong classified time series}}{\text{The number of time series in the test dataset}} \times 100\% \qquad (11)$$

As we known, the smaller the classification error is, the better the kernel transformation performance is.

SwedishLeaf Dataset. In the SwedishLeaf dataset, there are 500 training time series and 625 test time series. The length of each time series is 128. There are fifteen categories.

As shown in Fig. 2, ED-1NN and DTW-1NN are classified without any kernel mapping. DTW-1NN is better than ED-1NN, and has a CER of 20.8%. The traditional kernel transformation has reduced the classification error to 19.68% by GRBF-DTW. Our method can achieve the best performance, 11.68%. It is an amazing performance that AGDTW-1NN raises 8% classification performance. It mainly benefits from the import of warping path information so that the time series in the same class gather together, and those in the different classes separate from each other.

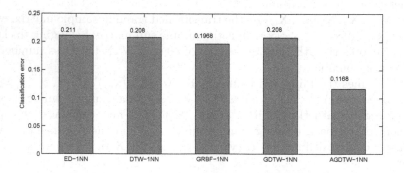

Fig. 2. 1NN classification performance on the Swedishleaf dataset

Table 1. The attributes of three time series dataset

Dataset	Class	Training	Test	Length
FaceFour	4	24	88	350
FISH	7	175	175	463
Gun-Point	2	50	150	150

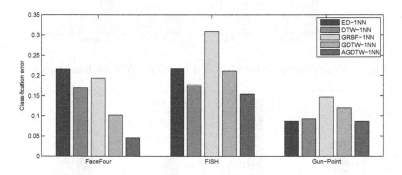

Fig. 3. Classification error obtained by using NN on three datasets

More Datasets. We validate the performance of the AGDTW kernel transformation performance on another three time series datasets. The more details of three datasets are shown in Table 1.

The details of transformation performance using different kernels based on NNC are shown in Table 2. The bold one in Table 2 denotes the best classification error. Figure 3 gives an intuitively contrast on different transformation combined with NNC. It is obvious that the AGDTW kernel transformation achieves remarkable progress. For instance, AGDTW-1NN acquires 4.55% CER on the FaceFour dataset, which indicates that the classification accuracy of AGDTW-1NN is twice as high as that of GDTW-1NN.

As for the FISH dataset, the AGDTW-1NN achieves 15.43% when the best performance of traditional method is 17.7% in DTW-1NN. The AGDTW-1NN also gets 8.67% when the best classification error of other methods is 8.7%.

4.2 Classification Performance

Besides the NNC framework, we also take SVM and kernel SRC as classification models for the sake of checking robust of our kernel.

SVM. As one of the state-of-the-art classifier, SVM achieves good results in many applications. Here, we mainly compare the classification performance of the GRBF, GDTW and AGDTW kernels using the SVM classifier, namely GRBF-SVM, GDTW-SVM and AGDTW-SVM. Figure 4 presents the result of classification.

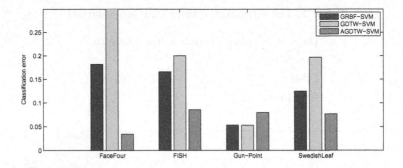

Fig. 4. Classification error obtained by using SVM on four datasets

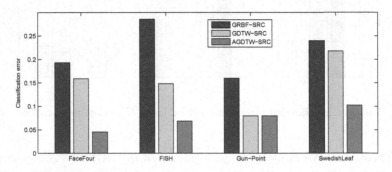

Fig. 5. Classification error obtained by using SRC on four datasets

As we can see from Fig. 4, AGTW-SVM is inferior to GRBF-SVM and GDTW-SVM only in Gun-Point dataset, but not too worse. AGTW-SVM obtains better performance on the FaceFour, FISH and SwedishLeaf datasets than the other two methods.

On the FaceFour dataset, AGDTW-SVM achieves 3.41% when GRBF-SVM obtains 18.18%. As for the FISH dataset, our method achieves 8.57% when GRBF-SVM obtains 16.57%. On the SwedishLeaf dataset, AGDTW-SVM achieves 7.68% when GRBF-SVM obtains 12.48%. It is clearly that GDTW-SVM cannot obtain a better performance compared with GRBF-SVM, partially because the GDTW kernel is not a positive semi-definite (PSD) kernel [17].

KSRC. Although little attention has been given to SRC for time series, we have to admit that SRC has attracted a good deal of attention for many applications. This experiment compares the classification result of different kernels with the kernel SRC framework, namely GRBF-KSRC, GDTW-KSRC and AGDTW-KSRC. The result is shown in Fig. 5.

As obviously shown in Fig. 5, AGDTW-KSRC has obtained the minimal classification error on all four dataset. Moreover, AGDTW-KSRC decreases the error from 15.91% to 4.55% using GDTW-KSRC on the FaceFour dataset, from 14.86% to 6.86% using GDTW-KSRC on the FISH dataset, from 21.76% to

10.24% using GDTW-KSRC on the SwedishLeaf dataset. On the Gun-Point dataset, it also achieves 8% which is same as GDTW-KSRC.

We can conclude from Figs. 2, 3, 4 and 5 that the transformation performance of the AGDTW kernel is better than that of GRBF and GDTW kernels in most cases. In addition, the classification results on different classifiers listed in Table 2 demonstrate the robustness of this new kernel mapping for time series. The bolded values are the best performance among different measurement ways for the same classifier.

Table 2. Classification error (%) details of different transformation combined with NNC, SVM, and SRC classifier

Dataset		FaceFour	FISH	Gun-Point	SwedishLeaf
1NN	ED	21.60	21.70	8.70	21.10
	DTW	17.00	17.70	9.30	20.80
	GRBF	19.32	30.86	14.67	19.68
	GDTW	10.23	21.14	12.00	20.80
	AGDTW	**4.55**	**15.43**	**8.67**	**11.68**
SVM	GRBF	18.18	16.57	**5.33**	12.48
	GDTW	82.95	20.00	**5.33**	19.68
	AGDTW	**3.41**	**8.57**	8.00	**7.68**
KSRC	GRBF	19.32	28.57	**8.00**	24.00
	GDTW	15.91	14.86	**8.00**	21.76
	AGDTW	**4.55**	**6.68**	16.00	**10.24**

5   Conclusion

In this paper, an altered Gaussian dynamic time warping (AGDTW) kernel has been proposed based on DTW and the GRBF kernel. The AGDTW kernel considering the information hidden in each of warping path generated by DTW may be conductive to the transformation of original time series. The homogeneous time series can be mapped closer and the heterogeneous one can have opposite effect. In addition, we prove that the AGDTW kernel is a positive semi-definite kernel. Benefit from it, existing classification classifiers can have significant progress in mapped dataset. Experimental results on four time series datasets indicate that the AGDTW kernel is better than GRBF and GDTW kernels in the transformation and classification performance.

This paper mainly gives a novelty thought that takes the warping path into kernel mapping and applies the constructed kernel to time series similar measures, such as ERP and TWED. Since AGDTW spends vast of time in kernel transformation, we plan to develop a method to speed up it.

Acknowledgments. This work was supported in part by the National Natural Science Foundation of China under Grant Nos. 61373093 and 61672364, by the Natural Science Foundation of Jiangsu Province of China under Grant No. BK20140008, and by the Soochow Scholar Project.

References

1. Chen, Z., Zuo, W., Hu, Q., Lin, L.: Kernel sparse representation for time series classification. Inf. Sci. **292**, 15–26 (2015)
2. Kai, N., Kortelainen, J., Seppänen, T.: Invariant trajectory classification of dynamical systems with a case study on ECG. Pattern Recogn. **42**(9), 1832–1844 (2009)
3. Lichtenauer, J.F., Hendriks, E.A., Reinders, M.J.: Sign language recognition by combining statistical DTW and independent classification. IEEE Trans. Pattern Anal. Mach. Intell. **30**(11), 2040–2046 (2008)
4. Fu, T.C., Law, C.W., Chan, K.K., Chung, F.L., Ng, C.M.: Stock time series categorization and clustering via SB-tree optimization. In: Wang, L., Jiao, L., Shi, G., Li, X., Liu, J. (eds.) FSKD 2006. LNCS, vol. 4223, pp. 1130–1139. Springer, Heidelberg (2006). doi:10.1007/11881599_141
5. Faloutsos, C., Ranganathan, M., Manolopoulos, Y.: Fast subsequence matching in time-series databases. In: ACM SIGMOD International Conference on Management of Data, vol. 23, pp. 419–429. ACM (2001)
6. Li, H.L., Guo, C.H.: Piecewise cloud approximation for time series mining. Control Decis. **26**(10), 1525–1529 (2011)
7. Li, H., Guo, C., Qiu, W.: Similarity measure based on piecewise linear approximation and derivative dynamic time warping for time series mining. Expert Syst. Appl. **38**, 14732–14743 (2011). Pergamon Press, Inc.
8. Zhang, L., Tao, Z.: Time series classification based on multi-codebook piecewise vector quantized approximation. In: IEEE 27th International Conference on Tools with Artificial Intelligence (ICTAI), pp. 385–390 (2015)
9. Cortes, C., Vapnik, V.: Support vector networks. Mach. Learn. **20**(3), 273–297 (1995)
10. Zhang, D., Zuo, W., Zhang, D., Zhang, H.: Time series classification using support vector machine with Gaussian elastic metric kernel. In: IEEE International Conference on Pattern Recognition, pp. 29–32 (2010)
11. Zhang, L., Zhou, W.D., Chang, P.C., Liu, J., Yan, Z., Wang, T., Li, F.Z.: Kernel sparse representation-based classifier. IEEE Trans. Signal Process. **60**(4), 1684–1695 (2012)
12. Gao, S., Tsang, I.W., Ma, Y.: Learning category-specific dictionary and shared dictionary for fine-grained image categorization. IEEE Trans. Image Process. **23**(2), 623–634 (2014)
13. Berndt, D.J., Clifford, J.: Using dynamic time warping to find patterns in time series. In: KDD workshop, vol. 10, pp. 359–370 (1994)
14. Vlachos, M., Hadjieleftheriou, M., Gunopulos, D., Keogh, E.: Indexing multidimensional time-series with support for multiple distance measures. In: Proceedings of the Ninth ACM SIGKDD International Conference on Knowledge Discovery and Data Mining, pp. 216–225, ACM (2003)
15. Chen, L., Ng, R.: On the marriage of lp-norms and edit distance. In: Proceedings of the Thirtieth International Conference on Very Large Data Bases, vol. 30, pp. 792–803. VLDB Endowment (2004)

16. Marteau, P.F.: Time warp edit distance with stiffness adjustment for time series matching. IEEE Trans. Pattern Anal. Mach. Intell. **31**(2), 306–318 (2009)
17. Pree, H., Herwig, B., Gruber, T., Sick, B., David, K., Lukowicz, P.: On general purpose time series similarity measures and their use as kernel functions in support vector machines. Inf. Sci. **281**(4), 478–495 (2014)
18. Shawe-Taylor, J., Cristianini, N.: Kernel Methods for Pattern Analysis. Cambridge University Press, Cambridge (2004)

An Interweaved Time Series Locally Connected Recurrent Neural Network Model on Crime Forecasting

Ke Wang[1], Peidong Zhu[1,2(✉)], Haoyang Zhu[1], Pengshuai Cui[1], and Zhenyu Zhang[1]

[1] College of Computer, National University of Defense Technology, Changsha, China
{wangke,pdzhu,zhuhaoyang,cuipengshuai,zhangzhenyu}@nudt.edu.cn
[2] Department of Electronic Information and Electrical Engineering, Changsha University, Changsha, China

Abstract. Forecasting events like crimes and terrorist activities is a vital important and challenging problem. Researches in recent years focused on qualitative forecasting of a single type event, such as protests or gun crimes. However, events like crimes usually have complicated correlations with each other, and a single type event forecasting cannot meet actual demands. In reality, a quantitative forecasting is more practical for policy making, decision making and police resources allocating. In this paper, we propose an interweaved time series and an interpretative locally connected Recurrent Neural Network model, which forecasts not only whether an event would happen but also how many it would be by each type. Using open source data from Crimes in Chicago provided by Chicago Police Department, we demonstrate our approach more accurately in forecasting the crime events than the existing methods.

Keywords: Crime forecasting · Interweaved time series · Interpretative locally connected RNN

1 Introduction

Crime forecasting is a significant problem which is related to social stability and people's livelihood. From the perspective of ordinary people, as residents or tourists, it is more attractive to tell them a qualitative forecasting result than a quantitative one. But in the view of police deployment, or decision maker, it is much more practical to treat the quantitative forecasting results, which are the most critical foundation of an intelligent decision.

Crime is a kind of human social activities, and crime forecasting belongs in the discussion of social event forecasting. Researches in recent years focus on forecasting the occurrence of large scale social events [5, 8, 13]. However, they are all qualitative researches, which provide only limited help.

As the third largest city in the USA, Chicago is famous for high crime rate not only in film and television works but also in the real world. It is a city which

© Springer International Publishing AG 2017
D. Liu et al. (Eds.): ICONIP 2017, Part V, LNCS 10638, pp. 466–474, 2017.
https://doi.org/10.1007/978-3-319-70139-4_47

attracts many attentions for crime research [6,7,9]. Anyhow, these researches still focus on forecasting a specific crime type.

Technically, crime forecasting belongs to time series prediction problems. In classic research fields such as stock forecasting, a time series is a sequence taken at successive equally spaced points in time. Traditional time series analysis models, such as Generalized Autoregressive Conditional Heteroskedasticity (GARCH) [1,14], can obtain good sequence prediction results, but deal with the complex relationships inefficiently. And the Recurrent Neural Network (RNN) [2,12] models show their advantages. In order to deal with the vanishing gradients and exploding gradients problems, a Long Short-Term Memory (LSTM) [4,10,15] is proposed. And there is another RNN model called Gated Recurrent Unit (GRU) [3,11], which is similar to LSTM but in simpler structure. To the best of our knowledge, RNNs have not been used in the field of crime forecasting.

In this paper, we propose an interweaved time series approach and a locally connected RNN structure, in which each time sequence has its own time interval without resampling and the multidimensional structures interpret the relationships between the crimes. The main contribution of our work can be summarized as below: (1) We introduce the RNNs into the quantitative crime forecasting for the first time; (2) Our interweaved time series deal with different time intervals without resampling; (3) Our locally connected RNN structure can not only implement the interweaved time series but also take into account the complicated relationships among various crimes types and districts in the prediction process.

2   Interweaved Time Series

Time series forecasting is a complex and interesting problem. We may find various of trends of one event, such as seasonal trends and cycles of volatility. In common ways, multiple time series should be resampled to the same time interval. As shown in Fig. 1(b) *Time interval 1*, the time series ts-1-2 and ts-1-3 should be transformed to re-ts-1-2 and re-ts-1-3 which have the same space of time as ts-1-1, and then we analyze the three time series together. In our opinion of crime forecasting, we care more about what and how many will happen, relative to the time span. Shown in Fig. 1(b) *Observation slice*, the time point is the central issue, where time series of different intervals interweave together.

Time series always have their own characters. When resampled, they may be more suitable for a mathematical hypothesis or formula. But the information in the data is more or less lost in the transform process. So we suppose that the time series with different spans have their own patterns. And if we ignore these patterns and project the time series to the same interval, we get the common perspective of time series, which means our point of view contains the existing ones.

We take morning run as an example in Fig. 1(a). There are three types of time series in the figure. The fist type, ts-1, records daily exercise data. The second type, ts-2, sets down the weight training data on every Tuesday. And the last one,

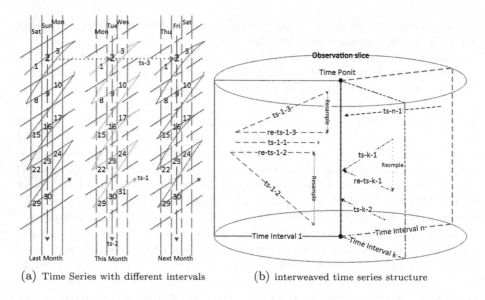

(a) Time Series with different intervals   (b) interweaved time series structure

Fig. 1. Perspective of interweaved time series

ts-3, chronicles the self test data of each month. Now our purpose is to forecast the data of 2nd in next month. In the day-by-day records, we maybe find that the runner was out of his state in recent days, and we may predict the runner's time to increase. In the week data, Friday is the last workday, and the runner is always happy for the coming weekend. So we infer that the runner will get a shorter time. And in each self test of the month, the runner all along tries his best, and we conclude that there will be a performance promotion. Obviously, the three types have different time intervals, and these intervals interpret the same event differently and reasonably. If we simply resample these time series, the explanation of the time series will be lost.

In the angle of our view, we interpret the time series problem into several time intervals, and resample the time series in the same interval only if necessary. Back to the example of last paragraph, the three time series should be in three time intervals, and the forecasting date is called time point, which is the intersection of the time series.

3   Interpretative Locally Connected Recurrent Neural Network

In our crime forecasting problem, the memory of the data history state is vitally important, so we choose the RNN cell structures for each node. To solve the vanishing gradients and exploding gradients problems, we use LSTM (Eq. (3.1)) and the GRU (Eq. (3.2)).

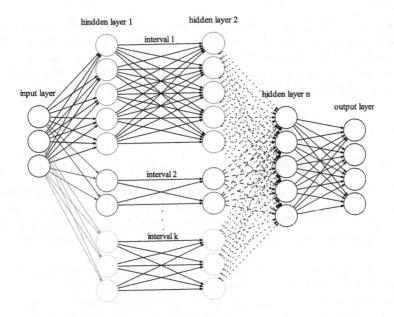

Fig. 2. Interpretative locally connected recurrent neural network

In the interest of putting the interweaved time series into practice and taking full advantage of the relations between different crime types, we propose an Interpretative Locally Connected Recurrent Neural Network (ILC-RNN), as Fig. 2 shown.

In this network structure, we lay stress on the interpretation of the interweaved time series. Diverse time series with distinct intervals have their own patterns. Trying to keep this inherent pattern of the series, we deal with them separately, which means each interval has its own full connected structure until the last hidden layer. In the first layer, the input layer, a full connected structure is built to process series with the minimum interval. And to other intervals, locally connected method is used to filter their own input data. Then in the hid-

$$f_t = \sigma(W_f h \bullet h_{t-1} + W_f x \bullet X_t + b_f) \quad (3.1a)$$

$$i_t = \sigma(W_i h \bullet h_{t-1} + W_i x \bullet X_t + b_i) \quad (3.1b)$$

$$\tilde{c}_t = tanh(W_c h \bullet h_{t-1} + W_c x \bullet X_t + b_c) \quad (3.1c)$$

$$c_t = f_t \times c_{t-1} + i_t \times \tilde{c}_t \quad (3.1d)$$

$$o_t = \sigma(W_o h \bullet h_{t-1} + W_o x \bullet X_t + b_o) \quad (3.1e)$$

$$h_t = o_t \times tanh(c_t) \quad (3.1f)$$

$$z_t = \sigma(W_z \bullet h_{t-1} + U_z \bullet X_t) \quad (3.2a)$$

$$r_t = \sigma(W_r \bullet h_{t-1} + U_r \bullet X_t) \quad (3.2b)$$

$$\tilde{h_t} = tanh(W \bullet (r_t \times h_{t-1}) + U \bullet X_t) \quad (3.2c)$$

$$h_t = (1 - z_t) \times h_{t-1} + z_t \times \tilde{h_t} \quad (3.2d)$$

$$h_{integrate} = W_{h1} \times h_{t1} + W_{h2} \times h_{t2} + \cdots + W_{hk} \times h_{tk} \quad (3.3)$$

den layers, for sake of unearthing the patterns adequately, fully connected way is applied to different intervals respectively, and each interval has its own hidden layer size. In the last hidden layer, we integrate the data from all the intervals together, as shown in Eq. (3.3), where h_{tk} represents the state of interval k. At last, in the output layer, we make a fine tuning to the results.

4   Experiment and Results

4.1   Experimental Data Analysis

With the development of data science, the Chicago Police Department provide an open source dataset[1], which reflects millions of reported incidents of crime that occurred in the City of Chicago from 2001 to present. Our experimental Data is extracted from January 1, 2001 to August 26, 2016, with 6,147,883 records and 22 features.

Figure 3 illustrates the number of crimes in different districts, in virtue of which we choose the data in 22 practical districts. Comparison and analysis are made on data in diverse types of crimes, and we get rid of the types where occurrence is less than 400 in nearly 16 years. Then we take 25 types in Fig. 4, where we display the data information as standard deviation, minimum value, the first quartile, median, mean value, the third quartile and maximum value.

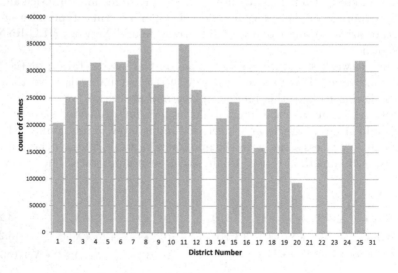

Fig. 3. The count of crimes in different districts

---

[1] https://data.cityofchicago.org/Public-Safety/Crimes-2001-to-present-Map/c4ep-ee
5m.

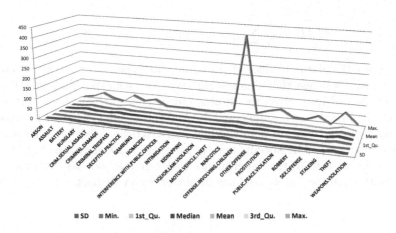

Fig. 4. Data information of different crime types

4.2 Experimental Results and Discussions

By the analysis of the data information, we find that the data has heteroscedasticity. So we choose GARCH as the baseline model. In order to verify the validity of the ILC-RNN structure, full connected LSTM and GRU are also selected as the models of comparison.

In evaluation, majority of forecasting experiments choose the Mean Squared Error (MSE, Eq. (4.1)), Root Mean Squared Error (RMSE, Eq. (4.2)) or Mean Absolute Error (MAE, Eq. (4.3)). In virtue of better describing the characteristics of the result, we choose both MSE and MAE. A toy example of the evaluation methodology is shown in Table 1. In this table, $pred_1$ and $pred_2$ have the same MAE, just as $pred_1$ and $pred_4$ have the same MSE, which are easy to compare the results. But when it comes to $pred_3$ and $pred_5$, the situation is different. Relative to $pred_1$, $pred_3$ has a smaller MAE and a bigger MSE, and $pred_5$ is on the contrary. In our experiments, we find that data sequences like $pred_3$ have more data points close to the real results but the outliers have bigger errors, while sequences like $pred_5$ are smoother but get more deviation points with small values. From the point of view of our forecasting problem, there is no absolutely better one between the two predictive results. But in the decision maker's opinion, there should be a better choice depending on the specific circumstances.

$$MSE = \frac{1}{N}\sum_{i=1}^{N}(pred\_result_i - true\_result_i)^2 \qquad (4.1)$$

$$RMSE = \sqrt{\frac{1}{N}\sum_{i=1}^{N}(pred\_result_i - true\_result_i)^2} \qquad (4.2)$$

$$MAE = \frac{1}{N} \sum_{i=1}^{N} |pred\_result_i - true\_result_i| \tag{4.3}$$

We group the data by district, integrate all types of crime data into one input data, and forecast the crimes in ten days by each type. The general effects are illustrated in Fig. 5. From the results, our methods achieve both better MSE and MAE than the three baseline models. And the ICL-LSTM gets the smoother results with more deviation points, while ICL-GRU has closer prediction curve with bigger outliers. The specific results by district are shown in Fig. 6(a) and (b). It can be seen from the experiments that our method has better accuracy and stability than the three baselines in almost every district.

Table 1. A Toy Example of MSE VS. MAE

col_name	col₁	col₂	col₃	col₄	MSE	MAE
true_data	2	2	2	2	–	–
pred₁	2	2	5	6	6.25	1.75
pred₂	2	2	2	9	12.25	1.75
pred₃	2	2	2	8	9	1.5
pred₄	2	2	2	7	6.25	1.25
pred₅	2	4	5	5	5.5	2

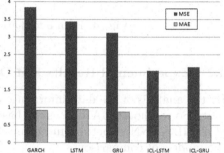

Fig. 5. Effects of different methods

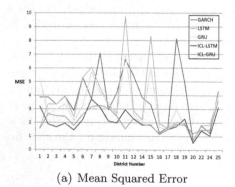

(a) Mean Squared Error

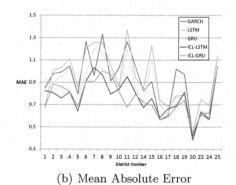

(b) Mean Absolute Error

Fig. 6. Experimental results of different districts

5   Conclusion and Future Work

In this paper, we studied the quantitative crime forecasting by RNN structure. We propose an interweaved time series method to deal with times series with diverse intervals, for sake of keeping the data fluctuation trends in various perspectives. Then we implement this method in two popular RNN models, LSTM and GRU, where we utilize a locally connected structure to interpret and handle different intervals. In experiments, we make comparisons between our methods and the three baseline models using real open source data from Chicago Police Department. Our method achieves better performance than the three baselines and could easily extend its scope of application. In the following work, we aim to explore interpretative method on more Neural Networks for crime forecasting.

Acknowledgements. This research has been supported by National Natural Science Foundation of China (No. 61572514) and (No. 61170285).

References

1. Bollerslevb, T.: Generalized autoregressive conditional heteroskedasticity. J. Econom. **31**(3), 307–327 (1986)
2. Chen, Y., Yang, J., Qian, J.: Recurrent neural network for facial landmark detection. Neurocomputing **219**(2017), 26–38 (2017)
3. Cho, K., van Merrienboer, B., Gulcehre, C., Bahdanau, D., Bougares, F., Schwenk, H., Bengio, Y.: Learning phrase representations using RNN encoder-decoder for statistical machine translation. In: Proceedings of the 2014 Conference on Empirical Methods in Natural Language Processing (EMNLP), pp. 1724–1734. Association for Computational Linguistics (2014)
4. Hochreiter, S., Schmidhuber, J.: Long short-term memory. Neural Comput. **9**(8), 1735–1780 (1997)
5. Hong, Q., Manrique, P., Johnson, D., Restrepo, E., Johnson, N.F.: Open source data reveals connection between online and on-street protest activity. EPJ Data Sci. **5**(1), 1–12 (2016)
6. Kieltyka, J., Kucybala, K., Crandall, M.: Ecologic factors relating to firearm injuries and gun violence in Chicago. J. Forensic Leg. Med. **37**(2016), 87–90 (2016)
7. Mohler, G.: Marked point process hotspot maps for homicide and gun crime prediction in Chicago. Int. J. Forecast. **30**, 491–497 (2014)
8. Ning, Y., Muthiah, S., Rangwala, H., Ramakrishnan, N.: Modeling precursors for event forecasting via nested multi-instance learning. In: Proceedings of the 22nd ACM SIGKDD International Conference on Knowledge Discovery and Data Mining, pp. 1095–1104. ACM (2016)
9. Smith, C.M., Papachristos, A.V.: Trust thy crooked neighbor: multiplexity in Chicago organized crime networks. Am. Sociol. Rev. **81**(4), 1–24 (2016)
10. Sutskever, I., Vinyals, O., Le, Q.V.: Sequence to sequence learning with neural networks. In: Advances in Neural Information Processing Systems, vol. 4, pp. 3104–3112 (2014)
11. Tang, Y., Huang, Y., Wu, Z., Meng, H., Xu, M., Cai, L.: Question detection from acoustic features using recurrent neural network with gated recurrent unit. In: IEEE International Conference on Acoustics, Speech and Signal Processing, pp. 6125–6129 (2016)

12. Williams, R.J., Zipser, D.: A learning algorithm for continually running fully recurrent neural networks. Neural Comput. **1**(2), 270–280 (1998)
13. Zhao, L., Sun, Q., Ye, J., Chen, F., Lu, C.T., Ramakrishnan, N.: Multi-task learning for spatio-temporal event forecasting. In: ACM SIGKDD International Conference on Knowledge Discovery and Data Mining, pp. 1503–1512 (2015)
14. Zhao, Y., Zou, X., Xu, H.: Improving forecasts of generalized autoregressive conditional heteroskedasticity with wavelet transform. Res. J. Appl. Sci. Eng. Technol. **5**(2), 649–653 (2013)
15. Zhu, W., Lan, C., Xing, J., Zeng, W., Li, Y., Shen, L., Xie, X.: Co-occurrence feature learning for skeleton based action recognition using regularized deep LSTM networks. In: The AAAI Conference on Artificial Intelligence, pp. 3697–3703 (2016)

Decouple Adversarial Capacities
with Dual-Reservoir Network

Qianli Ma[1,2(✉)], Lifeng Shen[1], Wanqing Zhuang[1], and Jieyu Chen[3]

[1] School of Computer Science and Engineering,
South China University of Technology, Guangzhou 510006, China
`qianlima@scut.edu.cn`, `scuterlifeng@foxmail.com`
[2] Guangdong Key Laboratory of Big Data Analysis and Processing,
Guangzhou 510006, China
[3] Linguistic Department, University of California, San Diego, CA 92093, USA
`jic387@ucsd.edu`

Abstract. Reservoir computing such as Echo State Network (ESN) and
Liquid State Machine (LSM) has been successfully applied in dynamical system modeling. However, there is an antagonistic trade-off between
the non-linear mapping capacity and the short-term memory capacity
in single-reservoir networks, especially when the input signals contain
high non-linearity and short-term dependencies. To address this problem, we propose a novel reservoir computing model called Dual-Reservoir
Network (DRN), which connects two reservoirs with an unsupervised
encoder such as PCA. Specifically, we allow these two adversarial capacities to be decoupled and enhanced in the dual reservoirs respectively. In
our experiments, we first verify DRN's feasibility on an extended polynomial system, which allows us to control the nonlinearity and short-term
dependencies of data. In addition, we demonstrate the effectiveness of
DRN on the synthesis and real-world time series predictions.

Keywords: Reservoir computing · Echo-state network · Short-term
memory · Non-linearity mapping · Time series prediction

1 Introduction

Reservoir Computing (RC) [9] is a popular framework of designing and training
recurrent neural networks (RNNs) due to its simplicity and effectiveness. A RC
network usually consists of three components: an input layer, a dynamic layer
called reservoir and an output layer. Weights in the input layer and the reservoir
are all fixed randomly, and output weights need to be adapted during training.
The reservoir is the core of the whole system as it can provide abundant dynamics
with its fixed, random and sparse recurrent connections.

Echo state network (ESN) is a main type of RC networks [7]. Given an ESN
with a N-size reservoir, we can define its state equation of ESN as follows:

$$\mathbf{x}(n) = (1-\gamma)\mathbf{x}(n-1)+\gamma f(\mathbf{z}) \tag{1}$$

© Springer International Publishing AG 2017
D. Liu et al. (Eds.): ICONIP 2017, Part V, LNCS 10638, pp. 475–483, 2017.
https://doi.org/10.1007/978-3-319-70139-4_48

where z is named as the working point within activation function and is formulated by

$$\mathbf{z} = \mathbf{W}\mathbf{x}(n-1) + IS \cdot \mathbf{W}_{in}\mathbf{u}(n) \tag{2}$$

where n is the time step, $\mathbf{u}$ is a T-length K-dim input signal, $\mathbf{x}$ is reservoir's echo state. IS denotes the input scaling. γ is a hyperparamer and it denotes the leaky rate of reservoir, and $\mathbf{W}_{in} \in \mathbb{R}^{N \times K}$ is the projection matrix. $\mathbf{W} \in \mathbb{R}^{N \times N}$ is the state transition matrix and is generated by

$$\mathbf{W} = \frac{\rho}{\mathbf{W}_0}\lambda_{max}(\mathbf{W}_0) \tag{3}$$

where $\lambda_{max}(\mathbf{W}_0)$ is the largest eigenvalue of matrix $\mathbf{W}_0$. The elements of $\mathbf{W}_0$ are sampled randomly from $[-0.5, 0.5]$. To ensure the echo state property (necessary stability condition) [6], the spectral radius ρ is suggested to be smaller than one.

From the perspective of dynamic modeling, ESN has two most important capacities: non-linear mapping capacity (NMC) and short-term memory capacity (MC). High NMC means that we can model non-linearity data well, while MC focuses on capturing the short-term dependencies of data. However, as pointed out in previous works [2,12], there is an antagonistic trade-off between NMC and MC in the single-reservoir networks.

In [12], Verstraeten investigated the interplay between these two capacities in reservoir over a simulation task, which allows an accurate control over NMC and MC within data. Their results showed that the overall performance of reservoir system is mostly dominated by memory requirements of a given task. In [2], Butcher analyzed this problem from the effects of the working point $\mathbf{z}$ in Eq. (2). As seen in (2) and (3), $\mathbf{z}$ is affected by the input scaling IS and the spectral radius ρ. As illustrated in Fig. 1, Butcher argued that given a small input signal, an IS smaller than one and ρ close to but smaller than one, the working point will generally be in the linear region of the activation function, which will create a linear reservoir with the highest MC, and the amount of non-linearity will be its minimum. When we enlarge IS and ρ (values over one), the memory capacity

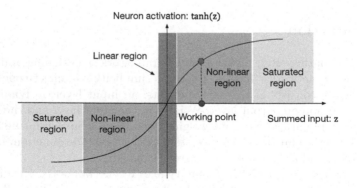

Fig. 1. Illustration of the working point and different regions of the tangent activation function.

of a reservoir will be weakened and its non-linearity will be enhanced as a result of moving working points into non-linear regions. From Butcher's analysis, this trade-off between NMC and MC is mainly due to the unicity of the working point location (linear region or nonlinear one). To resolve this conflict, Butcher introduced two static nonlinear Extreme Learning Machines (ELMs) [5] into an ESN and named the model as R^2SP [2,3]. Another similar work is Gallicchio's φ-ESN [4], which also added an ELM on the top of ESN and enhanced the nonlinearity of the whole system.

In this paper, we propose Dual-Reservoir Network (DRN) to break this antagonistic trade-off between the non-linear mapping capacity and the short term memory capacity. Our main idea is to construct two reservoirs to adjust these two capacities respectively, which is why it is termed as "DUAL". Inspired by the work of φ-ESN, we connect these two reservoirs in a pipeline. However, based on the idea of representation learning in deep learning, we also introduce an unsupervised encoder PCA between the two reservoirs. In this way, we can encoder the state information (with high nonlinearity or short-term dependencies) into an intermediary encoder, and then pass the information from the former reservoir to the latter one. The scales of two capacities in DRN are mainly determined by the hyperparameters IS and ρ and we adopt the genetic algorithm (GA) to optimize them. Finally, several simulations and real-world time series prediction experiments are used to analyze and demonstrate the effectiveness of our proposed DRN.

2   Dual-Reservoir Network

A simple illustration of our proposed Dual-Reservoir Network (DRN) is shown in Fig. 2.

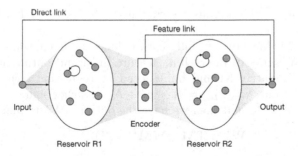

Fig. 2. A simple architecture illustration of the proposed Dual-Rerservoirs Network.

Compared with the general ESN, the defining characteristic of our DRN is in that the hidden layer is replaced by two dual-reservoir (they are denoted by R1 and R2 respectively) and an intermediary unsupervised encoder, where the encoder is selected as the Principal-Component Analysis (PCA) [13]. The reasons are two folds: (1) PCA is one of the most popular dimension reduction tools and has been used in many data-analysis applications. (2) The reservoir produces a relative high-dimensional state space and in this space, the hidden states in reservoir can be linearly separable with high probability. That also explains why we can use simple regressor to train ESN [7].

2.1   Training Methods

For the convenience of distinguishing parameters and states in the dual reservoirs, we use the superscript (1) to denote reservoir R1, and use superscript (2) to denote reservoir R2. Given T-length K-dim input signals $\{\mathbf{u}(n) \in \mathbb{R}^K\}$, n denotes the time step and $n = 1, 2, \ldots, T$, and let the size of the reservoir R1 and R2 be N_1 and N_2 respectively, the size of encoder be M and the dimension of the output be L, we can formulate the updating equation of the reservoir R1:

$$\mathbf{x}^{(1)}(n) = (1-\gamma^{(1)})\mathbf{x}^{(1)}(n-1)+\gamma^{(1)} f^{(1)}(\mathbf{W}^{(1)}\mathbf{x}^{(1)}(n-1)+IS^{(1)} \cdot \mathbf{W}^{(1)}_{in}\mathbf{u}(n)) \quad (4)$$

where $\mathbf{x}^{(1)} \in \mathbb{R}^{N_1}$ is the echo states of R1. $f^{(1)}$ denotes the activation function and usually is $tanh$ function. γ denotes the leaky rate. $\mathbf{W}^{(1)} \in \mathbb{R}^{N_1 \times N_1}$ denotes the transmission matrix, and $\mathbf{W}_{in} \in \mathbb{R}^{N_1 \times K}$ is the projection matrix. At the initial step, $\mathbf{x}^{(R1)}(0)$ is initialized as zeros.

After that, we obtain the echo states of R1 with specific $IS^{(1)}$ and $\rho^{(1)}$, which implies that R1's states have presented some non-linearity and short-term dependencies related with $IS^{(1)}$ and $\rho^{(1)}$. And then, encoding R1's states $\{\mathbf{x}^{(1)}(n)\}$ with PCA, we will get abstract representations $\{\mathbf{h}(n)\}$ from R1's hidden states.

Drive the $\{\mathbf{h}(n)\}$ into the reservoir $R2$, update its echo states and obtain $\{\mathbf{x}^{(2)}(n)\}$ by

$$\mathbf{x}^{(2)}(n) = (1-\gamma^{(2)})\mathbf{x}^{(2)}(n-1)+\gamma^{(2)} f^{(2)}(\mathbf{W}^{(2)}\mathbf{x}^{(2)}(n-1)+IS^{(2)} \cdot \mathbf{W}^{(2)}_{in}\mathbf{h}(n)) \quad (5)$$

where the meanings of the notions are similar to the Eq. (4), but with different inputs $\{\mathbf{h}(n)\}$.

Finally, we introduce direct and feature links from encoders to outputs shown in Fig. 2. Weights of direct link, feature links and output layer all will be collected into a matrix $\mathbf{M}$ and be adapted by regression technique, where $\mathbf{M} \in \mathbb{R}^{N_{collected} \times T}$, and $N_{collected} = K + M + N_2$. Teacher signals are collected into a matrix $\mathbf{T} \in \mathbb{R}^{L \times T}$, and then we have the optimal $\mathbf{W}^{\star}$ formulated by

$$\mathbf{W}^{\star} = \mathbf{TM}^T(\mathbf{MM}^T + \beta\mathbf{I})^{-1} \quad (6)$$

which is the well-known ridge regression solution and the β is the *Thikhonov* regularization term.

3   Experiments and Results

In this section, we conduct several experiments to analyze and evaluate the proposed Dual-Reservoir network (DRN). The experiments can be divided into two parts: (1) firstly, we test the decouple performance of DRN over an extended polynomial dataset with different nonlinearity and short-term dependency levels [2], which allows an accurate control of nonlinearity and short-term dependencies within data. (2) and then, we demonstrate the performance on the tasks of synthesis and real-world time series prediction.

The baseline models we selected here include the leaky ESN [8], R^2SP [2] and φ-ESN [4]. Apart from these baselines, we also introduce two types of encoders in DRN: the elm-based auto-encoder [10] and the random projection (RP) [1]. Hyperparameters of all above reservoir baselines are optimized by the genetic algorithm (GA) and these hyperparameters include IS, ρ and the leaky rate γ. The performance indicator is the normalized root mean squared error (NRMSE), which can be given by

$$NRMSE = \frac{\sqrt{\frac{1}{T}\sum_{t=1}^{T}[y(t) - \hat{y}(t)]^2}}{Var(\hat{y}(t))} \tag{7}$$

where T is the length of signals, $y(n)$ the outputs at time step n, and $y^{target}(n)$ the corresponding target signals.

As for our implementation details, the size of all the reservoirs of all the baselines is fixed to be 300, and the size of encoders in DRN is given by the cross-validation. The machine setup is the Matlab platform on an Intel Core i5-2410M, 2.30-GHz CPU 8-GB RAM.

3.1   Extended Polynomial Tasks

This dataset allows an accurate control over the nonlinearity and short-term dependency levels within data [2]. It can be viewed as a time series generated by a specific polynomial system, which can be formulated by

$$y(t) = \sum_{i=0}^{p} \sum_{j=0}^{p-i} c_{ij} u^i(t) u^j(t-d) \qquad s.t. \ i + j \leq p \tag{8}$$

where $\{y(t)\}$ denotes the target outputs. $\{u(t)\}$ is the input sequence drawn from an uniform distribution between -1 and +1, c_{ij} is a random number drawn from the same range as $\{u(t)\}$. The terms p and d are two important parameters in this system. The p denotes the order of the polynomial system and the d is the delay. p and d both can be adjusted to produce time series with different characteristics. Larger p means the more non-linear mapping capacity requirements, while larger d requires more short-term memory capacities. These changes are visualized in Fig. 3. In this experiment, we generate 3000 ordered points, and 64% is used for training, 16% for testing and the rest for testing. We do the one-step-ahead prediction over this dataset.

As shown in Table 1, we reported the performance (NRMSE) of our DRN and other baselines over this dataset with different orders p and delays d. From these results, we found that our DRN achieves the best performance among all

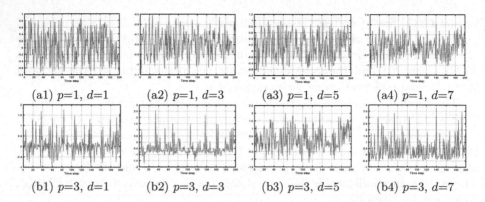

(a1) $p=1$, $d=1$ (a2) $p=1$, $d=3$ (a3) $p=1$, $d=5$ (a4) $p=1$, $d=7$

(b1) $p=3$, $d=1$ (b2) $p=3$, $d=3$ (b3) $p=3$, $d=5$ (b4) $p=3$, $d=7$

Fig. 3. Visualizations of the extended polynomial series with different p and d.

Table 1. Performance (NRMSE) for the extended polynomial tasks. Data with larger p performs more non-linear, and with larger d requires more memory capacity.

NRMSE	MODELS	d=1	d=3	d=5	d=7	d=9
p=1	ESN	1.07E-04	1.45E-03	2.62E-03	5.12E-02	2.10E-03
	Fai-ESN	6.90E-05	1.09E-03	1.41E-02	4.72E-02	8.32E-01
	R2SP	1.02E-04	6.10E-04	1.83E-03	4.54E-04	3.07E-03
	DRN	4.40E-05	2.20E-04	3.10E-04	2.85E-04	1.20E-03
p=3	ESN	5.97E-01	6.07E-01	4.76E-01	4.06E-01	7.58E-01
	Fai-ESN	8.38E-02	5.50E-01	2.93E-01	4.10E-01	7.56E-01
	R2SP	4.92E-01	6.03E-01	4.14E-01	4.21E-01	7.66E-01
	DRN	5.47E-02	5.43E-01	2.48E-01	4.03E-01	6.57E-01
p=5	ESN	6.82E-01	7.02E-01	5.66E-01	9.67E-01	9.89E-01
	Fai-ESN	1.56E-01	3.70E-01	5.16E-01	9.29E-01	9.33E-01
	R2SP	4.22E-01	6.04E-01	5.83E-01	9.62E-01	9.57E-01
	DRN	1.40E-01	3.61E-01	3.76E-01	9.22E-01	9.43E-01
p=7	ESN	4.12E-01	5.62E-01	9.37E-01	1.01E+00	7.09E-01
	Fai-ESN	1.50E-01	5.57E-01	7.53E-01	1.02E+00	8.80E-01
	R2SP	3.32E-01	4.98E-01	8.59E-01	8.84E-01	7.02E-01
	DRN	1.22E-01	4.44E-01	7.41E-01	9.12E-01	6.89E-01

the methods. In particular, we observed that all models tend to be worse when the nonlinearity is higher (increasing p), but our DRN still perform with a high prediction accuracy.

3.2   Time Series Predictions

We select three datasets to evaluate the practical performance of DRN:
(1) **Mackey-Glass System**: This is a classical time series for evaluating the performance of dynamical system identification methods. In discrete time, the Mackey-Glass delay differential equation can be formulated by

$$y(t+1) = y(t) + \delta \cdot (a\frac{y(t - \tau/\delta)}{1 + y(t - \tau/\delta)^n} - by(t)) \tag{9}$$

where the parameters δ, a, b, n usually are set to 0.1, 0.2, −0.1, 10. When $\tau > 16.8$, the system becomes chaotic, and in most previous work, τ is set to 17. Thus, we also let τ be 17 in this task. The model details can be found in the literature [7]. In this task, we adopt the 84 time steps ahead prediction to test our methods. In details, we simulate a 10000-length MGS time series, and split these 10000 points into three parts with length $T_{train} = 6400$, $T_{validate} = 1600$ and $T_{test} = 2000$. To avoid the effects of initial states, we discard a certain number of initial steps, $T_{washout} = 100$ for each reservoir.

(2) **NARMA System**: This is a difficult task to test the performance of recurrent networks. To model this dynamical system, high nonlinearity and strong memory capacity are required. The updated equation of 10th-order NARMA can be defined as follows:

$$y(t+1) = 0.3y(t) + 0.05y(t)\sum_{i=0}^{9} y(t-i) + 1.5u(t-9)u(t) + 0.1 \tag{10}$$

where $u(t)$ is a random value drawn from an uniform distribution between [0, 0.5] for each time step t, and the output signal $y(t)$ is initialized by zeros for the first ten steps. The total length we used is 4000 and $T_{train} = 2560$, $T_{validate} = 640$ and $T_{test} = 800$, respectively. The washout length is set to be 30 for each reservoir. In this task, we conduct the one-step-ahead prediction.

(3) **Daily Minimum Temperatures** in Melbourne, Australia is a real world time series dataset, recorded from January 1, 1981 to December 31, 1990 [11]. There are a total of 3650 sample points. We let $T_{train} = 2336$, $T_{validate} = 584$ and $T_{test} = 730$, respectively. The washout length for each reservoir is set to 30. Since this real world time series presents strong nonlinearity, we smooth it with a 5-step sliding window.

Table 2. Average prediction results (NRMSE) with standard deviation.

MODELS	MGS-84	NARMA	Temperatures
ESN	2.01E-01±2.91E-02	2.45E-01±2.00E-02	1.39E-01±1.02E-03
φ-ESN	3.96E-02±7.49E-03	1.69E-01±1.75E-02	1.41E-01±1.10E-03
R^2SP	1.25E-01±1.96E-02	1.81E-01±2.21E-02	1.37E-01±9.82E-04
DRN-RP	3.73E-02±4.59E-03	1.37E-01±1.89E-02	1.36E-01±1.06E-03
DRN-ELMAE	3.41E-02±7.74E-03	1.53E-01±2.14E-02	1.36E-01±8.00E-04
DRN-PCA	3.67E-02±4.78E-03	1.31E-01±9.50E-03	1.35E-01±4.31E-04

All average prediction results (NRMSE) are summarized in Table 2. As can be seen in Table 2, DRNs outperform other baselines significantly. Within the variants with different encoders, we found that the RP works the worst among three encoder types but still performs better than other baselines. The ELM-AE performs best over the MGS task, and the PCA also performs well. As for the NARMA and temperature tasks, the PCA achieves the best result.

4   Conclusions and Discussions

In this paper, we focus on the adversarial problems between the non-linear mapping capacity (NMC) and the short-term memory capacity (MC) in ESN. To address this problem, we proposed the dual-reservoir network (DRN), which is to allow the dual-reservoirs to adjust these two capacities respectively. Specifically, these two capacities are affected by the hyperparameters input scaling and spectral radius, which we adopt the genetic algorithm (GA) to optimize. In experiments, we used the extended polynomial dataset to verify the effectiveness of our models under different nonlinearity and memory requirements. We also test on the synthesis and real-world prediction tasks. It is worth noting that our DRN not only works in the adversarial problem of NMC and NC, but also can be developed into a hierarchical reservoir-computing framework.

Acknowledgment. This work is supported by the National Natural Science Foundation of China (Grant Nos. 61502174, 61402181), the Natural Science Foundation of Guangdong Province (Grant Nos. S2012010009961, 2015A030313215), the Science and Technology Planning Project of Guangdong Province (Grant No. 2016A040403046), the Guangzhou Science and Technology Planning Project (Grant Nos. 201704030051, 2014J4100006), the Opening Project of Guangdong Province Key Laboratory of Big Data Analysis and Processing (Grant No. 2017014), and the Fundamental Research Funds for the Central Universities (Grant No. D2153950).

References

1. Bingham, E., Mannila, H.: Random projection in dimensionality reduction: applications to image and text data. In: Proceedings of the Seventh ACM SIGKDD International Conference on Knowledge Discovery and Data Mining, pp. 245–250. ACM (2001)
2. Butcher, J., Verstraeten, D., Schrauwen, B., Day, C., Haycock, P.: Reservoir computing and extreme learning machines for non-linear time-series data analysis. Neural Netw. **38**, 76–89 (2013)
3. Butcher, J., Verstraeten, D., Schrauwen, B., Day, C., Haycock, P.: Extending reservoir computing with random static projections: a hybrid between extreme learning and RC. In: 18th European Symposium on Artificial Neural Networks (ESANN 2010), pp. 303–308. D-Side (2010)
4. Gallicchio, C., Micheli, A.: Architectural and markovian factors of echo state networks. Neural Netw. **24**(5), 440–456 (2011)
5. Huang, G.B., Zhu, Q.Y., Siew, C.K.: Extreme learning machine: theory and applications. Neurocomputing **70**(13), 489–501 (2006). Neural Networks Selected Papers from the 7th Brazilian Symposium on Neural Networks (SBRN 2004)
6. Jaeger, H.: The echo state approach to analysing and training recurrent neural networks-with an erratum note. German National Research Center for Information Technology GMD Technical report, Bonn, Germany, vol. 148(34), p. 13 (2001)

7. Jaeger, H., Haas, H.: Harnessing nonlinearity: predicting chaotic systems and saving energy in wireless communication. Science **304**(5667), 78–80 (2004)
8. Jaeger, H., Lukoeviius, M., Popovici, D., Siewert, U.: Optimization and applications of echo state networks with leaky-integrator neurons. Neural Netw. **20**(3), 335–352 (2007). Echo State Networks and Liquid State Machines
9. LukošEvičlus, M., Jaeger, H.: Reservoir computing approaches to recurrent neural network training. Comput. Sci. Rev. **3**(3), 127–149 (2009)
10. Tang, J., Deng, C., Huang, G.B.: Extreme learning machine for multilayer perceptron. IEEE Trans. Neural Netw. Learn. Syst. **27**(4), 809–821 (2016)
11. Time Series Data Library: daily minimum temperatures in Melbourne, Australia, 1981–1990. https://datamarket.com/data/set/2324/daily-minimum-temperatures-in-melbourne-australia-1981-1990
12. Verstraeten, D., Dambre, J., Dutoit, X., Schrauwen, B.: Memory versus nonlinearity in reservoirs. In: The 2010 International Joint Conference on Neural Networks (IJCNN), pp. 1–8. IEEE (2010)
13. Wold, S., Esbensen, K., Geladi, P.: Principal component analysis. Chemometr. Intell. Lab. Syst. **2**(1), 37–52 (1987). Proceedings of the Multivariate Statistical Workshop for Geologists and Geochemists

Tree Factored Conditional Restricted Boltzmann Machines for Mixed Motion Style

Chunzhi Xie, Jiancheng Lv$^{(\boxtimes)}$, Bijue Jia, and Lei Xia

Machine Intelligence Laboratory, College of Computer Science,
Sichuan University, Chengdu 610065, People's Republic of China
lvjiancheng@scu.edu.cn

Abstract. A factored conditional restricted Boltzmann machine (FCRBM) is an efficient, compact model for multi-class temporal data (e.g. multi-label human motion data). However, since all factors in FCRBM are linked to the labels directly, data generated by the model is heavily dependent on the learned tags. In this paper, we propose a tree-based FCRBM model in which the factors are tree-like connected and only part of the factors are directly connected to the labels. The proposed model can make the newly generated data have a variety of sports styles and achieve a smooth transition between the styles using little or even no labeled data.

Keywords: FCRBM · CRBM · RBM · Deep learning · Human motion style

1 Introduction

Temporal data has been the focus of research for decades such as video data, music data and human motion capture data [1,2]. Human motion capture data is high-dimensional, time-dependent and noisy. Modeling human motion is the basis of numerous applications such as tracking, activity identification and new motion data synthesis [3,4]. A restricted Boltzmann machine (RBM) has been successfully applied in many applications such as dimensionality reduction, feature extraction, and classification [5–7]. There have been a variety of related researches on improving RBM to extract the characteristics of temporal data, such as temporal RBM (TRBM) [8], recurrent temporal RBM (RTRBM) [9], structured recurrent temporal RBM (SRTRBM) [10], conditional RBM(CRBM) and FCRBM [11]. Among these RBM variants, CRBM and TRBM are the first models for sequence modeling. RTRBM and SRTRBM are improved models of TRBM, while FCRBM is an improved model of CRBM. A CRBM takes the previous temporal class features as the hidden bias and takes the individual characteristics of the previous moments as the visible bias, so that the newly generated data not only has the group features but also has temporal features. The difference between CRBM and TRBM is that the temporal features of TRBM come from previous moments, while those of CRBM come from the first few moments.

© Springer International Publishing AG 2017
D. Liu et al. (Eds.): ICONIP 2017, Part V, LNCS 10638, pp. 484–492, 2017.
https://doi.org/10.1007/978-3-319-70139-4_49

Via factors, FCRBM binds motion labels to CRBM, so that label features can be bound up with current and previous data and then applied to the hidden variables. Next, label features are once again bound up with the obtained hidden variables and previous data, and then applied to the current data. In this way, FCRBM only needs to be given the initial motion frame and the trained motion label to generate the corresponding motion style data when using captured style. However, FCRBM using label set in both the hidden and visible variables makes the generated data heavily dependent on the trained label. Furthermore, traditional FCRBM can not deal with unlabeled data, which are more common in many real-world applications [12].

This paper proposes a tree factored conditional restricted Boltzmann machine (TFCRBM) for mixed-style motion generation. We didn't change the number of the factors, and tied labels and previous features to RBM by factors just as FCRBM. Unlike FCRBM, we do not bind the past labels directly to the three factors, but only to one factor. Moreover, considering that the tree structure itself has the ability of learning temporal features [13, 14], we organize factors into a tree structure, which allows the model to learn smooth transformation of styles, even it no longer depends on the trained label. Experiments show that the proposed model can be adapted to capture the temporal features of unlabeled data.

2   Notations and Preliminaries

The CRBM is a modified RBM for time-series data (Fig. 1(a)). Because the frame data are continuous, the units in CRBM should be linear and real-valued variables with noise. FCRBM is based on CRBM which preserves the most important computational properties of CRBM. It contains multiplicative three-way interactions that efficiently allow interactive weight between two nodes be dynamically adjusted by the state of the third node (Fig. 1(b)). The energy function of FCRBM is given by [15]:

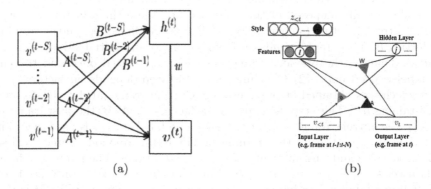

(a)   (b)

Fig. 1. Architecture of Conditional Restricted Boltzmann Machine (CRBM) and Factored Conditional Restricted Boltzmann Machine (FCRBM). (a) One-level model of CRBM. (b) One-level model of FCRBM.

$$E(v_t, h_t \mid v_{<t}, y_t, \theta) = \frac{1}{2} \sum_i (\hat{a}_{i,t} - v_{i,t})^2 - \sum_f \sum_{ijt} W_{if}^v W_{jf}^h W_{lf}^z v_{i,t} h_{j,t} z_{l,t} - \sum_j \hat{b}_{j,t} h_{j,t}, \quad (1)$$

Equation (1) corresponds to three sub-models. For each sub-model, the weight matrix is now replaced by three weight sets connected to the factor. The feature z_t is determined by a "hot" linear function, which is the encoding of the label y_t:

$$z_{l,t} = \sum_p R_{pl} y_{p,t} \, . \quad (2)$$

3   Motivation

In FCRBM, using factors makes it learn more and better styles, and sharing parameters can synthesize higher quality sports, as shown in Fig. 1(b). Training process is divided into three steps: (1) Getting h. In this step, initial value $v_{<t}$, v_t, and past frame label set $z_{<t}$ are contributed to h by factors B and W respectively. That is, the previous frame label set $z_{<t}$ is used twice in the process. (2) Obtaining current frame v^-. In this step, it depends on factors W, A and the h that was obtained at previous step to get v^-. The past frame label set $z_{<t}$ is used twice. (3) Obtaining current hidden unit h^-. This step depends on factors B, W, the previous frame set $v_{<t}$, $z_{<t}$ and the current frame v^-. The previous frame label set is used twice. That is, the past frame label set $z_{<t}$ is used six times throughout the process. Noticed that the number of previous frame labels used in FCRBM is $n_t (n_t >= 1)$. The total labels in the training is $6n_t$. Moreover, the obtained h and h^- heavily depend on the previous frame label set $z_{<t}$. As a result, label must be given in the later generation process, and value of the label must be a trained one. That means, FCRBM cannot flexibly achieve human motion style transition, let alone produce a new style.

Inspired by that tree structure can group similar values naturally, and that relevance between values is organized automatically and meaningfully, we presented the tree factored conditional restricted Boltzmann machine (TFCRBM). The model is still composed of three factors, but these three factors are organized as a tree structure (shown in Fig. 2). There are three steps in the training: (1) Getting h. This step depends on factors C, A, B and initial values $z_{<t}$, $v_{<t}$, z and v. That is, the previous frame label set $z_{<t}$ is used once and the current label is used once. (2) Obtaining v^-. This step depends on factors C, B and h which was obtained at previous step. Current frame label z is used once. (3) Obtaining h^-. This step depends on the factors C, A, B, the past frame set $v_{<t}$, $z_{<t}$, the current frame v^- and z. Past frame label set $z_{<t}$ is used twice and the z is used once. During the training of TFCRBM, previous frame label set $z_{<t}$ is used twice and current frame label z is used 3 times. The number of frame labels used is $2n_t + 3 \leq 5n_t$ $(n_t \geq 1)$. Note that in step (2), it only needs the current frame label z. Furthermore, the two-way interaction in FCRBM is multiplication interaction, while in our TFCRBM is the additive interaction, which makes TFCRBM have lower reconstruction error.

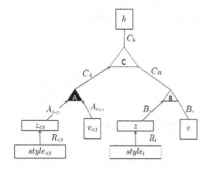

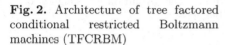

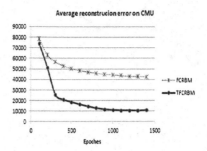

Fig. 2. Architecture of tree factored conditional restricted Boltzmann machines (TFCRBM)

Fig. 3. Evolution of reconstruction errors in 1400 epoches

4 Modeling of TFCRBM for Mixed Motion Style

To model multi-pose human motions, we use factorization, three-way interaction and mutual interaction. Figure 2 shows a different way of using three-way interactions of factorization to take real-valued style features. As standard FCRBM, the model defines a joint probability distribution between v_t and h_t, which is conditional on the past N observations, the style label $v_{<t}$, the past labels $style_{<t}$ and the current label $style_t$ to model the parameters θ.

4.1 TFCRBM Training Based on the Style

Note that in our model (shown in Fig. 2), there are 11 parameters that need to be updated, which are C_h, C_A, C_B, B_v, B_z, $A_{v<t}$, $A_{z<t}$, $R_{<t}$, R_t, $hidbias$ and $visbias$. The main idea of training RBM is applying Gibbs sampling on $h^{(0)}$, $h^{(k)}$, $v^{(0)}$ and $v^{(k)}$ and then getting the increments using CD algorithm. The TFCRBM training process is similar to that of RBM, but due to the additive three factors as well as the factor and factor which organized in a tree structure, the connection between one factor and another will have a cumulative effect on the third factor. We define an effect on a factor as a contribution, defining the effects of factor and factor as cross contributions on the third path.

Definition 1. (Contribution). The contribution of a factor on a way is the product of weight and scalar on the way. For example, $v \rightarrow B = v * B_v$.

Definition 2. (Cross contribution on the third way). The cross contribution of a factor on the third way is the sum contributions of the other two ways on this factor. For example, $z \rightarrow B \leftarrow v = (z \rightarrow B) + (v \rightarrow B)$.

Definition 3. (Gibbs sampling contribution on the third way). The Gibbs sampling contribution on the third way is the product of cross contribution and Gibbs sampling scalar on the way. For example, $C^{(k)}_{B-product} = (z \rightarrow B \leftarrow v)^{\top} * (p^{(k)}(h) \rightarrow C \leftarrow A)$. The training details are as follows:

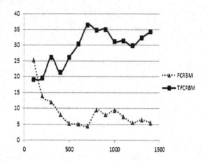

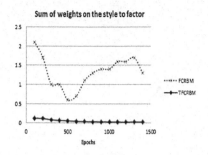

Fig. 4. The evolution of sum of weights which connects to hidden units directly.

Fig. 5. The evolution of weights of label class.

(1) Obtaining $h^{(0)}$.

$$p^{(0)}(h) = \frac{1}{1 + exp(-(B \to C \leftarrow A) * C_h^\top - hidbias)}, \tag{3}$$

where $B \to C \leftarrow A = (B \to C) + (A \to C)$, $B \to C = (z \to B \leftarrow v_t) * C_B$, $A \to C = (z_{<t} \to A \leftarrow v_{<t}) * C_A$, $z_{<t} \to A \leftarrow v_{<t} = (z_{<t} \to A) + (v_{<t} \to A)$.

(2) Calculating the Gibbs contribution when $k = 0$. There are 11 parameters that need to be updated in our model, which are:

$$C^{(0)}_{h-product} = p^{(0)}(h)^\top * (B \to C \leftarrow A), \tag{4}$$

$$C^{(0)}_{B-product} = (z \to B \leftarrow v)^\top * (p^{(0)}(h) \to C \leftarrow A), \tag{5}$$

$$C^{(0)}_{A-product} = (z_{<t} \to A \leftarrow v_{<t})^\top * (p^{(0)}(h) \to C \leftarrow B), \tag{6}$$

$$B^{(0)}_{v-product} = v^\top * (C \to B \leftarrow z), \tag{7}$$

$$B^{(0)}_{z-product} = (style_t \to z)^\top * (C \to B \leftarrow v), \tag{8}$$

$$A^{(0)}_{v_{<t}-product} = (v_{<t})^\top * (C \to A \leftarrow z_{<t}), \tag{9}$$

$$A^{(0)}_{z_{<t}-product} = (style_t \to z_{<t})^\top * (C \to A \leftarrow v_{<t}), \tag{10}$$

$$R^{(0)}_{t-product} = style_t^\top * ((C \to B \leftarrow v) * R_t^\top), \tag{11}$$

$$R^{(0)}_{<t-product} = (style_{<t})^\top * ((C \to A \leftarrow v_{<t}) * R_{<t}^\top), \tag{12}$$

$$v^{(0)}_{act} = sum(v, 1)_{1, numdims}, \tag{13}$$

$$h^{(0)}_{act} = sum(p^{(0)}(h), 1)_{1, numhid}. \tag{14}$$

(3) Obtaining $h^{(k)}, v^{(k)}$.

$$v^{(k)} = (C^{(k)} \rightarrow B \leftarrow z) * B_v^\top + visbias \tag{15}$$

$$p^{(k)}(h) = \frac{1}{1 + exp(-(B^{(k)} \rightarrow C \leftarrow A) * C_h^\top - hidbias)}. \tag{16}$$

where $C^{(k)} \rightarrow B \leftarrow z = (C^{(k)} \rightarrow B) + (z \rightarrow B)$, $C^{(k-1)} \rightarrow B = (p^{(k-1)}(h) \rightarrow C \leftarrow A) * C_B^\top$, $B^{(k)} \rightarrow C \leftarrow A = (B^{(k)} \rightarrow C) + (A \rightarrow C)$, $B^{(k)} \rightarrow C = (z \rightarrow B \leftarrow v^{(k)}) * C_B$.

(4) Calculating k steps Gibbs contributions of the 11 parameters. The contribution of each parameter is the same as step (2), except that $h^{(0)}, h^{(0)}$ is replaced by the corresponding $h^{(k)}, h^{(k)}$.

(5) Calculating the increment of each parameter.

$$\Delta f_r = f_r^{(0)} - f_r^{(k)}, \quad \Delta hidbias = h_{act}^{(0)} - h_{act}^{(k)}, \quad \Delta visbias = v_{act}^{(0)} - v_{act}^{(k)}, \tag{17}$$

where $f = \begin{cases} C & r \in \{h, A, B\}, \\ A & r \in \{style_{<t}, v_{<t}, C\}, \\ B & r \in \{style_t, v, C\}. \end{cases}$

5   Experiments

In our experiments, we used labeled data comes from [15], and the unlabeled data comes from all the motion styles of subject 35 in the CMU database (http://mocap.cs.cmu.edu/.) We preprocessed the motion data with labels according to the method in [15]. Motion styles of unlabeled data are 'jogging', 'running', 'navigating around obstacle' and 'walking'. All data in subject 35 in CMU was sampled at 40 Hz and the first few frames on each sequence were removed. The original frame order was reorganized randomly, dividing data into small batches. Thus, motions were spread around the training sequence, which means that the order of updates was randomly arranged.

Reconstruction Error and Weights to Hidden Units. The proposed TFCRBM model is consistent on labeled dataset in the following parameter settings: (1) Number of hidden units and factor units were 600, 200 respectively, and number of past frames that each frame relied on was $n_t = 12$; (2) Small batches were used (scale 100) to the training; (3) All parameters used a learning rate of 10^{-2}, except for those parameters that were connected to the factors and parameters $R_{<t}$ and R_t indicating the label feature. Figure 3 is the evolution of reconstruction errors in 1400 epochs. It can be seen that reconstruction error in the corresponding epoches of our TFCRBM is smaller than that of FCRBM and is significantly reduced as epoch increases. TFCRBM can characterize temporal features and store them in the corresponding factor weights, especially on the weight of factor C on top (Figs. 2 and 4). This time-dependent capture is benefited from the factorial connection of tree structure.

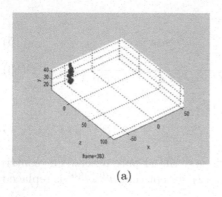

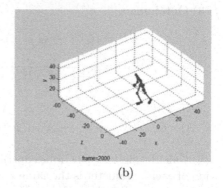

(a)   (b)

Fig. 6. The generated frames of TFCRBM. (a) The 383th frame of the newly generated 2000 frames in the case of selecting label 1('cat'). (b) The 2000th frame of the newly generated 2000 frames in the case of selecting label 14 (not in training labels).

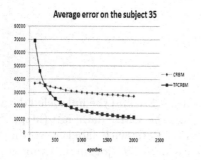

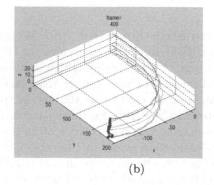

Fig. 7. Evolution of reconstruction errors in layer1 in 1400 epoches given the unlabeled data

Fig. 8. The generated frames by TFCRBM without training label.

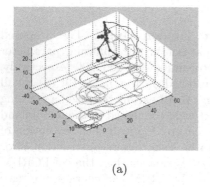

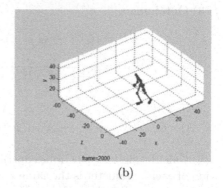

(a)   (b)

Fig. 9. Trajectories of generated frames by TFCRBM and CRBM. (a) TFCRBM. (b) CRBM.

Class Labels Learning. In order to verify that TFCRBM can learn class labels and not rely too much on the trained class labels, we carried out two groups of experiments. The first experiment used the labeled data from [15]. Figure 5 presents the evolution of labels by epochs. FCRBM learned labels from the beginning. As epoch increased, the weight of label classes arose. This indicated that FCRBM learned more labels than TFCRBM, but it also demonstrated data synthesis using FCRBM depended more on labels. For example, when generating data, FCRBM performs better when the given label is arbitrarily selected in 1, 2, ..., 10 but poorly when given label not in 1, 2, ..., 10. Compared to FCRBM, TFCRBM can: (1) If the given label is in the training labels category, the synthesized new frame will try to satisfy the given label style as much as possible. On the other hand, new frame is satisfied to be the deduction of the past frames. (2) If the given label is not in the training label category (e.g. the given label is 14), the synthesized new frame will try to be satisfied being the deduction of the past frames as much as possible, and it is necessary to satisfy the mixed label category (e.g. the given label 14 can the mixed by 1 and 4 label). Thus the newly generated data by TFCRBM has a stronger generalization. Figure 6(a) is the 383th frame in the newly generated 2000 frames by our TFCRBM in the case of selecting label 1 ('cat'). It can be seen that the frame does have 'cat' characteristics. And at about 800th frame, the generated data is displayed as 'graceful'. The whole 2000 frame data has a more obvious style ('cat') constraints. Figure 6(b) shows the new frame generated by our TFCRBM model in label 1 ('cat') and label 4 ('drunk'), respectively.

The second group of experiments used unlabeled data. The experimental configuration was as same as that in Sect. 5. In order to facilitate the use of unlabeled data, we set the labels in TFCRBM to 0, and disrupted the frames order. Compared with the traditional CRBM (same configuration), our TFCRBM used only one layer of RBM and had a lower reconstructed error (Fig. 7), resulting in a smoother transition between pose and pose (Fig. 8). Figure 9 shows the trajectories of generated frames by TFCRBM and CRBM using unlabeled data (consisting of six limbs-end data). It can be seen that TFCRBM learns more postures.

6   Conclusion

This paper presented a new model TFCRBM. Three factors are added to the traditional CRBM so that past data and label data can be used as conditions to learn. Moreover, the factors are organized in a tree structure, which makes only part of the TFCRBM factors connected labels directly. TFCRBM model can learn more temporal features rather than label features, can generate new data with both the given label in or not in the training labels. Moreover, TFCRBM model can learn temporal features in the case of no labels. Therefore the proposed model is more suitable for real-time data learning.

Acknowledgments. This work was supported by the National Science Foundation of China (Grant Nos. 61375065 and 61625204), partially supported by the State Key Program of National Science Foundation of China (Grant Nos. 61432012 and 61432014).

References

1. Gan, Z., Li, C., Henao, R., Carlson, D.E., Carin, L.: Deep temporal sigmoid belief networks for sequence modeling. In: Advances in Neural Information Processing Systems, pp. 2467–2475 (2015)
2. Zhou, F., De la Torre, F., Hodgins, J.K.: Hierarchical aligned cluster analysis for temporal clustering of human motion. IEEE Trans. Pattern Anal. Mach. Intell. **35**(3), 582–596 (2013)
3. Lehrmann, A.M., Gehler, P.V., Nowozin, S.: Efficient nonlinear Markov models for human motion. In: Proceedings of the IEEE Conference on Computer Vision and Pattern Recognition, pp. 1314–1321 (2014)
4. Li, K., Fu, Y.: Prediction of human activity by discovering temporal sequence patterns. IEEE Trans. Pattern Anal. Mach. Intell. **36**(8), 1644–1657 (2014)
5. Xie, C., Lv, J., Li, X.: Finding a good initial configuration of parameters for restricted Boltzmann machine pre-training. Soft Comput. **21**, 1–9 (2016)
6. Hinton, G.E., Salakhutdinov, R.R.: Reducing the dimensionality of data with neural networks. Science **313**(5786), 504–507 (2006)
7. Elaiwat, S., Bennamoun, M., Boussaid, F.: A spatio-temporal RBM-based model for facial expression recognition. Pattern Recogn. **49**, 152–161 (2016)
8. Sutskever, I., Hinton, G.: Learning multilevel distributed representations for high-dimensional sequences. In: Artificial Intelligence and Statistics, pp. 548–555 (2007)
9. Sutskever, I., Hinton, G.E., Taylor, G.W.: The recurrent temporal restricted Boltzmann machine. In: Advances in Neural Information Processing Systems, pp. 1601–1608 (2009)
10. Mittelman, R., Kuipers, B., Savarese, S., Lee, H.: Structured recurrent temporal restricted Boltzmann machines. In: Proceedings of the 31st International Conference on Machine Learning (ICML 2014), pp. 1647–1655 (2014)
11. Taylor, G.W., Hinton, G.E., Roweis, S.T.: Modeling human motion using binary latent variables. In: Advances in Neural Information Processing Systems, vol. 19, p. 1345 (2007)
12. Längkvist, M., Karlsson, L., Loutfi, A.: A review of unsupervised feature learning and deep learning for time-series modeling. Pattern Recogn. Lett. **42**, 11–24 (2014)
13. Zhang, P., Zhou, C., Wang, P., Gao, B.J., Zhu, X., Guo, L.: E-tree: an efficient indexing structure for ensemble models on data streams. IEEE Trans. Knowl. Data Eng. **27**(2), 461–474 (2015)
14. Li, J., Luong, M.T., Jurafsky, D., Hovy, E.: When are tree structures necessary for deep learning of representations? arXiv preprint (2015). arXiv:1503.00185
15. Taylor, G.W., Hinton, G.E.: Factored conditional restricted Boltzmann machines for modeling motion style. In: Proceedings of the 26th Annual International Conference on Machine Learning, pp. 1025–1032. ACM (2009)

A Piecewise Hybrid of ARIMA and SVMs for Short-Term Traffic Flow Prediction

Yong Wang, Li Li$^{(\boxtimes)}$, and Xiaofei Xu

School of Computer and Information Science, Southwest University,
Chongqing 400715, China
wy8654@email.swu.edu.cn, lily@swu.edu.cn, nakamura.desutini@gmail.com

Abstract. Short-term traffic flow is a variable affected by many factors. Thus, it is quite difficult to forecast accurately with only one model. The ARIMA model and the SVMs model have their own advantages in terms of linearity and nonlinearity. Therefore, making full use of the advantages of ARIMA model and SVMs model to predict traffic flow can significantly improve the overall effect. The current hybrid approach does not take full account of the characteristics of the data, which cause the effect of hybrid model is not always good. In this paper, first of all, we will use time series analysis and feature analysis to find the characteristics of data. Then, based on the analysis results, we decided to use the method of piecewise to fit the data and make the final prediction. The experiment shows that the piecewise hybrid model can give better play to the advantages of the two models.

Keywords: Traffic flow prediction · ARIMA model · SVMs model · Hybrid model

1 Introduction

It is very important to forecast the road traffic flow, which has great influence on the passenger, traffic department and business organization. Traffic flow prediction has become a key issue in intelligent transportation system study [1–3]. However, the traffic flow data based on time series are affected by many uncertain factors, such as abnormal weather, traffic accidents on roads, temporary maintenance of some roads. The existence of these problems increases the complexity of traffic flow prediction.

At present, many models have been proposed to solve the problem of traffic flow prediction [1,2,4]. They are divided into two categories according to their characteristics: the traditional methods represented by time series and the modern intelligent methods represented by SVMs. The traditional methods in which ARIMA is in the most common use have the advantages of mature technology and simple algorithm, but these methods are based on linear analysis. The modern intelligent methods are characterized by intelligent learning: including expert system [5], artificial neural networks [6], and support vector machines (SVMs)

© Springer International Publishing AG 2017
D. Liu et al. (Eds.): ICONIP 2017, Part V, LNCS 10638, pp. 493–502, 2017.
https://doi.org/10.1007/978-3-319-70139-4_50

[7], in which SVMs, owing to the great power to generalize, global optimal solution and fast calculation, have become the research hotspot of many subjects. In order to make full use of these two kinds of methods, in recent years, many combination forecasting methods that combine two or more models have been proposed [8, 13].

In this paper, we divide the work into two parts. First of all, based on the analysis of time series, we can find the factors which may influence the traffic flow effectively. We set these factors as specific feature, and then the importance of these feature will be analyzed. After that, we will use a piecewise hybrid model to predict traffic flow. Experiments show that the effect of prediction is improved obviously by the method of subsection mixing.

The rest of the paper is organized as follows: Sect. 2 briefly reviews related research work on the prediction of traffic flow and Sect. 3 presents the hybrid model. The experimental results are presented in Sect. 4, and the conclusions are offered in Sect. 5.

2   Related Work

Traffic flow prediction has been a popular research direction in intelligent transportation system. In the past few decades, researchers have used a wide variety of specifications to produce short-term traffic flow forecasts, such as BoxCJenkins autoregressive integrated moving average (ARIMA) models [9], Kalman filtering [3], and nonparametric statistical methods [10]. Among them, the ARIMA model is widely used, and has achieved good effect. This method makes predictions by extracting the correlation between sequences. Besides, Even in the case of a small data size, it can also get relatively good prediction results.

On the other hand, SVR (the algorithm of SVM in regression) has been successfully used to predict traffic parameters such as hourly flow and travel time. Wu et al. [11] presented the SVR for travel-time prediction and compared it with other baseline travel-time prediction methods using real highway traffic data. CastroNeto et al. [1] presented an OLSVR for the prediction of short-term freeway traffic flow under both typical and atypical conditions. In addition, on the basis of this article, Jeong et al. [12] modified the SVR model and put forward the OLWSVR model to predict the traffic flow. We find that SVR has been widely used in this problem. So we finally continue to choose to use this method to predict traffic flow data.

At present, the hybrid model is proposed to deal with the problem of traffic flow prediction [8, 13]. The current hybrid approach treats all the residual of the ARIMA model as a set of nonlinear data, and then SVR is used to fit these residuals. Actully, just as this paper analyses [14], the effect of this hybrid approach is not always satisfactory. Time series data tend to have different characteristics in different time periods, and it is unwise to consider all residuals as nonlinear data without consideration. This is the main reason for the instability of the current hybrid model. Based on the above situation, we combine the ARIMA model and the SVM model from another point of view and the piecewise hybrid model is applied to traffic flow prediction.

3   Hybrid ARIMA and SVM Model

The behavior of traffic flow data can not easily be captured. As we can see from Fig. 1, before the seventh data point, the data have a clear linear trend, while the nonlinear trend of the remaining data points is more obvious. Figure 2 has a similar situation, but the trend is opposite. This is also easy to explain in reality, and people travel more frequently at some period of time. So, the traffic situation is more complicated during that time. Therefore, a hybrid strategy that has both linear and nonlinear modeling abilities is a good alternative for forecasting future traffic flow. It is obviously that the ARIMA and the SVMs models have different capabilities to capture data characteristics in linear or nonlinear domains, so the hybrid model proposed in this study is composed of the ARIMA component and the SVMs component.

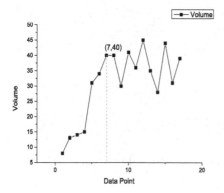 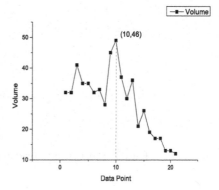

Fig. 1. From 6:00 a.m. to 10:00 a.m.   **Fig. 2.** From 2:00 p.m. to 6:00 p.m.

Figure 3 shows the process of the whole segmented mixture model. The hybrid model(Z_t) can be represented as follows:

$$Z_t = Y_t + N_t \qquad (1)$$

where Y_t is the linear part and N_t is the nonlinear part of the hybrid model. Both Y_t and N_t are estimated from the data set. We use $\hat{Y}_t$ to represent the prediction value of the ARIMA model at time t. Let ε_t represent the residual at time t as obtained from the ARIMA model. Then

$$\varepsilon_t = Z_t - \hat{Y}_t \qquad (2)$$

Now, let's divide the residuals into two parts. We use the SVMs model to fit the parts that are changing frequently. The other parts are no longer further fitted. We assume that the number we need to model by the SVMs model is n. The residual (ε_t') are modeled by the SVMs can be represented as follows

$$\varepsilon_t' = f(\varepsilon_{t-1}, \varepsilon_{t-2}, \cdots, \varepsilon_{t-n}) + \Delta_t \qquad (3)$$

where f is a nonlinear function modeled by the SVMs and Δ_t is the random error. Therefore, the combined forecast is

$$\hat{Z}_t = \begin{cases} \hat{Y}_t + \hat{N}_t & \text{the nonlinear part} \\ \hat{Y}_t & \text{others} \end{cases} \qquad (4)$$

Notably, $\hat{N}_t$ is forecast value of (3). We use the parameter x to represent the demarcation point between linear and nonlinear. The data point x determines the curve range to which the SVMs model is required to fit. We suggest that the data segments which change frequently should be jointly predicted by the two models and it will greatly enhance the effect of the hybrid model.

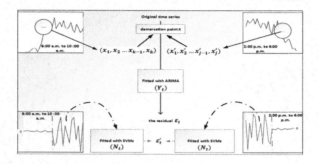

Fig. 3. Piecewise hybrid model procedure description

4   Experiments and Results

4.1   Experimental Data

We use a set of toll-gate traffic flow data from September 19th, 2016 to October 17th, 2016 provided by KDD cup. The data set includes the time for each vehicle to pass the toll-gate and types of vehicles from type 0 to type 7. In addition, the corresponding weather data sets are provided which incudes pressure, sea pressure, wind direction, wind speed, temperature, relative humidity and precipitation. As far as we know, this group of data is recorded somewhere in China, so we can clearly know the specific holiday time. For the purpose of prediction, we process the initial data into the standard traffic flow data and the time window size is set to 15 min.

4.2   Traffic Flow Time-Series Analysis

The time span of the entire dataset is close to a month, but our time window is set to 15 min. As we can see from the Fig. 4, the figure of the original time series changes too frequently during the busy period of traffic. Therefore, we use wavelet transform to de-noise the original sequence as shown in Fig. 5. It shows that the whole sequence has obvious periodicity. Besides, there will be a clearly up trend from about 6 a.m. until about 8 a.m. and a distinct downtrend in the afternoon every day. It is these acutely changing time quantum that we should focus on.

At the same time, in Fig. 5, we can found that the traffic flow was significantly higher than the other within a week of the time from the 1065th data point (September 30, 2016 at 2 a.m.) to 1829th data point (October 8, 2016 at 1 a.m.). The highest peak appears at the 1179th data point. In China, as we all know, this period is a statutory holiday. During this time, the travel volume will suddenly increase. Figure 6 shows the traffic flow corresponding to different types of vehicles. From this figure, we can see that the number of type 0 and type 1 is much larger than the others. Furthermore, during the holiday, the number of type 0 shot up. This may be due to the fact that people choose to travel by their own cars. In addition, we note that the traffic flow from 893rd data point to the 931st data point is higher than the other time (except holidays). We know that there is a continuous rainfall during this period by checking the weather dataset.

After the above analysis, we can find two interesting phenomena: First, the traffic flow has obvious periodicity, and the changing trend is basically the same at the same time every day. Second, the curve fluctuates frequently during certain periods of the day (for example, from 9 a.m. to 2 p.m.). In other time periods, the curves have obvious linear rising and downward trend.

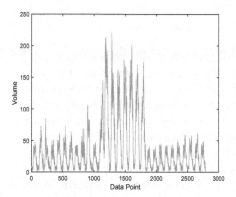

Fig. 4. Original sequence

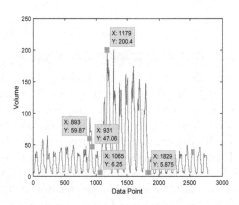

Fig. 5. De-noised sequence

4.3   Feature Analysis

From the previous analysis, we know that the change of traffic flow is closely related to time, vacation and weather. In addition, the weather dataset includes a number of features. In this section, we will use a specific approach to analyze the importance of each feature. From Fig. 5, the traffic flow in September 28th

Table 1. Feature list

Feature type	Feature example	Feature number
Time	Day	0
	Hour	1
	Minute	2
Weather	Pressure	3
	Sea pressure	4
	Wind direction	5
	Wind speed	6
	Temperature	7
	Relative humidity	8
	Precipitation (the amount of rainfall in a region)	9
	Duration	10
Holiday	0 or 1 (use 0 or 1 to indicate whether it is a holiday)	11

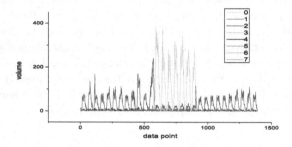

Fig. 6. Traffic flow of different vehicle types

was affected by the unusual weather. Although there will be rain at other time, their traffic flow is not affected. We found that the only difference in September 28th is a long period of rain. So, we assume that it may be that the duration of rainfall affects the traffic flow and we add the time span of rainfall as a feature. We finally extract 11 features from the traffic flow dataset to model its dynamics. Table 1 lists major features used in this study.

As we all know, feature selection is an essential task in machine learning. Since we need to use SVMs as the initial model, it is necessary to perform feature selection before modeling to achieve better results. Next, we use the method based on random forests to select features. Random forests are composed of multiple decision trees. Each node in the decision tree is a condition of a feature, and the data set is divided into two parts according to these conditions. We can use variance or least square fit to set these nodes (i.e., the optimal conditions). For a decision tree forest, we use the variance of the mean reduction as the value of feature selection.

In fact, there are only a few statutory holidays per year (except weekends). In order to prevent over fitting, according to the performance of the experiment, we do not take the holiday as a feature to train. As we can see from Fig. 7, the features related to time are the most important, while the importance of other features is not obvious. That is to say, we can ignore other features that are independent of time for the overall fitting effect. Of course, we can not simply treat abnormal weather data as outliers. In order to predict the traffic flow on rainy days accurately, we also analyze the feature importance of all the rainy days. As shown in Fig. 8, the duration of rainfall has become the most important factor. The results confirmed our previous conjecture: the real reason for traffic flow changes is not the amount of rainfall but the duration. Through feature analysis, we can find that traffic flow data has a strong time correlation, and many external features do not affect it significantly.

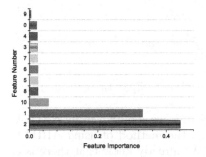

Fig. 7. Feature importance analysis

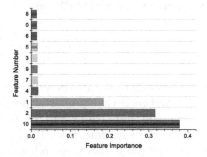

Fig. 8. Feature importance analysis in rainy day

4.4   Traffic Flow Prediction Using Hybrid Model

Based on the feature analysis above, the traffic data set was collected with traffic under two scenarios:

Scenario 1: Typical traffic conditions. In this scenario, no special occurrences that may significantly change the traffic pattern. In our experiments, the time window size was 15 min. we use two consecutive days of data points as historical data for model fitting when we use ARIMA model. After calculating the residual, in order to improve the effectiveness of the model, we use 21 days of data points as training data when we use the SVMs model. Finally, we use the two time quantum of the last day as the test data.

Scenario 2: Atypical traffic conditions. In scenario 2, the testing day (13th day) either was a special day of traffic (holiday) or had an unusual weather (continuous rainfall). Unlike the previous scenario, we use a cross validation approach to test our model because vacations and unusual weather are not common. In addition, in order to increase the accuracy of SVMs fitting, we increase the duration of rainfall as a feature in the case of rain.

To evaluate the prediction performance of the piecewise hybrid model, ARIMA, SVM and the ARIMA-SVMs model (current hybrid model) are selected

for experimental comparison. The method of using ARIMA to fit the daily traffic flow is as follows. Firstly, the load series is transferred into the stationary
time series by a periodic difference transformation and a first-order difference
transformation. We know that the periodic difference can eliminate the periodicity of the non-stationary sequence. In the case of the difference, the first-order
difference can eliminate the linear trend. As for the curve trend, we can use the
second to the third difference to eliminate it [10]. Secondly, the model of the raw
load series is confirmed through order determination, and the values of parameters are determined by parameter estimation. Finally, we use the confirmed
model to predict. The parameters of SVMs have a great effect on prediction
accuracy, so we use the grid search method [10] to find the optimal parameters.
In addition, we quantified the prediction performance of each model with mean
absolute percent error (MAPE) as

$$MAPE(\%) = \frac{1}{m} \sum_{t=1}^{m} \frac{|Z_t - \hat{Z}_t|}{Z_t} * 100 \tag{5}$$

where $\hat{Z}_t$ represents the predicted traffic flow at time t, Z_t is the actual traffic
flow data, and m is the number of predictions.

For example, from 6 a.m. to 10 a.m. in Scenario 1, we should use the ARIMA
model to fit the raw data first. In this Scenario, the method of using ARIMA
to forecast the daily load is as follows. Firstly, We already know that there is a
linear trend in the data sequence, so we use the first order difference to eliminate
it. At the same time, we use a periodic difference transformation to eliminate its
periodicity. Secondly, we confirm the parameter p, d, q of this model (ARIMA)
by order determination and parameter estimation. Finally, we use the confirmed
ARIMA (4, 2, 6) model to predict. After the model fitting process is over, we get
a set of residual sequences. We find that the nonlinear trend of the curve begins
at a certain point every day (usually between 7:30 a.m. and 8:30 a.m.). In other
words, after this time point, traffic has increased and the situation has become
complicated. There are four data points between 7:30 a.m. and 8:30 a.m., and
we use x_1, x_2, x_3, x_4 to represent these time points respectively. We use these
four points as the demarcation point, and then use the error of the test set to
find the right point. From Table 2, we can find that the overall error is minimal
when the time point is x_2 (that is, 7:45 a.m.). That is to say, we choose this time
point as the demarcation point.

Table 2. Determination of demarcation point

Time point	x_1	x_2	x_3	x_4
MAPE(%)	14.7	13.4	13.9	15.1

We sorted out residuals from 7:40 a.m. (cutoff point) to 10 a.m. and used the
new sequence to fit the SVMs model. We have extracted the necessary features

from the previous feature analysis. Now, we use these feature data to train the SVMs model. Then,through the grid search method, the proper values of them are determined as follows: $c = 20, \epsilon = 0, 13, \sigma = 3$. We apply the fitted piecewise hybrid model to the test set and obtain the final MPAE with a value of 13.4.

As for scenario 2, we selected one day during the holiday (October 4th) and a rainy day (September 28th) to do the experiment. The experimental procedure is similar to the previous one, and the difference is that it is necessary to add some features according to the previous analysis. As can be seen from Table 3, the overall prediction effect of the piecewise hybrid model is better than the others. Besides, despite the good performance of current hybrid model, there are still mixed results that are not as good as a single model (as described in [14]). Experiments show that we can get good results when we apply the idea of segmentation to the hybrid model.

Table 3. Experiment results

Time intervals		MAPE(%)			
		ARIMA	SVMs	ARIMA-SVMs	P-ARIMA-SVMs
Scenario 1	6 a.m. to 10 a.m.	17.5	16.8	15.8	**13.4**
	2 p.m. to 6 p.m.	18.9	15.4	16.7	**14.8**
Scenario 2.0 (holiday)	6 a.m. to 10 a.m.	16.2	15.6	13.9	**13.6**
	2 p.m. to 6 p.m.	17.7	15.3	13.5	**13.1**
Scenario 2.1 (rainy)	6 a.m. to 10 a.m.	18.5	16.2	16.7	**15.8**
	2 p.m. to 6 p.m.	17.2	15.4	14.8	**14.4**

5   Conclusions

This paper applies the piecewise hybrid model to predict. Time series analysis allows us to intuitively understand that traffic flow data has both linear and nonlinear characteristics. The process of feature analysis allows us to clearly understand that traffic flow data has a strong temporal correlation, and other external features have little impact. Experiments show that our consideration is meaningful, and the piecewise hybrid model does improve the prediction accuracy.

References

1. Castro-Neto, M., Jeong, Y.S., Jeong, M.K., Han, L.D.: Online-SVR for short-term traffic flow prediction under typical and atypical traffic conditions. Expert Syst. Appl. **36**, 6164–6173 (2009)
2. Jiang, X., Adeli, H.: Dynamic wavelet neural network model for traffic flow forecasting. J. Transp. Eng. **131**, 771–779 (2005)
3. Lee, W.H., Tseng, S.S., Tsai, S.H.: A knowledge based real-time travel time prediction system for urban network. Expert Syst. Appl. **36**, 4239–4247 (2009)

4. Sun, S., Xu, X.: Variational inference for infinite mixtures of Gaussian processes with applications to traffic flow prediction. IEEE Trans. Intell. Transp. Syst. **12**, 466–475 (2011)
5. Kandil, M.S., El-Debeiky, S.M., Hasanien, N.E.: Long-term load forecasting for fast developing utility using a knowledge-based expert system. IEEE Power Eng. Rev. **22**, 491–496 (2002)
6. Zhou, D.M., Guan, X.H., Sun, J., Huang, Y.: A short-term load forcasting system based on BP artificial neural network. Power Syst. Technol. **26**, 10–13 (2002)
7. Yuan-Cheng, L.I., Fang, T.J., Er-Keng, Y.U.: Study of support vector machines for short-term load forcasting. Proc. CSEE **5**, 654–659 (2003)
8. Nie, H., Liu, G., Liu, X., Wang, Y.: Hybrid of ARIMA and SVMs for short-term load forecasting. Energy Procedia **16**, 1455–1460 (2012)
9. Ahmed, M.S., Cook, A.R.: Analysis of Freeway Traffic Time-Series Data by Using Box-Jenkins Techniques (1979)
10. Smith, B.L., Williams, B.M., Oswald, R.K.: Comparison of parametric and non-parametric models for traffic flow forecasting. Transp. Res. Part C: Emerg. Technol. **10**, 303–321 (2002)
11. Wu, C.H., Wei, C.C., Su, D.C., Chang, M.H.: Travel time prediction with support vector regression. In: 2003 Proceedings of Intelligent Transportation Systems, vol. 2, pp. 1438–1442 (2004)
12. Jeong, Y.S., Byon, Y.J., Castro-Neto, M.M., Easa, S.M.: Supervised weighting-online learning algorithm for short-term traffic flow prediction. IEEE Trans. Intell. Transp. Syst. **14**, 1700–1707 (2013)
13. Zhu, Z., Sun, Y., Li, H.: Hybrid of EMD and SVMs for Short-Term Load Forecasting. In: IEEE International Conference on Control and Automation, pp. 1044–1047 (2007)
14. Pai, P.F., Lin, C.S.: A hybrid ARIMA and support vector machines model in stock price forecasting. Omega **33**, 497–505 (2005)

TMRCP: A Trend-Matching Resources Coupled Prediction Method over Data Stream

Runfan Wu, Yijie Wang$^{(\boxtimes)}$, Xingkong Ma, and Li Cheng

National Laboratory for Parallel and Distributed Processing College of Computer,
National University of Defense Technology, Changsha 410073, China
{wurunfan,wangyijie,maxingkong,chengli09}@nudt.edu.cn

Abstract. Resource prediction promotes dynamic scheduling and
energy saving in cloud computing. However, resource prediction becomes
a challenge with the diversity and dynamicity of the cloud environment. Existing methods merely focus on single specific resource and
ignore the correlation among resources, resulting in inaccurate predictions. Therefore, we propose a trend-matching resources coupled prediction method (TMRCP) based on incremental learning over data stream,
which consists of three algorithms. Firstly, to cope with the diversity of
the cloud environment, we propose a Resources Utilization Trend Matching algorithm (RUTM), which defines a new similarity measure for multidimensional sequences and takes the correlation among resources into
consideration. Secondly, we propose a dynamic prediction window adjustment algorithm that selects appropriate prediction length for different
resource utilization trends to overcome the disadvantage of fixed window.
Thirdly, in response to the sudden changes, we put forward a mixed synthesis algorithm to improve the robustness of the method. Experiments
on Google's cluster usage trace show that the Mean Absolute Percentage Error of TMRCP is 4.7%, 20% better than the state-of-the-art. In
addition, the TMRCP is still accurate in multi-step-ahead prediction.

Keywords: Multiple resources prediction · Resources correlation ·
Trend-matching

1 Introduction

Cloud computing is emerging as an increasingly popular computing paradigm
because of its elasticity of providing resources. As a result of the diversity of
the cloud environment, the resource utilization of data center changes dynamically. The maximum resources allocation of the application required results in
a waste of resources while insufficient allocation of resources leads to a Service
Level Agreement (SLA) violation. Currently, cloud data centers generally adopt
the maximum resources allocation to achieve a high quality of service (QoS).
Consulting firm McKinsey and Gartner estimated the resource utilization of
data centers at 6% to 12% [1], revealing waste of resources. Therefore, dynamic
resource allocation according to the needs of the application is the best solution.

© Springer International Publishing AG 2017
D. Liu et al. (Eds.): ICONIP 2017, Part V, LNCS 10638, pp. 503–513, 2017.
https://doi.org/10.1007/978-3-319-70139-4_51

Resource prediction is a vital step of on demand resource allocation. Accurate resource prediction is helpful to resource management and cost reduction of data centers. However, it is difficult to accurately predict multiple resources because of the following reasons:

Diversity of the cloud environment. Various applications with different types of resource demands are running in the cloud. Figure 1 shows some CPU utilization sequences extracted from the Google cluster [2]. Existing methods merely focused on the relation with their proximity and did not sufficiently mine the trends of history data. Furthermore, different resources correlate with each other and the overload of any single resource will cause a poor overall performance. For example, when CPU is overloaded, I/O operations would be slow even if the memory is idle.

Different prediction characteristics. Prediction characteristic is the relation between prediction and history values. In Fig. 1, curve (a) shows an overall trend, but curve (b) and (c) are not continuous trends, whose first few values have no effect on prediction. Sequences with different trends have different prediction characteristics and fixed window would cause an inaccurate prediction.

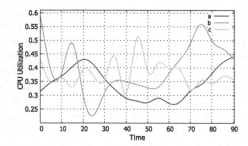

Fig. 1. Examples of CPU usage from Google cluster

Dynamicity of the cloud environment. The resource utilization of the application fluctuates dynamically with the frequent change of workloads. It includes not only the predictable workloads based on the fixed patterns, but also the unexpected workloads because of the sudden rising or falling.

To this end, we propose a trend-matching resources coupled prediction method based on incremental learning over data stream, called TMRCP. In the training phase, the TMRCP first identifies trends of the history data, then builds a Support Vector Regression (SVR) model with appropriate prediction window for each trend. In the prediction phase, the TMRCP first updates the model using the newly arrived monitor data, then matches the new sequence with the identified trends and predicts load based on the match result. The TMRCP contains three key algorithms and provides the following contributions:

1. To take advantage of the resource utilization trends in history data and the correlation among resources, we propose a Resources Utilization Trend Matching algorithm (RUTM). It defines a new similarity measure for multi-dimensional sequences and takes correlation among resources into consideration.
2. To overcome the disadvantage of the fixed window, we propose a dynamic prediction window adjustment algorithm, which selects appropriate prediction length for sequences with different resource utilization trends.
3. To respond to sudden workload changes, we propose a mixed synthesis algorithm, which improves the robustness of the method.

Experiments on Google's cluster usage trace show that the Mean Absolute Percentage Error (MAPE) of TMRCP is 4.7%, 20% better than the state-of-the-art. In addition, the MAPE is less than 10% within four-step ahead prediction, indicating the TMRCP has a 25% improvement in multi-step-ahead prediction.

The rest of this paper is organized as follows. In Sect. 2, we review the related work on the resource prediction. Section 3 presents the architecture of the TMRCP. Section 4 describes the TMRCP in detail. We evaluate the performance of the TMRCP in Sect. 5. Finally, we conclude the paper in Sect. 6.

2   Related Work

Recently, there has been considerable interest in resource prediction. The majority of existing studies are devoted to single specific resource, such as CPU and memory, which are classified into four main methods: time series models, machine learning methods, signal processing methods and hybrid methods.

Time series models are built by curve fitting and parameter estimation, including autoregressive (AR) models, the integrated (I) models, the moving average (MA) models and their combinations. Tang et al. in [3] used Auto Regression (AR) to predict the future memory demand and the parameter of each VM is updated frequently. Roy et al. in [4] proposed an ARMA method to forecast the number of users in the cloud and Xiao et al. in [5] proposed an improved EWMA algorithm to predict resource demand. They cannot adapt to the dynamicity because of their fixed parameters. ARIMA models were used by Niu et al. in [6] and Calheiros et al. in [7] to predict the server bandwidth and workload for SaaS providers, but finding the appropriate parameters is critical. Time series models are simple and low complexity, whose parameters are determined by experience and past observations. They have a good performance at the linear relationship, however, they cannot capture complex changes of the cloud.

The machine learning methods analyze history data from different views. Islam et al. in [8] presented empirical prediction models for adaptive resource prediction in a cloud environment based on Neural Network and Linear Regression to predict upcoming resource demands, the result showed NN performs well than LR. Shyam and Manvi in [9] proposed a Bayesian model to determine virtual resource requirement of the CPU/memory intensive applications on the basis of workload patterns. Support Vector Regression is suitable for the prediction of small sample nonlinear data and used to predict the fast change of the workloads by Liu et al. in [10]. The machine learning methods improve the prediction accuracy while increasing the computation complexity.

Signal processing approaches are based on harmonic analysis and filtering using the Fourier transform and spectral density estimation. Dabbagh et al. in [11] proposed a Wiener filter approach for future CPU or memory demands, however, they assumed the utilization is simple and weighted sum of the recently observed samples, resulting in low accuracy. Subbiah et al. in [12] presented a wavelet based resource prediction algorithm using a fixed size subset of the history data that is not adaptive to dynamicity of the cloud. Signal processing and filtering usually provide the accurate results for linear models. Due to the non-linear behavior and the dynamicity of the cloud, they cannot perform well.

The hybrid methods consider that single method cannot provide the most accurate results for different workloads [13]. Khan et al. in [14] developed a co-clustering technique to identify VMs that have similar workload patterns, then predict variations of patterns by Hidden Markov Modeling. Istin et al. in [15] proposed a linear-nonlinear method, which predicts with different steps by ARIMA and combines the results with neural network. The hybrid methods increase computation complexity, but effective combinations improve the accuracy.

3   Architecture

We summarize the complete procedure of the TMRCP in Fig. 2, which is divided into two phases: training and prediction.

In the training phase, we first split the historical monitor data to sequences by the sliding window and then normalize these sequences in the preprocessing step for the purpose of showing the trend over a period. Afterwards we adaptively identify the trends of preprocessed sequences in trend-matching step. Finally, we select the appropriate prediction window and train a SVR model for each trend in dynamic window adjustment step.

In the prediction phase, a new sequence is formed by the newly arrived monitor data and the data in the cache. Because the newly arrived load is the actual value of last prediction, we validate the last prediction and update model incrementally. Then we match the sequences with our identified trends in the trend-matching step. Finally, we predict workload in the mixed synthesis step.

4   Design of the TMRCP

4.1   Resources Utilization Trend Matching Algorithm

In this subsection, we propose a Resources Utilization Trend Matching algorithm. The algorithm identifies the historical resource utilization trends and predicts

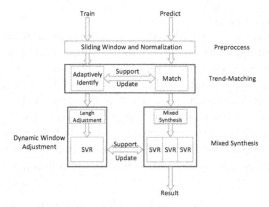

Fig. 2. Architecture of the TMRCP

the future resource utilization trend based on our newly defined similarity. Three key components are contained as follows.

Similarity Measure

Preprocessed sequences record the monitor data of the resources over the window time, such as CPU, cache, memory, disk and I/O time. Each sequence is treated as a K * N matrix, representing the trend of the resources utilization. The correlation among resources is taken into consideration when we study the trends of multiple resources utilization through the following defined similarity measure. First, we define the trends of various resources over the window time and the resources utilization of the same window time based on the Gaussian Function as follows:

$$TS = K - \sum_{i=0}^{k-1} Gauss\left(A\left(i\right), B\left(i\right)\right).$$ (1)

$$RS = N - \sum_{j=0}^{N-1} Gauss\left(A\left(:,j\right), B\left(:,j\right)\right).$$ (2)

Then, we introduce the variance to consider the influence of deviate and avoid the disadvantage that the translational sequences have the same similarity.

$$VA = \sum_{i=0}^{k-1}\sum_{j=0}^{N-1}\left(A\left(i,j\right) - B\left(i,j\right)\right).$$ (3)

We define the similarity of multi-dimensional sequences as follow. TS and RS represent the trend of each resource and system status, while VA represents the deviate of sequences. We normalize them and set coefficients for them in order to make the similarity at 0 to 1, while 0 represents similar and 1 represents different.

$$Similarity = \frac{1}{3K} * TS + \frac{1}{3N} * RS + \frac{1}{3*(4NK-4)} * VA.$$ (4)

Adaptive Identification

In the training phase, the trend-matching algorithm adaptively identifies the trends of the sequences. To achieve this, we improve the spectral clustering algorithm. The distance measure is replaced by the previously defined similarity (Algorithm 1, Line 4), so that the improved spectral clustering algorithm can process multi-dimensional sequences with consideration of the correlation among resources. Then we set a threshold to adaptively select the number of clusters (Algorithm 1, Line 5–6). The number of clusters will increase when similarity in the clusters is smaller than threshold, so that it can ensure that the trends within a cluster are similar and among clusters are different.

Matching

In the prediction phase, the trend-matching algorithm compares the new sequence with the last matching and updates the trend-matching model incrementally (Algorithm 1, Line 8). Then we evict the oldest value of the sequence and combine the newly arrived monitor data with the sequence (Algorithm 1, Line 9–10). Finally, we match the new sequence of $k-1$ sizes with the first $k-1$ elements of the identified trends in the training step (Algorithm 1, Line 11). The matching is based on the previously defined similarity. So we get the possibility distributions of various trends of the K-size sequence.

Algorithm 1: Resources Utilization Trend Matching Algorithm

```
Input:   Historical Monitor Data, Monitor Stream, threshold
Output:  Adaptive Clusters, Match Result
    1:   Sequences Set = Sliding Window (Historical Monitor Data)
    2:   For sequence in Sequences Do:
    3:       Normalization (sequence)
    4:   Spectral Clustering Using our Definition
    5:   While Existing Similarity < Threshold
    6:       Cluster Number++
    7:   While Monitor Stream arrives:
    8:       Update Trend-Matching Compared With the Last Matching
    9:       Evict the Oldest Value
   10:       New Sequence = Newly Arrived Monitor + Old Sequence
   11:       Match New Sequence with Trends
```

4.2 Dynamic Prediction Window Adjustment Algorithm

Applications with different characteristics are placed in cloud, resulting in different resource demand trends. Processing sequences of different trends with fixed window is not reasonable. In this subsection, we propose a dynamic prediction window adjustment algorithm, which selects appropriate prediction length for sequences with different trends. The algorithm is composed of the window scoring and model building.

The window scoring is implemented by cross validation with training set including sequences from each cluster. In our method, the last value of the

sequence is the predicted label. According to the accuracy by cross validation
and the complexity, we give a score to each prediction window (Algorithm 2,
Line 1–4). The lower the score is, the more appropriate the prediction window
is (Algorithm 2, Line 5). As the trend-matching algorithm has identified different trends, we only need to consider selecting appropriate prediction window
through traversal for each cluster previously obtained. Then we build Support
Vector Regression model using the selected prediction window for each trend
(Algorithm 2, Line 6). SVR has the applicability for small sample and its decision characteristic of support vector is suitable for data stream. When updating
model incrementally, we train the support vectors and the new sequence to
replace the old support vectors (Algorithm 2, Line 7–8).

Algorithm 2: Dynamic Prediction Window Adjustment Algorithm

```
Input:   Sequences of Each Trend
Output:  Window Length for Each Trend
    1:   For each trend:
    2:       For each prediction window:
    3:           Cross Validation
    4:           Window Scoring = Balance (accuracy, complexity)
    5:       Select the window of lowest score
    6:       Build SVR using the window
    7:   While updating the SVR
    8:       New Support Vectors = Train (SVs + New Sequence)
```

4.3 Mixed Synthesis Algorithm

The previous subsection introduces a trend-matching algorithm that predicts the
future trend. However, the workload changes frequently, increasing the probability of error matching. Prediction will deviate if the predicted trend is wrong.
In this subsection, we propose a mixed synthesis algorithm based on ensemble
learning to respond to sudden changes and improve the robustness.

The mixed synthesis algorithm chooses the appropriate learners by striking a
balance between the matching rate and the accuracy of the prediction (Algorithm
3, Line 1–3). Then we detect the possibility of sudden changes based on the best
L matching results and the proximity values (Algorithm 3, Line 4–6). For each
possible result, predict with their SVR model and combine them to the final
result (Algorithm 3, Line 7–9).

Algorithm 3: Mixed Synthesis Algorithm

```
Input:   Matching Result
Output:  Final Result
    1:   For sequence in Sequences:
    2:       Match sequence with Trends
    3:   L = Balance (matching rate, accuracy)
    4:   For sequence in Sequences:
```

```
5:        For {L1,L2..Ln} in Best L matching result:
6:            Train with {SVR(L1), SVR(L2)...SVR(Ln); Label}
7:   While matching possibility distributions arrives:
8:        Predict with L results
9:   Combine to Final Result
```

5 Experiments

We conduct a series of experiments to evaluate the accuracy of the TMRCP and
the affect of the proposed algorithms on Apache Storm. The experiments are
based on the Google's cluster trace dataset [2], which consists of the workload
on more than 12 k machines in Google's production cluster, recorded over 29
days in May of 2011. The recorded resources include CPU, Memory and other
usage data captured at 5 min intervals, producing 8352 data points per machine.

We evaluate the performance of TMRCP through three metrics: Mean
Absolute Percentage Error (MAPE), Root Mean Squared Error (RMSE) and
Mean Absolute Error (MAE).

5.1 Accuracy

In this subsection, we conduct experiments on the performance of the TMRCP
and the comparison among prediction methods.

We evaluate the accuracy of the TMRCP through predicting the workload
of Google trace. Figure 3 shows the results of CPU and memory synchronous
prediction by the TMRCP. As we can see, it is close between actual and predicted
load, which indicates the accuracy of the TMRCP.

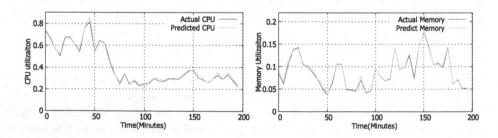

Fig. 3. CPU and memory usage predictions

Then we implement the AR, MA, ARMA, ARIMA [7] and Neural Network
[8] for comparison. As shown in Fig. 4, the TMRCP has significantly improved
the prediction. The MAPE, RMSE and MAE of the TMRCP are 4.7%, 0.056
and 0.018. The TMRCP adaptively identifies trends and dynamically selects
appropriate prediction window for each trend, making the TMRCP sensitive to
load changes and different prediction characteristic.

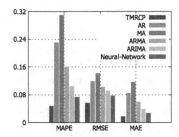

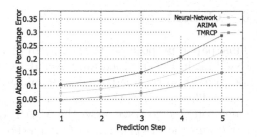

Fig. 4. Comparison of methods   **Fig. 5.** Multi-step ahead prediction

We further evaluate the multi-step-ahead prediction based on the one-step-ahead prediction in Fig. 5. The MAPE of the TMRCP is less than 0.1 within four step ahead prediction, which is 25% more accurate than the NN and ARIMA, indicating the TMRCP is suitable for multi-step-ahead prediction.

5.2 Effect of the Proposed Algorithms

In this subsection, we evaluate the effect of the proposed algorithms through measuring the MAPE changing of removing these algorithms.

The TMRCP-XX is the TMRCP method without the XX algorithm. We evaluate the effect of the proposed algorithms in Table 1. We first predict workload without considering the trend-matching. We train a SVR model with inputting the resources utilization data and outputting multiple resources prediction. The MAPE changes to 0.18. Next, we remove the correlation out of our method. The distance metric is replaced by the Gaussian Formula of vectors and the MAPE of CPU changes to 0.065. Then the MAPE decreases to 0.071 when we use fixed prediction window. Finally, we change the number of learners in the mixed synthesis algorithm and the MAPE first decreases and then increases in Fig. 6. It can be seen the TMRCP has more accurate result than removing anyone algorithm, which indicates the effect of our proposed algorithms.

Table 1. The effect of the algorithms

Method	MAPE
TMRCP	0.047
TMRCP-Trend	0.18
TMRCP-Correlation	0.065
TMRCP-Dynamic Window	0.071
TMRCP-Mixed Synthesis	0.115

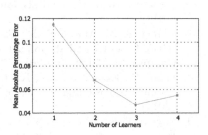

Fig. 6. MAPE with different learners

6   Conclusion

This paper proposes a trend-matching resources coupled prediction method based on incremental learning over data stream, called TMRCP. The TMRCP identifies history trends and dynamically selects prediction window for each trend. Then predicts the workload based on the matched future trend. Experiments on Google trace show the MAPE of the TMRCP is 4.7%, the RMSE is 0.056 and the MAE is 0.018. Our method has improved the accuracy by 20%. In addition, the TMRCP is still accurate within four step ahead prediction.

Acknowledgments. This work was supported by the National Natural Science Foundation of China (Grant No. 61379052), the National Key Research and Development Program (Grant No. 2016YFB1000101), the Natural Science Foundation for Distinguished Young Scholars of Hunan Province (Grant No. 14JJ1026), Specialized Research Fund for the Doctoral Program of Higher Education (Grant No. 20124307110015), the National Natural Science Foundation of China (Grant No. 61502513), the National Natural Science Foundation of China (Grant No. 61502513).

References

1. Plummer, D.C., Bittman, T.J., Austin, T., Cearley, D.W., Smith, D.M.: Cloud-computing: defining and describing an emerging phenomenon. Gartner (2008)
2. Reiss, C., Wilkes, J., Hellerstein, J.L.: Google cluster-usage traces: format+schema. Google Inc. (2011)
3. Tang, Z., Mo, Y., Li, K., Li, K.: Dynamic forecast scheduling algorithm for virtualmachine placement in cloud computing environment. J. Supercomput. **70**(3), 1279–1296 (2014)
4. Roy, N., Dubey, A., Gokhale, A.: Effcient autoscaling in the cloud using predictive models for workload forecasting. In: IEEE International Conference on CloudComputing, pp. 500–507 (2011)
5. Xiao, Z., Song, W., Chen, Q.: Dynamic resource allocation using virtual machines for cloud computing environment. IEEE Trans. Parallel Distrib. Syst. **24**(6), 1107–1117 (2013)
6. Niu, D., Liu, Z., Li, B., Zhao, S.: Demand forecast and performance predictionin peer-assisted on-demand streaming systems. In: IEEE INFOCOM, pp. 421–425 (2011)
7. Calheiros, R., Masoumi, E., Ranjan, R., Buyya, R.: Workload prediction using ARIMA model and its impact on cloud applications' QoS. IEEE Trans. Cloud Comput. **3**(4), 449–458 (2014)
8. Islam, S., Keung, J., Lee, K., Liu, A.: Empirical prediction models for adaptive resource provisioning in the cloud. Future Gener. Comput. Syst. **28**(1), 155–162 (2014)
9. Shyam, G.K., Manvi, S.S.: Virtual resource prediction in cloud environment: a Bayesian approach. J. Netw. Comput. Appl. **65**, 144–154 (2016)
10. Liu, C., Shang, Y., Duan, L., Chen, S., Liu, C., Chen, J.: Optimizing workload category for adaptive workload prediction in service clouds. In: Barros, A., Grigori, D., Narendra, N.C., Dam, H.K. (eds.) ICSOC 2015. LNCS, vol. 9435, pp. 87–104. Springer, Heidelberg (2015). doi:10.1007/978-3-662-48616-0_6

11. Dabbagh, M., Hamdaoui, B., Guizani, M., Rayes, A.: Efficient datacenter resource utilization through cloud resource overcommitment. In: Proceedings of IEEE INFOCOM 2015 (2015)
12. Subbiah, S., Wilkes, J., Gu, X.H., Nguyen, H., Shen, Z.: AGILE: elastic distributed resource scaling for infrastructure-as-a-service. In: International Conference on Autonomic Computing (2014)
13. Amiri, M., Mohammad-Khanli, L.: Survey on prediction models of applications forresources provisioning in cloud. J. Netw. Comput. Appl. **82**, 93–113 (2017)
14. Khan, A., Yan, X., Tao, S., Anerousis, N.: Workload characterization and prediction in the cloud: a multiple time series approach. In: 2012 IEEE Network Operations and Management Symposium (NOMS), pp. 1287–1294. IEEE (2012)
15. Istin, M., Visan, A., Pop, F., Cristea, V.: Decomposition based algorithm for stateprediction in large scale distributed systems. In: Ninth International Symposium on Parallel and Distributed Computing, pp. 17–24 (2010)

App Uninstalls Prediction: A Machine Learning and Time Series Mining Approach

Jiaxing Shang[1,2($\boxtimes$)], Jinghao Wang[1], Ge Liu[1], Hongchun Wu[1,2],
Shangbo Zhou[1,2], and Yong Feng[1,2]

[1] College of Computer Science, Chongiqng University, Chongqing, China
{shangjx,shbzhou,fengyong}@cqu.edu.cn,
{jinghao_wang,liuge1229}@foxmail.com, wuhc0217@163.com
[2] Key Laboratory of Dependable Service Computing in Cyber Physical Society,
Ministry of Education, Chongqing University, Chongqing, China

Abstract. Nowadays mobile applications (a.k.a. app) are playing unprecedented important roles in our daily life and their research has attracted many scholars. However, traditional research mainly focuses on mining app usage patterns or making app recommendations, little attention is paid to the study of app uninstall behaviors. In this paper, we study the problem of app uninstalls prediction based on a machine learning and time series mining approach. Our approach consists of two steps: (1) feature construction and (2) model training. In the first step we extract features from the dynamic app usage data with a time series mining algorithm. In the second step we train classifiers with the extracted features and use them to predict whether a user will uninstall an app in the near future. We conduct experiments on the data collected from AppChina, a leading Android app marketplace in China. Results show that the features mined from time series data can significantly improve the prediction performance.

Keywords: App uninstalls prediction · Time series · Machine learning · Data mining · Mobile application

1 Introduction

Nowadays, with the rapid growing and developing of mobile devices, mobile apps are playing unprecedented important roles in our daily life. The enormous apps from marketplace provide us with a variety of facilities. For example, one can use the Twitter/Facebook/WeChat apps to communicate and share interesting things with our friends. People use the Taobao/Amazon apps to buy all kinds of stuffs online. Other applications include news reading, shopping, travelling, fit keeping, having fun, vehicle sharing, etc. The research of mobile apps has attracted many scholars in recent years [1–3].

The current research of mobile apps mainly focuses on mining app usage patterns or making app recommendations [4,5]. However, little attention is paid to the research of people's app uninstall behavior. The motivations behind the

D. Liu et al. (Eds.): ICONIP 2017, Part V, LNCS 10638, pp. 514–522, 2017.
https://doi.org/10.1007/978-3-319-70139-4_52

problem are two folds. For app developers, knowing whether their product will be largely uninstalled by users can help them make quick response to the market (e.g. upgrade the product, fix the bugs, etc.). For app marketplace providers, if they know someone will give up an app in the near future, they may recommend alternative apps to the user in advance, providing better user experience.

Motivated by this, in this paper we study the problem of app uninstalls prediction based on a machine learning and time series mining approach. Our data is collected from AppChina[1], a leading Android app marketplace in China. Our approach consists of two steps: (1) feature construction and (2) model training. In the first step we extract features from the dynamic app usage data with a time series mining algorithm. In the second step we train classifiers with the extracted features and use them to predict whether a user will uninstall an app in the near future. Experimental results show that the features mined from time series data can significantly improve the prediction performance.

The rest of this paper is organized as follows. Section 2 introduces some related work. Section 3 gives detailed description about the data. We present our approach in Sect. 4. Section 5 shows the experimental results. Section 6 concludes this paper.

2   Related Work

A majority of the research works focus on app usage pattern mining and recommendation. For example, Pan et al. [4] developed a simple computational model to predict app installation by using a composite network. Shin et al. [5] studied the app usage pattern by predicting the next app to be used by a user and display it on the main screen. Tan et al. [6] proposed an algorithm which automatically determines and predicts each user's frequently-used applications in a fixed time slot. Liao et al. [7] proposed a framework to predict apps to be used regarding current device status. They designed a personalized feature selection algorithm based on minimum description length (MDL) and used kNN classifier for prediction. Xu et al. [8] developed an app usage prediction model that considers different factors such as user history, contextual information, etc. Kim and Mielikäinen [9] predicted apps that are mostly likely to be used in a given moment with a conditional log-linear model. Lu et al. [10] predicted app usage with considerations of both physical location moving paths and virtual application usage paths simultaneously. Srinivasan et al. [11] designed a middleware running on the phone and discovers frequent co-occurrence patterns indicating which context events frequently occuer together. Baeza-Yates et al. [12] proposed a machine learning model to predicted the next app to be used and display it to the user once the phone is unlocked. Li et al. [13] conducted an empirical analysis of app usage behaviors on an Android app marketplace and studied two types of user behaviors: app management activities and app network traffic.

Some research works focus on other aspects in studying mobile apps. Fu et al. [14] proposed WisCom: a systems that can analysis user ratings and comments at

---

[1] http://www.appchina.com/.

three different levels of detail. Their method is able to discover inconsistencies in reviews and identify reasons why users like or dislike a given app. Ferdous et al. [15] predicted stress levels at workplace based on smart phone app usage patterns. Ma et al. [16] proposed Monkeydroid which automatically pinpoints whether an app would leak sensitive information. Ding et al. [17] proposed a malware detection method which extracts structural features from Android app function call graph and uses the features to identify malware.

3   Data Collection

Our data is collected from AppChina, a leading Android marketplace in China. The architecture of AppChina is shown in Fig. 1.

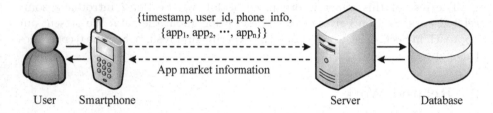

Fig. 1. The architecture of AppChina

After a user has installed the AppChina client on his/her smartphone, he/she can explore new apps, manage existing apps, get app market information from the AppChina server, etc. Meanwhile, each time the user launches the AppChina client, it will collect the user data and send the data to the server. The data contains the following information:

$$\{timestamp, user\_id, phone\_info, app\_list = \{app_1, app_2, \cdots, app_k\}\} \qquad (1)$$

where $timestamp$ is the time when the data is collected, $phone\_info$ contains the smartphone information including phone model, screen resolution, dpi, android version, network connection, etc. $\{app_1, app_2, \cdots, app_k\}$ is the app list installed on the user's smartphone at the current timestamp. For each app, we have crawled its information from the AppChina website and build an app profile with the following information:

$$\{app\_id, app\_name, \#ratings, avg\_rating, \#downloads, category\} \qquad (2)$$

where app_id and app_name are the id and name of the app, $\#ratings$ and $\#downloads$ are the number of app ratings and downloads, avg_rating is the average rating (ranging from 1 to 5) for the app.

4   Methodology

In this section we present our machine learning and time series mining approach, which consists of two steps: (1) feature construction and (2) model training.

4.1   Feature Construction

From the data as described in Sect. 3 we cannot directly tell which app is uninstalled or which one is newly installed. However, it is worthwhile to see that the app list $\{app_1, app_2, \cdots, app_k\}$ for each user is dynamic and changes over time, exhibiting time series nature [18]. By mining the time series data, we can infer the app uninstalls and installs. For example, if we recorded an app list $\{Faccbook, WeChat, Twitter\}$ for Tom at time t_1, and recorded another app lsit $\{WeChat, Twitter, Weibo\}$ for him at t_2 ($t_1 < t_2$), then it is natural to see that Tom has installed a new app ($Weibo$) and uninstalled an old one ($Facebook$) during the time $(t_1, t_2]$. Without loss of generality, we assume the installation and uninstallation behaviors are performed at time t_2.

Algorithm 1 shows the time series mining algorithm which takes the user app data as input and outputs five-tuples, where each five-tuple defines the relationship between a user and an app. The format of a five-tuple is:

$$\{user\_id, app\_id, t_1, t_2, flag\} \tag{3}$$

where t_1 is the time when the app firstly appears in the user's app list, t_2 is the time when the app was last seen or disappeared from the user's app list, $flag$ is a boolean variable indicating whether the app was uninstalled by the user.

After the five-tuples are generated by Algorithm 1, we can extract features by doing statistical analysis on them. For example, for a five-tuple $\{user\_id, app\_id, t_1, t_2, flag\}$, we can use $t_2 - t_1$ to evaluate the time that the user keeps the app on his/her smartphone. We divide these features into three categories, i.e., *user features*, *app features*, and *correlated feature*, as shown in Table 1.

4.2   Model Training

After the features are extracted, the next step is to build training and testing sets. We have collected 50 days data, from June 1st, 2012 to July 31st, 2012. The data of the former 40 days are used to train the classifiers. We then use the classifiers to prediction whether a user will uninstall an app from his/her current app list within the following 10 days. If a user uninstalles an app, the example is marked as positive, otherwise it is negative. The training set contains 220,789 examples (110,181 positive and 110,608 negative), while the testing set contains 222,466 examples (111,016 positive and 111,450 negative).

Features: We use two groups of features to train our classifiers. The first group includes 8 features: $\{\#apps, user\_avg\_time, \#ratings, avg\_rating, \#downloads, \#users, category, run\_time\}$, the second group includes 12 features: $\{\#apps,$

Algorithm 1. Time series data mining

Require:

Users' dynamic app usage data:

$\{timestamp, user\_id, phone\_info, app\_list = \{app_1, app_2, \cdots, app_k\}\}$

Ensure:

Five-tuples: $\{user\_id, app\_id, t_1, t_2, flag\}$ (indexed by $\{user\_id, app\_id\}$)

1: Initialize: five-tuple set $S = \Phi$
2: Group all users' dynamic app usage data according to $user\_id$.
3: **for** each user u **do**
4:     Sort u's app usage data according to $timestamp$.
5:     **for** each app usage data d of u **do**
6:         **for** each app a in $d.app\_list$ **do**
7:             **if** $\{u, a\}$ not in S **then**
8:                 Add five-tuple $\{u, a, d.timestamp, d.timestamp, 0\}$ to S.
9:             **else**
10:                Five-tuple $tp = S.find(u, a)$
11:                    $tp.t_2 \leftarrow d.timestamp$
12:            **end if**
13:        **end for**
14:    **end for**
15:    **for** each tp in $S.find(u)$ and $tp.flag = 0$ and $tp.app\_id \notin d.app\_list$ **do**
16:        $tp.t_2 \leftarrow d.timestamp$, $tp.flag \leftarrow 1$ //Mark the app as uninstalled
17:    **end for**
18: **end for**
19: **return** Five tuple set S

Table 1. Features extracted by time series data mining

Category	Feature	Description
User features	$model$	The model of the smartphone, e.g., Huawei
	$os\_version$	The version of Android operation system
	$resolution$	The screen resolution of the smartphone
	dpi	The dpi of the smartphone
	$\#apps$	The number of apps that the user has ever used
	$\#user\_uni$	The number of apps that the user has ever uninstalled
	$user\_un\_ratio$	The user's uninstallation ratio: $\#user\_uni/\#apps$
	$user\_avg\_time$	The average time that an app stays on the user's smartphone
App features	$\#ratings$	The number of ratings of the app
	avg_rating	The average rating of the app
	$\#downloads$	The number of downloads of the app
	$category$	The category of the app
	$\#users$	The number of users who have ever used the app
	$\#app\_uni$	The number of users who have ever uninstalled the app
	app_un_ratio	The app's uninstallation ratio: $\#app\_uni/\#users$
	app_avg_time	The average time that a user keeps the app
Correlated feature	run_time	The time that the app stays on the user's smart phone

$\#user\_uni$, $user\_un\_ratio$, $user\_avg\_time$, $\#ratings$, avg_rating, $\#downloads$, $\#users$, $\#app\_uni$, app_un_ratio, $category$, $run\_time\}$. The difference between the two feature groups is that the second feature group considers the uninstallation ratio while the first group does not.

Classifiers: We consider 8 classifiers in this paper, i.e., (1) Linear Regression (**LR**), (2) Support Vector Machine (**SVM**), (3) Naive Bayes (**NB**), (4) Decision Tree (**DT**), (5) k-Nearest Neighbors (**kNN**, (6) Random Forest (**RF**), and (7) Gradient Boosting Decision Tree (**GBDT**). All the classifiers are implemented with the Python scikit-learn [19] toolbox.

5   Experimental Results

In this section we will introduce the experimental study, including the evaluation metrics, experimental environment and the results.

Evaluation Metrics: We evaluate the prediction performance of our approach with four metrics, i.e., (1) **Precision**, (2) **Recall**, (3) **F1-Score**, (4) **Accuracy**.

Experimental Environment: The experiments are carried out on a computer with 2.5 GHz AMD A10-5750M CPU and 8 GB memory. Our code is implemented in Python 2.7 programming language.

Results of Different Classifiers: The prediction performance of different classifiers is shown in Fig. 2. Most of the classifiers perform better in terms of precision than recall. Specifically, the SVM classifier exhibits the best performance in precision on both the two feature groups, while the Naive Bayes performs the worst in that metric. When evaluated by the overall prediction performance, i.e., f1-score and accuracy, the GBDT classifier performs best in the 8 feature group while the GBDT, LR and SVM classifiers perform best on the 12 feature group. In general, the GBDT classifier exhibits better performance than other classifiers.

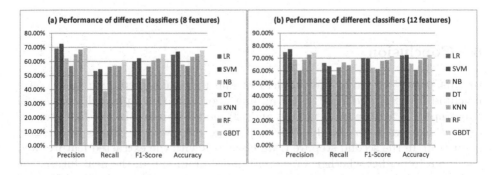

Fig. 2. The prediction performance comparison of different classifiers

Performance of Different Feature Groups: The prediction performance of different feature groups is shown in Fig. 3, from which we see that for all the classifiers and evaluation metrics, the 12 feature group significantly outperforms the 8 feature group. The most significant difference appears on Fig. 3(b), i.e., the performance evaluated by recall. For example, the recall value of the Naive Bayes classifier increase from 0.389 to 0.568, about 46% increment. Given that the 12 feature group includes the uninstallation ratio features which are mined through our time series data mining algorithm, the results indicate that our machine learning and time series mining approach can significantly improve the app uninstalls prediction performance.

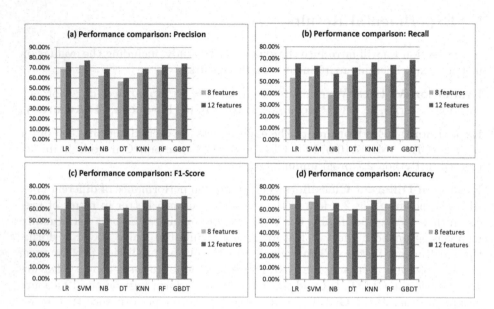

Fig. 3. The prediction performance comparison of different feature groups

6   Conclusion

In this paper, we study the problem of app uninstalls prediction based on a machine learning and time series mining approach. Our approach consists of two steps: (1) feature construction and (2) model training. In the first step we extract features from the dynamic app usage data with a time series mining algorithm. In the second step we train classifiers with the extracted features and use them to predict whether a user will uninstall an app in the near future. We conduct experiments on the data collected from AppChina, a leading Android app marketplace in China. Results show that the features mined from time series data can significantly improve the prediction performance.

Acknowledgements. This work was supported in part by National Natural Science Foundation of China (No. 61702059), China Postdoctoral Science Foundation (No. 2017M612913), Fundamental Research Funds for the Central Universities of China (No. 106112016CDJXY180003), Graduate Student Research and Innovation Foundation of Chongqing City (No. CYS17024), Frontier and Application Foundation Research Program of Chongqing City (No. cstc2017jcyjAX0340, cstc2015jcyjA40006), Social Undertakings and Livelihood Security Science and Technology Innovation Funds of Chongqing City (No. cstc2017shmsA20013).

References

1. Rehman, M., Liew, C., Wah, T.: Frequent pattern mining in mobile devices: a feasibility study. In: 6th IEEE International Conference on Information Technology and Multimedia (ICIMU), Putrajaya, pp. 351–356. IEEE Press (2014)
2. Rehman, M., et al.: Mining personal data using smartphones and wearable devices: a survey. Sensors **15**, 4430–4469 (2015)
3. Cao, H., Lin, M.: Mining smartphone data for app usage prediction and recommendations: a survey. Pervasive Mob. Comput. **37**, 1–22 (2017)
4. Pan, W., Nadav, A., Alex, P.: Composite social network for predicting mobile apps installation. In: 25th AAAI International Conference on Artificial Intelligence (AAAI), San Francisco, pp. 821–827. AAAI (2011)
5. Shin, C., Hong, J.H., Dey, A.K.: Understanding and prediction of mobile application usage for smart phones. In: 14th International Conference on Ubiquitous Computing (UbiCom), Pittsburgh, pp. 173–182. ACM (2012)
6. Tan, C., Liu, Q., Chen, E., Xiong, H.: Prediction for mobile application usage patterns. In: Nokia MDC Workshop, vol. 12 (2012)
7. Liao, Z.X., Li, S.C., Peng, W.C., Philip, S.Y., Liu, T.C.: On the feature discovery for app usage prediction in smartphones. In: 13th IEEE International Conference on Data Mining (ICDM), Dallas, pp. 1127–1132. IEEE Press (2013)
8. Xu, Y., Lin, M., Lu, H., Cardone, G., Lane, N., Chen, Z., Campbell, A., Choudhury, T.: Preference, context and communities: a multi-faceted approach to predicting smartphone app usage patterns. In: 17th ACM International Symposium on Wearable Computers (ISWC), Zurich, pp. 69–76. ACM (2013)
9. Kim, J., Mielikäinen, T.: Conditional log-linear models for mobile application usage prediction. In: Calders, T., Esposito, F., Hüllermeier, E., Meo, R. (eds.) ECML PKDD 2014. LNCS, vol. 8724, pp. 672–687. Springer, Heidelberg (2014). doi:10. 1007/978-3-662-44848-9_43
10. Lu, E.H.C., Lin, Y.W., Ciou, J.B.: Mining mobile application sequential patterns for usage prediction. In: IEEE International Conference on Granular Computing (GrC), Hokkaido, pp. 185–190. IEEE Press (2014)
11. Srinivasan, V., Moghaddam, S., Mukherji, A., Rachuri, K.K., Xu, C., Tapia, E.M.: Mobileminer: mining your frequent patterns on your phone. In: ACM International Joint Conference on Pervasive and Ubiquitous Computing (UbiCom), Seattle, pp. 389–400. ACM (2014)
12. Baeza-Yates, R., Jiang, D., Silvestri, F., Harrison, B.: Predicting the next app that you are going to use. In: 8th ACM International Conference on Web Search and Data Mining (WSDM), Shanghai, pp. 285–294. ACM (2015)
13. Li, H., Lu, X., Liu, X., Xie, T., Bian, K., Lin, F.X., Mei, Q., Feng, F.: Characterizing smartphone usage patterns from millions of android users. In: ACM Internet Measurement Conference (IMC), Tokyo, pp. 459–472. ACM (2015)

14. Fu, B., Lin, J., Li, L., Faloutsos, C., Hong, J., Sadeh, N.: Why people hate your app: making sense of user feedback in a mobile app store. In: 19th ACM SIGKDD International Conference on Knowledge Discovery and Data Mining (KDD), Chicago, pp. 1276–1284. ACM (2013)
15. Ferdous, R., Osmani, V., Mayora, O.: Smartphone app usage as a predictor of perceived stress levels at workplace. In: 9th International Conference on Pervasive Computing Technologies for Healthcare (PervasiveHealth), Istanbul, pp. 225–228. IEEE Press (2015)
16. Ma, K., Liu, M., Guo, S., Ban, T.: MonkeyDroid: detecting unreasonable privacy leakages of android applications. In: Arik, S., Huang, T., Lai, W.K., Liu, Q. (eds.) ICONIP 2015. LNCS, vol. 9491, pp. 384–391. Springer, Cham (2015). doi:10.1007/978-3-319-26555-1_43
17. Ding, Y., Zhu, S., Xia, X.: Android malware detection method based on function call graphs. In: Hirose, A., Ozawa, S., Doya, K., Ikeda, K., Lee, M., Liu, D. (eds.) ICONIP 2016. LNCS, vol. 9950, pp. 70–77. Springer, Cham (2016). doi:10.1007/978-3-319-46681-1_9
18. Yu, S., Abraham, Z.: Concept drift detection with hierarchical hypothesis testing. In: Proceedings of the 2017 SIAM International Conference on Data Mining, Texas, pp. 768–776. SIAM (2017)
19. Pedregosa, F., et al.: Scikit-learn: machine learning in Python. J. Mach. Learn. **12**, 2825–2830 (2011)

Time Series Forecasting Using GRU Neural Network with Multi-lag After Decomposition

Xu Zhang[1], Furao Shen[1(✉)], Jinxi Zhao[1], and GuoHai Yang[2]

[1] National Key Laboratory for Novel Software Technology,
Department of Computer Science and Technology,
Collaborative Innovation Center of Novel Software Technology and Industrialization,
Nanjing University, Nanjing, China
zhangxu0307@163.com, {frshen,jxzhao}@nju.edu.cn
[2] Nanjing Melangy Energy Science and Technology Co. Ltd., Nanjing, China
usedplaneandship@188.com

Abstract. Time series forecasting has a wide range of applications in society, industry, market, etc. In this paper, a new time series forecasting method (FCD-MLGRU) is proposed for solving short-term forecasting problem. First we decompose the original time series using Filtering Cycle Decomposition (FCD) proposed in this paper, secondly we train the Gated Recurrent Unit (GRU) Neural Network to forecasting the subseries respectively. In the process of training and forecasting, the multi-time-lag sampling and ensemble forecasting method is adopted, which reduces the dependence on the selection of time lag and enhance the generalization and stability of the model. The comparative experiments on the real data sets and theoretical analysis show that our proposed method performs better than other related methods.

Keywords: Time series forecasting · Gated Recurrent Unit Neural Network · Time series decomposition

1 Introduction

Large-scale time series data have widely emerged in the fields of economics, industry, education, society, etc. The forecasting of time series provides important guidance for decision-makers to take corresponding strategy. The forecasting problem of time series can be summed up as following: Build a forecasting model, which can capture the regularity of the history time series, so that it can predict value of the future which approximate the ground-truth. It can be formally written as:

$$x_{t+n}, x_{t+2}, ..., x_{t+1} = f(x_t, x_{t-1}, ..., x_{t-w}) \tag{1}$$

where f is forecasting model, x_t is time series data point at time step t, n is forecasting horizon, which means the number forecasting ahead, w is window

© Springer International Publishing AG 2017
D. Liu et al. (Eds.): ICONIP 2017, Part V, LNCS 10638, pp. 523–532, 2017.
https://doi.org/10.1007/978-3-319-70139-4_53

size which means the number of historical data used to forecast the future data, we also call it as time lag.

To solve forecasting problem, some classical forecasting methods including Exponential Smoothing, Trend Extrapolation, Moving Average Model (MA), Autoregressive Integrated Moving Average Model (ARIMA) [1], etc. was proposed. However, these methods are linear models, not suitable for the non-linear large-scale time series forecasting.

In recent years, the Artificial Neural Network (ANN), especially deep learning method, has been widely used in the time series forecasting problem. Due to the ability of ANN which can approximate the nonlinear function with arbitrary precision [2], it has strong power and robustness to fit the nonlinear data. Shen *et al.* leveraged the CDBN to forecasting the exchange rate [3]. Shi *et al.* combined the Convolutional Neural Network and LSTM network to forecasting the precipitation [4], Marino *et al.* studied energy forecasting by comparing the standard LSTM and sequence to sequence LSTM [5].

However, there are also two main problems with the above time series forecasting methods. First of all, in these forecasting methods, the time lag is generally fixed, it has a great influence on the final forecasting result but is difficult to determined [6,7]. More importantly, with time series data generating continually, time lag, as a important parameter, may drift with the time. Secondly, if a single forecasting model is setup on time series straightforward, it will be more susceptible to noise information, especially in large-scale time series.

In order to cope with the above-mentioned problems encountered in time series forecasting, we propose a new forecasting method FCD-MLGRU. The contribution of this paper is mainly shown as following. Firstly, we propose Filtering Cycle Decomposition (FCD) method to decompose the time series into trend, cycle and residual subseries. Secondly, because we train the Gated Recurrent Unit (GRU) Neural Network to forecasting the subseries respectively, in the process of training and forecasting, the multi-time-lag sampling and ensemble forecasting method can be adopted. It can reduce the dependence on the selection of time lag and enhance the generalization and stability of the model. The comparative experiments on the real data sets show that our proposed method performs better.

2   Proposed Method

The forecasting method of this paper mainly involves into three parts: Time series decomposition, GRU Neural Network training and forecasting, Multi-lag sampling and ensemble forecasting in process of GRU. The whole process of proposed method is shown in Fig. 1.

First, given a time series, we decompose the original time series into trend, cycle and residual subseries. Based on the traditional time series seasonal decomposition model X-11 [8], we propose a simplified version of this decomposition model, we refer it as Filter Cycle Decomposition (FCD).

Next we train forecasting model on trend and residual subseries after FCD decomposition by using GRU Neural Network. Cycle subseries is invariant in period, so it need't forecasting.

In the process of training and forecasting, different from the previous forecasting methods with fixed time lag, in this paper, we use the variable-length time lag sampling and multi-lag forecasting method. Because the GRU is a variant of the Recurrent Neural Network (RNN), it can handle variable-length sequences.

In the last step, the subseries forecasting results sum up to obtain the final forecasting result.

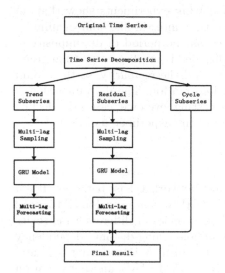

Fig. 1. Flowchart of proposed method   **Fig. 2.** Multi-lag sampling schematic

2.1   Time Series Decomposition

In the theory of time series analysis, time series can be considered as a combination of several subcomponents. A time series X can be decomposed as the following formula:

$$X\left(t\right) = T\left(t\right) + C\left(t\right) + R\left(t\right) \tag{2}$$

where $T(t)$ is trend time series, $C(t)$ is cyclic time series, $R(t)$ is residual time series. It is called addition model.

Another form of decomposition is the multiply model. In this paper, we use the additive model of time series decomposition, which means that we believe the subcomponents are completely independent of each other. This assumption will be an important basis for our later decomposition necessity analysis.

The Filtering Cyclic Decomposition (FCD) method is as following:

1. Select a fixed cyclic period m. Use $2 * m$ (m is even) or m (m is odd) window size to do Moving Average (MA) to estimate the trend subseries T.

2. Remove the trend information $y - T$.
3. Calculate the average of each moment in the cyclic period as cyclic m information C.
4. Obtain residual information $R = y - T - C$.

This is simple version of seasonal decomposition. Different from the traditional seasonal decomposition model like X-11 [8], the cycle period here is no longer the periodic frequency of the natural time like 24-h (Day), 7-day (Week) or 30-day (Month), but a fixed small given value (like 4 or 8, has not especially physics meaning). FCD is equivalent to a smooth filter which can extract periodic and residual characteristics. The following experiments show that, in the case of large data volumes at fine-grained time interval like hour or minute, if the cycle frequency of natural time is used as the period of decomposition, sometimes the magnitude of the residual subseries is much greater than trend time series. Residual time series is difficult to find the regularity, so that trend information is overwhelmed by the residual information. The decomposition is not conducive to improve the accuracy of forecasting, even will cause the decline of accuracy. We will show these detail results in the experiments of Sect. 4.

2.2 Gated Recurrent Unit (GRU)

We choose Gated Recurrent Unit (GRU) Neural Network as base regressor. It is a variation of Recurrent Neural Network (RNN). GRU was proposed in [9] to make each recurrent unit to adaptively capture dependencies of different time scales [10]. Compared with the vanilla RNN, GRU can hold a long-distance dependency because it reduces the problem about gradient vanish by introducing the gate. Unlike the Long-Short Term Memory Neural Network (LSTM), another variation of RNN which also can hold long-distance dependency, the GRU's network neural units architecture is much simpler, but its effectiveness is not reduced, sometimes even slightly better than LSTM [10].

As a variation of RNN, GRU Neural Network has a characteristic that more recent time step input will has greater influence on neural network [11], which is in consist with the characteristics of the time series forecasting, i.e. the information which is closer to the forecasting point is more important.

There are two gate structures in GRU: Update Gate z_t and Reset Gate r_t. Update Gate decides how much the neural unit updates, Reset Gate decide how much previously state the neural unit forgets. It can be calculate as following formula:

$$z_t = tanh\left(\boldsymbol{W_z} \cdot [\boldsymbol{h_{t-1}}, \boldsymbol{x_t}]\right) \tag{3}$$

$$r_t = tanh\left(\boldsymbol{W_r} \cdot [\boldsymbol{h_{t-1}}, \boldsymbol{x_t}]\right) \tag{4}$$

where $\boldsymbol{h_{t-1}}$ is last output and $\boldsymbol{x_t}$ is input, $\boldsymbol{W}$ represent the weight, $\cdot$ is element-wise multiply.

After calculating the gate, we must choose what information will be added in the neural units memory as $\tilde{h}_t$ and expose state as output h_t, it can be calculated as following formula:

$$\tilde{h}_t = tanh\ (\boldsymbol{W} \cdot [\boldsymbol{r_t} * \boldsymbol{h_{t-1}}, \boldsymbol{x_t}]) \tag{5}$$

$$\boldsymbol{h_t} = (\boldsymbol{1} - \boldsymbol{z_t}) * \boldsymbol{h_{t-1}} + \boldsymbol{z_t} * \tilde{\boldsymbol{h}}_t \tag{6}$$

More details about GRU Neural Networks can be referred to the [9].

2.3   Variable-Length Sampling and Multi-lag Ensemble Forecasting

In this paper we use the variable-length time lag sampling method in GRU training and forecasting to take full advantage of GRU which can handle variable length sequence. At the time of training, we forecast value of next time point x_{t+1} as ground-truth, we specify minimum length called L_{min} and a maximum length called L_{max}, then cut sequences in all possible length within $L_{min} \sim L_{max}$ prior to this point as regressor input, like $[x_t, x_{t-1}, ..., x_{t-L_{min}}]$, $[x_t, x_{t-1}, ..., x_{t-L_{min}+1}]$, ..., $[x_t, x_{t-1}, ..., x_{t-L_{max}}]$, so we get a group of samples-ground-truth pairs, then we can continuously slide to get next group. Note that the sequences whose length are less than L_{max} are padded with zero at the left end, so it can be operated by the GRU. The schematic diagram of multi-lag sampling is shown in Fig. 2.

At the time of forecasting, we use GRU to get different forecasting results in different time-lag, and finally take average of them as final result. This method can effectively suppress the instability of single-time-lag forecasting results and enhance the generalization ability of the model. It also reduce the dependence on selection of the time lag because we only need to specify a approximate range rather than a fixed accurate time lag. It can even maintain forecasting ability under the case when time lag drifts in data-stream.

3   Analysis

3.1   Why Decomposition

We make a semi-quantity explanation about the advantage of time series decomposition from the viewpoint of machine learning model error theory.

The square expectation error of any machine learning model consists of three parts, bias, variance and random noise [12]. As shown in the formula:

$$E\ (f\ (\boldsymbol{x}) - y')^2 = \sigma^2 + Var\ (\boldsymbol{x}) + Bias\ (\boldsymbol{x})^2 \tag{7}$$

where $\boldsymbol{x}$ is input data, y' is ground-truth value, σ is variance of the random noise, Var and $Bias$ represent variance and bias of the model.

After time series decomposition, the cyclic information can be completely retained in the final result. Assuming that under normal circumstances, the cyclic attributes will not change (if there is change, they all included in the residual information), so the cyclic subseries forecasting result bias and variance can be considered to be zero. In addition, without interference of cyclic information and residual information, the trend curve is very smooth, thus it can be easily fitted, so its bias and variance will be smaller. The residual subseries fluctuates greatly and its regularity is not obvious. It is the main source of the forecasting error. Therefore, it is necessary to reduce the magnitude of the residual time series in order to drop its effect to the final forecasting result.

According to the previous discussion, the subcomponents after the time series decomposition are independent to each other. Thus it means its covariance is 0. According to the attribution of the bias and variance, (7) can be overwritten as following:

$$
\begin{aligned}
E\left(f\left(\boldsymbol{x}\right)-\hat{y}\right)^2 &= \sigma^2 + Var\left(\boldsymbol{x}\right) + Bias\left(\boldsymbol{x}\right)^2 \\
&= Bias\left(T+C+R\right)^2 + Var\left(T+C+R\right) + \sigma^2 \\
&= \left(Bias\left(T\right) + Bias\left(C\right) + Bias\left(R\right)\right)^2 \\
&\quad + Var\left(T\right) + Var\left(C\right) + Var\left(R\right) + \sigma^2
\end{aligned}
\tag{8}
$$

where $\boldsymbol{x}$ is original series data, T,C,R represent trend, cyclic, residual data respectively after decomposition. According to the discussion above, we can see that $Bias\left(C\right) = 0$ and $Var\left(C\right) = 0$, other items should be reduced after decomposition. This is reason why we use FCD. That is to say, we need trend time series after smoothing filter is easier to fit, and the magnitude of residual time series is as small as possible to reduce the effect of residual forecasting on the final result.

Of course, if the sub-components are not independent, we also can analyze the final forecast results through the covariance changing. We will discuss this problem in the future.

3.2   Advantages of Multi-lag Sampling and Ensemble Forecasting

Many researchers have devoted to discuss how to choose appropriate time lag. Frank *et al.* uses the False Nearest Neighbour method and Heuristics for window size estimation to select the appropriate time lag [13]. Rahman *et al.* train several ANN as the base regressor under different time lag for ensemble learning [14]. Most of these methods use traditional neural networks or machine learning algorithms, and their input dimensions are generally fixed, so it is difficult to deal with situations of different time lag. Generally these methods are training completely independent models under different time lags or require a lot of cross validation experiments to determine the final time lags.

However, GRU neural network can cope with different time lag. Assume we specify max time lag as L_{max} and min time lag as L_{min}. L is the total length of a time series, according to the sampling method mentioned in Sect. 2.3, we can get max number of all possible training samples groups G is:

$$
G = L - L_{max}
\tag{9}
$$

every group has different time lag samples, so we can get max number of all possible training samples N:

$$
N = \left(L_{max} - L_{min} + 1\right) * G
\tag{10}
$$

So it can be seen that we get more training samples through this method. This is equivalent to finding the dependencies of all possible sequences within the minimum length to maximum length, which is used to train the GRU. Note that we use all these training samples with different time lag to train only one GRU neural network, not independent models under different time lags.

In forecasting process, we calculate the forecasting results under different time lag and finally take the average as the final result. Although there is only one GRU model on each subseries, but multi-time-lag forecasting are equivalent to ensemble learning method forecasting. According to the theory of ensemble learning, the variance of the forecasting model is reduced.

4 Experiment

4.1 Data

In order to verify the effectiveness of our proposed method, we conducted comparative experiments on several real time series data sets. The experimental data includes NSW 2013 and TAS 2016 annual electricity demand, the length is 17521, both of them data are collected from the Australian Energy Market Operator (AEMO) (https://www.aemo.com.au/). There is also a shared bicycle registration data record, the sequence length is 17380, collected from the UCI time series data set (https://archive.ics.uci.edu/ml/datasets.html).

We use the three most common metrics to evaluate the merits of these models: mean absolute error (MAE), mean root square error (MRSE) and symmetric mean absolute percentage error (SMAPE).

4.2 Results

GRU as Base Regressor. The single forecasting model comparative experiment results are shown in Table 1. It can be seen from the experimental results that the forecasting accuracy of LSTM and GRU is better than the other methods. Vanilla RNN can not memorize long sequences due to gradient vanish, thus the performance on the three data sets are not good enough. ANN and SVR are

Table 1. Base regressor used in forecasting straightforward

Method	NSW2013			Shared-Bike			TAS2016		
	MAE	RMSE	SMAPE	MAE	RMSE	SMAPE	MAE	RMSE	SMAPE
ANN	68.15	88.61	0.90	31.46	45.96	31.01	16.27	22.37	1.54
GRU	61.06	80.09	0.81	26.62	41.75	32.76	14.77	20.77	1.41
LSTM	65.12	83.82	0.88	29.22	47.69	30.17	14.75	20.74	1.40
RNN	70.54	91.71	0.93	33.91	48.05	34.60	24.23	30.37	2.31
SVR	66.85	87.12	0.89	35.97	55.58	36.30	14.74	21.03	1.41

slightly worse than RNN family, but they must require the fixed input dimension. Although the effect of LSTM and GRU is roughly close, the structure of GRU is more simple, thus it iterates and converges more faster than LSTM, thus we choose GRU as base regressor.

The Effectiveness of Time Series Decomposition. We compare the traditional seasonal decomposition (X-11) and the Filter Cycle Decomposition (FCD) method on NSW2013 and Shared-Bike, which have fine-grained record frequency (like hours or minutes) and larger data volume. The results of the two decomposition in Shared-Bike data set are shown as Figs. 3 and 4, it can be seen that the residual time series magnitude of FCD is lower than X-11, and cycle information is much more regular.

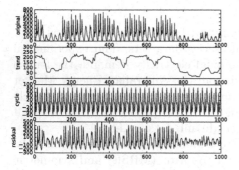

Fig. 3. Shared-Bike X-11 weekly seasonal decomposition (regional, peroid = 7 days)

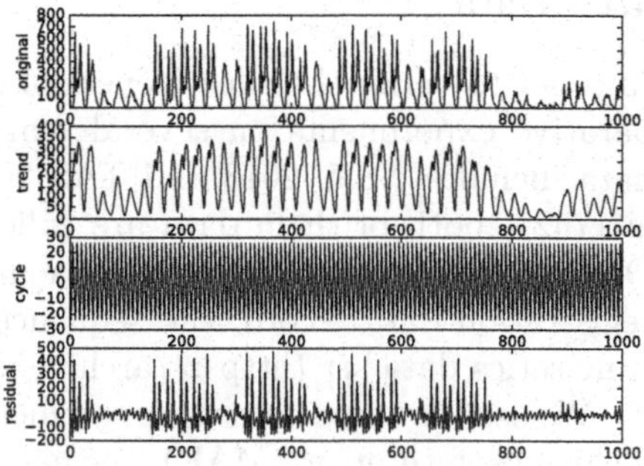

Fig. 4. Shared-Bike filter cycle decomposition (regional, peroid = 8)

In the experiment, we combined the two kinds of decomposition methods into the single forecasting model, the experimental results are shown in Table 2, obviously in most time, the second method is better. It is consistent with our previous theoretical analysis in Sect. 2.2.

Table 2. Forecasting using traditional time series decomposition and filter cycle decomposition

Method	NSW2013			Bike-Shared			Method	NSW2013			Bike-Shared		
	MAE	RMSE	SMAPE	MAE	RMSE	SMAPE		MAE	RMSE	SMAPE	MAE	RMSE	SMAPE
X-11ANN	70.51	91.78	0.92	33.14	48.69	38.12	FCD-ANN	80.93	99.44	1.05	15.69	21.53	24.06
X-11GRU	62.22	81.38	0.82	28.93	41.40	34.23	FCD-GRU	38.47	47.11	0.51	15.83	21.71	22.43
X-11LSTM	66.03	85.08	0.87	34.26	49.64	34.49	FCD-LSTM	41.23	51.62	0.53	13.68	19.61	17.98
X-11RNN	70.32	89.27	0.93	33.40	45.54	34.99	FCD-RNN	33.65	43.02	0.44	20.82	29.35	31.25
X-11SVR	59.52	76.66	0.78	35.56	53.72	38.97	FCD-SVR	45.83	59.02	0.63	17.04	28.58	21.38

Comparative Experiments. The above comparative experiments are carried out in the case of the same fixed time lag at 20. As discussed above, time lag itself is a difficult choice. For this reason, we use variable-length sampling to increase the number of training samples as discussed in Sect. 3.2, while using multi-lag ensemble forecasting method at forecasting phase. In the experiments, we conducted comprehensive comparative experiments on NSW2013, TAS2016 and Shared-Bike three data sets. In the experiment, we used the vanilla RNN, GRU, LSTM, ANN, SVR algorithm and these after decomposition version (FCD-RNN, FCD-GRU, FCD-LSTM, FCD-ANN, FCD-SVR) and the method which we propose in this paper (FCD-MLGRU). The experimental results are shown in Table 3.

Table 3. Comprehensive comparative experiments results

Method	NSW2013			Bike-Shared			TAS2016		
	MAE	RMSE	SMAPE	MAE	RMSE	SMAPE	MAE	RMSE	SMAPE
ANN	68.15	88.61	0.90	31.46	45.96	31.01	16.27	22.37	1.54
GRU	61.06	80.09	0.81	26.62	41.75	32.76	14.77	20.77	1.41
LSTM	65.12	83.82	0.88	29.22	47.69	30.17	14.75	20.74	1.40
RNN	70.54	91.71	0.93	33.91	48.05	34.60	24.23	30.37	2.31
SVR	66.85	87.12	0.89	35.97	55.58	36.30	14.74	21.03	1.41
FCD-ANN	80.93	99.44	1.05	15.69	21.53	24.06	14.59	20.75	1.39
FCD-GRU	38.47	47.11	0.51	15.83	21.71	22.43	10.38	14.65	0.98
FCD-LSTM	41.23	51.62	0.53	13.91	19.97	17.98	7.48	10.45	0.71
FCD-RNN	33.65	43.02	0.44	20.82	29.35	31.25	9.84	12.35	0.95
FCD-SVR	45.83	59.02	0.63	17.04	28.58	21.38	9.62	13.87	0.92
FCD-MLGRU	**25.36**	**33.38**	**0.34**	**13.82**	**19.81**	**17.33**	**6.87**	**9.81**	**0.65**

Through the results of the final comprehensive experiment, we can see that the forecasting error of proposed method is smaller than the remaining methods. On the one hand, almost all forecasting model after using decomposition performs better than single forecasting model, so it can be seen that the time series decomposition is important and necessary. On the other hand, after using variable-length sampling and multi-lag ensemble forecasting, the stability and generalization ability of the model are further strengthened, so the forecasting error is reduced further. In addition, we do not need to specify a time lag, but rather specify a range of time lag, which greatly reduces the dependence on the fixed time lag, reducing the burden of adjusting parameters. Even in the time series stream generation process, the model can also be adjusted adaptively to cope with time lag drift.

5   Conclusion

In this paper, a new time series forecasting method is proposed, which combines Filtering Cycle Decomposition (FCD), GRU Neural Network, variable length time lag sampling and multi-lag ensemble forecasting. Through the theoretical analysis, the necessity of time series decomposition is studied, and the advantages of variable length sampling and multi-lag ensemble forecasting are illustrated. The experimental results show that the method proposed in this paper performs better than other related methods on the real data sets.

Acknowledgments. This work is supported in part by the National Science Foundation of China under Grant Nos. (61373130, 61375064, 61373001), and Jiangsu NSF grant (BK20141319).

References

1. Box, G.E.P., Jenkins, G.: Time Series Analysis, Forecasting and Control. Holden-Day, Amsterdam (1976)
2. Hornik, K., Stinchcombe, M., White, H.: Multilayer feedforward networks are universal approximators. Neural Netw. **2**(5), 359–366 (1989)
3. Shen, F., Chao, J., Zhao, J.: Forecasting exchange rate using deep belief networks and conjugate gradient method. Neurocomputing **167**(C), 243–253 (2015)
4. Shi, X., Chen, Z., Wang, H., Yeung, D.Y., Wong, W., Woo, W.: Convolutional LSTM network: a machine learning approach for precipitation nowcasting. Computer Science (2015)
5. Marino, D.L., Amarasinghe, K., Manic, M.: Building energy load forecasting using deep neural networks (2016)
6. Chen, H., Yao, X.: Ensemble regression trees for time series prediscitions (2008)
7. Zhang, G.P., Berardi, V.L.: Time series forecasting with neural network ensembles: an application for exchange rate prediction. J. Oper. Res. Soc. **52**(6), 652–664 (2001)
8. Shiskin, J., Young, A.H., Musgrave, J.C.: The X-11 Variant of the Census Method II Seasonal Adjustment Program. U.S. Department of Commerce, Bureau of the Census, Suitland (1967)
9. Cho, K., Van Merrienboer, B., Bahdanau, D., Bengio, Y.: On the properties of neural machine translation: encoder-decoder approaches. Computer Science (2014)
10. Chung, J., Gulcehre, C., Cho, K.H., Bengio, Y.: Empirical evaluation of gated recurrent neural networks on sequence modeling. Eprint Arxiv arXiv:1412.3555 (2014)
11. Bahdanau, D., Cho, K., Bengio, Y.: Neural machine translation by jointly learning to align and translate. arXiv preprint arXiv:1409.0473 (2014)
12. Bishop, C.M.: Pattern Recognition and Machine Learning (Information Science and Statistics). Springer, New York (2006). p. 049901
13. Frank, R.J., Davey, N., Hunt, S.P.: Time series prediction and neural networks. J. Intell. Robot. Syst. **31**(1), 91–103 (2001)
14. Rahman, M.M., Islam, M.M., Murase, K., Yao, X.: Layered ensemble architecture for time series forecasting. IEEE Trans. Cybern. **46**(1), 270 (2016)

Position-Based Content Attention for Time Series Forecasting with Sequence-to-Sequence RNNs

Yagmur Gizem Cinar[1(✉)], Hamid Mirisaee[1], Parantapa Goswami[2], Eric Gaussier[1], Ali Aït-Bachir[3], and Vadim Strijov[4]

[1] Univ. Grenoble Alpes, CNRS, Grenoble INP, LIG, Grenoble, France
{yagmur.cinar,hamid.mirisaee,eric.gaussier}@imag.fr
[2] Viseo R&D, Grenoble, France
parantapa.goswami@viseo.com
[3] Coservit, Grenoble, France
a.ait-bachir@coservit.com
[4] Moscow Institute of Physics and Technology, Moscow, Russia
Strijov@ccas.ru

Abstract. We propose here an extended attention model for sequence-to-sequence recurrent neural networks (RNNs) designed to capture (pseudo-)periods in time series. This extended attention model can be deployed on top of any RNN and is shown to yield state-of-the-art performance for time series forecasting on several univariate and multivariate time series.

Keywords: Recurrent neural networks · Attention model · Time series

1 Introduction

Predicting future values of temporal variables is termed as *time series forecasting* and has applications in a variety of fields, as finance, economics, meteorology, or customer support center operations. Time series often display pseudo-periods, *i.e.* time intervals at which there is a strong correlation, positive or negative, between the values of the times series. In a forecasting scenario, the pseudo-periods correspond to the difference between the positions of the output being predicted and specific inputs. Pseudo-periods may be due to seasonality or to the patterns underlying the activities measured.

A considerable number of stochastic [1] and machine learning based [2] approaches have been proposed for this problem. A particular class of approaches that has recently received much attention for modelling sequences is based on sequence-to-sequence Recurrent Neural Networks (RNNs) [3], hereafter referred to as Seq-RNNs. In order to capture pseudo-periods in Seq-RNNs, one needs a *memory* of the input sequence, *i.e.* a mechanism to reuse specific (representations of) input values to predict output values. As the input sequence is usually

© Springer International Publishing AG 2017
D. Liu et al. (Eds.): ICONIP 2017, Part V, LNCS 10638, pp. 533–544, 2017.
https://doi.org/10.1007/978-3-319-70139-4_54

longer than the pseudo-periods underlying the time series, longer-term memories that store information pertaining to past input sequences (as described in e.g. [4–6]) are not required. A particular model of interest here is the content attention model proposed in [7] and described in Sect. 3. This model allows one to reuse the content of the input sequence to predict the output values. However, this model was designed for text translation and does not directly capture position-based pseudo-periods in time series. It has nevertheless been specialized in [8], under the name pointer network, so as to select the best input to be reused as the output. This model would be perfect for noise-free, truly periodic times series. In practice, however, times series are noisy and if the output is highly correlated to the input corresponding to the pseudo-period, it is not an exact copy of it. We propose in this paper extensions of the attention model that capture pseudo-periods and lead to state-of-the-art methods for time series forecasting.

The remainder of the paper is organised as follows: Sect. 2 discusses the related work. Section 3 presents the position-based content attention models for both univariate and multivariate time series. Experiments illustrating the behaviour of the proposed models are described in Sect. 4. Lastly, Sect. 5 concludes the paper.

2   Related Work

Various stochastic models have been developed for time series modeling and forecasting. Notable among these are autoregressive (AR) [9] and moving averages (MA) [10] models, that were combined in a more general and effective framework, known as autoregressive moving average (ARMA), or autoregressive integrated moving average (ARIMA) when the differencing is included in the model [11]. Vector ARIMA, or VARIMA [12], is the multivariate extension of the univariate ARIMA model. More recently, based on the development of statistical machine learning, time series prediction has been formulated as a regression problem typically solved with Support Vector Machines (SVM) [13] and, even more recently, with Random Forests (RF) [14,15]. RF have been in particular used for prediction in the field of finance [14] and bioinformatics [15], and have been shown to outperform ARIMA in different cases [16].

In this study, RNNs are used for modeling time series as they incorporate contextual information from past inputs and are thus an attractive choice for predicting sequence data, including time series [3]. Early work [17] has shown that RNNs (a) are a type of nonlinear autoregressive moving average (NARMA) model and (b) outperform feedforward networks and various types of linear statistical models on time series. Subsequently, various RNN-based models were developed for different time series, as noisy foreign exchange rate prediction [18], chaotic time series prediction in communication engineering [19] or stock price prediction [20]. A detailed review can be found in [21] for different time series prediction tasks.

RNNs based on LSTMs [22], that we consider here, alleviate the *vanishing gradient* problem of the traditional RNNs. They have furthermore been shown to outperform traditional RNNs on various temporal tasks [23,24]. Recently, they have been used for predicting the next frame in a video and for interpolating intermediate frames [25], for forecasting the future rainfall intensity in a region [26], or for modeling clinical data of multivariate time series [27]. The attention model in Seq-RNNs [3,7] has been studied very recently for time series prediction [28] and classification [29]. In particular, the study in [28] uses the attention to determine the importance of a factor for prediction.

None of the previous studies, to the best of our knowledge, investigated the possibility to capture pseudo-periods in time series via the attention model. This is precisely the focus of the present study, that introduces generalizations of the content based attention model to capture pseudo-periods and improve forecasting in time series.

3   Theoretical Framework

We first focus on univariate time series. As mentioned before, time series forecasting consists in predicting future values from past, observed values. The time span of the past values, denoted by T, is termed as *history*, whereas the time span of the future values to be predicted, denoted by T', is termed as *forecast horizon* (in multi-step ahead prediction, which we consider here, $T' > 1$). The prediction problem can be formulated as a regression-like problem where the goal is to learn the relation $\mathbf{y} = r(\mathbf{x})$ where $\mathbf{y} = (y_{T+1}, \ldots, y_{T+i}, \ldots, y_{T+T'})$ is the output sequence and $\mathbf{x} = (x_1, \ldots, x_j, \ldots, x_T)$ is the input sequence. Both input and output sequences are ordered and indexed by time instants. For clarity's sake, and without loss of generality, for the input sequence $\mathbf{x} = (x_1, \ldots, x_j, \ldots, x_T)$, the output sequence $\mathbf{y}$ is rewritten as $\mathbf{y} = (y_1, \ldots, y_i, \ldots, y_{T'})$.

3.1   Background

Seq-RNNs with memories rely on three parts: one dedicated to encoding the input, and referred to as *encoder*, one dedicated to generating the output, and referred to as *decoder*, and one dedicated to the *memory model*, the role of which being to provide information from the input to generate each output element. The encoder represents each input x_j, $1 \leq j \leq T$ as a *hidden state*: $\overrightarrow{\mathbf{h}}_j = F(x_j, \overrightarrow{\mathbf{h}}_{j-1})$, with $\overrightarrow{\mathbf{h}}_j \in \mathbb{R}^n$ and where the function F is non-linear transformation that takes different forms depending on the RNN considered. We use here LSTMs with peephole connections as described in [24]. The function F is further refined, in bidirectional RNNs [30], by reading the input both forward and backward, leading to two vectors $\overrightarrow{\mathbf{h}}_j = f(x_j, \overrightarrow{\mathbf{h}}_{j-1})$ and $\overleftarrow{\mathbf{h}}_j = f(x_j, \overleftarrow{\mathbf{h}}_{j+1})$. The final hidden state for any input x_j is constructed simply by concatenating the corresponding forward and backward hidden states, *i.e.* $\mathbf{h}_j = [\overrightarrow{\mathbf{h}}_j; \overleftarrow{\mathbf{h}}_j]^\mathsf{T}$, where now $\mathbf{h}_j \in \mathbb{R}^{2n}$.

The decoder parallels the encoder by associating each output $y_i, 1 \leq i \leq T'$ to a hidden state vector $\mathbf{s}_i$ that is directly used to predict the output:

$$y_i = \mathbf{W}_{\text{out}}\mathbf{s}_i + b_{\text{out}}, \ s_i = G(y_{i-1}, \mathbf{s}_{i-1}, \mathbf{c}_i)$$

with $\mathbf{s}_i \in \mathbb{R}^n$. $\mathbf{c}_i$ is usually referred to as a *context* and corresponds to the output of the memory model. In this study, the function G corresponds to an LSTM with peephole connections integrating a context [3].

The memory model builds, from the sequence of input hidden states $\mathbf{h}_j, 1 \leq j \leq T$, the context vector $\mathbf{c} = q(\{\mathbf{h}_1, \ldots, \mathbf{h}_j, \ldots, \mathbf{h}_T\})$ that provides a summary of the input sequences to be used for predicting the output. In its most simple form, the function q just selects the last hidden state [3]: $q(\{\mathbf{h}_1, \ldots, \mathbf{h}_j, \ldots, \mathbf{h}_T\}) = \mathbf{h}_T$. More recently, in [7], a content attention model is used to construct different context vectors (also called attention vectors) $\mathbf{c}_i$ for different outputs y_i $(1 \leq i \leq T')$ as a weighted sum of the hidden states of the encoder representing the input history:

$$e_{ij} = \mathbf{v}_a^\mathsf{T} \tanh(\mathbf{W}_a \mathbf{s}_{i-1} + \mathbf{U}_a \mathbf{h}_j), \alpha_{ij} = \mathrm{softmax}(e_{ij}), \mathbf{c}_i = \sum_{j=1}^T \alpha_{ij} \mathbf{h}_j \quad (1)$$

where softmax normalizes the vector e_i of length T to be the attention mask over the input. The weights α_{ij}, referred to as the attention weights, correspond to the importance of the input at time j to predict the output at time i. They allow the model to concentrate, or *put attention*, on certain parts of the input history to predict each output. Lastly, $\mathbf{W}_a$, $\mathbf{U}_a$ and $\mathbf{v}_a$ are trained in conjunction with the entire encoder-decoder framework.

We present below two extensions for univariate time series to integrate pseudo-periods.

3.2   Position-Based Content Attention Mechanism

We assume here that the pseudo-periods of a time series lie in the set $\{1, \ldots, T\}$ where T is the history size of the time series[1]. One can then explicitly model all possible pseudo-periods as a real vector, which we will refer to as $\boldsymbol{\pi}^{(1)}$, of dimension T, whose coordinate j encodes the importance of the input at position j in the input sequence to predict output at position i. From this, one can modify the weight of the original attention mechanism relating input j to output i as follows:

$$e_{ij} = \begin{cases} \mathbf{v}_a^\mathsf{T} \tanh(\mathbf{W}_a \mathbf{s}_{i-1} + <\boldsymbol{\pi}^{(1)}, \boldsymbol{\Delta}^{(i,j)} > \mathbf{U}_a \mathbf{h}_j) & \text{if } (i + T - j) \leq T \\ 0 & \text{otherwise} \end{cases}$$

where $<.,.>$ denotes the scalar product and $\boldsymbol{\Delta}^{(i,j)} \in \mathbb{R}^T$ is a binary vector that is 1 on dimension $(i + T - j)$ and 0 elsewhere. $\boldsymbol{\Delta}^{(i,j)}$ thus selects the coordinate of $\boldsymbol{\pi}^{(1)}$ corresponding to the difference in positions between input j and output i. This coordinate is then used to increase or decrease the importance of the hidden state h_j in e_{ij}. Note that, as the history is limited to T, there is no need to consider dependencies between an input j and an output i that are distant by more than T time steps (hence the test: $i + T - j \leq T$).

---

[1] This assumption is easy to satisfy by increasing the size of the history if the pseudo-periods are known or by resorting to a validation set to tune T.

The vector e_i can then be normalized using the softmax operator again, and a context be built by taking the expectation of the hidden states over the normalized weights. For practical purposes, however, one can simplify the above formulation by extending the vectors $\boldsymbol{\pi}^{(1)}$ and $\boldsymbol{\Delta}^{(i,j)}$ with T' dimensions that are set to 0 in $\boldsymbol{\Delta}^{(i,j)}$, and by considering a vector $\boldsymbol{\Delta}$ of dimension $(T + T')$ that has 1 on its first T coordinates and 0 on the last T' ones. The resulting position-based attention mechanism then amounts to:

$$\text{RNN-}\pi^{(1)} : \begin{cases} e_{ij} = \mathbf{v}_a^{\mathsf{T}} \tanh(\mathbf{W}_a \mathbf{s}_{i-1} + <\boldsymbol{\pi}^{(1)}, \boldsymbol{\Delta}^{(i,j)} > \mathbf{U}_a \mathbf{h}_j) \boldsymbol{\Delta}_{i+T-j} \\ \alpha_{ij} = \text{softmax}(e_{ij}), \mathbf{c}_i = \sum_{j=1}^{T} \alpha_{ij} \mathbf{h}_j \end{cases} \quad (2)$$

As one can note, $\boldsymbol{\pi}^{(1)}_{i+T-j}$ will either decrease or increase the hidden state vector $\mathbf{h}_j$ for output i. Since $\boldsymbol{\pi}^{(1)}$ is learned along with the other parameters of the Seq-RNN, we expect that $\boldsymbol{\pi}^{(1)}_{i+T-j}$ will be high for those values of $i+T-j$ that correspond to pseudo-periods of the time series. We will refer to this model as RNN-$\pi^{(1)}$. Lastly, note that the original attention mechanism can be recovered by setting $\boldsymbol{\pi}^{(1)}$ to $\mathbf{1}$ (a vector consisting of 1 on each coordinate).

In the above formulation, the position information is used to modify the importance of each hidden state in the input side. It may be, however, that some elements in $\mathbf{h}_j$ are less important than others to predict output i. It is possible to capture this by considering that, instead of having a scalar at each position relating the input to the output, one has a vector in $\mathbf{R}^{2n}$ that can now reweigh each coordinate of $\mathbf{h}_j$ independently. This leads to:

$$\text{RNN-}\pi^{(2)} : \begin{cases} e_{ij} = \mathbf{v}_a^{\mathsf{T}} \tanh(\mathbf{W}_a \mathbf{s}_{i-1} + \mathbf{U}_a((\boldsymbol{\pi}^{(2)} \boldsymbol{\Delta}^{(i,j)}) \odot \mathbf{h}_j)) \boldsymbol{\Delta}_{i+T-j} \\ \alpha_{ij} = \text{softmax}(e_{ij}), \mathbf{c}_i = \sum_{j=1}^{T} \alpha_{ij} \mathbf{h}_j \end{cases} \quad (3)$$

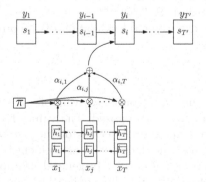

Fig. 1. The illustration of the proposed position-based content attention mechanism.

where $\odot$ denotes the Hadamard product (element wise multiplication) and $\boldsymbol{\pi}^{(2)}$ is a matrix in $\mathbf{R}^{2n \times (T+T')}$. $\boldsymbol{\Delta}^{(i,j)}$ and $\boldsymbol{\Delta}$ are defined as before. We will refer to this model as RNN-$\pi^{(2)}$.

Figure 1 illustrates the overall network in which $\boldsymbol{\pi}$ is a vector for RNN-$\pi^{(1)}$ and a matrix for RNN-$\pi^{(2)}$.

3.3 Multivariate Extensions

As each variable in a K multivariate time series can have its own pseudo-periods, a direct extension of the above approaches to multivariate time series is to consider that each variable k, $1 \leq k \leq K$, of the time series has its own encoder and attention mechanism. The context vector for the ith output of the kth variable is then defined by $\mathbf{c}_i^{(k)} = \sum_{j=1}^{T} \alpha_{ij}^{(k)} \mathbf{h}_j^{(k)}$, where $\mathbf{h}_j^{(k)}$ is the input hidden state at time stamp j for the kth variable and $\alpha_{ij}^{(k)}$ are the weights given by the attention mechanism of the kth variable. To predict the output while taking into account potential dependencies between different variables, one can simply concatenate the context vectors from the different variables into a single context vector $\mathbf{c}_i$ that is used as input to the decoder, the rest of the decoder architecture being unchanged:

$$\mathbf{c}_i = [\mathbf{c}_i^{(1)^T} \cdots \mathbf{c}_i^{(K)^T}]^\mathsf{T}$$

As each $\mathbf{c}_i^{(k)}$ is of dimension $2n$ (that is the dimension of the input hidden states), $\mathbf{c}_i$ is of dimension $2Kn$. This strategy can readily be applied to the original attention mechanism as well as the ones based on $\boldsymbol{\pi}^{(1)}$ and $\boldsymbol{\pi}^{(2)}$.

It is nevertheless possible to rely on a single attention model for all variables while having separate representations for them in order to select, for each output, specific hidden states from the different variables. To do so, one can simply concatenate the hidden states of each variable into a single hidden state ($\mathbf{h}_j = [\mathbf{h}_j^{(1)^T} \cdots \mathbf{h}_j^{(K)^T}]^\mathsf{T}$) and deploy the previous attention model on top of them. This leads to the multivariate model which we refer to as RNN-$\pi^{(3)}$ and is based on the same ingredients and equations as RNN-$\pi^{(2)}$, the only difference being that RNN-$\pi^{(3)}$ is now a matrix in $\mathbf{R}^{2Kn \times (T+T')}$.

We now turn to the experimental validation of the proposed models.

4   Experiments

We retained six widely used and publicly available [31] datasets, described in Table 1, to assess the models we proposed. The values for the history size were set so as they encompass the known periods of the datasets. They can also be tuned by cross-validation if one does not want to identify the potential periods by checking the autocorrelation curves. In general, the forecast horizon should reflect the nature of the data and the application one has in mind, with of course a trade off between long forecast horizon and prediction quality. For this purpose, the forecast horizons of these sets along the sampling rates are chosen as illustrated in Table 1. All datasets were split by retaining the first 75% of each dataset for training-validation and the last 25% for testing. For RNN-based methods, the training-validation sets were further divided by retaining the first 75% for training (56.25% of the data) and the last 25% for validation (18.75% of the data). For the baseline methods, we used 5-fold cross-validation on the training-validation sets to tune the hyperparameters. Lastly, linear interpolation was used whenever there are missing values in the time series[2].

Table 1. Datasets.

Name	Usage	#Instances	History	Forecast horizon	Sampling rate
Polish electricity (PSE)	Univariate	46379	96	4	2 h
Polish weather (PW)	Univariate	4595	548	7	1 days
Numenta benchmark (NAB)	Univariate	18050	72	6	5 min
Air quality (AQ)	Univ./Multiv	9471	192	6	1 h
Appliances energy pred. (AEP)	Univ./Multiv	19735	216	6	10 min
Ozone level detection (OLD)	Univ./Multiv.	2536	548	7	1 day

We compared the methods introduced before, namely RNN-$\pi^{(1/2/3)}$, with the original attention model (RNN-A) and several baseline methods, namely ARIMA, an ensemble learning method (RF) and the standard support vector regression methods. Among these baselines, we retained ARIMA and RF as these were the two best performing methods in our datasets. These methods, discussed in Sect. 2, have also been shown to provide state-of-the-art results on various forecasting problems (*e.g.* [16]). For ARIMA, we relied on the seasonal variant [32]. To implement the RNN models, we used theano[3] and Lasagne[4] on a Linux system with 256 GB of memory and 32-core Intel Xeon @2.60 GHz. All parameters are regularized and learned through stochastic backpropagation (the mini-batch size was set to 64) with an adaptive learning rate for each parameter [33], the objective function being the Mean Square Error (MSE) on the output. For tuning the hyperparameters, we used a grid search over the learning rate, the regularization type and its coefficient, and the number of units in the LSTM and attention models. The values finally obtained are 10^{-3} for the initial learning rate and 10^{-4} for the coefficient of the regularization, the type of regularization selected being L_2. The number of units vary among the set $\{128, 256\}$ for LSTMs and $\{256, 512\}$ for the attention models respectively. We report hereafter the results with the minimum MSE on the test set. For evaluation, we use MSE and

---

[2] We compared several methods for missing values, namely linear, non-linear spline and kernel based Fourier transform interpolation as well as padding for the RNN-based models. The best reconstruction was obtained with linear interpolation, hence its choice here.

[3] http://deeplearning.net/software/theano/.

[4] https://lasagne.readthedocs.io.

the symmetric mean absolute percentage error (SMAPE). MSE corresponds to the objective function used to learn the model. SMAPE presents the advantage of being bounded and represents a scaled L_1 error.

Overall Results on Univariate Time Series

For univariate experiments using multivariate time series, we chose the following variables from the datasets: for PW, we selected the *max temperature* series from the Warsaw metropolitan area that covers only one weather recording station; for AQ we selected *C6H6(GT)*; for AEP we selected the *outside humidity (RH6)*; for NAB we selected the *Amazon Web Services CPU usage* and for OLD we selected *T3*. Table 2 displays the results obtained with the MSE (left value) and the SMAPE (right value) as evaluation measures. Once again, one should note that MSE was the metric being optimized. For each time series, the best performance among all methods is shown in bold and other methods are marked with an asterisk if they are significantly worse than the best method according to a paired t-test with 5% significance level. Lastly, the last column of the table, *Selected-π*, indicates which method, among RNN-$\pi^{(1/2)}$, was selected as the best method on the validation set using MSE.

As one can note, except for AEP where the baselines are better than RNN-based methods, for all other datasets, the best results are obtained with RNN-$\pi^{(1)}$ and RNN-$\pi^{(2)}$, these results being furthermore significantly better than the ones obtained with RNN-A and baseline methods, for both MSE and SMAPE. The MSE improvement varies from one dataset to another: between 8% (PW) and 26% (NAB) w.r.t RNN-A. Compared to ARIMA, one can achieve an improvement ranging from 18% (15%), in OLD, to 94% (75%), in PSE, w.r.t MSE (SMAPE). In addition, the selected RNN-π method (column *Selected-π*) is the best performing method on three out of six datasets (AQ, PW, PSE) and the best performing RNN-π method on AEP. It is furthermore equivalent to the best performing method on OLD, the only dataset on which the selection fails being NAB (a failure means here that the selection does not select the best RNN-π method). However, on this dataset, the selected method is still better than the original attention model and the baselines. Overall, these results show that RNN-$\pi^{(1/2)}$ significantly improves forecasting in the univariate time series we considered, and that one can automatically select the best RNN-π method.

Table 2. Overall results for univariate case with MSE (left value) and SMAPE (right value).

Dataset	RNN-A	RNN-$\pi^{(1)}$	RNN-$\pi^{(2)}$	ARIMA	RF	*Selected-π*
AQ	0.282*/0.694*	0.257/**0.661**	**0.25**/0.669	0.546*/0.962*	0.299*/0.762*	$\pi^{(2)}$
OLD	0.319*/0.595*	**0.271/0.523**	0.275/0.586*	0.331*/0.619*	0.305*/0.606*	$\pi^{(2)}$
AEP	0.025*/0.085*	0.029*/0.101*	0.027*/0.095*	**0.021/0.066**	0.021/0.085*	$\pi^{(2)}$
NAB	0.642*/0.442*	**0.475/0.323**	0.54*/0.369*	1.677*/1.31*	0.779*/0.608*	$\pi^{(2)}$
PW	0.166*/0.558	**0.152**/0.547	0.162*/0.565*	0.213*/0.61*	0.156/**0.544**	$\pi^{(1)}$
PSE	0.034*/0.282*	**0.032**/0.264*	0.033*/**0.256**	0.623*/1.006*	0.053*/0.318*	$\pi^{(1)}$

Lastly, to illustrate the ability of RNN-π to capture pseudo-periods, we display in Fig. 2 (left) the autocorrelation plot for PSE, and in Fig. 2 (right) the average attention weights for the same time series obtained with RNN-A and RNN-$\pi^{(2)}$ (averaged over all test examples and forecast horizon points). As one can see from the autocorrelation plot, PSE has two main weekly and daily pseudo-periods. This two pseudo-periods are clearly visible in the attention weights of RNN-$\pi^{(2)}$ that gives higher weights to the four points located at positions *minus 7 days* and *minus 1 day* (these four points correspond to the four points of the forecast horizon). The attention weights of RNN-$\pi^{(1)}$ (not shown here for space reasons) are very similar. In contrast, the attention weights of the original attention model (RNN-A) follow a general increasing behaviour with more weights on the more recent time stamps. This model thus misses the pseudo-periods.

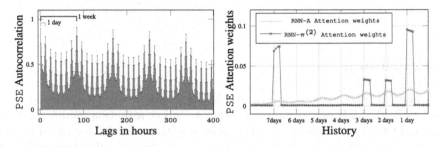

Fig. 2. Autocorrelation of PSE (left). RNN-A and RNN-$\pi^{(2)}$ attention weights (right).

Results on Multivariate Time Series

As mentioned in Table 1, we furthermore conducted multivariate experiments on AQ, AEP and OLD using the multivariate extensions described in Sect. 3. For AQ, we selected the four variables associated to real sensors, namely C6H6(GT), NO2(GT), CO(GT) and NOx(GT) and predicted the same one as the univariate case (C6H6(GT)). For AEP, we selected two temperature time series, namely T1 and T6, and two humidity time series, RH6 and RH8, and we predict RH6 as in the univariate case. For OLD, we trained the model using T0 to T3 and predicted T3, as we did on the univariate case. As RF outperformed ARIMA on five out of six univariate datasets and was equivalent on the sixth one, we retained only RF and RNN-A for comparison with RNN-$\pi^{(1/2/3)}$.

Table 3 shows the results of our experiments on multivariate sets with MSE (SMAPE, not displayed here for readability reasons, has a similar behaviour). As before, for each time series, the best result is in bold and an asterisk indicates that the method is significantly worse than the best method (again according to a paired t-test with 5% significance level). Similarly to the univariate case, the best results, that are always significantly better than the other results, are obtained with the RNN-π methods: for AQ and OLD datasets, RNN-π can respectively bring 24% and 18% of significant improvement over RNN-A. Similarly, for AEP, the improvement is significant over RNN-A (17% with RNN-$\pi^{(1)}$). Compared to

RF, one can obtain between 11% (AEP) and 40% (AQ) of improvement. As one can note, the selected method is always RNN-$\pi^{(3)}$. The selection is this time not as good as for the univariate case as the best method (sometimes significantly better than the one selected) is missed. That said, RNN-$\pi^{(3)}$ remains better than the state-of-the-art baselines retained, RNN-A and RF.

Table 3. Overall results for multivariate case with MSE.

Dataset	RNN-A	RNN-$\pi^{(1)}$	RNN-$\pi^{(2)}$	RNN-$\pi^{(3)}$	RF	*Selected-π*
AQ	0.352*	0.276*	**0.268**	0.3*	0.45*	$\pi^{(3)}$
OLD	0.336*	0.328*	0.327*	**0.274**	0.315*	$\pi^{(3)}$
AEP	0.029*	**0.024**	0.036*	0.026*	0.027*	$\pi^{(3)}$

5   Conclusion

We studied in this paper the use of Seq-RNNs, in particular the state-of-the-art bidirectional LSTMs encoder-decoder with a content attention model, for modelling and forecasting time series. If content attention models are crucial for this task, they were not designed for time series and currently are deficient as they do not capture pseudo-periods. We thus proposed three extensions of the content attention model making use of the (relative) positions in the input and output sequences (hence the term *position-based content attention*). The experiments we conducted over several univariate and multivariate time series demonstrate the effectiveness of these extensions, on time series with either clear pseudo-periods, as PSE, or less clear ones, as AEP. Indeed, these extensions perform significantly better than the original attention model as well as state-of-the-art baseline methods based on ARIMA and random forests.

In the future, we plan on studying formal criteria to select the best extension for both univariate and multivariate time series. This would allow one to avoid using a validation set that may be not large enough to properly select the best method. We conjecture that this is what happening on the multivariate time series we have retained.

References

1. De Gooijer, J.G., Hyndman, R.J.: 25 years of time series forecasting. Int. J. Forecast. **22**(3), 443–473 (2006)
2. Bontempi, G., Ben Taieb, S., Le Borgne, Y.-A.: Machine learning strategies for time series forecasting. In: Aufaure, M.-A., Zimányi, E. (eds.) eBISS 2012. LNBIP, vol. 138, pp. 62–77. Springer, Heidelberg (2013). doi:10.1007/978-3-642-36318-4_3
3. Graves, A.: Generating sequences with recurrent neural networks. arXiv preprint arXiv:1308.0850 (2013)

4. Weston, J., Chopra, S., Bordes, A.: Memory networks. CoRR abs/1410.3916 (2014)
5. Weston, J., Bordes, A., Chopra, S., Mikolov, T.: Towards AI-Complete question answering: A set of prerequisite toy tasks. CoRR abs/1502.05698 (2015)
6. Graves, A., Wayne, G., Reynolds, M., Harley, T., Danihelka, I., Grabska-Barwinska, A., Colmenarejo, S.G., Grefenstette, E., Ramalho, T., Agapiou, J., Badia, A.P., Hermann, K.M., Zwols, Y., Ostrovski, G., Cain, A., King, H., Summerfield, C., Blunsom, P., Kavukcuoglu, K., Hassabis, D.: Hybrid computing using a neural network with dynamic external memory. Nature **538**(7626), 471–476 (2016)
7. Bahdanau, D., Cho, K., Bengio, Y.: Neural machine translation by jointly learning to align and translate. arXiv preprint arXiv:1409.0473 (2014)
8. Vinyals, O., Fortunato, M., Jaitly, N.: Pointer networks. In: NIPS, pp. 2692–2700 (2015)
9. Walker, G.: On periodicity in series of related terms. In: Proceedings of Royal Society of London. Series A, Containing Papers of a Mathematical and Physical Character, vol. 131, no. 818, pp. 518–532 (1931)
10. Slutzky, E.: The summation of random causes as the source of cyclic processes. Econometrica: J. Econometr. Soc. **5**(2), 105–146 (1937)
11. Box, G.E., Jenkins, G.M.: Some recent advances in forecasting and control. J. Roy. Stat. Soc.: Ser. C (Appl. Stat.) **17**(2), 91–109 (1968)
12. Tiao, G.C., Box, G.E.: Modeling multiple time series with applications. J. Am. Stat. Assoc. **76**(376), 802–816 (1981)
13. Sapankevych, N.I., Sankar, R.: Time series prediction using support vector machines: a survey (2009)
14. Creamer, G.G., Freund, Y.: Predicting performance and quantifying corporate governance risk for Latin American ADRs and banks (2004)
15. Kusiak, A., Verma, A., Wei, X.: A data-mining approach to predict influent quality. Environ. Monit. Assess. **185**(3), 2197–2210 (2013)
16. Kane, M.J., Price, N., Scotch, M., Rabinowitz, P.: Comparison of arima and random forest time series models for prediction of avian influenza H5N1 outbreaks. BMC Bioinform. **15**(1), 276 (2014)
17. Connor, J., Atlas, L.E., Martin, D.R.: Recurrent networks and NARMA modeling. In: NIPS, pp. 301–308 (1991)
18. Giles, C.L., Lawrence, S., Tsoi, A.C.: Noisy time series prediction using recurrent neural networks and grammatical inference. Mach. Learn. **44**(1–2), 161–183 (2001)
19. Jaeger, H., Haas, H.: Harnessing nonlinearity: predicting chaotic systems and saving energy in wireless communication. Science **304**(5667), 78–80 (2004)
20. Hsieh, T.J., Hsiao, H.F., Yeh, W.C.: Forecasting stock markets using wavelet transforms and recurrent neural networks: an integrated system based on artificial bee colony algorithm. Appl. Soft Comput. **11**(2), 2510–2525 (2011)
21. Längkvist, M., Karlsson, L., Loutfi, A.: A review of unsupervised feature learning and deep learning for time-series modeling. Pattern Recogn. Lett. **42**, 11–24 (2014)
22. Hochreiter, S., Schmidhuber, J.: Long short-term memory. Neural Comput. **9**(8), 1735–1780 (1997)
23. Gers, F.A., Eck, D., Schmidhuber, J.: Applying LSTM to time series predictable through time-window approaches. In: Tagliaferri, R., Marinaro, M. (eds.) Neural Nets WIRN Vietri-01, pp. 669–676. Springer, London (2001). doi:10.1007/978-1-4471-0219-9_20
24. Gers, F.A., Schraudolph, N.N., Schmidhuber, J.: Learning precise timing with LSTM recurrent networks. J. Mach. Learn. Res. **3**(8), 115–143 (2002)

25. Ranzato, M., Szlam, A., Bruna, J., Mathieu, M., Collobert, R., Chopra, S.: Video (language) modeling: a baseline for generative models of natural videos. arXiv preprint arXiv:1412.6604 (2014)
26. Xingjian, S., Chen, Z., Wang, H., Yeung, D.Y., Wong, W.K., Woo, W.c.: Convolutional LSTM network: a machine learning approach for precipitation nowcasting. In: Advances in Neural Information Processing Systems, pp. 802–810 (2015)
27. Lipton, Z.C., Kale, D.C., Elkan, C., Wetzell, R.: Learning to diagnose with LSTM recurrent neural networks. arXiv preprint arXiv:1511.03677 (2015)
28. Riemer, M., Vempaty, A., Calmon, F.P., Heath III., F.F., Hull, R., Khabiri, E.: Correcting forecasts with multifactor neural attention. In: Proceedings of The 33rd International Conference on Machine Learning, pp. 3010–3019 (2016)
29. Choi, E., Bahadori, M.T., Sun, J., Kulas, J., Schuetz, A., Stewart, W.: RETAIN: an interpretable predictive model for healthcare using reverse time attention mechanism. In: NIPS, pp. 3504–3512 (2016)
30. Schuster, M., Paliwal, K.K.: Bidirectional recurrent neural networks. IEEE Trans. Sig. Process. **45**(11), 2673–2681 (1997)
31. Datasets: PSE: http://www.pse.pl. PW: https://globalweather.tamu.edu. AQ/OLD/AEP: http://archive.ics.uci.edu. NAB: https://numenta.com
32. Hyndman, R., Khandakar, Y.: Automatic time series forecasting: the forecast package for R. J. Stat. Softw. **27**(3), 1–22 (2008). https://www.jstatsoft.org/v027/i03
33. Kingma, D., Ba, J.: Adam: a method for stochastic optimization. arXiv preprint arXiv:1412.6980 (2014)

Deep Sequence-to-Sequence Neural Networks for Ionospheric Activity Map Prediction

Noëlie Cherrier[(✉)], Thibaut Castaings, and Alexandre Boulch

ONERA, The French Aerospace Lab, 91761 Palaiseau, France
noelie.cherrier@wanadoo.fr, {thibaut.castaings,alexandre.boulch}@onera.fr

Abstract. The ability to predict the ionosphere activity is of interest for several applications such as satellite telecommunications or Global Navigation Satellite Systems (GNSS). A few studies have proposed models able to predict Total Electron Content (TEC) values of the ionosphere locally over measuring stations, but not worldwide for most of them. We propose a method using Deep Neural Networks (DNN) to predict a sequence of global TEC maps consecutive to an input sequence of past TEC maps, by combining Convolutional Neural Networks (CNNs) with convolutional Long Short-Term Memory (LSTM) networks. The numerical experiments show that the approach provides significant improvement over methods implemented for benchmarking and is competitive with state-of-the-art methods while providing global TEC predictions. The proposed architecture can be adapted to any sequence-to-sequence prediction problem.

Keywords: Sequence prediction · Neural network · Forecasting · TEC · Ionosphere · Deep learning · CNN · LSTM

1 Introduction

Ionospheric activity is mainly measured by the Total Electron Content (TEC), which is the total number of electrons in the ionosphere integrated along a vertical path above a given location. It is expressed in TEC Units (1 TECU = 10^{16} el/m^2), usually ranging from a few units to one hundred TECU. During periods of high ionospheric TEC levels, the services provided by satellite telecommunication systems and Global Navigation Satellite Systems (GNSS) may be deteriorated due to changes in the paths of transionospheric radio waves inducing significant bitrate reduction and positioning errors [1,2]. As a consequence, forecasting TEC globally (*i.e.* worldwide) in the short term (up to two hours) or in the longer term (a few days) increases the ability of the users of these services to evaluate, as one example, data loss or mispositioning risks in operations planning.

Several services relying on measurements provided by GNSS ground networks [3] exist to address this issue at a global scale, *i.e.* forecasting TEC values worldwide. CTIPe, an experimental tool implementing complex physics models [4] developed by the US Space Weather Prediction Center, produces global

© Springer International Publishing AG 2017
D. Liu et al. (Eds.): ICONIP 2017, Part V, LNCS 10638, pp. 545–555, 2017.
https://doi.org/10.1007/978-3-319-70139-4_55

forecasts 30 min ahead of real-time. In Europe, the ESA Ionospheric Weather Expert Service Center combines products from different national services to provide global and regional 1-h TEC forecasts. However, the records of the input data and forecasts are not published.

In the literature, several methods have been published to predict TEC a few minutes to several days ahead above specific stations using time series analysis and statistical methods [5–9]. Among these, a few provide a reconstruction of a small area [10,11] with methods such as Bezier surface-fitting or Kriging. Several works are based on artificial neural networks [10,12,13], but they only focus on local stations. In order to make a prediction at a regional or global scale with these approaches, one model for each location must be adapted. In [14], a global analytical TEC model is proposed to address the global forecasting issue, using open source TEC data from the Center for Orbit Determination in Europe (CODE). This model is intended to apply to any temporal range, without relying on a record of TEC values.

In this paper, the objective is to predict global TEC maps 2 to 48 h ahead of real-time (Fig. 1). The proposed approach is entirely based on recurrent neural networks, taking advantage of convolutional and Long Short-Term Memory (LSTM) networks. Deep Neural Networks (DNN) have the advantage to enable complex modeling of large input data, such as global TEC maps in this case, with little or no prior knowledge.

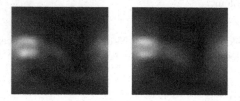

Fig. 1. Ground-truth (left) and predicted (right) TEC map example

The paper is organized as follows: Sect. 2 presents the approach used in this paper, including the data, the proposed network architecture and the training procedure; Sect. 3 discusses the results of our numerical experiments on TEC prediction; and finally the limits and perspectives of the proposed approach are discussed in Sect. 4.

2   Proposed Approach

The objective is to design a neural network architecture able to be fed with an input sequence of a number of maps and to output the next 48 h of TEC maps.

2.1 Datasets

Open source TEC data from the CODE is used in this study. Two datasets of TEC maps are available: Rapid TEC maps, which are accessible quicker, and Final TEC maps, more precise. In this study, final TEC maps are used with a $5° \times 2.5°$ resolution on longitude and latitude and 2 h temporal resolution, covering all latitudes and longitudes. One pixel in these maps represents the vertical TEC at this point.

Data from 1/1/2014 to 5/31/2016 is used for training, and data from 7/1/2016 to 12/31/2016 for testing. Splitting the data into sequences corresponding to five consecutive days allows for convincing results in terms of convergence and in the test procedure (see Sect. 3). Data is loaded as a sequence of 60 maps (one map every two hours): the first 36 maps (*i.e.* 3 days) are fed to the network, the last 24 maps (*i.e.* 48 h) being the prediction targets. This amount of input is sufficient to understand the context, and more information would lead to an overuse of the network's neurons (several neurons being dedicated to process out-of-date data).

2.2 Preprocessing

In order to fit the data to the input of the network and scale the dynamics of the TEC maps, the data undergoes feature normalization and is resized to 72×80 pixels. To prevent the algorithm from focusing on deterministic phenomena that can be easily isolated, the frame of reference is also changed from Fixed-Earth to Heliocentric, so that the effect of Earth's rotation is no longer visible to the DNN architecture. The transformed data is then the input of the network.

However, the resulting TEC maps appear to have a residual low spatial frequency, 24 h-periodicity due to the Earth's magnetic poles, distinct from the geographic poles. In order to avoid learning the residual periodicity, we choose the target T of the neural network to be the difference between the true TEC map I and a Gaussian filtered TEC map at $t - 48$ h $I_{t-48h}^{blurred}$, so that the subtracted component can always be computed from the input sequence. The Gaussian filter is applied in order to remove high frequency variations from the past map.

$$T_t = I_t - I_{t-48h}^{blurred} \tag{1}$$

2.3 Network Architecture

Deep Neural Networks (DNN) taking as input a sequence of images to output *e.g.* audio description or video labels have become a common topic in the past years [15]. However, sequence prediction problems in which both the inputs and targets are a sequence of images have barely been handled. In [16], the authors propose an architecture for precipitation nowcasting and introduce a convolutional LSTM structure.

To give an outline, the proposed approach uses Convolutional Neural Network (CNN) layers [17] to extract spatial features from an input TEC map. The resulting features are then fed to a convolutional LSTM [16], derived from the fully-convolutional LSTM [18]. This enables to handle spatially structured data. Deconvolutional layers are then fed with the output of the convolutional LSTM in order to generate the residual TEC map corresponding to the next time step.

Figure 2 presents the network architecture. To go into the details, we first explain the architecture designed to predict one TEC map (2 h ahead of real-time), represented as the dotted black rectangle in Fig. 2. The network is built as the repetition of a column composed of three modules: the **encoder** (four 3 × 3 convolutions with Rectified Linear Unit (ReLU) in between), a **convolutional LSTM cell** (with a 3 × 3 convolution operation and ReLU) and a **decoder** (four 3 × 3 deconvolutions and ReLU) in order to produce an output of the same size as the input. The columns are cascaded to enable the temporal informa-tion to go through the network, each column handling a single TEC map. The "many-to-one" architecture (dotted box in Fig. 2) is the starting point of the final architecture.

In order to generate a sequence of TEC maps, we derive the "many-to-one" architecture to obtain a "many-to-many" network able to output 48 h of TEC maps. The same columns as in the "many-to-one" part are reused. The predic-tion process is achieved by recursively feeding the next column of the network with the sum of the last prediction and the filtered TEC map at $t - 48$ h (see Subsect. 2.2). The cost function is finally summed over the differences produced by the successive prediction columns.

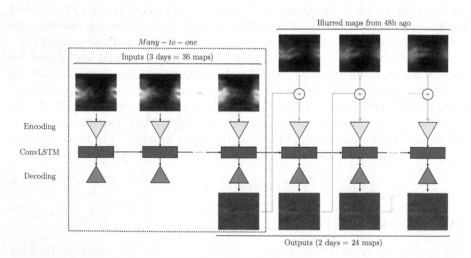

Fig. 2. Network architecture: encoders (*green*), convolutional LSTM cells (*yellow*), and decoders (*red*) (Color figure online)

This architecture is inspired from the Convolutional Autoencoders [19] with a LSTM inserted in between. It is preferred to other well-known architectures such as Fully Convolutional Autoencoders or Deep Belief Networks [20] which cannot handle 2D data. This approach is also preferred to three simpler methods for performance reasons:

- *Many-to-one:* the loss is computed only over one map and the sequence is recursively forecast from the previously predicted map. It leads to competitive results for 2 h prediction but diverges quickly (see the results in Subsection 3.1).
- *Direct TEC map prediction:* this approach performs poorly. As a comparison, taking the map corresponding to $t - 48\,\text{h}$ as a prediction leads to better performance (average RMS of 2.810 against 2.353 for the selected architecture).
- *Difference maps as input:* this method does not perform better than the selected approach (average RMS of 2.918 against 2.353 for the selected architecture).

Cost Function. The cost function for the DNN training is the pixel-wise ℓ_1-norm between the predicted maps and the targets, summed with a measure of relative error as a regularizing factor in order to improve the performance in areas where the TEC level is low. The ℓ_1-norm is chosen instead of ℓ_2: the ℓ_1-norm is known for capturing more details whereas the ℓ_2-norm tends to smooth the results. In this study, we obtained an average RMS of 2.524 with ℓ_2-norm and 2.353 with ℓ_1-norm for the selected architecture. The cost function $\mathcal{L}$ is given as:

$$\mathcal{L} = \sum_{t \in \mathcal{S}} w_t \sum_{i \in \mathcal{M}^t} \left(|P_i^t - T_i^t| + \frac{|P_i^t - T_i^t|}{T_i^t} \right) \tag{2}$$

with $\mathcal{S}$ the sequence of TEC maps, w_t the weight at time t, $\mathcal{M}^t$ the TEC map at t, P the predicted map and T the ground-truth map, where t indexes time and i is the map pixel index.

Different weight arrays $[w_t]$ are considered, corresponding to different learning strategies: *e.g.* putting more weight on the first predicted TEC maps, arguing that the accuracy of the following predictions depend on them. In this study three weight profiles are taken into account: uniform weights over the 48-h sequence; uniform weights on a 24 h-window; linear decreasing weights over the 48-h sequence. These weight arrays enable a better control over the temporal error and a possible improvement of performance at several prediction horizons.

Data Loading into the Network. The sequences are loaded chronologically from the training set, from 1/1/2014 to 5/31/2016, and reloaded again from the beginning of the dataset until convergence in training. This sequential sampling approach is more efficient than random sampling in this study (average RMS of 3.311 against 2.353 with the sequential sampling).

3   Results

Once it has been trained, the network can be fed with any 3-day sequence from the test set and produce 2-day forecasts consecutive to this sequence.

Baselines. Two simple prediction methods are implemented for benchmarking:

- *The 3-day mean prediction* (or constant prediction): the mean map computed over the three days of input data is the prediction result for the next 48 h of data.
- *The periodic prediction:* the predicted map at time t is exactly the map at $t - 48$ h. This exploits the 24 hour-periodicity of TEC maps (see Sect. 2.2).

The three alternative architectures presented in Subsect. 2.3 are also considered as baselines.

We finally compare our method to previous works [5, 8, 12, 14] in Subsect. 3.3.

Performance Criterion. The RMS error is used to compare the performance of the different methods. Other performance criteria may be taken into account, such as the Mean Absolute Percentage Error (MAPE) which is used in several related works to evaluate the forecasts [6, 9]. However, MAPE is not suited to this study since we predict global TEC maps, which include low-level TEC areas (*e.g.* Earth poles) where a small estimation deviation leads to very high MAPE (see discussion in Sect. 4).

3.1 Quantitative Results

The RMS errors averaged for each prediction horizon for the three weighting profiles and for the baselines are summarized in Table 1 for the period from 10/20/2016 to 12/20/2016. In any case, the proposed approach performs clearly better than the 3-day mean prediction, equal or better than the periodic prediction at all prediction horizons, and is also significantly better than the three alternative architectures presented in Subsect. 2.3.

The global generic model proposed in [14] achieves a mean RMS deviation of 7.5 TECU at any given time, whereas the approach proposed in this article performs significantly better for 48-h ahead forecasting (2.4 TECU in average for the uniform weighting).

Decreasing weighting is the best for the first prediction horizons, most likely because it has the bigger weights at the beginning. The weights emphasizing only the first 24 h provide good results until 24 h as well, and then the uniform weighting takes over the other two, since no other weight arrays focus on the last part of the forecasting range. Combining weight arrays enables a good overall optimization of the error.

Direct TEC map prediction does not perform as well as the selected architecture. The network may have to dedicate most of its weights to learn the periodic component and fails to catch transformations of interest. Similarly, giving the difference maps as input to the network does not improve the performance. The reason might be that a new temporal dependency is induced while subtracting the past maps, making it harder to extract the relevant phenomena.

Considering the record of TEC values in order to forecast 48 h of TEC maps, with any weighting, helps improving over a generic model such as [14] which does not take advantage of the last TEC data.

Table 1. Mean RMS of models at a given prediction horizon over period from 10/20/2016 to 12/20/2016

Method	2 h	6 h	12 h	18 h	24 h	30 h	36 h	42 h	48 h
Baselines									
3-day mean	3.072	3.102	3.121	3.140	3.173	3.217	3.241	3.260	3.291
Periodic (48 h)	2.518	2.516	2.514	2.514	2.510	2.509	2.509	2.508	2.505
Direct TEC map	2.712	2.639	2.601	2.575	2.728	3.001	3.003	2.963	3.142
Difference as input	2.447	2.800	2.904	2.898	2.915	2.958	3.007	3.041	3.041
Many-to-one	1.996	3.055	5.356	5.658	7.240	7.586	8.671	8.926	9.427
Selected architecture									
- Uniform weighting	2.050	2.265	2.254	2.256	2.331	**2.434**	**2.459**	**2.475**	**2.490**
- Decreasing weights	**1.987**	2.239	2.248	2.258	2.409	2.585	2.635	2.647	2.674
- Only first 24 h	2.078	**2.195**	**2.197**	**2.214**	**2.297**	2.486	2.513	2.500	2.498

3.2 Performance over the Test Period

The RMS value at 12 h ahead of the input sequence is evaluated for sequences in the test period from 10/01/2016 to 12/31/2016 (see Fig. 3).

RMS errors are stationary after 10/20/2016. They reach an upper point at 7-8 TECU around 10/15/2016 (see Fig. 3). However, during the day 10/17/2016, the proposed algorithm is not outperformed by the periodic prediction.

The algorithm is designed to predict residual differences between TEC maps separated by 48 h. In this case, there is a significant difference between the maps on 10/15/2016 and 10/13/2016. Such perturbations may be linked to external phenomena such as solar events or geomagnetic storms (discussed in Sect. 4).

3.3 Indicative Comparison to Local Prediction Methods

As an attempt to compare to previous works on TEC forecast, Table 2 presents a synthesis of the performance obtained in three related studies and an indicative comparison with the results presented in this paper. The comparison is only indicative since these works differ by their prediction horizons or by the considered areas and since several studies focus on one or a few specific measuring stations instead of producing a worldwide TEC prediction. The last column shows the RMS obtained with the proposed approach, computed as the mean of RMS values at the latitude(s) of the station(s) used by the cited papers.

The obtained results are competitive with state-of-the-art models. To put this in perspective, one should take into account the experimental periods of these studies. Periods from 2011 to 2016 are considered as an active period for the ionosphere due to solar activity, whereas 2008 is a very calm year. Good performance is easier to achieve in calm periods, as explained in [6]. The data used in this study is taken from 2014 to 2016, which makes the results from [12]

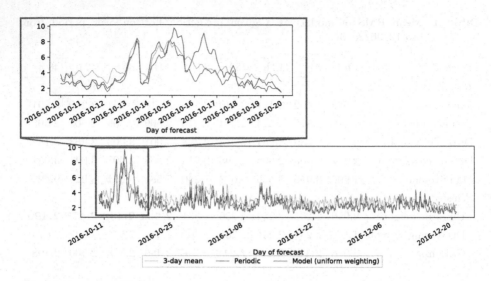

Fig. 3. RMS value for 12 h ahead forecast over whole test period (bottom) and zoom on a disturbed period (top)

Table 2. Results of previous works

Reference		Description	RMS (ref)	RMS (proposed)
[5]	Chunli and Jinsong	1 day forecast, 22° N in China region (2008 data)	1.45	2.049
[12]	Huang and Yuan	24 h forecast, three stations at latitudes 39.61° N, 30.53° N, 25.03° N (2011 data)	≤ 2	1.936
[8]	Niu *et al.*	Forecast of global mean TEC value (0 to 48 h ahead (2012–2013 data)	3.1	0.800

and [8] comparable. On the other hand, the better results of [5] are emphasized by the fact that the experiments are done during a calm year.

The proposed approach provides an accurate mean TEC value, significantly improving the results from [8]. In this study, giving more detailed data as input (*i.e.* a global TEC map) may indeed have helped to produce a better prediction than the one obtained by extrapolating from TEC mean values alone in [8].

4   Limits and Perspectives

The presented method overcomes the baseline methods and is competitive with state-of-the-art local approaches measuring the RMS error.

MAPE (*i.e.* relative error) is mentioned in Subsect. 3 as an alternative to RMS to evaluate the forecasts. Since global TEC maps are produced, the comparison with papers that use MAPE to evaluate local predictions is irrelevant as things stand. In this paper, we prioritized the achievement of a good global minimization with the ℓ_1-loss rather than a good relative error. In order to investigate further the proposed approach, we could consider minimizing the MAPE instead of the ℓ_1-loss. To avoid being over-influenced by ill-conditioned MAPE values, North and South poles (*i.e.* areas where TEC values are very small) could be cropped from TEC maps, so that the remaining TEC values are similar to the TEC values considered in [6,9], making fair comparison possible in future works.

During the implementation, trying to increase the complexity of the DNN by adding parameters and/or convolutional layers to the network did not improve the performance. Either another network is more appropriate to learn more complex dependencies, or there are dynamics in TEC evolution that can not be learned from TEC maps alone. For example magnetosphere or solar particles can significantly impact the ionosphere [21,22] and its evolution. As a consequence, two possibilities are offered to try to improve performance: train another network architecture such as U-net architecture [23], or include solar information in the input to investigate the capacity of a DNN architecture to learn the dependency between TEC levels and solar activity. Both options require learning from a larger training set (data from CODE is available from 1998 to today, whereas only data from 2014 to 2016 is used in this study) in order to have more data from perturbation periods for the network to infer the right behavior.

5   Conclusion

A promising model able to forecast global TEC maps 2 to 48 h ahead of real-time is proposed. This model takes advantage of Deep Neural Network architectures with CNNs and LSTM networks. The proposed approach is able to output a sequence of global TEC maps with comparable or smaller errors compared to single-station prediction models. In the specific case of [8], the global mean TEC value can also be significantly more accurately forecast.

First, the proposed approach forecasts global TEC maps as opposed to local TEC estimates proposed in other published works. Moreover, our method enables range forecasting from 2 to 48 h ahead of real-time, which is a further horizon than the predictions provided by existing operational services.

This method opens new possibilities for using Deep Neural Networks to forecast global TEC maps: the proposed approach can be adapted to other needs such as forecasting a specific region or minimizing another criterion. The presented architecture can also be easily extended to other sequence prediction problems in which both the inputs and prediction targets are sequences of images.

References

1. Datta-Barua, S., Lee, J., Pullen, S., Luo, M., Ene, A., Qiu, D., Zhang, G., Enge, P.: Ionospheric threat parameterization for local area GPS-based aircraft landing systems. J. Aircraft **47**(4), 1141–1151 (2010)
2. Lee, J., Datta-Barua, S., Zhang, G., Pullen, S., Enge, P.: Observations of low-elevation ionospheric anomalies for ground-based augmentation of GNSS. Radio Sci. **46**(6), 1–11 (2011)
3. Tulunay, E., Senalp, E.T., Cander, L.R., Tulunay, Y.K., Bilge, A.H., Mizrahi, E., Kouris, S.S., Jakowski, N.: Development of algorithms and software for forecasting, nowcasting and variability of TEC. Ann. Geophys. **47**(2/3), 1201–1214 (2004). (Supplement to Volume 47)
4. Millward, G.H., Muller-Wodarg, I.C.F., Aylward, A.D., Fuller-Rowell, T.J., Richmond, A.D., Moffett, R.J.: An investigation into the influence of tidal forcing on F region equatorial vertical ion drift using a global ionosphere-thermosphere model with coupled electrodynamics. J. Geophys. Res.: Space Phys. **106**(A11), 24733–24744 (2001)
5. Chunli, D., Jinsong, P.: Modeling and prediction of TEC in China region for satellite navigation. In: 2009 15th Asia-Pacific Conference on Communications. pp. 310–313, October 2009
6. Elmunim, N.A., Abdullah, M., Hasbi, A.M.: Improving ionospheric forecasting using statistical method for accurate GPS positioning over Malaysia. In: 2016 International Conference on Advances in Electrical, Electronic and Systems Engineering (ICAEES), pp. 352–355, November 2016
7. Li, X., Guo, D.: Modeling and prediction of ionospheric total electron content by time series analysis. In: 2010 2nd International Conference on Advanced Computer Control, vol. 2, pp. 375–379, March 2010
8. Niu, R., Guo, C., Zhang, Y., He, L., Mao, Y.: Study of ionospheric TEC short-term forecast model based on combination method. In: 2014 12th International Conference on Signal Processing (ICSP), pp. 2426–2430, October 2014
9. Zhenzhong, X., Weimin, W., Bo, W.: Ionosphere TEC prediction based on Chaos. In: ISAPE2012, pp. 458–460, October 2012
10. Tulunay, E., Senalp, E.T., Radicella, S.M., Tulunay, Y.: Forecasting total electron content maps by neural network technique. Radio Sci. **41**(4) (2006)
11. Wu, Y.W., Liu, R.Y., Jian-Ping, W., Wu, Z.S.: Ionospheric TEC short-term forecasting in CHINA. In: Proceedings of 9th International Symposium on Antennas, Propagation and EM Theory, pp. 418–421, November 2010
12. Huang, Z., Yuan, H.: Ionospheric single-station TEC short-term forecast using RBF neural network. Radio Sci. **49**(4), 283–292 (2014)
13. Senalp, E.T., Tulunay, E., Tulunay, Y.: Total electron content (TEC) forecasting by cascade modeling: a possible alternative to the IRI-2001. Radio Sci. **43**(4) (2008)
14. Jakowski, N., Hoque, M.M., Mayer, C.: A new global TEC model for estimating transionospheric radio wave propagation errors. J. Geodesy **85**(12), 965–974 (2011)
15. Donahue, J., Hendricks, L.A., Guadarrama, S., Rohrbach, M., Venugopalan, S., Saenko, K., Darrell, T.: Long-term recurrent convolutional networks for visual recognition and description. CoRR abs/1411.4389 (2014)
16. Xingjian, S., Chen, Z., Wang, H., Yeung, D.Y., Wong, W.K., Woo, W.C.: Convolutional LSTM network: a machine learning approach for precipitation nowcasting. In: Advances in NIPS, pp. 802–810 (2015)

17. Jarrett, K., Kavukcuoglu, K., Ranzato, M., LeCun, Y.: What is the best multi-stage architecture for object recognition? In: 2009 IEEE 12th ICCV, pp. 2146–2153, September 2009

18. Hochreiter, S., Schmidhuber, J.: Long short-term memory. Neural Comput. **9**(8), 1735–1780 (1997)

19. Masci, J., Meier, U., Cireşan, D., Schmidhuber, J.: Stacked convolutional auto-encoders for hierarchical feature extraction. In: Honkela, T., Duch, W., Girolami, M., Kaski, S. (eds.) ICANN 2011. LNCS, vol. 6791, pp. 52–59. Springer, Heidelberg (2011). doi:10.1007/978-3-642-21735-7_7

20. Hinton, G.E., Salakhutdinov, R.R.: Reducing the dimensionality of data with neural networks. Science **313**(5786), 504–507 (2006)

21. Webb, D.F.: Coronal mass ejections: origins, evolution, and role in space weather. IEEE Trans. Plasma Sci. **28**(6), 1795–1806 (2000)

22. Wells, H.W.: Effects of Solar activity on the ionosphere and radio communications. Proc. IRE **31**(4), 147–157 (1943)

23. Ronneberger, O., Fischer, P., Brox, T.: U-Net: convolutional networks for biomedical image segmentation. In: Navab, N., Hornegger, J., Wells, W.M., Frangi, A.F. (eds.) MICCAI 2015. LNCS, vol. 9351, pp. 234–241. Springer, Cham (2015). doi:10.1007/978-3-319-24574-4_28

Spatio-Temporal Wind Power Prediction Using Recurrent Neural Networks

Wei Lee Woon[1]([✉]), Stefan Oehmcke[2], and Oliver Kramer[2]

[1] Masdar Institute of Science and Technology (part of KUST),
Abu Dhabi, United Arab Emirates
wwoon@masdar.ac.ae
[2] Department of Computer Science, University of Oldenburg,
26111 Oldenburg, Germany
{stefan.oehmcke,oliver.kramer}@uni-oldenburg.de

Abstract. While wind is an abundant source of energy, integrating wind power into existing electricity grids is a major challenge due to its inherent variability. The ability to accurately predict future generation output would greatly mitigate this problem and is thus extremely valuable. Numerical Weather Prediction (NWP) techniques have been the basis of many wind prediction approaches, but the use of machine learning techniques is steadily gaining ground. Deep Learning (DL) is a sub-class of machine learning which has been particularly successful and is now the state of the art for a variety of classification and regression problems, notably image processing and natural language processing. In this paper, we demonstrate the use of *Recurrent Neural Networks*, a type of DL architecture, to extract patterns from the spatio-temporal information collected from neighboring turbines. These are used to generate short term wind energy forecasts which are then benchmarked against various prediction algorithms. The results show significant improvements over forecasts produced using state of the art algorithms.

1 Introduction

1.1 Motivations

The integration of wind power into existing grid infrastructure is a difficult challenge. The inherent variability in wind speeds and direction complicates efforts to ensure sustainable and safe provision (for e.g., Fig. 1(a)). NWP methods can be used to predict wind power production, but are computationally expensive and sensitive to complexities such as local topography and model stability. More recent approaches leverage advancements in machine learning, particularly with intelligent algorithms which can "learn" from complex data sets.

The approach proposed in this paper builds upon existing research published by some of the authors. The starting point is the "spatio-temporal" formulation presented in [1], which uses past measurements from both the target and neighboring turbines as inputs to the prediction models. The proximity of the neighboring turbines suggest that their wind power measurements could be closely

© Springer International Publishing AG 2017
D. Liu et al. (Eds.): ICONIP 2017, Part V, LNCS 10638, pp. 556–563, 2017.
https://doi.org/10.1007/978-3-319-70139-4_56

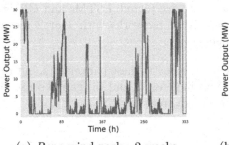

(a) *Reno* wind park - 2 weeks

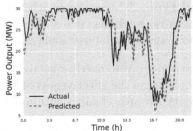

(b) *Casper* wind park - 1 day (actual and predicted values)

Fig. 1. Examples of wind power time series

related to those of the target turbine, though variations in the surrounding landscape and the relative positions of these turbines could make these relationships difficult to model. Machine learning methods are ideal in such circumstances as a way of extracting information from the expanded feature set and combining them to produce improved predictions.

1.2 Related Work

One problem with wind power prediction is the lack of a single standard challenge dataset. The results presented here will be benchmarked against two sets of experimental results, which were previously published in [2,3]. Both these papers used regression ensembles that are highly competitive in a range of different applications. In [2] ensembles of Support Vector Regressors (SVRs) were used in conjunction with evolutionary parameter tuning, while in [3] the performance of SVR and Decision Tree (DT) ensembles were studied and compared.

DL techniques have been used for wind prediction in the past. For example, in [4] future wind speeds were predicted using NWP features. However, this study used a three layer MLP, which doesn't leverage the benefits afforded by recurrent architectures. In [5] a mixture density RNN was used to predict wind power, again using NWP features. Predictions were for the 48 h time horizon, which is a very different problem, while both papers still depended on NWP features, with the attendant shortcomings. Finally, the spatio-temporal formulation mentioned above has shown great promise in previous studies using non DL algorithms, and is worth exploring in the context of RNNs. As such, it would appear that this area still has a lot of potential and we hope that this work helps to advance this promising research direction.

1.3 Objectives

The overall aim of the research described here is to explore and evaluate the effectiveness of RNNs when combined with the spatio-temporal formulation for short term wind power prediction. Two main dimensions are explored:

1. **Cell type** - The two main RNN variants are tested: LSTMs and GRUs
2. **Network architecture** - we deploy RNN networks with 1 or 2 reccurent layers, and with bidirectional connections

The aim is not to overly optimize with respect to these parameters, but rather to (i) determine the effectiveness of RNNs in general, and (ii) gain a sense of the sensitivity towards changes within the design space of RNNs.

2   Methods and Data

2.1   Recurrent Neural Networks (RNNs)

RNNs are a special class of neural network, where the outputs are concatenated with the input vector in the subsequent time step. This makes RNNs uniquely suited to modeling phenomena that are sequential in nature, where the order of data vectors carry some significance. In this paper, two main types of RNNs are studied:

Long Short Term Memory (LSTM) networks [6] are RNNs, which use a specialized processing unit known as the LSTM *cell* (shown in Fig. 2). LSTMs enable better distributions of gradients by remembering, which parts of the previous signal to keep, which ones to forget and which to output. This is done using a set of *gates*, each controlled by a neural network.

Gated Recurrent Networks (GRU) are RNNs which use a version of the LSTM cell [7] with the output gate removed. The resulting cells have fewer parameters but are reported to have similar performance to a regular LSTM. The network structure is the same as in Fig. 2.

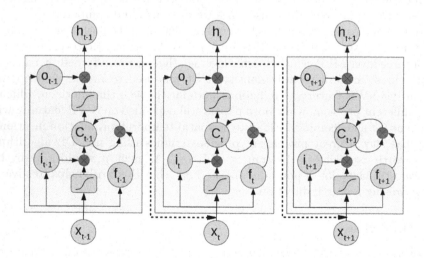

Fig. 2. RNN layer composed of LSTM cells

In addition, we also test *bidirectional* networks of RNN cells (for both LSTMs and GRUs) [8]. This is a more advanced architecture, where both forward and backward connections are employed (Fig. 3).

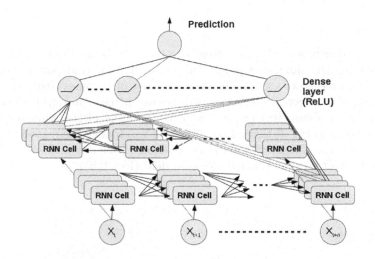

Fig. 3. Bidirectional RNN architecture (the RNN cells can be LSTM or GRU cells)

2.2 Benchmark Algorithms

State of the art ML algorithms were used to generate benchmark predictions. Some were implemented as part of this study to control for variations in implementation and pre-processing, while results for the ensemble predictions are obtained from the cited references (details in [2,3]). In the latter cases, our experimental conditions were carefully controlled to be as similar as possible to those in the benchmarks.

Random Forest - this is an ensemble based approach which uses "bagging" to combine the output of numerous CART regression trees. More details can be found in [9].

k-Nearest Neighbors - as a baseline method we employ k-NN regression. It averages the labels of the k nearest patterns in the training dataset. Nearest neighbor methods are comparatively simple yet powerful, and are useful for creating reasonable benchmarks.

Support Vector Regression (SVR) - Briefly, SVRs work by finding hyperplanes which minimize the following cost function:

$$\text{minimize } \frac{1}{2}||\mathbf{w}||^2 \tag{1}$$

$$\text{subject to } \begin{cases} y_i - \langle \mathbf{w}, \mathbf{x}_i \rangle - b \leq \epsilon \\ \langle \mathbf{w}, \mathbf{x}_i \rangle + b - y_i \leq \epsilon \end{cases} \tag{2}$$

where, $\mathbf{x}_i$ is the pattern, y_i is its corresponding label, $\mathbf{w}$ is the normal vector of the SVR, and ϵ defines a threshold of precision. Unfortunately, a more detailed description is not possible due to space constraints, so the interested reader is referred to [10].

2.3   Data

The data used here was obtained from the *Western Wind Resources Dataset*[1], which contains the simulated power outputs of 32,043 wind power stations in the US. The windparks used in this study consist of turbines situated within a 3 Km radius of 9 selected sites. The turbines are placed on a grid within these windparks, as is shown in Figs. 4(a) and (b). Samples of the wind power time series are shown in Fig. 1, taken from the *Reno* windpark for a period of 2 weeks (Fig. 1(a)) and the *Casper* windpark for a period of one day and with k-Nearest Neighbor (k-NN) generated predictions overlaid (Fig. 1(b)).

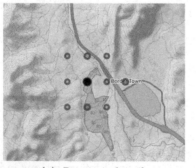
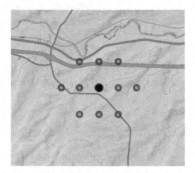

(a) *Reno* wind park   (b) *Casper* wind park

Fig. 4. Examples of turbine placements within wind parks. The filled circles in both cases are the "target" turbines, the outputs of which are to be predicted

The results presented in this paper were generated using the power outputs of two sets of five windparks:

Batch 1. Centered on *Cheyenne, Lancaster, Palm Springs, Vantage* and *Yucca Valley*, for the years 2004–2005. Matches data used in [2].
Batch 2. Centered on *Casper, Hesperia, Las Vegas, Reno* and *Vantage*, and for the years 2004–2006. Matches data used in [3].

To create the input data, feature vectors were created by concatenating the μ past outputs of the target turbine with other wind turbines in a radius of r kilometers. The dependent variable is the power output of the target turbine τ steps in the future. We set $\tau = 3$ and $r = 3$; for RNNs, we used $\mu = 1$, and $\mu = 3$ for the benchmark algorithms. The resulting data matches the data used in [2] exactly and is very close to the data format in [3].

---

[1] https://www.nrel.gov/grid/western-wind-data.html.

2.4 Implementation Details

All the experiments presented in this paper were implemented using the Python programming language. Most of the experiments and development were conducted on an Intel Quad Core i7 2.70 GHz computer. For the benchmark Machine Learning algorithms (k-NN and SVR), the widely used *Scikit-Learn* toolkit [11] was used[2], while the RNNs were implemented using *Keras*[3] and accelerated using an Nvidia Tesla K20m GPU.

The data used is publicly available from the website of the National Renewable Energy Laboratory (NREL). However, for convenience and reliability we used the *WindML* toolkit[4], which automates many of the data collection and feature construction steps.

3 Experimental Results

The results for the experiments described in the preceding sections are presented in Tables 1 and 2 for **Batch 1** and **Batch 2** windparks respectively. As in [2,3], the first half of the relevant period of time was used to generate training data and the second half was used for validation.

For each model, a small selection of hyperparameters were tested, as follows: **RNN**, layer 1 *#units* $\in \{64, 100, 32\}$; layer 2 *#units* $\in \{20, 32, 64, 100\}$; **Random Forest** *#estimators* $\in \{50, 100, 150\}$; **kNN** $k \in \{5, 10, 15\}$; **SVR** $c \in \{10000, 10000, 1000\}$. As stated in Sect. 1, this was not intended as a comprehensive parameter optimization process but just to check for sensitivity. Due

Table 1. Prediction error (MSE) for windparks in **Batch 1**. Top scores are in bold. Due to space constraints, in each row only the best score obtained over a small selection of model parameters is presented

Windparks	Cheyenne	Lancaster	Palm Springs	Vantage	Yucca Valley
Ensemble [2]	7.78	8.19	5.50	5.78	10.57
LSTM (2L)	7.67	**8.12**	5.46	**5.18**	**9.99**
GRU (2L)	7.65	8.22	5.46	5.25	10.02
LSTM (1L)	7.72	8.46	5.48	5.26	10.23
GRU (1L)	7.75	8.33	**5.42**	5.27	10.14
BiLSTM	7.77	8.36	5.44	5.36	10.18
BiGRU	7.91	8.49	5.47	5.35	10.15
RF	**7.59**	8.72	5.99	5.51	10.28
kNN	8.01	8.99	6.14	5.80	10.69
SVR	7.70	8.85	6.15	5.40	10.40

---

[2] http://scikit-learn.org.
[3] https://keras.io.
[4] http://www.windml.org.

Table 2. Prediction error (MSE) for windparks in **Batch 2**. Top scores are in bold. Due to space constraints, in each row only the best score obtained over a small selection of model parameters is presented

Windparks	Casper	Hesperia	Las Vegas	Reno	Vantage
Ensemble [3]	9.88	**7.25**	9.78	13.16	6.41
LSTM (2L)	**9.02**	7.28	**9.43**	12.77	**6.26**
GRU (2L)	9.20	7.33	9.49	**12.73**	6.34
LSTM (1L)	9.11	7.35	9.52	12.95	**6.26**
GRU (1L)	9.43	7.44	9.69	12.82	6.35
BiLSTM	9.39	7.47	9.54	13.08	6.42
BiGRU	9.31	7.46	9.55	13.14	6.57
RF	9.92	7.43	9.96	13.48	6.59
kNN	10.55	7.77	10.54	14.68	6.89
SVR	10.70	7.63	10.47	14.29	6.55

to space constraints, only the best score for each set is presented. The main observations were as follows:

1. Overall, the performance of the RNNs was very impressive. In 8 out of the 10 test cases presented here, RNNs had the best performance compared to existing forecasting techniques.
2. The top performing architecture was the two layered LSTM network, which had the lowest MSE score in 6 test cases, while single layered LSTMs and GRUs and two layered GRUs had the best scores in one case each (in the case of the *Vantage* windpark, the one and two layered LSTMs jointly achieved the top performance.
3. The performance of the RNNs was also very consistent across the range of architectures and cell types studied. We see that *all* RNN based predictors outperformed the top non- RNN predictor in 3 out of 5 windparks in both batches of windparks. In the other columns, they were still generally amongst the top predictors (and again, as noted above at least one RNN architecture achieved the top accuracy in an additional 3 windparks).

4 Conclusion

In summary, the results represented a significant improvement over those presented in [2,3], indicating that the use of RNNs is an extremely promising research direction.

In the future, we plan to extend our methods and improve on these experimental results by taking into account further wind time series, and to deploy parameter optimization techniques such as evolutionary optimization, and Bayesian optimization.

References

1. Kramer, O., Gieseke, F., Satzger, B.: Wind energy prediction and monitoring with neural computation. Neurocomputing **109**, 84–93 (2013)
2. Woon, W.L., Kramer, O.: Enhanced SVR ensembles for wind power prediction. In: 2016 International Joint Conference on Neural Networks (IJCNN), pp. 2743–2748. IEEE (2016)
3. Heinermann, J., Kramer, O.: Machine learning ensembles for wind power prediction. Renew. Energy **89**, 671–679 (2016)
4. Dalto, M., Matuško, J., Vašak, M.: Deep neural networks for ultra-short-term wind forecasting. In: 2015 IEEE International Conference on Industrial Technology (ICIT), pp. 1657–1663. IEEE (2015)
5. Felder, M., Kaifel, A., Graves, A.: Wind power prediction using mixture density recurrent neural networks. In: Poster Presentation gehalten auf der European Wind Energy Conference (2010)
6. Hochreiter, S., Schmidhuber, J.: Long short-term memory. Neural Comput. **9**(8), 1735–1780 (1997)
7. Chung, J., Gulcehre, C., Cho, K., Bengio, Y.: Empirical evaluation of gated recurrent neural networks on sequence modeling. arXiv preprint arXiv:1412.3555 (2014)
8. Baets, L.D., Ruyssinck, J., Peiffer, T., Decruyenaere, J., Turck, F.D., Ongenae, F., Dhaene, T.: Positive blood culture detection in time series data using a BiLSTM network. CoRR abs/1612.00962 (2016). http://arxiv.org/abs/1612.00962
9. Breiman, L.: Random forests. Mach. Learn. **45**(1), 5–32 (2001)
10. Hastie, T., Tibshirani, R., Friedman, J.: The Elements of Statistical Learning, vol. 2. Springer, New York (2009). doi:10.1007/978-0-387-84858-7
11. Pedregosa, F., Varoquaux, G., Gramfort, A., Michel, V., Thirion, B., Grisel, O., Blondel, M., Prettenhofer, P., Weiss, R., Dubourg, V., Vanderplas, J., Passos, A., Cournapeau, D., Brucher, M., Perrot, M., Duchesnay, E.: Scikit-learn: machine learning in Python. J. Mach. Learn. Res. **12**, 2825–2830 (2011)

Bayesian Neural Learning via Langevin Dynamics for Chaotic Time Series Prediction

Rohitash Chandra[1,2]([⊠]), Lamiae Azizi[1,2], and Sally Cripps[1,2]

[1] School of Mathematics and Statistics, The University of Sydney,
Sydney, NSW 2006, Australia
rohitash.chandra@sydney.edu.au
[2] Centre for Translational Data Science, The University of Sydney,
Sydney, NSW 2006, Australia

Abstract. Although neural networks have been very promising tools for chaotic time series prediction, they lack methodology for uncertainty quantification. Bayesian inference using Markov Chain Mont-Carlo (MCMC) algorithms have been popular for uncertainty quantification for linear and non-linear models. Langevin dynamics refer to a class of MCMC algorithms that incorporate gradients with Gaussian noise in parameter updates. In the case of neural networks, the parameter updates refer to the weights of the network. We apply Langevin dynamics in neural networks for chaotic time series prediction. The results show that the proposed method improves the MCMC random-walk algorithm for majority of the problems considered. In particular, it gave much better performance for the real-world problems that featured noise.

Keywords: Backpropagation · Gradient descent · MCMC algorithms · Chaotic time series · Neural networks

1 Introduction

The success of neural networks is partially due to the ability to train them on massive data sets with variants of the backpropagation algorithm [1,14]. Despite the successes, there are some challenges in training neural networks with back-propagation. Firstly, with large data sets, finding the optimal values of hyper-parameters (e.g. learning and momentum rate) and weights take a large amount of time [1]. Secondly, through canonical backpropagation, we can only obtain point estimates of the weights in the network. As a result, these networks make predictions that do not account for uncertainty in the parameters. However, in many cases, these weights may be poorly specified and it is desirable to produce uncertainty estimates along with predictions. A Bayesian approach to neural networks can potentially avoid some of the pitfalls of canonical backpropagation [9]. Bayesian techniques, in principle, can automatically infer hyperparameter values by marginalizing them out of the posterior distribution [10]. Furthermore, Bayesian methods naturally account for uncertainty in parameter estimates and can propagate this uncertainty into predictions. Finally, they are often more

© Springer International Publishing AG 2017
D. Liu et al. (Eds.): ICONIP 2017, Part V, LNCS 10638, pp. 564–573, 2017.
https://doi.org/10.1007/978-3-319-70139-4_57

robust to overfitting, since they average over parameter values instead of choosing a single point estimate. Bayesian neural networks (BNNs) use Markov Chain Monti-Carlo (MCMC) methods such as those based on e.g. the Laplace approximation [9], Hamiltonian Monte Carlo [12], expectation propagation [4] and variational inference [2]. However, these approaches have not seen widespread adoption due to their lack of scalability in both network architecture and data size. A number of attempts have been proposed combining MCMC techniques with the gradient optimization algorithms [6], the simulated annealing method [8], and evolutionary algorithms [5].

In the context of the non-linear time series prediction, there are a number of studies in the literature. Liang et al. presented an MCMC algorithm for neural networks for selected time series problems [7]. Bayesian techniques have also been used for controlling model complexity and selecting inputs in neural networks for the short term time series forecasting [3]. Moreover, recursive Bayesian recurrent neural networks [11] and evolutionary MCMC Baysiean neural networks have been used for time series forecasting [5]. Although these methods have been used, there has not been much emphasis on experimental evaluation of the method on chaotic time series benchmark problems. Langevin dynamics refer to a class of MCMC algorithms that incorporate gradients with Gaussian noise in parameter updates [13]. In the case of neural networks, the parameter updates refer to the weights of the network.

In this paper, we provide the synergy of MCMC random-walk algorithm and gradient descent via backpropagation neural network. This is referred as Langevin dynamics MCMC for training neural networks. We employ six benchmark chaotic time series problems to demonstrate the effectiveness of the proposed method. Furthermore, comparison is done with standalone gradient descent and the MCMC random-walk algorithm.

The rest of the paper is organised as follows. Section 2 presents the proposed method and Sect. 3 presents experiments and results. Section 4 concludes the paper with a discussion of future work.

2 Langevin Dynamics for Neural Networks

2.1 Model and Priors

Let y_t denote a univariate time series modelled by:

$$y_t = f(\mathbf{x}_t) + \epsilon_t, \text{ for } t = 1, 2, \dots, n \tag{1}$$

where $f(\mathbf{x}_t) = E(y_t|\mathbf{x}_t)$, is an unknown function, $\mathbf{x}_t = (y_{t-1}, \dots, y_{t-D})$ is a vector of lagged values of y_t, and ϵ_t is the noise with $\epsilon_t \sim \mathcal{N}(0, \tau^2) \; \forall t$.

In order to use neural networks for time series prediction, the original dataset is constructed into a state-space vector through Taken's theorem [15] which is governed by the embedding dimension (D) and time-lag (T).

Define

$$\mathcal{A}_{D,T} = \{t; t > D, \mod (t - (D+1), T) = 0\} \tag{2}$$

Let $\mathbf{y}_{\mathcal{A}_{\mathcal{D},\mathcal{T}}}$ to be the collection of y_t's for which $t \in \mathcal{A}_{\mathcal{D},\mathcal{T}}$, then, $\forall t \in \mathcal{A}_{D,T}$, we compute the $f(\mathbf{x}_t)$ by a feedforward neural network with one hidden layer defined by the function

$$f(\mathbf{x}_t) = g\left(\delta_o + \sum_{h=1}^{H} v_j g\left(\delta_h + \sum_{d=1}^{D} w_{dh} y_{t-d}\right)\right) \tag{3}$$

where δ_o and δ_h are the bias weights for the output o and hidden h layer, respectively. V_j is the weight which maps the hidden layer h to the output layer. w_{dh} is the weight which maps y_{t-d} to the hidden layer h and g is the activation function, which we assume to be a sigmoid function for the hidden and output layer units.

Let $\boldsymbol{\theta} = (\tilde{\mathbf{w}}, \mathbf{v}, \boldsymbol{\delta}, \tau^2)$, with $\boldsymbol{\delta} = (\delta_o, \delta_h)$, denote $L = (DH + (2 * H) + O + 1)$ vector of parameters that includes weights and biases, with O number of neurons in output layer. H is the number of hidden neurons required to evaluate the likelihood for the model given by (1), with $\tilde{\mathbf{w}} = (\mathbf{w}'_{1.}, \ldots, \mathbf{w}'_{D.})'$, and $\mathbf{w}_{d.} = (w_{d1}, \ldots, w_{dH})'$, for $d = 1, \ldots, D$.

To conduct a Bayesian analysis, we need to specify prior distributions for the elements of θ which we choose to be

$$v_h \sim \mathcal{N}(0, \sigma^2) \text{ for } h = 1, \ldots, H,$$
$$\delta_0 \sim N(0, \sigma^2)$$
$$\delta_h \sim N(0, \sigma^2)$$
$$w_{dj} \sim \mathcal{N}(0, \sigma^2) \text{ for } h = 1, \ldots, H \text{ and } d = 1, \ldots, D,$$
$$\tau^2 \sim \mathcal{IG}(\nu_1, \nu_2) \tag{4}$$

where H is the number of hidden neurons. In general the log posterior is

$$\log\left(p(\boldsymbol{\theta} \mid \mathbf{y})\right) = \log\left(p(\boldsymbol{\theta})\right) + \log\left(p(\mathbf{y} \mid \boldsymbol{\theta})\right)$$

In our particular model the log likelihood is

$$\log\left(p(\mathbf{y}_{\mathcal{A}_{\mathbf{D},\mathcal{T}}} \mid \boldsymbol{\theta})\right) = -\frac{n-1}{2}\log(\tau^2) - \frac{1}{2\tau^2}\sum_{t \in \mathcal{A}_{\mathcal{D},\mathcal{T}}}(y_t - E(y_t \mid \mathbf{x}_t))^2 \tag{5}$$

where $E(y_t \mid \mathbf{x}_t)$ is given by (3). We further assume that the elements of θ are independent *apriori* so that the log of the prior distributions is

$$\log\left(p(\boldsymbol{\theta})\right) = -\frac{HD + H + 2}{2}\log(\sigma^2) - \frac{1}{2\sigma^2}\left(\sum_{h=1}^{H}\sum_{d=1}^{D} w_{dh}^2 + \sum_{h=1}^{H}(\delta_h^2 + v_h^2) + \delta_o^2\right)$$
$$-(1 + \nu_1)\log(\tau^2) - \frac{\nu_2}{\tau^2} \tag{6}$$

2.2   Algorithm

As discussed earlier, principled way to incorporate uncertainty during learning is to use Bayesian inference [9,12]. This requires obtaining samples from the posterior distribution $p(\boldsymbol{\theta} \,|\, \mathbf{y}_{\mathcal{A}_{\mathbf{D},\mathbf{T}}})$. The proposed MCMC algorithm consists of a single Metropolis Hasting step with proposals formed using Langevin dynamics that employs gradients. In particular, we propose a new value of $\boldsymbol{\theta}$ from

$$\boldsymbol{\theta}^p \sim \mathcal{N}(\bar{\boldsymbol{\theta}}^{[k]}, \Sigma_\theta), \text{ where} \tag{7}$$

$$\bar{\boldsymbol{\theta}}^{[k]} = \boldsymbol{\theta}^{[k]} + r \times \nabla E_{\mathbf{y}_{\mathcal{A}_{\mathbf{D},\mathbf{T}}}}[\boldsymbol{\theta}^{[k]}], \tag{8}$$

$$E_{\mathbf{y}_{\mathcal{A}_{\mathbf{D},\mathbf{T}}}}[\boldsymbol{\theta}^{[k]}] = \sum_{t \in \mathcal{A}_{\mathbf{D},\mathbf{T}}} (y_t - f(\mathbf{x}_t)^{[k]})^2,$$

$$\nabla E_{\mathbf{y}_{\mathcal{A}_{\mathbf{D},\mathbf{T}}}}[\boldsymbol{\theta}^{[k]}] = \left(\frac{\partial E}{\partial \theta_1}, \dots, \frac{\partial E}{\partial \theta_L} \right)$$

r is the learning rate, $\Sigma_\theta = \sigma_\theta^2 I_L$ and I_L is the $L \times L$ identity matrix. So that the newly proposed value of $\boldsymbol{\theta}^p$, consists of 2 parts:

1. An gradient descent based weight update given by Eq. (8).
2. Add an amount of noise, from $\mathcal{N}(0, \Sigma_\theta)$.

Hereafter, we refer to the proposed Langevin dynamics for neural networks as LD-MCMC. This combined update is used as a proposal in a Metropolis-Hastings step, which accepts the proposed value of $\boldsymbol{\theta}^p$ with the usual probability α, where

$$\alpha = \min \left\{ 1, \frac{p(\boldsymbol{\theta}^p \,|\, \mathbf{y}_{\mathcal{A}_{D,T}}) q(\boldsymbol{\theta}^{[k]} \,|\, \boldsymbol{\theta}^p)}{p(\boldsymbol{\theta}^{[k]} \,|\, \mathbf{y}_{\mathcal{A}_{D,T}}) q(\boldsymbol{\theta}^p \,|\, \boldsymbol{\theta}^{[k]})} \right\} \tag{9}$$

where $p(\boldsymbol{\theta}^p \,|\, \mathbf{y}_{\mathcal{A}_{D,T}})$ and $p(\boldsymbol{\theta}^{[k]} \,|\, \mathbf{y}_{\mathcal{A}_{D,T}})$ can be computed using Eqs. (5) and (6). $q(\boldsymbol{\theta}^p \,|\, \boldsymbol{\theta}^{[k]})$, is given by Eq. (7) and $q(\boldsymbol{\theta}^{[k]} \,|\, \boldsymbol{\theta}^p) \sim N(\bar{\boldsymbol{\theta}}^p, \Sigma_\theta)$, with $\bar{\boldsymbol{\theta}}^p = \boldsymbol{\theta}^p + r \times \nabla E_{\mathbf{y}_{\mathcal{A}_{\mathbf{D},\mathbf{T}}}}[\boldsymbol{\theta}^p]$, thus ensuring that the detailed balance condition holds and the sequence $\boldsymbol{\theta}^{[k]}$ converges to draws from the posterior $p(\boldsymbol{\theta} \,|\, \mathbf{y})$.

For testing, given a test input $\tilde{\mathbf{x}}$ (with missing label $\tilde{y}$), the uncertainty learned in training is transferred to prediction, yielding the following posterior distribution:

$$p(\tilde{y} \,|\, \tilde{x}, \mathbf{y}_{\mathcal{A}_{D,T}}) = \mathbb{E}_{p(\boldsymbol{\theta} \,|\, \mathbf{y}_{\mathcal{A}_{D,T}})}[p(\tilde{y} \,|\, \tilde{x}, \boldsymbol{\theta})] = \int_\theta p(\tilde{y} \,|\, \tilde{x}, \boldsymbol{\theta}) p(\boldsymbol{\theta} \,|\, \mathbf{y}_{\mathcal{A}_{D,T}}) d\boldsymbol{\theta} \tag{10}$$

The predicted distribution of $\tilde{y}$ can be viewed in terms of model averaging across parameters, based on the learned $p(\boldsymbol{\theta} \,|\, \mathbf{y}_{\mathcal{A}_{D,T}})$; this is contrasted with learning a single point estimate of $\boldsymbol{\theta}$ based on $\mathbf{y}_{\mathcal{A}_{D,T}}$.

Each iteration of the LD-MCMC algorithm requires refinement of the proposal drawn through gradient descent on the entire dataset for one iteration via backpropagation. The LD-MCMC algorithm for feedforward neural networks (FNNs) is given in Algorithm 1. The major difference of this algorithm from MCMC random-walk lies in the additional step where gradient descent is used to update the location of the proposal distribution. This is helpful as gradients help in better proposals when compared to a random-walk MCMC algorithm.

Langevin dynamics through gradient descent is possible for the selected time series problems given that the size of the datasets are relatively low when compared to big data problems. The implementation of Algorithm 1 in Python is also given online[1]. The MCMC random-walk for neural networks is also given online[2].

Algorithm 1. Langevin Dynamics for neural networks

Data: Univariate time series $\mathbf{y}$
Result: Posterior of weights and biases $p(\boldsymbol{\theta} \,|\, \mathbf{y})$
Step 1: State-space reconstruction $\mathbf{y}_{\mathcal{A}_{D,T}}$ by Eq. 2
Step 2: Define feedforward network as given in Eq. 3
Step 3: Define $\boldsymbol{\theta}$ as the set of all weights and biases
Step 4: Set parameters σ^2 ,ν_1, ν_2 for prior given in Eq. 6

for *each k until max-samples* **do**
  1. Compute gradient $\Delta\boldsymbol{\theta}^{[k]}$ given by Eq. 8
  2. Draw $\boldsymbol{\eta}$ from $\mathcal{N}(0, \Sigma_\eta)$
  3. Propose $\boldsymbol{\theta}^* = \theta^{[k]} + \Delta\boldsymbol{\theta}^{[k]} + \boldsymbol{\eta}$
  4. Draw from uniform distribution $u \sim \mathcal{U}[0, 1]$
  5. Obtain acceptance probability α given by Eq. 9
  if $u < \alpha$ **then**
    | $\theta^{[k+1]} = \boldsymbol{\theta}^*$
  end
  else
    | $\theta^{[k+1]} = \theta^{[k]}$
  end
end

---

3   Experiments and Results

This section presents experimental evaluation of LD-MCMC algorithm (Algorithm 1) for six benchmark one-step-ahead chaotic time series prediction problems. The comparison is done with MCMC random-walk algorithm and gradient descent (backpropagation).

3.1   Problem Description

The benchmark problems employed are Mackey-Glass, Lorenz, Sunspot, and Laser, Henon, and Rossler time series. Takens' embedding theorem [16] is applied

---

[1] https://github.com/rohitash-chandra/LDMCMC_timeseries.
[2] https://github.com/rohitash-chandra/MCMC_fnn_timeseries.

with selected values as follows. $D = 4$ and $T = 2$ is used for reconstruction of the respective time series into state-space vector. All the problems used first 1000 data points of the time series from which 60% was used for training and 40% for testing the method. The prediction performance is measured using the root mean squared error (RMSE)

$$RMSE = \sqrt{\frac{1}{N} \sum_{i=1}^{N} (y_i - \hat{y}_i)^2} \tag{11}$$

where $y_i, \hat{y}_i$ are the observed data and predicted data, respectively. N is the length of the observed data. These results are also compared with related methods from the literature.

3.2 Experimental Design

We provide details of the experimental design followed by experimentation and results. The FNN employs one hidden layer and sigmoid activation function for all the respective problems. The first 10% was discarded for "burn-in" which is standard procedure for MCMC based algorithms. In the LD-MCMC algorithm, each sample drawn was further refined using gradient descent for $n = 1$ iterations. Note that the FNN used 4 neurons in the input and 5 neurons in the hidden layer for all the respective problems. In the MCMC methods, the random-walk for the respective weights and biases θ was drawn using a normal distribution with mean $\mu = 0$ and standard deviation $\sigma = 0.02$. A normal distribution was also used to draw samples that define η with mean $\mu = 0$ and standard deviation $\sigma = 0.2$. These values gave acceptable results in trial runs and hence were used in all the problems. In the case of GD, fixed learning rate of 0.1 and momentum rate of 0.01 was used for all the problems. The parameters for the prior given in Eq. 6 was set: $\sigma^2 = 25$, $\nu_1 = 0$, $\nu_2 = 0$.

3.3 Results

The results for the three methods for the six benchmark problems are given in Table 1. Note that (*) represents early convergence or termination criteria which is 2000 epochs for GD*. On the other hand, GD convergence is defined by 80 000 iterations (epochs) in order to compare with MCMC random-walk algorithm which draws 80 000 samples. LD-MCMC draws 40 000 samples and each of them are refined by GD for 1 iteration which makes the computation time similar in terms of number of fitness evaluation given by the loss function. Note that GD method employed 30 independence experimental runs, while MCMC and LD-MCMC results are for a single experimental run. The mean and standard deviation (std) of the results are given.

We evaluate the performance achieved by the respective algorithms for six benchmark chaotic time series problems which include Lazer, Sunspot, Mackey-Glass, Lorenz, Rossler and Henon. We note that Lazer and Sunspot are real-world problems that contain noise while the rest are simulated time series. We find that backpropagation via GD outperforms the rest of the methods for all the problems. Moving on, we observe that LD-MCMC outperforms MCMC random-walk algorithm for all the problems except for the Mackey-Glass problem. Generally, LD-MCMC accepted more solutions when compared to MCMC which could be due to the refinement of samples through gradient descent. In the case of Lorenz problem, they achieve similar results. The MCMC methods generally gave better performance when compared to early convergence instance of gradient descent (GD*). The results for early convergence was shown taking into consideration typical training time used for related problems in the literature.

Table 1. Results for the respective methods

Problem	Method	Train (mean)	Train (std)	Test (mean)	Test (std)	% Accepted
Lazer	GD*	0.02191	0.00129	0.02675	0.00085	-
	GD	0.01035	0.00162	0.01732	0.00142	-
	MCMC	0.02549	0.00837	0.02643	0.00718	9.70
	LD-MCMC	0.01658	0.00211	0.02280	0.00351	57.86
Sunspot	GD*	0.02117	0.00160	0.02359	0.00209	-
	GD	0.00775	0.00026	0.00975	0.00031	-
	MCMC	0.01466	0.00174	0.01402	0.00178	4.55
	LD-MCMC	0.01155	0.00167	0.01090	0.00146	45.89
Mackey	GD*	0.00590	0.00026	0.00669	0.00029	-
	GD	0.00286	0.00025	0.0033	0.00026	-
	MCMC	0.00511	0.00058	0.00520	0.00058	1.74
	LD-MCMC	0.00615	0.00091	0.00627	0.00091	30.35
Lorenz	GD*	0.01570	0.00056	0.01608	0.00052	-
	GD	0.00400	0.00024	0.00460	0.00026	-
	MCMC	0.00813	0.00174	0.00713	0.00150	2.74
	LD-MCMC	0.00890	0.00211	0.00821	0.00207	26.95
Rossler	GD*	0.01570	0.00056	0.01608	0.00052	-
	GD	0.00281	0.00029	0.00462	0.00045	-
	MCMC	0.01371	0.00291	0.01355	0.00297	3.69
	LD-MCMC	0.00722	0.00121	0.00692	0.00120	35.53
Henon	GD*	0.01366	0.00033	0.01778	0.00025	-
	GD	0.00555	0.00029	0.00604	0.00025	-
	MCMC	0.03256	0.03920	0.03127	0.03850	8.71
	LD-MCMC	0.00948	0.00112	0.00912	0.00114	27.17

The results for a selected problem (Sunspot) is given in Fig. 1 which shows uncertainty quantification using LD-MCMC. The results show the uncertainty (through 5th and 95th percentile) and prediction on the training and the test datasets. The posterior distributions for the set of weights and biases are given as box-plot in Fig. 2 that could be further visualised as probability distributions.

The results in general have shown that LD-MCMC further improves the performance of MCMC random-walk for training FNNs for majority of the

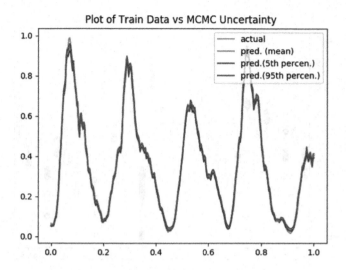

(a) Prediction and uncertainty quantification over training data

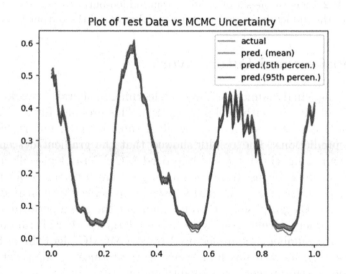

(b) Prediction and uncertainty quantification over test data

Fig. 1. Results for Sunspot time series using LD-MCMC algorithm. Note that the x-axis represents the time in year while y-axis gives the Sunspot index.

problems. This improvement has been through the incorporation of gradients in weight updates via Langevin dynamics. This has been beneficial for improving the proposals drawn by MCMC random-walk algorithm. In the case for larger datasets, stochastic gradient version of Langevin dynamics can be implemented [17].

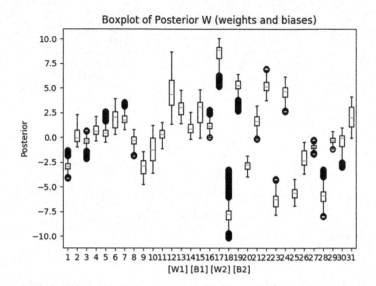

Fig. 2. Posterior of weights and biases. W1 refers to the set of weights from input-hidden layer, W2 refers to the set of weights from hidden-output layer, B1 refers to the set of biases of the hidden layer and B2 refers to the bias of the output layer.

4   Conclusions and Future Work

We applied Langevin dynamics for Bayesian learning in neural networks that featured synergy of MCMC with backpropagation. This provides further advantage to conventional neural network training algorithms through uncertainty quantification in predictions. The results showed that the gradient descent weight updates though Langevin dynamics improved MCMC random-walk algorithm for majority of the problems. In particular, it gave much better performance for real-world problems such as Lazer and Sunspot time series that featured noise. A limitation is that the proposed method cannot get the same performance when compared to gradient decent alone when given a large number of training epochs that match with number of samples used by the MCMC algorithms. However, gradient decent on its own has limitations as it produces a one-point solution which does not feature uncertainty quantification.

Future work would focus on Langevin dynamics based Bayesian learning for other neural network architectures. The incorporation of Langevin dynamics with hyper-parameters such as momentum and learning rate could be further explored. In addition, uncertainty quantification could also be provided in terms of number of hidden neurons and layers of the network. Moreover, the method can be directly used for real-world time series applications that includes climate change problems such as rainfall, storms and cyclones.

References

1. Bottou, L.: Large-scale machine learning with stochastic gradient descent. In: Proceedings of COMPSTAT 2010, pp. 177–186. Springer, Heidelberg (2010). doi:10. 1007/978-3-7908-2604-3_16
2. Hinton, G.E., Van Camp, D.: Keeping the neural networks simple by minimizing the description length of the weights. In: Proceedings of 6th Annual Conference on Computational Learning Theory, pp. 5–13. ACM (1993)
3. Hippert, H.S., Taylor, J.W.: An evaluation of bayesian techniques for controlling model complexity and selecting inputs in a neural network for short-term load forecasting. Neural Netw. 23(3), 386–395 (2010)
4. Jylänki, P., Nummenmaa, A., Vehtari, A.: Expectation propagation for neural networks with sparsity-promoting priors. J. Mach. Learn. Res. 15(1), 1849–1901 (2014)
5. Kocadağlı, O., Aşıkgil, B.: Nonlinear time series forecasting with Bayesian neural networks. Expert Syst. Appl. 41(15), 6596–6610 (2014)
6. Li, C., Chen, C., Carlson, D., Carin, L.: Preconditioned stochastic gradient Langevin dynamics for deep neural networks. In: 30th AAAI Conference on Artificial Intelligence, pp. 1788–1794 (2016)
7. Liang, F.: Bayesian neural networks for nonlinear time series forecasting. Stat. Comput. 15(1), 13–29 (2005)
8. Liang, F.: Annealing stochastic approximation Monte Carlo algorithm for neural network training. Mach. Learn. 68(3), 201–233 (2007)
9. MacKay, D.J.: A practical Bayesian framework for backpropagation networks. Neural Comput. 4(3), 448–472 (1992)
10. MacKay, D.J.: Hyperparameters: optimize, or integrate out? In: Heidbreder, G.R. (ed.) Maximum entropy and Bayesian Methods, pp. 43–59. Springer, Dordrecht (1996). doi:10.1007/978-94-015-8729-7_2
11. Mirikitani, D.T., Nikolaev, N.: Recursive Bayesian recurrent neural networks for time-series modeling. IEEE Trans. Neural Netw. 21(2), 262–274 (2010)
12. Neal, R.M.: Bayesian Learning for Neural Networks. Springer, New York (1996)
13. Neal, R.M.: MCMC using Hamiltonian dynamics. In: Brooks, S., Gelman, A., Jones, G.L., Meng, X.-L. (eds.) Handbook of Markov Chain Monte Carlo, pp. 113–162. Chapman & Hall/CRC Press (2010)
14. Rumelhart, D.E., Hinton, G.E., Williams, R.J.: Learning representations by backpropagating errors. Nature 323, 533–536 (1986). doi:10.1038/323533a0
15. Takens, F.: Detecting strange attractors in turbulence. In: Rand, D.A., Young, L.S. (eds.) Dynamical Systems and Turbulence, Warwick 1980. LNM, pp. 366–381. Springer, Heidelberg (1981). doi:10.1007/BFb0091903
16. Takens, F.: On the numerical determination of the dimension of an attractor. In: Braaksma, B.L.J., Broer, H.W., Takens, F. (eds.) Dynamical Systems and Bifurcations. LNM, vol. 1125, pp. 99–106. Springer, Heidelberg (1985). doi:10.1007/ BFb0075637
17. Welling, M., Teh, Y.W.: Bayesian learning via stochastic gradient Langevin dynamics. In: Proceedings of 28th International Conference on Machine Learning (ICML-2011), pp. 681–688 (2011)

Causality Analysis Between Soil of Different Depth Moisture and Precipitation in the United States

Hui Su[1], Sanqing Hu[1], Tong Cao[1], Jianhai Zhang[1(✉)],
Yuying Zhu[1], Bocheng Wang[2], and Lan Jiang[3]

[1] College of Computer Science, Hangzhou Dianzi University,
Hangzhou 310018, Zhejiang, China
suhui08@foxmail.com, {sqhu,jhzhang}@hdu.edu.cn,
tonneyc@126.com, xjjhzyy@126.com
[2] Zhejiang University of Media & Communications,
Hangzhou 310018, Zhejiang, China
baixibao@gmail.com
[3] Zhejiang Environmental Monitoring Center, Hangzhou 310018, Zhejiang, China

Abstract. Previously the stronger coupling between soil moisture and precipitation in the land-atmosphere interaction have widely been studied. However, few work discusses the causality between them. In this paper, we use Granger causality (GC) and New causality (NC) to detect the causality between soil of different depth moisture and precipitation. Our results demonstrate that the causality between shallow soil moisture and precipitation is greater than that between deep soil moisture and precipitation. And the results also demonstrate that the NC method is much clearer to reveal the causal influence between soil moisture and precipitation than GC method in the time domain.

Keywords: Granger causality · New causality · Soil moisture · Precipitation

1 Introduction

The relationship between soil moisture and precipitation is an important issue for the study of climate change and water cycle. Soil moisture influences latent and sensible heat fluxes at the land surface, thus altering precipitation profiles by moisture evaporation. Roberts [1] confirmed that the heat fluxs exchanges between the ocean and atmosphere were fundamental components of these balances about the Modern Era Retrospective-Analysis for Research and Applications (MERRA). Precipitation is the primary connection in the water cycle that provides the delivery of atmospheric water to the Earth. For example, Wu [2] demonstrated that precipitation play an important role in regulating both interannual and spatial variations of soil water. Soil water content can modify atmospheric processes on a range, potentially leading to cloud formation

© Springer International Publishing AG 2017
D. Liu et al. (Eds.): ICONIP 2017, Part V, LNCS 10638, pp. 574–580, 2017.
https://doi.org/10.1007/978-3-319-70139-4_58

and precipitation. In addition, Li [3] used 19-year Chinese soil moisture data, including ERA40, NCEP/NCAR reanalysis (R-1) and NCEP/DOE reanalysis 2 (R-2), and demonstrated that the research on soil moisture provided important information for other studies of climate change, water cycle and weather model evaluation. So studying the relationship between soil moisture and precipitation has great significant in the land-atmosphere interaction.

So far, most of researchers by proposing a variety of methods discuss the correlation relationship between soil moisture and precipitation. However, they have not reached a consensus. For example, Findell [4] revealed that soil moisture positively correlated with precipitation by analyzing 14-year data of the Illinois. Paolo [5] also analyzed this relationship in Illinois and found soil moisture to be correlated with precipitation. However, Salvucci [6] applied the GC method to analyze the relationship and concluded that there is no significant influence from soil moisture to subsequent precipitation in the same regions. In addition, Duerinck [7] analyzed this relationship through different time scales in the same place, and found that there is the positive correlation between them on the monthly to seasonal time scale but no relation on the daily to weekly time scale.

There are few work to discuss causality between soil moisture and precipitation. For example, employing GC method, Salvucci found that there is no causal influence from soil moisture to precipitation. In addition, Koster [8] and Hurk [9] demonstrated there is causal influence between them by observing data changes and reanalyzing models. For example, Koster confirmed that soil moisture variations affect precipitation by observing the 50-year the United States precipitation dataset. Hurk revealed that soil moisture affects precipitation through model analysis. They do not use any causal methods (such as GC method), only through observation data and model analysis to prove that there is a causal relationship between them. Observation data and model analysis do not completely reveal the causal mechanisms between them. In this paper, we apply that our recently proposed New causality (NC) method [10] which was shown to be much better reveal true causality of the underlying system than the widely used GC by many illustrative examples [11].

2   GC and NC

Consider two stochastic time series which are assumed to be jointly stationary. Individually, under fairly general conditions, each time series admits an autoregressive representation

$$
\begin{cases}
X_{1,t} = \sum_{j=1}^{m} \mathbf{a}_{1,j} X_{1,t-j} + \epsilon_{1,t} \\
X_{2,t} = \sum_{j=1}^{m} \mathbf{a}_{2,j} X_{1,t-j} + \epsilon_{2,t}
\end{cases}
\tag{1}
$$

and their joint representations are described as

$$
\begin{cases}
X_{1,t} = \sum_{j=1}^{m} a_{11,j} X_{1,t-j} + \sum_{j=1}^{m} a_{12,j} X_{2,t-j} + \eta_{1,t} \\
X_{2,t} = \sum_{j=1}^{m} a_{21,j} X_{1,t-j} + \sum_{j=1}^{m} a_{22,j} X_{2,t-j} + \eta_{2,t}
\end{cases}
\tag{2}
$$

where $\mathbf{a}_{1,j}, \mathbf{a}_{2,j}$ and $a_{11,j}, a_{12,j}, a_{21,j}, a_{22,j} (j = 1, 2, \cdots, m)$ represent the coefficients of the autoregressive model, m represents the order of the model, $t = 0, 1, \cdots, N$, N is the size of the sampled data, ϵ_i and η_i have zero means and variances of $\sigma_{\epsilon_i}^2$, and $\sigma_{\eta_i}^2$, $i = 1, 2$. The covariance between η_1 and η_2 is defined by $\sigma_{\eta_1 \eta_2} = cov(\eta_1, \eta_2)$. For a practical system, a general approach for determining the order of the model is the AIC–Akaike Information Criterion [12].

From the Eqs. (1) and (2), if $\sigma_{\eta_1}^2$ is less than $\sigma_{\epsilon_1}^2$ in some suitable sense X_2 is said to have a causal influence on X_1. Otherwise, if $\sigma_{\eta_1}^2 = \sigma_{\epsilon_1}^2$, X_2 is said to have no causal influence on X_1. In this case, two equalities are same. Such kind of causal influence, called GC, is defined by

$$
F_{X_2 \to X_1} = \ln \frac{\sigma_{\epsilon_1}^2}{\sigma_{\eta_1}^2}
\tag{3}
$$

Obviously, $F_{X_2 \to X_1} = 0$ when there is no causal influence from X_2 to X_1 and $F_{X_2 \to X_1} > 0$ when there is. Similarly, the causal influence from X_1 to X_2 is defined by

$$
F_{X_1 \to X_2} = \ln \frac{\sigma_{\epsilon_2}^2}{\sigma_{\eta_2}^2}
\tag{4}
$$

Based on the first equality in (2), we can see contributions to $X_{1,t}$, which include $\sum_{j=1}^{m} a_{11,j} X_{1,t-j}, \sum_{j=1}^{m} a_{12,j} X_{2,t-j}$ and the noise term $\eta_{1,t}$ where the influence from $\sum_{j=1}^{m} a_{11,j} X_{1,t-j}$ is causality from X_1's own past values. Each contribution plays an important role in determining $X_{1,t}$. If $\sum_{j=1}^{m} a_{12,j} X_{2,t-j}$ occupies a larger portion among all those contributions, then X_2 has stronger causality on X_1, or vice versa. Thus, a good definition for causality from X_2 to X_1 in time domain should be able to describe what proportion X_2 occupies among all these contributions. So based on this general guideline NC from X_2 to X_1 is defined as

$$
n_{X_2 \to X_1} = \frac{\sum_{t=m}^{N} \left(\sum_{j=1}^{m} a_{12,j} X_{2,t-j} \right)^2}{\sum_{h=1}^{2} \sum_{t=m}^{N} \left(\sum_{j=1}^{m} a_{1h,j} X_{h,t-j} \right)^2 + \sum_{t=m}^{N} \eta_{1,t}^2}
\tag{5}
$$

Similarly, NC in time domain from X_1 to X_2 is defined by

$$n_{X_1 \to X_2} = \frac{\sum\limits_{t=m}^{N} \left(\sum\limits_{j=1}^{m} a_{21,j} X_{1,t-j} \right)^2}{\sum\limits_{h=1}^{2} \sum\limits_{t=m}^{N} \left(\sum\limits_{j=1}^{m} a_{2h,j} X_{h,t-j} \right)^2 + \sum\limits_{t=m}^{N} \eta_{2,t}^2} \tag{6}$$

3   Experimental Method

In this paper, in order to analyze the causality between precipitation and soil moisture, we choose data which covers the period from June 19, 2002 to June 19, 2011 from the United States. Three datasets are used for this analysis: precipitation data is gridded gauge observations from the North American Land Data Assimilation System (NLDAS), soil moisture contains the University of Montana (UMT) (≤ 2 cm soil depth) and the Noah-3.3 land-surface (NOAH) (≤ 10 cm soil depth). We first average these data to 0.25 degree spatial resolution and to a daily timestep, and then take preprocessing for soil moisture and the precipitation by subtracting mean values, finally we calculate the GC and NC values in the time domain for soil moisture and precipitation.

4   Results

We now analyze the causality between NLDAS precipitation (X_P) and soil moisture in two datasets UMT (X_{SM1}) and NOAH (X_{SM2}). We apply GC to calculate $F_{X_{SM1} \to X_P}$, $F_{X_P \to X_{SM1}}$, $F_{X_{SM2} \to X_P}$, $F_{X_P \to X_{SM2}}$, and meanwhile we apply NC to calculate $n_{X_{SM1} \to X_P}$, $n_{X_P \to X_{SM1}}$, $n_{X_{SM2} \to X_P}$, $n_{X_P \to X_{SM2}}$. Each GC and NC values is of significance ($p < 0.05$). The results are shown in Table 1 where each number is the average value of the causal values in the United States.

From Table 1 one can see that (i) NC values from precipitation to UMT soil moisture (≤ 2 cm soil depth) are greater than that from soil moisture to precipitation. NC values from precipitation to NOAH soil moisture (≤ 10 cm soil depth) are less than that from soil moisture to precipitation. However, all GC values from precipitation to soil moisture are greater than that from soil moisture to precipitation. (ii) NC and GC values from UMT soil moisture (≤ 2 cm soil

Table 1. Causality values between precipitation (X_P) and soil moisture (X_{SM}) with GC and NC methods ($p < 0.05$)

Datasets	Soil depth	Causality methods			
		GC		NC	
		$F_{X_{SM} \to X_P}$	$F_{X_P \to X_{SM}}$	$n_{X_{SM} \to X_P}$	$n_{X_P \to X_{SM}}$
UMT	≤ 2 cm	0.0113	0.0399	0.0144	0.0299
NOAH	≤ 10 cm	0.0099	0.0801	0.0132	0.0116

depth) to precipitation are greater than that NOAH soil moisture (≤ 10 cm soil depth) to precipitation. (iii) NC and GC values from precipitation to UMT soil moisture (≤ 2 cm soil depth) are greater than that from precipitation to NOAH soil moisture (≤ 10 cm soil depth).

To give a more clear visual observation of the distribution of all areas having significant causality values, all areas not having significant are omitted. The areas having causality values are marked in Figs. 1 and 2.

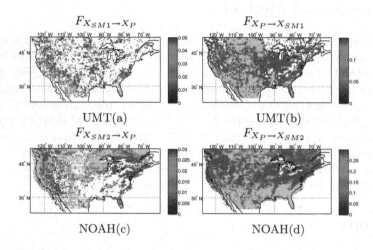

Fig. 1. Maps of the causality between precipitation (X_P) and soil moisture with GC method. (a), (b) the causality between UMT (X_{SM1}) soil moisture and precipitation. (c), (d) the causality between NOAH (X_{SM2}) soil moisture and precipitation. The white area of boundary line inside denote the absence of a significant influence (p < 0.05), and the white area of boundary line outside are not tested.

From Figs. 1 and 2 one can see that (i) Comparing (a) and (c) of Fig. 2, NC results show that the causality from UMT soil moisture (≤ 2 cm soil depth) to precipitation is greater than that NOAH soil moisture (≤ 10 cm soil depth) to precipitation. The shallow soil can cause the surface temperature rising, thus leading to greater evaporation in the west. (ii) Comparing (b) and (d) of Fig. 2, NC results show that the causality from precipitation to UMT soil moisture (≤ 2 cm soil depth) is greater than that from precipitation to NOAH soil moisture (≤ 10 cm soil depth). From Fig. 1, GC results can draw the same conclusion. As the surface water content is close to saturation, precipitation is difficult to change deep soil moisture content. However, since the shallow soil is dry and contain less water, precipitation is more likely to change the soil moisture. (iii) By calculating the significance in Table 1, the proportion of significant causality from soil moisture to precipitation is 64.57%. The proportion of significant causality from precipitation to soil moisture is 85.52%. The results illustrate that the impact from precipitation to shallow soil moisture is much evident than that from shallow soil moisture to precipitation. This can be observed clearly from

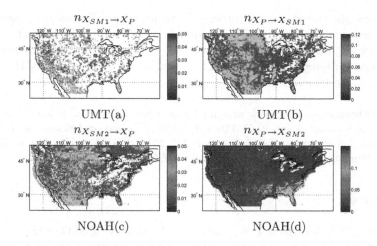

Fig. 2. Maps of the causality between precipitation (X_P) and soil moisture with NC method. (a), (b) the causality between UMT (X_{SM1}) soil moisture and precipitation. (c), (d) the causality between NOAH (X_{SM2}) soil moisture and precipitation. The white area of boundary line inside denote the absence of a significant influence $(p < 0.05)$, and the white area of boundary line outside are not tested.

Fig. 2. (iv) Compare (c), (d) of Fig. 1 and (c), (d) of Fig. 2, NC results show that the causality from NAOH soil moisture to precipitation is greater than that from precipitation to soil moisture. As soil depth deepens, penetration of rainwater into the soil decrease, which leading to the decrease of the influence from precipitation to soil moisture. However, GC results show the opposite conclusion. It means that NC method is more likely to reveal the causal relationship between soil moisture and precipitation than GC method.

5   Conclusion

In this paper, we used GC and NC methods to analyze the causality between soil of different depth moisture and precipitation in the United States. The results from our study show that there is causal influence between soil of different depth moisture and precipitation. This means that there is an interdependent relationship between surface water content and precipitation, and that soil moisture change may be caused by precipitation, and vice versa. And the results also demonstrate that the causality between shallow soil moisture and precipitation is greater than that between deep soil moisture and precipitation. In the shallow soil layer, precipitation is easier to change the soil moisture content. Therefore, a feedback between soil moisture and precipitation could increase the duration of severe dry and wet periods (droughts and floods). The GC method can not draw the same conclusion in the deep soil moisture. Thus, the conclusions demonstrate that the NC method is much clearer to reveal the causal influence between soil moisture and precipitation than GC method in the time domain.

We only use the bivariate autoregressive model to discuss the causal relationship between the two important climatic properties in the land-atmosphere cycle. Since NC may be defined for multiple climate attributes, our future work will focus on applying the NC method to reveal the causal influence of the various attributes in the climate model.

Acknowledgments. This work was funded by National Natural Science Foundation of China under Grants (Nos. 61473110, 61633010), International Science and Technology Cooperation Program of China, Grant No. 2014DFG12570, Key Lab of Complex Systems Modeling and Simulation, Ministry of Education, China.

References

1. Roberts, J.B., Robertson, F.R., Clayson, C.A., Bosilovich, M.G.: Characterization of turbulent latent and sensible heat flux exchange between the atmosphere and ocean in MERRA. J. Clim. **25**(3), 821–838 (2012)
2. Wu, C., Chen, J.M., Pumpanen, J., Cescatti, A., Marcolla, B., Blanken, P.D.: An underestimated role of precipitation frequency in regulating summer soil moisture. Environ. Res. Lett. **7**(2), 33–40 (2012)
3. Li, H., Robock, A., Liu, S., Mo, X., Viterbo, P.: Evaluation of reanalysis soil moisture simulations using updated chinese soil moisture observations. J. Hydrometeorol. **6**(2), 180–193 (2005)
4. Findell, K.L., Eltahir, E.A.B.: An analysis of the soil moisture-rainfall feedback, based on direct observations from Illinois. Water Resour. Res. **33**(4), 725–735 (1997)
5. D'Odorico, P., Porporato, A.: Preferential states in soil moisture and climate dynamics. Proc. Natl. Acad. Sci. USA **101**(24), 8848 (2004)
6. Salvucci, G.D., Saleem, J.A., Kaufmann, R., Miller, C.T., Parlange, M.B., Hassanizadeh, S.M.: Investigating soil moisture feedbacks on precipitation with tests of granger causality. Adv. Water Resour. **25**(8), 1305–1312 (2002)
7. Duerinck, H.M., Ent, R.J.V.D., Giesen, N.C.V.D., Schoups, G., Babovic, V., Yeh, J.F.: Observed soil moisture-precipitation feedback in Illinois: a systematic analysis over different scales. J. Hydrometeorol. (2014)
8. Koster, R.D., Suarez, M.J., Higgins, R.W., Dool, H.M.V.D.: Observational evidence that soil moisture variations affect precipitation. Geophys. Res. Lett. **30**(5), 45–41 (2003)
9. Hurk, B.V.D., Doblas-Reyes, F., Balsamo, G., Koster, R.D., Seneviratne Jr., S.I., Camargo, H.: Soil moisture effects on seasonal temperature and precipitation forecast scores in europe. Clim. Dyn. **38**(1–2), 349–362 (2012)
10. Hu, S., Dai, G., Worrell, G.A., Dai, Q., Liang, H.: Causality analysis of neural connectivity: critical examination of existing methods and advances of new methods. IEEE Trans. Neural Netw. **22**(6), 829–844 (2011)
11. Hu, S., Wang, H., Zhang, J., Kong, W., Cao, Y., Kozma, R.: Comparison analysis: granger causality and new causality and their applications to motor imagery. IEEE Trans. Neural Netw. Learn. Syst. **27**(7), 1429–1444 (2016)
12. Taylor, C.C.: Akaike's information criterion and the histogram. Biometrika **74**(3), 636–639 (1987)

Fix-Budget and Recurrent Data Mining for Online Haptic Perception

Lele Cao[1,2]([⊠]), Fuchun Sun[1], Xiaolong Liu[1], Wenbing Huang[1],
Weihao Cheng[2], and Ramamohanarao Kotagiri[2]

[1] Department of Computer Science and Technology, Tsinghua University,
Beijing 100084, China
{caoll12,liuxl12,huangwb12}@mails.tsinghua.edu.cn,
fcsun@mail.tsinghua.edu.cn
[2] Department of Computing and Information Systems, The University of Melbourne,
Melbourne, VIC 3010, Australia
weihaoc@student.unimelb.edu.au, kotagiri@unimelb.edu.au

Abstract. Haptic perception is to identify different targets from haptic input. Haptic data have two prominent features: sequentially real-time and temporally correlated, which calls for a fixed-budget and recurrent perception procedure. Based on an efficient-robust spatio-temporal feature representation, we handle the problem with a bounded online-sequential learning framework (MBS-ESN), and incorporates the strength of batch-regularization bootstrapping, bounded recursive reservoir, and momentum-based estimation. Experimental evaluations show that it outperforms the state-of-the-art methods by a large margin on test accuracy; and its training performance is superior to most compared models from aspects of computational complexity and storage efficiency.

Keywords: Haptic perception · Echo state network · Online learning · Recurrent neural network · Fixed-budget learning

1 Introduction

Haptic sensor is a device for repeatedly measuring spatial and temporal property of a physical contact event; and haptic object perception is to identify target objects from haptic sensory readings/sequences. When a haptic-enabled robotic hand grasps an object, its haptic sensors output haptic frames that have 2 properties: (a) sequentially real-time and (b) temporally correlated. The first property calls for algorithms with strictly bounded computational complexity and storage capacity (i.e. fixed-budget), which are able to continuously adapt to the unpredictable incoming sensory readings and accordingly make proper predictions as soon as possible. The second property requests creating an recurrent internal state pool (i.e. reservoir) that allows exhibiting dynamic temporal correlations embodied in haptic sequences. So in this work, we plan to address fixed-budget and recurrent haptic object perception.

© Springer International Publishing AG 2017
D. Liu et al. (Eds.): ICONIP 2017, Part V, LNCS 10638, pp. 581–591, 2017.
https://doi.org/10.1007/978-3-319-70139-4_59

The recent advances of the fixed-budget learning approaches have propelled the emergence of several models in this league, such as Projectron++ [20], budget stochastic gradient descent (BSGD) [29], online independent SVM (OISVM) [19], online sequential extreme learning machine (OS-ELM) [15], fixed-budget kernel recursive least-squares (FB-KRLS) [28]; but none of them explicitly considers the temporal correlations between consecutive samples. In the community of time-series data modeling, recurrent neural networks (RNNs) become increasingly dominant. The application of RNNs is however not always feasible because of the high training costs and possible fading gradient [1]. Hence a lightweight alternative, reservoir computing [16,17], has been proposed to obtain the optimum weights: train the output weights directly (with a specific classifier or regressor) while leaving the internal connection weights and input weights unchanged. Among various reservoir models, echo-state network (ESN) [13,22] is an efficient RNN framework that has yielded satisfying results on many problems of time-series prediction [12]. However, up till this date, there are merely a few attempts [12,14,26,27] in converting the standard ESN into an online learning schema; and none of them has been applied to haptic perception tasks.

To realize online fixed-budget ESN training, Jaeger [12] proposed finding the output weights using a "vanilla" version [8] of the recursive least squares (RLS) algorithm; yet it appeared challenging to balance between the factors of computational cost and numerical stability. Authors of [26] incorporated Bayesian online learning for Gaussian process (GP) [6] with ESN, creating the model of online echo-state GP (OESGP); however, the truncation strategy of reservoir states (to bound the algorithm by maintaining L states at most) might significantly affect its convergence and prediction performance. To avoid constructing an explicit reservoir for OESGP, Soh and Demiris [27] further developed the online infinite echo-state GP (OIESGP) by employing recursive kernels with automatic relevance detection [18]; unfortunately, optimization of the hyper-parameters requires gradient computations on the order of $O(L^2)$ and periodic matrix inversions of order $O(L^3)$, which obstructs its straightforward application in many real-time machine learning tasks. To our best knowledge, OIESGP/OESGP is the approach evaluated towards incremental haptic sensing in [24,25].

In this paper, we explore fix-budget and recurrent data mining for online haptic perception by proposing a simple-yet-effective method: momentum batch-sequential ESN (MBS-ESN), which is a lightweight reservoir RNN behaving in an online-sequential style. We largely follow [4] to carry out the unsupervised haptic feature encoding, since it is capable of extracting highly over-complete, discriminative, robust, and fault-tolerant feature representations efficiently. Our approach has the following advantages over previous work:

- *mini-batch bootstrapping:* it can initialize a reasonable model with prediction capability from the very first frame using fixed-budget batch regularization;
- *bounded recursive reservoir:* upon memory overflow, it can seamlessly switch to a sequential training mode with bounded computational and storage cost;
- *momentum-based estimation:* perception stability and continuity is guaranteed by a recursive linear combiner of previous and current predictions.

2   Spatio-Temporal Haptic Feature

Recently in [4], Cao et al. suggested to decompose haptic data into **spatial and temporal components**, from which the spatio-temporal feature descriptors are extracted using the randomized tiling convolution network (RTCN). Let haptic frames for training formalized as $\aleph = \{(\mathbf{x}_t, \mathbf{y}_t) \mid \mathbf{x}_t \in \mathbb{R}^{d \times d}, \mathbf{y}_t \in \mathbb{R}^m, t = 1, \ldots, N\}$, where $\mathbf{x}_t$ is the t-th frame in $d \times d$ dimension; and its corresponding label is denoted as a vector $\mathbf{y}_t$ of size m representing m object classes in total. Simply put, the **spatial component** is each individual haptic frame $\mathbf{x}_t$, which is fed to an RTCN with F feature maps to generate a joint pooling activation $\mathbf{p}_{\text{spatial}}(\mathbf{x}_t) = [p_1(\mathbf{x}_t), \ldots, p_F(\mathbf{x}_t)]$.

For **temporal component**, we calculate the inter-frame haptic flow following [11], perform mean flow subtraction [23], and obtain the two-dimensional vector field $\vec{\mathbf{u}}(\mathbf{x}_t)$ of haptic flow between frames $\mathbf{x}_t$ and $\mathbf{x}_{t+1}$. The horizontal and vertical components of $\vec{\mathbf{u}}(\mathbf{x}_t)$ are noted as $\mathbf{u}_{\text{h}}(\mathbf{x}_t)$ and $\mathbf{u}_{\text{v}}(\mathbf{x}_t)$ respectively. To represent force motion across longer time period, we stack two consecutive vector fields together as is empirically recommended in [4]; hence the stacked haptic flow $\mathbf{U}_t$ for frame $\mathbf{x}_t$ is formulated as $\mathbf{U}_t = \{\mathbf{u}_{\text{h}}(\mathbf{x}_t), \mathbf{u}_{\text{v}}(\mathbf{x}_t), \mathbf{u}_{\text{h}}(\mathbf{x}_{t+1}), \mathbf{u}_{\text{v}}(\mathbf{x}_{t+1})\}$, each element of which is treated as a channel in RTCN, constituting a temporal feature space $\mathbf{p}_{\text{temporal}}(\mathbf{x}_t) = [p_1(\mathbf{U}_t), \ldots, p_F(\mathbf{U}_t)]$. Finally, the spatial and temporal haptic features are concatenated to compose a joint spatio-temporal feature representation $\mathbf{p}_t$ for the t-th frame:

$$\mathbf{p}_t = [p_1(\mathbf{x}_t), \ldots, p_F(\mathbf{x}_t), p_1(\mathbf{U}_t), \ldots, p_F(\mathbf{U}_t)]. \tag{1}$$

3   Momentum Batch-Sequential Echo State Net

The key of any fundamental ESN is a randomly-generated fixed RNN (i.e. reservoir), which essentially contains a large number of randomly and sparsely connected neurons [16]. The overall mechanism is to drive the reservoir using the input signal (e.g. training samples) and derive the output (e.g. predicted labels) via certain combination of the reservoir units [27]; hence, instead of tuning all network weights, we only need to train the output weights. To begin with, the state of the reservoir ϕ is updated as follows:

$$\phi_{t+1} = h(\mathbf{W}_{\text{res}} \cdot \phi_t + \mathbf{W}_{\text{in}} \cdot \mathbf{p}_{t+1}), \tag{2}$$

where $\mathbf{p}_{t+1}$ is the extracted haptic feature descriptor (Eq. 1) at time point $t + 1$; $\phi_t = [\phi_{t,1}, \ldots, \phi_{t,l}]$ denote the t-th reservoir state; variable l indicates the size of reservoir (i.e. the number of ESN neurons); $h(\cdot)$ is the activation function that is infinitely differentiable in any given interval; $\mathbf{W}_{\text{in}}$ and $\mathbf{W}_{\text{res}}$ represent the input weight matrix and the internal connection weight matrix of the reservoir, respectively. We can compute the predicted outputs $\hat{\mathbf{y}}_t$ via $\phi_t \mathbf{W}_{\text{out}}$, where $\mathbf{W}_{\text{out}} = [\mathbf{w}_1, \ldots, \mathbf{w}_l]_{l \times m}^{\top}$ is the linear output weight matrix. Obviously, the "agile" determination of $\mathbf{W}_{\text{out}}$ becomes the core of this study; and we will start tackling this problem from a fixed-budget solver of batch regularization.

3.1   Naive Fixed-Budget Batch Regularization for $\mathbf{W}_{\text{out}}$

To discard the influence of the first b $(\ll N)$ reservoir transitions and only consider no more than L (usually expected to be as large as possible, e.g. $L > l$, to enable traceability of much older temporal relations) reservoir states at any time t $(> b)$, we define the t-th $(t \in \{b+1, \ldots, N\})$ valid set of reservoir states $\mathbf{\Phi}_t$ as

$$\mathbf{\Phi}_t = \begin{cases} [\phi_{b+1}^{\top}, \ldots, \phi_t^{\top}]_{(t-b)\times l}^{\top}, & t - b \leq L \\ [\phi_{t-L+1}^{\top}, \ldots, \phi_t^{\top}]_{L\times l}^{\top}, & t - b > L \end{cases}, \tag{3}$$

and matrix $\mathbf{Y}_t \in \mathbb{R}^{\min\{t-b,L\}\times l}$ as true labels of the corresponding training samples in scope. Thereupon in the t-th round, we are interested in finding a $\mathbf{W}_{\text{out}}$ that minimizes the training error $\|\mathbf{\Phi}_t\mathbf{W}_{\text{out}} - \mathbf{Y}_t\|_{\text{F}}^2$ as well as the norm of the output weights $\|\mathbf{W}_{\text{out}}\|$. Instead of the standard optimization approach, a closed form solution can be derived by utilizing Moore–Penrose generalized inverse [21] and Tikhonov regularization [10]:

$$\mathbf{W}_{\text{out}} = \begin{cases} \mathbf{\Phi}_t^{\top}(\dfrac{\mathbf{I}}{C} + \mathbf{\Phi}_t\mathbf{\Phi}_t^{\top})^{-1}\mathbf{Y}_t, \forall t \leq l+b < L+b, & (4) \\[2mm] (\dfrac{\mathbf{I}}{C} + \mathbf{\Phi}_t^{\top}\mathbf{\Phi}_t)^{-1}\mathbf{\Phi}_t^{\top}\mathbf{Y}_t, \forall t > l+b, & (5) \end{cases}$$

where C is the regularization parameter to balance the quality and complexity of the approximation function; and we practically stipulate $L > l$.

3.2   Sequential Train $\mathbf{W}_{\text{out}}$ After Memory-Overflow

From the $(l + b + 1)$-th haptic frame, Eq. (5) is applied under the assumption of the availability of L most-recent reservoir states. Additionally, the first-in-first-out reservoir window $\mathbf{\Phi}_t$ might constantly throw away discriminative reservoir states and lead to fluctuating and unstable perception performance. In order to avoid (or at least alleviate) those drawbacks, we sequentialize Eq. (5) when the memory overflows (i.e. $t - b > L$), therefrom creating a *sequential fixed-budget ESN* (i.e. a specific implementation of "agile-recurrent learning"), which is primed on one haptic frame and progressively update $\mathbf{W}_{\text{out}}$ as each new frame arrives.

Assume that the output weight matrix attained at the critical point $(\hat{t} = L + b)$ is $\mathbf{W}_{\text{out}}^{(\hat{t})}$, We immediately have $\mathbf{W}_{\text{out}}^{(\hat{t})} = (\mathbf{I}/C + \mathbf{\Omega}_{\hat{t}})^{-1}\mathbf{\Phi}_{\hat{t}}^{\top}\mathbf{Y}_{\hat{t}}$ where $\mathbf{\Omega}_{\hat{t}} = \mathbf{\Phi}_{\hat{t}}^{\top}\mathbf{\Phi}_{\hat{t}}$; and suppose now that the next haptic frame-label doublet $(\mathbf{x}_{\hat{t}+1}, \mathbf{y}_{\hat{t}+1})$ is received, then following a similar approach as [5, 15], we can derive and generalize that the output weight can be sequentially computed via

$$\mathbf{W}_{\text{out}}^{(t+1)} = \mathbf{W}_{\text{out}}^{(t)} + (\dfrac{\mathbf{I}}{C} + \mathbf{\Omega}_t + \phi_{t+1}^{\top}\phi_{t+1})^{-1} \cdot \phi_{t+1}^{\top} \cdot (\mathbf{y}_{t+1} - \phi_{t+1}\mathbf{W}_{\text{out}}^{(t)}), \tag{6}$$

which is a recursive process, and whose convergence ability can be proved following a similar technique as [5]; and the predicted probability vector and the object category are respectively

$$\hat{\mathbf{y}}_{t+1} = \boldsymbol{\phi}_{t+1} \mathbf{W}_{\text{out}}^{(t+1)} \quad \text{and} \quad Label(\mathbf{x}_{t+1}) = \arg \max_{y \in \{1,\dots,m\}} (\hat{\mathbf{y}}_{t+1}). \tag{7}$$

3.3 Apply Momentum Sequential Estimation

Relating to the ground truth label vector $\mathbf{y}$ of each haptic frame, the prediction label $\hat{\mathbf{y}}$ could be approximately expressed as $\hat{\mathbf{y}} = \mathbf{y} + \boldsymbol{\epsilon}$, where $\boldsymbol{\epsilon}$ is the error probably caused by certain noise or measurement inaccuracy. In the context of haptic perception, prediction **continuity** and **consistency** turn out to be salient characteristics, because the frames in the same haptic sequence are obtained via manipulating the same object. As a consequence, we suggest to treat the estimated object labels in a haptic data stream as a series of states (noted as $\mathbf{z}_t$, $t = 1, \dots, N_k$ where N_k is the index of the last frame in the k-th sequence), each of which is a linear function of previous estimation and current prediction:

$$\mathbf{z}_0 = \hat{\mathbf{y}}_0 = 0, \quad \mathbf{z}_{t+1} = \lambda \cdot \mathbf{z}_t + \hat{\mathbf{y}}_{t+1}, \quad \forall t \in \{0, 1, \dots, N_k\}, \tag{8}$$

where $\lambda \in [0, 1)$ is the momentum factor that determines the proportion between historical and current prediction; a greater value of λ indicate a **stronger** and **longer** impact of historical estimations, and vice versa; if $\lambda = 0$, Eq. (8) falls back to Eq. (7). By unfolding Eq. (8), we get a more explicit form of $\mathbf{z}_{t+1} = \sum_{i=0}^{t+1} \lambda^{t-i+1} \hat{\mathbf{y}}_i$, which is essentially the sum of $\hat{\mathbf{y}}_i$'s times a decreasing parameter over time. Further noticing that $\hat{\mathbf{y}}_i = \mathbf{y}_i + \boldsymbol{\epsilon}_i$ and $\mathbf{y}_0 = \mathbf{y}_1 = \dots = \mathbf{y}_i = \dots = \mathbf{y}_{N_k}$, we can therefore represent $\mathbf{z}_{t+1}$ in the following way:

$$\frac{1}{\sum_{i=0}^{t+1} \lambda^{t-i+1}} \mathbf{z}_{t+1} = \mathbf{y}_{N_k} + \frac{\sum_{i=0}^{t+1}(\lambda^{t-i+1} \boldsymbol{\epsilon}_i)}{\sum_{i=0}^{t+1} \lambda^{t-i+1}}, \tag{9}$$

where the right-most term $\sum_{i=0}^{t+1}(\lambda^{t-i+1} \epsilon_i)/\sum_{i=0}^{t+1} \lambda^{t-i+1}$ is the error of momentum-based estimation. Since "arg max" operation is scale-invariant, we have

$$Label(\mathbf{x}_{t+1}) = \arg \max_{y \in \{1,\dots,m\}} \frac{1}{\sum_{i=0}^{t+1} \lambda^{t-i+1}} \mathbf{z}_{t+1}. \tag{10}$$

To understand the advantage of using momentum-based estimation $\mathbf{z}_{t+1}$ instead of original prediction $\hat{\mathbf{y}}_{t+1}$, we compare the variance of momentum error term, i.e. $Var(\sum_{i=0}^{t+1}(\lambda^{t-i+1} \epsilon_i)/\sum_{i=0}^{t+1} \lambda^{t-i+1})$, with that of the original error term $Var(\epsilon_{t+1})$:

$$Var\left(\frac{\sum_{i=0}^{t+1}(\lambda^{t-i+1} \epsilon_i)}{\sum_{i=0}^{t+1} \lambda^{t-i+1}}\right) = \frac{\sum_{i=0}^{t+1} \lambda^{2(t-i+1)}}{(\sum_{i=0}^{t+1} \lambda^{t-i+1})^2} Var(\epsilon) < Var(\epsilon); \tag{11}$$

Note that we have $Var(\epsilon) = Var(\epsilon_{t+1})$, because ϵ is independent and identical distributed random variable. To this end, introducing momentum factor λ on top of the sequential frame-wise prediction makes the estimation $\mathbf{z}_{t+1}$ more stable than raw prediction $\hat{\mathbf{y}}_{t+1}$, and thus improves the accuracy and consistency of agile-recurrent haptic perception.

4   Evaluations on Haptic Perception Tasks

In this section, we evaluate MBS-ESN on several benchmark datasets (cf. Table 1): SD10 [2,3], SPr7 [7], BDH5 [30], and HCs10 [4], which represent different haptic perception tasks with various complexities.

Table 1. The dataset specifications (left part) and parameter settings (right part)

Dataset [Ref.]	Frame size	# Obj. classes	#Seq.	Avg. seq. length	# Train frames	(F_s, k_s, r_s, s_s) (F_t, k_t, r_t, s_t)	b C	l L	λ^*
SD10 [2,3]	13 × 18 (units)	10	100	349 (frames)	25,001 (avg.)	(16, 4, 6, 1) (16, 4, 6, 2)	3 10^4	200 400	0.8
SPr7 [7]	8 × 16	7	70	405	19,827	(12, 4, 6, 1) (48, 3, 4, 1)	5 10^2	200 400	0.9
BDH5 [30]	8 × 9	2	100	267	18,452	(16, 4, 6, 2) (48, 4, 6, 4)	5 10^5	500 10^3	0.7
HCs10 [4]	4 × 4	10	180	37	4,720	(52, 3, 4, 1) (52, 3, 4, 1)	2 10^3	400 600	0.9

*During the search of optimal λ, we empirically discovered that large $\lambda \in [0.7, 0.9]$ is favored when the end/beginning of any sequence is known, while a moderate value is preferred (e.g. $\lambda \in [0.4, 0.6]$) if the sequence segmentation is unknown to MBS-ESN.

Table 2. The comparison of the average online classification ratio (on testing set after 8 epochs) between M/BS-ESN and the state-of-the-arts. (**best**, 2nd best)

Dataset	MBS-ESN	BS-ESN	Proj.++	BSGD	OS-ELM	FB-KRLS	OISVM	OIESGP
SD10	**81.388**	78.519	77.050	71.936	78.070	76.310	<u>81.016</u>	74.456
SPr7	**87.748**	85.959	84.980	83.619	85.420	<u>86.850</u>	82.070	83.160
BDH5	**87.268**	<u>83.921</u>	80.704	79.060	81.082	79.867	79.484	82.630
HCs10	**89.024**	<u>84.420</u>	82.950	76.365	84.090	80.110	81.710	77.600

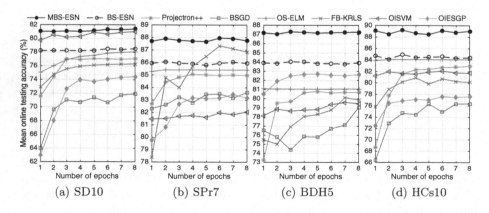

Fig. 1. The average test accuracy as a function of epochs.

The statistical results presented hereafter were, without exception, averaged over 10 trials, each of which used a 7:3 train/testing split. We compare MBS-ESN (together with BS-ESN that works in the same manner except that it has no momentum-based estimation) with the state-of-the-art models for fixed-budget online learning, which include Projectron++ [20], OISVM [19], BSGD [29], OS-ELM [15], FB-KRLS [28], and OIESGP [26,27]. Concerning haptic feature extraction, we used the same set of parameters (i.e. F, k, r, s) as in [4]; the model specific parameters (e.g. for MBS-ESN, they include C, l, L, and λ listed in Table 1) were determined by a combination of grid search and manual search [9] on validation sets.

4.1   Accuracy on Testing Sets

We ran 8 epochs for all datasets and reported the perception performance (Fig. 1) on a separate test set as a function of the number of epochs. It is observed that the best results of MBS-ESN outperform all compared models on all four tasks. We also noticed that M/BS-ESN and OISVM seem much less sensitive to the number of epochs than other models in our scope, the reasons of which varies: M/BS-ESN essentially employs a closed-form solution while OISVM might slowly improve its test accuracy by using a smaller η (i.e. allowing more samples to become SV and severely jeopardizing the training efficiency). Since a high accuracy can already be achieved after only one epoch, MBS-ESN turns out to be more advantageous that no extra storage is needed for supporting more training epochs. Cao et al. [4] reported the best frame-wise batch prediction accuracy for those datasets as SD10: 89.25%, SPr7: 90.48%, BDH5: 92.8%, HCs10: 79.8%. From Table 2, MBS-ESN and a few other models managed to approach or even surpass (e.g. HCs10) those batch learning scores within only 8 epochs, which is probably attributed to the attempt of learning temporal correlations via recurrent states and momentum-based estimations. Interestingly, OISVM and FB-KRLS performed even slightly better (after 8 epochs) than BS-ESN on SD10 and SPr7 benchmarks respectively, indicating complex reservoirs might be unnecessary for certain perception tasks.

4.2   Analysis of Time Complexity

Comparing computational costs of both training and testing (cf. Fig. 2), M/BS-ESN and OS-ELM remained the overall fastest algorithms in terms of CPU time; it could ascribe to the improved computational scalability of Eq. (6) on the order of $O(l^2)$ that is irrelevant to the number of total samples N or bootstrap samples L. Because of the update-skipping mechanism demonstrated by Fig. 5 in [20], Projectron++ also achieved low training costs. The high training cost of BSGD and OISVM on SD10, SPr7, and BDH5 datasets also said, during our tests, we had to carry out parameter search of those models, which may took days of calculation under certain parameterizations. In view of the minimum sampling frequency (15 Hz) for most haptic sensors, we can accordingly define a threshold

CPU-time (≈ 0.067 s, i.e. the dashed line in the upper part of Fig. 2) to help telling the level of difficulty for a model to be applied to real-time applications.

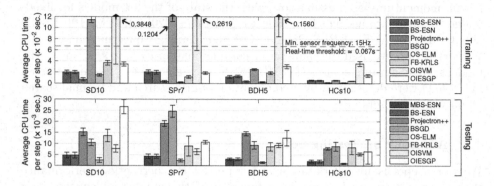

Fig. 2. The average CPU time per step for testing (upper) and training (lower).

4.3   Comparison of Storage Efficiency

M/BS-ESN and OS-ELM are based on the covariance matrices (of $l \times l$ dimensions), while the rest of the models use Gram matrices, whose dimensions depend on the number of input patterns (not feature dimensionality) and increase upon the inclusion of new data. We use the term "size of the active vector set" to measure the memory consumption required by each model; it is referred to as the dimension of covariance matrices or Gram matrices. The 3D plots in Fig. 3

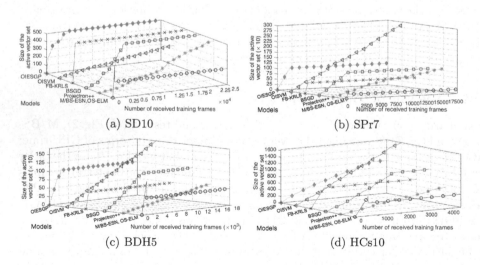

Fig. 3. Storage efficiency: # active vectors in relation to # training samples.

show the growth of the active vector set as a function of the number of training samples. While the growth for M/BS-ESN and OS-ELM is clearly bounded by l, it is linear/sub-linear for OISVM and Projectron++. When Projectron++ is trained over multiple epochs, its active vector set will reach a maximum size and then cease further growing. The trend of OISVM seems to be proportional to the number of training data in the first epoch, but provided an appropriate choice of η (for trading off between compactness of the solution and optimal performance), it is capable of finding a plateau in memory over more epochs [19]. There is a hard-limit of vector size configured for OIESGP, FB-KRLS, and BSGD, hence their kernel update strategy will eventually make differences to classification accuracy under the same storage limit.

5 Conclusions and Perspectives

This paper presented and evaluated a fixed-budget recurrent data mining framework MBS-ESN in the application domain of haptic object perception. MBS-ESN is intrinsically a lightweight randomized RNN that strives to learn temporal correlations embodied in sequential data using a flow of reservoir states. MBS-ESN is bootstrapped by Tikhonov regularization; once the reservoir memory overflows, it switches seamlessly to a regularized sequential training mode with bounded computational and storage requirement; the prediction stability and continuity are further guaranteed by momentum-based estimation. The experimental evaluations show that our approach outperformed several state-of-the-art methods by a large margin on testing accuracy, which is achieved by using significantly fewer epochs; furthermore, the training performance of MBS-ESN is superior to most compared models from the perspectives of computational complexity and storage efficiency.

Acknowledgments. This work is supported by National Natural Science Foundation of China with grant number 041320190.

References

1. Atiya, A.F., Parlos, A.G.: New results on recurrent network training: unifying the algorithms and accelerating convergence. IEEE Trans. Neural Netw. **11**(3), 697–709 (2000)
2. Bekiroglu, Y., Kragic, D., Kyrki, V.: Learning grasp stability based on tactile data and HMMs. In: Proceedings of 19th International Conference on RO-MAN, pp. 132–137. IEEE, Viareggio (2010)
3. Bekiroglu, Y., Laaksonen, J., Jorgensen, J.A., Kyrki, V., Kragic, D.: Assessing grasp stability based on learning and haptic data. IEEE Trans. Robot. **27**(3), 616–629 (2011)
4. Cao, L., Kotagiri, R., Sun, F., Li, H., Huang, W., Aye, Z.M.M.: Efficient spatio-temporal tactile object recognition with randomized tiling convolutional networks in a hierarchical fusion strategy. In: Proceedings of 30th AAAI, pp. 3337–3345. AAAI Press, Phoenix (2016)

5. Chong, E.K., Zak, S.H.: An Introduction to Optimization, vol. 76. Wiley, Hoboken (2013)
6. Csató, L., Opper, M.: Sparse on-line Gaussian processes. Neural Comput. **14**(3), 641–668 (2002)
7. Drimus, A., Kootstra, G., Bilberg, A., Kragic, D.: Design of a flexible tactile sensor for classification of rigid and deformable objects. Robot. Auton. Syst. **62**(1), 3–15 (2014)
8. Farhang-Boroujeny, B.: Adaptive Filters: Theory and Applications. Wiley, Hoboken (2013)
9. Hinton, G.E.: A practical guide to training restricted Boltzmann machines. In: Montavon, G., Orr, G.B., Müller, K.-R. (eds.) Neural Networks: Tricks of the Trade. LNCS, vol. 7700, 2nd edn, pp. 599–619. Springer, Heidelberg (2012). doi:10. 1007/978-3-642-35289-8_32
10. Hoerl, A.E., Kennard, R.W.: Ridge regression: biased estimation for nonorthogonal problems. Technometrics **12**(1), 55–67 (1970)
11. Horn, B.K., Schunck, B.G.: Determining optical flow. Artif. Intell. **17**, 185–203 (1981)
12. Jaeger, H.: Adaptive nonlinear system identification with echo state networks. In: Advances in Neural Information Processing Systems, pp. 593–600 (2002)
13. Jaeger, H., Haas, H.: Harnessing nonlinearity: predicting chaotic systems and saving energy in wireless communication. Science **304**(5667), 78–80 (2004)
14. Kountouriotis, P., Obradovic, D., Goh, S.L., Mandic, D.P.: Multi-step forecasting using echo state networks. In: International Conference on Computer as a Tool, EUROCON 2005, vol. 2, pp. 1574–1577. IEEE (2005)
15. Liang, N.Y., Huang, G.B., Saratchandran, P., Sundararajan, N.: A fast and accurate online sequential learning algorithm for feedforward networks. IEEE Trans. Neural Netw. **17**(6), 1411–1423 (2006)
16. LukošEvičIus, M., Jaeger, H.: Reservoir computing approaches to recurrent neural network training. Comput. Sci. Rev. **3**(3), 127–149 (2009)
17. Maass, W., Natschläger, T., Markram, H.: Real-time computing without stable states: a new framework for neural computation based on perturbations. Neural Comput. **14**(11), 2531–2560 (2002)
18. Neal, R.M.: Bayesian Learning for Neural Networks, vol. 118. Springer Science & Business Media, New York (2012). doi:10.1007/978-1-4612-0745-0
19. Orabona, F., Castellini, C., Caputo, B., Jie, L., Sandini, G.: On-line independent support vector machines. Pattern Recogn. **43**(4), 1402–1412 (2010)
20. Orabona, F., Keshet, J., Caputo, B.: Bounded kernel-based online learning. J. Mach. Learn. Res. **10**, 2643–2666 (2009)
21. Rao, C.R., Mitra, S.K.: Generalized Inverse of Matrices and Its Applications, vol. 7. Wiley, New York (1971)
22. Shi, Z., Han, M.: Support vector echo-state machine for chaotic time-series prediction. IEEE Trans. Neural Netw. **18**(2), 359–372 (2007)
23. Simonyan, K., Zisserman, A.: Two-stream convolutional networks for action recognition in videos. In: Advances in Neural Information Processing Systems, pp. 568–576 (2014)
24. Soh, H., Demiris, Y.: Incrementally learning objects by touch: online discriminative and generative models for tactile-based recognition. IEEE Trans. Haptics **7**(4), 512 (2014)
25. Soh, H., Su, Y., Demiris, Y.: Online spatio-temporal Gaussian process experts with application to tactile classification. In: Proceedings of 25th IROS, pp. 4489–4496. IEEE/RSJ, Algarve (2012)

26. Soh, H., Demiris, Y.: Iterative temporal learning and prediction with the sparse online echo state Gaussian process. In: Proceedings of 25th IJCNN, pp. 1–8. IEEE, Brisbane (2012)
27. Soh, H., Demiris, Y.: Spatio-temporal learning with the online finite and infinite echo-state Gaussian processes. IEEE Trans. Neural Netw. Learn. Syst. **26**(3), 522–536 (2015)
28. Van Vaerenbergh, S., Santamaría, I., Liu, W., Príncipe, J.C.: Fixed-budget kernel recursive least-squares. In: 2010 IEEE International Conference on Acoustics Speech and Signal Processing (ICASSP), pp. 1882–1885. IEEE (2010)
29. Wang, Z., Crammer, K., Vucetic, S.: Breaking the curse of kernelization: budgeted stochastic gradient descent for large-scale SVM training. J. Mach. Learn. Res. **13**(1), 3103–3131 (2012)
30. Yang, J., Liu, H., Sun, F., Gao, M.: Tactile sequence classification using joint kernel sparse coding. In: Proceedings of 28th IJCNN, pp. 1–6. IEEE, Killarney (2015)

Arterial Coordination for Dedicated Bus Priority Based on a Spectral Clustering Algorithm

Shuhui Zheng[1], Xiaoming Liu[2], Chunlin Shang[2(✉)], Guorong Zheng[2], and Guifang Zheng[2]

[1] Beijing Information Technology College, Beijing, China
[2] Beijing Key Lab of Urban Road Traffic Intelligent Technology,
North China University of Technology, Beijing, China
itsshangchunlin@gmail.com

Abstract. The current method of dedicated bus arterial coordination priority is mostly based on the arterial coordination control scheme of social vehicle, which makes the dedicated bus arterial coordination priority has many limitations. This paper compare social vehicle traffic flow data with bus traffic flow data which obtained from survey to determine the weighted proportion between them by using spectral clustering (SC) method. And then design multi-period division program for intersection by using Piecewise Aggregate Approximation (PAA). At last we get new arterial coordination control scheme by using graphic method. This paper selects per capita delays as efficiency indicator to measure intersection traffic efficiency. After VISSIM simulation we find out that the new-control-methods outstanding performance on bus traffic efficiency which can decrease the per capita delays reach with public transit-oriented purposes. *abstract* environment.

Keywords: Arterial coordination · Dedicated bus priority · Spectral clustering · Piecewise Aggregate Approximation

1 Introduction

The key to improve urban road traffic condition is solving intersections traffic problems, and multi-period control scheme is the main way to improve intersection capacity [1]. Due to this, the first thing for arterial coordination dedicated bus priority is to solve multi-period control problem. Paper [2] takes capacity, delay, number of stops and other efficiency indicators as the optimization target, in accordance with changes of traffic intersections set multi-period division control methods under different traffic flow. Paper [3] takes optimal bus running state as a goal, establish coordination speed and optimization priority methods to improve traffic efficiency and service level of intersection public transport. Paper [4] comprehensive analysis of the bandwidth of social vehicles arterial coordination and dedicated bus arterial coordination, finally obtain active bus priority control method. Paper [5] chooses arterial coordination cycle optimization, phase difference, green ratio as the upper layer algorithm, and bus priority algorithm which is bounded by the green wave bandwidth as lower coordination

© Springer International Publishing AG 2017
D. Liu et al. (Eds.): ICONIP 2017, Part V, LNCS 10638, pp. 592–599, 2017.
https://doi.org/10.1007/978-3-319-70139-4_60

method. However, the current division of multi-period mostly based social vehicle, while the intersection fixed cycle for bus priority designs are mostly based on peak hours. There is no research result that combining bus priority and fixed cycle multi-period division control method now. Those researches on bus priority are mostly based on active bus priority, which have high requirements on monitoring equipment and actual situation of the road, without considering bus priority control method under multi-period of the day.

Aiming at current intersection multi-period division methods for fixed cycle and arterial coordination control scheme are mostly based on social vehicles without considering the specificity of dedicated bus. Firstly, based on flow data of intersection and supported by PPA algorithm, we will get the multi-period division scheme which mainly to social vehicles, and verify the correctness of PAA by VISSIM simulation. Then analyzing the characteristics of bus traffic, which is collected by actual investigation. We obtain the weighted values of bus in different traffic conditions by using the passenger capacity of social vehicles and buses as the data input of spectral clustering. And then we will get the intersection multi-period division method which reflects the requirements of dedicated bus priority. At last we get new arterial coordination control scheme by using graphic method which analyzed and verified through simulation.

2   Algorithm Description

2.1   Algorithm Description for Spectral Clustering

Divide data groups into several classes or clusters by clustering, and data in the same cluster has high degree of similarity, but has a large degree in difference in different clusters. Cluster analysis is an effective method widely used in data mining, data analysis, computer vision, VLSI design, and machine learning. Traditional clustering algorithms like K-means clustering, EM algorithm, they are all based on the convex spherical sample space, but when the sample space is not convex, these algorithm may just reach local optimization. In order to cluster data group in any shape and converge to the global optimal solution, scholars start to find a new clustering algorithm named spectral clustering [6].

Spectral clustering algorithm is based on the theory of spectral division, the basic idea is to use of the sample data similar matrix (Laplace matrix) of the feature vectors obtained after decomposition clustering feature.

This paper use social vehicle passenger capacity and buses passenger capacity of the intersection as the input of spectral clustering algorithm and one social vehicle passenger capacity is 3 men, one bus passenger capacity is 30–50 men [7]. The input matrix constitute by social vehicle passenger capacity and buses passenger capacity. Write MATLAB control program and follow the algorithm steps of spectral clustering above. And finally get the weighting factor at different times.

Defined spectral clustering algorithm is: given a data construct an undirected weighted graph (similarity matrix) $G = \{V, E\}$, it represents a symmetric matrix $W = [W_{ij}] \, n \times n$, and W_{ij} is the weight of the point i and the point j. W_{ij} is the set of vertices. The set $E = \{W_{ij}\}$ is the similarity between two points that based on some similarity metric computation [8]. Steps like follow:

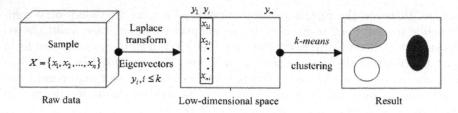

Fig. 1. Schematic model of spectral clustering: from data group $X = \{x_1, x_1, \ldots, x_n\}$, after Laplace transform we get eigenvectors y_i forming matrix $Y = [y_1, y_2, \ldots, y_m]$, and then we get the final result after K-means clustering.

Step1: Structure similarity matrix W according to the data given;

Step2: Calculation matrixD, its diagonal correspond with the sum of matrix W row (or column), others is 0,

$$D\left(i, i\right) = \sum_j A\left(i, j\right). \tag{1}$$

Step3: Calculation Laplace matrix

$$L = D - W. \tag{2}$$

Step4: Normalized matrix

$$L' = D^{-\frac{1}{2}} L D^{-\frac{1}{2}} = I - D^{-\frac{1}{2}} W D^{-\frac{1}{2}}. \tag{3}$$

Step5: Calculate the minimum eigenvalues and corresponding eigenvectors of normalized matrix L, and put these together to make up a $N \times K$ matrix named Q

Step6: After K-means clustering of feature matrix Q, we get a N-dimensional vector C, and this is the final clustering results.

2.2   Algorithm Description for PAA

The main method to improve the adaptability of intersection signal control is multi-period division. But previous multi-period division method just takes peak and off-peak period into account, and this method often resulting in not take advantage of the green time that may cause intersection unnecessary delays and traffic jams, and also may cause environmental pollution and economic losses. Paper [9] use PAM clustering method to divide multi-period, but it ignore the ordering of the traffic flow and this may cause noncontiguous period divided together, but the actual situation is that signal controller is fixed should not jumping, it is difficult to be applied in the real world. For the problems above and considering the particularity of periods division, this paper decide use PAA algorithm to design intersection multi-period division methods for fixed cycle signal control.

PAA is a basic method for data stream processing based on time series segmentation algorithm, set the data in the window as the analysis object and mean value calculation within the window of data. As the window continues to receive new data arrives from the data stream, the data within the window constantly updated, and then we get new Mean window value. And we finally divided multi-period by comparing of different data we got.

According to these, we get multi-period division result of intersections. Keogh and Yi proposed time-series algorithm Piecewise Aggregate Approximation (PAA). The basic idea of PAA: divided the data flow into a length and record the average length of these parts to approximate the original data [10]. PAA can divide into four steps:

Step1: n data length named as C, like $C = \{c_1, c_2, \ldots, c_n\}$;

Step2: Set a parameter ω ($1 \leq \omega \leq n$) as the width size of the slide window;

Step3: Make a N ($1 \leq N \leq n$) data length named as C' , and

$$c_i' = 1/\omega \sum_{j-\omega(i-1)+1}^{\omega i} c_i, (\omega = [n/N]). \tag{4}$$

Step4:Use C' to replace C .

2.3   Model of Arterial Coordination for Dedicated Bus

Green wave band is one of the main methods of traffic signal adjustment in urban arterial roads. It has the advantages of obvious control effect and simple realization. The core of it is to make as many vehicles as possible through the various intersections on the arterial road. However, in the traditional arterial coordination control, there is no consideration of dedicated bus priority. No matter in the process of signal optimization for single intersection, public cycle selection or green wave design of arterial control. It is designed based on social vehicles by using only a simple equivalent conversion to consider the bus and proposing a new control strategy. From signal optimization for single intersection, public cycle selection and multi-period division to make the arterial coordination design of dedicated bus transit. The main control ideas are as follows:

Step1: The number of intersections of arterial road is defined as n. Every half hour to collect the flow of social vehicles and dedicated bus of every direction of intersection i ($0 < i \leq n$). Make the matrix A_i with the traffic flow within 24 h.

$$A_i = \begin{pmatrix} a_{i11} & a_{i12} & \cdots & a_{i1m} & b_{i11} & b_{i12} & \cdots & b_{i1m} \\ a_{i21} & a_{i22} & \cdots & a_{i2m} & b_{i21} & b_{i22} & \cdots & b_{i2m} \\ \vdots & \vdots & \ddots & \vdots & \vdots & \vdots & \ddots & \vdots \\ a_{ij1} & a_{ij2} & \cdots & a_{ijm} & b_{ij1} & b_{ij2} & \cdots & b_{ijm} \end{pmatrix} \tag{5}$$

m is the number of entrance direction of the i th intersection. j is the number of data sets obtained within 24 h.

Make the matrix $L = [L_1 \ L_2 \ \cdots \ L_{i-1}]$ with the distance of every intersection.

Step2: According to the spectral clustering weighting analysis method of 2.2, the weighting efficient β_i of the i th intersection in the j th statistic time-interval.

$$\beta_i = (\beta_{i1} \ \beta_{i2} \ \cdots \ \beta_{ij}). \tag{6}$$

Step3: According to the weighting efficient, traffic flow for every direction of the i th intersection is weighted to make the new weighted traffic data matrix D_i:

$$D_i = \begin{pmatrix} d_{i11} & d_{i12} & \cdots & d_{i1m} \\ d_{i21} & d_{i22} & \cdots & d_{i2m} \\ \vdots & \vdots & \ddots & \vdots \\ d_{ij1} & d_{ij2} & \cdots & d_{ijm} \end{pmatrix}. \tag{7}$$

$$d_{ijm} = a_{ijm} + \beta_{ij} \times b_{ijm}. \tag{8}$$

Step4: According to the PAA multi-period division model of 2.1, the intersections are divided into a number of periods one by one. Because the arterial flow has certain continuity, so the difference of multi-period division of each intersection can be ignored. Set periods, and the p th period of the p th intersection is P_{ip}.

Step5: The selection of the public cycle depends on the time period. During the period p, selecting the key intersection from intersections with the maximum cycle principle. Using the Webster algorithm to calculate the time of cycle, and record the maximum cycle of this period as $C_{pi\max}$

$$C_{\max} = \begin{pmatrix} C_{1i\max} & C_{2i\max} & \cdots & C_{pi\max} \end{pmatrix}. \tag{9}$$

Step6: Calculating the offset and green ratio of each intersection with the regular algorithm for the maximum cycle $C_{pi\max}$ of each period, distance data matrix L, weighted traffic data matrix D_i.

3   Simulation Verification

3.1   Effectiveness Analysis for PAA

Referring to Fig. 2 and combined with traffic characteristics of each period, use Webster calculation method to the signal time of every period to verify the validity of this algorithm partitions PAA.

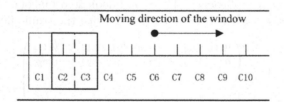

Fig. 2. Schematic model of PAA method: it should be noted that more than one data must be included within the selected window, and window moving follow the setting direction. Form Fig. 2 we find that the first stage window include data group C_1 and C_2, and then the second stage window include data group C_2 and C_3. This paper use the traffic data group collected every half hour to constitute sequence C, set half hour as sliding step, one hour as sliding window size. And then calculate mean value of the window.

These are not many differences between the new method and the old one in the night period form Fig. 3 we can find, but in the daytime the difference is obviously. This is because the vehicle number is small that caused low utilization of green light. It shows that the new method performed well.

3.2 Effect Verification of Multi-period Arterial Coordination

In order to verify whether the arterial coordination algorithm is to optimize the running state of dedicated bus., We choosing the travel time and number of stops of bus as indexes to verify it, and choosing delay per person and delay per

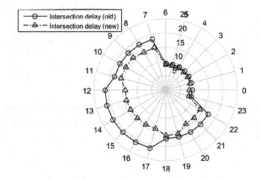

Fig. 3. Intersection delay comparison: the new method worked well in the daytime compared with the old one, especially at the off-peak period, it has 30% decrement in time delay.

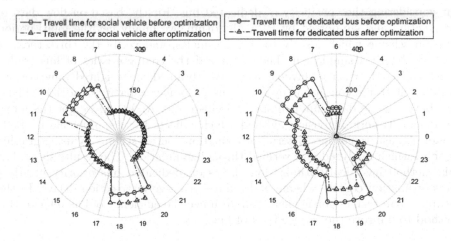

Fig. 4. Before and after comparison of travel time: since the new optimal control scheme has some effect on the social vehicles, the travel time of social vehicle has increased by almost 10% in the peak hour. But after using this method the travel time of dedicated bus has a optimization of 22.23%

vehicle as indexes to verify whether the method will have a negative impact on the running state of social vehicle. The simulation results are as follows:

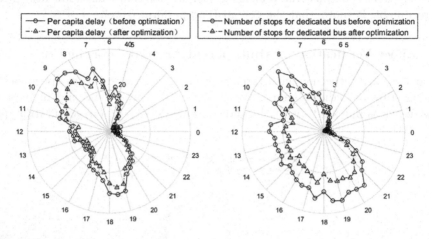

Fig. 5. Before and after comparison of delay per person and stops of bus: further found that the delay per person of arterial road has a optimization of 18.43%. And the stops of dedicated bus have a 35.67% optimization

4   Conclusion

Aiming at current intersection multi-period division methods for fixed cycle and arterial coordination control scheme are mostly based on social vehicles without considering the specificity of dedicated bus. Firstly, based on flow data of intersection and supported by PPA algorithm, we finally get the multi-period division scheme which mainly to social vehicles, and verify the correctness of PAA by VISSIM simulation. Then analyzed the characteristics of bus traffic, which collected by actual investigation. We obtained the weighted values of bus in different traffic conditions by using the passenger capacity of social vehicles and buses as the data input of spectral clustering. And then we got the intersection multi-period division method which reflects the requirements of TOD model. At last we got new arterial coordination control scheme by using graphic method which analyzed and verified through simulation. The results show that this method has ideal optimization effects on dedicated bus priority in arterial coordination control scheme and reduces average intersection delays. In the future, urban traffic management can optimize the efficiency of bus refer to this method to improve the service level of bus.

Acknowledgments. This work is partially supported by National Natural Science Foundation (61374191) and by the Great Wall Scholar Program (15038).

References

1. Yang, G.X.: Research on multi - time signal timing of single intersection. HIT (2005). (in Chinese)
2. Tang, Q.L., Li, Y.: Research on dynamic multi-time traffic based on assignment model. J. Technol. Econ. Areas Commun. 1(12), 18–20 (2010). (in Chinese)
3. Ma, W.J., Wu, M.M., Han, B.X., et al.: Bus signal priority control method for isolated intersection based on dynamic variable speed adjustment. Chin. J. Highw. Transp. 2(26), 127–133 (2013). (in Chinese)
4. Feng, L.: Research on Coordinated Control Theory of Transit Signal Priority. Jilin University School of Transportation, Chang Chun (2009). (in Chinese)
5. Li, F.: Bus signal prioritya useful technology for Beijing public transportation development. J. Highw. Transp. Res. Dev. 28(7), 26–30 (2011). (in Chinese)
6. Cai, X.Y., Dai, G.Z., Yang, L.B.: Survey on spectral clustering algorithms. Comput. Sci. 7(35), 14–18 (2008). (in Chinese)
7. Zhang, W.H., Lu, H.P., Shi, Q.: Optimal signal-planning method of intersections based on bus priority. J. Traffic Transp. Eng. 4(3), 49–53 (2004). (in Chinese)
8. Xu, T.S.: Research on spectral clustering. Comput. Knowl. Technol. 16(8), 3948–3950 (2012). (in Chinese)
9. Li, Y., Li, W., Wang, H.C.: An application of cluster analysis algorithm in traffic control. Syst. Eng. 22(2), 66–68 (2004). (in Chinese)
10. Su, L., Zhou, Y.X., Zhang, G.Q.: A piecewise aggregate approximation algorithm of data stream based on variable sliding window. Sci. Technol. Eng. 9(14), 211–214 (2014). (in Chinese)

Multi-resolution Selective Ensemble Extreme Learning Machine for Electricity Consumption Prediction

Hui Song[1(⊠)], A.K. Qin[2], and Flora D. Salim[1]

[1] Computer Science and Information Technology, School of Science,
RMIT University, Melbourne, VIC 3000, Australia
{hui.song,flora.salim}@rmit.edu.au
[2] Computer Science and Software Engineering,
School of Software and Electrical Engineering, Swinburne University of Technology,
Hawthorn, VIC 3122, Australia
kqin@swin.edu.au

Abstract. We propose a multi-resolution selective ensemble extreme learning machine (MRSE-ELM) method for time-series prediction with the application to the next-step and next-day electricity consumption prediction. Specifically, at the current time stamp, the preceding time-series data is sampled at different time intervals (i.e. resolutions) to constitute the time windows used for the prediction. The value at each sampled point can be certain statistics calculated from its associated time interval. At each resolution, multiple extreme learning machines (ELMs) with different numbers of hidden neurons are first trained. Then, sequential forward selection and least square regression are used to select an optimal set of trained ELMs to constitute the final ensemble model. The experimental results demonstrate that the proposed MRSE-ELM outperforms the best single ELM model across all resolutions. Compared to three state-of-the-art prediction models, MRSE-ELM shows its superiority on the next-step and next-day electricity consumption prediction tasks.

Keywords: Multi-resolution · Extreme Learning Machine · Least Square · Ensemble

1 Introduction

A rapidly increase in world energy use has caused issues of supply difficulties, exhaustion of energy resources and adverse environmental impacts. The prediction of electricity consumption in buildings has the followings benefits: it helps to improve energy monitoring and use in buildings; it plays a significant role in improving electricity performance, with the aim of achieving energy consumption conservation and reducing environmental impact [1]; it can play a vital role in decision-making and future planning that rely on prediction accuracy; and it

© Springer International Publishing AG 2017
D. Liu et al. (Eds.): ICONIP 2017, Part V, LNCS 10638, pp. 600–609, 2017.
https://doi.org/10.1007/978-3-319-70139-4_61

is an indispensable part of easing the conflict between supply and demand based on the analysis of existing electricity usage [2]. Therefore, the reliable prediction of electricity consumption is important and requires more attention.

Various techniques have been proposed to solve electricity usage prediction problems. In the early stage, statistical methods such as time series approaches and regression models were applied [3]. With the development of artificial intelligence techniques, artificial neural networks (ANN) and support vector machine (SVM) [4] have been successfully utilized to predict electricity consumption. Moreover, many hybrid prediction models have been developed to take the advantage of single models [5]. Recently, Extreme Learning Machine (ELM) [6] has attracted more attention given its super fast computational ability.

It has been shown that ensemble models have greater accuracy and robustness compared with single models. Several ensemble approaches have been developed for load/electricity prediction [7–11]. Among all these approaches, simple averaging is usually used for combining single predictors [8]. Generating useful instances for ensemble learning is of significant importance for improving prediction accuracy. In addition, in practice some predictors in the ensemble are better than others, therefore the ensemble output should select the individual outputs to generate the best prediction performance.

In this paper, we aim to predict next-step and next-day electricity usage in a university building. The data is at 15-min to 1-day intervals collected from the smart meter in building 80 in the city campus of RMIT University, Melbourne. Temporal resolutions such as Wavelet Transform [11], usually needs to be decided a priori, whereas in here it does not need to. Here, at the current time stamp, the preceding time-series data is sampled at different resolutions to constitute the time windows used for the prediction. The value at each sampled point can be certain statistics calculated from its associated time interval. Then the instances for each resolution are generated with different hidden neuron settings of ELM. Finally, Sequential Forward Selection (SFS) [12] and Least Square Regression (LS) is proposed to perform selective ensemble (SE) to obtain the optimal subset of instances for improving prediction performance of next-step and next-day electricity consumption. Therefore, the main contributions of the designed MRSE-ELM are as follows:

- A new multi-resolution mechanism is proposed as the combination of different resolutions will capture more time information for improving prediction performance;
- Selective ensemble based on SFS and LS is proposed to obtain the optimal subset of instances which can lead to the best prediction performance;
- MRSE-ELM is compared to three state-of-the-art prediction models and shows its superiority on our prediction problems.

The rest of this paper is organized as follows: Sect. 2 gives the background of electricity consumption prediction and ELM related prediction; the proposed MRSE-ELM will be introduced in Sect. 3; Sect. 4 mainly focuses on data description, experimental settings and related results; conclusions and future work will be presented in Sect. 5.

2 Background

2.1 Electricity Consumption Prediction

Electricity consumption prediction as a hot topic has a history of more than 20 years. There are several review articles about energy consumption forecasting, including electricity consumption forecasting [1,3,5,13,14], which introduce the energy demand forecasting from different viewpoints. Suganthi and Samuel [5] summarized the models as traditional methods and new techniques. Traditional methods, such as time series, regression, econometric, Autoregressive integrated moving average (ARIMA) and soft computing techniques have been applied for electricity consumption prediction. Newer methods, also called integrated models, include SVR, Ant Colony and PSO. Zhao and Magoulès [3] categorized the energy consumption prediction models into five different kinds: Engineering methods, statistical methods, ANN, SVM [4] and Grey Models. In [1], the authors did not only discuss the physical models for building thermal behavior modeling, but also review the prediction models from limitations to applications, which contain Multiple Linear Regression (MLR), Genetic algorithm (GA), ANN, SVR and hybrid models. However, it is very difficult to say which one outperforms others without complete comparison under the same circumstances because each of them is still being developed [3].

2.2 ELM Related Prediction

Extreme Learning Machine (ELM), proposed by Huang *et al.* [6], is based on Single hidden Layer Feedforward (SLNF) and has input layer, hidden layer and output layer. The hidden bias and the weight for connecting the input layer and hidden layer are generated randomly and maintained through the whole training process.

Assuming dataset $(\mathbf{x}_i, \mathbf{y}_i)$ with a set of M distinct samples, satisfy $\mathbf{x}_i \in \mathcal{R}^{d1}$ and $\mathbf{y}_i \in \mathcal{R}^{d2}$, so a SLNF with N hidden neurons can be formulated as:

$$\sum_{i=1}^{N} \beta_i \mathbf{f}(\mathbf{w}_i^T \mathbf{x}_j + b_i), 1 \leq j \leq M \tag{1}$$

where f is the activation function; $\mathbf{w}_i$ represents the weight for connecting input layer and hidden layer; b_i is bias and β_i is the output weight.

In ELM, the structure perfectly approximates to the given output data:

$$\sum_{i=1}^{N} \beta_i \mathbf{f}(\mathbf{w}_i^T \mathbf{x}_j + b_i) = \mathbf{y}_j, 1 \leq j \leq M \tag{2}$$

which can be written as $\mathbf{HB} = \mathbf{Y}$, the matrix $\mathbf{H}$ can be represented as:

$$\mathbf{H} = \begin{pmatrix} f(\mathbf{w}_1^T \mathbf{x}_1 + b_1) & \cdots & f(\mathbf{w}_N^T \mathbf{x}_1 + b_N) \\ \cdots & \cdots & \cdots \\ f(\mathbf{w}_1^T \mathbf{x}_M + b_1) & \cdots & f(\mathbf{w}_N^T \mathbf{x}_M + b_N) \end{pmatrix} \tag{3}$$

$\mathbf{B} = (\beta_1^T, \beta_2^T, \ldots, \beta_N^T)^T$ and $\mathbf{Y} = (y_1^T, y_2^T, \ldots, y_M^T)^T$.

The output weight $\mathbf{B}$ is calculated by $\mathbf{B} = \mathbf{H}^{+}\mathbf{Y}$, and $\mathbf{H}^{+}$ is a Moore-Penrose generalized inverse of $\mathbf{H}$ [15]. Theoretical proofs and a more thorough presentation of the ELM algorithm are detailed in the original paper [6].

The only task for ELM applications is to select a suitable activation function and set the number of hidden neurons. Moreover, compared with conventional learning approaches, it removes the challenges of setting up a model like learning rates, learning epochs, stop criteria and local optima [16], all of which makes it easier to be applied in load/electricity prediction [17–20] issues. All the advantages motivate us to utilize it as a basic prediction model for our problems.

3   Multi-resolution Selective Ensemble Extreme Learning Machine

3.1   Instances Generation Based on Multi-Resolution and ELMs

The general framework of the proposed MRSE-ELM will be introduced in detail here. Figure 1 describes the training procedure of the selective ensemble prediction based on multi-resolution and multiple ELMs. The total training data D is split into D_1 to D_k, which is used for cross validation in order to train the model, with k being the number of folders. In Fig. 1, the construction of multi-resolution is illustrated in detail.

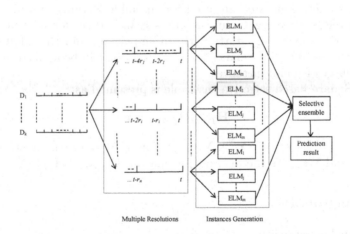

Fig. 1. The training procedure of MRSE-ELM

Multiple resolutions are constituted of different time intervals. At the current time stamp, the preceding time-series data is sampled at different resolutions, denoted as $r_i, i = 1, 2, \ldots n$. r_1 is the minimal time interval and n is the number of resolutions. The potential relationship between r_1 and r_i is: $r_i = ir_1$. For example, for the current resolution at the current time t, the previous time stamp $t - r_i$ means that there are i minimal time intervals between two joint time intervals. The value at $t - r_i$ is the mean value from $t - ir_1$ to $t - r_1$. By using this approach, multiple resolutions are constructed for next-step prediction using minimal interval. The daily electricity consumption is represented by the sum of all minimal intervals in one day and the multiple resolutions are constructed similarly with next-step prediction.

As ELM is the basic prediction model, the only parameter in ELM is the number of hidden neuron settings. With respect to each resolution, there will be m different hidden neuron settings in ELM. Therefore, with n resolutions, there will be $m \cdot n$ instances generated with ELM, as seen in the instances generation in Fig. 1.

3.2 Selective Least Square Regression

For a single ELM, the last hidden layer output matrix $\mathbf{H}$ and output $\mathbf{Y}$ can be perfectly approximated through $\mathbf{HB} = \mathbf{Y}$, which is actually the procedure of Least Square. Therefore, the output weight $\mathbf{B}$ can be calculated from $\mathbf{B} = \mathbf{H}^+\mathbf{Y}$.

From Fig. 1, it can be seen there are $m \cdot n$ different ELMs, all of which have the same output $\mathbf{Y}$, along with $m \cdot n$ different predictors. The respective output weight matrix in the instances generation in Fig. 1 is $\mathbf{B}^l = \mathbf{H}^{l+}\mathbf{Y}, l = 1, \ldots, m \cdot n$ and the predicted outputs of each ELM is $\hat{\mathbf{Y}}^l = \mathbf{H}^l\mathbf{B}^l$, where $\hat{\mathbf{Y}}^l$ is the predicted value for each of the ELM corresponding to the real value $\mathbf{Y}$.

If all instances generated are used for ensemble learning, it will cause overfitting, therefore selecting useful instances is important for efficient ensemble prediction. Here, SFS [12] starts with an empty set, and the addition of one feature with the best performance for each iteration will be applied to select the optimal subset of useful instances.

Least Square based selective ensemble is presented as:

$$\mathbf{H}_{final}\mathbf{B}_{final} = \mathbf{Y} \tag{4}$$

where $\mathbf{H}_{final} = (\mathbf{Y}^1, \ldots, \mathbf{Y}^N), N \leqslant mn$ and N is the number of selected instances, and $\mathbf{B}_{final}$ is the final output matrix, $\mathbf{B}_{final} = \mathbf{H}_{final}^+\mathbf{Y}$.

4 Experiments

4.1 Data Description

As the data is collected with different time intervals, we focus on predicting next interval, which is the minimal interval, and next day, which is the maximal interval. Here we define the minimal time interval (15 min) as interval electricity data while the maximal interval (one day, 96 minimal intervals) is called daily electricity data. The dataset for interval electricity is from 01.03.2017 to 31.05.2017 with 8352 samples. Daily electricity data is from 01.01.2013 to 31.05.2017 with 1573 samples at the maximal time interval of 24 h. Both are normalized to $[0.15, 1]$ except for the prediction targets in the training (70%) and testing (30%) sets.

The time window for interval electricity data is set at 48 steps for purpose of symmetry, even for different resolutions. The number of resolutions is set at 10. For daily data, the window size is set at 7 days for 7 different resolutions with the lengths of 7, 14, 21, 28, 42, 56 and 84 respectively. During the process of training, the training dataset is split into 5 folders in order to adjust the parameters. Once the parameters leading to the best performance have been found, the performance will be evaluated on the testing dataset by Root Mean Square Error (RMSE).

4.2   Comparison Models and Experimental Settings

Three state-of-the-art regression models will be compared: SVR, H-ELM [21] and Gated Recurrent Units (GRUs) [22]. There are two different kinds of SVRs, called ϵ-SVR model and nu-SVR model. In this paper, ϵ-SVR model is implemented using the LibSVM library [23] for SVR. H-ELM, proposed by Tang et al. [21], has unsupervised and supervised stages, where unsupervised stage focuses on feature extraction and supervised stage aims for classification or regression. GRUs is a simpler variant of Long Short Term Memory (LSTM) [24]. GRUs can be used to learn long term dependencies, similar to LSTM. The parameters for the proposed model and comparison models are presented as:

- **ELM (RBF):** Number of Hidden Neurons (100–500 with interval 100 for interval data and 50–500 with interval 50 for daily data)
- **ϵ-SVR (RBF):** C (Cost, 2^{-5} 1 2^5 2^{10} 2^{15}); ϵ (0.01 0.03 0.05 0.1 0.15 0.2)
- **H-ELM:** Number of Hidden Neurons (10–300 with interval 20 for three different hidden layers, two for unsupervised stage and one for supervised stage); C (regularized least square calculation, 10^{-10} to 10^{10} with interval 10^2)
- **GRUs:** Number of Hidden Neurons (1 to 10 with interval 1).

4.3   Results

In this experiment, there are two problems to be addressed:

- How can it be proven that selective ensemble will improve prediction performance?
- How does the performance of MRSE-ELM compare with state-of-the-art models?

Result for Problem 1. Since there are 5 and 10 different hidden neuron settings, with 10 and 7 different resolutions for interval and daily electricity data respectively, there are 50 and 70 instances generated from ELM on each dataset. Figure 2 shows the best training performance at each resolution without ensemble learning on two different datasets for different models.

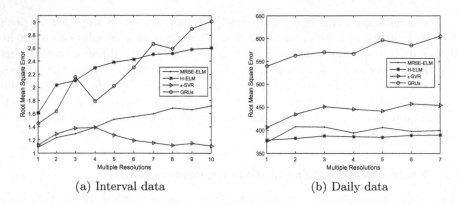

(a) Interval data   (b) Daily data

Fig. 2. The performance of multiple resolutions on different datasets

Figure 3a and b show the dynamic change of the number of selected instances N used for ensemble learning with respect to interval and daily electricity data (previous 50 shown because of the increasing trend) respectively. $N = 1$ represents only one instance selected and there is no ensemble learning. Further, $N = 1$ is the best one among all the generated instances. When the selected instances N is more than 1, the performance will improve until the prediction accuracy reaches the best level, which are labeled by red circles ($N = 19$ and $N = 4$, for interval and daily data respectively). Then the performance will decrease with more instances selected, which can be seen from Fig. 3a and b. Therefore, Fig. 3 does not only illustrate that the proposed selective ensemble performs better than the single best, but also shows when the selective ensemble will reach the best performance for different datasets.

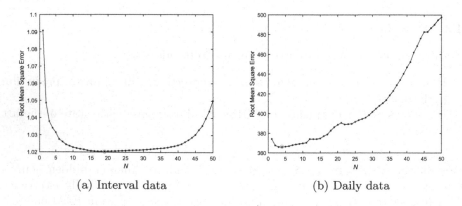

(a) Interval data   (b) Daily data

Fig. 3. The performance of MRSE-ELM on different datasets

Result for Problem 2. The comparison experiment for MRSE-ELM is based on the optimal subset ($N = 19$ and $N = 4$) of generated instances for selective ensemble in Fig. 3 for both interval data and daily data. The result presented was achieved by the best parameter settings after adjustment. Table 1 presents the results of two different prediction tasks using MRSE-ELM and comparison models. All the results presented are mean and standard deviation values with 10 runs. Wilcoxon rank sum test is used to evaluate the differences between MRSE-ELM and comparison models. The returned result 1 indicates a rejection of the null hypothesis, while 0 indicates a failure to reject the null hypothesis at the 5% significance level.

Table 1. RMSE performance on testing data

Datasets		Models			
		ELM	H-ELM	ϵ-SVR	GRUs
Interval data	Mean (std)	**1.0021** (0.0063)	1.6411 (0.0982)	1.1025 (0)	1.3056 (0.0835)
	h	-	1	1	1
Daily data	Mean (std)	**385.5006** (4.5142)	428.0085 (5.8491)	411.9080 (0)	534.5920 (26.1012)
	h	-	1	1	1

The result in Table 1 shows the performance with respect to MRSE-ELM and comparison models. The proposed MRSE-ELM outperforms the comparison prediction models on average values 1.0021 and 385.5006 (labeled bold) for next-step and next-day prediction respectively. Moreover, compared the statistical performance with the state-of-the-art prediction models H-ELM, ϵ-SVR and GRUs, MRSE-ELM shows superiority because of '$h = 1$' for all comparison models.

5 Conclusions and Future Work

In this paper, we proposed a multi-resolution selective ensemble extreme learning machine (MRSE-ELM) method for time-series prediction with the application to next-step and next-day electricity consumption. Firstly, the optimal subset of instances used for ensemble prediction was analyzed on the training data for the interval and daily data. After that, the performance of MRSE-ELM was evaluated by the testing datasets with the selected instances and compared with the state-of-the-art models H-ELM, SVR and GRUs. Experimental result does not only demonstrate that the proposed MRSE-ELM performs better than the best single ELM, but also show its superiority on improving prediction accuracy compared to the state-of-the-art models. The proposed multi-resolution selective ensemble mechanism is not limited in being applied with ELM. In the future, we will utilize it in the state-of-the-art models to further improve prediction performance.

Acknowledgments. This work is supported by Buildings Engineered for Sustainability research project, funded by RMIT Sustainable Urban Precincts Program.

References

1. Foucquier, A., Robert, S., Suard, F., Stéphan, L., Jay, A.: State of the art in building modelling and energy performances prediction: a review. Renew. Sustain. Energy Rev. **23**, 272–288 (2013)
2. Yalcintas, M., Akkurt, S.: Artificial neural networks applications in building energy predictions and a case study for tropical climates. Int. J. Energy Res. **29**, 891–901 (2005)
3. Zhao, H.X., Magoulès, F.: A review on the prediction of building energy consumption. Renew. Sustain. Energy Rev. **16**(6), 3586–3592 (2012)
4. Dong, B., Cao, C., Lee, S.E.: Applying support vector machines to predict building energy consumption in tropical region. Energy Build. **37**, 545–553 (2005)
5. Suganthi, L., Samuel, A.A.: Energy models for demand forecasting–a review. Renew. Sustain. Energy Rev. **16**(2), 1223–1240 (2012)
6. Huang, G.B., Zhu, Q.Y., Siew, C.K.: Extreme learning machine: theory and applications. Neurocomputing **70**, 489–501 (2006)
7. Taylor, J.W., Buizza, R.: Neural network load forecasting with weather ensemble predictions. IEEE Trans. Power Syst. **17**(3), 626–632 (2002)
8. Abdel-Aal, R.: Improving electric load forecasts using network committees. Electr. Power Syst. Res. **74**(1), 83–94 (2005)
9. De Felice, M., Yao, X.: Short-term load forecasting with neural network ensembles: a comparative study [application notes]. IEEE Comput. Intell. Mag. **6**(3), 47–56 (2011)
10. Zhang, R., Dong, Z.Y., Xu, Y., Meng, K., Wong, K.P.: Short-term load forecasting of australian national electricity market by an ensemble model of extreme learning machine. IET Gener. Transm. Distrib. **7**(4), 391–397 (2013)
11. Li, S., Goel, L., Wang, P.: An ensemble approach for short-term load forecasting by extreme learning machine. Appl. Energy **170**, 22–29 (2016)
12. Reunanen, J.: Overfitting in making comparisons between variable selection methods. J. Mach. Learn. Res. **3**, 1371–1382 (2003)
13. Fumo, N.: A review on the basics of building energy estimation. Renew. Sustain. Energy Rev. **31**, 53–60 (2014)
14. Ghalehkhondabi, I., Ardjmand, E., Weckman, G.R., Young II, W.A.: An overview of energy demand forecasting methods published in 2005–2015. Energy Syst. **8**, 1–37 (2016)
15. Rao, C.R., Mitra, S.K.: Generalized Inverse of Matrices and Its Applications, vol. 7. Wiley, New York (1971)
16. Zhang, R., Lan, Y., Huang, G.B., Xu, Z.B.: Universal approximation of extreme learning machine with adaptive growth of hidden nodes. IEEE Trans. Neural Netw. Learn. Syst. **23**, 365–371 (2012)
17. Li, S., Wang, P., Goel, L.: Short-term load forecasting by wavelet transform and evolutionary extreme learning machine. Electr. Power Syst. Res. **122**, 96–103 (2015)
18. Sajjadi, S., Shamshirband, S., Alizamir, M., Yee, L., Mansor, Z., Manaf, A.A., Altameem, T.A., Mostafaeipour, A.: Extreme learning machine for prediction of heat load in district heating systems. Energy Build. **122**, 222–227 (2016)

19. Wan, C., Xu, Z., Pinson, P., Dong, Z.Y., Wong, K.P.: Probabilistic forecasting of wind power generation using extreme learning machine. IEEE Trans. Power Syst. **29**(3), 1033–1044 (2014)
20. Song, H., Qin, A.K., Salim, F.D.: Multivariate electricity consumption prediction with extreme learning machine. In: International Joint Conference on Neural Networks (IJCNN), pp. 2313–2320. IEEE (2016)
21. Tang, J., Deng, C., Huang, G.B.: Extreme learning machine for multilayer perceptron. IEEE Trans. Neural Netw. Learn. Syst. **27**(4), 809–821 (2016)
22. Cho, K., Van Merriënboer, B., Bahdanau, D., Bengio, Y.: On the properties of neural machine translation: encoder-decoder approaches. arXiv preprint arXiv:1409.1259 (2014)
23. Chang, C.C., Lin, C.J.: LIBSVM: a library for support vector machines. ACM Trans. Intell. Syst. Technol. (TIST) **2**, 1–27 (2011)
24. Hochreiter, S., Schmidhuber, J.: Long short-term memory. Neural Comput. **9**(8), 1735–1780 (1997)

Dow Jones Index is Driven Periodically by the Unemployment Rate During Economic Crisis and Non-economic Crisis Periods

Tong Cao[1], Sanqing Hu[1], Yuying Zhu[1], Jianhai Zhang[1(✉)], Hui Su[1], and Bocheng Wang[2]

[1] College of Computer Science, Hangzhou Dianzi University,
Hangzhou 310018, Zhejiang, China
`tonneyc@126.com`, {`sqhu,jhzhang`}`@hdu.edu.cn`, `xjjhzyy@126.com`,
`suhui08@foxmail.com`
[2] Zhejiang University of Media and Communications,
Hangzhou 310018, Zhejiang, China
`baixibao@gmail.com`

Abstract. Previous researchers have made some causality hypotheses: the change of stock index causing volatility of economic data or short-run impact of anticipated unemployment rate on stock price. However, they have not reached a consensus. In this article we apply New Causality (NC) method to investigate the causality between Dow Jones Index and the unemployment rate. The results demonstrate stock market is periodically driven by the unemployment rate during all periods, and the causal direction during one ECP and on-going NECP together is uncertain because there may exist two different causal mechanisms in two periods. In this point of view, we conclude that anticipated unemployment rate change results in Dow Jones Index fluctuation in each period. Our conclusion is consistent with the phenomenon that Dow Jones Index was pushed to historical high level after Donald Trump came into power.

Keywords: New causality · Stock index · Unemployment rate · Economic crisis

1 Introduction

It is well known that stock market is generally considered as the economy "weather glass" of a country's economic condition. Some studies have shown that the volatility of macroeconomic variables can forecast stock market, including the growth rate of industrial production, orders inflow and so on [1,2]. For example, national unemployment rate has the correlation with ups and downs of stock market. When there exists stock market volatility or instability, the economy will be affected, and the unemployment rate will fluctuate subsequently. In these researches, generally, it is believed that stock market can be used to predict the economy in advance and affects the unemployment rate. Besides, many

© Springer International Publishing AG 2017
D. Liu et al. (Eds.): ICONIP 2017, Part V, LNCS 10638, pp. 610–617, 2017.
https://doi.org/10.1007/978-3-319-70139-4_62

researchers report stock market volatility correlates to business cycle fluctuations [3]. The causality between stock market and economic variables is essential for stock investors, economists, politicians, and it is helpful to take more measures and have a better understanding of economy. Furthermore, it is also essential to make monetary policy stable when the economic crisis breaks out [4]. However, the debate about direction of causality between stock market and the unemployment rate has been going on for a long time.

Farsio and Fazel [5] argued that it would be a mistake to rely on the unemployment rate data to make investment decision in stock market and there is no stable long-term causality relationship and even no causal influence from the unemployment rate to stock prices. His analysis focused on USA, China and Japan covering the 1970–2011 period through Granger Causality tests [6]. Farmer [7,8] also established that there had been a high correlation between unemployment and stock market in the USA data since 1929 and stock market Granger caused the unemployment rate. However, For example, Gonzalo and Taamouti [9] found that only anticipated unemployment rate had a strong impact on stock prices which was contrary to general findings in the literature by using nonparametric Granger causality and regression based tests.

Previous researchers have made some hypotheses including either the change of stock index causes the volatility of economic data or the economic data causes the volatility of stock market. They do not take long-term or short-term into account either. Therefore, they have not come to consensus. There are two possible reasons for this phenomenon: (i) the used methods such as Granger Causality tests may not be suitable; (ii) The analyzed data may involve different causal mechanisms. In this article we apply our recently proposed NC method [10] which is shown to be much better reveal true causality of the underlying system than the widely used Granger Causality by many illustrative examples [10,11]. We study causal relationship between stock index (Dow Jones Industrial Average Index) and the unemployment rate. We choose data in USA covering the period from January 1962 to March 2017, which includes 6 NECPs and 5 ECPs. The results revealed by NC method demonstrate that (i) stock market is periodically driven by the unemployment rate during all 6 NECPs and 5 ECPs, which cannot be drawn by GC results, and the causal direction between two variables during one ECP and on-going NECP together is uncertain because there may exist two different causal mechanisms in two periods (for example, both the unemployment rate and Dow Jones Index rise in ECP, and the unemployment rate decreases and Dow Jones Index rises in on-going NECP). (ii) We can always derive the undirectional significant causal influence flow from the unemployment rate to stock market only if the data in NECP or ECP is enough longer. The results present strong evidence to confirm that the unemployment rate may have a significant causal impact on stock market and thus provide important useful information for market traders, economists and politicians of the capitalist countries.

2   Data and Methods

According to the information about the USA economic expansion and contraction periods and some other references that introduce the development of the capitalist countries [12], we make appropriate modifications based on the specific situations. As a result, the period from January 1962 to March 2017 is divided into 6 NECPs and 5 ECPs displayed in Table 1, and the related data are plotted in Fig. 1 for NECPs and Fig. 2 for ECPs respectively. All the unemployment rate datasets and the Dow Jones Industrial Average index datasets are preprocessed by using logarithmic method and then subtracting its mean value from the data.

Firstly, the Granger Causality method is applied to analyze and calculate the causal relationship between the Dow Jones Industrial Average index and the USA monthly unemployment rate during various non-economic crisis periods. Granger Causality is a popular method to measure the causal relationship between two time series proposed by Granger [6]. Granger Causality is widely used in economics, neuro-science, meteorology and other fields. According to Granger Causality, a variable X_1 "Granger causes" a variable X_2 if information in the past of X_1 helps predict the future of X_2 better than when considering only information in the past of X_2 itself.

The auto-regression model of Granger Causality can be defined as follow:

$$\begin{cases} X_{1,t} = \sum_{j=1}^{m} \mathbf{a}_{1,j} X_{1,t-j} + \epsilon_{1,t} \\ X_{2,t} = \sum_{j=1}^{m} \mathbf{a}_{2,j} X_{2,t-j} + \epsilon_{2,t} \end{cases} \tag{1}$$

where, $a_{1,j}$, $a_{2,j}$ are the coefficient of the auto regressive model, $\epsilon_{1,t}$, $\epsilon_{2,t}$ are denoted as the error term. At this point each error item depends on the value of the past related variable itself. The joint regression model can be expressed as follows:

$$\begin{cases} X_{1,t} = \sum_{j=1}^{m} a_{11,j} X_{1,t-j} + \sum_{j=1}^{m} a_{12,j} X_{2,t-j} + \eta_{1,t} \\ X_{2,t} = \sum_{j=1}^{m} a_{21,j} X_{1,t-j} + \sum_{j=1}^{m} a_{22,j} X_{2,t-j} + \eta_{2,t} \end{cases} \tag{2}$$

where, $a_{11,j}$, $a_{12,j}$, $a_{21,j}$, $a_{22,j}$ are the coefficient of the joint regression model, $\eta_{1,t}$, $\eta_{2,t}$ are the prediction error and they are relevant when considering the time series. m means the order of the model which is determined by the AIC-Akaike Information Criterion. Then causality can be defined as follows:

$$G_{X_2 \to X_1} = \ln \frac{\sigma_{\epsilon_1}^2}{\sigma_{\eta_1}^2} \tag{3}$$

Obviously, $F_{X_2 \to X_1} = 0$ means there is no causal influence from X_2 to X_1 and $F_{X_2 \to X_1} > 0$ means X_2 Granger causes X_1. Similarly, the causal influence from X_1 to X_2 is defined by

$$G_{X_1 \to X_2} = \ln \frac{\sigma_{\epsilon_2}^2}{\sigma_{\eta_2}^2} \tag{4}$$

Furthermore, the NC method [10] is applied to calculate the causality direction. The NC method takes all the error items into account, and has proved that the NC is better than GC to reveal the true causal relationship. NC between variable X_1 and X_2 can be calculated by:

$$N_{X_2 \to X_1} = \frac{\sum\limits_{t=m}^{N} \left(\sum\limits_{j=1}^{m} a_{12,j} X_{2,t-j} \right)^2}{\sum\limits_{h=1}^{2} \sum\limits_{t=m}^{N} \left(\sum\limits_{j=1}^{m} a_{1h,j} X_{h,t-j} \right)^2 + \sum\limits_{t=m}^{N} \eta_{1,t}^2} \tag{5}$$

$$N_{X_1 \to X_2} = \frac{\sum\limits_{t=m}^{N} \left(\sum\limits_{j=1}^{m} a_{21,j} X_{1,t-j} \right)^2}{\sum\limits_{h=1}^{2} \sum\limits_{t=m}^{N} \left(\sum\limits_{j=1}^{m} a_{2h,j} X_{h,t-j} \right)^2 + \sum\limits_{t=m}^{N} \eta_{2,t}^2} \tag{6}$$

3   Results

From Fig. 1 one can see that generally the unemployment rate decreases and Dow Jones index rises during each NECP except (a) and (b). From Fig. 2 one can see that the unemployment rate always rises during all ECPs, but Dow Jones index fluctuates depending on the specific economic crisis property.

From Table 1 one can see that (i) For five economic crisis periods, if we set the threshold of 0.003 then all NC values from Dow Jones index to the unemployment rate are less than the threshold and thus causal influence from Dow Jones index to the unemployment rate can be ignored. However, all NC values from the unemployment rate to Dow Jones index are greater than the threshold and thus there are significant causal influence from the unemployment rate to Dow Jones index. Therefore, there are significant causal influence from the unemployment rate to Dow Jones index during five economic crisis periods by NC method. On the contrary, GC cannot draw such a conclusion. (ii) For six non-economic crisis periods, if we set the threshold of 3.0e−04 then all NC values from the unemployment rate to Dow Jones index are greater than the threshold and thus all causal influence from the unemployment rate to Dow Jones index are significant. However, there are two NC values (0.0053, 5.2647e−04) from Dow Jones index to the unemployment rate are greater than the threshold which imply significance and there are four NC values from Dow Jones index to the unemployment rate are less than the threshold which imply no significance. Since 0.0308 > 0.0053 and 0.0148 > 5.2647e−04, the whole causal influence flow for two NECPs (1975.08–1978.10 and 1982.08–1988.12) are from the unemployment rate to Dow Jones index. Therefore, there are significant causal influence flow from the unemployment rate to Dow Jones index during six NECPs by NC method. On the contrary, GC cannot draw such a conclusion. In summary, we can

Table 1. Causality values between the unemployment rate (Unem) and Dow Jones Index (DJI) by two causality methods where all bold numbers imply significance ($p <$ 0.05)

Economic state	Time periods	Causality methods			
		GC		NC	
		Unem→DJI	DJI→Unem	Unem→DJI	DJI→Unem
NECPs	1962.01–1972.12	0.1641	**0.1729**	**3.0797e−04**	5.9472e−05
	1975.08–1978.10	**0.5711**	**0.8700**	**0.0308**	**0.0053**
	1982.08–1988.12	0.0502	**0.0185**	**0.0114**	**5.2647e−04**
	1993.01–2000.02	**0.1322**	0.0011	**0.0148**	**5.1657e−05**
	2002.04–2007.06	**0.5859**	**0.2979**	**3.1813e−04**	2.5413e−04
	2009.07–2017.03	0.1504	**0.0230**	**0.0022**	1.0756e−04
ECPs	1973.01–1975.07	0.1005	**0.1104**	**0.0067**	**1.6903e−04**
	1978.11–1982.07	**0.0893**	0.0511	**0.0053**	2.3029e−05
	1989.01–1992.12	**0.2531**	0.0131	**0.0042**	**7.4205e−04**
	2000.03–2002.03	**0.7782**	**0.6548**	**0.6580**	3.0777e−06
	2007.07–2009.06	**0.5954**	**0.9838**	**0.1012**	**0.0025**
ECPs + NECPs	1973.01-1978.10	**0.2001**	0.0301	**0.0544**	5.0387e−04
	1978.11–1988.12	**0.0451**	**0.0785**	2.3585e−04	**2.2036e−04**
	1989.01–2000.02	**0.1592**	**0.1725**	**3.8485e−04**	0.0026
	2000.03–2007.06	**0.6593**	**0.3574**	**0.0084**	5.8040e−05
	2007.07–2017.03	0.0476	**0.1921**	2.0080e−05	**8.9877e−04**

conclude that there are significant causal influence flow from the unemployment rate to Dow Jones index during six NECPs and five ECPs by NC method. However, GC method cannot draw such a conclusion.

For ECP and on-going NECP together it is noted that from Table 1 one cannot always draw above conclusion for both NC and GC methods. This fact can be understandable. For example, for both ECP 1989.01–1992.12 and on-going NECP 1993.01-2000.02 there are significant causal influence from the unemployment rate to Dow Jones index. However, from (c) of Fig. 2 and (d) of Fig. 1 one can see that the causal influence from the unemployment rate to Dow Jones index for ECP 1989.01–1992.12 is positive (NC result) and the causal influence from the unemployment rate to Dow Jones index for on-going NECP 1993.01–2000.02 is negative (NC result). As a result there is no longer the causal influence flow from the unemployment rate to Dow Jones index for ECP 1989.01–1992.12 and on-going NECP 1993.01–2000.02 together. Therefore, different causal mechanisms in ECP and on-going NECP result in conclusion that there is no longer the causal influence from the unemployment rate to Dow Jones index for ECP and on-going NECP together although there are causal influence from the unemployment rate to Dow Jones index for both ECP and NECP.

Since there are significant causal influence from the unemployment rate to Dow Jones index for all six NECPs and five ECPs based on NC results, once the data in each ECP or NECP is long enough, there must exist significant causal influence from the unemployment rate to Dow Jones index for this data. An example of 2009.07–2017.03 is shown in Table 2 from which one can see that we can always obtain significant causal influence from the unemployment rate to Dow Jones index (NC results) as new data are added and GC results cannot reveal such a conclusion.

Gonzalo and Taamouti [9] pointed out that anticipated unemployment rate can cause change of stock price. Our conclusion is consistent with this finding in each NECP and ECP. According to length of anticipated time, it may include short-run and long-run. For example, in short-run case, the unemployment rate will be announced tomorrow, stock price may react today. Such kind of causal influence belongs to short-run influence from unemployment rate to stock price. In long-run case, some policies made by politicians and economists having long-time important effect on unemployment rate which further significantly affect stock market. For example, the new president of the United States named Donald Trump has taken some stimulus measures, such as largely increasing investment in infrastructure and thus create a large number of work opportunities, which result in largely decreasing unemployment rate. Therefore, since he came into power, we see the phenomenon that Dow Jones Index creates continuously historical high level. We believe that Dow Jones Index will keep historical high level in couple of years.

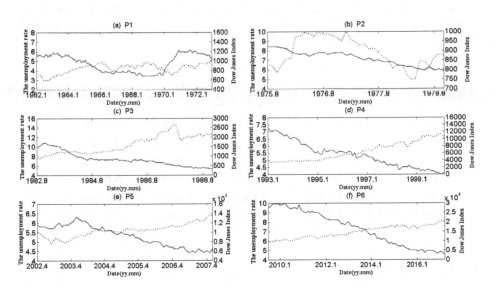

Fig. 1. Monthly recorded data of Dow Jones Index (red dotted line) and the unemployment rate (blue solid line) during six NECPs. (a) The first NECP. (b) The second NECP. (c) The third NECP. (d) The forth NECP. (e) The fifth NECP. (f) The sixth NECP. (Color figure online)

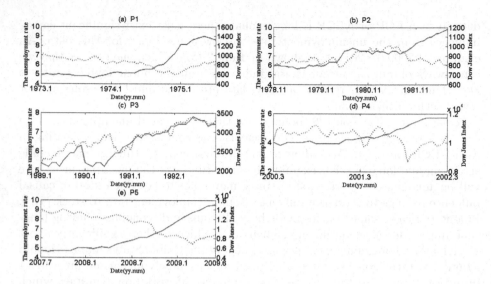

Fig. 2. Monthly recorded data of Dow Jones Index (red dotted line) and the unemployment rate (blue solid line) during five ECPs. (a) The first ECP. (b) The second ECP. (c) The third ECP. (d) The forth ECP. (e) The fifth ECP. (f) The sixth NECP. (Color figure online)

Table 2. Causality values of different time periods of the sixth NECP between the unemployment rate (Unem) and Dow Jones Index (DJI) by two causality methods where all bold numbers imply significance ($p < 0.05$)

Time periods	Causality methods			
	GC		NC	
	Unem→DJI	DJI→Unem	Unem→DJI	DJI→Unem
2009.07–2012.06	**0.5002**	**1.0909**	0.1894	**0.0366**
2009.07–2013.06	0.4015	**1.3089**	0.1776	**0.0097**
2009.07–2014.06	0.5544	**0.3928**	0.0893	**0.0104**
2009.07–2015.06	**0.3729**	0.1680	**0.0048**	**0.0023**
2009.07–2016.06	0.2802	**0.0302**	**0.0039**	9.8330e−05
2009.07–2017.03	0.1504	**0.0230**	**0.0022**	1.0756e−04

4   Conclusions

In this study, we always obtain that Dow Jones Index is caused by unemployment rate from the causal analysis of above six NECPs and five ECPs. This imply Dow Jones Index change may be caused by anticipated the unemployment rate. When the economic crisis ends in NECP, it is known that unemployment rate starts falling in a longer time. This kind of anticipation will leads to Dow Jones Index rising in a longer time. Accordingly, the unemployment rate will rise in near future in ECP when the economic crisis breaks out. This kind of anticipation will leads to Dow Jones Index falling. Since economic crisis breaks out periodically, that is ECP and NECP exchange continuously. Thus, violation of stock index is driven periodically by anticipated unemployment rate change.

Acknowledgments. This work was funded by National Natural Science Foundation of China under Grants (Nos. 61473110, 61633010), International Science and Technology Cooperation Program of China, Grant No. 2014DFG12570, Key Lab of Complex Systems Modeling and Simulation, Ministry of Education, China.

References

1. Brahmasrene, T., Jiranyakul, K.: Cointegration and causality between stock index and macroeconomic variables in an emerging market. Acad. Account. Finan. Stud. J. **3**, 17–30 (2007)
2. Pierdzioch, C., Döpke, J., Hartmann, D.: Forecasting stock market volatility with macroeconomic variables in real time. J. Econ. Bus. **60**, 256–276 (2005)
3. Hamilton, J.D., Lin, G.: Stock market volatility and the business cycle. J. Appl. Econometrics **11**, 573–593 (1996)
4. Weller, C.E.: Whose bank is it anyway? The importance of unemployment and the stock market for monetary policy. Rev. Radical Political Econ. **34**, 303–310 (2002)
5. Farsio, F., Fazel, S.: The stock market/unemployment relationship in USA, China and Japan. Int. J. Econ. Finance **5**(3) (2013)
6. Granger, C.W.J.: Investigating causal relations by econometric models and cross-spectral methods. Econometrica **37**, 424–438 (1969)
7. Farmer, R.: The stock market crash of 2008 caused the great recession. In: Meeting Papers (2012)
8. Farmer, R.E.A.: The stock market crash really did cause the great recession. Oxford Bull. Econ. Stat. **77**, 617–633 (2015)
9. Gonzalo, J., Taamouti, A.: The reaction of stock market returns to anticipated unemployment. In: Uc3m Working Papers Economics (2011)
10. Hu, S., Dai, G., Worrell, G.A., Dai, Q., Liang, H.: Causality analysis of neural connectivity: critical examination of existing methods and advances of new methods. IEEE Trans. Neural Netw. **22**, 829–844 (2011)
11. Hu, S., Wang, H., Zhang, J., Kong, W., Cao, Y., Kozma, R.: Comparison analysis: granger causality and new causality and their applications to motor imagery. IEEE Trans. Neural Netw. Learn. Syst. **27**, 1429–1444 (2016)
12. Orlov, P.A.: Trial of economies of the developed capitalist countries and countries with economies in transition by world economic crisis. Studia Universitatis Vasile Goldis Arad Seria Stiinte Economice **42**, 5 (2012)

Dynamic Cyclone Wind-Intensity Prediction Using Co-Evolutionary Multi-task Learning

Rohitash Chandra[(✉)]

Centre for Translational Data Science, The University of Sydney,
Camperdown, NSW 2006, Australia
rohitash.chandra@sydney.edu.au

Abstract. A new category called dynamic time series prediction is introduced to address robust "on the fly" prediction needed in events such as natural disasters. A co-evolutionary multi-task learning algorithm is presented which incorporates features from modular and multi-task learning. The algorithm is used for prediction of tropical cyclone wind-intensity. This addresses the need for a robust and dynamic prediction model during the occurrence of a cyclone. The results show that the method addresses dynamic time series effectively when compared to conventional methods.

Keywords: Backpropagation · Modular network design · Multi-task learning · Modular pattern classification

1   Introduction

In prediction for climate extremes [1, 15, 21, 25], it is important to develop models that can make predictions dynamically, i.e. robust "on the fly" prediction given minimal information about the event. In time series prediction, it is typical to reconstruct the time series into a state-space vector defined by fixed embedding dimension and time lag. The embedding dimension defines the minimal timespan or number of past data-points needed to make a prediction. This can be seen as a limitation since data from the past cyclones mostly considered readings every 6 hours [14]. Therefore, the model should have the feature to make timely prediction as soon as the cyclone has been detected. *Dynamic time series prediction* is defined as the ability of a model to make prediction for different number of input features or set of values of the embedding dimension. It has been highlighted in recent work [9] that recurrent neural networks trained with a predefined embedding dimension can only provide effective prediction for the same embedding dimension which makes dynamic time series prediction a challenging problem. A way to address such categories of problems is to develop robust models that can make prediction given different values of embedding dimension that define the number of features in the input. The overlapping information in different groups of features can harnessed through shared knowledge representation.

© Springer International Publishing AG 2017
D. Liu et al. (Eds.): ICONIP 2017, Part V, LNCS 10638, pp. 618–627, 2017.
https://doi.org/10.1007/978-3-319-70139-4_63

Multi-task learning employs shared representation knowledge for learning multiple instances from the same problem [3,10,23,24]. In the case of time series, multi-task learning can consider different a set of embedding dimensions as tasks that have shared knowledge representation. Neuro-evolution refers to the use of evolutionary algorithms for training neural networks [2]. Cooperative coevolution [19] has been a prominent neuro-evolution methodology where the neural network is decomposed into modules [18]. The modules are implemented as sub-populations and been used for time series prediction [4,7] Modular neural networks are motivated from repeating structures in nature [8,13]. Although neuro-evolution has been successfully applied for training neural networks, multi-task learning for enhancing neuro-evolution has not been fully explored. Evolutionary multi-task learning has been used for enforcing modularity in neural networks that can be beneficial when certain modules are perturbed or damaged [6]. There has not been any investigation that explores the embedding dimension of a time series as subtasks for multi-task learning which can be beneficial for dynamic time series prediction.

In this paper, a co-evolutionary multi-tasking learning algorithm is proposed that provides a synergy between multi-task learning and co-evolutionary algorithms. It enables neural networks to feature shared knowledge representation while retaining modularity for robust "on the fly" prediction considered to be dynamic time series. The original time series is reconstructed with a set of values for the embedding dimension that defines the subtasks for multi-task learning. Here, the tasks are referred as subtasks to avoid confusion with typical multi-task learning where a set of datasets define the tasks. The proposed method is used for tropical cyclone wind-intensity prediction and addresses the problem of dynamic prediction.

The rest of the paper is organised as follows. Section 2 gives details of the co-evolutionary multi-task learning method for dynamic time series prediction. Section 3 presents the results with a discussion. Section 4 presents the conclusions and directions for future research.

2   Co-Evolutionary Multi-task Learning

2.1   Dynamic Time Series Prediction

In state-space reconstruction, the original time series is divided using overlapping windows at regular intervals that can be used for one-step-ahead prediction. Taken's theorem expresses that the state-space vector reproduces important characteristics of the original time series [22]. Hence, given an observed time series $x(t)$, an embedded phase space $Y(t) = [(x(t), x(t-T), ..., x(t-(D-1)T)]$ can be generated, where, T is the time delay, D is the embedding dimension (window), $t = (D-1)T, DT, ..., N-1$, and N is the length of the original time series. The optimal values for D and T must be chosen in order to efficiently apply Taken's theorem [11]. Taken's proved that if the original attractor is of dimension d, then $D = 2d + 1$ will be sufficient to reconstruct the attractor [22]. In the case of using feedforward neural networks, D is the a number of input neurons. As defined earlier, dynamic time series prediction refers to the ability of a model to make prediction for different number of input features or set of values of the embedding dimension. Therefore, one step ahead prediction $\hat{y}_{t+1}$ for subtask Ω_m that feature the respective embedding dimension can be given by

$$\Omega_m = x_t, x_{t-1}, ..., x_{t-m}$$
$$\hat{y}_{t+1} = f(\Omega_m) \tag{1}$$

where $f(.)$ is a model such as a feedforward neural network, m is defines the length of the input features, where $m = 1, 2, ..., M$, and M is the number of subtasks.

2.2   Algorithm

The proposed algorithm is inspired by the strategies used in dynamic programming where a subset of the solution is used as the main building block for the optimisation problem. It enables neural networks to feature shared knowledge representation while retaining modularity for robust "on the fly" prediction considered to be dynamic time series. The original time series is reconstructed with a set of values for the embedding dimension that defines the respective subtasks. The problem is in sequentially learning the modules of a neural network that link with their subtasks. The base network module features lowest number of input features and hidden neurons. The base network module is part of larger cascaded neural network architecture that expands in size as more subtasks are considered. Additional network modules are required for the subtasks that have shared knowledge representation, hence, the method is called *co-evolutionary multi-task learning* (CMTL). The algorithm considers learning cascaded network module θ_m which is composed of network module Φ_m. The input time series is defined by embedding dimension that is referred as a subtask Ω_m. The cascaded network architecture can also be viewed as an ensemble of neural networks that feature distinct topologies as shown in Fig. 1. The output layer employs one neurons in all the respective subtasks since the overall problem is one-step ahead prediction. The input-hidden layer ω_m weights and the hidden-output layer υ_m weights are combined for the respective network module Φ_m. The base module is given as $\Phi_1 = [\omega_1, \upsilon_1]$. Multi-task learning is used via coevolution to update the respective network module Φ_m given subtask Ω_m. Note that the cascaded network module θ_m of subtask m is constructed by combining with current Φ_m and previous network module Φ_{m-1} as follows.

$$\Phi_1 = [\omega_1, \upsilon_1]; \quad \theta_1 = (\Phi_1)$$
$$\Phi_2 = [\omega_2, \upsilon_2]; \quad \theta_2 = [\theta_1, \Phi_2]$$
$$\vdots$$
$$\Phi_M = [\omega_M, \upsilon_M]; \quad \theta_M = [\theta_{M-1}, \Phi_M]$$

$$\tag{2}$$

The list of network modules considered for training or optimisation is therefore
$\Phi = (\Phi_1, \ldots, \Phi_M)$.

$$y_1 = f(\theta_1, \Omega_1)$$
$$y_2 = f(\theta_2, \Omega_2)$$
$$\vdots$$
$$y_M = f(\theta_M, \Omega_M)$$

(3)

The loss L for subtask m can be calculated by root mean squared error.

$$L_m = \sqrt{\frac{1}{N} \sum_{i=1}^{N} (\hat{y}_{m_i} - y_i)^2}$$

(4)

where y is the observed time series and $\hat{y}_m$ is the prediction given by subtask m, and N is the number of samples.

Cooperative coevolution breaks down a problem into modules that are implemented as sub-population. Through the problem formulation via multi-task learning, it is natural to use sub-populations to optimise the given network modules Φ_m. The sub-populations are given by $S_1, S_2, ..S_M$, where M is number of subtasks. The individuals that make up the sub-populations that consist of respective network module m, $S_m = \Phi_m$, where $\Phi_m = [\omega_m, \upsilon_m]$. The sub-populations $S_{m_{ij}}$ consist of a pool of P individuals that are referenced by i. j is used to reference the elements in the individuals that consist of weights and biases in the network module. The individuals of the sub-population are also referred as genotype while the corresponding network module are referred as the phenotype.

Algorithm 1 gives further details which begins by initialising all the components which include the sub-populations S_m. The sub-populations S_m are initialised with real values in a range $[-\alpha, \alpha]$ drawn from uniform distribution where α defines the range. Once this has been done, the algorithm moves into the evolution phase where each subtask is evolved for a fixed number for generations defined by depth of search, β. The major issues of concern here is the way the phenotype is mapped into genotype where a group of weight matrices given by $\Phi_m = [\omega_m, \upsilon_m]$ that makes up subtask θ_m are converted into vector X_m. Next, the notion of utilizing knowledge from previous subtasks through multi-task learning is executed. In the case if the subtask is a base problem ($m == 1$), then the subtask solution X_m is utilised in a conventional matter where knowledge from other subtasks or modules are not needed. Given that the subtask is not a base problem, the current subtask individual X_m is appended with best individuals from the previous subtasks, therefore, $X_m = [B_1, ..., B_{m-1}, V_m]$ where B is the best individual from previous subtask and V is the current individual that needs to be evaluated. The algorithm then makes subtask prediction

---

Algorithm 1. Co-evolutionary multi-task learning

Data: Reconstructed state-space vector for the respective subtasks Ω_m.
Result: Prediction error for the respective subtasks Ω_m.
initialisation
for *each subtask Ω_m* **do**
  | 1. Define different sub-populations S_m using network module Φ_m
  |
  | 2. Initialise all individuals in sub-population S_m
end
while *each phase until termination* **do**
  | **for** *each sub-population S_m* **do**
  | | **for** *each generation until depth β* **do**
  | | | **for** *each i individual in sub-population $S_{m_{ij}}$* **do**
  | | | | **for** *each j in individual $S_{m_{ij}}$* **do**
  | | | | | Assign individual $V_m = S_{m_{ij}}$
  | | | | **end**
  | | | | **if** *m == 1* **then**
  | | | | | 1. $Z_m = [V_m]$
  | | | | | 2. Fitness evaluation via Eq. 4 by encoding Z_m in the cascaded network module θ_m
  | | | | **end**
  | | | | **else**
  | | | | | 1. Append to best Individual B_{m-1} of previous sub-population: $Z_m = [V_m, B_{m-1}]$
  | | | | | 2. Fitness evaluation via Eq. 4 by encoding Z_m in the cascaded network module θ_m
  | | | | **end**
  | | | **end**
  | | | **for** *each i individual in sub-population $S_{m_{ij}}$* **do**
  | | | | * Select and create new offspring via evolutionary operators:
  | | | | 1. Selection, 2. Crossover, and 3. Mutation
  | | | **end**
  | | **end**
  | **end**
end
for *each subtask Ω_m* **do**
  | 1. Get best solution B_m from sub-population S_m
  | 2. Load test data
  | 3. Report prediction performance given by loss L_m.
end

$\hat{y}_m = f(\theta_m, \Omega_m)$ and evaluates it though the loss function given in Eq. 4. This procedure is executed for every individual in the sub-population and repeated for every sub-population until the termination condition is satisfied. The termination condition can be either the maximum number of function evaluations or a minimum fitness value (loss) from the training or validation dataset. Figure 1 shows an ensemble of the network modules, however, they are part of one cascaded network architecture. An implementation in Matlab has been given online[1].

---

[1] https://github.com/rohitash-chandra/CMTL_dynamictimeseries.

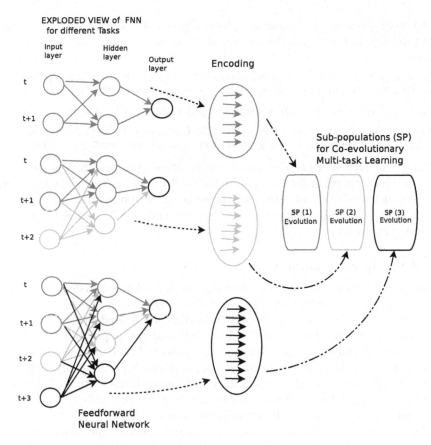

Fig. 1. Subtasks in co-evolutionary multi-task learning. Subtask 1 employs a network topology with 2 hidden neurons while the rest of the subtasks add extra input and hidden neurons.

3   Simulation and Results

This section presents an experimental study that compares the performance of CMTL with conventional evolutionary (single task learning) methods such as cooperative neuro-evolution (CNE) and an evolutionary algorithm (EA) for dynamic time series prediction. Tropical cyclones from South Pacific and South Indian Ocean are considered in order to address the minimal timespan problem as previously identified [9].

3.1   Cyclone Time Series

The Southern Hemisphere tropical cyclone best-track data from Joint Typhoon Warning Center recorded every 6-h is used as the main source of data [14]. The austral summer tropical cyclone season (November to April) from 1985 to 2013 is considered in the current study as data prior to the satellite era is not reliable due to inconsistencies and missing values. The original data of tropical cyclone wind intensity in the South Pacific was divided into training and testing set as follows:

- Training Set: Cyclones from 1985–2005 (219 Cyclones with 6837 data points)
- Testing Set: Cyclones from 2006–2013 (71 Cyclones with 2600 data points)

In the case for South Indian Ocean, the details are as follows:

- Training Set: Cyclones from 1985–2001 (285 Cyclones with 9365 data points)
- Testing Set: Cyclones from 2002–2013 (190 Cyclones with 8295 data points)

Note that although the cyclones are separate events, the cyclones are combined in a consecutive order as given by their data of occurrence. The time series is reconstructed with set of values for the embedding dimensions, $D_m = \{3, 5, 7\}$ that refers to the respective subtasks Ω_m. The time lag $T = 2$ is used for the respective subtasks.

3.2   Experimental Design

In the case of cooperative neuro-evolution, neuron level problem decomposition is applied for training feedforward networks [5] on the given problems. Covariance matrix adaptation evolution strategies (CMAES) [12] is used as the evolutionary algorithm in sub-populations of CMTL, CNE and the population of EA. The training and generalisation performances are reported for each case given by the different tasks in the respective time series problems. Note that only CMTL can be used to approach dynamic time series problem with the feature of multi-task learning, however, the results for single task learning methods (CNE and EA) are given in order to provide baseline performance.

The respective neural networks used sigmoid units in the hidden and output layer for all the different problems. The loss given by root mean squared error (RMSE) in Eq. 4 is used as the main performance measure. Each neural network architecture was tested with different numbers of hidden neurons. The depth of search, $\beta = 5$ generations is used in the sub-populations of CMTL and CNE. Note that all the sub-populations evolve for the same depth of search. The population size of the CMAES in all the respective methods is given by $P = 4 + floor(3 * log(\gamma))$, where γ is the total number of weights and biases in the cascaded network architecture. The termination condition is fixed at 30 000 function evaluations for each task, hence, CMTL employs 120 000 function evaluations while single task learning methods use 30 000 for each of the respective subtasks for all the problems. Note that since there is a fixed training time, there was no validation set used to stop training.

3.3   Results

The results are presented for the performance of the given methods on the two selected cyclone problems, which focus on South Pacific and South Indian ocean as shown in Figs. 3 and 2, respectively. In the case of the South Pacific ocean, the results show that CMTL provides the best generalisation performance when compared to CNE and EA. This is also observed for the South Indian ocean.

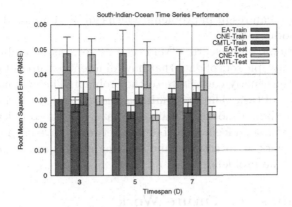

Fig. 2. Performance given by EA, CNE, CMTL for South Indian Ocean

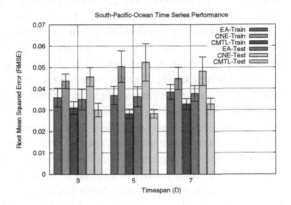

Fig. 3. Performance given by EA, CNE, CMTL for South Pacific Ocean

3.4 Discussion

The goal of the experiments was to evaluate if the proposed algorithm can deliver similar results when compared to single task learning methods for the dynamic time series problems. Therefore, the comparison was to ensure that the approach does not lose quality in terms of generalisation performance when compared to single-task learning methods. The results have shown that CMTL not only addresses dynamic time series, but also improves the performance when compared to single-task learning. This could be seen as a form of developmental learning [17,20] which enables modularity for dynamic time series prediction as demonstrated through the case of cyclones. The feature of modularity was implemented through the network modules that are combined as the nature or complexity of the problem increases. Modularity helps in making a prediction given knowledge from base network module in cases if the information in the other subtasks are not available. Modularity is important for the design of neural networks tailored for hardware implementation [16]. The disruptions in certain synapse(s) can result in problems which can be eliminated by persevering knowledge as modules [8].

Being evolutionary in nature, CMTL can be seen as a flexible method that can be used for multiple sets of data that have different features of which some are overlapping and distinctly contribute to the problem. The common features can be captured as a subtask. Through multi-task learning, the overlapping features can be used as building blocks to learn the nature of the problem through the model at hand. Although feedforward neural networks have been used in CMTL, other neural network architectures, and learning models can be used depending on the nature of the problem.

4   Conclusions and Future Work

A novel algorithm has been presented that provides a synergy between neuro-evolution and multi-task learning for dynamic time series problems. Co-evolutionary multi-task learning is needed for tropical cyclones where robust prediction needs to be given as soon as the event takes place. In the cyclones studied, each point in a given timespan represented six hours. Therefore, the proposed algorithm implemented a model that can work even when a single point is implemented as a subtask. As more points of data are given to the subtask, more predictions can be made which makes the model dynamic and robust. The results show that the proposed algorithm not only addressed dynamic time series, but also improved prediction performance when compared to conventional methods.

In future work, the proposed algorithm can be used for other time series problems that can be broken into multiple subtasks. It can also be extended to transfer learning problems that can include both heterogeneous and homogeneous domain adaptation cases. In case of tropical cyclones, which is a multivariate problem, the different subtasks can be redefined with information such as cyclone tracks, seas surface temperature, and humidity.

References

1. Ali, M.M., Jagadeesh, P.S.V., Lin, I.I., Hsu, J.Y.: A neural network approach to estimate tropical cyclone heat potential in the Indian ocean. IEEE Geosci. Remote Sens. Lett. **9**(6), 1114–1117 (2012)
2. Angeline, P., Saunders, G., Pollack, J.: An evolutionary algorithm that constructs recurrent neural networks. IEEE Trans. Neural Netw. **5**(1), 54–65 (1994)
3. Caruana, R.: Multitask learning. Mach. Learn. **28**(1), 41–75 (1997)
4. Chandra, R.: Competition and collaboration in cooperative coevolution of Elman recurrent neural networks for time-series prediction. IEEE Trans. Neural Netw. Learn. Syst. **26**, 3123–3136 (2015)
5. Chandra, R., Frean, M., Zhang, M.: On the issue of separability for problem decomposition in cooperative neuro-evolution. Neurocomputing **87**, 33–40 (2012)
6. Chandra, R., Gupta, A., Ong, Y.-S., Goh, C.-K.: Evolutionary multi-task learning for modular training of feedforward neural networks. In: Hirose, A., Ozawa, S., Doya, K., Ikeda, K., Lee, M., Liu, D. (eds.) ICONIP 2016. LNCS, vol. 9948, pp. 37–46. Springer, Cham (2016). doi:10.1007/978-3-319-46672-9_5

7. Chandra, R., Zhang, M.: Cooperative coevolution of Elman recurrent neural networks for chaotic time series prediction. Neurocomputing **186**, 116–123 (2012)

8. Clune, J., Mouret, J.B., Lipson, H.: The evolutionary origins of modularity. In: Proceedings of the Royal Society of London B: Biological Sciences, vol. 280, no. 1755 (2013)

9. Deo, R., Chandra, R.: Identification of minimal timespan problem for recurrent neural networks with application to cyclone wind-intensity prediction. In: International Joint Conference on Neural Networks (IJCNN), pp. 489–496, July 2016

10. Evgeniou, T., Micchelli, C.A., Pontil, M.: Learning multiple tasks with kernel methods. J. Mach. Learn. Res. **6**(Apr), 615–637 (2005)

11. Frazier, C., Kockelman, K.: Chaos theory and transportation systems: instructive example. Transp. Res. Rec.: J. Transp. Res. Board **20**, 9–17 (2004)

12. Hansen, N., Müller, S.D., Koumoutsakos, P.: Reducing the time complexity of the derandomized evolution strategy with covariance matrix adaptation (CMA-ES). Evol. Comput. **11**(1), 1–18 (2003)

13. Happel, B.L., Murre, J.M.: Design and evolution of modular neural network architectures. Neural Netw. **7**(6–7), 985–1004 (1994)

14. JTWC: Joint Typhoon Warning Center - tropical cyclone best track data site. http://www.usno.navy.mil/NOOC/nmfc-ph/RSS/jtwc/ (2017). Accessed 16 Aug 2017

15. Knaff, J.A., DeMaria, M., Sampson, C.R., Gross, J.M.: Statistical, five-day tropical cyclone intensity forecasts derived from climatology and persistence. Weather Forecast. **18**, 80–92 (2003)

16. Misra, J., Saha, I.: Artificial neural networks in hardware: a survey of two decades of progress. Neurocomputing **74**(1–3), 239–255 (2010). Artificial Brains

17. Morse, A.F., De Greeff, J., Belpeame, T., Cangelosi, A.: Epigenetic robotics architecture (ERA). IEEE Trans. Auton. Mental Dev. **2**(4), 325–339 (2010)

18. Potter, M.A., De Jong, K.A.: Cooperative coevolution: an architecture for evolving coadapted subcomponents. Evol. Comput. **8**(1), 1–29 (2000)

19. Potter, M.A., De Jong, K.A.: A cooperative coevolutionary approach to function optimization. In: Davidor, Y., Schwefel, H. P., Männer, R. (eds.) PPSN 1994. LNCS, vol. 866, pp. 249–257. Springer, Heidelberg (1994). doi:10.1007/3-540-58484-6_269

20. Prince, C., Helder, N., Hollich, G.: Ongoing emergence: a core concept in epigenetic robotics. In: Proceedings of the Fifth International Workshop on Epigenetic Robotics: Modeling Cognitive Development in Robotic Systems, pp. 63–70. Lund University Cognitive Studies (2005)

21. Stiles, B.W., Danielson, R.E., Poulsen, W.L., Brennan, M.J., Hristova-Veleva, S., Shen, T.P., Fore, A.G.: Optimized tropical cyclone winds from QuikSCAT: a neural network approach. IEEE Trans. Geosci. Remote Sens. **52**(11), 7418–7434 (2014)

22. Takens, F.: Detecting strange attractors in turbulence. In: Rand, D., Young, L.-S. (eds.) Dynamical Systems and Turbulence, Warwick 1980. LNM, vol. 898, pp. 366–381. Springer, Heidelberg (1981). doi:10.1007/BFb0091924

23. Zeng, T., Ji, S.: Deep convolutional neural networks for multi-instance multi-task learning. In: 2015 IEEE International Conference on Data Mining (ICDM), pp. 579–588, November 2015

24. Zheng, H., Geng, X., Tao, D., Jin, Z.: A multi-task model for simultaneous face identification and facial expression recognition. Neurocomputing **171**, 515–523 (2016)

25. Zjavka, L.: Numerical weather prediction revisions using the locally trained differential polynomial network. Expert Syst. Appl. **44**, 265–274 (2016)

Social Networks

Layer-Prioritized Influence Maximization
in Social Networks

Qianwen Zhang, Yuzhu Wu, and Jinkui Xie[(⊠)]

Department of Computer Science and Technology, East China Normal University,
Shanghai 200062, China
zqw1005@126.com, cstxpxz@163.com, jkxie@cs.ecnu.edu.cn

Abstract. Influence maximization, first proposed by Kempe, is the
problem of finding seed nodes that maximizes the number of affected
nodes. However, not only influenced number, but also influence layer
is a crucial element which may play an important role in viral marketing. In this paper, we design a new framework, *layer-prioritized influence
maximization* (LPIM), to address the problem of influence maximization
with an emphasis on influence layer. The proposed framework is mainly
composed of three parts: (1) graph clustering. (2) key node selection.
(3) seed node detecting. We also demonstrate the effective and efficient
of our proposed framework by experiments on large collaboration networks and complexity analysis respectively.

Keywords: Social networks · Influence maximization · Layer-
prioritized

1 Introduction

In recent years, various online social networks have emerged. Many social networks, such as Facebook, Google+, Flickr, Weibo, and Youtube, help strengthen
individuals' relationships online, and make it easy to propagate information via
word-of-mouth effect. This phenomenon has been found useful for viral marketing, social influence maximization, etc. For example, to promote a film, a
company may give some free tickets to influential users, hoping they will recommend to their friends. The problem of finding individuals who can trigger
maximum adoptions was defined as *Influence Maximization Problem* (IM problem) by Kempe [1], attracting a lot of research interest [2–5].

While modeling the process of influence propagation, there are two basic diffusion models: *Linear Threshold Model* (LT model) proposed by Granovetter
and Schelling [6,7], and *Independent Cascade Model* (IC model) proposed by
Goldenberg et al. [8,9]. Several studies aim at addressing the IM problem under
these models. In Ref. [10], Kimura and Saito presented shortest-path based on IC
model for finding sets of influential nodes. In Ref. [11], Chen et al. further improved
the greedy algorithm, greatly decreasing the time while keeping closed influence
spread. They continued to propose PMIA model in Ref. [12], maintaining good

© Springer International Publishing AG 2017
D. Liu et al. (Eds.): ICONIP 2017, Part V, LNCS 10638, pp. 631–638, 2017.
https://doi.org/10.1007/978-3-319-70139-4_64

balance between efficiency and effectiveness, which is a popular algorithm to select seed sets. Purohit et al. proposed influence-based coarsening for networks, and obtained greatly speed-up for tackling IM problem [13]. In Ref. [14], Chen et al. selected influential individuals by exploring the community structures, improving efficiency and scalability with almost no compromise of effectiveness.

The above work mainly focus on the number of affected individuals and time consumption for solving the problem. However, besides the affected number, affected range is also a key target. Consider the following scenario as a motivating example. To issue some alert, like severe weather forecasting or disease prevention, concerned organizations may particularly inform several people, with the hope to spread widely. In this situation, it is reasonable to consider influence range besides the total number of influenced people. In this paper, we introduce influence layer as an indicator to evaluate affected range. Furthermore, to address the problem of influence maximization with an emphasis on influence layer, we design a new framework, *layer-prioritized influence maximization* (LPIM), which comprises three phase: (1) graph clustering, (2) key node selection, (3) seed sub-graph(node) detecting. Since we hope the chosen individuals spread out as much as possible, we adopt spectral clustering methods in phase (1). The two remaining questions how to select key node and detect seed subgraph will be addressed in Sect. 2.

Organization. In Sect. 2, we describe the research problem, further detail the LPIM framework and associated algorithms. In Sect. 3, we present the experiments on several real datasets. We conclude the paper in Sect. 4.

2   Layer-Prioritized Influence Maximization

In this section, we first provide a review of the influence maximization problem, including its definition and popular IC model. Then we present the proposed layer-prioritized influence maximization framework and our approaches dealing with the arising issue.

2.1   Influence Maximization: Review

We consider a social network to be an influence graph $G = (V, E)$, where each vertex $v \in V$ represents a user, and $e \in E$ represents the link between them. For every edge $(i, j) \in E$, p_{ij} denotes the probability that j is activated by i through the edge after i is activated.

The Independent Cascade model is a popular diffusion model used to model the influence propagation. Given a seed set S, the IC model works as follows. Let S_n be the set of vertices that are activated in the nth round, with $S_0 = S$. In the $n + 1$th round, each newly activated vertex v_i may activate its neighbor v_j which is not yet activated with an independent probability of p_{ij}. This process is repeated until S_n is empty. Note that each activated vertex only has one chance to activate its neighbors. Use $\sigma(S)$ to denote the expected number of active vertices when the process finishes, which we call *influence spread*.

Let X represents a set of vertices, $f(X)$ denotes the vertices activated by X. During the random process of propagation in the IC model, given a seed set S, the influence spread can be written as follows:

$$\begin{cases} I^0(S) = \emptyset, \quad I^1(S) = S, \\ I^{n+2}(S) = I^{n+1}(S) \cup f(I^{n+1}(S) \backslash I^n(S)) \quad \text{for } n \geq 1. \end{cases}$$

where $I^n(S)$ denotes the set of activated vertices at nth round by the seed set S. The expected influence spread $\sigma(S)$ is $|I^\infty(S)|$. Based on the above definition, we formalize the IM problem as below:

Problem 1 (IM Problem). For a graph $G = (V, E)$ and a positive integer k, compute the seed set

$$S^* = \underset{S \subset V \wedge |S| = k}{\arg\max} \ \sigma(S).$$

2.2 Layer-Prioritized Influence Maximization Framework

Original algorithms of influence maximization target at finding individuals that maximize the influence spread. However, the chosen individuals may gather under specific situation. For example, two individuals have a common social circle, and both of them are influential among their friends. Some previous work may choose both of them, because of their high-impact, while ignoring the distribution of the chosen seeds. Based on the above observation, we define *influence layer* to evaluate the range of influenced area as below:

Definition 1 (influence layer). *Given seed nodes S, the influence layer $\phi(S)$ refers to the sum of length of distinct longest path influenced by each seed node under influence cascade model.*

Our proposed LPIM framework, emphasizing on influence layer, aims at finding seeds with appropriate distribution and less compromise of influence spread. Figure 1 shows the overview of the LPIM framework, including three phases: (1) graph clustering. Since we want the chosen individuals spread out as much as possible, we adopt spectral clustering methods to obtain subgraphs; (2) key node selection. For each subgraph, we select an influential node; (3) seed subgraph detecting. Among all subgraphs who have different status in social graph, we detect the most influential subgraphs, whose quantity equals to the number of seed nodes defined in IM problem.

Key Node Selection. Given subgraphs, phase (2) of PLIM is to select an influential node for each subgraph. But how to identify the high-impact node in subgraph is a problem. In this part, we give two strategies of selecting key node. One is to regard the node with highest degree as key node. Another strategy is based on the observation of influence propagation, we consider the node who have close connection with others to be influential nodes. Specifically, we adopt the thought of random walk, choose a few nodes to be initial points, and let them propagate influence for T steps. By repeating the process, the node who has been affected most times is considered to be influential.

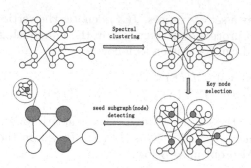

Fig. 1. Overview of the LPIM framework

Formally speaking, since each vertex has a chance to influence its neighbor under IC model, we define simulate influence probability $q_{ij}(t)$ from jth vertex to ith vertex at tth step as

$$q_{ij}(t) = \begin{cases} 1 & \text{if } rand() \leq p_{ji} \\ 0 & \text{otherwise} \end{cases},$$

then we denote by $x_i(t)$ whether the ith vertex is newly influenced at step t as

$$x_i(t) = \begin{cases} 1 & \text{if } \sum_j q_{ij}(t)x_j(t-1) \geq 1 \\ 0 & \text{otherwise} \end{cases},$$

with the initial value where $x_i(0) = 1$ if ith vertex chosen to be initial point. Then, we have

$$Y = \sum_{t=1}^{T} Q(t) \cdot X(t-1), \tag{1}$$

in which

- $Q(t) = \{q_{ij}(t)\}_{n \times n}$ is a matrix consisting of simulate influence probability at tth step,
- $X(t)$ is the column vector consisting of each $x_i(t)$, viz. $X(t) = [x_1(t), x_2(t), \ldots, x_n(t)]^{\text{T}}$.
- Y is the column vector consisting of each y_i, which represent how many times the ith vertex has been influenced during the whole process.

As mentioned, we repeat the process several times, then regard the node with biggest y_i as key node.

Seed Subgraph Detecting. In phase (3), we aim at detecting important subgraph among all clusters in phase (1). Since finding nodes with as much influence as possible is one of our purpose, we decide using PMIA [12] which is a fast and popular algorithm among existing influence maximization algorithms to identify seed subgraphs. The problem is how to define the weight between each subgraph which we consider as a vertex.

Notice in phase (2), we have selected key node for each subgraph, the procedure of influence propagation between each subgraph could be treated as the procedure between each key node of subgraph from the overall framework.

For a shortest path $SP = <u = a_1, a_2, \cdots, a_d = v>$, we define propagation probability, $P(SP_{u,v})$ as

$$P(SP_{u,v}) = \prod_{i=1}^{d-1} p(a_i, a_i + 1)$$

For two subgraphs c_i and c_j, let v_i and v_j to denote chosen seed node respectively. For an edge connecting subgraphs with vertices $u_1 \in c_i$ and $u_2 \in c_j$, we define propagation probability of the effective path, $P(EP_{v_i,v_j})$ as

$$P(EP_{v_i,v_j}) = P(SP_{v_i,u_1}) \cdot p(u_1, u_2) \cdot P(SP_{u_2,v_j}) \tag{2}$$

Figure 2 shows how we define the weight between two subgraphs. Since we have selected the key node of each subgraph, we convert the subgraph to a tree with the chosen node as root. According to the form of (2), we further calculate the propagation probability of the effective path, e_1 and e_2 in this example. Then we abstract the relationship between two subgraphs as average propagation probability of all effective paths between them.

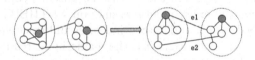

Fig. 2. Subgraph transform

Then we use PMIA algorithm on the new graph, whose vertices are key nodes we selected in phase (2). Once we detect the seed subgraphs, key nodes of these subgraphs are final seed nodes we target at.

Based on three phases, the algorithm for our proposed LPIM framework is shown in Algorithm 1.

Algorithm 1. LPIM(G, k)

Input: Graph of a social network $G(V, E)$; number of seeds set k
Output: k seeds
1: $C_s = \{c_1, c_2, \cdots, c_m\} \leftarrow$ spectral clustering
2: **for** each $c_i \in C_s$ **do**
3:     $v_i \leftarrow$ key node selection
4:     $c'_i \leftarrow$ converted rooted tree of c_i
5: **end for**
6: update weight between each pair of v_i
7: $V' = \{v_1, v_2, \cdots, v_m\}$
8: $S \leftarrow$ PMIA on $G'(V', E')$
9: **return** S;

---

3 Experiments

We conduct experiments on several real-life networks. Our experiments aim at illustrating the effectiveness of our proposed LPIM.

Datasets. We use three real social network datasets. Two collaboration networks NetHEPT and NetPHY are obtained from arXiv.org in the High Energy Physics Theory and Physics domains respectively. Another dataset is DBLP, which is an academic collaboration network. In these datasets, each vertex in the network represents an author, and an edge between a pair of vertices represents their co-authorship. These datasets are commonly used in the literature of influence maximization [1,11,15,16].

Since our algorithm base on general IC model in which weighted cascade model is usually adopted to obtain the probabilities [1]. We set $p_{ij} = 1/d(j)$ for an edge e_{ij}, where $d(j)$ is the in-degree of jth vertex.

We compare our LPIM algorithm with several other methods. LPIM(Degree) is our algorithm in which selects highest degree node in phase (2) for LPIM model. LPIM(RandomWalk) is our algorithm with another strategy of choosing key node. PMIA [12], proposed by Chen et al. is a heuristic algorithm which is a fast and popular solution to influence maximization problem. Pagerank is a well known algorithm for ranking web pages [17]. We also adopt Random algorithm that selects k random vertices in the graph as a baseline comparison.

Figure 3(a) shows the experimental results on NETHEPT, The x-axis indicates number of seed nodes and y-axis indicates influence spread, namely the number of affected nodes. In Fig. 3(b), the y-axis indicates influence layer. For graph NETHEPT, different algorithms other than random method have similar performance, which may result from the structure of graph. Figure 3(c) and (d) show the result on NETPHY dataset. It is clearly that LPIM(Degree) perform best among these methods on both influence spread and influence layer. LPIM(RandomWalk) is slightly better than PMIA with regard to influence layer, but simulated influence spread is worse than LPIM(Degree) and PMIA method. The results on DBLP dataset is similar to the NETPHY dataset.

Based on the experimental results, we find LPIM(Degree) is better than LPIM(RandomWalk). The phenomenon may be explained as that when we use the idea of random walk to select key node, top p nodes perform stable while we only choose one node for each subgraph. So we adopt selecting key node with highest degree in phase (2), which is better and faster.

Time Complexity. In this part, we present an analysis of the time complexity of the algorithm LPIM. Given a network with n vertices and m edges, we first adopt a spectral clustering algorithm CASP [18] for graph clustering, whose time complexity is $O(p^3) + O(nlogn)$, where p is the number of clusters and $(p << n)$. While selecting key node of each subgraph by choosing the node with highest degree, time complexity is $O(n)$. In phase(3), the time complexity of calculating the weight between connected subgraphs is $O(mlogm)$ whereas using PMIA to choose k influential vertices on construct graph cost $O(kp^2logp)$. So, the total time complexity is within $O(nlogn) + O(mlogm) + O(p^3)$.

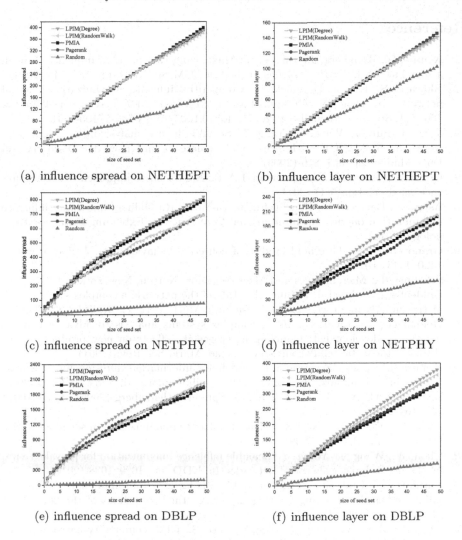

(a) influence spread on NETHEPT

(b) influence layer on NETHEPT

(c) influence spread on NETPHY

(d) influence layer on NETPHY

(e) influence spread on DBLP

(f) influence layer on DBLP

Fig. 3. Experimental results

4   Conclusion

In this paper, we have proposed the LPIM framework to solve the IM problem emphasizing on influence layer which we defined to evaluate the range of influence propagation. Experiments showed our algorithm performed well on both influence layer and influence spread compared to other methods.

For further work, to study the influence graph more realistically, we are interested at designing a clustering algorithm to obtain better clusters. Furthermore, we also intend to study the strategy of choosing seed nodes whose influence propagation could achieve maximal influence layer.

References

1. Kempe, D., Kleinberg, J., Tardos, E.: Maximizing the spread of influence through a social network. In: Proceedings of the 9th ACM SIGKDD, pp. 137–146 (2003)
2. Bharathi, S., Kempe, D., Salek, M.: Competitive influence maximization in social networks. In: Deng, X., Graham, F.C. (eds.) WINE 2007. LNCS, vol. 4858, pp. 306–311. Springer, Heidelberg (2007). doi:10.1007/978-3-540-77105-0_31
3. Tang, J., Sun, J., Wang, C., Yang, Z.: Social influence analysis in large-scale networks. In: ACM SIGKDD International Conference on Knowledge Discovery and Data Mining, pp. 807–816 (2009)
4. Goyal, A., Bonchi, F., Lakshmanan, L.V.S.: A data-based approach to social influence maximization 5(1) (2011)
5. Aslay, C., Barbieri, N., Bonchi, F., Baeza-Yates, R.: Online topic-aware influence maximization queries. In: International Conference on Extending Database Technology (2014)
6. Granovetter, M.: Threshold models of collective behavior. Am. J. Sociol. 83(6), 1420–1443 (1978)
7. Schelling, T.: Micromotives and Macrobehavior. Norton, New York (1978)
8. Goldenberg, J., Libai, B., Muller, E.: Talk of the network: a complex systems look at the underlying process of word-of-mouth. Mark. Lett. 12(3), 211–223 (2001)
9. Goldenberg, J., Libai, B.: Using complex systems analysis to advance marketing theory development: modeling heterogeneity effects on new product growth through stochastic cellular automata. Acad. Mark. Sci. Rev. (2001)
10. Kimura, M., Saito, K.: Tractable models for information diffusion in social networks. In: Fürnkranz, J., Scheffer, T., Spiliopoulou, M. (eds.) PKDD 2006. LNCS (LNAI), vol. 4213, pp. 259–271. Springer, Heidelberg (2006). doi:10.1007/11871637_27
11. Chen, W., Wang, Y., Yang, S.: Efficient influence maximization in social networks. In: KDD, pp. 199–208 (2009)
12. Chen, W., Wang, C., Wang, Y.: Scalable influence maximization for prevalent viral marketing in large-scale social networks. In: KDD, pp. 1029–1038 (2010)
13. Purohit, M., Prakash, B.A., Kang, C., Zhang, Y., Subrahmanian, V.S.: Fast influence-based coarsening for large networks. In: Proceedings of the 20th ACM SIGKDD, pp. 1296–1305 (2014)
14. Chen, Y., Zhu, W., Peng, W., Lee, W., Lee, S.: CIM: community-based influence maximization in social networks. ACM Trans. Intell. Syst. Technol. 5(2), 25 (2014)
15. Lei, S., Maniu, S., Mo, L., Cheng, R., Senellart, P.: Online influence maximization. In: Proceedings of the 20th ACM SIGKDD, pp. 645–654 (2014)
16. Zhang, Q., Huang, C.-C., Xie, J.: Influence spread evaluation and propagation rebuilding. In: Hirose, A., Ozawa, S., Doya, K., Ikeda, K., Lee, M., Liu, D. (eds.) ICONIP 2016. LNCS, vol. 9948, pp. 481–490. Springer, Cham (2016). doi:10.1007/978-3-319-46672-9_54
17. Brin, S., Page, L.: The anatomy of a large-scale hypertextual Web search engine. In: International Conference on World Wide Web, pp. 107–117 (1998)
18. Zhu, M., Meng, F., Zhou, Y., Yuan, G.: An approximate spectral clustering for community detection based on coarsening networks. Int. J. Adv. Comput. Technol. 4(4), 235–243 (2012)

Design of Traffic Signal Controller Based on Network

Xiaoming Liu, Yulin Tian$^{(\boxtimes)}$, Chunlin Shang, Peizhou Yan, and Lu Wei

Beijing Key Lab of Urban Road Traffic Intelligent Technology,
North China University of Technology, Beijing, China
tian_yulin@yeah.net

Abstract. In the paper, a traffic signal controller based on network was designed after analyzing the development and actuality of traffic signal controllers. The controller consisted of Server, Network Bus and IP Nodes. Power line carrier communication module was designed as the medium of network communication, which could meet the requirements of traffic control network for network bandwidth and communication distance. The software based on Firework computing paradigm can achieve efficient data processing and improve the coordination and optimization ability of traffic control system. With precise actual control effect, the controller can reduce equipment cost, lower difficulties of upgrading and maintenance, and provide complete data support for collaborative optimization of traffic network.

Keywords: Network · Traffic signal controller · Firework

1 Introduction

With the combination of informatization and industrialization, traffic signal control systems are developing toward network and intelligent system [1], traffic intersections gradually become the data conversion nodes in the traffic information network, traffic operation systems are shifting from "automatic control based on traffic signal" to "intelligent computing based on traffic data" [2].

In current traffic control systems, most of the peripheral devices are still unintelligent, each device needs a set of independent wires for communication or being controlled. For example, each group of traffic lights need 4 wires to complete its assembly, more than 60 wires are needed to complete all the lights assembly in a traffic intersection, which leads to the high cost of wires and difficulties in construction, maintenance and promotion. The traffic signal controllers carry out basic logical control command to the underlying devices by predefined rules [3]. Most of the control methods are fixed-rule control models which have limited control ability of coordinated optimization. The traffic controllers can't store and reuse important traffic data, and the network can't transmit large amounts of data to the center because of centralized network structure. This leads to the traffic

© Springer International Publishing AG 2017
D. Liu et al. (Eds.): ICONIP 2017, Part V, LNCS 10638, pp. 639–646, 2017.
https://doi.org/10.1007/978-3-319-70139-4_65

data which is crucial to the optimization of the intersection signal can't be utilized effectively, and hampered the optimal control of traffic signals seriously [4].

A traffic signal controller based on network, designed in this paper, was made up of Server, Network Bus and IP Nodes. The "Server" was the Integrated Traffic Control Server (ITCS) with powerful capability of computing, network communication and data storage, which could achieve remote processing and prompt storage of traffic intersection real-time data. "IP Nodes" were IP-based peripheral traffic devices which could achieve network communication, state detection, device driving and data computing, which made the traditional traffic peripherals become the Intelligent IP Nodes (IIPU) in the traffic network. "Network Bus" achieved the network connection between the ITCS and the IIPUs by using power line communication technology, which could meet the requirements of communication bandwidth and communication distance in traffic signal network control system. The system topology structure could not only reduce cost and difficulty of construction and maintenance, but also improve the intelligent level of equipment and the efficiency of system operation (Fig. 1).

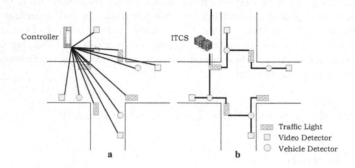

Fig. 1. The comparison of traditional traffic control system (a) and network control system (b) in engineering.

2   Hardware Design

The hardware design of the traffic signal controller based on network included two parts, ITCS and IIPU. ITCS was the core of the controller which was composed of CPU module, power-supply module, network switching module, data storage module, indicating module and network bus interface module. The design of IIPU consisted of two parts: the design of network interface for connection with network bus and the drive circuit for driving traditional traffic facilities. The network bus was constructed by power line carrier communication module which could achieve network communication between IIPUs and ITCS (Fig. 2).

The CPU module was the data processing and signal control center of the controller, and responsible for the unified and coordinated control of each module in the controller. The power-supply module included AC and DC power supply

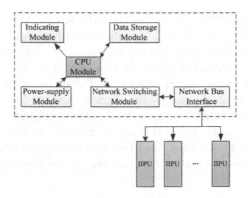

Fig. 2. The overall composition of the hardware.

sections, two sections generate AC and DC (24 V, 12 V, 5 V) outputs for other devices and modules. The storage module was designed for saving important traffic data. The indicating module was used to achieve the real-time display of the traffic signal state. The network switching module provided network support for ITCS and IIPUs in the controller, and provided the external network interface at the same time. Network bus interface module provided the network communication bus for external equipment, and could make it a networked control system by connecting IIPUs with controller. The IIPUs could accept control instructions from ITCS through the network bus, at the same time upload the traffic data such as vehicle flow data, traffic phase state and running state of IIPUs, thus the controller could provide traffic data support for the whole traffic network control system, and make the traffic signal control system a more efficient system.

2.1 CPU Module

The CPU module needed to monitor running state of other modules, generate traffic signal control schemes, and process the traffic data. Therefore, the module should have a powerful ability of computing. The CPU module used the quad core high performance industrial chip T1040 produced by Freescale as processor chip which could receive, analyze and process large number of traffic real-time data. T1040 has four cores built on Power Architecture technology, and each core has a private 256 KB L2 cache. The CPU module had an integrated 8-port Gigabit Ethernet switch which support dynamic learning of MAC addresses and aging. The module had a rich peripheral interface, such as 2 high-speed USB 2.0 controllers with integrated PHY, 4 I2C controllers and 1 Enhanced Serial peripheral interface (eSPI).

2.2 Power Line Carrier Communication Module

The design used the power line carrier communication technology to construct the network bus when compared several conventional network communication media. The communication distance of the Ethernet cable is about 100 m which can't meet the requirement (500 m) of traffic control system for network communication. Fiber Optical Communication supports long distance network communication, but star-type network made the topologic structure of the network very complicated and increased engineering cost. Power line carrier communication technology modulates and couples the signals through Orthogonal Frequency Division Multiplexing (OFDM), then loads the signal onto the power line for transmission [5]. When the signal is coupled, filtered and demodulated, the original information can be restored, and the data communication can be achieved after a similar inverse process at the other end of the power line (Fig. 3).

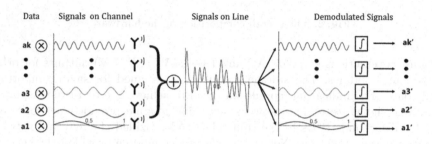

Fig. 3. The schematic diagram of carrier communication based on OFDM.

OFDM is a multicarrier modulation method, the main principle is as the follows. Firstly, divide the signal channel into x orthogonal sub channels, then encode high-speed serial data into binary and convert to x strands parallel low speed data stream, finally, modulate them into each sub channel for transmission. The multi-channel signals are equal in intervals and orthogonal to each other in frequency domain. This method can not only recover the original signal without distortion but also improve the utilization of the frequency-band by utilizing the mutual orthogonality of each signal [6].

Based on the above principles, the waveform of the signal loaded onto each sub channel can be described as follow:

$$x_k(t) = B_k \cos(2\pi f_k t + \varphi_k), k = 0, 1, \cdots, X - 1. \tag{1}$$

Then the sum of the signals loaded onto the channels was:

$$S(t) = \sum_{k=0}^{X-1} x_k(t) = \sum_{k=0}^{X-1} B_k \cos(2\pi f_k t + \varphi_k). \tag{2}$$

After a Laplace transformation, the signals on the power line can be described as follow:

$$S(t) = \sum_{k=0}^{X-1} x_k(t) = \sum_{k=0}^{X-1} \overrightarrow{B_k} e^{i2\pi f_k t + \varphi_k}. \tag{3}$$

Based on the above formula, OFDM can be implemented by IFFT/FFT or IDFT/DFT. The original signal can be restored as follow when integrate the result of the received signals multiplied by the corresponding source carrier signal.

$$B_k = \frac{1}{T} \int_{t_s}^{t_s+T} e^{-i2\pi f_k t + \varphi_k} \bullet \sum_{k=0}^{X-1} \overrightarrow{B_k} e^{i2\pi f_k t + \varphi_k} \bullet dt. \tag{4}$$

Power line carrier communication module was designed by SSC1660 chip in this paper. The performance of network communication media is mainly determined by throughput, packet loss rate and jitter, so we built a local area network using power line carrier communication modules to test the indicators. Figure 4 shows the average speed of network transmission is 110.68 Mbps, the average jitter is 0.48 ms, and the packet loss rate is 0%, which could meet the actual needs of traffic network communication.

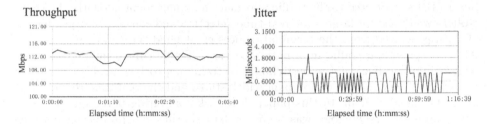

Fig. 4. The results of throughput and jitter.

2.3 IIPU

The IIPUs could achieve external network communication and its own state detection, and execute the corresponding control instructions through the power line carrier communication network bus when install the peripheral driving module on traditional non-intelligent traffic field equipment. Traffic field devices have different types, such as traffic lights, vehicle detectors and traffic cameras, their driving modes and interface types are also different. In the design of peripheral driver module, the upward and downward compatibility of module were taken into account in this paper. As the Fig. 5 shows, the peripheral driving module used the universal power line carrier communication bus interface to achieve the communication with ITCS, and designed multiple interfaces, such as high voltage electric devices drive interface and data communication interface, which could meet the needs of different traffic field equipment.

The peripheral driving module could drive 1 group of traffic lights, a set of video equipment with RJ45 interface, 1 detector device with 32 GPIO interface, and 1 third-party device with RS485 interface. Meanwhile, the peripheral driving module was equipped with the running state detection circuit and the identification circuit, which could help upgrade traditional traffic equipment to IIPUs rapidly and improve the level of intelligence, reduce the difficulty of upgrading and maintenance.

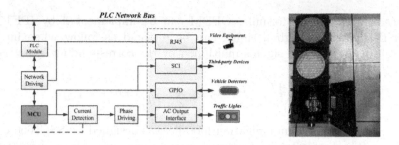

Fig. 5. The structure diagram and physical picture of IIPU.

3 Software Design

The design adopted all IP-based network control mode, there were a large number of IIPUs connected to the traffic information management network when controllers were used for large-scale real-time detection and control. A large amount of data generated from the terminal devices challenged the processing capability while providing traffic information services. Traffic control model based on edge computing became the best choice for traffic network control system when considering the communication bandwidth states and the real-time requirement of traffic signal control. In this paper, Firework computing paradigm based on edge computing technology was used for data sharing and processing between ITCS and IIPUs [7]. Figure 6 shows the composition diagram of Firework.

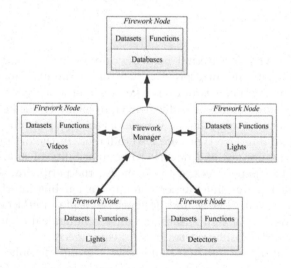

Fig. 6. The composition diagram of firework.

By pushing the data processing as close as to data producers, Firework aims to avoids data movement from the edge of the network to the center of system and reduce the response latency [8]. The Firework computing paradigm

included Fireworks Manager and Fireworks Nodes which were designed for facil-
itating data sharing and processing in such collaborative edge environment [9].
So we design the software of traffic cooperative optimization control according
to the model structure and data processing methods. In this model, the ITCS
was Firework Manager while the IIPUs were Firework Nodes, and each device
achieved coordinated optimization control through the network bus.

Figure 7 shows the software running process based on the Firework. The
IIPUs stored the configuration data timely needed by the program through
built-in flash memory. When IIPUs were powered on, they initialized the program configuration by reading the data about network configuration, device ID,
and default center IP from flash memory. The IIPUs sent register messages and
registered as the member of the network equipment after they got the IP address
from ITCS. Then the IIPUs began to detect the current communication state so
as to achieve the stability and reliability of network communication. When the
network state was abnormal, the IIPUs dealt with it accordingly. In the normal
network condition, the IIPUs executed the received control commands and report
the running states in real time. The IIPUs generated heartbeat data periodically,
and sent it to the network by multicast, in order to maintain communication for-
wardly with other devices. At the same time, the IIPUs also supported firmware
update, program configuration, program reboot and other online configuration
operations, to ensure the IIPUs could work steadily.

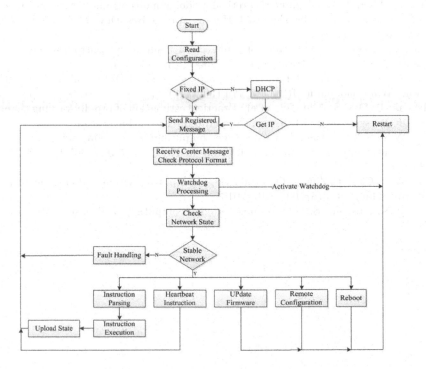

Fig. 7. The software execution process of IIPU.

4   Conclusions

After completing the specific design of each part, we carried out the electrical safety and environmental test in the laboratory environment, moreover we tested the controllers at several standard junctions and they worked well. The design could improve the integration level of traffic controller, provide complete data support for the optimization of road network with lower cost, reduce the difficulty of construction, maintenance and upgrade. The design was technical exploration and application practice based on edge computing in the field of traffic signal control system. It provided a feasible reference for the development of traffic signal controller in the direction of network and intelligence.

Acknowledgments. This work is partially supported by National Natural Science Foundation (61374191) and by the Great Wall Scholar Program (15038).

References

1. Wang, F.Y.: Agent-based control for networked traffic management systems. IEEE Intell. Syst. **20**(5), 92–96 (2005)
2. Lu, X.M., Wang, X.: Transport strategy of world metropolis. Urban Plan. Overseas **2001**(5), 17–19 (2001). (in Chinese)
3. Li, Z.L., Chen, D.W.: A game theoretical model of multi-agents in area coordination and optimization of traffic signals. J. Highw. Transp. Res. Dev. **21**(1), 85–88 (2004). (in Chinese)
4. Liu, X.M., Wang, F.Y.: Study of intersection traffic flow control on the basis of agents. J. Syst. Simul. **16**(4), 854–857 (2004). (in Chinese)
5. Mostofi, Y., Cox, D.C.: ICI mitigation for pilot-aided OFDM mobile systems. IEEE Trans. Wirel. Commun. **4**(2), 765–774 (2005)
6. Hrycak, T., Das, S., Matz, G., et al.: Practical estimation of rapidly varying channels for OFDM systems. IEEE Trans. Commun. **59**(11), 3040–3048 (2011)
7. Shi, W.S., Sun, H., Cao, J., et al.: Edge computing: an emerging computing model for the internet of everything era. J. Comput. Res. Dev. **54**(5), 907–924 (2017). (in Chinese)
8. Shi, W.S., Cao, J., Zhang, Q., et al.: Edge computing: vision and challenges. IEEE Internet Things J. **3**(5), 637–646 (2016)
9. Shi, W.S., Dustdar, S.: The promise of edge computing. J. Comput. **49**(5), 78–81 (2016)

Motif Iteration Model for Network Representation

Lintao Lv[1], Zengchang Qin[1(✉)], and Tao Wan[2(✉)]

[1] Intelligent Computing and Machine Learning Lab, School of Automation Science and Electrical Engineering, Beihang University, Beijing 100191, China
zcqin@buaa.edu.cn
[2] School of Biological Science and Medical Engineering, Beihang University, Beijing 100191, China
taowan@buaa.edu.cn

Abstract. Social media mining has become one of the most popular research areas in Big Data with the explosion of social networking information from Facebook, Twitter, LinkedIn, Weibo and so on. Understanding and representing the structure of a social network is a key in social media mining. In this paper, we propose the *Motif Iteration Model* (MIM) to represent the structure of a social network. As the name suggested, the new model is based on iteration of basic network motifs. In order to better show the properties of the model, a heuristic and greedy algorithm called *Vertex Reordering and Arranging* (VRA) is proposed by studying the adjacency matrix of the three-vertex undirected network motifs. The algorithm is for mapping from the adjacency matrix of a network to a binary image, it shows a new perspective of network structure visualization. In summary, this model provides a useful approach towards building link between images and networks and offers a new way of representing the structure of a social network.

Keywords: Motif Iteration Model (MIM) · Vertex Reordering and Arranging (VRA)

1 Introduction

Over the past a few years, there has been an explosion of interests in social media (network) mining with the increase popularity of online social networking services like Facebook, Twitter, Weibo and so on. In the study of social networks, the most fundamental idea in social network research is that a node's position in a network determines in part the opportunities and constraints that it encounter [1]. From transcriptional regulation networks, computer networks to electrical circuits networks, network motifs is regarded as recurring circuits of interactions from which the networks are built. Network motifs, depend on a small set of recurring regulation patterns, inspire more research on the networks [2–5].

© Springer International Publishing AG 2017
D. Liu et al. (Eds.): ICONIP 2017, Part V, LNCS 10638, pp. 647–656, 2017.
https://doi.org/10.1007/978-3-319-70139-4_66

Adjacency matrix is a square matrix representing the graph of a finite network. The binary value in adjacency matrix indicates whether the vertices are connected or not. On the other hand, the largest eigenvalue of the network adjacency matrix has emerged as a key value for study of a variety of dynamical networks. This allows the degree of a vertex to be easily found by taking the sum of the values in either its respective row or column in the adjacency matrix. Manipulating the adjacency matrix is the most direct way to study the complex network. In literatures, there are two major kinds of models for learning network structures: (1) count-based motifs methods, such as Generative Model Selection for Complex Networks (GMSCN) [6] and (2) Global adjacency matrix methods, such as Structural Perturbation Method (SPM) for link predictability of complex networks [7]. Currently, both models suffer significant drawbacks. While methods like GMSCN efficiently leverage statistical information, they do relatively poorly on the globe structure. SPM may do better on the analogy task of globe structure, but they poorly utilize the statistics of the graphlets.

In this work, we analyze the both model properties to produce linear mapping between the social network structure and the adjacency matrix binary (amb) image. We also introduce the Motifs Iteration Model (MIM) for globe network structure and use an algorithm to indicate the relevance of the network structure and the amb image. Under certain constraints, it can be a good representation of the real network structure, including the local motifs unit information and the overall frame information. After discussing the relationship between the adjacency matrix and the network motifs, we present a heuristic algorithm VRA. This method consists of two important parts, including reordering of the complex network vertex and replacing the sorted results symmetrically.

2   Related Work

Considering the structure, there are mainly four types of well-studied networks: random networks (Erdös-Rényi model) [8], nearest-neighbor coupled networks (NCN model) [9], small world networks (Watts-Strogatz model) [10] and scale-free networks (Barabási-Albert model) [11].

(1) Given a random network with N vertices, there can be C_2^N edges, and the network from which we randomly connect M edges is called random network. The links between the vertices in the network are random, the characteristics of the network is also random. (2) For the nearest-neighbor coupled networks, each vertex in the network is linked to a fixed number of vertices around the vertex. NCN network has a stable degree distribution. (3)The small-world networks is another classic network. It has a small path length and high clustering properties, started with a regular network, such as NCN, and we can obtain a small-world network by re-routing some edges randomly. (4) A typical feature of the scale-free network is that most vertices in a network are connected to only a few vertices, and very few vertices are connected to a very large number of vertices. The existence of such a critical vertex makes the scale-free network a strong ability to withstand unexpected failures, but in the face of collaborative attack is vulnerable.

In the field of complex networks, Janssen et al. [12] tries to classify the multiple network models by using a broad array of features, include the frequency counts of small subgraphs as well as features capturing the degree distribution and small world property, in order to make a better judgment on the real network. In the study of compressing the large-scale network, Liakos et al. [13] improves the state-of-the-art method for graph compression by exploiting the locality of reference observed in social network graphs. They apply the Layered Label Propagation (LLP) [14] algorithm on the origin social network graphs. More complex a network is, more accurate results will be. But they can not visualize the good performance of the model, only from theoretical and experimental verification of the superiority of the method. Although the network motifs and adjacency matrix in the application of the social networks were achieved good results, the use of combination of the two studies to social networks is relatively unexplored.

3   Topology of Networks

3.1   Motifs Iteration Model

Most study about complex networks mainly analyze their global statistical characteristics, such as small world features, scale-free features. However, in addition to these global features, the characteristics of the basic elements of each type of network is also very important. The network motifs are the key patterns of interaction in complex networks, which are more common in complex networks than in random networks [15]. Network motifs, as one of the best conditions of complex network, reveals the basic information of structure or basic building blocks of most complex networks in the real-world.

In different types of network structures, the number and type of different motifs vary widely. In this paper, to simplify the problem, we mainly study the network structure of undirected graphs. We will select the graphlets of the three-node, four-node undirected graphs in our study. Since the network element can only express local information in the network structure, we want to amplify the representation of the network motifs. The most basic element in the network motifs is a vertex, and we replace it with another network motifs. As each member in the family will eventually form their own family, this alternative approach is a typical iterative approach. The replaced network motifs is defined as a second-level network motifs. We then will obtain a higher level of network motifs. This is referred to as Motifs Iteration Model (MIM).

We only need to ensure that each layer of the network motifs are consisted of node, three-node motifs and four-node motifs. The details of the MIM are shown in Fig. 1. From left to right are the network graph with different node number, and each node in the network connected with the adjacent node. Black node as a first-level network motifs, red node with red dashed ellipse as second-layer network motifs. The black and red connections represent the element links of the first-layer and second-layer network motifs, respectively. Finally, we can use MIM of network to form any complex network structure easily. Prove as follows:

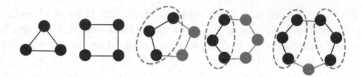

Fig. 1. Displays the Motifs Iteration Model of the networks under different number of nodes. There is an NCN network with parameter $k = 2$. Black nodes and segments are denoted as first-level motifs, red nodes and line segments represent second-layer motifs. The link between the second-layer motifs is called the second-link. (Color figure online)

Proof. Let n be the size of the networks. We omit the situation of $n < 6$, because this is clearly held. And define vertices as V, and the basic network motifs as $M(n) = (x, y, z)$ or $M(n) = (x, y, z, p)$. Where x, y, z, p can be a single vertex or motifs. So $M(7) = (M(3), M(3), 1) = (M(4), 1, 1, 1)$.

We assume that the number of vertices in a complex network is k, we can use the above formula. Which is:

$$M(k) = (\alpha, \beta, \gamma) | (\alpha, \beta, \gamma, \delta) \tag{1}$$

So we can introduce the network vertex count for the case of $k + 1$. Here, we will discuss two cases. If $M(k) = (\alpha, \beta, \gamma)$, so $M(k + 1) = (\alpha, \beta, \gamma, 1)$. If $M(k) = (\alpha, \beta, \gamma, \delta)$, $M(k + 1)$ will become $((\alpha, \beta, \gamma), \delta, 1)$. It is worth noting that: α, β, γ stand for the elements randomly. Based on this derivation, it is proved that all networks will be represented by the MIM.   □

3.2   Standard Motifs Iteration Model

In the process of building the model based on MIM, we will still face the choice, including the motifs level and node placement. For medium-sized networks, there will be tens of thousands of cases. In order to better represent the network structure, we define Standard Motifs Iteration Model (SMIM) that reducing the choice of building a network structure using MIM. The biggest difference from the previous model is that the SMIM adds additional constraints.

– The nodes in adjacent locations are preferentially linked.
– The network motifs at the same level are preferentially linked.

The first constraint is to ensure that the elements of the motifs in each network are close, so that the local information of the network structure is better preserved. The other constraint is for having good uniformity in the adjacency matrix binary*amb* image. Based on this model, we need to re-evaluate the capabilities of SMIM for the network. We can prove this as following:

Proof. When $M(k) = (\alpha, \beta, \gamma)$ is true, $M(k+1) = (\alpha, \beta, \gamma, 1)$ is true at the same time. When $M(k) = (\alpha, \beta, \gamma, \delta)$ is true, since k is greater than 7, there must be an element in $\alpha, \beta, \gamma, \delta$ that is not a single vertex.

- If $M(k) = (\alpha, 1, 1, 1)$, so $M(k+1) = (\alpha, M(3), 1)$;
- If $M(k) = (\alpha, \beta, 1, 1)$, so $M(k+1) = (\alpha, \beta, M(3))$;
- If $M(k) = (\alpha, \beta, \gamma, 1)$, so $M(k+1) = ((\alpha, \beta, \gamma), 1, 1)$;
- If $M(k) = (\alpha, \beta, \gamma, \delta)$, so $M(k+1) = ((\alpha, \beta, \gamma), \delta, 1)$; □

Through the study of the proof, we can find: the most advanced motifs in the complex network is showing the three nodes, the four nodes change alternately. And in the later work, the symmetry of this property in the network motifs and network adjacency matrix combination process will play an important role. In order to specify the model, the following describes the network model with a vertex number of 13:

$$M(13) = ((1,1,1), ((1,1,1), (1,1,1), (1,1,1)), 1) \tag{2}$$

3.3 Adjacency Matrix of Motifs

The structure of the undirected network decides adjacency matrix symmetry. Each element in the adjacency matrix corresponds to a single pixel, the adjacency matrix can correspond to an $n \times n$ binary image. In the adjacency matrix, if there is a connection between the two vertices, the corresponding element is 1, otherwise 0. Element 1 corresponds to white plot, and element 0 is a black plot. Therefore, based on the adjacency matrix, we can plot an *adjacency matrix binary (amb)* image. For a fixed network motif, we hope to find the unique representation of the *amb* image. If we can solve the mapping problem of the undirected motifs and the adjacency matrix binary image, we can observe the complex network from the network adjacency matrix binary image of the real network.

In the SMIM, three network motifs are used as essential components. In other words, the state of more than four motifs is only regarded as generated structure from basic three-vertices. However, even with a simple three-vertices network of the same structure, different vertex indexing will result in different *amb* images. For example, see Fig. 2.

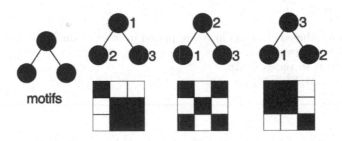

Fig. 2. This is an example of a three-node motifs with its corresponding *amb* image. The first line is the same motifs structure with different node indices. The second line is network's *amb* image. Though the structure of the network is identical, but the *amb* images are different, we need one unique image for the same network structure regardless node indices.

In Fig. 2, the left side shows the network motifs we used, in order to illustrate the importance of the vertex number, we did not use fully connected three-node network motifs. Because when the network with fully connected, the vertex indexing will not have any affect on the *amb* image. There are three cases are named (a), (b) and (c), from the left to right. All the cases have the same structure, but for their *amb* images, they are not symmetrical. The *amb* image of case (b) is symmetric and symmetrical with the standard multi-layer network motifs. Based on the advantages of motifs, we will examine the correspondence between the network motifs and the *amb* image of case (b).

3.4   Vertex Reordering and Arranging

Inspired by the motifs of the three-node example, we can arrange the important vertices in the middle, and the secondary vertex is in the adjacent position. In accordance with this idea, propose the following Vertex Reordering and Arranging (VRA) algorithm.

Algorithm 1. Vertex reordering and arranging

Input: Adjacency matrix M of the given network.
Output: New adjacency matrix $\hat{M}$.
 1: Calculate the degree d_i of ith vertex.
 2: Compute degree list $\hat{d}_i$.
 3: Define empty list X and $max = max(\hat{d}_i) - 1$.
 4: **for all** $\hat{d}_i = max$ **do**
 5:     all possible values are denoted as Ω.
 6:     Select a in Ω randomly.
 7:     **for all** $len(a) - 1$ **do**
 8:         Append a in X list.
 9:         **if** b in Ω but not in X and b satisfies the priority **then**
10:             $a = b$ and break.
11:         **end if**
12:         Select a in $(\Omega - X)$ randomly and append a in X.
13:     **end for**
14:     $max = max - 1$
15: **end for**
16: The odd-numbered elements in X are placed in front of the list in turn.
17: $X \times X$ constitute the $\hat{M}$.
18: **for all** (i, j) in vertexes **do**
19:     $\hat{M}[i][j] = M[X[i]][X[j]]$
20: **end for**

In the above algorithm, more choices are provided to satisfy the priority order. The implementation of the code in this article is based on the interconnection of nodes to be ordered firstly. VRA is an algorithm based on sorting and special placement, and it is a heuristic algorithm essentially.

In Fig. 3, it shows that the algorithm has good performance in identifying the network structure. With a rough look, two images are completely different. However, two *amb* images represent the same network structure. The example shown in Fig. 3 is a WS network with parameters $(200, 20, 0.1)$. Representing 200 vertices in the network, each of which is connected to 20 vertices adjacently to the surroundings, and the presence of 0.1 probability fluctuations in the connection. The image on the left-hand side is the *amb* image corresponding to the number of the vertex of the network, and the image on the right-hand side is the *amb* image after the VRA processing.

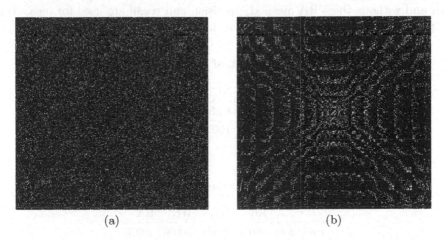

(a)   (b)

Fig. 3. An illustration of the VRA effect for small-world networks. (a) and (b) represent the *amb* image after random generation and VRA processing respectively.

It is easy to see that the right-hand side image has a clear pattern. This is because in the SMIM, the adjacent three vertices are small groups, and the small group of elements in the algorithm is adjacent, so the main diagonal is regular. On the other hand, left and right symmetry of small groups are closely related, so the same data on the diagonal also have the law. And the white point that is almost parallel to the image is the display form of the relationship between the small group and the large group. It can also be said that VRA method can better discover the nature of the network. Now we have established a network and the corresponding relationship between the image. Combined with SMIM and VRA, we will be able to apply the method of image processing to the application of social networks.

4   Experimental Studies

In order to test the validity of our new model to capture the properties of complex networks, we classify the *amb* image into different network classes based on the convolutional neural network (CNN) [17] method. We select the ER network, NCN network, WS network and BA network as our algorithm in the simulation of network classification of test data. The network is represented by an *amb* image after applying the VRA algorithm, and the network type is used as the label of the image as the input of the CNN. By training the neural network model, we try to find a good network type classifier.

In the data set, different parameters are set for each network. The number of vertices in the network here is fixed with 100. In order to test the rendering ability of for complex network's *amb* image, we selected the experiments without VRA as a baseline methods. In order to reflect the network type classification with and without the VRA method, precision and recall are used for measuring the performance. The results are shown in Tables 1 and 2, respectively.

Table 1. Precision and recall of the classifier without VRA.

	ER	NCN	WS	BA
Precision	97.5%	84.3%	98.7%	100%
Recall	99.0%	100%	79.0%	100%

Table 2. Precision and recall of the classifier with VRA.

	ER	NCN	WS	BA
Precision	100%	100%	100%	100%
Recall	100%	100%	100%	100%

The VRA processed network classifier shows a strong classification capability. In the processing of the model training, 100% accuracy can be achieved. The difference between two groups of experiments is that the resolution ability of random networks and some small-world networks. With the change of probability, the small world network's *amb* image is very similar with that of the random network. However, network with VRA can be very easy to avoid this problem.

In the real large-scale social network category identification processing, we found that a social network is not simply generated by one standard network. In order to better describe the reality of the network, we propose a rough solution. Based on standards community detection [18], each community need to identify what kind of network it is. Finally, the network can be described as a combination of multiple basic models.

$$\mathbf{D}(\mathbf{x}) = 0.5\mathbf{T}(\mathbf{x}) + 0.5\sum_{i=1}^{K} \lambda_i \mathbf{T}(\mathbf{x}_i) \tag{3}$$

where $\mathbf{x}$ is the entire network, $\mathbf{x}_i$ represents each small community network. The number of communities found after the standard network community is k, and

λ_i represents the weight. The $\mathbf{D}()$ function and the $\mathbf{T}()$ function denote the functions describing the network and the network type function, respectively.

For example, the Zacharys Karate Club is a social network of friendships between 34 members of a karate club at a US university in the 1970s [16]. Need to pay attention to the process of dealing with the real network, the use of the classifier is through the number of vertex 50 training from the network. When dealing with a network with fewer than 50 network points, we need to put the *amb* image corresponding to the network to be tested in the center and the rest to use the black supplement. The decomposition of this network is as follows: scale-free networks and small-world networks, and a small number of nearest-neighbor coupled networks.

$$\mathbf{D(Zachary)} = 0.50\mathbf{BA} + 0.43\mathbf{WS} + 0.07\mathbf{NCN} \tag{4}$$

5   Conclusions

Although either network motif and adjacency matrix is well used in studies of social networks and had achieved good results, how to use both of them is relatively unexplored. Through the study of network structure, by using motifs, we propose MIM and SMIM models. Through theoretical exposition and experimental verification, the networks using VRA can present best patterns of adjacency matrix images. Based on an image representation of the network structure, some image processing or pattern recognition (e.g. convolution neural networks) can be used to study network structures. This leaves a lot of possibilities for our future research.

Acknowledgments. This work is supported by the National Science Foundation of China Nos. 61401012 and 61305047.

References

1. Borgatti, S.P., Mehra, A., Brass, D.J., Labianca, G.: Network analysis in the social sciences. Science **323**(5916), 892–895 (2009)
2. Alon, U.: Network motifs: theory and experimental approaches. Nat. Rev. Genet. **8**(6), 450–461 (2007)
3. Alon, U.: Introduction to Systems Biology: Design Principles of Biological Circuits. CRC, Boca Raton (2006)
4. Davidson, E.H.: The Regulatory Genome: Gene Regulatory Networks in Development and Evolution. Academic, Burlington (2006)
5. Levine, M., Davidson, E.H.: Gene regulatory networks for development. Proc. Natl. Acad. Sci. U.S.A. **102**(14), 4936 (2005)
6. Motallebi, S., Aliakbary, S., Habibi, J.: Generative model selection using a scalable and size-independent complex network classifier. Chaos Interdiscip. J. Nonlinear Sci. **23**(4), 043127 (2014)
7. Lv, L., Pan, L., Zhou, T., Zhang, Y.-C., Stanley, H.E.: Toward link predictability of complex networks. Proc. Natl. Acad. Sci. U.S.A. **112**(8), 2325–2330 (2015)

8. Erdos, P., Renyi, A.: On random graphs. Publ. Math. **6**(4), 290–297 (1959)
9. Yu, Y., Wang, X., Liu, C.: Synchronization in a nearest-neighbor coupled network and motif dynamics. J. Ningxia Univ. **31**(1), 44–48 (2010)
10. Strogatz, S.H.: Exploring complex networks. Nature **410**(6825), 268 (2001)
11. Barabsi, A.L., Albert, R.: Emergence of scaling in random networks. Science **286**(5439), 509 (1999)
12. Janssen, J., Hurshman, M., Kalyaniwalla, N.: Model selection for social networks using graphlets. Internet Math. **8**, 338–363 (2012)
13. Liakos, P., Papakonstantinopoulou, K., Sioutis, M.: On the effect of locality in compressing social networks. In: de Rijke, M., Kenter, T., de Vries, A.P., Zhai, C.X., de Jong, F., Radinsky, K., Hofmann, K. (eds.) ECIR 2014. LNCS, vol. 8416, pp. 650–655. Springer, Cham (2014). doi:10.1007/978-3-319-06028-6_71
14. Boldi, P., Marco, R., Santini, M., Vigna, S.: Layered label propagation: a multiresolution coordinate-free ordering for compressing social networks. Comput. Sci. **133**(6), 587–596 (2011)
15. Milo, R., Shenorr, S., Itzkovitz, S., Kashtan, N., Chklovskii, D., Alon, U.: Network motifs: simple building blocks of complex networks. Science **298**, 824–827 (2002)
16. Börner, K., Sanyal, S., Vespignani, A.: Network science. Annu. Rev. Inf. Sci. Technol. **41**(1), 537–607 (2007)
17. Sermanet, P., Chintala, S., Lecun, Y.: Convolutional neural networks applied to house numbers digit classification. In: CVPR, pp. 3288–3291 (2012)
18. Fortunato, S.: Community detection in graphs. Phys. Rep. **486**(3–5), 75–174 (2010)

Inferring Social Network User's Interest Based on Convolutional Neural Network

Yanan Cao[1], Shi Wang[2(✉)], Xiaoxue Li[1], Cong Cao[1], Yanbing Liu[1], and Jianlong Tan[1]

[1] Institute of Information Engineering,
Chinese Academy of Sciences, Beijing, China
{caoyanan,lixiaoxue,caocong,
liuyanbing,tanjianlong}@iie.ac.cn
[2] Institute of Computing Technology, Chinese Academy of Sciences,
Beijing, China
wangshi@ict.ac.cn

Abstract. Learning microblog users' interest has important significance for constructing more precise user profile, and can be useful for some commercial applications such as personalized advertisement, or potential customer analysis. Existing works generally utilize text mining or label propagation methods to solve this problem, which leverage either the user's publicly available comments or the user's social links, but not both. As we will show, these learning methods achieve limited precision rates. To address this challenge, we consider the interest inference task as a multi-value classification problem, and solve it using a convolutional neural network architecture. We innovatively present an ego social-attribute network model which integrates the target users' attributes, social links and their comments, and represent the ego SA network as the input fed to CNN. As a result, we assign each microblog user one or more interest labels (such as "loving sports"), which is different from previous approaches using non-uniform interest keywords (such as "basketball", "tennis", etc.). Experimental results on SMP CUP and Zhihu dataset showed that the precision rate of user interest inference reached 77.9% at best.

Keywords: Social-attribute network · Convolutional neural network · User interest inference

1 Introduction

As an important social media service, microblog is a wonderful platform where people share their thoughts, status and even their personal information. As the continuously increasing of microblog users, the analysis of users' attributes, relations and behaviors has received more and more attention both in academic and industry. Specifically, microblog users' interests can reflect users' preference and also have a close relationship with users' other attributes such as gender, age and occupation. Therefore, modeling users' interests has important significance for getting more precise user profile, and can be useful for commercial applications such as personalized advertisement, or potential customer analysis. At present, the miss rate of users' registration

© Springer International Publishing AG 2017
D. Liu et al. (Eds.): ICONIP 2017, Part V, LNCS 10638, pp. 657–666, 2017.
https://doi.org/10.1007/978-3-319-70139-4_67

interest tags is higher than 70% [1], which means that most users' implicit interests should be learned. As important data sources, microblog contents and social links involve personal preferences which directly reflect user interests.

To mine user interests from microblog contents, existing studies have proposed two major methods including TextRank [2, 3, 5] and Topic Model [6, 7]. Although these methods are considered the state-of-the-art unsupervised keyword extraction and clustering methods, they face two challenges. On one hand, extracted interest keyword may provide an ambiguous representation of the topic. On the other hand, the topic model can obtain potential topics in texts, but explicit topic semantic labels are not given.

To make use of social links, label-propagation-based works [16–19] propagate missing attribute values from label nodes to unlabel nodes. The foundation of label-propagation-based work is homophily, which means that two linked users share similar attributes. Main label-propgation algorithms include MV, GSSL and CP are used in different kinds of attribute value inference, such as school, location, interest, etc. The average precision reached 60% to 70% and computing cost is typically high.

To deal with these above problems, we propose a novel method to infer user interest based on both users' attributes, social links and comments. Here, we consider the interest inference task as a multi-value classification problem, and solve it using the convolutional neural network architecture. We innovatively present an ego social-attribute network model which integrates the target users' attributes, social links and their comments, and represent the ego SA network as the input fed to CNN. Our CNN architecture contains two layers of convolution, which capture latent relations between the target user and his neighbor nodes, and the output is the probability distribution over interest labels. During the pre-processing stage, we mine and cluster frequent interest phrases, which have clearer semantic information than keywords, from users' comments based on the method proposed in [20]. For each user, we select the top N topics his interest phrases belong to as important attributes.

We evaluate our method both on the SMP CUP 2016 dataset and Zhihu dataset, which contains about 20 thousand Sina microblog users and 30 million contents in Chinese. Experimental results show that CNN architecture can achieve better results than traditional classifier models on precision, and performs well on time complexity. The precision rate of user interest inference reached 77.9% at best.

The main contributions of this paper are as follows:

- We present an ego social-attribute network model which integrates the target users' attributes, social links and their comments.
- We design a convolutional neural network architecture under ego SA network to infer users' implicit interests, which specially performs well on multi-valued interest inference.

2   Related Works

Existing interest inference works can be roughly classified into two categories, text mining based method and label propagation based one.

Text Mining Based Methods: Researchers tried to use TextRank to build a word-based graph and to use PageRank [8] to get top n candidate keywords as users' interest keywords, which gained 31.2% precision and recall rate of 43.1% [5]. Some researchers describe users' interests by using a set of tuples of content directives (categories to which user interests belong) and action indicators (actions related to interest categories), which can effectively exploit the real-time interests of microblog users [10]. Others consider the time distribution of microblog contents and use the time series to classify users' contents [3]. The precision rate of the classification was increased to 67%. These methods make use of the statistical properties or semantic information of words in text. They have made some effects in mining the interest information of microblog users, but they can't make use of statistical features in documents and between documents, and can't solve the ambiguity problem of interest words either.

The topic model performs better in this respect. Zhang used LDA to extend the text feature space, and then used words' frequency to extract the hot topic, which makes the hot topic rank higher [14]. Ramage also use the aggregated information to train the LDA model, and the experimental results show that the model is more suitable for the modeling of "author-feature topic" [15]. Weng proposed Twitter-LDA to filter the non-hot topic words and compared them with the distribution of hot topics in traditional media [6]. They find that most of the topics in microblog contents are about the daily life of users, which more reflect users' personal interests. These studies showed that the topic model can efficiently mine interests from the sparse and short text such as microblog contents by using the distribution of words and topics in the text and the distribution of topics and documents. However, in the existing works, the semantic information of topics and the categories of user interests are not clearly identified.

Label Propagation Based Methods: To make use of social links, label-propgation-based works [16–18] propagate missing interest values from labeled nodes to unlabeled nodes. Li present a hidden factor in social connections-relationship type and propose a co-profile users' attributes and relationship types base on this development [16]. Through iteratively profiles attributes by propagation via certain types of connections, and profiles types of connections based on attributes and the network structure, their algorithm profiles various attributes accurately. Dong design different strategies for computing the relational weights between users' attributes and social links and used a graph-based semi-supervised learning (GSSL) algorithm to infer attributes [17]. Dougnon proposed a new lazy algorithm PGPI to infer user profiles by using rich information (such as group memberships) without training [18]. These methods are used in different kinds of attribute value inference, such as school, location, interest, etc. The average precision reached 60% to 70%. Interest inference using social structure and attributes could also be solved by a social recommender system in [19]. However, such approaches have higher computational complexity than above methods.

To address above shortages of existing research, we consider the interest inference task as a multi-value classification problem, and solve it using a convolutional neural network architecture. And we present an ego social-attribute network model which integrates the target users' attributes, social links and their comments.

3 Problem Definition

In this section, we will formally introduce the definition of user interests' inference problem. We start by describing our social-attribute network model, which integrates social structure and user attributes in a unified framework. Here, user attributes not only involve gender, status, country, etc., but also contain the top N topical phrases extracted from user comments, which may reflect the user interests in a certain extent.

Definition 1 (Social-Attribute Network). We denote a Social-Attribute network (SA network for short) as $G = (V, E, t)$, where V is the set of nodes, E is the set of links, and t a function that maps a node to its node type, i.e., t_u is the node type of u. Nodes corresponding to users, attributes and interests are respectively called *social nodes*, *attribute nodes* and *interest nodes*, which are represented as S, A and I. Links between social nodes are called *social links*, links between social nodes and attribute nodes are called *attribute links*, and links between social nodes and interest nodes are called *interest links*. Additionally, for a given node u in the SA network, we denote by $\Gamma_{u,S}$, $\Gamma_{u,A}$, $\Gamma_{u,I}$, respectively the sets of all social neighbors, attribute neighbors and interests of u, and the neighbor nodes of u is $\Gamma_u = \Gamma_{u,S} \cup \Gamma_{u,A} \cup \Gamma_{u,I}$.

Definition 2 (ego SA network). The ego social-attribute network of a target user u is represented as a graph $EG_u = (V', E', t)$, where $V' = \bigcup_{v \in u \cup \Gamma_{u,s}} (\Gamma_{v,A} \cup \Gamma_{v,I}) \cup u \cup \Gamma_{u,s}$, $E' = \{(u', v') | u', v' \in V')\}$, and t a function that maps a node to its node type. Later in this article, we call the target user in an SA ego network an ego user.

Figure 1 illustrates an example SA network, in which the social node set is $S=\{v_1, v_2, v_3, v_4, v_5\}$, attribute node set is $A=\{status, gender, age, country\}$ and interest set is $I=\{sports, music, drawing\}$. For a given ego user v_1, its social neighbor set is $\Gamma_{u,s}=\{v_2, v_3\}$, its attribute neighbor set is $\Gamma_{u,A}=\{student, China\}$, and its interest node set is $\Gamma_{u,I} = \phi$. Next, we aim to predict $\Gamma_{u,I}$ in the ego network EGv_1, which contains social nodes $\{v_1, v_2, v_3\}$, their attribute nodes, interest nodes and links between these.

More formally, the problem of interest inference of a target user u in an ego network G is to predict the interest link of u based on his attributes and neighbor nodes' information in the ego network EG_u. We represent these information as an $n*k$ matrix and feed it to a CNN model to implement the multi-value classification.

4 Proposed Architecture

The model architecture, shown is Fig. 2, is a slight variant of the CNN architecture of [21]. The input is an $n * k$ matrix which represents an ego SA network. The second layer contains several convolution operations, and we take multiple convolution kernels of different sizes to extract features of input data.

We use a notation $f(\cdot)$ to represent a neural network. Any feed-forward neural network with P layers can be seen as a composition of functions f, corresponding to each layer p.

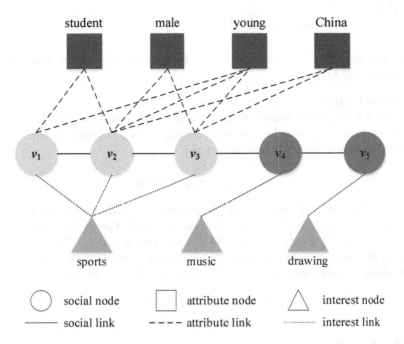

Fig. 1. An example of a social-attribute network and an ego SA network of v_1. All nodes and edges belong to a full social-attribute network and an ego-network of v_1 is represented as orange nodes and their attributes (Color figure online)

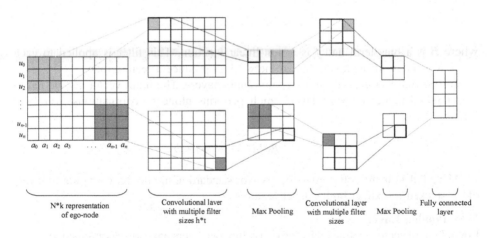

Fig. 2. CNN architecture for social network user's interest inference

$$f_\theta(\cdot) = f_\theta^P(f_\theta^{P-1}(\cdots f_\theta^1(\cdot)\cdots) \tag{1}$$

And initially, $f_\theta^0(\cdot) = F_i$ where $i \in (1, n)$. In the following, we could introduce our architecture layer by layer.

Input Feature Map

Let $u_1 \in R^k$ be the k-dimensional attribute vector corresponding to the i-th social node in the ego-network and u_0 is the ego-node specially. An ego-network of size n (padded or pruned where necessary) is represented as

$$u_{1:n} = u_1 \oplus u_2 \oplus \ldots \oplus u_n \tag{2}$$

where $\oplus$ is the concatenation operator. In general, $u_{i:i+j}$ refer to the concatenation of social nodes $u_i, u_{i+1}, \ldots, u_{i+j}$. Each attribute value of u_i is first passed through the lookup table layer, producing a numeric vector ML_i of the same size as $L_{i\_}$. The feature can be viewed as the initial input of the standard convolution neural network. More formally, the initial input feature map fed to the convolution layer can be written as

$$f_\theta^P(\cdot) = ML_i = LTF(F_i) \tag{3}$$

Convolution Layer

A convolution operation involves a filter w, which is applied to a window with size $h * r$ to produce a new feature. For example, a feature is generated from a window of social nodes $u_{i:i+h-1}$ and a window of attribute nodes $a_{j:j+r-1}$ by

$$c_i = f(w \cdot u_{i:i+h-1} a_{j:j+r-1} + b) \tag{4}$$

where b is a bias term and f is a non-linear function. This filter is applied to each possible window of nodes in the ego-network to produce a feature map.

In our architecture, we use two convolution layers. The input vector can be fed to the standard neural network layer which performs affine transformations over their inputs

$$f_\theta^P(\cdot) = ReLU(wf_\theta^{P-1}(\cdot) + b) \tag{5}$$

Here ReLU is the active function. As for standard affine layers, convolution layers often stacked to extract higher level features.

Max Pooling Layer

Local feature vectors extracted by the convolutional layers have to be combined to obtain a global feature vector, with a fixed size independent of the $L_{i\_}$, in order to apply subsequent standard affine layers. Then, we apply a max pooling over the feature map and take the maximum value as the feature corresponding to this particular filter. The idea is to capture the most important feature with the highest value for each feature map. This pooling scheme naturally deals with variable matrix size. Formally, given a matrix $f_\theta^{P-1}(\cdot)$ output by a convolution layer $p-1$, the max pooling layer output a vector $f_\theta^{P-1}(\cdot)$

$$[f_\theta^p]_i = \max_i [f_\theta^{p-1}]_{i,t} \tag{6}$$

where t is the number of layer $p-1$ output. The fixed size global feature vector can be then fed to the standard affine network layers.

Fully Connected Softmax Layer

We have described the process by which one feature is extracted from one filter. The model uses multiple filters to obtain multiple features. These features from the penultimate layer and are passed to a fully connected softmax layer whose output is the probability distribution over interest labels.

5 Experiments and Results

In this section, we start with the introduction of the datasets and experimental setting. We then describe comparative methods and the evaluation on precision.

5.1 Datasets

There is no public benchmark in social network user interest inference problem. So, we use a dataset provided from SMP CUP 2016 (a microblog user profile contest held by Sina) which contains more than 2000 users and 230,000 microblog contents. This dataset is divided from a real Sina microblog dataset in Chinese, which contains about 46,000 users, more than 30,000,000 microblog contents. SMP CUP dataset provides plenty of user contents but inadequate user attributes. So, we constructed another dataset containing 20000 users, 50 contents and 13 dimensions of attributes for each user, in which the user attribute, user link and user comments are crawled from the online social network website Zhihu. The interest label for each user is extracted semi-automatically and verified annually.

User Attribute Selection: We select 8 attributes (gender, age, status, major, university, i.e.) which are commonly used information in social media platforms. In our dataset, there are some missing attribute value and noisy information. For example, some user's status is 'loving money'. In this case, we use specific tag to represent missing attribute values and noisy data.

Topic Mining: We preprocess users' contents by deleting specific symbols and removing duplication. Chinese stop words are also removed for phrase mining and topic modeling steps. Then, we extract users' interest candidate phrases using an effective topical phrase mining method [19]. In order to identify the topics users interested in, we semi-automatically construct a hierarchical topic knowledge base and utilize it to identify users' topical phrases. For each user, his top 5 topical phrases are used as user attributes in the input matrix.

After these preprocessing procedures, above datasets are divided to training dataset and test dataset according to the proportion of 2:1.

5.2   Experimental Setting

We compared the performance of the proposed algorithm with four state-of-the-art algorithms: Linear Regression, Naïve Bayes classifiers [9], Graph Semi-Supervised Learning and Majority Voting. These four algorithms predict the value of target user's gender, status and major respectively.

Linear Regression (LR): we construct a linear function by using our training dataset, and predict the missing values using this function.

Naïve Bayes (NB) Classifiers: NB infer user profiles strictly based on correlation between attributes values which is as well as our UPS model.

Graph Semi-Supervised Learning (GSSL) [11] and Majority Voting (MV) [6] infer user's profiles by using the social structures which is the same as our CNN model. For algorithms which need specific parameters, we use empirical value to achieve the state-of-art results.

Both the Linear Regression and Naïve Bayes classifiers perform well on binary classification. Because inferring user's potential interest is a problem of multi-classification, we also use a general neural network (NN) algorithm as a comparative method. Note that, these three methods don't use user links in the interest inference procedure, which is different from GSSL and MV.

Contrast experiments were performed on a computer with a fourth generation 64-bit Core i5 processor running Ubuntu 14.5 and 16 GB of RAM.

5.3   Experimental Results

We evaluate the accuracy of CNN and the comparative models on user interest inference. Table 1 shows the evaluation results on SMP CUP dataset and Zhihu dataset. We can see that, on both datasets, CNN significantly outperforms other methods, which achieved 52.6% and 77.9% precision respectively. This result demonstrates that CNN has great advantages in dealing with multiple classification problems. Besides, the effectiveness of NN and GSSL (which make good use of multiple dimensions of user attributes and user links respectively) is also acceptable. MV is the worst model in user interest inference, because its strategy is too simple to deal with this problem. Compared with NN, CNN improved the accuracy significantly for its model architecture and the use of social links in the neural network, which reflects that CNN could learn the hidden relation between the ego node and its social neighbors.

Table 1. Accuracy of different algorithms on two datasets

Dataset\ Accuracy	CNN	NN	NB	LR	GSSL	MV
SMP CUP dataset	52.6%	48.2%	44.3%	48.4%	49.0%	40.2%
Zhihu dataset	77.9%	72.3%	67.4%	70.6%	71.30%	52.50%

We also evaluate the impact of different training dataset size on above classification models, including CNN, NN, NB, LR. The evaluation result is demonstrated in Fig. 3. We can see that the effectiveness of these classifiers was generally improved as the training data increases. When the training set contains less than 5,000 users, the effectiveness of CNN and NN is almost equivalent. Because under this situation, users' links are sparse and unbalanced, which conducts a negative impact on the effectiveness of CNN.

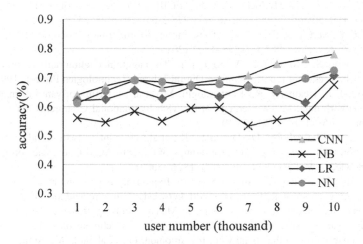

Fig. 3. The accuracy of four algorithms on different dataset size (Zhihu)

6   Conclusion and the Future Work

In contrast to other classifier models, CNN performs better on predicting multi-valued interest for online social network users. Our experimental results showed that it is possible to learn users' potential interests based on user links and user profiles. Our CNN architecture is just a single channel model. In the future work, we would like to use a multichannel architecture which may make better use of attributes and topical phrases. Furthermore, we will evaluate the impact of users' topical phrases on our algorithms.

Acknowledgement. This work was supported by the National Natural Science Foundation of China grants (NOs. 61403369, 61602466), the National Key Research and Development program of China (Nos. 2016YFB0801304, 2016YFB0800303).

References

1. Ding, Y.X., Xiao, X., Wu, M.J.: Predicting users' profiles in social network based on semi-supervised learning. J. Commun. **35**(8), 15–22 (2014)
2. Vu, T., Perez, V.: Interest mining from user tweets. In: Proceedings of ACM International Conference on Information and Knowledge Management, San Francisco, CA, USA (2013)
3. Yang, T., Lee, D.W., Yan, S.: Steeler nation, 12th man, and boo birds: classifying Twitter user interests using time series. In: Proceedings of the 2013 IEEE/ACM International Conference on Advances in Social Networks and Mining, pp. 684–691 (2013)
4. He, L., Jia, Y., Han, W.H., Ding, Z.H.Y.: Mining user interest in microblogs with a user-topic model. China Commun. **11**, 131–144 (2014)

5. Mihalcea, R., Tarau, P.: Textrank: bringing order into texts. In: Proceedings of 2004 Conference on Empirical Methods in Natural Language Processing, pp. 404–411 (2004)
6. Zhao, W.X., Jiang, J., Weng, J., He, J., Lim, E.-P., Yan, H., Li, X.: Comparing Twitter and traditional media using topic models. In: Clough, P., Foley, C., Gurrin, C., Jones, G.J.F., Kraaij, W., Lee, H., Mudoch, V. (eds.) ECIR 2011. LNCS, vol. 6611, pp. 338–349. Springer, Heidelberg (2011). doi:10.1007/978-3-642-20161-5_34
7. Zhang, C.Y., Sun, J.L., Ding, Y.Q.: Topic mining for microblog based on MB-LDA model. J. Comput. Res. Dev. **48**(10), 1795–1802 (2011)
8. Page, L., Brin, S., Motwani, R., Winograd, T.: The pagerank citation ranking: bringing order to the web. Technical report, Stanford Digital Library Technologies Project (1998)
9. Lindamood, J., Heatherly, R., Kantarcioglu, M., et al.: Inferring private information using social network data. In: Proceedings of the 18th International Conference on World Wide Web, pp. 1145–1146. ACM (2009)
10. Banerjee, N., Chakraborty, D., Dasgupta, K., et al.: User interests in social media sites: an exploration with micro-blogs. In: Proceedings of the 18th ACM Conference on Information and Knowledge Management, pp. 1823–1826 (2009)
11. Hu, X., Sun, N., Zhang, C., Chua, T.S., et al.: Exploiting internal and external semantics for the clustering of short texts using world knowledge. In: Proceedings of the 18th ACM Conference on Information and Knowledge Management, pp. 919–928 (2009)
12. Abel, F., Gao, Q., Houben, G.-J., Tao, K.: Semantic enrichment of Twitter posts for user profile construction on the social web. In: Antoniou, G., Grobelnik, M., Simperl, E., Parsia, B., Plexousakis, D., De Leenheer, P., Pan, J. (eds.) ESWC 2011. LNCS, vol. 6644, pp. 375–389. Springer, Heidelberg (2011). doi:10.1007/978-3-642-21064-8_26
13. Musat, C.C., Velcin, J., Trausan-Matu, S., Rizoiu, M.A., et al.: Improving topic evaluation using conceptual knowledge. In: Proceedings of the Twenty-Second International Joint Conference on Artifical Intelligence-Volume, vol. 3, pp. 1866–1871 (2011)
14. Zhang, S., Luo, J., Liu, Y., Yao, D., et al.: Hotspots detection on microblog. In: 2012 Fourth International Conference on Multimedia Information Networking and Security(MINES), pp. 922–925. IEEE (2012)
15. Ramage, D., Hall, D., Nallapati, R., et al.: Labeled LDA: a supervised topic model for creditattribution in multi-labeled corpora. In: Proceedings of the 2009 Conference on Empirical Methods in Natural Language Processing, pp. 248–256 (2009)
16. Li, R., Wang, C., Chang, K.C.C.: User profiling in an ego network: co-profiling attributes and relationships. In: Proceedings of the 23rd International Conference on World Wide Web, pp. 819–830 (2014)
17. Dong, Y., Tang, J., Wu, S., Tian, J., et al.: Link prediction and recommendation across heterogeneous social networks. In: 2012 IEEE 12th International Conference on Data Mining (ICDM), pp. 181–190. IEEE (2012)
18. Dougnon, R.Y., Fournier-Viger, P., Nkambou, R.: Inferring user profiles in online social networks using a partial social graph. In: Barbosa, D., Milios, E. (eds.) CANADIAN AI 2015. LNCS, vol. 9091, pp. 84–99. Springer, Cham (2015). doi:10.1007/978-3-319-18356-5_8
19. Ye, M., Liu, X., Lee, W.C.: Exploring social influence for recommendation a probabilistic generative model approach. In: SIGIR (2012)
20. El-Kishky, A., Song, Y., Wang, C., Voss, C.R., Han, J.W.: Scalable topical phrase mining from text corpora. PVLDB **8**(3), 305–316 (2015). Also, In: Proceedings of 2015 International Conference on Very Large Data Bases (VLDB 2015), Kohala Coast, Hawaii, September 2015
21. Krizhevsky, A., Sutskever, I., Hinton, G.E.: ImageNet classification with deep convolutional neural networks. In: Proceedings of NIPS (2012)

Enhanced Deep Learning Models
for Sentiment Analysis in Arab Social Media

Mariem Abbes[✉], Zied Kechaou, and Adel M. Alimi

REsearch Groups in Intelligent Machines (REGIM),
National School of Engineers (ENIS), University of Sfax,
BP 1173, 3038 Sfax, Tunisia
{mariem.abbes,zied.kechaou,adel.alimi}@ieee.org

Abstract. Over the last few years, the amount of Arab sentiment rich data as appearing on the web has been marked with a rapid surge, owing mainly to the remarkable increase noticed in the number of social media users. In this respect, various companies are now turning to online forums, blogs, and tweets with the aim of getting reviews of their products, as drown from customers. Hence, sentiment analysis turns out to lie at the heart of social media associated research, targeted towards detecting people opinion as embedded within the wide range of texts while attempting to capture their pertaining polarities, whether positive or negative.

While research associated with English sentiment analysis has already achieved significant progress and success, a remarkable efforts have been made to extend the focus of interest to cover the Arabic language domain. Indeed, most of the Arabic sentiment analysis systems tend to still rely on costly hand-crafted features, where features representation seems to rest on manual pre-processing procedures for the intended accuracy to be achieved. This is mainly due to the Arabic language morphological complexity, linguistic specificities and lack of the resources. For this purpose, deep learning (DL) techniques for Sentiment Analysis turn out to be very versatile and popular. It is in this context that the present paper can be set, with the major focus of the interest being laid on proposing a novel automated information processing systems based DL. The experiment result show that RNN outperforms DNN in term of precision.

Keywords: Sentiment analysis · Deep learning · Arabic language · DNN · RNN · LSTM

1 Introduction

Recognizing the other's opinion usually stands as an important source of information as far as to the decision-making process is concerned. In the past, however, wherever a decision needs to be taken with respect to a certain phenomenon topic or item, people used to carry out particular enquiries, questioners and investigations regarding the typical opinions of friends, relatives and family, etc. Similarly, whenever an organization is faced with the need to retrieve and recognize the general public's views, attitudes and prospects concerning particular services and/or product ranges, it would

© Springer International Publishing AG 2017
D. Liu et al. (Eds.): ICONIP 2017, Part V, LNCS 10638, pp. 667–676, 2017.
https://doi.org/10.1007/978-3-319-70139-4_68

certainly conduct investigations and center the focus of interests on specific target groups. Hence, organizations tend most often to incur noticeable costs for hiring consultants and conducting surveys in a bid to construct reliable source of information, useful for maintaining an equitable background whereby public opinions can be maintained as to their products. Similarly, people are generally interested in the others' opinions concerning particular ranges of products, services and events helping them in making and reaching the most appropriate choices. In this context, opinion mining, or sentiment analysis, process stands as critical mechanism whereby measures of people's opinions can be effectively and automatically drawn on the basis of digital data. Thus, owing to the surging wave of the 2.0 Web technology, a remarkable transformation have been noticed to predominate worldwide and uses turnout and end up by being the data source generators, culminating in the emergence of the so-called "Big Data" [1, 2].

Nowadays, people can mail, publish opinions or product reviews on merchant sites while publishing and expressing their proper views regarding almost anything by means of discussion forums, blogs, and social networking sites. As a matter of facts, such comments appear to be highly complex, massive and diverse that an automated system can hardly process, particularly if one is to define the most optimally appropriate product in the batch, on the basis of such comments. This highlights the need for a reliable sentiment-classification system likely help to digest such a huge repository of reviews and capable of deciphering the hidden hints and opinions.

While research on English concerning the English opinion mining area has already made great steps achieving significant progress and success, research works dealing with opinion mining as associated with the Arabic linguistic and cultural context remains still lagging. This modest situation is mainly due to the morphological complexity [3–7] and lack of opinion resources relevant to the Arabic context.

For the purpose of narrowing such a remarkable gap, Deep Learning (DL) techniques closely associated with the Sentiment Analysis area are discovered to be very popular [8]. They have helped a participated greatly in providing automatic feature extraction measures, along with the establishment of both richer representation capabilities and rather effective performance with respect to the traditional feature based techniques. Indeed, DL, also known as hierarchical learning, has emerged as a new research area involving machine learning research and classical classification techniques, worth citing among which are the SVM, Naïve Bayes, etc. In this work, a DL approach is put forward useful for coping with the sentiment classification problem concerning Arabic text scrutinization. Actually, two relevant architectures are suggested to deal with the deep neural network and recurrent neural network domains.

The present research is organized as follows. Section 2 is devoted to highlight the major related works. Section 3 details our advanced approach and deals with the construction of word vector model for Arabic, along with the data extraction and pre-processing procedures. It is focused on implementing the Deep Neural Networks (DNN) and Recursive Neural Network (RNN) models on our study dataset. Section 4 is dedicated to highlight our reached experimental results. As for the conclusion, along with the prospects for a potential work prospective make subject of Sect. 5.

2   Related Works

More recently, a wide range of research has been conducted to deal with the area of sentiment analysis, as related to the English language user-generated content [9–13]. Noteworthy, however, very few are those studies devoted to address the issue of sentiment analysis with respect to the Arabic language. In fact, the majority of Arabic opinions modeling approaches appear to predominantly rely on training machine learning classifiers through implementation of various feature engineering options. Still, only few studies have been devoted to analyze Arabic specific social media.

In [14], for instance, the focus of interest has been laid on movie and product reviews, while the authors in [15] implements a genetic algorithm for sentiment detection with respect to both English and Arabic Web forums at document level. Even though both of the textual syntactic and stylistic features are exploited, no reference has been made to the associated morphological features. In [16] a special study in conducted to treat the pre-processing procedure's impact on the sentiment analysis of Egyptian dialect tweets. In [17], a semantic approach is advanced to depict the user related attitudes and business insights from social media formulated in Arabic, both standard and dialects. In doing so, they introduce the initial Arabic Sentiment Ontology version, enclosing different word classes that express feelings and the extent to which the relevant word ranges help in conveying the feelings of different Twitter feeds on different topics.

Recently, DL has been proposed in a bid to solve various Natural Language Pre-processing (NLP) related tasks [18]. Contrary to supervised learning, the major advantage of this approach in the fact that it does not require manually tuned features based on expert knowledge and available linguistic resources. DL is an emergent machine learning area that provides special learning feature representation methods, that operate in either supervised or unsupervised modes within a hierarchy. DL models have recently gained popularity, and are potentially applicable in the sentiment analysis field. These models include DNN, Convolutional Neural Networks (CNN) [19], and Deep Belief Networks (DBN) with fast inference of the model parameters [20], and RNN [21]. In [21], the authors introduce the Recursive Neural Tensor Netpoel and data culminates in the setting up of single sentence sentiment detection system that further enhances the state of the art framework by 5.4% with respect to the positive/negative sentence classification. Still, Socher [22] suggests a novel machine learning framework based on recursive auto-encoders relevant to sentence-level prediction of sentiment label distributions. He also introduces a new dataset encompassing distributions over a wide range of human emotions. As for the authors in [23], they put forward a new deep convolutional neural network that helps exploit character-to-sentence-level information to apply the sentiment analysis proceeding to short texts. In [24], a hybrid method, integrating the bilingual text sentiment DL features, and shallow learning features, is proposed. Models such as recurrent neural networks (RNNs) with long-short-term memory (LSTM), Naïve Bayes Support Vector Machine (NB-SVM), word vectors and bag-of-words are thoroughly explored. The processed data are based on NLPCC 2014 sentiment classification task; including positive and negative reviews in English and Chinese respectively.

In what follow is an overview of the most popular approaches devised to deal with opinion mining in Arabic, along with a series of recently appearing opinion models based on DL techniques, and opinion lexical resources developed for Arabic language analyses purposes. The authors in [25] explore four different architectures relating to DL of text sentiment classification concerning the Arabic language. Three among them appear to be based on DBN and Deep Auto Encoders (DAE), in which the input data model rests on the ordinary Bag-of-Words, with features being based on the recently developed Arabic Sentiment Lexicon as jointly combined with other standard lexicon features. As for the fourth model, it relies on the Recursive Auto Encoder (RAE), and is proposed to cope with the lack of context handling procedures as persistent in the first three models. The evaluation process is carried out using Linguistic Data Consortium Arabic Tree Bank (LDC ATB) dataset [26] and the results achieved turn out to reveal remarkable improvement of the fourth model over the state of the art ones, with a recorded advantage of using no lexicon resources that prove to be scarce and costly in terms of development.

In [27], a crawling scheme for a large multi-domain corpus is devised with the aim of contracting an Arabic word embedding model. To this end, the authors have provided a short practical and empirically informed procedure that sounds useful for exploring the Arabic word embedding process, along with the CNN for sentiment classification purposes. In [28], Several Machine Learning algorithms (Naïve Bayes, Support Vector Machine and Logistic Regression) alongside Deep and Convolutional Neural Networks are applied as effective experimental sentiment analysis tool relevant to scrutinizing the Arabic health care domain associated dataset. To our knowledge, and following an exhaustively intense biographical search, the latter turns out to stand as the first and unique work elaborated to apply the RNN methodology for sentiment classification purposes, as oriented to deal with the Arabic social media context.

3   Deep Learning Models for Sentiment Analysis in Arabic

The proposed approach, as envisaged to investigate the social media sentiment analysis domain involves a number of explanatory steps illustrated through a flowchart as appearing in Fig. 1.

In a first stage, data are gathered and collected, with the major source being book reviews. In a second stage (the preprocessing step), the NPL techniques are applied, prior to the implementation of our proposed lexicon model. The third phase (the feature extraction step), consists in deriving the lexicon-based relevant features from the stored data. The ultimate step consists in a sentence level to-class categorization procedure (i.e. a positive and negative sentence classification stage)

In this section, we highlight of our system architecture along with the two applied models namely DNN and RNN.

3.1   Data Collection

With regard to the Arabic language, there is a little amount of existing corpus which is freely accessible for downloading, which sounds practically not large enough for

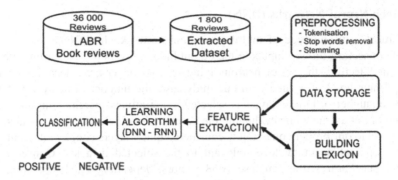

Fig. 1. The proposed sentiment analysis processing architecture.

artificial neural network training. Our dataset consists of book reviews which have been selected out of a large scale dataset of publication reviews published in Arabic language (LABR) [29]. To note, the LABR content is written in Modern Standard Arabic (MSA) as well as in Egyptian dialect. It involves about 63 k book reviews, each of which is rated from 1 to 5. At the end of the data collection stage, a compilation of 1800 Arabic book review dataset, written in MSA, have been gathered, loaded with either of two sentiment polarities: positive and negative.

Data Preprocessing

Data pre-processing is an important stage in the sentiment analysis process. The major preprocessing steps involved in this stage are, mainly, the sentence tokenization step, used to divide sentences into lists of words. Followed by the dataset parsing step including the deletion of all stop-words, punctuation marks, blank space, tab space, etc., while the final step involves converting each sentence containing words into their original word stem. Note that concerning the preprocessing stage, the NLTK [30] platform is applied to harmoniously fit and jointly operate with human language data.

Word Embedding and Lexicon

At this level, the word embedding technique as applied for text feature selection is presented. It actually consists in implementing the bag-of-word model. As our data is in linguistic format, rather than numerical format, it need be converted into a vector of features mode, since the entirety of the feature sets are required to have an identical length for the training process to take place. In a first step, our lexicon englobing all the unique words appearing in the dataset must be setup formed by means of Term Frequency-Inverse Document Frequency (TF-IDF) representation [31]. TF increases the weight of the terms (words) occurring more frequently in the document. IDF diminishes the weight of the terms figuring in all the dataset documents, and similarly increases the weight of the terms that occur in rare documents across the dataset. For the empirical study purposes 1068 unique words have been selected as size of vectors. In the second Step, a binary vector is implemented. Feature vectors are established for every word appearing in the sample sentence, thus 1 is accorded if the word appears to exist in the lexicon and 0 otherwise.

3.2   Deep Neural Networks (DNN)

As the most common type of multilayer neural network, the DNN stand as a feed forward connection linking inputs to the output hidden layers [8]. Regarding the DNN specific architecture, the set of neurons lying on each layer are selected to yield the most optimum level of accuracy. In our study case, the number of neurons is 500 per layer. The number of hidden layers is selected by iteratively incrementing the number of layers, one at a time while evaluating the accuracy level relevant to every increment. With respect to our proper context, it is considered that 4 layers could well help to yield the most optimal accuracy level relevant to the selected data set. Concerning the applied input features, they enclose 1068 vectors, depending on their recurrent frequency in the dataset (between 5 and 100 times) to be classified according to the model final output version into either positive or negative. The resulting architecture of the DNN model is shown in Fig. 2.

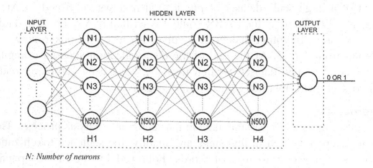

Fig. 2.   The DNN architecture.

3.3   Recurrent Neural Network (RNN)

In its complete form, the RNN represents a type of neural network that encompasses directed loops. The latter stand as the propagation of activations to future inputs in a sequence. Actually, the RNN is increasingly applied to NLP because it helps account for the words order relevant to each sentence, which could positively impact the expected results [32]. Noteworthy, also, is that the RNN architectures may take various forms, e.g. Long-Short Term Memory (LSTM), Gated Recurrent Unit (GRU), etc. In this regard, the LSTM stands as the most widely used models applied to DL in the context of NLP procedure, as it is capable of leaning long-term dependencies. Information in LSTM can be stored, written or read from a cell, much like data in a computer memory. In this context, the number of units determined to persist within the LSTM cells is equal to 256 units, with two hidden layer being applied. The resulting DBN model architecture is depicted in Fig. 3.

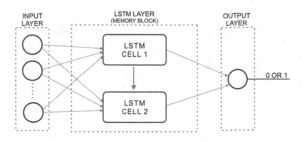

Fig. 3. The RNN architecture.

4 Experimentation

For the system's assessment purposes, the following accuracy rate evaluation method is applied, defined as: $Accuracy = (TP + TN)/(TP + TN + FP + FN)$. Where (TP) denotes the number of true positive sentiment items; (TN) stands for the number of true negatives; (FP) signifies the number of false positives, while (FN) denotes the number of false negatives. The entire dataset is divided into 80% (1440 sentences) destined to undergo the training process and 20% (360 sentences) oriented to undergo the DNN and the LSTM tested process. To note the back propagation proceeding is implemented for the sake of minimizing the error term of the network output in respect of the true sentiment class label regarding each single training case. Figure 4 illustrate the accuracy result as reached through the DNN applied experiment, attaining 64.4% in 200 epochs with batches size equal to 100 (Fig. 5).

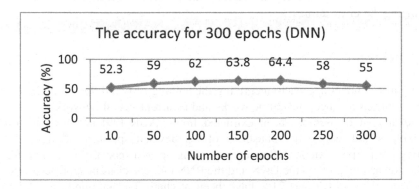

Fig. 4. Accuracy results per 300 epochs for DNN

As can be revealed through the implemented experiments, we have managed to achieve highly satisfactory results via LSTM applied model relevant to the existing deep neural networks, of the order of words in sentences and also he makes communication between LSTM layers for minimize error. Our approach provides an absolute accuracy improvement of 9.9 over the DNN approach proposed by Al-Sallab in [25].

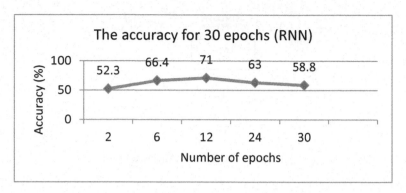

Fig. 5. Accuracy results per 30 epochs for RNN

Table 1 summarizes the performance of the proposed models in terms of accuracy, precision, recall and F1-score. F1-score = 2 * ((recall * precision)/recall + precsion).

Table 1. The results of two classifiers

	Accuracy	Precision	Recall	F1-score
DNN	64.4%	61.1%	75.3%	67.5%
DNN [25]	55.5%	–	–	44.5%
RNN (LSTM)	71%	68.3%	77%	72.4%

Due mainly to the fact that the applied model (LSTM) proves to helps greatly in accounting not only for the words' order at sentence level but also for the maintaining communication among the LSTM various layers, which contributes remarkably in minimizing the error term.

5   Conclusion

In this paper, a novel approach integrating the training of two different modes of deep neural architectures: deep neural networks and recurrent neural networks.

The applied dataset has been extracted from LABR [29] and used to test the effectiveness and assess the robustness of our devised approach, along with constructing our proper lexicon, following the setting up of a special vector of words. On test the attained dataset via the DNN and the RNN models, classification accuracy rates of respectively 64, 4% and 71% have been reached. To our mind, and following exhaustive biographical search, the present work appears to stand as a novel pioneering-study conducted to apply the RNN architecture to the context of Arabic sentiment classification area as appearing in social media. As a potential work perspective, we consider experimenting with larger dataset range useful for training other more highly thorough and deep architectures, along with attempting to devise architectural combinations in a bid to achieve even more remarkably improved accurate results.

Acknowledgement. The research leading to these results has received funding from the Ministry of Higher Education and Scientific Research of Tunisia under the grant agreement number LR11ES48.

References

1. Ravi, K., Ravi, V.: A survey on opinion mining and sentiment analysis: tasks, approaches and applications. Knowl.-Based Syst. **89**, 14–46 (2015)
2. Agerri, R., Artola, X., Beloki, Z., Rigau, G., Soroa, A.: Big data for natural language processing: a streaming approach. Knowl.-Based Syst. **79**, 36–42 (2015)
3. Moussa, S.B., Zahour, A., Benabdelhafid, A., Alimi, A.M.: New features using fractal multi dimensions for generalized Arabic font recognition. Pattern Recogn. Lett. **31**, 361–371 (2010)
4. Boubaker, H., Kherallah, M., Alimi, A.M.: New algorithm of straight or curved baseline detection for short Arabic handwritten writing. In: International Conference on Document Analysis and Recognition, ICDAR, p. 778. IEEE (2009)
5. Slimane, F., Kanoun, S., Hennebert, J., Alimi, A.M., Ingold, R.: A study on font-family and font-size recognition applied to Arabic word images at ultra-low resolution. Pattern Recogn. Lett. **34**, 209–218 (2013)
6. Elbaati, A., Boubaker, H., Kherallah, M., Alimi, A.M., Ennaji, A., Abed, H.E.: Arabic handwriting recognition using restored stroke chronology. In: International Conference on Document Analysis and Recognition, ICDAR, p. 411. IEEE (2009)
7. Kechaou, Z., Kanoun, S.: A new-arabic-text classification system using a hidden Markov model. KES J. **18**(4), 201–210 (2014)
8. Collobert, R., Weston, J.: A unified architecture for natural language processing: deep neural networks with multitask learning. In: International Conference on Machine Learning, ICML, pp. 160–167. ACM (2008)
9. Kechaou, Z., Wali, A., Ammar, M.B., Karray, H., Alimi, A.M.: A novel system for video news sentiment analysis. J. Syst. Inf. Technol. **15**(1), 24–44 (2013)
10. Kechaou, Z., Ammar, M.B., Alimi, A.M.: A multi-agent based system for sentiment analysis of user-generated content. Int. J. Artif. Intell. Tools **22**(2), 1350004 (2013)
11. Kechaou, Z., Ammar, M.B., Alimi A.M.: Improving e-learning with sentiment analysis of users' opinions. In: Global Engineering Education Conference, EDUCON, pp. 1032–1038. IEEE (2011)
12. Kechaou, Z., Ammar, M.B., Alimi A.M.: A new linguistic approach to sentiment auto-matic processing. In: Cognitive Informatics, ICCI, pp. 265–272. IEEE (2010)
13. Kechaou, Z., Wali, A., Ammar, M.B., Alimi A.M.: Novel hybrid method for sentiment classification of movie reviews. In: International Conference on Data Mining, DMIN, pp. 415–421. IEEE (2010)
14. Abdul-Mageed, M., Diab, M., Kuebler, S.: SAMAR: subjectivity and sentiment analysis for Arabic social media. Comput. Speech Lang. **28**, 20–37 (2014)
15. Abbasi, A., Chen, H., Salem, A.: Sentiment analysis in multiple languages: feature selection for opinion classification in web forums. ACM Trans. Inf. Syst. **26**(3), 12:1–12:34 (2008)
16. Shoukry, A., Rafea, A. Sentence-level Arabic sentiment analysis. In: Proceedings of Collaboration Technologies and Systems, CTS, pp. 546–550. IEEE (2012)
17. Tartar, A., Abdul-Nabi, I.: Semantic sentiment analysis in Arabic social media. J. King Saud Univ. Comput. Inf. Sci. **29**, 229–233 (2016)

18. Collobert, R., Weston, J., Bottou, L., Karlen, M., Kavukcuglu, K., Kuksa, P.: Natural language processing (almost) from scratch. J. Mach. Learn. Res. **12**, 2493–2537 (2011)
19. LeCun,Y., Bengio, Y.: Convolutional networks for images, speech, and time series. In: The handbook of brain theory and neural networks, pp. 255–258. ACM (1995)
20. Hinton, G.E., Osindero, S., Teh, Y.: A fast learning algorithm for deep belief nets. Neural Comput. **18**, 1527–1554 (2006)
21. Socher, R., Perelygin, A., Wu, J.Y., Chuang, J., Manning, C.D., Ng, A.Y., Potts, C.: Recursive deep models for semantic compositionality over a sentiment Treebank. In: Conference on Empirical Methods in Natural Language Processing, EMNLP, pp. 1631–1642. ACL (2013)
22. Socher, R., Pennington, J., Huang, E.H., Ng, A.Y., Manning, C.D.: Semi-supervised recursive autoencoders for predicting sentiment distributions. In: Conference on Empirical Methods in Natural Language Processing, EMNLP, pp. 151–161. ACM (2011)
23. Dos-Santos, C.N., Zadrozny, B.: Learning character-level representations for part-of-speech tagging. In: International Conference on Machine Learning, ICML, pp. 1818–1826. ACM (2014)
24. Liu, G., Xu, X., Deng, B., Chen, S., Li, L. A hybrid method for bilingual text sentiment classification based on deep learning. In: Software Engineering, Artificial Intelligence, Networking and Parallel/Distributed Computing, SNPD, pp. 93–98. IEEE (2016)
25. Al-Sallab, A., Baly, R., Badaro, G., Hajj, H., El-Hajj, W., Shaban, KB.: Deep learning models for sentiment analysis in Arabic. In: The Second Workshop on Arabic Natural Language Processing, ANLP, pp. 26–31. ACL (2015)
26. LDC homepage. https://catalog.ldc.upenn.edu/LDC2005T20
27. Dahou, A., Xiong, S., Zhou, J., Haddoud, M.H., Duan, P. Word embeddings and convolutional neural network for Arabic sentiment classification. In: International Conference on Computational Linguistics, COLING, pp. 2418–2427. ACL (2016)
28. Alayba, A.M., Palade, V., England, M., Iqbal, R.: Arabic language sentiment analysis on health services. CoRR abs/1702.03197 (2017)
29. Aly, M., Atiya, A.: Large-scale Arabic Book Reviews Dataset. Association of Computational Linguistics, ACL (2013)
30. Bird, S., Loper, E., Klein, E.: Natural language processing with Python. O'Reilly Media Inc., Sebastopol (2009)
31. Manning, A.H., Raghavan, C., Schütze, P.: Introduction to Information Retrieval, 1st edn. Cambridge University Press, New York (2008)
32. Mikolov, T., Karafiat, M., Burget, L., Cernocky, J., Khudanpur, S.: Recurrent neural network based language model. In: Interspeech, pp. 1045–1048. ISCA (2010)

Collective Actions in Three Types of Continuous Public Goods Games in Spatial Networks

Zimin Xu, Qiaoyu Li, and Jianlei Zhang[✉]

Department of Automation, College of Computer and Control Engineering,
Nankai University, Tianjin 300071, China
jianleizhang@nankai.edu.cn

Abstract. Collective action in the provision of pubic goods is analyzed in the framework of three kinds of public goods dilemmas routinely encountered in real-life situations. We study the evolution of cooperation in structured populations within three PGG models: the traditional public goods game (PGG), complementary public goods game (PPGG) and containable public goods game (TPGG), differing in supplying patterns of public goods. In addition, we extend the combination of dual strategy (cooperation and defection) to a portfolio of multiple strategies. We reveal that, is a fundamental property promoting cooperation in groups of selfish individuals, irrespective of which social dilemma applies. For a parallel comparison, it is found that the system in PGG and PPGG can perform comparatively better than TPGG, which reduces the provision of the public goods. Our study can be helpful in effectively portraying the characteristics of cooperative dilemmas in real social systems.

Keywords: Evolutionary game · Multi-agent systems · Complex networks

1 Introduction

In multi-agent systems, collective dilemmas are situations in which the optimal decision of an individual contrasts with the optimal decision of the group, which is very common in the real world. Rational agents, who attempt to maximize their In multi-agent systems, collective dilemmas are situations in which the optimal decision of an individual contrasts with the optimal decision of the group, which is very common in the real world [1,2]. Rational agents, who attempt to maximize their own benefits, may thus attempt to free-ride on the others - benefiting from the contributions of others without offering their own to the group. In investigating this problem the often-used tool for modeling the dilemma is evolutionary game theory [3–5].

The popular archetype model of reciprocal altruism, such as the Prisoner's Dilemma Game (PDG), is a two-person game. The game represents the simplest possible form of competitive environment, since cooperation is only meaningful

© Springer International Publishing AG 2017
D. Liu et al. (Eds.): ICONIP 2017, Part V, LNCS 10638, pp. 677–688, 2017.
https://doi.org/10.1007/978-3-319-70139-4_69

with at least two persons [6–8]. However, multiple agents instead of two individuals are usually involved in many real-world events or projects, and thus they naturally constitute genuine N-person problems instead of two-person problems. To simulate the interactions and cooperations among a group of agents, many researchers treated these N-person problems as a summation of many two-person problems. So while the PDG is unrivaled in popularity when it comes to studying the evolution of cooperation through pairwise interactions, the closely related public goods game (PGG) accounts for group interactions as well. Similarly, the N-person battle of sexes game [9] and the N-person evolutionary snowdrift game or the stag hunt game [10,11] are N-person extension of the corresponding games, which constitute powerful metaphors to describe conflicting situations often encountered in natural and social sciences [12,13].

Emergence of altruism under social dilemmas can be explained by various mechanisms, such as kin selection, direct reciprocity, indirect reciprocity and group selection [14–17]. In the last decade altruism is also promoted by the viscosity of populations, where players are aligned on a complex structured system which goes beyond the regularity of simple lattices [18–21]. Comparing with the well-mixed or fully-connected case, a typical setup is the following: agents are assigned to the nodes of a network, which can be a regular lattice or more complex structure. Each individual occupies a vertex and is constrained to play with its immediate neighbors along the edges.

The PGG, in particular, has proven itself times and again as the classic paradigm that succinctly captures the essential social dilemma that emerges as a consequence of group and individual interests being inherently different [22–24]. In its simplest form, the game has two strategies: cooperate and defect. Cooperators pay a cost for contributing to the public goods, whereas defectors refrain from doing so. After all individuals are given the chance to contribute to the public goods, the accumulated contribution is multiplied by an enhancement factor. Then the result is distributed equally among all members of the group. Since defectors can always exploit the benefits of living in a group without contribution, being free-riders (defection) is the dominant strategy.

Although most of the relevant literatures have focused on the N-person PDG in the form of provision of PGGs, there are also other collective dilemmas routinely encountered in social sciences. Here we introduce other two special cases of public goods provision, which differ in the decisive patterns of the private supply from the common adopted PGG extended from traditional PDG.

(i) *Complementary public goods game* (PPGG). It describes such a scene that the individual contributions of participants are complementary. It involves a group of N individuals, who can contribute any cost to the public goods. After all the participants are given the chance to contribute, the lowest contribution among group members will be found out. Then the final contribution is a multiplied amount of the lowest contribution by an enhancement factor $N\eta$. Finally all individuals in the group will benefit from the common pool equally, irrespective of individual contribution. There are many such social dilemmas in many real-world events. For example, when heavy rain signals the possibility of floods which threat the safety of a dam, this public construction project requires the coordinated action of villagers living in the low-lying terrain. If each villager is assigned to construct the part of dam in front of it. In this case, once the flood waters washed off the lowest part of the dam, the whole village will be submerged. The effectiveness of flood protection depends on the lowest hight of the dam construction. Thus, a feature of these examples which is captured in the framework of PPGG is the fact that the final public goods depends on the minimum contribution among the members of the construction group.

(ii) *Containable public goods game* (TPGG), meaning that the contribution of individual players can be contained by others. Assuming as usual, N players participate this many-person game which will be beneficial to the entire group. Each one can contribute any cost to perform the given task. After all the participants are given the chance to contribute, the highest contribution among group members will be found out. Then the final contribution is a multiplied amount of the highest contribution by an enhancement factor $N\eta$. Clearly, the assumption is that all individuals in the group will equally enjoy the same benefit, irrespective of individual contribution. The collective action captured in the framework of TPGG is also common in real-life situations. Taking the intelligence answering competition for example, the result of a group is decided by the best performance of its cooperating members. The relatively poor performance of members in this group does not affect the final result brought by the best players. Notably, the achieved public goods is determined by the highest contribution of group members, and shared evenly by all group members.

Furthermore, agents usually have various choices to invest any part of their wealth according to their own personal situations, instead of just none or all. Thus, instead of the feasible choices between two alternatives (cooperate and defect), each player has a variety of possible options, and the diversity of strategies is denoted by multiple contributions of players in our study. A variation on the PGG, in which each individual can choose not only full cooperation or no cooperation, but also intermediate cooperation levels in the form of contributing funds to a collectively advantageous group project.

The scope of the present paper is to model the three different PGG models and analyze the strategic behavior of the participators by means of game theory. In the next section we describe the employed spatial PGGs and other details of the evolutionary process. Section 3 is devoted to the presentation of main findings, whereas in the last section conclusions are summarized.

2   Preliminaries of Evolutionary Game Theory

Evolutionary game model description: As mentioned, evolutionary game theory provides an interdisciplinary framework to embody several relevant features of the spread of strategy choices. Strategy is one of the key factors in the game playing and the payoff calculation. Generally, for any player in the multi-agent system, her provided strategy profile is denoted by $\mathcal{S} := \{A, B, \ldots\}$. Among these, for example A or B, will be the strategy each player can play with in the multi-agent system.

The two-strategy payoff matrix: Here, we consider a game contested by players who can make an option from two strategies, i.e. A and B. Players choose strategies from a binary set $\mathcal{S} := \{A, B\}$. In general, a A-player interacting with another A-player receives the benefits of a. If she interacts with a B-player, she obtains the payoff os b. Similarly, the B-player receives c from the A-player and d from other B-players. The payoff gained by each player depends on the following payoff matrix,

$$\begin{array}{c} \quad A \quad B \\ \begin{array}{c} A \\ B \end{array} \left(\begin{array}{cc} a & b \\ c & d \end{array} \right) \end{array}$$

Payoff Calculation in networked system: In a networked game, the states of nodes can be called strategies. The fitness of an individual player (node) is acquired by playing games with its neighbors.

Let $\mathcal{G}(\mathcal{V}, \epsilon)$ denote an undirected network whose node set $V = 1, \ldots, n$ corresponds to agents who participate in the two-strategy two-player games. Assuming that the strategy profile is $\mathcal{S} := \{A, B, \ldots\}$. After game playing individuals accumulate their payoffs according to the payoff matrix.

At each time t, we denote the strategy state of the population by $x(t) = [x_1(t), \ldots, x_n(t)]^T$, each belongs to the strategy profile of $\mathcal{S} := \{A, B, \ldots\}$. In the two-strategy of games, for example, the cooperation or defection. Total payoffs are given by

$$y_i(t) = \omega_i \Sigma_{j \in \mathcal{N}_i} \tag{1}$$

Updating Rules: Evolutionary game dynamics generally involve how players update their strategies as time evolves. Several strategy update rules are customary in evolutionary game theory. Many updating rules are based on replication or imitation. For example, the payoffs of the focal player i and a random neighbor are determined, according to their characters and their local groups. Then, the probability that player i will adopt the strategy of one of its randomly chosen neighbors s_j depends on the payoffs p_i and p_j of both players in the light of

$$q(x_i(t) \to x_j(t)) := \max\{0, \min[1, \frac{\lambda}{|\mathcal{N}_i|}(y_j(t) - y_i(t))]\} \tag{2}$$

3   Model Settings

A diluted well-mixed $N = 100 \times 100$ population is modeled by the lattice in which each player has four neighbors. Participators contribute to the collective welfare at a random personal cost, a mark that the strategy set is extended to the continuous strategy distribution. In one generation, each agent interacts, i.e. participates in the PGG with all its neighbors. Here, the games staged on a lattice grid are discussed below respectively.

(i) *Traditional public goods game* (PGG). In this model, individuals have the opportunity to contribute any amount of money ranging from 0 to c under full anonymity. For simplicity, the costs c are set to unity in the remainder of the text. It is played in interaction groups of size N. Players must decide simultaneously whether they wish to contribute to the common pool or not. In this case, all contributions of N members are summed and multiplied by an amplification factor η. Hence, the final collected investment is

$$P_{sum} = \eta \sum_{i=1}^{N} c_i; \tag{3}$$

where c_i denotes the contribution amount of player i.

Then, the resulting amount is divided equally among all N members irrespective of their initial decision. Thus, player i obtains the following net payoff

$$P_i = \frac{P_{sum}}{N} - c_i \tag{4}$$

It is easy to see that none contribution is the better choice irrespective of the opponent's selection, so defection is expected to be dominating, though mutual cooperation improves the average payoff.

(ii) *Complementary public goods game* (PPGG). In an N-person game, an agent competes with a group of $N-1$ other agents. There will be a task to be done and N individuals have the opportunity to contribute any amount of money ranging from 0 to 1 under full anonymity. Players must decide simultaneously whether they wish to contribute to the common pool or not. It is different from the traditional PGG where the common goods is typically evaluated by summing up the contributions of all members. Here, the final investment depends on the smallest contribution amount in the interacting group, and then are multiplied to take into account synergetic effects of cooperation. Hence,

$$P_{sum} = \eta N \times Min(c_i), i = 1, 2, ...N; \tag{5}$$

where c_i denotes the contribution amount of player i in the proposed PPGG. $Min(c_i)$ is the smallest contribution amount among the all N participating members.

Then, the obtained contribution is equally distributed among all group members irrespective of their initial decision. Thus, if the agent under consideration contributes c_i, then it has a net payoff

$$P_i = \frac{P_{sum}}{N} - c_i = \eta \times Min(c_i) - c_i \tag{6}$$

If the payoff determines reproductive fitness, evolution will lead to the spreading of contributing nothing. However, the payoff for mutual defection is smaller than the payoff for mutual cooperation and thus creates a dilemma.

(iii) *Containable public goods game* (TPGG), is illustrated as a situation: N individuals have the opportunity to disburse any amount of contribution ranging from 0 to 1 under full anonymity. Players must decide simultaneously whether they wish to contribute to the common pool or not. With N agents in the group, the total contributions $\eta N \times Max(c_i)$ are shared evenly among all members in the group, where η is called the public goods multiplier. Then the total contribution equally benefits all members and is given by

$$P_{sum} = \eta N \times Max(c_i), i = 1, 2, ...N. \tag{7}$$

where c_i denotes the contribution amount of player i, N is the number of all participants. $Max(c_i)$ is the largest contribution among all the participants.

Then the resulting public goods are distributed equally amongst all the group members irrespective of their initial contributions. Thus, a given player i obtains the following net payoff

$$P_i = \frac{P_{sum}}{N} - c_i = \eta \times Min(c_i) - c_i \tag{8}$$

Starting from an equal fraction of cooperators and defectors, forward iteration is performed according to the following elementary steps. First, a randomly selected player x acquires its payoff P_x by playing the game with its neighbors. Next, one randomly chosen neighbor y, also acquires its payoff P_y by playing the game with its neighbors. Last, player y tries to enforce its strategy s_y on player x according to the probability

$$W_{s_x \to s_y} = \frac{1}{1 + exp[(p_x - p_y)/K]}, \tag{9}$$

where K characterizes the magnitude of noise involving many different effects (fluctuations in payoffs, errors in decision, individual trials, etc.). $K = 0$ and $K \to \infty$ denote the completely deterministic and completely random selection of the y's strategy s_y respectively. While for any finite positive values, K incorporates the uncertainties in the strategy update, where the better one's strategy is definitely imitated, but there is a probability of selecting the worst ones. Herein, we only consider the simple situation for individuals' selection probability and simply set $K = 0.1$ in this paper. $0 < K \to 1$ implies that the better performing player is readily adopted, whilst it is not completely impossible to adopt the strategy of a worse performing player. After every such iteration cycle, we set s_x and s_y for $\forall i, j$.

4   Dynamic Results and Discussion

We will now address by the use of Monte Carlo simulations how the dynamic model evolves in time, and the central question is the steady state. Here, we track the evolution of strategies by iterating the model for a maximus of 10^6 time steps. Here, a stationary state is one in which no further change in strategy distribution is possible. Initially, the strategies of the agents are randomly assigned and distributed in $[0, 1]$.

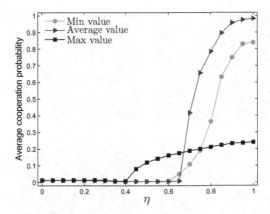

Fig. 1. (Color online) The average cooperation frequency as a function of η in the framework of three kinds of PGGs respectively, each drawn with a different color. It can be observed that the cooperation can be better promoted in traditional PGG than TPGG and PPGG. The interaction graph is a lattice graph with the average degree $\langle k \rangle = 4$. Equilibrium frequencies of cooperators are averaged over 10^4 time steps after a transient of 10^5 time steps. Curves are averages over 100 independent realizations of both the networks and the initial conditions for a population of size 10^4. Lines are just guides for the eye.

We start by comparing results obtained with the presently introduced evolutionary models, by visually inspecting the stationary average cooperation probability f_c obtained versus the synergy factor η (see Fig. 1). As can be deduced from results presented in Fig. 1, in strongly defection-prone environments, cooperators are outperformed by defectors. Our explanation is related with the fact that large temptation to defect enhances the amount of defection as one would expect. Notably, f_c is a monotonically increasing function as synergy factor η, irrespective of the adopted PGGs. More remarkably, the minimally required η for cooperative behavior to survive is different in the three kinds of PGGs. The critical value of η where cooperators emerge is about $\eta = 0.4$ in the TPGG, much lower than the $\eta > 0.6$ in the traditional setup of PGG. Additional insight is provided in Fig. 1 that the maximum values of f_c are closely related with the studied PGG models. More precisely, when $\eta \approx 1$, a small fraction of surviving cooperators $f_c < 0.3$ in the context of TPGG, $f_c > 0.8$ in PPGG, and $f_c \approx 1$ in the traditional PGG. This is a straightforward consequence of different provision forms of the proposed PGGs.

On a more practical level, we are interested in exploring the the instantaneous snapshots of individual contributions in the framework of the adopted models respectively. In Fig. 2, we first present the snapshots of cooperation frequency distributions of the whole population in PPGG. As expected based on results presented in Fig. 1, the majority of the population adopt low contribution amounts. As Fig. 2 illustrates, the few individuals with high contribution amounts (red) are scattered among the majority population with low contribution amounts (blue). In this case, cooperators that are disadvantageous in

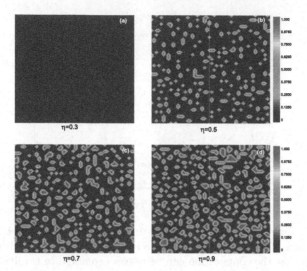

Fig. 2. (Color online) Characteristic snapshots of cooperation frequency distributions of the whole population playing TPGG when starting from a random initial state. The snapshots were take at $\eta = 0.3$, $\eta = 0.5$, $\eta = 0.7$ and $\eta = 0.9$ respectively. All panels are snapshots of the full 100×100 lattices. The color code in the right indicates the value of cooperation probability, where the color's change from blue to red corresponds to the increment of ρ_c from 0 to 1.

the dilemma benefit more from the formation of clusters on the square lattice and so protect themselves against the exploitation by defectors. Once clusters of cooperators have formed, selection for strong game partners can effectively shield clusters of cooperators from the invasion of defectors. Strategy updating promotes a local assortment of strategies cooperators 'breed' cooperators.

In addition, cooperators within the cluster attract interactions with individuals at the cluster boundary. As demonstrated in Fig. 2, individuals at the boundary of clusters (red) contribute less (e.g., green) than that in the clusters. In this way payoffs of cooperator nodes inside the cluster and at its fringe are enhanced and interactions with defectors are avoided. What is important is to note that the adopted strategies are relatively monotonous and do not show much diversity. One possible explanation is closely related with the fact that the final contribution is a multiplied amount of the highest contribution by an enhancement factor. All individuals in the group will equally enjoy the same benefit, irrespective of individual contribution. From the perspective of a single player choosing to contribute nothing always yields a higher payoff regardless of the action of the group members. Further, the total contribution is directly affected by the largest individual contribution amount in TPGG. Therefore, the actual contributions less than maximum contribution do not affect the collected amount. According to this, the payoff difference among individuals was reduced in this social dilemma, thus, leading to the final state of low cooperation level.

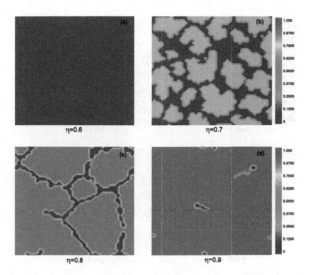

Fig. 3. (Color online) Typical snapshots of cooperation frequency distributions of the whole population playing traditional PGGs from a random initial state. The snapshots were take at $\eta = 0.3$, $\eta = 0.5$, $\eta = 0.7$ and $\eta = 0.9$ respectively. All panels are snapshots of the full 100×100 lattices.

In addition, increasing the value of η is beneficial for the survival of cooperation, however, significant changes did not occur with the increment of η.

Figure 3 summarizes the dependence of the average cooperation level as a function of the synergy factor η in the traditional PGGs. It appears that clusters of higher contribution amount also occur in the PGG. Clearly, the nature of the game does not induce a wide diversity in the emerging strategy distributions. As depicted in Fig. 3, the small number of clusters assemble the majority of the population. The cluster sizes the numbers are both smaller than that in Fig. 2. As the synergy factor η approaches to 1, the cluster numbers decrease and the colors turned to be red finally. In this case, the highest cooperation level is attained, though the emergence is difficult.

Figure 4 sheds more light on the dependence of the fraction of cooperators with respect to the synergy factor η in the framework of PPGG. Fascinatingly, clusters did not occur in the framework of PPGG. The individual contributions tend to be very consistent, that is, the contribution differences between the whole population is trivial. The possible explanation is also closely related with the fact that the final contribution is a multiplied amount of the lowest contribution by an enhancement factor. All individuals in the group will then equally enjoy the same benefit. From the perspective of a single player, choosing to contribute nothing always yields a higher payoff regardless of the action of the group members. Further, the total contribution is directly affected by the largest individual contribution amount in TPGG.

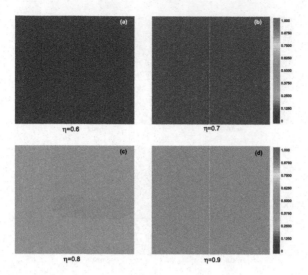

Fig. 4. (Color online) Typical snapshots of cooperation frequency distributions of the whole population playing PPGGs from a random initial state. The snapshots were take at $\eta = 0.3$, $\eta = 0.5$, $\eta = 0.7$ and $\eta = 0.9$ respectively. All panels are snapshots of the full 100×100 lattices.

5   Conclusions

In summary, we extend the public goods game strategy set from a common combination of cooperation and defection to the continuous strategy distribution. We quantify, model and simulate three kinds of collective behaviors under conflicts between individuals and public interests. Simulation results show that cooperation behavior depends on the underlying social dilemma. Cooperation behavior emerges more easily in the framework of traditional PGG than other two types of PGGs. Taken together, our results cast prospects for modeling the collective action and dilemmas among the structured population in a hopeful light. The parameter settings and adjustment differ in the framework of different collective dilemmas. These models also demonstrate the importance of establishing effective models for describing the collective actions in multi-agent systems.

Acknowledgments. This work was supported by the National Natural Science Foundation of China (Grant Nos. 61603199, 61603201 and 61573199), and the Foundation of Key Laboratory of Machine Intelligence and Advanced Computing of the Ministry of Education (Grant No. MSC-201709A).

References

1. Smith, J.: Evolution and the Theory of Games. Cambridge University Press, Cambridge (1982)

2. Nowak, M.A., Sigmund, K.: Phage-lift for game theory. Nature **399**, 367–368 (1999)
3. Cleary, A.S., Leonard, T.L., Gestl, S., Gunther, E.: Tumour cell heterogeneity maintained by cooperating subclones in Wnt-driven mammary cancers. Nature **508**(7494), 113–117 (2014)
4. Xu, Z., Zhang, J., Zhang, C., Chen, Z.: Fixation of strategies driven by switching probabilities in evolutionary games. EPL (Europhys. Lett.) **116**(5), 58002 (2017)
5. Ramazi, P., Riehl, J., Cao, M.: Networks of conforming or nonconforming individuals tend to reach satisfactory decisions. Proc. Natl. Acad. Sci. **113**(46), 12985–12990 (2016)
6. Choi, W., Yook, S.H., Kim, Y.: Percolation in spatial evolutionary prisoner's dilemma game on two-dimensional lattices. Phys. Rev. E **92**(5) (2015)
7. Zhang, J., Zhang, C., Chu, T.: Cooperation enhanced by the survival of the fittest' rule in prisoner's dilemma games on complex networks. J. Theor. Biol. **267**, 41–47 (2010)
8. Ghoneim, A., Abbass, H., Barlow, M.: Characterizing game dynamics in two-player strategy games using network motifs. IEEE Trans. Syst. Man Cybern. **38**, 682–690 (2008)
9. Zhao, J., Szilágyi, M., Szidarovszky, F.: An n-person battle of sexes game. Phys. A **387**, 3669–3677 (2008)
10. Chan, C.H., Yin, H., Hui, P.M., Zheng, D.F.: Evolution of cooperation in well-mixed n-person snowdrift games. Phys. A **387**, 2919–2925 (2008)
11. Pacheco, J.M., Santos, F.C., Souza, M.O., Skyrms, B.: Evolutionary dynamics of collective action in n-person stag hunt dilemmas. Proc. R. Soc. Lond. B **276**, 315–321 (2009)
12. Vamvoudakis, K.G., Hespanha, J.P.: Online optimal operation of parallel voltage-source inverters using partial information. IEEE Trans. Industr. Electron. **64**(5), 4296–4305 (2017)
13. Wang, J., Hipel, K.W., Fang, L., Xu, H., Kilgour, D.M.: Behavioral analysis in the graph model for conflict resolution. IEEE Trans. Syst. Man Cybern.: Syst. (2017)
14. Wedekind, C., Milinski, M.: Cooperation through image scoring in humans. Science **288**, 850–852 (2000)
15. Nowak, M.A.: Five rules for the evolution of cooperation. Science **314**, 1560–1563 (2006)
16. Groot, N., De Schutter, B., Hellendoorn, H.: On systematic computation of optimal nonlinear solutions for the reverse stackelberg game. IEEE Trans. Syst. Man Cybern.: Syst. **44**(10), 1315–1327 (2014)
17. Shahrivar, E.M., Sundaram, S.: The strategic formation of multi-layer networks. IEEE Trans. Netw. Sci. Eng. **2**(4), 164–178 (2015)
18. Nowak, M.A., May, R.M.: Evolutionary games and spatial chaos. Nature **359**, 826–829 (1992)
19. Weitz, J.S., Eksin, C., Paarporn, K., Brown, S.P., Ratcliff, W.C.: An oscillating tragedy of the commons in replicator dynamics with game-environment feedback. Proc. Natl. Acad. Sci. **113**(47), E7518–E7525 (2016)
20. Zhang, C., Zhang, J., Xie, G., Wang, L.: Coevolving agent strategies and network topology for the public goods games. Eur. Phys. J. B **80**, 217–222 (2011)
21. Zhang, J., Zhang, C., Cao, M., Weissing, F.: Crucial role of strategy updating for coexistence of strategies in interaction networks. Phys. Rev. E **91**(4), 042101 (2015)

22. Hauert, C., De Monte, S., Hofbauer, J., Sigmund, K.: Volunteering as Red Queen mechanism for cooperation in public goods game. Science **296**, 1129–1132 (2002)
23. Kümmerli, R., Burton-Chellew, M.N., Ross-Gillespie, A., West, S.A.: Resistance to extreme strategies, rather than prosocial preferences, can explain human cooperation in public goods games. Proc. Natl. Acad. Sci. USA **107**, 10125–10130 (2010)
24. Santos, F.C., Santos, M.D., Pacheco, J.M.: Social diversity promotes the emergence of cooperation in public goods games. Nature **454**, 213–216 (2008)

Analysing the Evolution of Contrary Opinions on a Controversial Network Event

Qu Liu[1], Yuanzhuo Wang[2], Chuang Lin[1(✉)], and Guoliang Xing[2]

[1] Institute of Computer Science, Tsinghua University,
Haidian District, Beijing, China
qu-liu13@mails.tsinghua.edu.cn, chlin@tsinghua.edu.cn
[2] Institute of Computing Technology, Chinese Academy of Science,
Haidian District, Beijing, China
{wangyuanzhuo,xingguoliang}@ict.ac.cn

Abstract. With the growing popularity of social networking services, network public opinion gradually plays an important role in social life. When a controversial network event happens, what people concern about is which opinion in the contrary opinions will be widely accepted by people and how long this event lasts. To solve this problem, we propose a social evolutionary game model based on Hawk-Dove game to simulate how contrary opinions evolve in social network. The effectiveness of our model is validated by actual data. This model can be used to estimate the potential dominant opinion group and the time length of the controversy. Besides, our simulation reveals some special features of the evolution process and results. This study may be useful for network public opinion supervision and market research.

Keywords: Social network · Opinion dynamics · Evolutionary game · Controversial network event

1 Introduction

Unquestionably, Social Networking Services (SNS) is essential to people's daily life nowadays. Everyday people chat, view information, post their ideas and repost messages they are interested in through SNS. Thanks to the open environment of SNS, people feel freer to express their opinions and the will of expressing themselves is becoming stronger and stronger. If one saw a twitter expressing the same opinion with his, his opinion would be enriched and his belief in this opinion would be enhanced. When people holding different opinions meet together, a heated argument possibly happens. In social network, controversial network event could happen in almost any area you can imagine. People may hold diverse opinions towards a new policy, a new movie, a sports match, a new product, an academic viewpoint and so on. People holding different opinions may debate, argue, even quarrel over a specific topic. The argument often result in no winner or loser but this process changes both sides' understanding of the topic. Contrary opinions diffuse and evolve in social network, influencing network public

© Springer International Publishing AG 2017
D. Liu et al. (Eds.): ICONIP 2017, Part V, LNCS 10638, pp. 689–698, 2017.
https://doi.org/10.1007/978-3-319-70139-4_70

opinion. It goes without saying that network public opinion matters to many people, who may benefit or get hurt from it. For instance, the opinion shown in people's discussion on SNS may affect the decision of a new policy, the reputation of a public figure, the assessment of a new product, the president campaign, etc. Therefore, researching on the evolution process of contrary opinions in social network is of great significance.

Opinion evolution is the process that individuals holding different opinions interact with each other and their opinions change as time goes on. Opinion diffusion and evolution has been studied since the early 2000s. Sznajd-Weron proposed Sznajd model [1], in which the individuals are easily affected by others and then abandon their own opinion. This model describes the phenomenon that opinion diffuses from the group's inside to outside. Bounded confidence model [2,3] describes the social phenomenon that people won't even interact with the people holding opinion a far cry from them. The CODA [4] (Continuous Opinions and Discrete Actions) updating rule emphasizes the phenomenon that though people make the same choices, their extent of how they believe in this opinion may be different. This model also explains how individuals with extreme opinions emerge. These early works of opinion dynamics provide some basic concepts and methods which are instructive to the future works in this research area. However, these models are not designed for the opinion evolution happened in online social networks these days.

Recent studies including [5–9] provide different research prospectives on the topic. Xu et al. focus on modeling the emergence of public opinion in network forum [5] while Zhang emphasizes the evolution of two different opinions [6]. However, they didn't adopt game theory in their research. Zheng et al. proposed a bargaining model with game theory to study opinion dynamics [8]. However, the studies mentioned above are not aimed at the social network prevalent nowadays, like twitter and weibo, since their models are not built on a topology network. Furthermore, these studies fail to consider the evolution of network topology. The fact is people in social networks are possibly change their neighbors from time to time. Jiang et al. mentioned the influence of network topology when information diffusing over social network [7] and Yu et al. proposed a social evolutionary game model to study the dissemination process of competitive information [9], but they didn't study the diffusion of two contrary opinions in the context of controversial network event.

In this paper, a contrary opinion evolution model based on social evolutionary game is presented. This model takes Hawk-Dove game as its game model, focusing on modeling the evolution of people's opinion and network topology. The effectiveness of the model is validated through comparison with actual data. We carry out the simulation of this model and several meaningful conclusions are summarized. The rest of the paper is organized as follows. Section 2 interprets what is social evolutionary game. In Sect. 3, our social evolutionary game model is specifically introduced. Section 4 shows the simulation of the model and our analysis of the results. Finally, our conclusion is drawn in Sect. 5.

2   Social Evolutionary Game

Evolutionary game theory [10] is originally used to study the evolution of species in nature world. The interactions of people in social network can be regarded as game behaviors. In an evolutionary game model, people may take different actions when interacting with other people in order to increase their utilities, which is a better hypothesis about people's behavior than everybody obeying fixed interaction rules. Social evolutionary game [11] is a modeling method used to investigate the evolution process of social network.

There is only one type of game played in a social evolutionary game model. The utility, the agent's short-term concern, is its cumulative payoffs obtained from his opponents. The reputation, the agents' long-term concern, is his parters' evaluation on his historical behavior. Agents independently update their game strategy and adjust their partnerships with others as time goes on so as to improve his short-term utility and long-term reputation. This modeling method proves its effectiveness in many research works including [9,12–14].

Social evolutionary game is taken as our modeling framework. Naturally, the model ought to be modified according to the scenario of controversial network event discussed in this paper. Despite considering taking Hawk-Dove game as the game model, plenty of details including the set of opinion threshold, various interaction rules, the definition of reputation, the rules of strategy update and partnership adjustment are all modified compared to the original model in [11]. These modifications with certain originality makes our model more specific and closer to reality.

3   The Model

The interaction rules, strategy updating rules and partnership adjusting rules are the basic components of a social evolutionary game model.

To begin with, a topological graph $G = \{V, E\}$ is set to represent the social network, where the nodes are expressed by $V = \{v_1, v_2, \ldots, v_n\}$, representing the individuals, and the edges are expressed by $E = \{e_{ij} | 1 \leq i \leq n, 1 \leq j \leq n\}$, representing the partnerships between them, for instance, one follows another in Twitter. By convention, an opinion value O is assigned to each individual. Positive or negative value of O represents contrary opinions, denoted by α and β, and the absolute value of O i.e. $|O|$ represents how one believes in the viewpoint. O_M is the predefined maximum value of $|O|$. The meaning of one's $|O|$ reaching O_M is that he has completely faith in this viewpoint. A positive number Tr is defined as the opinion threshold. When $|O| \leq Tr$, it is regarded as an individual with no opinion to the topic. Apart from this, the individual holds α opinion when $O > Tr$ or β opinion when $O < -Tr$. Only a few individuals are selected to be the opinionated individuals whose opinions are extreme. Their opinion value O is assigned by O_M or $-O_M$ before the simulation starts. Most individuals don't hold any opinion to the topic at the beginning.

3.1 Interaction Rules

The interaction rules are the core of our social evolutionary game model. Hawk-Dove game [15] is applied to one of the interaction rules. Hawk-Dove game describes the antagonistic behaviors between two sides. In this game, one may hold 'Hawk' strategy, to fight for benefit, or 'Dove' strategy, to avoid conflict. A 'Dove' gets nothing in the game, yet it reduces losses to some extent. A 'Hawk' gets some benefit if its opponent is a 'Dove', however, it will suffer serious losses if the opponent is also a 'Hawk'. Obviously, the Hawk-Dove game doesn't encourage cooperation. Individuals in the game only care about how to get more benefits and avoid losses. This game model fits our presupposed situation: controversial network event where people holding contrary opinions collide in social networks. In our model, St denotes the strategy held by an individual, the value of which may be H, for Hawk strategy, or D, for Dove strategy. [16–18] are some of the previous studies on Hawk-Dove games in evolutionary game.

Table 1. Hawk Dove game payoff matrix

	H	D
H	P, P	T, S
D	S, T	A, A

The payoff matrix of Hawk-Dove game is shown in Table 1. In this payoff matrix, P denotes punishment, S denotes surrender, T denotes temptation and A denotes avoidance. There are some constraints of these parameters according to the definition of Hawk-Dove game.

$$\begin{cases} P < S < 0 \\ T > A \geq 0 \end{cases} \tag{1}$$

We assume that all the individuals are willing to strengthen their own opinions, so they regard the increase of their own opinion value as their benefit, except the ones with maximum opinion value who think they ought to change their opponents' mind. The interaction rules in our model is more complicated than a single Hawk-Dove game, since there are different situations of interactions:

(1) Interactions between individuals with the same opinions: they exchange their knowledge on the topic, both of their opinion values increase by R.
(2) Interactions between individuals with different opinions: they two play a Hawk-Dove game, the increment of their opinion values is the payoffs they get in the game.
   - If they both hold Hawk strategy, a heated debate may happen between them. After obtaining enough information on the opinion they disapprove, their beliefs on their own opinion will decrease, i.e. their $|O|$ decreases by $|P|$.

- If one of them holds Hawk strategy and the other one holds Dove strategy, the person holding Hawk strategy expresses his opinion thoroughly while the other person holding Dove strategy, though disapproving the viewpoint, chooses not to reply to his opponent. Receiving no objection encourages the belief of the person holding Hawk strategy, resulting in his $|O|$ increases by T. The other person holding Dove strategy receives his opponent's opinion passively so that his opinion value $|O|$ decreases by $|S|$.
- Suppose they both hold Dove strategy, then they both refuse to discuss the divisive topic. Maybe they consider convincing people holding another opinion is a waste of time. The result is their opinion values remain the same. Accordingly, A is set to be 0.

(3) An individual with opinion interact with an individual without opinion: the one with opinion gives the other one a lesson on the topic, trying to persuade the ignorant individual to believe in his own opinion. Both of their opinion values increase by Te.

The formula of the interaction rules are defined in Eq. 2, in which a and b denotes the two interacting individuals.

$$
\begin{cases}
|O_a| > Tr, |O_b| > Tr : \begin{cases}
O_a O_b > 0 : O_a = O_a + \dfrac{O_a}{|O_a|}R, \ O_b = O_b + \dfrac{O_b}{|O_b|}R \\[2mm]
O_a O_b < 0 : \begin{cases}
St_a = H, St_b = H : O_a = O_a + \dfrac{O_a}{|O_a|}P, \ O_b = O_b + \dfrac{O_b}{|O_b|}P \\[2mm]
St_a = H, St_b = D : O_a = O_a + \dfrac{O_a}{|O_a|}T, \ O_b = O_b + \dfrac{O_b}{|O_b|}S \\[2mm]
St_a = D, St_b = H : O_a = O_a + \dfrac{O_a}{|O_a|}S, \ O_b = O_b + \dfrac{O_b}{|O_b|}T \\[2mm]
St_a = D, St_b = D : O_a = O_a + \dfrac{O_a}{|O_a|}A, \ O_b = O_b + \dfrac{O_b}{|O_b|}A
\end{cases}
\end{cases} \\[2mm]
|O_a| > Tr, |O_b| \le Tr : O_a = O_a + \dfrac{O_a}{|O_a|}Te, \ O_b = O_b + \dfrac{O_a}{|O_a|}Te \\[2mm]
|O_a| \le Tr, |O_b| > Tr : O_a = O_a + \dfrac{O_b}{|O_b|}Te, \ O_b = O_b + \dfrac{O_b}{|O_b|}Te
\end{cases}
\tag{2}
$$

There are some special rules for the individuals whose $|O| = O_M$ in the model. Firstly, their $|O|$ won't increase by any means. Secondly, we consider the individuals with extreme opinions to be aggressive, i.e. their $St = H$ if only their $|O| = O_M$. Meanwhile, they regard arguing with the individuals holding opinion contrary to them as a lucrative behavior even though their O won't increase any more. This rule makes the ones with extreme opinions eager to spread their thoughts, which is reasonable in the real world.

3.2 Strategy Updating Rules

Individuals actively update their strategies regularly in order to get more payoffs in the evolutionary game. The time cycle of strategy update is C_S. According to [11], the action of strategy update is triggered by probability p. The procedure of strategy update is:

(1) Sum up the game payoffs one individual gets between he and all his neighbors who hold different opinions with him.
(2) Compare the summation under different strategies and choose the strategy that brings better payoffs.
(3) Update the chosen strategy.

The calculation formula of individual a's expected game payoff from the interaction with his neighbor b is:

$$\text{Payoff} = \begin{cases} P(1 - r_b) + T \cdot r_b & St_a = H \\ S(1 - r_b) + A \cdot r_b & St_a = D \end{cases} \tag{3}$$

r refers to the reputation of an individual. The concept of reputation refers to the expected probability that one individual chooses Dove strategy. The value of one's reputation is updated every C_R time steps. σ is the decay factor of reputation. The reputation of an individual at time t is:

$$\begin{cases} r(t) = \sigma r(t - C_R) + (1 - \sigma)\Delta r(t) \\ \Delta r(t) = \begin{cases} 1 & St = D \\ -1 & St = H \end{cases} \end{cases} \tag{4}$$

3.3 Partnership Adjusting Rules

The purpose of partnership adjusting is to find the appropriate neighbors who may help him get more payoffs in the game. Each individual attempts to adjust his partnership in every C_P time steps by probability q. The individual seeks appropriate partners in his neighbors' neighbors. Meanwhile, he checks his neighbors and find the person who causes the maximum losses to his payoffs. After considering the potential payoff variation caused by the partnership adjustment, the decision of whether to break off relation with this individual or build relation with that individual will be made. The specific procedure of partnership adjustment is described below:

(1) An empty set M is defined.
(2) Calculate the payoffs generated from the interactions between this individual and his neighbors' neighbors respectively. These values binding with its node number are then added to set M.
(3) Calculate the payoffs generated from the interactions between this individual and his neighbors respectively. Add the negative value of these results to set M, these values are bound with node number, too.
(4) Choose the maximum value in M and check its node number. If this node is the individual's neighbor, cut off the edge between them. If this node is the individual's neighbor's neighbor, then link up the edge between them.

In addition, the quantity of one's neighbor is no more than M_Nei.

4 Simulation Results and Analysis

The simulation of this model is realized by python codes. Several libraries are used including igraph, matplotlib and numpy. The simulation runs on an ordinary PC, with Intel i3 CPU (3.4 GHz) and 4G RAM, windows system.

The influential parameters are described below. The initial ratio of the opinionated individuals to the whole group is denoted as α_r and β_r, referring to the two contrary opinions α and β. The type of topology network is denoted as NET, alternative values are: $Powerlaw, Random$ and $Smallworld$. N is the total number of the nodes in the network.

The process of a simulation round is described below. Firstly, assign 0 to the opinion value O of each node. Then, select $\lfloor \alpha_r \cdot N \rfloor$ and $\lfloor \beta_r \cdot N \rfloor$ nodes in the network and assign O_M or $-O_M$ to them respectively. Their strategy St is automatically assigned with H. Next, the clock starts to run. In each time step, every individual who has opinion randomly finds one of his neighbors and interacts with him. Individuals with no opinion won't initiatively interact with others. Individuals update their strategies in every C_S steps, update their reputation in every C_R steps and adjust their partnerships with neighbors in every C_P steps.

Finally, there are two possible terminating conditions of the simulation:

(1) The proportion of one opinion group to the whole group exceeds 90%.
(2) The proportions of the opinion groups remain the same over 30 time steps.

These regulations mean that the simulation ends when one opinion group dominates the whole network or the proportions of the opinion groups remain steady. All of the simulations we've done terminate on either of these two conditions.

4.1 Model Validation

A set of data containing tens of thousands of microblogs on event "Death of Wei Zexi" [19] from Sina Weibo is obtained from the Internet. Then we identified the sentiment of 30000 microblogs posted in 7 days. Due to the immaturity of automatic sentiment analysis technic, all the identification tasks are completed manually. The analysis result is shown in Fig. 1a. The length of the bars in the upper subfigure represents the amount of the people who participate in the discussion of the topic. The color of the bars differentiate one opinion group from another. The curve in the lower subfigure shows the ratio of the *pro* opinion group to the whole group. Figure 1b is the simulation result obtained from our model.

In both of these figures, there is a burst of participants at first and then the quantity of the participants goes down. Meanwhile, the ratio of one opinion group to the whole group is roughly monotonous. Compared with the actual data, the trend of the simulation result shows the same characteristics. This result validates the effectiveness of our model to a certain extent.

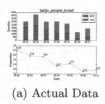

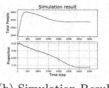

(a) Actual Data   (b) Simulation Result

Fig. 1. Actual data and simulation result of event "Death of Wei Zexi" (Color figure online)

4.2   Simulations and Analysis

It is obvious that α_r and β_r, the ratio of the initial opinionated individuals, matters to the result of the simulation, i.e. which opinion group wins the controversy and the time length of the simulation. Therefore, a specific research is done to the relationships between (α_r, β_r) and the simulation result. Meanwhile, we concern about the influence caused by the topology structure of the network. Studies show that social networks have both small-world [20] and power-law [21] properties. Thus this simulation is carried out on three different topology networks: power-law network, small-world network and random network.

Each of the networks has 1000 nodes and 2000 edges approximately. The *power* of the power-law network [22,23] is 2. The small-world network is a WS small-world network [24], the randomized reconnection probability of which is 0.5. The probability of the emergence of the edges is 0.0043 in the random network [25]. The parameters above ensure the generated networks with different topologies have about the same amount of nodes and edges.

We take $0.01, 0.02, \ldots, 0.09, 0.1$ as the values for the experimental parameters α_r and β_r. Hence there are $10 \times 10 = 100$ value pairs of α_r and β_r. Opinion ratio is the result of the simulation, referring to the proportion of the group holding opinion α to the whole group at the end of the simulation. For each pair of α_r and β_r, we run 20 rounds of simulations, the average value of the results of these simulations is denoted as *average opinion ratio* which is shown in Fig. 2a.

The three pictures in Fig. 2b show that the *average opinion ratio* increases as α_r increases and decreases as β_r increases. This phenomenon indicates that having more initial opinionated members is an important competitive advantage to the opinion groups in the controversial network event. Meanwhile, the figure also shows that having more initial opinionated members won't guarantee the opinion group's victory.

The influence of topology network is shown in Fig. 2b. The curves show that the simulation result in power-law network is more stable than that in random network and small-world network. This phenomenon indicates that it's harder for the opinion group with fewer initial members to win the controversy in a power-law network than in other networks.

The average time length of the simulations is shown in Fig. 2c. We observe that the time length increases when α_r is close to β_r and decreases when they are apart from each other. The reason is individuals spending more time arguing in the network if the scale of the two groups close to each other. The range of the average time length is much wider in power-law network than other networks.

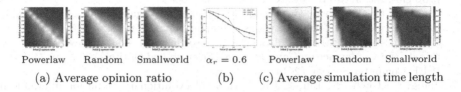

Powerlaw   Random   Smallworld $\alpha_r = 0.6$ Powerlaw   Random   Smallworld

(a) Average opinion ratio   (b)   (c) Average simulation time length

Fig. 2. Simulation results

5 Conclusions

In order to analyse the evolution process of people's opinion on a network event, an opinion evolution model based on social evolutionary game is proposed in this paper. This model simulates the diffusion and evolution process of two contrary opinions in social network. The Hawk-Dove game appropriately describes the interactions between people holding different opinions and the effectiveness of the model is validated by actual data. The opinion group with more initial members is more likely to dominate the whole network. The time length of the controversy is longer if the initial scale of the two opinion groups are close to each other. In power-law network, the result of the controversy is more stable and the duration of the controversy is longer than that in other topology networks. It confirms that network topology has certain influence on the opinion evolution process.

Our model is a powerful tool for the study of opinion dynamics in social network, especially when contrary opinions emerge at the same time. This model can be used to predict the trend of network public opinion, which facilitates many tasks related with analysis of network public opinion such as forecasting the result of a president campaign, estimating people's attitudes towards a new policy, calculating the sales forecast of a new product and so on.

We welcome further studies on opinion evolutionary model and opinion dynamics based on our work. The influence of other parameters of this model is still unknown. The evolution of three or more kinds of opinions and the opinion evolution happening on more than one social network platforms at the same time are also attractive scenarios for further studies.

The appendix of this paper including values of the parameters and the source codes of the simulation is posted online at http://pan.baidu.com/s/1o8mby2m.

References

1. Sznajd-Weron, K., Sznajd, J.: Opinion evolution in closed community. Int. J. Mod. Phys. C **11**(6), 0000093 (2000)
2. Amblard, F., Deffuant, G., Weisbuch, G.: How can extremism prevail? A study based on the relative agreement interaction model. J. Artif. Soc. Soc. Simul. **5**(4), 1 (2002)

3. Hegselmann, R., Krause, U.: Opinion dynamics and bounded confidence models, analysis and simulation. J. Artif. Soc. Soc. Simul. **5**(3), 2 (2002)
4. Martins, A.C.R.: Continuous opinions and discrete actions in opinion dynamics problems. Int. J. Mod. Phys. C **19**, 0801233 (2008)
5. Xu, H., Cai, W., Chen, G., Wang, J.: Opinion propagation and public opinion formation model for forum networks. Comput. Sci. **5**, 150–152 (2013)
6. Zhang, Y.: The evolution of public opinion in social simulation. In: Seventh International Joint Conference on Computational Sciences and Optimization, pp. 343–345. IEEE Computer Society (2014)
7. Jiang, C., Chen, Y., Liu, K.J.R.: Graphical evolutionary game for information diffusion over social networks. IEEE J. Sel. Topics Signal Process. **8**(4), 524–536 (2014)
8. Zheng, X., Lu, X., Chan, F.T.S., Deng, Y., Wang, Z.: Bargaining models in opinion dynamics. Appl. Math. Comput. **251**, 162–168 (2015)
9. Yu, J., Wang, Y., et al.: Analysis of competitive information dissemination in social network based on evolutionary game model. In: Third International Conference on Cloud & Green Computing, vol. 90, pp. 748–753. IEEE (2013)
10. Maynard-Smith, J., Price, G.R.: The logic of animal conflict. Nature **246**, 15–18 (1973)
11. Yu, J., Wang, Y., Jin, X., Cheng, X.: Social evolutionary games. In: International Conference on Game Theory for Networks, pp. 1–5. IEEE (2014)
12. Li, J., Xing, G., Wang, Y., Ren, Y.: Training opinion leaders in microblog: a game theory approach. In: Second International Conference on Cloud and Green Computing, pp. 754–759. IEEE Computer Society (2012)
13. Qiu, W., Wang, Y., Yu, J.: A game theoretical model of information dissemination in social network. In: International Conference on Complex Systems, vol. 229, pp. 1–6. IEEE (2013)
14. Yu, J., Wang, Y., Jin, X., Cheng, X.: Identifying interaction groups in social network using a game-theoretic approach. In: IEEE/WIC/ACM International Joint Conferences on Web Intelligence, vol. 2, pp. 511–518. IEEE Computer Society (2014)
15. Sugden, R.: The Economics of Rights, Cooperation and Welfare, 2nd edn. Palgrave Macmillan, Basingstoke (2005). P. 132
16. Liu, R., Wang, Y.: Evolutionary coordinative foraging algorithm based on Hawk-Dove game. In: Second International Symposium on Intelligent Information Technology Application, vol. 2, pp. 629–633. IEEE Computer Society (2008)
17. Tomassini, M., Luthi, L., Giacobini, M.: Hawks and Doves on small-world networks. Phys. Rev. E Stat. Nonlinear Soft Matter Phys. **73**(2), 016132 (2006)
18. Voelkl, B.: The Hawk-Dove game and the speed of the evolutionary process in small heterogeneous populations. Games **1**(2), 103–116 (2010)
19. Death of Wei Zexi-Wikipedia. https://en.wikipedia.org/wiki/Death_of_Wei_Zexi
20. Milgram, S.: The small world problem. Psychol. Today **2**, 185–195 (1967)
21. Barabasi, A.L., Bonabeau, E.: Scale-free networks. Sci. Am. **288**(5), 60 (2003)
22. Goh, K.I., Kahng, B., Kim, D.: Universal behavior of load distribution in scale-free networks. Phys. Rev. Lett. **87**(27 Pt 1), 278701 (2001)
23. Cho, Y.S., Kim, J.S., Park, J., Kahng, B., Kim, D.: Percolation transitions in scale-free networks under the Achlioptas process. Phys. Rev. Lett. **103**(13), 135702 (2009)
24. Watts, D.J., Strogatz, S.H.: Collective dynamics of 'small-world' networks. Nature **393**(6684), 440 (1998)
25. Erdos, P., Renyi, A.: On random graphs. Publ. Math. **6**(4), 290–297 (1959)

Category Prediction of Questions Posted in Community-Based Question Answering Services Using Deep Learning Methods

Qing Ma[1]([✉]), Reo Kato[1], and Masaki Murata[2]

[1] Department of Applied Mathematics and Informatics, Ryukoku University,
Kyoto, Japan
qma@math.ryukoku.ac.jp
[2] Department of Information and Electronics, Tottori University, Tottori, Japan

Abstract. This paper presents methods of predicting categories of questions posted in community-based question answering (CQA) services using deep learning methods, which are implemented with stacked denoising autoencoders (SdA), as well as deep belief networks (DBN). We compare them with conventional machine learning methods, i.e., multi-layer perceptron (MLP) and support vector machines (SVM). We also compare their performance when using dropout regularization. The experimental results indicate that (1) the proposed methods reach much higher prediction precision than that provided by CQA services, (2) deep learning with dropout has higher prediction precision than the conventional machine learning methods, whether or not the dropout regularization is used, i.e., DBN with dropout reaches the highest precision and SdA with dropout reaches the next highest precision among all the methods in general, and the SdA with dropout in a specific case reaches the highest precision across all experiments, (3) increasing the dimensions of feature vectors representing the questions is an effective measure for improving the prediction precision, (4) prediction precision can be further improved using titles in addition to the actual questions and by improving the quality of the corpus used for training.

Keywords: CQA · Q&A website · Category prediction · Deep learning · DBN · SdA · Dropout

1 Introduction

Community-based question answering (CQA) services, i.e., Q&A websites such as "OKWave", "Yahoo!知恵袋" (Yahoo!Answers), and "教えて!goo" (tell me!goo), each with several ten millions of users, have become popular tools for information seeking due to advantages, such as allowing people to ask natural language questions and providing more personalized answers, over conventional information retrieval methods. When posting questions on CQA sites, questioners must choose adequate categories provided by the CQA services to which the

© Springer International Publishing AG 2017
D. Liu et al. (Eds.): ICONIP 2017, Part V, LNCS 10638, pp. 699–709, 2017.
https://doi.org/10.1007/978-3-319-70139-4_71

question content should belong. For answering questions at these sites, answerers usually find the answerable questions from the categories they are familiar with. That is, it is important to determine correct question categories for obtaining appropriate answers.

However, it is difficult for questioners to choose categories because the numbers of categories are very large in these sites, e.g., there are respectively 19, 200, and 1,000 large, middle, and small classifications (categories) in OKWave. In these sites, services to help questioners determine categories, i.e., to predict and present category candidates of the submitted questions, are provided. However, by examining 20 questions that have already been posted on these sites and that have correct categories, we find that the prediction precisions of these three Q&A websites are extremely low. Specifically, the prediction precisions of "OKWave", "Yahoo!Answers", and "tell me!goo" are respectively 0.55, 0.45, 0.40 from only judging whether the top category of the candidate list presented by each of these sites is the correct one, and 0.75, 0.85, 0.70 from judging whether the candidate list presented by each of the three sites includes the correct category. It is therefore preferable to develop a method for predicting categories of questions posted on CQA sites with high accuracy.

Several studies on CQA have been conducted. Blooma et al. [1], Wang et al. [2], Agichtein et al. [3], and Liu et al. [4] aimed at finding the best answers or high-quality content. Jurczyk and Agichtein [5], Riahi et al. [6], and Yokoyama et al. [7] attempted to discover authoritative (expert) users for answering in CQA services. There have been studies in Japanese on question classifications [8,9]. However, the questions were either classified into two types, i.e., "information retrieval" and "social survey" (the former is questions to ask for objective facts or information and the later is the questions to ask for personal advice, comments, and experiences) or into five types, i.e., "fact", "evidence", "experience", "proposal", and "opinion". However, neither directly predicts categories provided by the CQA services.

This paper presents methods of predicting categories of questions posted in CQA services using deep learning methods, which are implemented with stacked denoising autoencoders (SdA), as well as deep belief networks (DBN). We compare them with conventional machine learning methods, i.e., multi-layer perceptron (MLP) and support vector machines (SVM). We also compare their performance when using dropout regularization. The experimental results indicate that (1) the proposed methods reach much higher prediction precision than that provided by CQA services, (2) deep learning with dropout has higher prediction precision than the conventional machine learning methods, whether or not the dropout regularization is used, i.e., DBN with dropout reaches the highest precision and SdA with dropout reaches the next highest precision among all the methods in general, and the SdA with dropout in a specific case reaches the highest precision across all experiments, (3) increasing the dimensions of feature vectors representing the questions is an effective measure for improving prediction precision, (4) prediction precision can be further improved using titles of questions, which are created by questioners in addition to the actual questions, and by improving the quality of the corpus used for training, which is done by integrating the questions with similar meanings but classified into different categories.

2 Data

2.1 Corpus

We focus on OKWave, the largest CQA service with about 30 million users in Japan to predict categories of questions posted to the site. For training and testing, a labeled data set consisting of pairs of inputs and their responses (or correct answers) is needed. In our case, we use pairs of questions posted in OKWave and their categories assigned by questioners. OKWave contains hierarchically organized categories in which there are 19, 200, and 1,000 large, middle, and small categories, respectively, and questioners manually assign (determine) specific categories when posting their questions.[1] Since the data is too large, we first confine the data to the ten middle categories under the large category called *digital life*, i.e., *audio visual systems, mobile phones, social networking service, on-line shopping, virus countermeasure, windows, macintosh, PC parts and peripherals, software,* and *multimedia*. We therefore obtain a corpus that consists of 33,000 questions with 10 categories and call it the **original corpus**.

By examining the corpus, however, we find that the question content of the small category of *digital camera* under the middle category of *audio visual systems* is similar to that of the middle category of *multimedia*. We believe that using these similar questions with two different middle categories (i.e., *audio visual systems* and *multimedia*) as training data will deteriorate category prediction. Therefore, we move the questions of the small category of *digital camera* under the middle category of *audio visual systems* to the middle category of *multimedia*. We therefore obtain a new corpus and call it the **improved corpus**[2].

2.2 Feature Vectors

We transform questions in the corpora into feature vectors, which are needed in machine learning as inputs, using the bag-of-words model. Since questions consist of titles and texts, we extract words, used as features, from the corpora in two ways. One involves extracting the words only from texts and the other involves extracting them from both titles and texts. By examining the corpora, we find cases in which there are only content words in titles, such as below, and believe that using words in titles is effective for improving prediction performance.

---

[1] Some questioners might use the category prediction service provided by the OKWave site and refer to their prediction results. However, to which categories the questions should belong are finally determined by the questioners, not the site.

[2] The total numbers of questions and categories in the corpus do not change.

title: method of copying TV programs in an external HDD to a DVD
text: described as in the title, and the individuals who are knowledgeable about
   this please tell me.

We extract words from texts or texts&titles in steps (1)–(4) and construct feature vectors by further adding steps (5) and (6), as follows. (1) Unify two-byte and one-byte characters (i.e., transform one-byte Japanese characters into two-byte ones and two-byte alphanumeric characters into one-byte ones) and transform lower-case letters into upper-case letters for English words; (2) conduct morphological analysis on texts or texts&titles of questions that are used for training and extract all nouns, including proper nouns, verbal nouns (nouns forming verbs by adding the word "する" (do), and general nouns; (3) connect the nouns successively appearing as single words; (4) extract the words whose appearance frequency in the text or text&title of each question is ranked in the top M; and (5) use the words obtained in the above steps as the vector elements with binary values, taking 1 if a word appears and 0 if not; and (6) morphologically analyze all data (i.e., all texts or texts&titles of questions) and construct the feature vectors in accordance with step (5).[3]

2.3   Data Sets

We use two data sets called **data set 1** and **data set 2**, respectively. Data set 1 is the set of feature vectors that are created by using words extracted only from the texts in questions of the original corpus. Data set 2 is the set of feature vectors that are created by using words extracted from both titles and texts in questions of the improved corpus.

3   Deep Learning

Deep learning consists of unsupervised learning for pre-training to extract features and supervised learning for fine-tuning to output labels. Deep learning can be implemented using two typical approaches: deep belief networks (DBN) (e.g., [10,11]) and stacked denoising autoencoders (SdA) (e.g., [11,12]). The same supervised learning method can be used with both of approaches; i.e., both approaches can be implemented with a single-layer or multi-layer perceptron or other techniques (linear regression, logistic regression, etc.), while a different unsupervised learning method is used, i.e., a DBN is formed by stacking restricted Boltzmann machines (RBM), and an SdA is formed by stacking denoising autoencoders (dA) using a greedy layer-wise training algorithm. In this study, we use SdA as well as DBN, both of which use a multi-layer perceptron for supervised learning.

Figure 1 shows an example of deep neural networks composed of three RBM or dA for pre-training and a supervised learning device for fine-tuning. Naturally the number of RBM/dA is changeable as needed. As shown in the figure, the hidden layers of the earlier RBM/dA become the visible layers of the new RBM/dA.

---

[3] The vectors created in this way are also called one-hot vectors.

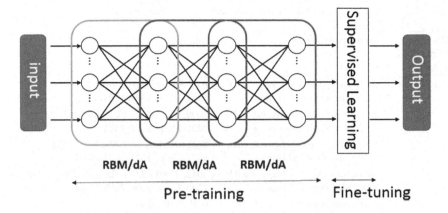

Fig. 1. An example of deep neural networks consisting of DBN or SdA.

A novel regularization method called dropout [13] was proposed to prevent overfitting and improve generalization performance. The key point is to train different models for each piece of a training data set by randomly removing units with probability p from the input and hidden layers. At test time, all units are always present (not removed) and their weights are multiplied by 1-p.

4 Experiments

4.1 Experimental Setup

We use data sets 1 and 2 described in Subsect. 2.3. From each data set, we use 28,000 questions, which are selected so that each category will have the same number of questions. That is, each category has 2,800 questions, which are randomly divided into 2,000, 400, and 400 questions for training, validation, and testing, respectively. In constructing feature vectors, we set M, the number of words with top appearance frequencies in the text or the text&title of each question, to 300 and 500, respectively. As a result, the dimensions of the feature vectors constructed from the training data of data set 1 in accordance with the steps in Subsect. 2.2 are 1,276 and 2,018, respectively, and those of data set 2 are 1,309 and 2,083, respectively.

The optimal hyperparameters of the machine learning methods used are determined by a grid search using validation data[4]. Table 1 lists the hyperparameters for grid search. To avoid unfair bias toward DBN/SdA during validation due to them having more hyperparameters than MLP and SVM, we divide the MLP and SVM hyperparameter grids more finely than those of DBN/SdA so that they have the same or more hyperparameter combinations (hyperparameter

---

[4] Instead of using cross-validation, we use a single validation data set due to its extremely high time consumption.

Table 1. Hyperparameters for grid search.

Machine learning methods	Hyperparameters	Values
DBN/SdA	Structure (hidden layers) with 1,276 dimensional inputs	638, 851-425, **957-638-319**[a], 1531, 1531-1531, 1531-1531-1531
	Structure (hidden layers) with 2,018 dimensional inputs	1000, 1345-672, 1513-1009-504, 2421, 2421-2421, 2421-2421-2421
	Structure (hidden layers) with 1,309 dimensional inputs	655, 872-436, 982-655-327, 1571, 1571-1571, 1571-1571-1571
	Structure (hidden layers) with 2,083 dimensional inputs	1042, 1389-694, 1562-1042-521, 2500, 2500-2500, 2500-2500-2500
	ϵ of pre-training	0.01, 0.05, 0.01
	ϵ of fine-tuning	0.1, 0.01
	Epoch of pre-training	100, 300, 500
	Epoch of fine-tuning	500
	Activation function	ReLU, Sigmoid
MLP	Structure (hidden layers) with 1,276 dimensional inputs	638, 851-425, 957-638-319, 1531, 1531-1531, 1531-1531-1531
	Structure (hidden layers) with 2,018 dimensional inputs	1009, 1345-672, 1513-1009-504, 2421, 2421-2421, 2421-2421-2421
	Structure (hidden layers) with 1,309 dimensional inputs	655, 872-436, 982-655-327, 1571, 1571-1571, 1571-1571-1571
	Structure (hidden layers) with 2,083 dimensional inputs	1042, 1389-694, 1562-1042-521, 2500, 2500-2500, 2500-2500-2500
	ϵ	10 divisions between 0.1–0.01 and 9 divisions between 0.009–0.001
	Epoch of training	500
	Activation function	ReLU, Sigmoid
SVM (linear)	C	225 divisions between 10^{-4}–10^4 in a logarithmic scale
SVM (RBF)	C	15 divisions between 10^{-4}–10^4 in a logarithmic scale
	γ	15 divisions between 10^{-4}–10^4 in a logarithmic scale

[a] As an example, the structure (hidden layers) **957-638-319**, shown in bold in the table, refers to DBN/SdA with a 1276-**957-638-319**-10 structure, where 1,276 and 10 respectively refer to the dimensions of the input and output layers. These figures are not set in an arbitrary manner. The first three structures are in decreasing size (pyramid-like), and the last three structures were set in accordance with the recommendations of [14], i.e., using the same size (1.2 times larger than that of the input layer) works generally as well as or better than using a decreasing (pyramid-like) or increasing (upside-down pyramid) size.

sets) as those of DBN/SdA. We therefore have 216, 228, and 225 hyperparameter sets in total for DBN/SdA, MLP, and SVM, respectively. For DBN/SdA, the dropout is only used in fine-tuning and the probability p is set to 0.8 in the input layers and 0.5 in the middle layers for both DBN/SdA and MLP.

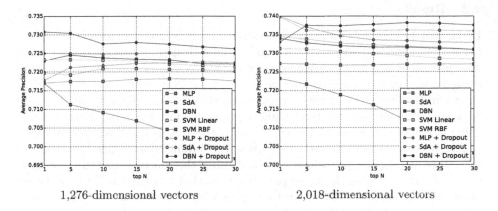

1,276-dimensional vectors   2,018-dimensional vectors

Fig. 2. Average precisions obtained using data set 1.

Table 2. Precisions obtained using data set 1.

Machine learning methods	1,276-dimensional vectors	2,018-dimensional vectors
MLP	0.717	0.727
SdA	0.720	0.735
DBN	0.723	0.734
SVM linear	0.723	0.731
SVM RBF	0.717	0.723
MLP + dropout	0.718	**0.740**
SdA + dropout	0.725	**0.740**
DBN + dropout	**0.731**	0.733

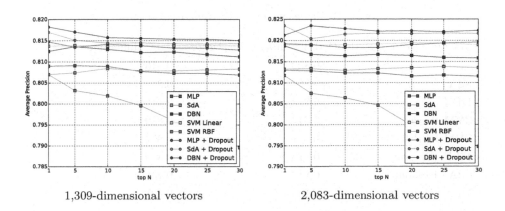

1,309-dimensional vectors   2,083-dimensional vectors

Fig. 3. Average precisions obtained using data set 2.

4.2   Results

Data Set 1: Figure 2 compares the testing data precision obtained using different machine learning methods with and without dropout when using data set 1. The precisions are averages when using the top N sets of the hyperparameters in ascending order of the validation errors, with N varying from 1 to 30. As shown in the figure, DBN with dropout generally has the highest precision and SdA with dropout has the next highest precision among all the methods when using different dimensional vectors. Using 2,018-dimensional vectors produces higher precision than that achieved using 1,276-dimensional ones for all methods. Table 2 lists the precisions obtained using the machine learning methods when N = 1, i.e., using the hyperparameter set that has the smallest validation error. We can see from the table that DBN with dropout has the highest precision among all the methods when using 1,276-dimensional vectors, and SdA with dropout, tied with the MLP with dropout, has the highest precision among all the methods when using 2,018-dimensional vectors.

Data Set 2: Figure 3 compares the testing data precision obtained using different machine learning methods with and without dropout when using data set 2. DBN with dropout generally has the highest precision and SdA with dropout has the next highest precision among all the methods when using different dimensional vectors, the same as with data set 1. Using 2,083-dimensional vectors produces higher precision than that achieved using 1,309-dimensional ones for all the methods, the same as with data set 1. Table 3 lists the precisions obtained using these methods when N = 1. We can see from the table that DBN with dropout has the highest precision among all methods when using 1,309-dimensional vectors, and SdA with dropout has the highest precision among all methods when using 2,083-dimensional vectors. In addition, SdA with dropout in this case has the highest precision in all cases, including different methods, different data sets, and different dimensional vectors.

Table 3. Precisions obtained using data set 2.

Machine learning methods	1,309-dimensional vectors	2,083-dimensional vectors
MLP	0.809	0.813
SdA	0.807	0.813
DBN	0.813	0.819
SVM Linear	0.814	0.819
SVM RBF	0.807	0.812
MLP+Dropout	0.815	0.819
SdA+Dropout	0.817	**0.824**
DBN+Dropout	**0.818**	0.821

Significance Test: By performing t-tests on the above results we find that the difference between DBN/SdA with dropout and MLP/SVM, whether or not the dropout is used, is significant at the 0.05 level in most cases.

Data Set 1 vs. Data Set 2: By comparing the above figures and tables we can see that the precisions obtained using data set 2 are higher than those obtained using data set 1. Table 4 lists the average precisions of questions in each category predicted using SdA with dropout with the hyperparameter set that has the smallest validation errors.[5] This table shows that the precision of any category when using data set 2 is higher than that of any category when using data set 1. These results indicate that prediction precision can be improved using data set 2, which is created from the improved corpus and uses both titles and texts in questions, instead of data set 1.

CQA vs. Our Methods: The precisions obtained with the proposed methods are higher than 0.8 as shown in Table 3, whereas the prediction precisions provided by CQA services are between 0.4–0.55 as described in Sect. 1.

Table 4. Precisions of categories obtained using data set 1 and data set 2.

Category	Data set 1	Data set 2
audio visual systems	0.703	0.840
mobile phones	0.748	0.820
social networking service	0.853	0.892
on-line shopping	0.915	0.930
virus countermeasure	0.818	0.890
Windows	0.637	0.698
Macintosh	0.608	0.873
PC parts and peripherals	0.762	0.848
software	0.667	0.708
multimedia	0.688	0.738

5 Conclusion

We proposed methods of predicting categories of questions posted in CQA services using deep learning methods, DBN and SdA. We compared them with conventional machine learning methods. The experimental results showed that (1) the proposed methods reach much higher prediction precision than that provided by CQA services, (2) deep learning with dropout has higher prediction

---

[5] We choose SdA with dropout in this experiment because it has the highest precision in all cases, as described above.

precision than the conventional machine learning, whether or not the dropout regularization is used, i.e., DBN with dropout reaches the highest precision and SdA with dropout reaches the next highest precision among the all methods in general, and SdA with dropout in a specific case reaches the highest precision across all experiments, (3) increasing the dimensions of feature vectors representing the questions is an effective measure for improving prediction precision, (4) prediction precision can be further improved using titles in addition to the actual questions and by improving the quality of the corpus used for training.

For future work, we plan to further improve the prediction performance of our proposed methods, scale up the experiment, and start developing practical systems that can predict all categories provided by CQA services.

Acknowledgments. This work was supported by JSPS KAKENHI Grant Number 25330368.

References

1. Blooma, M.J., Chua, A.Y.K., Goh, D.H.L.: A predictive framework for retrieving the best answer. In: Proceedings of 2008 ACM Symposium on Applied Computing, SAC 2008, pp. 1107–1111. ACM, New York (2008)
2. Wang, X.J., Tu, X., Zhang, L.: Ranking community answers by modeling question-answer relationships via analogical reasoning. In: Proceedings of the 32nd International ACM SIGIR Conference on Research and Development in Information Retrieval, SIGIR 2009, pp. 179–186. ACM, New York (2009)
3. Agichtein, E., Castillo, C., Donato, D., Gionis, A., Mishne, G.: Finding high-quality content in social media. In: Proceedings of the 2008 International Conference on Web Search and Data Mining, WSDM 2008, pp. 183–194. ACM, New York (2008)
4. Liu, Y., Li, S., Cao, Y., Lin, C.Y., Han, D., Yu, Y.: Proceedings of the 22nd International Conference on Computational Linguistics, COLING 2008, pp. 497–504, Stroudsburg, PA, USA. Association for Computational Linguistics (2008)
5. Jurczyk, P., Agichtein, E.: Discovering authorities in question answer communities by using link analysis. In: Proceedings of the Sixteenth ACM Conference on Conference on Information and Knowledge Management, CIKM 2007, pp. 919–922. ACM, New York (2007)
6. Riahi, F., Zolaktaf, Z., Shafiei, M., Milios, E.: Finding expert users in community question answering. In: Proceedings of the 21st International Conference on World Wide Web, WWW 2012 Companion, pp. 791–798. ACM, New York (2012)
7. Yokoyama, Y., Hochin, T., Nomiya, H.: Improvement of obtaining potential appropriate respondents to questions at Q&A sites by considering categories of answer statements. Int. J. Affect. Eng. **16**(2), 63–73 (2017). Japan Society of Kansei Engineering, Japan
8. Omori, Y., Morita, K., Fuketa, M., Aoe, J.: Question classification of Q&A sites using pseudo training data. In: Proceedings of the 21st Annual Meeting of the Japan Association for Natural Language Processing, pp. 489–492 (2015) (in Japanese)
9. Watanabe, N., Shimada, S., Seki, Y., Kando, N., Satoh, T.: A study for questions classification based on questioner demands in QA communities. In: Proceedings of the 3rd Forum on Data Engineering and Information Management, DIEM 2011, pp. 1–6 (2011) (in Japanese)

10. Hiton, G.E., Osindero, S., Teh, Y.: A fast learning algorithm for deep belief nets. Neural Comput. **18**, 1527–1554 (2006). MIT Press
11. Bengio, Y., Courville, A., Vincent, P.: Representation learning: a review and new perspectives. IEEE Trans. Pattern Anal. Mach. Intell. **35**(8), 1798–1828 (2013). IEEE Press
12. Vincent, P., Larochelle, H., Lajoie, I., Bengio, Y., Manzagol, P.A.: Stacked denoising autoencoders: learning useful representations in a deep network with a local denoising criterion. J. Mach. Learn. Res. **11**, 3371–3408 (2010)
13. Srivastava, N., Hinton, G., Krizhevsky, A., Sutskever, H., Salakhutdinov, R.: Dropout: a simple way to prevent neural networks from overfitting. J. Mach. Learn. Res. **15**, 1929–1958 (2014)
14. Bengio, Y.: Practical recommendations for gradient-based training of deep architectures. In: Montavon, G., Orr, G.B., Müller, K.-R. (eds.) Neural Networks: Tricks of the Trade. LNCS, vol. 7700, 2nd edn, pp. 437–478. Springer, Heidelberg (2012). doi:10.1007/978-3-642-35289-8_26

LCE: A Location Category Embedding Model for Predicting the Category Labels of POIs

Yue Wang[1], Meng Chen[2], Xiaohui Yu[1,2], and Yang Liu[1(✉)]

[1] School of Computer Science and Technology, Shandong University,
Jinan 250101, China
wysdu15@gmail.com, {xyu,yliu}@sdu.edu.cn
[2] School of Information Technology, York University, Toronto, ON M3J 1P3, Canada
mchen16@yorku.ca

Abstract. The proliferation of location-based social networks, makes it possible to record human mobility using an array of points-of-interest (POIs). Exploring the semantic meanings of POIs can be of great importance to many urban computing applications, e.g., personalized route recommendation and user trajectory clustering. Nonetheless, such information is not always available in practice. This paper aims at predicting the category labels, which will provide a succinct summarization of POIs. In particular, we first propose a Location Category Embedding (LCE) model, which projects user POIs and their associated category labels into the same vector space, and then identify the POIs' most related category labels according to their similarities. To capture the influence that might affect users' moving behavior, LCE considers sequential pattern, personal preference, and temporal influence, and further models the connection between the POIs and the three factors. Experimental results on two real-world datasets prove the effectiveness of the proposed method.

Keywords: Location Category Embedding · Points-of-interest · Category label prediction · Embedding learning

1 Introduction

The increasing prevalence of GPS-enabled smartphones and location-based social networks (LBSNs), such as Foursquare and Gowalla, allows users to explore new places and share their locations through *check-ins*. Each check-in represents a user's visit to a POI, such as a restaurant or a museum, at a specific time. In LBSNs, some POIs have been labeled with semantic categories, for example, Foursquare uses 9 categories (e.g., Food and Education) as the first level of semantic labels. With these semantic categories, we are able to better understand human's mobility patterns [3,4,18] and further improve the performance of future navigation recommendations [2,6,17]. However, our analysis of data on Foursquare reveals that about 30% of POIs lack the semantic category information. This semantic absence problem is also reported in previous studies [5,15].

D. Liu et al. (Eds.): ICONIP 2017, Part V, LNCS 10638, pp. 710–720, 2017.
https://doi.org/10.1007/978-3-319-70139-4_72

Current methods of category label prediction mainly construct a multi-label classifier with several previously defined features (e.g., visiting frequency [1], stay duration [1,5] and distribution of check-in time [7,15]). Nevertheless, the performance depends heavily on its way of feature selection. As there is no principled method to follow, it can be extremely daunting and time-consuming to determine a set of proper features. In this paper, we seek for a fire-new strategy to model the check-ins in LBSNs and further predict the category labels of POIs based on the discovered patterns.

In particular, when modeling the check-ins, we project the POIs and the category labels into the same low-dimensional latent space, in which the POIs with the same category label are closely clustered around the category. In light of recent advances in distributed representation [8,13], we explore the use of embedding learning. However, the following unique characteristics of check-in data make it unsuitable to directly apply existing methods [9,11] to our problem:

- **Sequential pattern.** The check-in POI is often related to the contextual POIs and their category labels. For example, users may check in sequentially at "Home → Subway → Office" when they go to work from home.
- **Personal preference.** Each user has its own moving preference, e.g., a student is more likely to check in at a library, whereas an artist may often visit a theatre.
- **Temporal influence.** Users tend to visit various POIs during different time periods. For instance, people may go to the restaurants at noon but visit the bars at night.

To take into account the above factors, we propose a novel Location Category Embedding (LCE) model, which considers the correlation between each POI and its multiple contexts in a check-in sequence, including user information, current time, the previous K and the successive K POIs together with their category labels. Furthermore, LCE minimizes the distances between the POIs and their category labels in the latent space. At forecast time, given an unlabeled POI, we compute the distance between the POI and each possible category label in the latent space, and select the one with the minimum distance as the predicted.

The major contributions of this paper are summarized as follows:

- We propose a novel Location Category Embedding (LCE) model for predicting the category labels of POIs, in which we project the POIs and their category labels into the same latent space.
- We not only model the connection between the POIs and their contexts by considering multiple factors, but also ensure the correlation between the POIs and their category labels in the LCE.
- We conduct extensive experiments on two real check-in datasets, and compare LCE with baselines in the task of category label prediction. Experimental results show the effectiveness of LCE.

2   Preliminaries

We present some definitions which are required for the subsequent discussion.

Definition 1 (Check-in record). *A check-in record is a quadruple $<u, l, c, t>$, which describes a user u visiting a POI l with the category label c at time t. Specifically, if the POI l is unlabeled, its category is a **null** value.*

Definition 2 (Check-in sequence). *Given a user u, its check-in sequence is a set of chronologically ordered records, denoted as $S_u = \{<l_1, c_1, t_1>, <l_2, c_2, t_2>, \cdots, <l_n, c_n, t_n>\}$, where $t_1 \leq t_2 \leq \cdots \leq t_n$.*

To develop a more robust model, we discretize the time span into equi-sized time bins and map all check-in records to the bins according to their timestamps. In what follows, we represent t with the id of the bin that it belongs to.

3 Location Category Embedding Model

In this section, we first present how to construct LCE from scratch, followed by parameter estimation with a SGD-based strategy. Assisted with the patterns discovered by LCE, we finally introduce our method of category label prediction.

3.1 Location Category Embedding

Considering the characteristics of people's check-ins, we propose a novel LCE model for predicting the category labels of POIs, which projects the POIs and the category labels into the same low-dimensional latent space. In the latent space, the POIs with the same category label are closely clustered around the category. To achieve this goal, LCE considers the correlation between each POI and its context in a check-in sequence, and minimizes the distances between the contextual POIs and their category labels in the latent space. Accordingly, the construction of LCE can be divided into 3 parts:

1. Modeling POIs and their contexts,
2. Modeling POIs and their category labels, and
3. Combining part 1 and 2 to form the integrated objective function of LCE.

Modeling POIs and Their Contexts. Based on our analysis of check-in sequences in LBSNs, we observe that the generation of a check-in POI is related to multiple contextual features, namely, user information, current time, the preceding K and the successive K POIs together with their category labels. For a single check-in sequence $S_u = \{<l_1, c_1, t_1>, <l_2, c_2, t_2>, \cdots, <l_n, c_n, t_n>\}$ of length N_u, we aim to model the correlation between each POI and its context; therefore, the objective function is to maximize the probability of each POI l_j given its corresponding contextual information:

$$\mathcal{L}(S_u) = \sum_{j=1}^{N_u} P(l_j | \text{context}(l_j)). \tag{1}$$

Formally, we represent each POI l_j with a d-dimensional embedding vector $v_{l_j} \in \mathbb{R}^d$ and embed each contextual feature into the same latent vector space. Given a POI l_j, we average the embedding vectors of multiple contexts to compute a contextual vector $\bar{v}_{l_j}$. To handle unlabeled POIs in its contexts, we introduce an indicated variable α_{j+k} with

$$\alpha_{j+k} = \begin{cases} 1, & \text{if } l_{j+k} \text{ is labeled;} \\ 0, & \text{otherwise.} \end{cases}$$

Therefore, the contextual vector $\bar{v}_{l_j}$ of POI l_j can be estimated by

$$\bar{v}_{l_j} = \frac{1}{2K+S+2} \left\{ \sum_{-K \leq k \leq K, k \neq 0} (v_{l_{j+k}} + \alpha_{j+k} v_{c_{j+k}}) + (v_u + v_{t_j}) \right\}, \quad (2)$$

where K is the context window size and S is the number of non-zero α_{j+k}. v_u, v_t, v_c are the embedding vectors of users, time bins, and category labels respectively. A softmax probability function is adopted to generate a check-in POI based on its contextual vector:

$$P(l_j | \text{context}(l_j)) = \frac{\exp(\bar{v}_{l_j}^\top \cdot v_{l_j})}{\sum_l \exp(\bar{v}_{l_j}^\top \cdot v_l)}. \quad (3)$$

To reduce the computational cost, the strategy of negative sampling is adopted. When we train the vector of l_j, we can get a corresponding negative sample set $NEG(l_j)$. If $l_x \in NEG(l_j)$, then $l_x \neq l_j$. Mikolov et al. [10] explain how to find negative samples in detail. Let $L(l_x)$ represents the label of a POI, we have

$$L^{l_j}(l_x) = \begin{cases} 1, & l_x = l_j; \\ 0, & l_x \neq l_j. \end{cases}$$

Given the contextual features of POI l_j, we want to maximize the occurrence of l_j and meanwhile minimize the probability of negative samples $l_x \in NEG(l_j)$. The objective function for the target POI l_j therefore becomes

$$\begin{aligned} g_1(l_j) &= P(l_j | \text{context}(l_j)) \prod_{l_x \in NEG(l_j)} [1 - P(l_x | \text{context}(l_j))] \\ &= \sigma(\bar{v}_{l_j}^\top \theta^{l_j}) \prod_{l_x \in NEG(l_j)} [1 - \sigma(\bar{v}_{l_j}^\top \theta^{l_x})] \\ &= \prod_{l_x \in \{l_j\} \cup NEG(l_j)} [\sigma(\bar{v}_{l_j}^\top \theta^{l_x})]^{L^{l_j}(l_x)} \cdot [1 - \sigma(\bar{v}_{l_j}^\top \theta^{l_x})]^{1 - L^{l_j}(l_x)}, \end{aligned} \quad (4)$$

where $\sigma(z) = (1 + \exp(-z))^{-1}$ is the sigmoid function and θ^{l_x} is an auxiliary vector corresponding to the POI l_x.

Modeling POIs and Their Category Labels. To ensure the correlation between the POIs and their category labels, we further expect to minimize the distances between the labeled POIs and their category labels in the latent vector space. Therefore, given a POI l_j, the objective function is to minimize the distances between the labeled POIs in its context and their category labels:

$$g_2(l_j) = \frac{1}{S} \sum_k \alpha_{j+k} \left\| v_{l_{j+k}} - v_{c_{j+k}} \right\|^2. \tag{5}$$

In particular, Euclidean distance is adopted in this work, and other distance metrics such as the Kullback-Leibler divergence can also be applied.

Integrated Objective Function. So far, both the contextual features and the correlation between the POIs and their category labels are taken into consideration in LCE. That is, the first part maximizes the probability of each POI given its context, and the second part minimizes the distances between the contextual POIs and their category labels in the latent vector space. To effectively model the POIs, we combine Eqs. (4) and (5) defined in each part respectively to construct the integrated objective function for all check-in sequences as follows:

$$\mathcal{L} = \sum_{u \in U} \sum_{j=1}^{N_u} \left\{ \log g_1(l_j) - \log g_2(l_j) \right\}$$

$$= \sum_{u \in U} \sum_{j=1}^{N_u} \left\{ \sum_{l_x} \left\{ L^{l_j}(l_x) \cdot \log[\sigma(\bar{v}_{l_j}^\top \theta^{l_x})] + [1 - L^{l_j}(l_x)] \cdot \log[1 - \sigma(\bar{v}_{l_j}^\top \theta^{l_x})] \right\} \right.$$

$$\left. - \log \left\{ \frac{1}{S} \sum_k \alpha_{j+k} \left\| v_{l_{j+k}} - v_{c_{j+k}} \right\|^2 \right\} \right\}, \tag{6}$$

where $l_x \in \{l_j\} \cup NEG(l_j)$ and $k \in \{k| -K \le k \le K, k \ne 0\}$.

3.2   Parameter Learning

We represent all the parameters with Θ, and learn the LCE using maximum a posterior (MAP):

$$\Theta = \arg\max \sum_{u \in U} \sum_{j=1}^{N_u} \left\{ \log g_1(l_j) - \log g_2(l_j) \right\} - \lambda \left\| \Theta \right\|^2, \tag{7}$$

where $\lambda \left\| \Theta \right\|^2$ is the regularization term.

Here, we apply the Stochastic Gradient Descent(SGD) algorithm to estimate the parameters. During the training process, the algorithm iterates over the POIs of all sequences until it converges. At each time, a target POI l_j with its context is used for update. For a certain context(l_j), the POI l_j is a positive sample, and we randomly sample m unobserved POIs as negative samples. The update procedure is shown in Eq. (8), where η denotes the learning rate.

$$\theta^{l_x} \leftarrow \theta^{l_x} + \eta\{[L^{l_j}(l_x) - \sigma(\bar{v}_{l_j}^{\top}\theta^{l_x})]\bar{v}_{l_j} - 2\lambda\theta^{l_x}\}, \quad l_x \in \{l_j\} \cup NEG(l_j)$$

$$v_u \leftarrow v_u + \eta\{\sum_{l_x}[L^{l_j}(l_x) - \sigma(\bar{v}_{l_j}^{\top}\theta^{l_x})]\theta^{l_x} - 2\lambda v_u\}$$

$$v_{t_j} \leftarrow v_{t_j} + \eta\{\sum_{l_x}[L^{l_j}(l_x) - \sigma(\bar{v}_{l_j}^{\top}\theta^{l_x})]\theta^{l_x} - 2\lambda v_{t_j}\}$$

$$v_{l_{j+k}} \leftarrow v_{l_{j+k}} + \eta\{\sum_{l_x}[L^{l_j}(l_x) - \sigma(\bar{v}_{l_j}^{\top}\theta^{l_x})]\theta^{l_x} - \frac{2\alpha_{j+k}(v_{l_{j+k}} - v_{c_{j+k}})}{\sum_k \alpha_{j+k}\left\|v_{l_{j+k}} - v_{c_{j+k}}\right\|^2} - 2\lambda v_{l_{j+k}}\}$$

$$v_{c_{j+k}} \leftarrow v_{c_{j+k}} + \eta\{\sum_{l_x}[L^{l_j}(l_x) - \sigma(\bar{v}_{l_j}^{\top}\theta^{l_x})]\theta^{l_x} + \frac{2\alpha_{j+k}(v_{l_{j+k}} - v_{c_{j+k}})}{\sum_k \alpha_{j+k}\left\|v_{l_{j+k}} - v_{c_{j+k}}\right\|^2} - 2\lambda v_{c_{j+k}}\}.$$

$$(8)$$

3.3 Category Label Prediction

At the stage of prediction, given an unlabeled POI l, we measure the distance between the target POI and each possible category label in the latent space by employing the Euclidean distance

$$distance(l, c_i) = ||v_l - v_{c_i}||^2, c_i \in C, \tag{9}$$

and select the one with the minimum distance as the predicted result.

4 Experiments

4.1 Datasets

Two publicly available check-in datasets [14] are adopted in the experiments. They are collected from Foursquare, one from New York and the other from Tokyo. Each check-in record contains four types of properties, namely, user ID, POI ID, category label, and timestamp. Foursquare organizes the categories of POIs with a hierarchical structure. Here, we adopt the first-level categories, including Entertainment, Education, Food, Nightlife, Outdoor, Professional, Residence, Shop&Service and Transportation, as the category labels of POIs.

In order to effectively model personal preference, we filter out inactive users who have provided less than three check-ins per week following Yang et al. [14]. Then check-in sequences are constructed based on Definition 2. For discretization, we divide one day into 24 equi-sized time bins. Considering that people tend to demonstrate periodical behaviors by weeks, e.g., *Tom* devotes to working on weekdays and takes leisure on weekends, we differentiate weekdays from weekends. Thus we can get 48 time bins in total, 24 for weekdays and 24 for weekends. After preprocessing, the statistical properties of the two datasets are shown in Table 1, where ♯Users, ♯POIs, ♯Check-ins are the number of users, POIs, and check-ins respectively. Further, we use a 10-fold cross validation to evaluate the prediction performance of different methods.

Table 1. Data statistics

Dataset	♯Users	♯POIs	♯Check-ins
New York	1,083	38,333	227,428
Tokyo	2,293	61,858	573,703

4.2 Experimental Setup

We compare the proposed LCE model with four baselines for predicting the category labels of POIs.

- SAP [15]: SAP explores some explicit properties of each POI (e.g., its number of check-ins, number of users and distribution of visiting time), as well as implicit relatedness among similar POIs. These extracted features are combined to develop a SVM classifier for category label prediction.
- word2vec [9]: In text mining, word2vec can effectively model the words' contextual correlations in sentences. We follow the same methodology by applying word2vec to learn vectors for POIs, and taking each POI as a word and the check-in sequence of a user as a sentence.
- MC-TEM [18]: MC-TEM exploits multiple contexts containing user-level, trajectory-level, location-level and temporal features, but it does not minimize the distances between the POIs and their category labels in the latent space.
- DeepCity [11]: DeepCity proposes a task-specific random walk to define each POI's neighbors and utilizes the network embedding method (i.e., the Skip-gram model) to learn different features. Then, one-vs-rest logistic regression is adopted to infer a POI's category.

The default parameters for our LCE model are as follows: the number of iterations is $I = 15$, the dimensionality of the embedding vector is $d = 50$, the context window size is $K = 5$, the number of negative samples is $m = 30$, the learning rate is $\eta = 0.01$ and the regularization parameter is $\lambda = 0.001$. For the baseline methods, the system parameters are optimized by a grid search. To evaluate the results, we adopt four well-known metrics, namely, accuracy, average precision (over different categories), average recall, and average F1-measure.

4.3 Performance of Methods

The prediction performances of different methods are presented in Table 2. We experiment with different combinations of features, and report the best result of SAP. Although it receives a decent result with the current setting, i.e., both explicit patterns and neighbor POIs' properties are considered to infer a POI's category; however, it is still not comparable with our proposed LCE. Word2vec only considers the sequential patterns in check-in sequences, and it performs the worst. MC-TEM takes the user and temporal information into account, and it performs better than *word2vec*. DeepCity introduces the location category biased values of temporal-users which further boost the prediction performance.

Our proposed LCE outperforms all the baselines significantly on both datasets. For instance, compared with DeepCity that has the best performance among the baselines, LCE gains about 31% and 28% improvements with average F1-measure on New York and Tokyo check-in datasets respectively, which is due to two reasons. On one hand, we model the connection between one POI and its context by considering multiple factors, including sequential pattern, personal preference, and temporal influence. On the other hand, we minimize the distances between the POIs and their category labels in the latent space, resulting in the POIs with the same category label are clustered around the category.

Table 2. Performance comparison

Method	New York				Tokyo			
	Accuracy	Precision	Recall	F1	Accuracy	Precision	Recall	F1
SAP	0.53	0.42	0.33	0.37	0.58	0.48	0.32	0.39
word2vec	0.30	0.20	0.22	0.21	0.34	0.18	0.23	0.20
MC-TEM	0.42	0.27	0.21	0.24	0.43	0.31	0.24	0.27
DeepCity	0.51	0.43	0.36	0.39	0.55	0.45	0.41	0.43
LCE	**0.62**	**0.53**	**0.49**	**0.51**	**0.70**	**0.59**	**0.52**	**0.55**

4.4   Parameter Settings and Tuning

We measure the effect of four major parameters in LCE, including the number of iterations I, the dimensionality of the embedding vector d, the context window size K, and the number of negative samples m. When studying the effect of one parameter, we fix the other parameters to their default values. After tuning those parameters on New York and Tokyo datasets, we discover that the trends for all the four parameters are very similar on both datasets. Due to space limitation, we only demonstrate the results on Tokyo dataset.

As shown in Fig. 1, we observe that (1) the performance improves gradually when we increase I, and becomes relatively stable after about 10 iterations; (2) the prediction performance improves as d increases from 10 to 50, and starts to drop slightly when d is greater than 100; (3) as we increase K, the performance of LCE improves significantly at first and varies little after $K = 5$; (4) with the increase of m, the prediction performance steadily rises and remains constant when m is greater than 30.

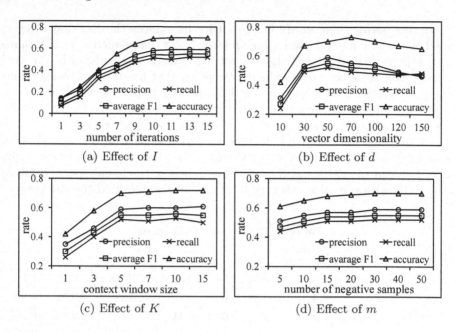

(a) Effect of I

(b) Effect of d

(c) Effect of K

(d) Effect of m

Fig. 1. Tuning parameters for LCE on Tokyo dataset.

5 Related Work

5.1 Location Category Prediction

Existing methods of category label prediction mainly cast it as a classification problem which first extract a set of features for each POI, and then train a multi-label classifier with some features. For example, Ye et al. [15] explore explicit properties of each POI (e.g., its number of check-ins), as well as implicit relatedness among similar POIs, to develop a SVM for prediction. Falcone et al. [5] identify POI semantics based purely on spatial-temporal features, such as POI popularity and stay duration. Chang et al. [1] aim at exploiting human's mobility patterns and capturing four types of temporal features about each POI for the semantic annotation. However, as selecting proper features requires extensive domain knowledge of the data, the performance may vary a lot in practice.

The work by Pang and Zhang [11] explores to train distributed vectors for POIs in LBSNs, which is more related to ours. They propose a feature learning framework called DeepCity, which uses a task-specific random walk to define each POI's neighbors, and then employs the Skip-gram model to preserve learned features. Different from DeepCity, we project the POIs and the category labels into the same low-dimensional latent space, in which a POI is close to its category label. Subsequently, we make prediction by finding the nearest category label for a given POI instead of using a classifier.

5.2   Embedding Learning

Embedding learning aims at modeling objects as continuous variables in a low-dimensional vector space. In NLP, *word2vec* proposed by Mikolov et al. [9,10] can well explain the words' co-occurrence relation in sentences and generate high-quality distributed vector representations of words. Due to its effectiveness in capturing the correlations of objects, *word2vec* has been successfully applied to various applications, such as network embedding [11], POI recommendation [6,18], music recommendation [13], and user modeling [12,16].

To effectively model the check-ins, Zhou et al. [18] propose a Multi-Context Trajectory Embedding Model (MC-TEM), which uses the framework of *word2vec* directly, and takes into consideration various contextual features, including user-level, trajectory-level, location-level and temporal contexts. The difference between our proposed LCE and MC-TEM lies in that we further minimize the distances between the labeled POIs and their category labels in the latent space.

6   Conclusion

In this paper, we have proposed a novel Location Category Embedding (LCE) model for predicting the category labels of POIs, which projects the POIs and the category labels into the same low-dimensional latent space. To effectively model the POIs, we not only capture the correlation between the POIs and their contexts, but also minimize the distances between the contextual POIs and their category labels in the latent space. We evaluate the performance of the proposed LCE on two real datasets, and experimental results show that LCE outperforms the state-of-the-art baselines significantly.

Acknowledgments. This work was partially supported by the National Basic Research 973 Program of China under Grant No. 2015CB352502, the National Natural Science Foundation of China under Grant Nos. 61272092 and 61572289.

References

1. Chang, C., Fan, Y., Wu, K., Chen, A.L.P.: On the semantic annotation of daily places: a machine-learning approach. In: LocWeb, pp. 3–8 (2014)
2. Chen, M., Liu, Y., Yu, X.: NLPMM: a next location predictor with markov modeling. In: Tseng, V.S., Ho, T.B., Zhou, Z.-H., Chen, A.L.P., Kao, H.-Y. (eds.) PAKDD 2014 Part II. LNCS (LNAI), vol. 8444, pp. 186–197. Springer, Cham (2014). doi:10.1007/978-3-319-06605-9_16
3. Chen, M., Yu, X., Liu, Y.: Mining moving patterns for predicting next location. Inf. Syst. **54**, 156–168 (2015)
4. Cheng, H., Ye, J., Zhu, Z.: What's your next move: user activity prediction in location-based social networks. In: SDM, pp. 171–179 (2013)
5. Falcone, D., Mascolo, C., Comito, C., Talia, D., Crowcroft, J.: What is this place? Inferring place categories through user patterns identification in geo-tagged tweets. In: MobiCASE (2014)

6. Feng, S., Li, X., Zeng, Y., Cong, G., Chee, Y.M., Yuan, Q.: Personalized ranking metric embedding for next new POI recommendation. In: IJCAI (2015)
7. Krumm, J., Rouhana, D.: Placer: semantic place labels from diary data. In: Ubi-Comp, pp. 163–172 (2013)
8. Le, Q.V., Mikolov, T.: Distributed representations of sentences and documents. In: ICML, pp. 1188–1196 (2014)
9. Mikolov, T., Chen, K., Corrado, G., Dean, J.: Efficient estimation of word representations in vector space. CoRR abs/1301.3781 (2013)
10. Mikolov, T., Sutskever, I., Chen, K., Corrado, G.S., Dean, J.: Distributed representations of words and phrases and their compositionality. In: NIPS (2013)
11. Pang, J., Zhang, Y.: DeepCity: a feature learning framework for mining location check-ins. In: ICWSM, pp. 652–655 (2017)
12. Tang, D., Qin, B., Liu, T., Yang, Y.: User modeling with neural network for review rating prediction. In: IJCAI, pp. 1340–1346 (2015)
13. Wang, D., Deng, S., Zhang, X., Xu, G.: Learning music embedding with metadata for context aware recommendation. In: ICMR, pp. 249–253 (2016)
14. Yang, D., Zhang, D., Zheng, V.W., Yu, Z.: Modeling user activity preference by leveraging user spatial temporal characteristics in LBSNs. IEEE Trans. Syst. Man Cybern.: Syst. **45**(1), 129–142 (2015)
15. Ye, M., Shou, D., Lee, W., Yin, P., Janowicz, K.: On the semantic annotation of places in location-based social networks. In: KDD, pp. 520–528 (2011)
16. Yu, Y., Wan, X., Zhou, X.: User embedding for scholarly microblog recommendation. In: ACL (2016)
17. Zhao, S., Zhao, T., Yang, H., Lyu, M.R., King, I.: STELLAR: spatial-temporal latent ranking for successive point-of-interest recommendation. In: AAAI (2016)
18. Zhou, N., Zhao, W.X., Zhang, X., Wen, J., Wang, S.: A general multi-context embedding model for mining human trajectory data. TKDE **28**(8), 1945–1958 (2016)

Knowledge Graph Based Question Routing for Community Question Answering

Zhu Liu[1], Kan Li[1(✉)], and Dacheng Qu[2]

[1] School of Computer Science and Technology, Beijing Institute of Technology,
Beijing 100081, China
{paradiser,likan}@bit.edu.cn
[2] School of Software, Beijing Institute of Technology, Beijing 100081, China
qudc@bit.edu.cn

Abstract. Community-based question answering (CQA) such as Stack Overflow and Quora face the challenge of providing unsolved questions with high expertise users to obtain high quality answers, which is called question routing. Many existing methods try to tackle this by learning user model from structure and topic information, which suffer from the sparsity issue of CQA data. In this paper, we propose a novel question routing method from the viewpoint of knowledge graph embedding. We integrate topic representations with network structure into a unified Knowledge Graph Question Routing framework, named as KGQR. The extensive experiments carried out on Stack Overflow data suggest that KGQR outperforms other state-of-the-art methods.

Keywords: Community question answering · Question routing · Knowledge graph · Embedding

1 Introduction

Community-based question answering (CQA) enables users to ask different kinds of questions and provide answers on questions to share their own opinions, which meets the needs of users to obtain and share knowledge. Some popular CQA sites such as Stack Overflow, Quora and Zhihu have collected large scale of metadata associated with questions, answers and users. However, most CQA sites have a common issue: there exist many long-term unresolved questions and the number of them is increasing rapidly. It's due to the growing number of newly asked questions every day with the stable numbers of users. The answerers are unable to meet with the questions they are interested in and the askers can't obtain the information they want. One fundamental task to tackle this issue is question routing which aims to provide a list of answerers who are most likely to post satisfying answers for the questions in the shortest possible time, which has been studied by lots of work [5,12,13,15,16]. This kind of recommendation in CQA sites is also known as expert finding [16] or question recommendation [10].

In order to route questions to proper answerers, some related work [1,6,9,14] tried to utilize structure information of CQA data. Jurczyk and Agichtein [6] and

© Springer International Publishing AG 2017
D. Liu et al. (Eds.): ICONIP 2017, Part V, LNCS 10638, pp. 721–730, 2017.
https://doi.org/10.1007/978-3-319-70139-4_73

Zhang et al. [14] built asker-replier networks based on users and user interactions when they play the role of askers or answerers. This kind of networks are inspired by link analysis (e.g., PageRank, HITS) which are quite qualified for measuring centrality of web pages. All these work only consider using structure features but ignore topic features of post contents.

Although most of existing question routing methods [8,10,11,13] have shown promising performance by capturing deeper advanced topic features and addressing the issue of the lexical gap between the user profiles and posted questions, it's undeniable that they fail to overcome the sparsity of CQA data. They learn user-topic model from past user interaction history including posting questions and answering questions. On average, only a few users will show their opinions for each question in CQA sites and it's costly to construct a sparse user-question matrix for some latent topic models including LDA and PLSA. To tackle the sparsity problem, it's necessary to full leverage both structure and topic features.

Recently, knowledge graph has gradually attracted wide attention since Google Knowledge Graph[1] was constructed. Knowledge graph combines semantic data with graph structure by regarding objects in various domains such as people, music, movies, books, events as entities and connecting them through different relationships. Depending on heterogeneity of knowledge graph, people develop insights on problems which are difficult to uncover with other simple data.

In this paper, we propose a novel question routing method from the viewpoint of knowledge graph embedding to integrate topic representations from CQA data with network structure into a unified Knowledge Graph Question Routing framework, named as KGQR. We first construct the initial knowledge graph using collected CQA data. Then, we introduce the thoughts similar to TransR [7], a state-of-the-art method in knowledge graph embedding, to represent our graph. When there comes a new question, KGQR will give a ranked list of users who are more suitable for it based on our trained model. The main contributions of this paper are summarized as the following:

- To the best of our knowledge, this is the first work applying knowledge graph embedding model to integrate structure features and topic features into a unified framework for question routing in CQA sites.
- Based on the dataset from the well-known CQA site Stack Overflow, we assess the performance of our methods with extensive experiments. The results suggest that our method outperforms other baseline methods.
- Our knowledge graph is scalable for large-scale CQA data, and it is capable of addressing other CQA problems such as answer ranking, question quality and question retrieval.

2   Related Work

In this section, we briefly review some related work on the problem of question routing and knowledge graph embedding.

---

[1] http://www.google.com/insidesearch/features/search/knowledge.html.

Question Routing. A body of literature exists around question routing in CQA sites. They can be mainly divided into two groups: structure-based methods and topic-based methods.

The structure-based question routing methods utilize past question-answering activities in CQA sites for link analysis. Jurczyk and Agichtein [6] proposed a method based on HITS to estimate the ranking expertise of the users by constructing question answering activity graphs. Zhu et al. [17] measured the relevance of questions and select a list of candidate users by ranking user authority in an expanded category graph. This kind of methods fail to employ topic features which questions belong to in CQA sites. Different from them, we consider question routing problem from the viewpoint of knowledge graph embedding. We employ questions topic features which they fail to use in our framework.

The topic-based question routing methods are based on PLSA and LDA, which focus on modeling and evaluating users. Yang et al. [13] introduced a Topic Expertise Model (TEM) to jointly model topics and expertise by integrating content, tags and answer quality information. They proposed a CQARank framework that combines user topical expertise estimation from the TEM model and user authority derived from link analysis in a CQA graph. Zhao et al. [16] proposed a unified Ranking Metric Network Learning framework (RMNL) for question routing. They combined recurrent neural networks with a random-walk based learning method to learn semantic representations of questions and users. However, in this paper, we apply knowledge graph embedding model to integrate structure features and topic features into a unified framework.

Knowledge Graph Embedding. Knowledge graph embedding focus on embedding entities and relations into a continuous vector space and model the semantics of structure in that space. TransE [2] represented a relation by a translation vector so that the head entity and tail entity of embedded entities in a triplets can be connected with low error. However, TransE fails to deal with N-to-N relations in a heterogeneous knowledge graph. To solve the issue of TransE, TransR [7] stepped to propose that entities and relations are completely different objects. They modeled entities and relations in distinct spaces and performed translation in relation space by using a translation matrix. In this paper, we employ translation similar to TransR to represent our CQA graph, which full leverage the structural knowledge in CQA sites.

3   Knowledge Graph Question Routing

In this section, we first briefly introduce the overview of the method, then we give a detailed introduction to the composition and learning method of the model and finally we explain the way we route questions to users.

3.1 Method Overview

We provide an overview of our method referred as KGQR, which is shown in Fig. 1.

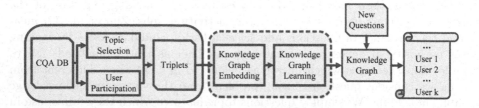

Fig. 1. The overview of proposed method for question routing.

We first crawl enough posts from a well-known CQA site to construct a large scale of dataset. Here the posts refer to all the questions including related data in a period of time. Secondly, based on a classic community detection algorithm called CNM [4], we select the top frequent topics and active users to filter raw data and construct triplets to build a heterogeneous CQA network. Thirdly, we use a knowledge graph embedding approach to represent our CQA network. Note that, our method can be regarded as off-line processing. We learn training and testing questions representation together. Then, the learning process of KGQR is carried out by stochastic gradient descent (SGD). Finally, for each question in testing set, we provide a list of ranked users based on our trained KGQR.

3.2 Knowledge Graph Embedding

We now introduce knowledge graph embedding in detail. We first use an directed graph $G = (V, E)$ to denote the CQA network, where $V = \{v_1, \ldots, v_{|v|}\}$ is a set of vertices referring to various entities and E is a set of directed edges referring to relation between head entities and tail entities. We represent entities and relations in different semantic spaces which are connected by relation-specific matrices, while other methods set entity and relation embeddings within the same space R^k. In our method, v_h and v_t are used to indicate head entity and tail entity respectively and r is the relation connecting them. For each triplet (v_h, r, v_t) in the graph, entities embeddings are set as vectors $v_h, v_t \in R^k$ and relation embedding is set as $r \in R^d$. Note that, k and d are not necessarily the same. We set a projection matrix $M_r \in R^{k \times d}$ for each relation r to projects entities from entity space to relation space. With the projection matrix, entity embeddings are defined as

$$v_h^r = v_h M_r \ , v_t^r = v_t M_r. \tag{1}$$

All the entities have different vector representations in different relation spaces. The score function of this triplet is relatively defined as

$$f_r(v_h, v_t) = ||v_h^r + r - v_t^r||^2. \tag{2}$$

Hence, we can assume that triplet (v_h, r, v_t) is correct when $f_r(v_h, v_t)$ is lower, and wrong otherwise.

3.3   Knowledge Graph Learning

In this subsection, we present the details of our knowledge graph learning method and summarize the main training process in Algorithm 1.

Algorithm 1. Knowledge graph learning.

Input: CQA network $G = (V, E)$, triplets $S = (v_h, r, v_t)$, embedding size k, d, batch size b, margin γ, learning rate λ, number of iterations T.

Output: v_h, v_t, r, M_r.

1: Initialize $v_h, v_t, r \leftarrow$ Results of TransE[2], $M_r \leftarrow \frac{M_r}{||M_r||}$
2: **for** $t = 1$ to T **do**
3:     $S_{batch} \leftarrow Sample(S, b)$
4:     **for** $(v_h, r, v_t) \in S_{batch}$ **do**
5:         $(v_h', r, v_t') \leftarrow Sample(S')$
6:         Accumulate $\mathcal{L}$ by Equation (3)
7:         Update v_h, v_t, r, M_r by SGD
8:     **end for**
9: **end for**

We define the following margin-based hinge loss function for learning knowledge graph representations

$$\mathcal{L} = \sum_{(v_h, r, v_t) \in S} \sum_{(v_h', r, v_t') \in S'} max(0, f_r(v_h, v_t) + \gamma - f_r(v_h', v_t')), \tag{3}$$

where S is the set of real triplets, S is the set of fake triplets and γ is the margin. In our training sets, there only contain real triplets. Consequently, we need to construct fake triplets $(v_h', r, v_t') \in S'$ by replacing entities in real triplets $(v_h, r, v_t) \in S$. When corrupting the real triplets, we follow the sampling method called "unif" in [7]. We utilize stochastic gradient descent (SGD) to minimize our margin-based hinge loss function.

3.4   Question Routing

In this paper, all the objects in CQA sites are mapped to entities in knowledge graph, termed as entities. So our heterogeneous CQA network with multiple types of entities and relations are represented by a knowledge graph. For a question session, entities including question, users (askers and answerers) and tags (question topics) are connected by some relations (ask, belong to and so on). Meanwhile, tags are treated as some kind of relations. When a question q and a user u are connected by a tag t, it means that t is one of the topics of q and u can provide high quality answers for q. So, we intend to calculate Eq. (2) for each (q, u) pair to evaluate whether u is an answerer for q or not. The final question routing for q is obtained according to Eq. (4).

$$list : u_1 > u_2 > ... > u_n \rightarrow f_{q_t}(v_q, v_{u_1}) < f_{q_t}(v_q, v_{u_2}) < ... < f_{q_t}(v_q, v_{u_n}). \quad (4)$$

4   Experiments

In this section, we conduct several experiments on the focused CQA site Stack Overflow to show the high performance of our approach KGQR for the problem of question routing.

4.1   Data Description

The experimental dataset is based on a data-dump[2] of the focused CQA site Stack Overflow. For our experiments, we crawl all the posts from February, 2015 to August, 2015 as our dataset. In terms of user participation, low participation levels are organic to CQA sites, where most of the answers are posted by a minority of contributors. To evaluate the impact of user participation level on question routing, we split the dataset into 6 subsets denoted by $Q_1, Q_2, ..., Q_6$ according to users who answered no less than x questions, $x = \{10, 20, 30, 40, 50, 60\}$. Then, we split first 80%, next 10% and remaining 10% of these subsets into a training set, a validation set and a testing set according to the timestamp when the question is posted. Table 1 shows the details of our subsets. We take subset Q_4 as an example and construct an undirected network as input of CNM [4] for topic extraction and CNM automatically extract 8 topics. Table 2 shows top frequent tags of extracted 8 topics in Q_4 and the first tag is regarded as representation of each topic. We assume that when a new question is queried, we will get its tag information and learn its representation together with training set.

4.2   Evaluation Metrics and Baseline Methods

We employ ranking evaluation metrics to compare our method with different baseline approaches. Note that the actual answerers for each test question are viewed as the ground truth and we sort users for each question in decreasing order of voting scores as our target rank to predict.

We use **Precision@5** to measure the number of users provided by model, which are in the top 5 positions of the target rank. **nDCG@5** is an extension of *Precision*@5, which allows for multiple relevant values. We use **Accuracy** in [15] to evaluate the ranking quality of the best answerer and **Mean Reciprocal**

---

[2] http://data.stackexchange.com.

Table 1. Detailed statistics of 6 datasets in stackoverflow.com. All datasets are divided by x mentioned above.

Dataset	Question number	Answer number	User number	Tag number
Q_1	148,143	334,859	12,458	18,651
Q_2	139,419	287,022	6,987	18,029
Q_3	132,123	252,080	4,680	17,485
Q_4	125,144	222,546	3,296	16,980
Q_5	118,641	199,667	2,505	16,553
Q_6	113,248	181,781	1,977	16,164

Table 2. Top frequent tags of different topics discovered by CNM.

Topic1	Topic2	Topic3	Topic4	Topic5	Topic6	Topic7	Topic8
c#	subjective	c++	java	php	sql	python	iphone
.net	language-agnostic	c	best-practices	javascript	sqlserver	beginner	objective-c
asp.net	programming	windows	eclipse	html	database	ruby	cocoa
visual-studio	not-programming-related	linux	design	css	mysql	svn	iphone-sdk
vb.net	career-development	algorithm	gui	jquery	sqlserver2005	git	xcode

Rank (MRR) to calculate the average of the Reciprocal Rank for a set of questions.

To evaluate the performance of KGQR, we compare against some related work including structure-based methods, topic-based models and mixture methods combining both. **AuthorityRank (AR)** [3] computes user expertise by considering the number of best answers he has provided, which is an in-degree method. **ExpertsRank (ER)** [14] finds expert users by using question answering relations to construct CQA graph via standard PageRank algorithm. **TSPM** [5] is a topic-sensitive probabilistic model, which learns question representation for expert finding in CQA sites via LDA-based model. **CQARank** [13] combines user topical expertise estimation from their Topic Expertise Model and user expertise derived from link analysis in CQA graph. **KGQR-NTE** is a version of our method which neglecting topic extraction. We regard this version as a baseline method to evaluate the effect of topic extraction on KGQR.

4.3 Performance Comparison Under Different Parameters

We care about four essential parameters: entity and relation embedding k, d, batch size b for SGD, margin γ and learning rate λ for SGD. Figure 2(a) to (d) report the *precision*@5 of different values of k, d, b, γ and λ in validation sets, respectively. According to the results, we can observe that:

1. The performance trend of our method becomes stable after the size of embedding k, d larger than 100.
2. The performance of our method tend to be stable when batch size b are increasing. Meanwhile, time consumption is also increasing.

3. When margin γ equals 2 and learning rate λ equals 0.0001, we obtain the best results in validation sets.

Considering both performance and time consumption, the optimal configurations we choose are: $k, d = 100, b = 120, \gamma = 2$ and $\lambda = 0.0001$.

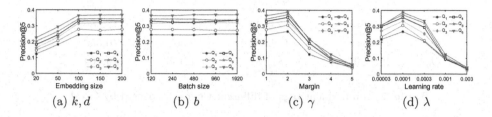

(a) k, d (b) b (c) γ (d) λ

Fig. 2. *Precision*@5 under different parameters.

4.4 Performance Comparison with Different Methods

Figure 3(a) to (d) present the comparisons of different methods on *Precision*@k, *nDCG*@k, *Accuracy* and *MRR* in testing sets, respectively. These experiments reveal a number of interesting points:

1. Compared with ER, CQARank performs better on different evaluation metrics, which shows the advantage of considering tag information for question routing. The results of CQARank are better than TSPM, which proves the effectiveness of combining topic features with link structure to improve question routing. Overall, mixture methods get better performance than single feature based method, which confirms our intuition.

2. KGQR significantly outperforms other baseline methods including KGRQ-NTE, which indicates that leveraging topic and structure features from the viewpoint of knowledge graph embedding can further enforce question routing. Specifically, KGQR improves 19.8% on *Precision*@5, 28.7% on *nDCG*@5, 14.6% on *Accuracy* and 11.3% on *MRR* compared with the state-of-the-art method CQARank. We can see detailed results in Table 3.

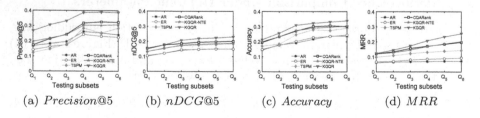

(a) *Precision*@5   (b) *nDCG*@5   (c) *Accuracy*   (d) *MRR*

Fig. 3. Performance comparison with different methods.

3. We evaluate all the baseline methods with subsets that have different user participation levels. We can observe that with the growth of user participation level, all the methods improves a lot on different evaluation metrics. However, we should notice that this kind of growth tends to level off or even be negative when user participation level is too high.
4. In term of MRR, the best value of KGQR in our experiments is 0.2545, which indicates that new questions should be routed to 4 users on average to get answered. However, AR and ER should route it to the top 14 and 11 users, the state-of-the-art methods TSPM and CQARank should consider top 7 and 5 users.

Table 3. Detailed performance comparison with different methods in subset Q_6.

	Precision@5	nDCG@5	Accuracy	MRR
AR	0.2413	0.1496	0.2379	0.0714
ER	0.2257	0.1502	0.2457	0.0926
TSPM	0.2851	0.1847	0.2911	0.1439
CQARank	0.3233	0.2019	0.3014	0.1977
KGQR-NTE	0.3072	0.1892	0.3044	0.2022
KGQR	**0.3877**	**0.2313**	**0.3388**	**0.2545**

5 Conclusion

In this paper, we introduce a novel knowledge graph based question routing method integrating structural features and topic features to tackle the sparsity data issues and enhance the satisfaction of askers in CQA sites. The extensive experiments conducted on a real world CQA dataset from Stack Overflow suggest that our method outperforms other baselines on different evaluation metrics.

Some meaningful future work could be continued. We plan to introduce other useful features including users social relations (e.g., Quora with Twitter) to improve our method. In addition, we intend to tackle other important problems such as answer ranking and question retrieval by the method we propose.

Acknowledgments. The research was supported in part by National Basic Research Program of China (973 Program, No. 2013CB329605) and National Natural Science Foundation of China (NSFC No. 61370136).

References

1. Aslay, Ç., O'Hare, N., Aiello, L.M., Jaimes, A.: Competition-based networks for expert finding. In: Proceedings of the 36th International ACM SIGIR Conference on Research and Development in Information Retrieval, pp. 1033–1036. ACM (2013)
2. Bordes, A., Usunier, N., Garcia-Duran, A., Weston, J., Yakhnenko, O.: Translating embeddings for modeling multi-relational data. In: Advances in Neural Information Processing Systems, pp. 2787–2795 (2013)

3. Bouguessa, M., Dumoulin, B., Wang, S.: Identifying authoritative actors in question-answering forums: the case of yahoo! answers. In: Proceedings of the 14th ACM SIGKDD International Conference on Knowledge Discovery and Data Mining, pp. 866–874. ACM (2008)
4. Clauset, A., Newman, M.E., Moore, C.: Finding community structure in very large networks. Phys. Rev. E **70**(6), 066111 (2004)
5. Guo, J., Xu, S., Bao, S., Yu, Y.: Tapping on the potential of q&a community by recommending answer providers. In: Proceedings of the 17th ACM conference on Information and knowledge management, pp. 921–930. ACM (2008)
6. Jurczyk, P., Agichtein, E.: Discovering authorities in question answer communities by using link analysis. In: Proceedings of the Sixteenth ACM Conference on Conference on Information and Knowledge Management, pp. 919–922. ACM (2007)
7. Lin, Y., Liu, Z., Sun, M., Liu, Y., Zhu, X.: Learning entity and relation embeddings for knowledge graph completion. In: AAAI, pp. 2181–2187 (2015)
8. Liu, D.R., Chen, Y.H., Kao, W.C., Wang, H.W.: Integrating expert profile, reputation and link analysis for expert finding in question-answering websites. Inf. Process. Manag. **49**(1), 312–329 (2013)
9. Liu, J., Song, Y.I., Lin, C.Y.: Competition-based user expertise score estimation. In: Proceedings of the 34th International ACM SIGIR Conference on Research and Development in Information Retrieval, pp. 425–434. ACM (2011)
10. San Pedro, J., Karatzoglou, A.: Question recommendation for collaborative question answering systems with RankSLDA. In: Proceedings of the 8th ACM Conference on Recommender systems. pp. 193–200. ACM (2014)
11. Van Dijk, D., Tsagkias, M., De Rijke, M.: Early detection of topical expertise in community question answering. In: Proceedings of the 38th International ACM SIGIR Conference on Research and Development in Information Retrieval, pp. 995–998. ACM (2015)
12. Xu, F., Ji, Z., Wang, B.: Dual role model for question recommendation in community question answering. In: Proceedings of the 35th International ACM SIGIR Conference on Research and Development in Information Retrieval, pp. 771–780. ACM (2012)
13. Yang, L., Qiu, M., Gottipati, S., Zhu, F., Jiang, J., Sun, H., Chen, Z.: CQArank: jointly model topics and expertise in community question answering. In: Proceedings of the 22nd ACM International Conference on Information and Knowledge Management, pp. 99–108. ACM (2013)
14. Zhang, J., Ackerman, M.S., Adamic, L.: Expertise networks in online communities: structure and algorithms. In: Proceedings of the 16th International Conference on World Wide Web, pp. 221–230. ACM (2007)
15. Zhao, Z., Wei, F., Zhou, M., Ng, W.: Cold-start expert finding in community question answering via graph regularization. In: Renz, M., Shahabi, C., Zhou, X., Cheema, M.A. (eds.) DASFAA 2015. LNCS, vol. 9049, pp. 21–38. Springer, Cham (2015). doi:10.1007/978-3-319-18120-2_2
16. Zhao, Z., Yang, Q., Cai, D., He, X., Zhuang, Y.: Expert finding for community-based question answering via ranking metric network learning. In: Proceedings of the Twenty-Fifth International Joint Conference on Artificial Intelligence, pp. 3000–3006. AAAI Press (2016)
17. Zhu, H., Chen, E., Xiong, H., Cao, H., Tian, J.: Ranking user authority with relevant knowledge categories for expert finding. World Wide Web **17**(5), 1081–1107 (2014)

Exploiting Non-visible Relationship in Link Prediction Based on Asymmetric Local Random Walk

Chunlong Fan, Dong Li$^{(\boxtimes)}$, Yiping Teng, Dongwan Fan,
and Guohui Ding

College of Computer Science, Shenyang Aerospace University,
Shenyang, Liaoning, China
zgr_mzd@163.com, 1d1259802213@163.com, typ@sau.edu.cn,
dongwan_fan@sina.com, dinggh.sau@gmail.com

Abstract. Link prediction is an important aspect of complex network evolution analysis. In the existing link prediction algorithms, the sparseness and scale of the target network have a great influence on the prediction results, and the link prediction algorithm based on local random walk is better in solving this problem. However, the existing local random walk link prediction algorithm simplifiy the definition of random walk process between nodes as symmetrical relationship, and ignore the influence of non-visible factors on the relationship of information diffusion between nodes. In this paper, for the first time, we introduce asymmetry and non-visible relationship of the network to the link prediction problem. Exploiting the unequal diffusion weights in different directions resulted from different degrees, we propose an asymmetric local random walk (ALRW) algorithm. In addition, with non-visible relationship to calculate of the similarity index, we propose a grounded asymmetric local random walk (GALRW) algorithm on the basis of ALRW. Compared with existing advanced link prediction algorithms, thorough experiments on typical datasets show that GALRW achieves better performance in prediction accuracy.

Keywords: Link prediction · Complex networks · Random walk · Asymmetric · Non-visible relationship

1 Introduction

Link prediction, as an important branch of complex network analysis, is widely used in the fields of protein network prediction, friend and commodity recommendation [1, 2]. Link prediction mainly deals with the restoration and prediction of error information or missing information of complex network [3], and generalizes the characteristics of network such as network node attributes and topology. It can find missing or incorrect links in the network, or predict the connections that will appear in the network [4, 5].

To achieve a better prediction effect, diffusion works in the definition of the similarity index between two nodes, which plays an essential role in the link prediction based on the structural similarity. Since the structure of real networks is always of asymmetry, for each pair of nodes, different degrees will result unequal diffusion

© Springer International Publishing AG 2017
D. Liu et al. (Eds.): ICONIP 2017, Part V, LNCS 10638, pp. 731–740, 2017.
https://doi.org/10.1007/978-3-319-70139-4_74

weights in the different diffusion directions, so that the similarity index between them can be computed in an asymmetric way. However, existing diffusion-based link prediction algorithms [6, 9, 19] do not consider the asymmetry of different diffusion directions in calculation of the similarity index, which hardly take advantage of the asymmetric structural information of networks in link prediction.

In addition, the network structure describes the visible relationship of the network, but the non-visible relationship are always ignored. Non-visible relationship represent the hidden relations among nodes in the network. For example, given a social network U, it includes two communities, b and c. *Bob* is a member of community b and has a friend *Alice* who is not in U. *Alice* recommends a topic of community c to *Bob*, therefore *Bob* joins community c due to *Alice*'s suggestion. Here *Alice*'s suggestion can be regarded as a non-visible relationship, which increases the probability to establish a connection between B and members in community c. However, in real networks, since the non-visible relationship can hardly be mined and difficult to describe, few of existing works [6–11, 19] take such relationship into account. Therefore, it calls for an effective link prediction algorithm which leverages the asymmetry of the network structure and non-visible relationship among nodes in the network.

In this paper, for the first time, we introduce asymmetry and non-visible relationship of the network to the link prediction problem. Exploiting the unequal diffusion weights in different directions resulted from different degrees, we propose an asymmetric local random walk (ALRW) algorithm, where different balance factors are set to an asymmetric local random walk process for different directions between two nodes. In addition, with non-visible relationship to calculate of the similarity index, we propose a grounded asymmetric local random walk (GALRW) algorithm on the basis ALRW, where a ground node, connecting to all nodes, is added in the target network [6]. Compared with existing advanced link prediction algorithms CN [7], AA [8], RA [9], Salton [10], Jaccard [11], LRW [19] and LGH [6], thorough experiments on typical datasets show that GALRW achieves a better performance in prediction accuracy.

In this paper, the main contributions are summarized as follows,

(1) To the best of our knowledge, for the first time, we introduce asymmetry and non-visible relationship of networks to the link prediction problem.
(2) Exploiting the unequal diffusion weights in different directions resulted from different degrees, we first propose an asymmetric local random walk (ALRW) algorithm. With non-visible relationship to calculate of the similarity index, we further propose a grounded asymmetric local random walk (GALRW) algorithm.
(3) Compared with existing advanced link prediction algorithms, GALRW achieves a better performance in prediction accuracy through the thorough experiments on typical datasets.

The other part of this paper is organized as follows: Sect. 2 introduces the problem description and related work, Sect. 3 describes the proposed algorithm, Sect. 4 verifies the proposed algorithm on different real datasets, and finally, Sect. 5 of the paper we conduct summarize and propose the future works.

2 Related Work

Link prediction based on structural similarity has three different research directions, local-information-based, path-based and random-walk-based link prediction. The link prediction based on local information mainly focuses on common neighbors, node degrees and other factors. The similarity index of link prediction algorithm based on local information mainly includes CN, AA, RA, Salton and Jaccard. This type of algorithms has the advantage of low computational complexity. Path-based link prediction algorithms use node paths to compute node similarity. The similarity index of path-based link prediction algorithms includes local path (LP) [12], Katz [13], and LHN-II [14]. Compared with these two types of link prediction algorithms mentioned above, the link prediction algorithms based on random walk benefit by fully exploiting network structure information, and have better prediction performance.

The link prediction based on random walk is usually based on two types of random walk algorithms, global random walk and local random walk. The similarity index in link prediction based on global random walk includes average commute time [15], Cos+ [16], random walk with restart [17], SimRank index [18]. This type of link prediction algorithm can exploit the network structure more adequately, and has good prediction performance, while its time complexity is high. To this end, Lv et al. [19] propose link prediction algorithm based on local random walk, which reduces the time complexity effectively. Recently, Zhou et al. [6] propose LGH based on local random walk and heat conduction, which is improved in diversity and accuracy. Considering asymmetry and non-visible relationship of networks can increase the utilization of network information in link prediction, however few of existing link prediction algorithms based on random walk take into account the asymmetry and non-visible relationship of the network.

3 Problem Statement and Preliminaries

3.1 Problem Statement

Define the target network as an undirected graph G (V,E), where V represents a set containing all the nodes in G, and E represents the set of edges. Note that multi-edge and self-loop are not considered. For each pair of nodes $x, y \in V$, we calculate a similarity value S_{xy}. The links with the highest S_{xy} are most likely to exist.

In this paper, the proposed link prediction algorithm is based on local random walk. Local random walk index (LRW) only considers the limited number of random walk process. Probabilistic evolution equation is as follows,

$$\pi_x(t+1) = P^T \pi_x(t),\qquad(1)$$

where p is the transition probability matrix. Assuming that the initial resource configuration of a node is q, the similarity based on the t-step random walk is defined as follows,

$$S_{xy}^{LRW}(t) = q_x\,\pi_{xy}(t) + q_y\,\pi_{yx}(t)\,. \tag{2}$$

3.2 Preliminaries

To test the performance of link prediction, the ensemble set E is divided into the training set E^T and the test set E^P, such that $E^T \cup E^P = E$, and $E^T \cap E^P = \emptyset$. The training set E^T is used to calculate the similarity between nodes and the test set E^P is used to test the accuracy of the prediction algorithm. U is defined as the edge set of the complete graph corresponding to G, and U^n is defined as a set of edges that $U^n \in U$, and $U^n \cup E = U$.

We use two standard metrics, AUC (Area Under the receiver operating characteristic Curve) [20] and P (Precision) [21], to quantify the accuracy of prediction algorithms. AUC can be regarded as the probability that the similarity value of an edge randomly selected in E^P is higher than that randomly selected in U^n. The equation is denoted as follows,

$$AUC = \frac{n_1 + n_2}{n}, \tag{3}$$

where n represents the number of independent sampling comparisons, n_1 represents the times of the sampling comparisons that the similarity value of the edge in E^P is higher than that in U^n, and n_2 represents the times that the similarity values are equal for E^P and U^n. If all scores are generated from an independent and identical distribution, AUC = 0.5. Thus, when AUC is higher than 0.5, it indicates a better performance of the algorithm.

P (Precision) focuses on the correct ratio in the prediction list. Given ranked non-observed links, select the previous L links as the predicted list. Then, P is the predicted percentage of the predicted list. The equation is denoted as follows,

$$P = \frac{m}{L}, \tag{4}$$

where m represents the number of edges in the predicted list that belong to E^P.

4 Algorithm

Exploiting the unequal diffusion weights in different directions resulted from different degrees, we propose an asymmetric local random walk (ALRW) algorithm, where different balance factors are set to an asymmetric local random walk process for different directions between two nodes. In addition, with non-visible relationship to calculate of the similarity index, we propose a grounded asymmetric local random walk (GALRW) algorithm on the basis of ALRW where a ground node, connecting to all nodes, is added in the target network.

4.1 ALRW Algorithm

According to the information spreading in the target network in a random walk way, we build a new LRW similarity index. First, A represents the adjacency matrix of the target network. If node x and y have edges, the element $a_{xy} = 1$, otherwise $a_{xy} = 0$. Assuming that the initial information resource of the target node x is 1, and the remaining node information resource is 0, each iteration information resource is propagated in random walk. Then, the amount of information resource for a node after $t + 1$ iterations is as follows,

$$S_x^{LRW}(t+1) = \sum_{y=1}^{M} \frac{a_{xy} S_y(t)}{k_y}, \tag{5}$$

where M is the number of nodes in the network, and k_y is the degree of node y. The similarity index based on local random walk can be expressed as follows,

$$S_{xy}^{LRW}(t) = S_x^{LRW}(t) + S_y^{LRW}(t). \tag{6}$$

The similarity of x and y is obtained by superimposing the bi-directional information propagation process based on local random walk. The information is spread in the form of random walk in the network. For each node, the information propagated to the neighboring nodes is negatively correlated with the degree of the node. In the real network, the importance of the node is different, so is the influence of the node. In order to reflect the asymmetry and the influence of the nodes, we set the balance factor of the random walk process from x to y as $k_y/(k_x + k_y)$ and that from y to x as $k_x/(k_x + k_y)$. Then, we get the asymmetric local random walk index (ALRW) as follows,

$$S_{xy}^{ALRW}(t) = \frac{k_y}{k_x + k_y} S_x^{LRW}(t) + \frac{k_x}{k_x + k_y} S_y^{LRW}(t). \tag{7}$$

Let us give an example, where CE is chosen as the target network. We select the first nine nodes, where Node 3 is selected as the target node. We calculate the similarity between Node 3 and other nodes. In Table 1, S_{3y}^{LRW} is the node similarity calculated by the LRW index between Node 3 and other nodes, and S_{3y}^{ALRW} is the node similarity calculated by the ALRW index between Node 3 and other nodes.

Table 1. Similarity values for different similarity indicators

Node y	1	2	4	5	6	7	8	9
Degree	11	29	52	54	14	36	18	20
S_{3y}^{LRW}	0.0089	0.0106	0.0156	0.0161	0.0088	0.0101	0.0087	0.0104
S_{3y}^{ALRW}	0.0139	0.0127	0.0161	0.0165	0.0129	0.0113	0.0119	0.0138

We sort the sets of similarity values obtained from the two similarity indices respectively in Table 1. It can be found that the similarity between Node 3 and Node 1 is ranked sixth in the LRW method and to the third place in the ALRW method. From the results we get, adding the balance factor makes the low degree nodes improve the competitiveness, which is in line with our expectation.

4.2   GALRW Algorithm

The propagation of information in the network is not only affected by the visible network structure in the target network, but also the non-visible relationship between nodes. In addition, the diffusion step is also a key problem in the propagation of information. For sparse networks, the shortest distance between two nodes is relatively large result the information is difficult to spread in limited number of iterative steps, so the similarity index based on local random walk cannot accurately evaluate the similarity between nodes in finite iterative steps. In order to reflect the non-visible relation between nodes and shorten the average shortest distance of the network, we add a ground node g in the target network. g is connected with all the nodes in the target network. We get the local random walk iteration process as follows,

$$S_x^{GLRW}(t+1) = \sum_{y=1}^{M} \frac{a_{xy} S_y(t)}{k_y} + S_g(t). \tag{8}$$

After adding the ground node, it participates in the process of information propagation. The ground node accepts and transmits part of the information resources in each step of random walk. After the random walk process, the ground node will occupy part of the information resources. However, the ground node is a virtual node which should not have information resources. To this end, we allocate the information resources held by the ground node evenly to other nodes after each step of random walk. This process is as follows,

$$S_x^{GLRW}(t+1) = \sum_{y=1}^{M} \frac{a_{xy} S_y(t)}{k_y} + \frac{N+1}{N} S_g(t). \tag{9}$$

Accordingly, adding a ground node on the basis of the ALRW index, we present the grounded asymmetry random walk (GALRW) index as follows,

$$S_{xy}^{GALRW}(t) = \frac{k_y}{k_x + k_y} S_x^{GLRW}(t) + \frac{k_y}{k_x + k_y} S_y^{GLRW}(t). \tag{10}$$

5   Experiments

5.1   Datasets

We take four representative datasets from different fields. (1) US Air Transport System Network (USAir), which includes 322 airports and 2126 routes. (2) Nematode neural network C.elegans (CE), containing 297 neurons and 2148 connections. (3) Jazz musician network (JAZZ), including 198 musicians and 2742 relationship. (4) Football (FB, network of American football games between Division IA colleges), including 115 teams and 613 games match. Table 2 summarizes five statistics that describe the characteristics of the network. N represents the number of nodes. M represents the number of edges. C is the clustering coefficient. K represents the average degree. D represents the network diameter.

Table 2. Statistic of network characteristics

Network	N	M	C	K	D
USAir	322	2126	0.749	12.81	7
CE	297	2148	0.308	14.46	5
JAZZ	198	2742	0.618	27.7	6
FB	115	613	0.403	10.661	4

5.2   Experimental Result

In this paper, the proposed algorithm is compared with the following algorithm: CN, AA, RA, Salton, Jaccard, LRW and LGH. In order to test the prediction effect of the algorithm, we divide the data set into training set and test set, the ratio is 9: 1, which means that the training set contains 90% of the edges, and the test set contains 10% of the edges. The training set is used for link prediction, and the test set is used for testing. We have two evaluation indicators: accuracy AUC and precision P. All the results are calculated on a different data set 100 times independently.

Table 3: Evaluate the accuracy of different prediction algorithms on four datasets using AUC. The maximum AUC value for each column is marked with bold.

Table 3. AUC of different link prediction algorithms

Network	CE	FB	JAZZ	USAir
CN	0.8484	0.8440	0.9546	0.9543
AA	0.8628	0.8467	0.9640	0.9664
RA	0.8697	0.8416	0.9722	0.9724
Salton	0.7978	0.8545	0.9657	0.9253
Jaccard	0.7927	0.8589	0.9635	0.9138
LRW	0.8993	0.8676	0.9632	0.9587
LGH	0.8973	0.8760	0.9670	0.9542
ALRW	0.9111	0.8733	0.9671	0.9635
GALRW	**0.9238**	**0.8996**	**0.9756**	**0.9800**

Table 4: Evaluate the accuracy of different prediction algorithms on four datasets using P. The maximum P value for each column is marked with bold.

The AUC and P of the different prediction algorithms on all data sets are shown in Tables 3 and 4, with the highest values in each column being highlighted in bold.

Table 4. P of different link prediction algorithms

Network	CE	FB	JAZZ	USAir
LRW	0.1413	0.4132	0.3038	0.3473
LGH	0.2311	0.3993	0.6437	0.4121
ALRW	0.2302	0.3797	0.3555	0.2914
GALRW	**0.2457**	**0.4134**	**0.6655**	**0.4253**

ALRW is an abbreviation for asymmetry local random walk algorithm. GALRW is an abbreviation for the ground asymmetry local random walk algorithm.

Obviously, in the selected four networks, WLRW algorithm is superior to the classic benchmark algorithms (CN, Salton, Jaccard, AA and RA), but also slightly better than LRW and LGH. ALRW algorithm increases the diversity of prediction results by adding asymmetric relationship to the prediction effect, and improves the prediction probability of important nodes with low degree of importance. By comparing the results of ALRW and GALRW, it can be found that adding the ground node improves the prediction accuracy to a certain extent. For example, the AUC of the USAir dataset increased from 0.9735 to 0.9800, and P increased from 0.2915 to 0.4253. Adding a ground node to the network actually increases the additional transfer probability from the ground node to the other node. Thus, each node receives the same information from the ground node in each iteration step, and then passes the information evenly to the neighbor nodes, thereby increasing the information of large degree node. The improvement of the prediction of the important nodes of the large degree makes the prediction accuracy improved.

Figure 1: The correlation between the AUC and local random walk steps on all data sets are shown in Fig. 1. The crosses, oblique crosses, and squares respectively

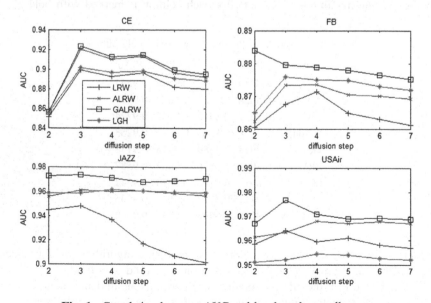

Fig. 1. Correlation between AUC and local random walk steps

represent the best AUC for the LRW, ALRW and GALRW algorithms corresponding to different random walk steps.

In addition, all diffusion-based approaches are limited by network connectivity. Liu W et al. have shown that the optimal iteration steps are positively correlated with the average shortest distance [19]. We make the average shortest distance of the network smaller and the network connection more closely by adding a ground node in the target network. As shown in Fig. 1, it is clear that GALRW can get the best prediction result on four data sets in three steps, especially for two-step iterations of FB datasets, So, compared with LRW, LGH and ALRW, GALRW has a better performance in time complexity. The experimental results show that the proposed algorithm has better prediction effect, and also verifies the rationality of the proposed viewpoint.

6   Conclusion

In this paper, a new link prediction algorithm - ground asymmetry optimization local random walk algorithm (GALRW) is proposed to deal with the asymmetry of information and the non-visible relationship among nodes. Based on the local random walk algorithm, the balance factor is used to reflect the asymmetry of information propagation between nodes, and the proportion of similarity between nodes is optimized. By adding the ground node to reflect the non-visible relationship between nodes, the proposed algorithm improves the connectivity strength of the network and shortens the average shortest distance of the network. Compared with the six typical prediction algorithms on four real networks, GALRW has better prediction effect.

Future research is mainly carried out from two aspects. First, we will examine how to reflect the asymmetry between nodes during random walk more accurately. Secondly, because the non-visible relationship between nodes in the network is not negligible to the prediction effect of the prediction algorithm, the non-visible relationship in the network will be further excavated.

References

1. Guimerà, R., Salespardo, M.: Missing and spurious interactions and the reconstruction of complex networks. In: Proceedings of the National Academy of Sciences of the United States of America. vol. 106, no. 52, pp. 22073–22078 (2010)
2. Zeng, A., Cimini, G.: Removing spurious interactions in complex networks. Phys. Rev. E Stat. Nonlinear Soft Matter Phys. 85(3 Pt 2), 036101 (2012)
3. Neuman, M.E.J.: The structure and function of complex networks. Siam Rev. 45(1–2), 40–45 (2003)
4. Getoor, L., Diehl, C.P.: Link mining: a survey. ACM SIGKDD Explor. Newsl. 7(2), 3–12 (2005)
5. Lü, L.Y., Zhou, T.: Link prediction in complex networks: a survey. Physica A Stat. Mech. Appl. 390(6), 1150–1170 (2010)
6. Liu, J.H., Zhu, Y.X., Zhou, T.: Improving personalized link prediction by hybrid diffusion. Physica A Stat. Mech. Appl. 447, 199–207 (2015)
7. Kossinets, G.: Effects of missing data in social networks. Soc. Netw. 28, 247–268 (2006)

8. Adamic, L.A., Adar, E.: Friends and neighbors on the web. Soc. Netw. **25**, 211–230 (2003)
9. Zhou, T., Lü, L., Zhang, Y.C.: Predicting missing links via local information. Eur. Phys. J. B **71**(4), 623–630 (2009)
10. Salton, G., McGill, M.J.: Introduction to Modern Information Retrieval. MuGraw-Hill, Auckland (1983)
11. Jaccard, P.: Etude comparative de la distribution florale dans une portio n des alpes et des jura. Bull. Soc. Vaudoise Sci. Nat. **37**, 547–579 (1901)
12. Lü, L., Jin, C.H., Zhou, T.: Similarity index based on local paths for link prediction of complex networks. Phys. Rev. E **80**, 046122 (2009)
13. Katz, L.: A new status index derived from sociometric analysis. Psychometrika **18**(1), 39–43 (1953)
14. Leicht, E.A., Holme, P., Newman, M.E.: Vertex similarity in networks. Phys. Rev. E **73**, 026120 (2006)
15. Klein, D.J., Randic, M.: Resistance distance. J. Math. Chem. **12**(1), 81–95 (1993)
16. Fouss, F., Pirotte, A., Renders, J.M., et al.: Random-walk computation of similarities between nodes of a graph with application to collaborative recommendation. IEEE Trans. Knowl. Data Eng. **19**(3), 355–369 (2007)
17. Brin, S., Page, L.: The anatomy of a large-scale hypertextual web search engine. Comput. Netw. ISDN Syst. **30**(1–7), 107–117 (1998)
18. Jeh, G., Widom, J.: SimRank: a measure of structuralcontext similarity. In: Proceedings of the ACM SIGKDD 2002, pp. 538–543, ACM Press, New York (2002)
19. Liu, W., Lu, L.: Link prediction based on local random walk. EPL. **89**(5), 58007–58012(6) (2010)
20. Hanely, J.A., McNeil, B.J.: The meaning and use of the area under a receiver operating characteristic (ROC) curve. Radiology **143**, 29–36 (1982)
21. Herlocker, J.L., Konstann, J.A., Terveen, K., et al.: Evaluating collaborative filtering recommender systems. ACM Trans. Inf. Syst. **22**(1), 5–53 (2004)

Ciphertext Retrieval Technology of Homomorphic Encryption Based on Cloud Pretreatment

Changqing Gong[✉], Yun Xiao, Mengfei Li, Shoufei Han, Na Lin,
and Zhenzhou Guo

College of Computer Science, Shenyang Aerospace University,
Shenyang 110136, Liaoning, China
gongchangqing@sau.edu.cn

Abstract. Ciphertext retrieval in cloud computing environments requires both security and retrieval efficiency. This paper proposes a Ciphertext Retrieval based on Cloud Pretreatment (CRBCP) based on cloud preprocessing. The scheme divides the cloud into the file server and the index server. Firstly, the program uploads the ciphertext document set to the file server and index server. A lot of preprocessing work is done in the index server and generate an inverted index table. Then, the program uploads the ciphertext retrieval item to the index server. Term Frequency-Inverse Document Frequency (TF-IDF) is used the to get the weight vector of the ciphertext document and the ciphertext retrieval item. Finally, the index server calculates the similarity and returns the result to the client. The simulation results show that the efficiency of encryption and decryption time in the algorithm is obviously higher than that of DjikGentryHaleviVaikuntanathan (DGHV) and Based Vector space model and Homomorphism ciphertext retrieval scheme (BVH). The overall efficiency of ciphertext retrieval in the program is superior than others. In the protection of user data privacy and security under the premise, CRBCP scheme preprocesses the ciphertext in the index server. This will not only greatly improve the efficiency of ciphertext retrieval and reduce the computational pressure on the client, but also fully embody the concept and advantages of cloud computing.

Keywords: Cloud computing · Homomorphic encryption · Term Frequency-Inverse Document Frequency (TF-IDF) · Ciphertext retrieval · Cloud Pretreatment

1 Introduction

Cloud computing is an innovative service model that allows users to obtain near-infinite computing power and a wide variety of information services over the Internet. It is the evolution of distributed computing, parallel computing and grid computing [1]. With the rapid development of cloud computing and information exchange, a large number of confidential information is concentrated in the cloud [2]. In the cloud computing environment, homomorphic encryption technology can ensure that the data are not decrypted. In the absence of open plaintext data, homomorphic

© Springer International Publishing AG 2017
D. Liu et al. (Eds.): ICONIP 2017, Part V, LNCS 10638, pp. 741–751, 2017.
https://doi.org/10.1007/978-3-319-70139-4_75

encryption technology can meet the user's data security requirements and retrieval requirements. Algebraic encryption is the ability to arbitrarily calculate the ciphertext without knowing the key, that is, for any valid f and plaintext m, the property f (Enc (m)) = Enc (f (m)). In 2009, Gentry [3, 4] proposed an integer ring homomorphic encryption algorithm based on the rational lattice. In 2010, van Dijk et al. [5] proposed a homomorphic encryption scheme DjikGentryHaleviVaikuntanathan (DGHV) based on simple algebraic computation. It based on the idea of rational lattice and integer ring. In 2013, Gentry et al. [6] proposed a homogeneous cryptographic scheme Gentry Sahai Waters (GSW) based on the learning with errors (LWE). In 2014, Brakerski et al. [7] proposed a homomorphic encryption scheme without Bootstrapping. In 2015, Cheon and Kim [8] proposed a hybrid homomorphic encryption technique based on SomeWhat encryption theory and the public key cryptosystem. After the data in the cloud are encrypted, the user needs to find the information that best meets his requirements in a large number of ciphertext information. It needs to be realized by homomorphic cryptography. Huang et al. [9] through the analysis of existing encryption information retrieval algorithm to ensure the accuracy of quasi-search while checking both the sorting problem and accuracy problem. This paper proposed a search method for homomorphic encryption in cloud storage applications. Song [10] proposed a homology encryption ciphertext retrieve algorithm that can be used in cloud storage environment. The client does not have to pass the key to the cloud during the retrieval process. The cloud retrieves the data ciphertext directly. Zhang [11] proposed a homomorphic encryption based on the retrieval system. She used homomorphic encryption can be achieved on the ocean ciphertext data range search. Based on the analysis and comparison of the existing homomorphic encryption algorithm, she chose a homomorphic encryption algorithm that is appropriate for the system to effectively perform range searches.

Based on the scheme of Wei [12], this paper designs a Ciphertext Retrieval based on Cloud Pretreatment (CRBCP) scheme based on the previous research. It divides the cloud into the file server and the index server, and moves the ciphertext retrieval operation to the index server of the cloud as much as possible. The file server saves the ciphertext document set for the user to download. We use the vector space model, the retrieval term and the vector to be retrieved can be expressed as the vector in the multidimensional space. And the correlation is determined by calculating the angles of the two vectors. Finally, the sorting result is obtained. CRBCP is in the cloud index server to handle the ciphertext to be indexed. It can take full advantage of the cloud's powerful computing power to reduce the pressure on the client and improving the efficiency of cloud data retrieval.

The remainder of this paper is organized as follows. Section 2 introduces the homomorphic encryption algorithm (DGHV) briefly. The details of our proposed algorithms are described in Sect. 3. The experimental evaluations of our proposed algorithm are presented in Sect. 4. Finally, Sect. 5 concludes this paper.

2   The Basic Homomorphic Encryption Algorithm

The DGHV scheme is as follows.

The scheme first constructs a symmetric encryption scheme and then turns it into an asymmetric encryption scheme.

The symmetric encryption scheme is as follows.

Kengen: We select the η bit length prime p as a key, where $p \in [2^{\eta-1}, 2^{\eta})$.

Encrypt (p, m): Let any 1bit plaintext $m \in \{0, 1\}$, ciphertext $c = m + 2r + pq$, where integers q and r are randomly selected within the specified time interval, q is a Large integer, r is a small integer, and satisfies $|2r| < p/2$.

Decryption (p, c): Plaintext $m = (c \bmod p) \bmod 2$.

In the above scheme, $c \bmod p = m + 2r$, $m + 2r$ is called noise. If we want to decrypt correctly, we must meet the noise $m + 2r < p/2$, and the noise of the fresh ciphertext must be less than $p/2$. So, the above program can certainly be decrypted correctly.

Now, we convert its symmetric encryption scheme into an asymmetric encryption scheme.

Kengen (η): We select the prime $p \in [2^{\eta-1}, 2^{\eta})$ of η bit length as the private key; the public key pk is the set $\{x_i = 2r_i + pq_i\}$, and randomly select a subset S of the set.

Encrypt (pk, m): Let any 1 bit plaintext $m \in \{0, 1\}$, ciphertext $c = m + Sum(S)$, where S is a random subset of the set $\{x_i; x_i = pq_i + 2r_i\}$. The integer q and r are randomly selected within the specified time interval, q is a large integer, r is a small integer, and satisfies $|2r| < p/2$.

Decrypt (p, m): Plaintext $m = (c \bmod p) \bmod 2$.

From the above can be drawn, encryption process will produce noise. With the increase of ciphertext noise, the decryption may not get the correct plaintext. So, we can use Gentry's homomorphic encryption framework to improve the above scheme to get a homomorphic encryption scheme. It can be seen that each encryption must be selected from the collection of a subset of the encryption as the public key to encrypt. It is not only very complex in the specific implementation, but also reduce the efficiency of homomorphic encryption algorithm.

For the above reasons, (1) We can eliminate the set S in the algorithm to increase the random number method to guarantee the algorithm safety. (2) Retrieve algorithm: Retrieval(c): Retrieval = $((c_i - c_{index}) \bmod p) \bmod 2$, if Retrieval = 0, $m_i - m_{index} = 0$. It can be noted that the user must send the key p to the server when the user wants to retrieve the keyword in the cloud. Obviously, this will make the user stored in the server on the encrypted data completely exposed to the server. Therefore, the algorithm is not practical when the server is not trusted.

3 The Improved Program(CRBCP)

We use the homomorphism framework proposed by Gentry to construct a homomorphism scheme. It is more suitable for ciphertext retrieval in cloud storage.

3.1 Improved Homomorphic Encryption Algorithm

Keygen: We select a randomly generated P-bit secure large prime number as the key p, where $p \in [2^{\eta-1}, 2^{\eta})$.

Encrypt (m): We select a Q-bit secure large prime q, where $q \in [2^{Q-1}, 2^Q)$, P > Q > the length of the plaintext grouping. The client randomly generates a random number r. M is grouped $M = m_1 m_2 m_3 ... m_t$ (The length of m_i is L), and the client calculates the ciphertext $c = m + \lambda q^2 + (p + q)^2 r$, where $\lambda: = 1 \parallel \{0, 1\}^{256}$.

Decrypt (c): The decrypted plaintext is $m = c \mod (p + q)^2 \mod \lambda$.

Homoclinic analysis: With two clear m_1, m_2, the corresponding ciphertext were c_1, c_2, and $c_1 = m_1 + \lambda q^2 + (p + q)^2 r_1$, $c_2 = m_2 + \lambda q^2 + (p + q)^2 r_2$.

Additive homomorphism analysis is as follows.

$$c_1 + c_2 = (m_1 + m_2) + 2\lambda q^2 + (p+q)^2 (r_1 + r_2).$$

Since $(c_1 + c_2) \mod (p + q)^2 \mod \lambda = m_1 + m_2$, the algorithm satisfies the additive homomorphism.

Multiplication homomorphism analysis is as follows.

$$c_1 * c_2 = (m_1 * m_2) + (m_1 + m_2)\lambda q^2 + (p + q)^2 (m_1 r_2 + m_2 r_1) \\ + \lambda q^2 (p + q)^2 (r_1 + r_2) + \lambda^2 q^4 + (p + q)^4 r_1 r_2.$$

Since $(c_1 * c_2) \mod (p + q)^2 \mod \lambda = m_1 * m_2$, the algorithm satisfies multiplicative homomorphism.

Improved Retrieval Process. We use the improved program on the keyword m_{index} retrieval process is as follows.

(1) We use the above-mentioned improved homomorphic encryption scheme in the client to encrypt the keyword and geting the corresponding ciphertext $c_{index} = m_{index} + \lambda q^2 + (p + q)^2 r$. Then uploading them to the cloud to retrieve.

(2) After we receive the cipher key keyword c_{index}, the retrieve is performed by Eq. 1, where $N = (p + q)^2$.

$$\begin{aligned} Retrieval &= (c_i - c_{index}) \mod N \\ &= ((m_i - m_{index}) + (p+q)^2 (r_1 - r_2)) \mod N \end{aligned} \tag{1}$$

If Retrieval $= 0$, then $m_i - m_{index} = 0$. Say that, when $m_i = m_{index}$, $(p + q)^2 (r_1 - r_2) \mod N = 0$. If $N = (p + q)^2$, the cloud can not get the user's key p. The server can directly retrieve the user's ciphertext and other operations. Because the user's data is completely unknowable for the server.

DGHV's retrieve algorithm needs to send the key p to the server, it cause the key to be exposed directly to the cloud. Based Vector space model and Homomorphism ciphertext retrieval scheme (BVH) [13] and the scheme of Tan [14] need to upload N and qr_i to the cloud, but cannot ensure that q will be leaked to cause the key p to be leaked. In contrast, the program's search algorithm only need to send $N = (p + q)^2$ to the cloud, you can achieve the search for ciphertext keywords.

3.2   The Proposed Algorithm (CRBCP)

CRBCP is based on the scheme of Wei [12] to improve the ciphertext retrieval program and get a cloud-based vector space model isotactic ciphertext retrieval program.

The overall architecture of the CRBCP scheme consists of a client and a cloud (server), which in turn includes a file server and an index server. The main job of the client is to store the original document set and encrypt the document set and the retrieval items. Then it can generate the ciphertext document and the ciphertext retrieval item on the client and upload them to the file server and index server respectively. The main work of the cloud is divided into two parts: one part is the file server in the untrusted conditions need to store the ciphertext document set; the other part of the index server in the fully credible conditions, the ciphertext document set and the ciphertext retrieval item are preprocessed: the word segmentation filter, the establishment of the inverted index, the document vector set, calculate the similarity between ciphertext and ciphertext and return the result to the user.

The program is described in the following order. The overall flow chart of the program is shown in Fig. 1.

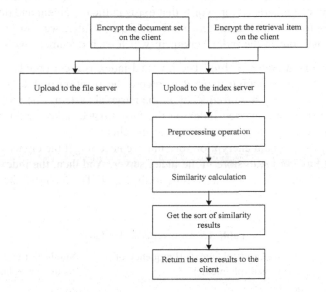

Fig. 1. The overall flow chart of the program

3.2.1   Encrypted Upload

In the client, the client encrypts each document content, document name and retrieval item through the homomorphic encryption algorithm on the improved integer. Then, the client uploads them to the file server and index server respectively to save and operate.

3.2.2 Preprocessing of Document Set

The client uploads the encrypted document set to the index server, and the index server preprocesses the ciphertext document set. The pretreatment is divided into three stages: segmentation filtering, establishment of inverted index, and generation of weight sets of document set.

Word Segmentation Filtering. In the information retrieval, there are many keywords that do not have any meaningful meaning to the search, which are known as the stop words. If the system encounters them during the text processing, the process is immediately stopped and discarded. These words that was thrown away can reduce the amount of index, and increase the efficiency of the retrieval, and usually improve the effect of the retrieval.

The process is a query process: for each keyword to determine whether it exists in the stop word list. If it exists, we must remove it from the document. Because the homomorphic encryption technique encrypts the same data twice, the ciphertext is different in the case of encryption. The CRBCP scheme uses the following method to match the query: First, according to the client encryption formula to get the ciphertext stop words c_i and cipher text of the ciphertext keyword c_{index}. Then, according to the Eq. 1 to query the ciphertext stop words that exists in the document and delete. Finally, all the queries to the ciphertext stop words delete and filter, we can generate a new document set to store in the index server, in which each document with d_j said.

Create an Inverted Index Table. The inverted index is a commonly used indexing method that uses the mapping method to record the position of the keyword in the document. It used a very high frequency in the retrieval system. The program will put the establishment of an inverted index on the fully trusted index server. We abandon some methods to create an inverted index in the client.

According to the client encryption formula, we need to get the cipher key c_{index} and the ciphertext keyword c_i to store in the index server. And then, the index server carry out the keyword matching query according to the Eq. 1. The recorded data is shown in Table 1.

Table 1. An inverted index table

Keyword (ciphertext)	Document id (plaintext) [Frequency of occurrence (plaintext)]	Number of existing documents (plaintext)
c_{index}	$d_j[f_{ij}]$	n_i

Generate a File Vector Set. When a user searches for a keyword, the document set to be retrieved actually corresponds to the weight vector set of the document set. The Term Frequency-Inverse Document Frequency (TF-IDF) algorithm is proposed by Salton and Yu [15], which takes into account the importance of keywords and the relationship between documents in the entire document set. So, CRBCP scheme uses TF-IDF weight calculation framework. Given the keyword weight is more representative, so that a good search results can be achieved. TF represents the word frequency of the keyword, IDF represents the inverse document frequency of the keyword. The value of the weight vector used in this paper is shown in Eq. 2.

$$\omega_{ij} = \begin{cases} (1 + \log f_{ij}) \times \log \frac{N}{n_i} & f_{ij} > 0 \\ 0 & else \end{cases} \tag{2}$$

where w_{ij} represents the TF-IDF weight of the ciphertext keyword k_i for the document d_j. f_{ij} represents the frequency of the keyword k_i appearing in the document d_j, that is the number of occurrences. N represents the total number of documents in the document set. n_i represents the number of documents in the document set that contain the keyword k_i. N/n_i represents the inverse document frequency of the keyword k_i.

There are t ciphertext keywords k_i, and they are independent of each other. We define a ciphertext document d_j is a vector on a t-dimensional space. The weight value of each ciphertext key k_i in the ciphertext document d_j is obtained according to the above formula. The index server generate the ciphertext weight of the document weight vector, in which each ciphertext keyword k_i obtained weight values are plaintext. Then the value of d_j is shown in Eq. 3.

$$d_j = (\omega_{1j}, \omega_{2j}, \ldots, \omega_{ij}, \ldots, \omega_{tj}) \tag{3}$$

3.2.3   Document Retrieval

After the user authorization is successful, it can enter the retrieval system. First, the user can enter the plain text retrieval item, the client will encrypt the plaintext retrieval item to the index server. Then, ciphertext retrieval item does the same preprocessing work as the ciphertext document set in the index server. Finally, the plaintext weight vector of the ciphertext retrieval term is generated. The formula for the weighting of the plaintext is shown in Eq. 4.

$$\omega_{iq} = (1 + \log f_{iq}) \times \log \frac{N}{n_i} \tag{4}$$

The weight vector q of the retrieved item is shown in Eq. 5.

$$q = (\omega_{1q}, \omega_{2q}, \ldots, \omega_{iq}, \ldots, \omega_{tq}) \tag{5}$$

According to the definition of the vector space model, we assume that the similarity is sim, the similarity formula is Eq. 6.

$$sim(\vec{d_j}, \vec{q}) = \frac{\vec{d_j} \cdot \vec{q}}{|\vec{d_j}| \cdot |\vec{q}|} = \frac{\sum_{i=1}^{t} \omega_{ij} \cdot \omega_{iq}}{\sqrt{\sum_{i=1}^{t} (\omega_{ij})^2} \cdot \sqrt{\sum_{i=1}^{t} (\omega_{iq})^2}} \tag{6}$$

where $|d_j|$ and $|q|$ are the plaintext vector module of the ciphertext document and the plaintext vector module of the ciphertext retrieval item. They are calculated by the index server. In the index server, using the Eq. 6 to calculate the ciphertext keywords and each ciphertext document similarity calculation results. The results of similarity calculation of ciphertext documents are sorted by similarity degree. The high degree of similarity of the document is arranged in the front. It is beneficial to the user to find. Finally, the index server returns the similarity sort results to the client for viewing.

3.2.4   Download the Document

First, the user enters the name of the document to be downloaded on the client. The client uses the encryption formula to get the cipher file name c_{index} to be downloaded, and the client uploads c_{index} to the cloud file server. And then we match the search operation in the file server, through the Eq. 1 can be found the document to be downloaded in the file server. If Retrieval = 0, then $c_i - c_{index} = 0$, it means that the ciphertext document has been queried in the ciphertext document. Finally, the client is homomorphic decrypted to get the correct plaintext.

4   Experiments and Results

4.1   Analysis of Efficiency

We randomly selected from a different number of five groups of documents to test the experimental data.

As shown in Fig. 2(a) and (b), compared with the classical DGHV integer homomorphic encryption algorithm and the BVH encryption algorithm, the improved integer homomorphic encryption algorithm has not increased the encryption time. Its encryption time not only did not increase, but also greatly reduced. And the decryption time compared to BVH and DGHV also greatly reduced. So, the efficiency of encryption and decryption has been improved. As shown in Fig. 2(c), the retrieval efficiency comparison chart of CRBCP and BVH shows that the CRBCP retrieval scheme is greatly reduced when compared with the time of BVH under the same number of retrieved documents. It can be seen that CRBCP homomorphic encryption algorithm can greatly improve the efficiency of encryption time and decryption time. Retrieval efficiency is also much higher than BVH.

4.2   Analysis of Result

As shown in Fig. 3(a) and (b), the maximum similarity is 10, BVH retains one decimal place, CRBCP retains three decimal places. It can be seen that the similarity value obtained by CRBCP is more specific than the value obtained by BVH, and the accuracy is higher and the value is more accurate.

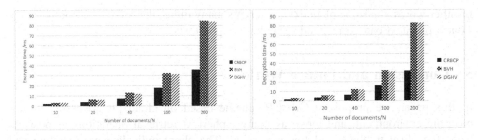

(a) Encryption time   (b) Decryption time

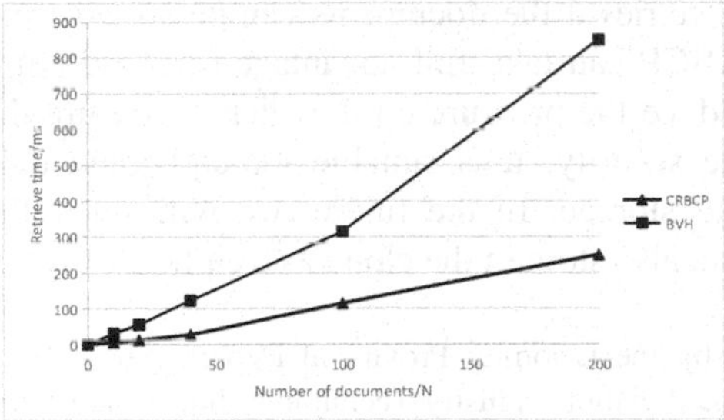

(c) Retrieve time

Fig. 2. Analysis of efficiency

```
E:\files\test9.txt=2.0          E:\files\test8.txt=0.1991
E:\files\test8.txt=2.0          E:\files\test9.txt=0.1539
E:\files\test12.txt=1.0         E:\files\test7.txt=0.1467
E:\files\test14.txt=1.0         E:\files\test12.txt=0.1271
E:\files\test7.txt=1.0          F:\files\test14.txt=0.0999
E:\files\test11.txt=1.0         E:\files\test11.txt=0.0522
E:\files\test4.txt=0.0          E:\files\test4.txt=0.0
E:\files\test2.txt=0.0          E:\files\test2.txt=0.0
E:\files\test6.txt=0.0          E:\files\test6.txt=0.0
E:\files\test3.txt=0.0          E:\files\test3.txt=0.0
E:\files\test5.txt=0.0          E:\files\test5.txt=0.0
E:\files\test1.txt=0.0          E:\files\test1.txt=0.0
E:\files\test13.txt=0.0         E:\files\test13.txt=0.0
E:\files\test10.txt=0.0         E:\files\test10.txt=0.0
```

(a) Similarity results of BVH   (b) Similarity results of CRBCP

Fig. 3. Analysis of result

4.3 The Indistinguishable of Ciphertext

The most basic security of the homomorphic encryption scheme is semantic security. Intuitively, ciphertext does not reveal any express information. Even if m_1, m_2 and p are known, a very important amount is used in the encryption algorithm: random number r. Even if the attacker gets the encryption result, the encryption algorithm is executed once for m_1 or m_2, and the same ciphertext data as the given ciphertext is not obtained.

In other words, the same plaintext m encrypts many times, the results are not the same. That a plaintext can correspond to multiple ciphertexts.

5 Conclusion and Future Works

In this paper, we propose a CRBCP scheme based on cloud preprocessing homomorphic environment, and move as many ciphertext retrieval operations as possible from the client to the cloud for processing. The client only deal with storage and encryption operations. The index server is responsible for processing and retrieving large amounts of data. And the results of retrieval the documents can be sorted from large to small for user to find better. CRBCP can take full advantage of cloud computing's powerful computing power to reduce the pressure on the client to ensure the security of data transmission and storage security. It is suitable for ciphertext data processing and user retrieval under cloud storage. In the future, we will study the inverted index table statistics out of the data also store in the cloud as a ciphertext form.

Acknowledgement. This work is supported by the Liaoning Provincial Department of Education Science Fund Project (L2013064), China Aviation Industry Technology Innovation Fund Project (Basic Research) (2013S60109R).

References

1. Ren, F.L., Zhu, Z.X., Wang, X.: A cloud computing security solution based on fully homomorphic encryption. J. Xi'an University of posts and telecommunications **18**(3), 92–95 (2013)
2. Duan, G.H., Ju, R., Wang, Y.B., Liu, Y.: Efficient ciphertext retrieval protocol in cloud computing environment. Netinfo Secur. **9**, 26–29 (2013)
3. Gentry, C.: Fully homomorphic encryption using ideal lattices. In: Proceedings of the 41st Annual ACM Symposium on Theory of Computing, vol. 9, no. 4, pp. 169–178. ACM Press, New York (2009)
4. Regev, O.: On lattices, learning with errors, random linear codes, and cryptography. J. ACM **56**(6), 84–93 (2005)
5. van Dijk, M., Gentry, C., Halevi, S., Vaikuntanathan, V.: Fully homomorphic encryption over the integers. In: Gilbert, H. (ed.) EUROCRYPT 2010. LNCS, vol. 6110, pp. 24–43. Springer, Heidelberg (2010). doi:10.1007/978-3-642-13190-5_2
6. Gentry, C., Sahai, A., Waters, B.: Homomorphic encryption from learning with errors: conceptually-simpler, asymptotically-faster, attribute-based. In: Canetti, R., Garay, Juan A. (eds.) CRYPTO 2013. LNCS, vol. 8042, pp. 75–92. Springer, Heidelberg (2013). doi:10.1007/978-3-642-40041-4_5
7. Brakerski, Z., Gentry, C., Vaikuntanathan, V.: Fully homomorphic encryption without bootstrapping. ACM Trans. Comput. Theory **6**(3), 1–36 (2014)
8. Cheon, J.H., Kim, J.: A hybird scheme of public-key encryption and somewhat homomorphic encryption. IEEE Trans. Inf. Forensics Secur. **10**(5), 1052–1063 (2017)
9. Huang, Y.F., Zhang, J.L., Li, X.: Encrypted storage and its retrieval in cloud storage applications. ZTE Commun. **16**(5), 33–35 (2010)

10. Song, D.J.: The design and implement of cloud storage system based on homomorphic encryption. Beijing University of Posts and Telecommunications (2013)
11. Zhang, X.J.: Ciphertext retrieval system for cloud storage system based on homomorphic encryption over integers. Ocean University of China (2013)
12. Wei, R.Q.: The design and implement of a ciphertext retrieval model based on homomorphic encryption. Xidian University (2014)
13. Lv, W.B.: Application and research of homomorphic encryption in cloud storage security. Shenyang Aerospace University (2016)
14. Tan, K.L.: Research of security communication in flight record data cloud storage. Shenyang Aerospace University (2014)
15. Salton, G., Yu, C.T.: On the construction of effective vocabularies for information retrieval. In: Proceedings of the 1973 Meeting on Programming Languages and Information Retrieval, vol. 10, pp. 48–60. ACM, New York (1973)

A Linear Time Algorithm for Influence Maximization in Large-Scale Social Networks

Hongchun Wu[1,2], Jiaxing Shang[1,2(✉)], Shangbo Zhou[1,2], and Yong Feng[1,2]

[1] College of Computer Science, Chongqing University, Chongqing, China
wuhc0217@163.com, {shangjx,shbzhou,fengyong}@cqu.edu.cn
[2] Key Laboratory of Dependable Service Computing in Cyber Physical Society,
Ministry of Education, Chongqing University, Chongqing, China

Abstract. Influence maximization is the problem of finding k seed nodes in a given network as information sources so that the influence cascade can be maximized. To solve this problem both efficiently and effectively, in this paper we propose LAIM: a linear time algorithm for influence maximization in large-scale social networks. Our LAIM algorithm consists of two parts: (1) influence computation; and (2) seed nodes selection. The first part approximates the influence of any node using its local influence, which can be efficiently computed with an iterative algorithm. The second part selects seed nodes in a greedy manner based on the results of the first part. We theoretically prove that the time and space complexities of our algorithm are proportional to the network size. Experimental results on six real-world datasets show that our approach significantly outperforms other state-of-the-art algorithms in terms of influence spread, running time and memory usage.

Keywords: Influence maximization · Social networks · Linear time algorithm · Computational complexity · Data mining

1 Introduction

Influence maximization (IM) is the problem of finding k "best" seed nodes in a network as information sources so that the maximum number of individuals can be influenced through the "word-of-mouth" effect [1]. Because of its various applications such as viral marketing, virus controlling, etc. [2], the study of influence maximization has attracted many researchers in recent years. The problem was firstly formatted by Kempe et al. [3] as follows: given a graph $G(V, E)$ modeling a network where V is set of nodes and the E is the set of edges, under a predefined propagation model D, the influence maximization problem aims to find a subset of nodes $S \subseteq V$ with $|S| = k$, so that the objective function $\sigma(S)$ is maximized, i.e.

$$S^* = \arg \max_S \left\{ \sigma(S) \big| |S| = k \right\} \tag{1}$$

Kempe et al. not only proved the NP-hardness of the optimization problem, but also showed that the objective function $\sigma(S)$ is submodular under several

© Springer International Publishing AG 2017
D. Liu et al. (Eds.): ICONIP 2017, Part V, LNCS 10638, pp. 752–761, 2017.
https://doi.org/10.1007/978-3-319-70139-4_76

diffusion models. Based on the properties of submodular functions, they further proposed a greedy algorithm to solve this problem with guaranteed accuracy. However, the greedy algorithm requires a large number of Monte-Carlo simulations to approximate the objective function, which severely affects its time efficiency and scalability.

To tackle the influence maximization problem, many research works have been proposed in recent years. Some works take advantage of the submodular functions, such as the CELF (Cost-Effective Lazy Forward) algorithm proposed by Leskovec et al. [4] and the CELF++ algorithm proposed by Goyal et al. [5]. These methods improve the time efficiency by maintaining the marginal gain of each node in influence evaluation. However, they still cannot handle large-scale networks. There are also node centrality based heuristic algorithms which simply select the top k nodes with the highest centrality values, such as the single discount (SD) algorithm based on discounted degree centrality [6], the Shapely centrality based algorithm [7], etc. Though these algorithms run quickly on large networks, without consideration of diffusion parameters they cannot provide any guarantee to the solution accuracy. To reach a balance between efficiency and effectiveness, researchers proposed path-based algorithms, which assume that influence can only spread along the network through some specific paths. These approaches include the local arborescence based algorithm PMIA [8], the independent path algorithm IPA [9], etc. These methods usually require huge amount of memory to store the influence paths, limiting their scalability on large-scale networks. Very recently scholars also proposed some other methods, such as the community-based algorithm [10], the multi-selector based algorithm [11], the reverse influence sampling (RIS) based algorithms [12,13], the learning based influence function study [14], etc.

As we mentioned above, existing influence maximization methods either require too much time or memory to handle large-scale networks, or generate poor solutions with low accuracy. To tackle the problem both efficiently and effectively, in this paper we propose **LAIM**: a **L**inear time **A**lgorithm for **I**nfluence **M**aximization in large-scale networks. Our algorithm consists of two parts: (1) influence computation; and (2) seed nodes selection. The first part uses the local influence of any node v to approximate its influence on the whole network with an iterative algorithm. The second part selects seed nodes with a greedy approach based on the results of the first part. We theoretically prove that the time and space complexity of the LAIM algorithm is proportional to the network size. Experimental results on six real world networks show that our approach significantly outperforms other state-of-the-art algorithms in terms of influence spread, running time and memory usage.

2   Linear Time Algorithm for Influence Maximization

2.1   Diffusion Model

In this paper, we consider the independent cascade (IC) diffusion model, where each node is in one of two states: active or inactive. Active nodes are those who have been influenced and will further influence their neighbors. Inactive nodes are those who have not been in touch with any active node or reject to be influenced. Initially all nodes are inactive, then k seed nodes are selected to be activated and propagation starts from the k nodes. At step t, for an active node u, it will try to activate each of its inactive neighbor v, and succeed with probability p_{uv}. Node u has only one chance to activate v, whether succeed or not, u will make no further attempt to activate v in the future. If v was successfully activated, then from step $t + 1$, v will be active and try to activate its inactive neighbors. If no node is activated as step T, the diffusion process will stop.

2.2 Preliminaries

Without loss of generality, we describe the directed network[1] $G(V, E)$ with an adjacency matrix A, where:

$$A_{i,j} = \begin{cases} 1 & \text{if } i \text{ and } j \text{ are connected} \\ 0 & \text{otherwise} \end{cases} \tag{2}$$

For any node v, its outgoing neighbor set is:

$$N(u) = \{v | A_{u,v} > 0\} \tag{3}$$

Before introducing our LAIM algorithm, we give the following definitions:

Definition 1 *(γ-th layer neighbor). Given a network, for any node u and non-negative integer $\gamma \geq 0$, we define u's γ-th layer neighbor set as:*

$$N_\gamma(u) = \begin{cases} \bigcup_{v \in N_{\gamma-1}(u)} N(v) & \gamma > 0 \\ \{u\} & \gamma = 0 \end{cases} \tag{4}$$

From definition 1 it is easy to see that $N_0(u) = \{u\}$ and $N_1(u) = N(u)$. Intuitively, the γ-th layer neighbor indicates the set of nodes whose distance from u is γ.

Based on the definition of $N_\gamma(u)$, we define u's γ neighborhood as $N_{\leq\gamma}(u) = \bigcup_{\ell=0}^{\gamma} N_\ell(u)$. Intuitively, u's γ neighborhood represents the set of nodes whose distance from u is less than or equal to γ.

Definition 2 *(Node influence). For any node $u \in V$, its influence on network $G(V, E)$ is defined as $inf_G(u)$. Similarly, u's influence on network $G - v$ is defined as $inf_{G-v}(u)$, where $G - v$ is the subgraph induced from G by V's subset $V \setminus \{v\}$.*

Definition 3 *(Local influence). For any node $u \in V$, its influence within its γ neighbourhood $N_{\leq\gamma}(u)$ on network G and $G - v$ are defined as $inf_G^{\leq\gamma}(u)$ and $inf_{G-v}^{\leq\gamma}(u)$, respectively. Similarly, u's influence to its γ-th layer neighbor $N_\gamma(u)$ on network G and $G - v$ are defined as $inf_G^\gamma(u)$ and $inf_{G-v}^\gamma(u)$, respectively.*

---

[1] Our approach is also applicable to undirected networks.

2.3   Linear Time Algorithm

Our LAIM algorithm consists of two parts: (1) influence computation; and (2) seed nodes selection. In the following part we will give detailed description about our approach.

(i) Influence computation: The computation of influence is motivated the following intuition: for any seed node u, its will first influence its first layer neighbors $N(u)$, and then its $2, 3, \cdots, n$-th layer neighbors $N_2(u), N_3(u), \cdots, N_n(u)$, until the influence stops propagating. From the IC model we know that influence generally cannot propagate far away from u, which means we may use the influence of u within its local neighbourhood (local influence) to approximate its influence on the overall network.

Based on the above idea, given network $G(V, E)$, for any node $u \in V$ and nonnegative integer $\gamma \geq 0$, we approximate u's overall influence on G by:

$$inf_G(u) \simeq inf_G^{\leq \gamma}(u) = \sum_{\ell=0}^{\gamma} inf_G^{\ell}(u) \tag{5}$$

Since u must first influence its first layer neighbor $N(u)$ in order to influence others nodes in $N_{\leq \gamma}(u)$, and from Eq. 4 we see that u's γ-th layer neighbors are included in the $\gamma - 1$-th layer neighbors of $N(u)$, so we may represent $inf_G^{\gamma}(u)$ (the influence of u to its γ-th layer neighbors) as a function of $N(u)$'s influence to their $\gamma - 1$-th layer neighbors:

$$inf_G^{\gamma}(u) = f\big(inf_G^{\gamma-1}(v_1), inf_G^{\gamma-1}(v_2), \cdots, inf_G^{\gamma-1}(v_q)\big) \tag{6}$$

where $v_1, v_2, \cdots, v_q \in N(u)$. In the IC diffusion model node u will successfully activate each of its inactive neighbor v with probability p_{uv}, so we can further formulate $inf_G^{\gamma}(u)$ as:

$$inf_G^{\gamma}(u) = \sum_{v \in N(u)} p_{uv} \cdot inf_G^{\gamma-1}(v) \tag{7}$$

Given that the $\gamma-1$-th layer neighbors of different nodes in $N(u)$ may overlap, the above formula is an approximate evaluation. In IC model, each node can be activated only once (no repeated activations allowed), so we should eliminate u from v's $\gamma - 1$-th layer neighbor set when computing $inf_G^{\gamma-1}(v)$ for any $v \in N(u)$ to avoid the re-computation of influence of u. As a result, $inf_G^{\gamma}(u)$ can be further represented as:

$$inf_G^{\gamma}(u) = \sum_{v \in N(u)} p_{uv} \cdot inf_{G-u}^{\gamma-1}(v) \tag{8}$$

where $\gamma \geq 1$. New we prove that Eq. 8 can be effectively approximated by a recursive formula.

Theorem 1. *Eq. 8 can be approximated by the following recurvise formula:*

$$inf_G^{\gamma}(u) = \begin{cases} 0 & \gamma = -1 \\ 1 & \gamma = 0 \\ \sum_{v \in N(u)} p_{uv}\left(inf_G^{\gamma-1}(v) - p_{vu}inf_G^{\gamma-2}(u)\right) & \gamma \geq 1 \end{cases} \quad (9)$$

Proof. We give the proof by considering $\gamma = 0, \gamma = 1$ and $\gamma > 1$ separately.

(1) If $\gamma = 0$, since $N_0(u) = u$, node u only influence itself, so its influence is a unit.

(2) If $\gamma = 1$, according to Eq. 9, it is easy to have: $inf_G^0(v) = inf_{G-u}^0(v) = 1$, so we have:

$$\begin{aligned} inf_G^1(u) &= \sum_{v \in N(u)} p_{uv}\left(inf_G^0(v) - p_{vu}inf_G^{-1}(u)\right) \\ &= \sum_{v \in N(u)} p_{uv} \cdot inf_G^0(v) = \sum_{v \in N(u)} p_{uv} \cdot inf_{G-u}^0(v) \end{aligned} \quad (10)$$

Equation 8 satisfies. Intuitively, when $\gamma = 1$, we have $N_1(u) = N(u)$, so node u only influences its first layer neighbors, and activate each neighbor $v \in N(u)$ with probability p_{uv}.

(3) If $\gamma > 1$, according to Eq. 9, we have:

$$inf_G^{\gamma}(u) = \sum_{v \in N(u)} p_{uv}\left(inf_G^{\gamma-1}(v) - p_{vu}inf_G^{\gamma-2}(u)\right) \quad (11)$$

By substituting $inf_G^{\gamma-1}(v)$ according to Eq. 9 again, we further have:

$$\begin{aligned} inf_G^{\gamma}(u) &= \sum_{v \in N(u)} p_{uv}\left[\sum_{z \in N(v)\backslash\{u\}} p_{vz}\left(inf_G^{\gamma-2}(z) - p_{zv}inf_G^{\gamma-3}(v)\right) \right. \\ &\quad \left. + p_{vu}\left(inf_G^{\gamma-2}(u) - p_{uv}inf_G^{\gamma-3}(v)\right) - p_{vu}inf_G^{\gamma-2}(u) \right] \\ &\simeq \sum_{v \in N(v)} p_{uv}\left[\sum_{z \in N(v)\backslash\{u\}} p_{vz}\left(inf_G^{\gamma-2}(z) - p_{zv}inf_G^{\gamma-3}(v)\right) \right] \\ &= \sum_{v \in N(v)} p_{uv} \cdot inf_{G-u}^{\gamma-1}(v) \end{aligned} \quad (12)$$

where the third line of the formula is achieved by ignoring the smaller item $p_{vu}p_{uv} \cdot inf_G^{\gamma-3}(v)$, so Eq. 8 satisfies.

In sum, Eq. 8 can be approximated by Eq. 9.

Based on Theorem 1, we give the following iterative algorithm to recursively compute the local influence of all nodes.

(ii) Seed nodes selection: Based on the above influence computation algorithm, we further propose the **LAIM** algorithm which selects top k seed nodes in

---

Algorithm 1. Influence computation

Require:

    Network $G(V, E)$, diffusion parameters $\{p_{uv}\}$, iterative parameter γ

Ensure:

    Local influence: $inf_{G}^{\leq\gamma}(u), \forall u \in V$

1: Initialize: Let $inf_{G}^{-1}(u) = 0, inf_{G}^{0}(u) = 1, inf_{G}^{\leq\gamma}(u) = 0, \forall u \in V$.

2: **for** $\ell = 1$ to γ **do**

3:    **for** $u \in V$ **do**

4:        $inf_{G}^{\ell}(u) = \sum_{v \in N(u)} p_{uv}\left(inf_{G}^{\gamma-1}(v) - p_{vu}inf_{G}^{\gamma-2}(u)\right)$

5:        $inf_{G}^{\leq\gamma}(u) \leftarrow inf_{G}^{\leq\gamma}(u) + inf_{G}^{\ell}(u)$

6:    **end for**

7: **end for**

8: **return** $inf_{G}^{\leq\gamma}(u), \forall u \in V$

---

a greedy manner. Specifically, we first use Algorithm 1 to compute the local influence of all nodes, then select the node with the maximum $inf_{G}^{\leq\gamma}(v)$ value and add it to the seed set S. After that we remove the node from V and re-execute Algorithm 1 to update the local influence of the rest nodes. The LAIM algorithm stops when k seed nodes are selected. Algorithm 2 shows the pseudocode of the LAIM algorithm. To further accelerate the algorithm, we provide the simple fast LAIM (**FastLAIM**) algorithm which directly selects top k nodes with the maximum values of $inf_{G}^{\leq\gamma}(v)$ computed in the first part.

Algorithm 2. The LAIM algorithm

Require:

    Network $G(V, E)$, diffusion parameters $\{p_{uv}\}$, iterative parameter γ

Ensure:

    Seed set S

1: Initialize: Let $S \leftarrow \Phi$

2: **for** $i = 1$ to k **do**

3:    Compute $inf_{G}^{\leq\gamma}(u), \forall u \in V$ with Algorithm 1

4:    $v = \arg_{u \in V} \max\{inf_{G}^{\leq\gamma}(u)\}$

5:    $S \leftarrow S \bigcup\{v\}$

6:    $V \leftarrow V \setminus \{v\}$

7: **end for**

8: **return** S

Time Complexity Analysis: We first show the time complexity of local influence computation. As show in Algorithm 1, the initialization in line 1 requires $O(n)$ time, where $n = |V|$ is the number of nodes. In each iteration, for any node $u \in V$, we need to traverse all the neighbors of u in order to compute $inf_{G}^{\gamma}(u)$ according to formula 9. Traversing the neighbors of u requires $O(|N(u)|)$ time. So the time complexity of one round of iteration is $O(\sum_{u \in V} |N(u)|) = O(m)$, where m is the number of edges. Since Algorithm 1 contains γ round of iterations, its time complexity is $O(\gamma m)$. From Algorithm 2 it is easy to see that the LAIM algorithm will repeat Algorithm 1 for k times to select the best k seed nodes. So the overall time complexity of the LAIM algorithm is $O(k\gamma m)$, which is linear to the network size.

Space Complexity Analysis: If we use adjacency list to represent the network, we need $O(m)$ space. Since Algorithm 1 requires γ rounds of iterations, and each round of iteration needs $O(n)$ space to maintain the $inf_G^\gamma(u)$ information of all nodes, the space complexity of Algorithm 1 is $O(\gamma n)$. So the overall space complexity of the LAIM algorithm is $O(m+\gamma n)$ which is necessarily $O(m)$ when $m \gg \gamma n$.

3   Evaluation

Datasets: We conduct experiments on six real-world datasets, as summarized in Table 1. The largest network has 3 million nodes and more that 100 million edges. We obtain these datasets from [6] and the SNAP website[2]. Readers are encouraged to go to the SNAP website for detailed description of the datasets. We treat all the edges as undirected ones, i.e., each edge has two directions. Table 1 also provides the values of the iterative parameter γ for the FastLAIM and LAIM algorithms on each dataset.

Table 1. Statistical properties of six real world datasets

Dataset	NetHEPT	NetPHY	Amazon	Pokec	LiveJournal	Orkut
# Ndoes	15K	37K	335K	1.6M	4M	3.1M
# Edges	31K	174K	926K	22.3M	35M	117M
Max. Degree	64	178	290	14,854	14,724	33,313
Avg. Degree	4.12	9.38	4.34	27.22	17.01	76.17
γ (LAIM)	4	5	4	6	7	5
γ (FastLAIM)	4	7	5	6	7	5

Baseline Algorithms: We compare our methods with five baseline algorithms: **IPA** [9], **TIM+** [12], **IMM** [13], **SD** (Single Discount) [6], and **Degree**, where the Degree algorithm simply select top k nodes with the highest degree values. The TIM+ and IMM algorithms have been shown to have good performance in dealing with large-scale networks. Due to the limit of space, we encourage readers to refer to the references for detailed introduction about these methods.

---

[2] http://snap.stanford.edu/data/.

Evaluation Metrics: We employ **(i) influence spread**, **(ii) running time**, and **(iii) memory usage** as the metrics to evaluate performance of an algorithm. Influence spread indicates the number of nodes influenced after information propagates through a given diffusion model. Similar to [3], we use the average number of active nodes with 10,000 Monte-Carlo simulations to approximate the influence spread for any given seed set S. Running time indicates how much time an algorithm takes to find k seed nodes. Memory usage is the bytes of memory used when an algorithm runs. They are important metrics to evaluate whether an algorithm is scalable to large-scale networks.

Experimental Environment: All the algorithms are carried out in single process and single thread, on a computer with 2.5 GHz Intel Xeon E5-2640 CPU and 32 GB memory. All algorithms are implemented in C/C++.

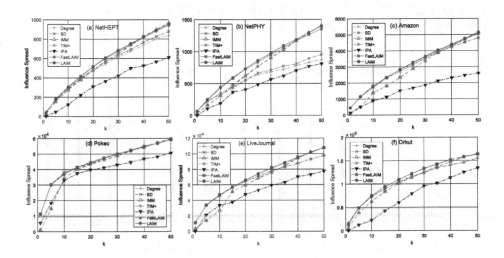

Fig. 1. Influence spread on six real-world networks

4   Results

Influence Spread: We first compared the influence spread of different algorithms, as shown in Fig. 1. The IPA algorithm gives the worst performance on all the datasets. The node centrality based heuristic algorithms (SD and Degree) cannot provide high quality solutions, as shown in Fig. 1(a),(b). Though the TIM+ and IMM algorithms give competitive results with $k = 50$, our LAIM/FastLAIM algorithms show better performances across different values of k. It has been reported in previous work [10] that the TIM+ and IMM algorithms cannot provide any accuracy guarantee for any $k < 50$. In summary, our algorithm always effectively and robustly finds high quality solutions as compared with other state-of-the-art algorithms.

Running Time & Memory Usage: Fig. 2 shows the running time and memory usage of different algorithms in finding $k = 50$ seed nodes on six real-world datasets. From Fig. 2(a) we see that the FastLAIM algorithm significantly outperforms three state-of-the-art algorithms (IPA, TIM+, and IMM). Even for the most competitive benchmark, i.e., the IMM algorithm, our FastLAIM algorithm is about an order of magnitude faster than it. On the Orkut dataset, the largest one with 3.1M nodes and 117M edges, the FastLAIM algorithm requires only 40 seconds to find the seed nodes, more than 7 times faster than the IMM algorithm and about 50 times faster than the TIM+ algorithm. From 2(b) we see LAIM/FastLAIM algorithms exhibit high scalability on mega-scale networks. Our algorithm requires almost no extra memory as compared with the simple node centrality based heuristic algorithms (SD and Degree). For example, on the NetHEPT dataset, the memory usage of our algorithm is 7M, only about 7.4% and 1.6% to that of the IMM and TIM+ algorithms respectively. On large-scale networks (Pokec, LiveJournal, Orkut), the differences in memory usage of different algorithms become non-significant.

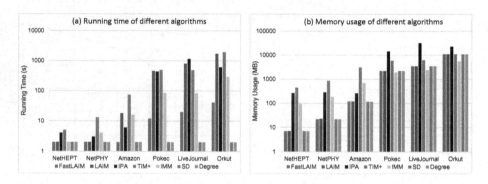

Fig. 2. Running time and memory usage on six real-world networks

5 Conclusion

In this paper, we propose a linear time algorithm for influence maximization. Our LAIM consists of two parts: (1) influence computation, and (2) seed nodes selection. We theoretically prove that the time and space complexities of our algorithm are proportional to the network size. Experimental results indicate that our algorithm significantly outperforms state-of-the-art algorithms in terms of influence spread, running time and memory usage. In sum, our work makes significant contributions to the design of new influence maximization algorithms.

Acknowledgments. This work was supported in part by National Natural Science Foundation of China (No. 61702059), Graduate Student Research and Innovation Foundation of Chongqing City (No. CYS17024), Fundamental Research Funds for the

Central Universities of China (No. 106112016CDJXY180003), China Postdoctoral Science Foundation (No. 2017M612913), Frontier and Application Foundation Research Program of Chongqing City (Nos. cstc2017jcyjAX0340, cstc2015jcyjA40006), Social Undertakings and Livelihood Security Science and Technology Innovation Funds of Chongqing City (No. cstc2017shmsA20013).

References

1. Domingos, P., Richardson, M.: Mining the network value of customers. In: ACM 7th SIGKDD International Conference on Knowledge Discovery and Data Mining, pp. 57–66. ACM, San Francisco (2001)
2. Goldenberg, J., Libai, B., Muller, E.: Talk of the network: a complex systems look at the underlying process of word-of-mouth. Market. Lett. **12**, 211–223 (2001)
3. Kempe, D., Kleinberg, J., Tardos, É.: Maximizing the spread of influence through a social network. In: ACM 9th SIGKDD International Conference on Knowledge Discovery and Data Mining, pp. 137–146. ACM, Washington DC (2003)
4. Leskovec, J., et al.: Cost-effective outbreak detection in networks. In: ACM 13th SIGKDD International Conference on Knowledge Discovery and Data Mining, pp. 420–429. ACM, San Jose (2007)
5. Goyal, A., Lu, W., Lakshmanan, L.: CELF++: optimizing the greedy algorithm for influence maximization in social networks. In: ACM 20th International Conference Companion on World Wide Web (WWW), pp. 47–48. ACM, Hyderabad (2011)
6. Chen, W., Wang, Y., Yang, S.: Efficient influence maximization in social networks. In: ACM 15th SIGKDD International Conference on Knowledge Discovery and Data Mining, pp. 199–208. ACM, Paris (2009)
7. Suri, N., Narahari, Y.: Determining the top-k nodes in social networks using the Shapley value. In: ACM 7th International Joint Conference on Autonomous Agents and Multiagent Systems (AAMAS), pp. 1509–1512. ACM, Estoril (2008)
8. Chen, W., Wang, C., Wang, Y.: Scalable influence maximization for prevalent viral marketing in large-scale social networks. In: ACM 16th SIGKDD International Conference on Knowledge Discovery and Data Mining, pp. 1029–1038. ACM, Washington DC (2010)
9. Kim, J., Kim, S.K., Yu, H.: Scalable and parallelizable processing of influence maximization for large-scale social networks. In: IEEE 29th International Conference on Data Engineering (ICDE), pp. 266–277. IEEE, Brisbane (2013)
10. Shang, J., et al.: CoFIM: a community-based framework for influence maximization on large-scale networks. Knowl.-Based Syst. **117**, 88–100 (2017)
11. Shang, J., Wu, H., Zhou, S., Liu, L., Tang, H.: Effective influence maximization based on the combination of multiple selectors. In: Ma, L., Khreishah, A., Zhang, Y., Yan, M. (eds.) WASA 2017. LNCS, vol. 10251, pp. 572–583. Springer, Cham (2017). doi:10.1007/978-3-319-60033-8_49
12. Tang, Y., Xiao, X., Shi, Y.: Influence maximization: near-optimal time complexity meets practical efficiency. In: ACM SIGMOD International Conference on Management of Data, pp. 75–86. ACM, Snowbird (2014)
13. Tang, Y., Shi, Y., Xiao, X.: Influence maximization in near-linear time: a martingale approach. In: ACM SIGMOD International Conference on Management of Data, pp. 1539–1554. ACM, Melbourne (2015)
14. Zhang, Q., Huang, C.-C., Xie, J.: Influence spread evaluation and propagation rebuilding. In: Hirose, A., Ozawa, S., Doya, K., Ikeda, K., Lee, M., Liu, D. (eds.) ICONIP 2016. LNCS, vol. 9948, pp. 481–490. Springer, Cham (2016). doi:10.1007/978-3-319-46672-9_54

Bioinformatics, Information Security and Social Cognition

Thyroid Nodule Classification Using Hierarchical Recurrent Neural Network with Multiple Ultrasound Reports

Dehua Chen[✉], Cheng Shi, Mei Wang, and Qiao Pan

School of Computer Science and Technology, Donghua University,
Shanghai 201620, China
{chendehua,wangmei,panqiao}@dhu.edu.cn, 2151535@mail.dhu.edu.cn

Abstract. Precise thyroid nodule classification is a key issue in endocrine clinic domain, which can enhance a patient's chance for survival. The reports of type-B ultrasound examination are important data source for thyroid nodule classification, and patients with thyroid nodules normally undergo several periodic ultrasound examinations during the process of diagnosis and treatment. However, most of the existing methods rely on feature engineering of single ultrasound reports and they did not take into consideration the historical records of the patients. In this paper, we propose a Hierarchical Recurrent Neural Network (HRNN) for thyroid nodule classification using historical ultrasound reports. HRNN consists of three layers of Long Short-Term Memory (LSTM) Neural Networks. Each LSTM layer is trained to produce the higher-level representations. We evaluate HRNN on real-world thyroid nodule ultrasound reports. The experiment results show that HRNN outperforms the baseline models with ultrasound reports.

Keywords: Deep learning · Text classification · Medical document · Thyroid nodule

1 Introduction

Thyroid cancer, malignant thyroid nodule, is the most common endocrine malignancy [1]. The incidence of thyroid cancer has rapidly increased in the worldwide, especially in Asian countries like China [2,3]. Thyroid cancer has become the most common cancer among Chinese women whose ages are below 30 years old [2].

Ultrasound is a common way that uses medical imaging to diagnose thyroid nodules, thus large quantities of ultrasound textual reports have been generated which are valuable for thyroid nodule classification. The existing thyroid nodule classification models are based on machine learning algorithms such as support vector machines (SVMs) [4], naive Bayes classifier [5], random forests [6]. However, these models rely on feature extracting methods based on natural language

© Springer International Publishing AG 2017
D. Liu et al. (Eds.): ICONIP 2017, Part V, LNCS 10638, pp. 765–773, 2017.
https://doi.org/10.1007/978-3-319-70139-4_77

processing (NLP) systems [7,8]. There are three major problems in existing models. Firstly, the feature engineering approach is effort intensive and nonadaptive to varying medical data. Automated feature representation method based on bag-of-words (BoW) is scalable, while it ignores the relationship between words, thus it cannot learn good representation of reports. Secondly, the medical document is a hierarchical structure, which can be divided into several sentences and words. Existing data-driven approaches fail to capture hierarchical regularities and dependencies. Finally, periodic ultrasound is important for patients with thyroid nodules, thus patients with thyroid nodules always have multiple historical ultrasound reports. Existing clinical text classification algorithms only take use of single document, ignoring the historical records of the patients which are the representations of disease states.

To address the above problems, we propose a deep learning based algorithm, Hierarchical Recurrent Neural Network (HRNN), an end-to-end classification algorithm which hierarchically builds textual representations for ultrasound reports sequences classification, shown in Fig. 1. HRNN does not require manual feature engineering, able to learn the hierarchical regularities and dependencies automatically and take into account the patient's historical ultrasound reports in order to improve the performance of classification. HRNN consists of three layers of Long Short-Term Memory (LSTM) Neural Networks [9] with average pooling and a sigmoid layer for classification. Each LSTM layer is trained to learn the hierarchical regularities and dependencies and produce the higher-level representations from the ultrasound reports.

We evaluate the proposed HRNN on a collection of real-world ultrasound reports extracted from the EMR system at Hospital. Experimental results show that our algorithm outperforms several strong baselines. Furthermore, we investigate the effects of model parameters for classification.

The remaining part of this paper is organized as follows. Section 2 discusses related work. In Sect. 3, we describe the architectures of HRNN. Section 4 shows the experimental results.

2   Related Work

Thyroid nodule classification problem is actually a clinical text classification issue which can be addressed by machine learning algorithms. These approaches rely on manual feature engineering based on clinical natural language processing systems [7,8]. Besides, there are also some methods which use automated feature extraction such as bag-of-words which do not rely on manual feature engineering. The bag-of-words can extract features automatically, but it fails to capture hierarchical regularities and dependencies in clinical text.

Deep learning is one of the machine learning approaches which is designed for producing end-to-end systems that learn from raw data without feature engineering. Deep learning uses multiple processing layers to learn representation of data with multiple levels of abstraction. Deep learning approaches have dramatically improved the state-of-the-art in multiple domains such as speech recognition, visual object recognition, object detection. Some recent papers use deep learning-based methods to build models for clinical text classification. Hughes et al. build a clinical text classification model with convolutional neural network to represent complex features [10]. Miotto et al. use denoising autoencoders to capture hierarchical regularities and dependencies in clinical data, and further use the random forest and SVM classifiers to build models to predict the future of patients [11]. These methods extract features automatically with single text, while they did not take into account the historical records of patients.

3 Model

In this section, we introduce the proposed Hierarchical Recurrent Neural Network (HRNN) for multiple historical ultrasound reports as shown in Fig. 1. HRNN is an end-to-end deep learning model which extract hierarchical features without feature engineering and use the features for classification. Our proposed model consists of four layers. In the first layer, our proposed model computes the vector representation for each sentence in reports with the word vectors. Then, in the second layer, the calculated sentence vectors are treated as input to produce report representations. After that, the report representations are used to calculate the representation for report sequences. In the last layer, report sequence representations are further used as features for thyroid nodule classification.

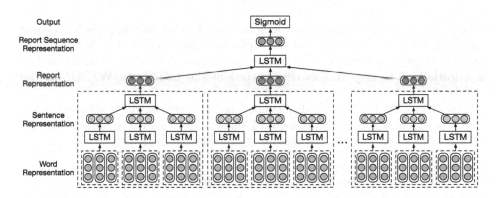

Fig. 1. The architecture of Hierarchical Recurrent Neural Network (HRNN). The sentence representations are computed with word representations. The report representations are calculated with sentence vectors. The report sequence representations are produced with report representations, and the obtained report sequence representations are used for thyroid nodule classification.

3.1 A Brief View of Long Short-Term Memory (LSTM) Networks

Recurrent Neural Network (RNN) is a type of neural network architecture particularly suited for modeling arbitrary-length sequential input. At each time step t, an RNN takes the input vector $\mathbf{x}_t \in \mathbb{R}^n$ at time t and the hidden state vector $\mathbf{h}_{t-1} \in \mathbb{R}^m$ at time $t-1$ and produces the next hidden state $\mathbf{h}_t$ at time t by applying the following recursive operation:

$$\mathbf{h}_t = f\left(\mathbf{W}\mathbf{x}_t + \mathbf{U}\mathbf{h}_{t-1} + \mathbf{b}\right) \tag{1}$$

where $\mathbf{W} \in \mathbb{R}^{m \times n}$, $\mathbf{U} \in \mathbb{R}^{m \times m}$, $\mathbf{b} \in \mathbb{R}^m$ are parameters of the model and f is an element-wise nonlinearity. In theory, the RNN can summarize all historical information up to time t with the hidden state $\mathbf{h}_t$. However, in practice, the traditional RNN suffers from the exploding or vanishing gradient problems, learning long-range dependencies with the traditional RNN is very difficult.

To address the problem of learning long range dependencies, Long Short-Term Memory (LSTM) networks augment the traditional RNN with a memory cell vector $\mathbf{c}_t \in \mathbb{R}^n$ at each time step. Concretely, at each time step t, the LSTM takes as input $\mathbf{x}_t$, $\mathbf{h}_{t-1}$, $\mathbf{c}_{t-1}$ and produces $\mathbf{h}_t$, $\mathbf{c}_t$ via the following equations:

$$
\begin{aligned}
\mathbf{i}_t &= \sigma\left(\mathbf{W}^i\mathbf{x}_t + \mathbf{U}^i\mathbf{h}_{t-1} + \mathbf{b}^i\right) \\
\mathbf{f}_t &= \sigma\left(\mathbf{W}^f\mathbf{x}_t + \mathbf{U}^f\mathbf{h}_{t-1} + \mathbf{b}^f\right) \\
\mathbf{o}_t &= \sigma\left(\mathbf{W}^o\mathbf{x}_t + \mathbf{U}^o\mathbf{h}_{t-1} + \mathbf{b}^o\right) \\
\mathbf{g}_t &= \tanh\left(\mathbf{W}^g\mathbf{x}_t + \mathbf{U}^g\mathbf{h}_{t-1} + \mathbf{b}^g\right) \\
\mathbf{c}_t &= \mathbf{f}_t \odot \mathbf{c}_{t-1} + \mathbf{i}_t \odot \mathbf{g}_t \\
\mathbf{h}_t &= \mathbf{o}_t \odot \tanh\left(\mathbf{c}_t\right)
\end{aligned} \tag{2}
$$

where $\sigma\left(\cdot\right)$ and $\tanh\left(\cdot\right)$ are the element-wise sigmoid and hyperbolic tangent functions, $\odot$ is the element-wise multiplication operator, and $\mathbf{i}_t$, $\mathbf{f}_t$, $\mathbf{o}_t$ denote the input gate, forget gate and output gate separately. At $t = 1$, $\mathbf{h}_0$ and $\mathbf{c}_0$ are initialized to zero vectors. Parameters of the LSTM are $\mathbf{W}^j$, $\mathbf{U}^j$, $\mathbf{b}^j$ for $j \in \{i, f, o, g\}$.

Memory cells in the LSTM alleviate the gradient vanishing problem, while gradient exploding is still an issue. LSTMs have been shown to outperform traditional RNNs on many tasks, including on language modeling.

3.2   Hierarchical Recurrent Neural Network

For words contained in each sentence of reports, they are represented as low dimensional and real-valued vectors, known as word embedding [12]. These word vectors can be either randomly initialized [13] or pre-trained from text corpus [14]. In our proposed model, we learn word vectors with word2vec tool and use them as the inputs of the model.

We use long short-term memory (LSTM) to compute the vector representation for each sentence in reports with the word vectors. LSTM are able to capture the regularities and dependencies and learn fixed-length vectors for sentences of varying length.

For the sentence consists of n words $[w_1, w_2, .., w_n]$, each word w_i is mapped to its word vectors $e_i \in \mathbb{R}^d$, we use LSTM to produce the hidden state $h_i \in \mathbb{R}^k$

for i-th word, where k is the memory cell number. We then calculate the average of the hidden states for all the words contained in sentence,

$$h = \frac{1}{n} \sum_{i=1}^{n} h_i \qquad (3)$$

where n is the length of the sentence. The averaging hidden state is the calculated sentence representation. This method is illustrated in Fig. 2.

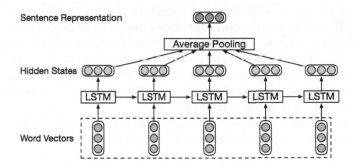

Fig. 2. Sentence representation calculated by long short-term memory (LSTM) network with word vectors.

The obtained sentence representations are then used for calculating the report representations. Let the report consists of n sentences $[s_1, s_2, ..., s_n]$. For each sentence, we obtain the sentence vectors $[v_1, v_2, ..., v_n]$ as in the previous subsection. The sentence vectors contain the semantics information for each sentence. We use the LSTM to combine the semantics and produce the report representation.

The LSTM produce the hidden state for each sentence $h_i \in \mathbb{R}^k$ for i-th sentence, where k is the memory cell number. We use the averaging hidden state as the report representation.

Ultrasound reports are stored for communication and documentation of diagnostic imaging, and they represent the disease states of patients at the check time. We take into account the historical ultrasound reports of the patients in order to further improve the performance of thyroid nodule classification.

Let the input of our model is a report sequence of n reports ordered by check time: $[r_1, r_2, ..., r_n]$, where the r_1 is the report when patients first visit the hospital. We use the LSTM to calculate the hidden state for each report and use the averaging hidden state as the report sequence representation.

The calculated report sequence representations can be regarded as features of patients with multiple ultrasound reports for thyroid nodule classification without feature engineering. We use the sigmoid function to produce the probability of thyroid cancer with the report representations. For patients with probability larger than 0.5, we predict that they have malignant thyroid nodules. Otherwise, we predict that the thyroid nodules are benign.

For model training, we use the binary-entropy function as the loss function. We calculate the derivative of loss function through back-propagation and update parameters through stochastic gradient descent over shuffled mini-batches with the Adadelta update rule.

4   Experiments

4.1   Datasets and Evaluation Metrics

We evaluated the proposed HRNN on a collection of ultrasound reports extracted from the EMR system at Hospital. The data consists of 13592 patients who have thyroid pathology diagnoses with multiple retrieved ultrasound reports from Jan 1, 2011 to Nov 30, 2015 in sequence of check time. The labels were extracted from thyroid pathology diagnoses. The dataset consists of 8785 malignant 4807 benign cases. All models are trained on 80% of the data and tested on 10%. The remaining 10% is used as the validation set.

We use the accuracy, precision, recall, F1-score and Area Under roc Curve (AUC) to measure the performance of classification.

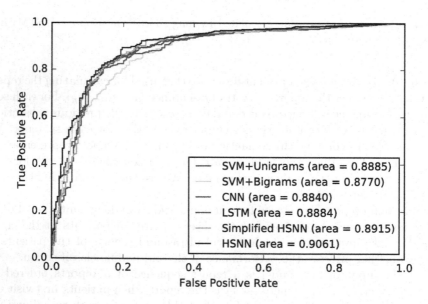

Fig. 3. Receiver operating characteristic (ROC) of HRNN and baselines.

4.2   Baseline Methods and Experiment Setup

We compare our methods (HRNN) with the following baseline methods.

(1) We use SVM baseline with linear kernel and the features of bag-of-unigrams (SVM + Unigrams) and bag-of-bigrams (SVM + Bigrams).
(2) We implement a convolutional neural network (CNN) baseline with 100 convolution filters each for window sizes of 3, 4, 5.
(3) We implement a long short-term memory (LSTM) neural network baseline with 128 memory cells and average pooling.
(4) We also implement a simplified version of HRNN (Simplified HRNN) as baseline, which exclude the report sequence representation layer. It can be used for single ultrasound classification.

We use the 300-dimensional word embedding pretrained by using the text corpus from the EMR system with word2vec tool. Our best performing HRNN used 128 memory cells in LSTMs and dropout of probability 0.5 on AUC.

4.3 Classification Performance

Table 1 and Fig. 3 presents the performance of HRNN compared to the baseline models. As shown in the table, we can see that simplified HRNN performs the best among all the baselines. This could be explained by that simplified HRNN can learn the hierarchical regularities and dependencies, while other baselines only learn flat features. SVM with unigrams features, CNN and LSTM are all effective methods for thyroid nodule classification. LSTM performs slight better than CNN in modeling sentence representations. HRNN take use of multiple reports and take into consideration the historical records of patients, thus it outperforms all the baseline methods.

Table 1. Experiment results of HRNN compared with other models. Performance is measured in accuracy, precision, recall, F1-score and AUC.

Model	Accuracy	Precision	Recall	F1-score	AUC
SVM + Unigrams	0.8493	0.8787	0.8896	0.8841	0.8885
SVM + Bigrams	0.8221	0.8567	0.8703	0.8634	0.8777
CNN	0.8426	0.8674	0.8931	0.8800	0.8840
LSTM	0.8485	0.8768	0.8908	0.8837	0.8884
Simplified HRNN	0.8588	0.8838	0.8999	0.8918	0.8915
HRNN	**0.8816**	**0.8989**	**0.9204**	**0.9095**	**0.9061**

4.4 Effect of Parameters

We investigate how model parameters influence classification performance. In this study, we focus on parameters related to model architectures including LSTM memory cells number and whether using dropout.

Table 2 presents the effects of the parameters on classification performance, including the LSTM memory cells number and whether using dropout. As shown in Table 2, HRNN models with dropout performs slightly better than those without dropout. However, the increasing of performance is limited, this could be explained by that overfitting is not the main reason for wrong classification especially with text data. Compared with different memory cells number in LSTM, we find that with the increasing of memory cells number, the accuracy increases. However, when we choose larger memory cell number, larger number of training time and computer memory are needed. In addition, comparing the HRNN with 128 cell number with dropout and 256 cell number with dropout, HRNN with 128 cell number with dropout even performs better than HRNN with 256 cell number with dropout on AUC.

Table 2. Experiment results of HRNN with different parameters.

Model	Accuracy	Precision	Recall	F1-score	AUC
64 cells	0.8610	0.8842	0.9033	0.8936	0.8985
128 cells	0.8603	0.8849	0.9010	0.8929	0.8976
256 cells	0.8750	0.8943	0.9147	0.9044	0.9041
64 cells + Dropout	0.8699	0.8900	0.9113	0.9005	0.9039
128 cells + Dropout	0.8816	0.8989	**0.9204**	0.9095	**0.9061**
256 cells + Dropout	**0.8824**	**0.9017**	0.9181	**0.9098**	0.9049

5   Conclusion and Future Work

In this paper, we present Hierarchical Recurrent Neural Network (HRNN), an end-to-end deep learning model that learns to extract features hierarchically and classify the thyroid nodules automatically from historical ultrasound reports. We show that it is effect to build the thyroid nodule classification model with the historical ultrasound reports, and it performs better than the methods only take use of single ultrasound report.

In future, we wish to implement our technique on other diseases and at a much larger scale. We acknowledge that the deep learning-based models are lack of interpretability when applied to complex medical problems. We are developing methods to interpret the representation learned by HRNN in order to better discriminate the patterns of malignant and benign to clinical users.

Acknowledgments. This work was supported by the Shanghai Innovation Action Project of Science and Technology (15511106900), the Science and Technology Development Foundation of Shanghai (16JC1400802), and the Shanghai Specific Fund Project for Information Development (XX-XXFZ-01-14-6349).

References

1. Burman, K., Wartofsky, L.: Clinical practice. Thyroid nodules. N. Engl. J. Med. **373**(24), 2347 (2015)
2. Chen, W., Zheng, R., Baade, P.D., Zhang, S., Zeng, H., Bray, F., Jemal, A., Yu, X.Q., He, J.: Cancer statistics in china, 2015. CA: Cancer J. Clin. **66**(2), 115–132 (2016)
3. Haugen, B.R., Alexander, E.K., Bible, K.C., Doherty, G.M., Mandel, S.J., Nikiforov, Y.E., Pacini, F., Randolph, G.W., Sawka, A.M., Schlumberger, M., et al.: 2015 American Thyroid Association management guidelines for adult patients with thyroid nodules and differentiated thyroid cancer: the American Thyroid Association guidelines task force on thyroid nodules and differentiated thyroid cancer. Thyroid **26**(1), 1–133 (2016)
4. Joachims, T.: Text categorization with support vector machines: learning with many relevant features. In: Nédellec, C., Rouveirol, C. (eds.) ECML 1998. LNCS, vol. 1398, pp. 137–142. Springer, Heidelberg (1998). doi:10.1007/BFb0026683
5. McCallum, A., Nigam, K., et al.: A comparison of event models for naive bayes text classification. In: AAAI 1998 Workshop on Learning for Text Categorization, vol. 752, pp. 41–48. Madison, WI (1998)
6. Breiman, L.: Random forests. Mach. Learn. **45**(1), 5–32 (2001)
7. Alghoson, A.M.: Medical document classification based on mesh. In: 47th Hawaii International Conference on System Sciences, HICSS 2014, Waikoloa, HI, USA, 6–9 January 2014, pp. 2571–2575 (2014)
8. Tran, T., Luo, W., Phung, D.Q., Gupta, S.K., Rana, S., Kennedy, R., Larkins, A., Venkatesh, S.: A framework for feature extraction from hospital medical data with applications in risk prediction. BMC Bioinform. **15**, 6596 (2014)
9. Hochreiter, S., Schmidhuber, J.: Long short-term memory. Neural Comput. **9**(8), 1735–1780 (1997)
10. Hughes, M., Li, I., Kotoulas, S., Suzumura, T.: Medical text classification using convolutional neural networks (2017). CoRR abs/1704.06841
11. Miotto, R., Li, L., Dudley, J.T.: Deep learning to predict patient future diseases from the electronic health records. In: Ferro, N., Crestani, F., Moens, M.-F., Mothe, J., Silvestri, F., Di Nunzio, G.M., Hauff, C., Silvello, G. (eds.) ECIR 2016. LNCS, vol. 9626, pp. 768–774. Springer, Cham (2016). doi:10.1007/978-3-319-30671-1_66
12. Bengio, Y., Ducharme, R., Vincent, P., Janvin, C.: A neural probabilistic language model. J. Mach. Learn. Res. **3**, 1137–1155 (2003)
13. Socher, R., Perelygin, A., Wu, J.Y., Chuang, J., Manning, C.D., Ng, A.Y., Potts, C., et al.: Recursive deep models for semantic compositionality over a sentiment treebank. In: Proceedings of the Conference on Empirical Methods in Natural Language Processing (EMNLP), vol. 1631, p. 1642 (2013)
14. Mikolov, T., Sutskever, I., Chen, K., Corrado, G.S., Dean, J.: Distributed representations of words and phrases and their compositionality. In: Advances in Neural Information Processing Systems 26: 27th Annual Conference on Neural Information Processing Systems 2013 and Proceedings of a meeting held 5–8 December 2013, Lake Tahoe, Nevada, United States. pp. 3111–3119 (2013)

Prediction of Stroke Using Deep Learning Model

Pattanapong Chantamit-o-pas[✉] and Madhu Goyal

Centre for Artificial Intelligence, Faculty of Engineering and Information
Technology, University of Technology Sydney, PO BOX 123,
Broadway, NSW 2007, Australia
Pattanapong.Chantamit-o-pas@student.uts.edu.au,
Madhu.Goyal-2@uts.edu.au

Abstract. Many predictive techniques have been widely applied in clinical decision making such as predicting occurrence of a disease or diagnosis, evaluating prognosis or outcome of diseases and assisting clinicians to recommend treatment of diseases. However, the conventional predictive models or techniques are still not effective enough in capturing the underlying knowledge because it is incapable of simulating the complexity on feature representation of the medical problem domains. This research reports predictive analytical techniques for stroke using deep learning model applied on heart disease dataset. The atrial fibrillation symptoms in heart patients are a major risk factor of stroke and share common variables to predict stroke. The outcomes of this research are more accurate than medical scoring systems currently in use for warning heart patients if they are likely to develop stroke.

Keywords: Deep learning · Predictive techniques · Stroke

1 Introduction

Data mining has become an essential instrument for researchers and clinical practitioners in medicine and numerous reports using these techniques are available. However, one of the biggest problems in data mining in medicine is that medical data is voluminous, heterogeneous and complex. The need for algorithms with very high accuracy is required as medical diagnosis is considered quite significant task that needs to be carried out precisely and efficiently.

Most common techniques that used to induce predictive models from data set are naïve Bayesian classifier and the decision tree. Bayesian classifier is one of the simplest yet a fairly accurate predictive data mining method [1]. A clinical decision support system [2] is used for prediction and diagnosis in heart disease. This approach is able to extract hidden pattern and relationships among medical data for prediction of heart disease using major risk factors. It applied genetic algorithms and neural networks and is called 'hybrid system'. It used a genetic algorithm feature for initialization of neural network weights.

For stroke, the predictive techniques range from simple models to more complex predictive models. Risk factors of stroke are complex and applied for finding different

© Springer International Publishing AG 2017
D. Liu et al. (Eds.): ICONIP 2017, Part V, LNCS 10638, pp. 774–781, 2017.
https://doi.org/10.1007/978-3-319-70139-4_78

complexities of disease and uncertainty from direct and/or non-direct risk factor sources. The analysis of stroke patients who were admitted in the TOAST study was done using stepwise regression methods [3]. This research used 1,266 stroke patients from database who had suffered in a transient ischemic attack (TIA) or recurrent stroke within 3 months after stroke, and selection 20 clinical variable for finding performance and evaluation.

The prognostic significance of blood pressure for stroke risk was examined by using the Cox proportional hazards regression model, which was adjusted for possible confounding factors. The results from a number of measurements showed that the predictive value of home blood pressure increased progressively. The initial home blood pressure values (one measurement) showed a significantly greater relation with stroke risk than conventional blood pressure values [4].

The Cox proportional hazards model and machine learning approach have been compared for stroke prediction on the Cardiovascular Health Study (CHS) dataset [5]. Specially, they considered the common problems of prediction in medical dataset, feature selection, and data imputation. This research proposes the use of an innovative algorithm for automatic feature selection - which chooses robust features based on heuristic: conservative mean. This algorithm was applied in combination with Support Vector Machines (SVMs). The feature selection algorithm achieves a greater area under the ROC curve (AUC) in comparison with the Cox proportional hazards model and L1 regularized Cox model. The method was also applied to clinical prediction of other diseases - where missing data are common, and risk factors are not well understood.

The Bayesian Rule Lists generated stroke prediction model employing the Market Scan Medicaid Multi-State Database (MDCD) with Atrial Fibrillation (AF) symptom [6]. The database categorised 12,586 patients on the basis of AF diagnosis. The observation was divided into two phases: a 1-year observation prior to the diagnosis; and 1-year observation after the diagnosis. The result found that 1,786 patients had a stroke within a year after suffering the atrial fibrillation. With regards to evaluation, the Bayesian Rule List (BRL) point is estimated by constructing a receiver operating characteristic (ROC) curve and measuring area under the curve (AUC) for each fold.

The classification algorithms i.e. Neural Network, Naïve Bayes, and Decision Tree are used for predicting the presence of stroke with various related attributes. The principle component analysis algorithm is commonly used for reducing the dimensions. Also, it is used for determining more relevant attributes towards the prediction of stroke and predicting whether the patient is suffering from stroke or not [7].

The conventional models are incapable of detecting fundamental knowledge because they fail to simulate the complexity and feature representation of medical problem domains. Researchers attempt to apply a deep model to overcome this weakness. Several applications of deep learning model to medical data analysis have been reported in recent years, for instance, an image analysis system for histopathological diagnosis on the images. Liang et al. [8] suggested the application of deep belief network for unsupervised feature extraction, and then perform supervised learning through a standard SVM. The results confirm the advantage of deep model towards knowledge modelling for data from medical information systems such as Electronic Medical Record (EMR) and Hospital Information System (HIS). Thus, predictive analytical techniques for stroke using deep learning techniques are potentially significant and beneficial.

The rest of this paper is organized as follows. Section 2 reviews deep learning technique in healthcare sector. Section 3 discusses the back propagation algorithm. Section 4 discusses application of deep learning model on heart disease dataset and the conclusion and future work are presented in Sect. 5 of this paper.

2 Deep Learning

Recently researchers [8–10] have been using deep learning technique for prediction. This technique employs learning from data with multiple level of abstraction by computational models that are associated with multiple processing layers. This method intended to discover complex structure in big data set by using the back propagation algorithm to predict the result. The machine can learn from source and change its internal parameters by computing the representation in each layer to form the representation in the previous layer [11].

Deep Learning architecture can be illustrated in form of different concept levels that uncover hidden layers in a problem domain. Deep learning models represent dataset in a multi-layer form. Each layer derives from the computation of node and weight of connections among nodes, and each transform represents one level, which will be the input for the next layer (see Fig. 1).

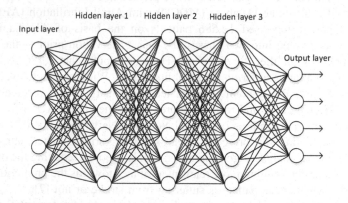

Fig. 1. A structure of deep neural network

In healthcare studies, researchers [10] developed health seekers by searching information needed for predicting possible disease, and there were able to show an evidence of symptoms for further analysis. Their research proposed the use of deep learning technique to find the possible disease based on the question of health seekers. The main discussion of research included two keys component. The first component is the detachment of the discriminant medical signatures form raw features. The next component is the input of raw features and their signatures into the first layers and hidden nodes. Meanwhile, the inter-relations in each layer learned between pre-training and pseudo-labelled data. In hidden layers, the raw features served the abstract signature mining. Therefore, the repetition and increase of these two components build a deep architecture to connect with three hidden layers.

Basically, a stroke is where an area of brain gets deprived of its blood supply, and it can happen by a vessel being blocked, which is then called an *"ischemic stroke"*. Three quarters of ischemic stroke result in a case of *"hemorrhagic stroke"* where a blood vessel can burst.

In acute ischemic stroke treatment, the prediction of tissue survival outcome plays a fundamental role in the clinical decision-making process as it can be used to assess the balance of risk and possible benefit when considering endovascular clotretrieval intervention. For the first time, Stier, Vincent, Liebeskind and Scalzo [9] constructed a deep learning model of tissue fate based on randomly sampled local patches from the hypoperfusion (Tmax) feature observed in MRI immediately after symptom onset. They evaluated the model with respect to the ground truth established by an expert neurologist four days after intervention. The results show the superiority of the proposed regional learning framework versus a single-voxel-based regression model. The previous research reveal to kernel of the deep learning techniques can be applied to healthcare sector as regulariser at the output layers or a part of model.

3 Deep Learning Architecture for Heart Disease Dataset

Features selection and classification analysis can be done by algorithms such as, Naïve Bayes, Decision Tree, SVM and many other techniques. These techniques extract knowledge from large databases for crucial decision support. A supervised learning technique is applied to healthcare applications for supporting medical diagnosis, improve patient care, and decision-making treatment [12]. This technique relies on a training dataset when constructing a prediction model for a target task. Each training is described various variables such as weights, classes, and attributes. The training set will compare between previous cases and new cases. However, diagnosis of disease is difficult problem because the number of risk factors are increase and complex. So, it need to improve prediction accuracy.

Backpropagation algorithm (BP) is applied to the main deep model as for learning in multi-layer networks in this study. This model is supervised concept extractor for the original dataset samples. A backpropagation is treated as a multi-layer feedforward neural networks with hidden layers or multiple-layer neural network (see Fig. 1). Each hidden unit can be considered as multiple outputs perceptron network, so we can be considered a soft-threshold linear combination of the hidden units, which are equivalent to the output unit perceptron.

We need to consider multiple output unit for multi-layer networks. $Let(x, y)$ be a single sample with its desired output labels $y = \{y_1, \ldots, y_i\}$. The error at the output unit is just $y - h_W(x)$, and it can use this to adjust the weights between the hidden layers and the output layers, so this process that produce the error at the hidden layers in terms of equivalent to the error at the hidden layers. This is subsequently used for update the weights between the input units and the hidden layers as in algorithm (see Fig. 2).

---

function BP-LEARNING **returns** a neural network
inputs: *examples*, a set of data, each with input vector **x** and output vector **y** *network*, a multi-layer network with L layers, weights $W_{j,i}$, activation function g
repeat
    for each *e* **in** examples **do**
        for each node *j* **in** the input layer **do** $a_j \leftarrow x_j[e]$
        for $\ell = 2$ **to** *M* **do**
            $in_i \leftarrow \Sigma_i W_{j,i} a_j$
            $a_i \leftarrow g(in_i)$
        for each node *i* **in** the output layer **do**
            $\Delta_j \leftarrow g'(in_j) \times (y_i[e] - a_i)$
        for $\ell = M - 1$ **to** 1 **do**
            for each node *j* in layer ℓ **do**
                $\Delta_j \leftarrow g'(in_j) \Sigma_i W_{j,i} \Delta_i$
            for each node *i* in layer $\ell + 1$ **do**
                $W_{j,i} \leftarrow W_{j,i} + \alpha \times a_j \times \Delta_i$
until some stopping criterion is satisfied
return NEURAL NETWORK

Fig. 2. The backpropagation algorithm for learning in multi-layer networks [13].

After different feature representations have been introduced to the network, the weight parameters automatically tune themselves to a supervisee feature. Augments of the previous steps become input to the next layer. The larger layers are, the faster model can compute. Either a classification or regression model is required for the hidden layers, and by virtue of it, the hidden layers can serve in filtering final output. However, it is required to decide again whether the outputs are valid.

4 Model Evaluation

4.1 Data Source

This research has applied deep learning model on heart disease dataset (available at UCI Machine learning website). It has 899 records and 76 attributes per record. It contains Patient Number, Social Security Number, Age, Gender, Blood pressure, type of chest pain, Cigarettes, Family history, Hypertension, Cholesterols, Years, EKG (day/month/year), Heart rate, Nitrates, and calcium channel blocker, and so on. It covers four hospitals at medical center in Hungarian, Switzerland, Cleveland and Long Beach Virginia. From this dataset, we selected ten attributes such as Age, Gender, Blood Pressure, Chest Pain, Cigarettes, Family History, Hypertension, Cholesterols, Heart Rate, and blood vessels. These attributes are related to stroke's risk factor used for prediction as described in AHA guideline [14, 15].

4.2 Evaluation

This research also did comparison of three models: Naïve Bayes; Support Vector Machine (SVM); and deep learning. Algorithms Naïve Bayes and SVM are wildly used in prediction. In Naïve Bayes, we used the classifier for discrete predictors that assumed independence of the predictor variables, and Gaussian distribution for metric prediction. The attributes with missing values and the corresponding table entries were omitted for prediction. In order to test, we calculated the values of the A-priori probabilities of 0.4784 (Non-Stroke) and 0.5215 (Stroke). Next, SVM method was used for sampling method in 10-flod cross validation for prediction, and SVM-Kernel was used for linear analysis. The cost of constraints or 'C'-constant of the regularization term is 0.1, and also their best performance is 0.465. Therefore, the prediction values of Naïve Bayes and SVM of 0 (Non-stroke) or 1 (Stroke). This result indicates that the patients are suffering from stroke or not.

In terms of deep learning, we used deep Neural Network model by using a feed-forward multi-layer artificial neural network. The computation shows that Mean Square Error (MSE) is 0.2596. This value indicates the confidence has best performance for prediction of stroke. The Mean Value and Standard Deviation are different from each technique, so the percentage of predict stroke with three models are showed in Table 1. The result shows that during training procedure, the mean values for prediction in deep

Table 1. Comparison of three techniques used for prediction of stroke.

Techniques	Mean value	Standard deviation
Naïve Bayes	49%	0.038
Support Vector Machine	47.78%	0.1106
Deep learning	36.73%	0.084

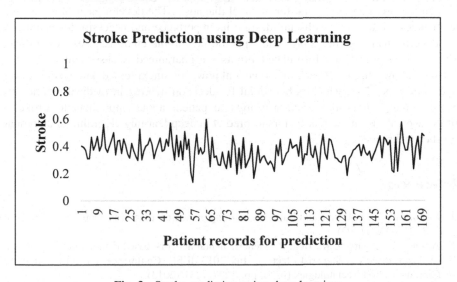

Fig. 3. Stroke prediction using deep learning.

learning is higher than two techniques. For the result prediction of deep learning technique, it shows in percentage that different from Naïve Bayes and SVM which shows a chance of stroke. The number of edges with deep learning are plotted (see Fig. 3). An edge of weight value in each layer is the predictor with risk factors. So, a prediction of stroke implied some concepts of a medical domain that preferred for good performance and explanation.

5 Conclusion and Future Work

In this paper, we compared three techniques: the deep learning technique; Naïve Bayes; and Support Vector Machine for stroke prediction. All techniques used data from heart patients for risk factors identification and disease prediction. The results are valid for warning to patients and significantly favorable to decision-making by a medical practitioner. We compared the use of deep learning with deep neural network, Naïve Bayes for discrete predictor, and linear performance in SVM. The result of Naïve Bayes and SVM show that patients are suffering from stroke or not deep learning technique shows in percentage of a chance of stroke. This confirmed that deep learning technique is most suitable for generating the heart dataset for predictive analysis in stroke.

Deep learning is widely used in prediction of diseases, especially in the prognosis and data analyses in healthcare sector. It employs the use of data obtained from patient health records and a comparison between previous cases, observation, or inspection. Stroke has complex risk factors. Therefore, the concepts or decision-making techniques cannot be directly extracted from a source that requires the involvement of a number of human experts.

However, a high volume of medical data, heterogeneity, and complexity have become the biggest challenges in diseases prediction. Algorithms with very high level of accuracy are, therefore, vital for medical diagnosis. The development of algorithms, nevertheless, still remains obscure despite its importance and necessity for healthcare. Good performance comes along with specific favourable circumstances, for instance, when well designed and formulated inputs are guaranteed. Consequently, the deep learning allows the disclosure of some unknown or unexpressed knowledge during prediction procedure, which is beneficial for decision-making in medical practice and provide useful suggestions and warnings to patient about unpredictable stroke. In future, we will use more risk factors to predict in deep learning algorithm and compare to other techniques.

References

1. Kononenko, I.: Inductive and Bayesian learning in medical diagnosis. Appl. Artif. Intell. Int. J. **7**, 317–337 (1993)
2. Amin, S.U., Agarwal, K., Beg, R.: Genetic neural network based data mining in prediction of heart disease using risk factors. In: 2013 IEEE Conference on Information and Communication Technologies (ICT), pp. 1227–1231 (2013)

3. Leira, E.C., Ku-Chou, C., Davis, P.H., Clarke, W.R., Woolson, R.F., Hansen, M.D., Adams Jr., H.P.: Can we predict early recurrence in acute stroke? Cerebrovasc. Dis. **18**, 139–144 (2004)
4. Ohkubo, T., Asayama, K., Kikuya, M., Metoki, H., Hoshi, H., Hashimoto, J., Totsune, K., Satoh, H., Imai, Y.: How many times should blood pressure be measured at home for better prediction of stroke risk? ten-year follow-up results from the Ohasama study. J. Hypertens. **22**, 1099–1104 (2004)
5. Khosla, A., Cao, Y., Lin, C.C.-Y., Chiu, H.-K., Hu, J., Lee, H.: An integrated machine learning approach to stroke prediction. In: Proceedings of the 16th ACM SIGKDD International Conference on Knowledge Discovery and Data Mining, pp. 183–192. ACM, Washington, DC, USA (2010)
6. Letham, B., Rudin, C., McCormick, T.H., Madigan, D.: Interpretable classifiers using rules and Bayesian analysis: building a better stroke prediction model. Ann. Appl. Stat. **9**, 1350–1371 (2015)
7. Sudha, A., Gayathri, P., Jaisankar, N.: Effective analysis and predictive model of stroke disease using classification methods. Int. J. Comput. Appl. **43**, 26–31 (2012)
8. Liang, Z., Zhang, G., Huang, J.X., Hu, Q.V.: Deep learning for healthcare decision making with EMRs. In: 2014 IEEE International Conference on Bioinformatics and Biomedicine (BIBM), pp. 556–559 (2014)
9. Stier, N., Vincent, N., Liebeskind, D., Scalzo, F.: Deep learning of tissue fate features in acute ischemic stroke. In: 2015 IEEE International Conference on Bioinformatics and Biomedicine (BIBM), pp. 1316–1321 (2015)
10. Nie, L., Wang, M., Zhang, L., Yan, S., Zhang, B., Chua, T.S.: Disease inference from health-related questions via sparse deep learning. IEEE Trans. Knowl. Data Eng. **27**, 2107–2119 (2015)
11. LeCun, Y., Bengio, Y., Hinton, G.: Deep learning. Nature **521**, 436–444 (2015)
12. Jiang, F., Jiang, Y., Zhi, H., Dong, Y., Li, H., Ma, S., Wang, Y., Dong, Q., Shen, H., Wang, Y.: Artificial intelligence in healthcare: past, present and future. Stroke Vasc. Neurol. (2017)
13. Russell, S.J., Norvig, P.: Artificial Intelligence: A Modern Approach (International Edition). Pearson, London (2010)
14. Goldstein, L.B., Adams, R., Alberts, M.J., Appel, L.J., Brass, L.M., Bushnell, C.D., Culebras, A., DeGraba, T.J., Gorelick, P.B., Guyton, J.R., Hart, R.G., Howard, G., Kelly-Hayes, M., Nixon, J.V., Sacco, R.L.: Primary prevention of ischemic stroke: a guideline from the American heart association/American stroke association stroke council: cosponsored by the atherosclerotic peripheral vascular disease interdisciplinary working group; cardiovascular nursing council; clinical cardiology council; nutrition, physical activity, and metabolism council; and the quality of care and outcomes research interdisciplinary working group: the American academy of neurology affirms the value of this guideline. Stroke **37**, 1583–1633 (2006)
15. Goldstein, L.B., Adams, R., Becker, K., Furberg, C.D., Gorelick, P.B., Hademenos, G., Hill, M., Howard, G., Howard, V.J., Jacobs, B., Levine, S.R., Mosca, L., Sacco, R.L., Sherman, D.G., Wolf, P.A., del Zoppo, G.J.: Members: primary prevention of ischemic stroke. Circulation **103**, 163–182 (2001)

A Method of Integrating Spatial Proteomics and Protein-Protein Interaction Network Data

Steven Squires$^{(\boxtimes)}$, Rob Ewing, Adam Prügel-Bennett, and Mahesan Niranjan

University of Southampton, Southampton, UK
{ses2g14,rob.ewing,apb,mn}@soton.ac.uk

Abstract. The increase in quantity of spatial proteomics data requires a range of analytical techniques to effectively analyse the data. We provide a method of integrating spatial proteomics data together with protein-protein interaction (PPI) networks to enable the extraction of more information. A strong relationship between spatial proteomics and PPI network data was demonstrated. Then a method of converting the PPI network into vectors using spatial proteomics data was explained which allows the integration of the two datasets. The resulting vectors were tested using machine learning techniques and reasonable predictive accuracy was found.

Keywords: Bioinformatics · Spatial proteomics · Machine learning

1 Introduction

Proteins can only perform their function in direct physical contact with other proteins or parts of the cell, therefore knowing the location of a protein (known as spatial proteomics) can aid in understanding its function. It has also been shown that there is a direct connection between diseases and subcellular protein localisation [1], consequently understanding certain diseases and cellular function depends on a reliable and accurate knowledge of protein localisation.

Protein-protein interactions (PPIs) have been studied for many years due, in part, to their importance in understanding cellular function. Interactions between pairs or groups of proteins, or proteins and other parts of the cell, have significant consequences for cell functionality including links to disease [2]. PPI networks chart these known or predicted interactions.

There is considerable interest in combining multiple sources of high throughput biological measurements. Examples include the integration of spatial and temporal patterns of gene expression [3], combining sequence and secondary structure of proteins [4], and the integrated analysis of the transcriptome and proteome [5].

Spatial proteomics and PPI networks should have significant similarities. A pair of proteins can only physically interact if they are in the same spatial location at the same time, hence we would expect that there would be a link

© Springer International Publishing AG 2017
D. Liu et al. (Eds.): ICONIP 2017, Part V, LNCS 10638, pp. 782–790, 2017.
https://doi.org/10.1007/978-3-319-70139-4_79

between proteins that interact and those that share a spatial location. In principle, accurate PPI networks might be able to predict which proteins co-localise. Conversely, spatial proteomics cannot on its own specify whether an interaction exists, but if two proteins are in the same compartment it may be more likely due to the increased likelihood that they share a function. In addition, proteins that never exist in the same spatial location cannot directly interact.

There has been work conducted which uses the PPI networks to make predictions on protein localisation. In particular, a recently published paper used PPI networks together with sequence predictors to classify proteins into spatial locations [6]. In contrast, we use the spatial proteomics profiles themselves and integrate them together with the PPI network data. In doing so we propose to aid the development of analytical tools for the analysis of proteomics data.

Our contributions are, first, to demonstrate the strong relationship between spatial proteomics and PPI network data. We then provide a mechanism to integrate the two datasets along with some useful visualisation techniques. We demonstrate that prediction of spatial localization from fractionation profiles can potentially be enhanced by the inclusion of information taken from PPI interactions and that interactions themselves are somewhat predictable from spatial profiles.

This paper is structured as follows: in Sect. 2 we discuss the spatial proteomics and PPI datasets along with the methods we use; in Sect. 3 we demonstrate the strong correlation between PPI and spatial proteomics data, the visualisation benefits of our technique, and the predictive power of the datasets. Finally, in Sect. 4 we provide a brief discussion of our results.

2 Methods

2.1 Spatial Proteomics and PPI Network Datasets

Spatial proteomics data is obtained from experiments which separate the contents of the cell into fractions and measure the relative abundance of each protein within each fraction. Proteins with a similar profile of fractional abundances are believed to occupy the same spatial location [7]. These proteins are then mapped to organelles by using marker proteins with similar profiles whose location is known [8,9]. The marker proteins tend to be extracted from literature and need to be highly reliable as they set the mapping from profiles to locations. In this study we used two sets of data obtained from Arabidopsis thaliana [10] and Drosophila melanogaster [11]. Spatial proteomics data is generally of a fairly low dimensionality (usually under 10 fractions) and the datasets contain 689 and 888 proteins for Arabidopsis and Drosophila respectively. We consider two marker sets for each organism, the same as the authors use [10,11]. The first, which we call the original marker set, are those proteins with known location extracted from literature. There are 27 for Arabidopsis and 55 for Drosophila. The authors then use the spatial proteomics datasets to assign previously unknown proteins to an organelle. We use the same assigned proteins which are named as extended marker sets in this paper.

Two PPI datasets were used: STRING [12] and BIOGRID [13]. These datasets have different methodologies to extract PPIs and present the results differently but results we have gained from both, where comparable, are consistent.

2.2   Combining Spatial Proteomics and PPI Network Data

Spatial proteomics data is produced in a format suitable for applying standard machine learning classification techniques as there is a matrix with m dimensions, n datapoints and each datapoint is within a class. In contrast the PPI data is presented as pairs of known interactions and needs to be converted into a form suitable for applying machine learning methods.

We create a simple fixed dimensional representation to capture information held in interaction networks and apply standard machine learning techniques. We do not consider more sophisticated techniques such as graph kernels [14] in this work because the amount of experimental data relating to subcellular measurements is small. Consider three organelles α, β and γ and a protein of interest, A. The PPI data for A is transformed into a three-dimensional vector by calculating the number of interactions between protein A and the marker proteins within α, β and γ respectively. Protein A is then represented as a vector with each dimension associated with an organelle. We normalise the vector by dividing by the number of proteins within each organelle and then dividing each protein individually by the sum of the vectors across each organelle. We then scale up each protein components so the sum across the PPI vector equals one.

3   Results

3.1   Correlation Between Spatial Proteomics and PPI Network Data

For the integration of spatial proteomics and PPI network data to be a valuable analytical technique it should first be demonstrated that they are structurally similar. A key structural similarity is the relationship between the spatial location and the chance of an interaction occurring. In the STRING database approximately 10% and 7% of proteins have links for Drosophila and Arabidopsis respectively; proteins located in the same organelle should, in general, have a higher likelihood of interacting. The ratio of interactions to potential interactions was measured for each protein in the extended marker set. Figure 1(a) shows the fraction of interactions occurring within the same organelle against the fraction to proteins in other organelles. The dots are the proteins for Drosophila and crosses for Arabidopsis, the stars are the averages for each organelle and the black line is the average expected if there was no correlation between the datasets. Any protein above the line has a higher fraction of links to proteins within its organelle than to proteins in other organelles. A majority of proteins have significantly more links within the organelles than between them.

The probability of these results occurring by chance from an uncorrelated PPI network (P-values) was calculated using the hypergeometric distribution.

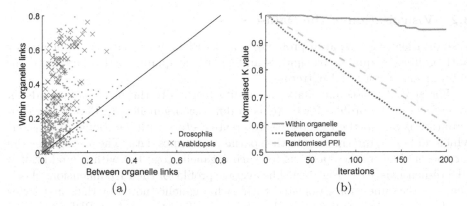

Fig. 1. (a) The fraction of interactions for each extended marker protein within the organelle is plotted against the fraction of interactions to proteins in different organelles. The black line is the expectation if there is no correlation and the stars show the averages for the organelles. The vast majority of proteins lie above the line. (b) The clustering algorithm iteratively removes links, preferentially removing PPI links which connect clusters together rather than those links that are within a cluster. The fall in number of links between proteins within the same organelle (solid line) is slower than those links between proteins in different organelles (dotted line) or a randomised PPI network (dashed line).

Within an individual organelle, the probability of the number of interactions observed occurring by chance ranges from $P = 10^{-12}$ to $P = 10^{-404}$. We can reasonably conclude that the marker proteins in the two datasets are correlated.

If similar clusters of proteins are formed in the PPI networks as in the organelle groupings in spatial proteomics data it would provide further evidence of the correlation between the data types. A clustering algorithm was developed based upon previous work [6,15] which removes links sequentially based on the structure of the PPI network. The first links to be removed are those considered to be joining separate clusters together. If the structures of the datasets are similar we would expect the number of links between proteins in different organelles to fall off much faster than between proteins within the same organelle. The normalised K value [15] gives a measure of how well connected a group of proteins are, if it falls fast then these proteins are unlikely to be in a cluster together. The average number of links between proteins within the same organelle, K_{in}, and the equivalent number of links between proteins in different organelles, K_{out}, were calculated and normalised. In Fig. 1(b) we show that the fall in K_{out} (dotted line) is far faster than for K_{in} (solid line) or for a randomised PPI network (dashed line) demonstrating that the clusters formed in the PPI data are similar to those in spatial proteomics data.

3.2   Visualisation of the Datasets

Visualisation of the spatial proteomics and PPI network data is important for both gaining a qualitative understanding of the quality of the data and for observation of potential patterns.

For spatial proteomics data the ability to classify the proteins depends on differences in the profiles for proteins in different organelles [8]. Part of our contribution is to add the PPI vectors to create vectors with additional dimensions which add extra information to the spatial proteomics data. The average vectors for the original marker proteins for each organelle together with the additional PPI dimensions are in Fig. 2(a). The average profile for the PPI dimensions shows peaks at the dimension associated with each organelle (noted by their first letter on the plots). Any protein not showing a significant peak in a PPI dimension while being well classified by the spatial proteomics profile may be of particular biological interest. The organelles shown are the endoplasmic reticulum (ER), mitochondrion (Mito), plasma membrane (PM), Golgi apparatus (Gol) and the vacuole (Vac).

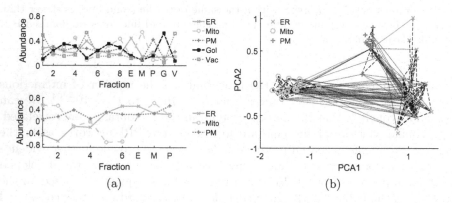

Fig. 2. (a) The average profiles for the original marker proteins for Arabidopsis (top) and Drosophila (bottom) are shown. The first eight (for Arabidopsis) and six (Drosophila) fractions are from the spatial proteomics data with the remaining created from the PPI data with labels associated with the first letter of the relevant organelle. (b) The original marker protein profiles for Drosophila were projected onto their principal components and PPI network links added. Proteins that have strong connections outside of their organelle can be investigated.

Visualisation of the PPI network data is difficult as the number of links are large. Here, an example of what the network looks like for the small number of original marker proteins for Drosophila is shown in Fig. 2(b). The spatial proteomics profiles were projected onto their principal components and known interactions from the STRING database are shown. It may be useful to inspect the links between individual proteins and the markers in this manner to gain insight into how each protein is linked. While there are many links between proteins in different organelles there are significantly more to proteins within the same organelle. It should also be noted that with only 55 proteins the network plots are already difficult to inspect in this format.

While the PPI networks may be difficult to visualise, the PPI vectors are somewhat easier when two components of the vector are compared. In Fig. 3(a) the PM component was plotted against the Mito component for all the Drosophila expanded marker proteins. The PPI vectors were also used to create a Gaussian probabilistic model, with means (shown as stars) and covariances extracted from the vectors. Contours of equal probability are then plotted which aids with understanding the separation and structure of the data. For example, the Mito vectors are much tighter bound than the PM vectors which visually represents that the Mito proteins are more closely connected to other Mito proteins than to proteins in other organelles than the PM proteins are. The structure of the plot is exactly as would be expected with the PM and Mito proteins tending to reside high up the PM and Mito axes respectively and the ER proteins following a fairly isotropic distribution near the origin.

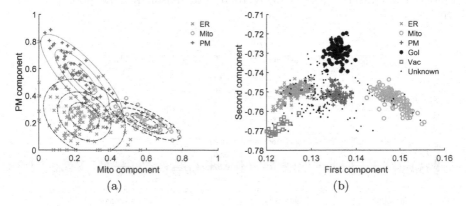

(a)   (b)

Fig. 3. (a) The PPI vectors for the PM and Mito components for Drosophila marker proteins are plotted. The PM proteins tend to cluster high up the PM axis and the Mito proteins along the Mito axis. The ER proteins are based near the origin. The contours of equal probability are centred on the average of each organelle. (b) The application of a Fisher Linear Discriminant to the combined spatial proteomics and PPI vectors for Arabidopsis allows for effective partition of the extended marker proteins into their respective organelles.

We also show the effect of projection using Fisher discriminant directions for the combined vector for Arabidopsis in Fig. 3(b). Most of the extended marker proteins are well separated into their organelles, the combined vector is able to effectively partition the proteins into different compartments.

3.3   Predictive Power

We have demonstrated the similarity of the two datasets and some visualisation techniques. Now we will show that there is considerable predictive power in the method of combining datasets.

First we show the PPI networks can predict spatial location. As we have converted the PPI network into a vector we can apply standard machine learning techniques. The protein extended marker data was partitioned randomly into two and a support vector machine [16] (SVM) was trained on half of the proteins with the PPI vector as input and the organelles as output classes. The trained SVM was then tested on the remainder of the randomised data. The process, with different random partitions, was then repeated two hundred times. The classification accuracies, sensitivities (true positive rate) and specificities (true negative rates) are shown as boxplots in Fig. 4(a). Generally, the SVM trained on the PPI data was able to predict the location of the proteins approximately 70% of the time. The most notable exception was the vacuole where very poor predictions are made. There are only small numbers of vacuole proteins in the extended marker set which is likely to be the reason for the poor predictive ability.

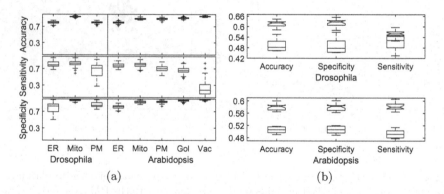

Fig. 4. (a) The fraction of extended marker proteins predicted correctly from two hundred runs of an SVM classifier based on the PPI vectors as input and the organelle markers as output classes. The classifier is able to correctly predict the organelle around 70% of the time. (b) The spatial proteomics data can make weak predictions on the existence of PPI links. The notched plots show the accuracy, specificity and sensitivity for Drosophila (top) and Arabidopsis (bottom) while the square plots are the results for a randomised PPI network. There is a clear (but faint) signal from the spatial proteomics data.

The inverse problem is more challenging. To attempt to use the spatial proteomics data to estimate whether a PPI exists between two proteins, first each pair of proteins were merged together to create a combined vector by multiplying each component of each vector by all the components of the other. The six dimensional Drosophila vector, for example, was transformed to a 36 dimensional combined vector. A two-class SVM for interactions and non-interactions was then trained. The new dataset is the combination of all the protein pairs so contains 393,828 protein pairs for Drosophila and 237,016 for Arabidopsis. The data was split into training sets of protein pairs and, as the data is highly skewed towards non-interactions, the sampling was biased to force the training set to contain 50% interactions. The process was repeated twenty times. The accuracy, sensitivity and specificity of the predictions are shown in the notched plots of Fig. 4(b). The equivalent SVM was applied to a randomised PPI network and shows what would be expected if there was no signal available (the non-notched plots). While the signal from the real data is small, it is consistently larger than the results from the randomised PPI network and can make some predictions about the PPI network.

4   Discussion

In this paper, we show how sub-cellular proteomics measurements can be combined with information contained in protein-protein interaction networks. Our work shows that there is significant correlation between spatial protein expression in cells and protein interaction information. Using a simple representation of interaction data in a fixed dimensional space, we show that predictions can be made in both directions between spatial proteomics and PPI networks.

There are many potential benefits from using the combined datasets. Differences in classification between the datasets may be of particular interest as it may imply interesting cases such as proteins that exist in multiple compartments or false data that should be re-evaluated. Confidence in conclusions can also be increased if the same conclusion is drawn using two separate datasets. Inspection of the data using some of the visualisation techniques discussed may also be useful for increasing understanding of data quality and building intuition.

References

1. Park, S., Yang, J.S., Shin, Y.E., Park, J., Jang, S.K., Kim, S.: Protein localization as a principal feature of the etiology and comorbidity of genetic diseases. Mol. Syst. Biol. **7**(1), 494 (2011)
2. Ideker, T., Sharan, R.: Protein networks in disease. Genome Res. **18**(4), 644–652 (2008)
3. Samsonova, A.A., Niranjan, M., Russell, S., Brazma, A.: Prediction of gene expression in embryonic structures of Drosophila melanogaster. PLoS Comput. Biol. **3**(7), e144 (2007)
4. Wieser, D., Niranjan, M.: Remote homology detection using a kernel method that combines sequence and secondary-structure similarity scores. Silico Biol. **9**(3), 89–103 (2009)
5. Gunawardana, Y., Fujiwara, S., Takeda, A., Woo, J., Woelk, C., Niranjan, M.: Outlier detection at the transcriptome-proteome interface. Bioinformatics **31**(15), 2530–2536 (2015)

6. Du, P., Wang, L.: Predicting human protein subcellular locations by the ensemble of multiple predictors via protein-protein interaction network with edge clustering coefficients. PloS One **9**(1), e86879 (2014)

7. De Duve, C., Beaufay, H.: A short history of tissue fractionation. J. Cell Biol. **91**(3), 293 (1981)

8. Gatto, L., Breckels, L.M., Burger, T., Nightingale, D.J., Groen, A.J., Campbell, C., Mulvey, C.M., Christoforou, A., Ferro, M., Lilley, K.S.: A foundation for reliable spatial proteomics data analysis. Mol. Cell. Proteomics **13**, mcp–M113 (2014)

9. Itzhak, D.N., Tyanova, S., Cox, J., Borner, G.H.: Global, quantitative and dynamic mapping of protein subcellular localization. Elife **5**, e16950 (2016)

10. Dunkley, T.P., Hester, S., Shadforth, I.P., Runions, J., Weimar, T., Hanton, S.L., Griffin, J.L., Bessant, C., Brandizzi, F., Hawes, C., et al.: Mapping the Arabidopsis organelle proteome. In: Proceedings of the National Academy of Sciences. vol. 103(17), pp. 6518–6523 (2006)

11. Tan, D.J., Dvinge, H., Christoforou, A., Bertone, P., Martinez Arias, A., Lilley, K.S.: Mapping organelle proteins and protein complexes in Drosophila melanogaster. J. Proteome Res. **8**(6), 2667–2678 (2009)

12. Jensen, L.J., Kuhn, M., Stark, M., Chaffron, S., Creevey, C., Muller, J., Doerks, T., Julien, P., Roth, A., Simonovic, M., et al.: String 8—a global view on proteins and their functional interactions in 630 organisms. Nucleic Acids Res. **37**(suppl_1), D412–D416 (2008)

13. Stark, C., Breitkreutz, B.J., Reguly, T., Boucher, L., Breitkreutz, A., Tyers, M.: BioGRID: a general repository for interaction datasets. Nucleic Acids Res. **34**(suppl_1), D535–D539 (2006)

14. Kondor, R.I., Lafferty, J.: Diffusion kernels on graphs and other discrete input spaces. In: ICML. vol. 2, pp. 315–322 (2002)

15. Radicchi, F., Castellano, C., Cecconi, F., Loreto, V., Parisi, D.: Defining and identifying communities in networks. In: Proceedings of the National Academy of Sciences of the United States of America. vol. 101, no. 9, pp. 2658–2663 (2004)

16. Vapnik, V.: The Nature of Statistical Learning Theory. Springer Science and Business Media, Berlin (2013). doi:10.1007/978-1-4757-3264-1

Tuning Hyperparameters for Gene Interaction Models in Genome-Wide Association Studies

Suneetha Uppu$^{(\boxtimes)}$ and Aneesh Krishna

Department of Computing, Curtin University, Perth, Australia
suneetha.uppu@postgrad.curtin.edu.au,
A.Krishna@curtin.edu.au

Abstract. In genetic epidemiology, epistasis has been the subject of several researchers to understand the underlying causes of complex diseases. Identifying gene-gene and/or gene-environmental interactions are becoming more challenging due to multiple genetic and environmental factors acting together or independently. The limitations of current computational approaches motivated the development of a deep learning method in our recent study. The approach trained a multilayered feedforward neural network to discover interacting genes associated with complex diseases. The models are evaluated under various simulated scenarios and compared with the previous methods. The results showed significant improvements in predicting gene interactions over the traditional machine learning techniques. This study is further extended to maximize the predictive performance of the method by tuning the hyperparameters using Cartesian grid and random grid searching. Several experiments are conducted on real datasets to identify higher-order interacting genes responsible for diseases. The findings demonstrated randomly chosen trials are more efficient than trials chosen by grid search for optimizing hyperparameters. The optimal configuration of hyperparameter values improved the model performance without overfitting. The results illustrate top 30 gene interactions responsible for sporadic breast cancer and hypertension.

Keywords: Gene interactions · Grid search · Random grid search · Hyperparameters · Deep feedforward neural network

1 Introduction

In this era of genetic epidemiology, genome-wide association studies (GWAS) have potential to identify genetic variants that are associated with a disease. These studies predominantly focuses on single-locus approaches, leading to the problem of "missing heritability" [1]. In reality, there are number of factors acting independently and/or together for underlying causes of a disease manifestation. As a step forward, genome-wide association interaction studies (GWAIS) have emerged to discover epistasis (gene-gene interactions) in understanding the biology behind complex diseases [2]. However, detecting these gene interactions remains a big challenge in high dimensional genome due to genetic heterogeneity, absence of marginal effects and computational limitations [1, 3]. These challenges have been addressed partially by a

© Springer International Publishing AG 2017
D. Liu et al. (Eds.): ICONIP 2017, Part V, LNCS 10638, pp. 791–801, 2017.
https://doi.org/10.1007/978-3-319-70139-4_80

number of data mining and machine learning approaches. A detailed review on the current methods and related software packages to detect gene-gene interactions are reviewed in the previous study [4]. Some of the pioneering works reviewed in the paper are MDR [5], MB-MDR [6], RJ [7], SNPHarvester [8], BOOST [9], PILINK [10], LogicFS [11], BEAM [12], GESVM [13] and GPNN [14]. The review also addressed the issues to be considered while designing the associated research methods. The paper further addresses the achievements in data simulation by evaluating the performance of the designed methods. In addition, the paper suggested few possible avenues to be explored for developing new methodologies by combining biological knowledge with statistical analysis.

Despite substantial advances in GWAIS, the progress in identifying multi-locus gene-gene interactions is still limited. Deep learning is a new era of machine learning, which allows the systems to learn representation of data at multiple levels of abstraction using sophisticated algorithms, and parallel computation [15]. Their applications into image processing, speech recognition, natural language processing, and bioinformatics have produced promising results [16]. Hence, this research focuses on the application of deep learning in GWAIS for discovering potential interacting genes responsible for complex diseases. A deep learning method was proposed in the previous study to detect higher-order gene-gene interactions [17]. The method was extended for unsupervised feature learning by discovering the anomalies in the reduced representation of the original data. The method was validated under different simulated scenarios, and the performances of the models were observed in terms of accuracy. The results showed remarkable improvements in predicting gene interactions over some of the existing methods, such as MDR, RF, SVM, NN, Naïve Bayes', classification based on predictive association rules (CPAR), logistic regression (LR), and Gradient Boosted Machines (GBM). This study is further extended to maximize the predictive performance of the models by tuning the hyperparameters using Cartesian grid and random grid searching. Several experiments were performed by varying number of parameters that can have impact on the model accuracy. The findings improved the model performance without overfitting on real datasets. The results illustrated top 30 gene interactions responsible for breast cancer and hypertension.

Rest of this paper is arranged as following: Sect. 2 gives a brief introduction to the chronological overview of the previously proposed deep learning method. The section further introduces hyperparameter tuning. Sections 3 elaborates and discusses several experimental evaluations performed on the proposed method. Finally, conclusion and future works are included in Sect. 4.

2   Methods

2.1   Deep Neural Networks (DNN)

The overview of the proposed extended method to detect higher-order gene-gene interactions is illustrated in Fig. 1. The trained predictive model is based on multi-layered feedforward neural network [17, 18]. It comprises of an input layer, hidden layers, and an output layer. The information transfers in forward direction from

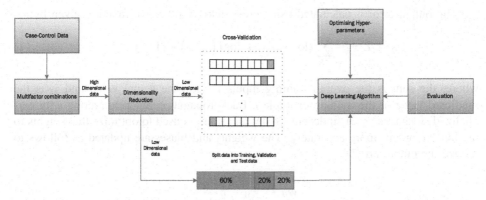

Fig. 1. Overview of the deep learning method (updated to previously proposed method [17]).

the input layer to the output layer. The output of the input layer serves as the input to the hidden layers, and the output of hidden layers serves as the input to the output layer. The computation in each layer transforms the representation of data into more abstract.

Figure 2 is an illustrative example of three layered feedforward neural network with one input layer, two hidden layers, and an output layer. The basic computational units of the network are neurons that are inspired from human brain. Each neuron represents the summation of the weighted inputs with a bias. The generalized parametric linear equation of a neuron is expressed as: $\alpha = \sum_{i=1}^{n} w_i I_i + B$, where w is the weight vector of input I, and B is bias that represents neuron's threshold. Bias is added to the network to improve the prediction accuracy by shifting the decision boundary along y axis. Each neuron is triggered by a non-linear activation function f, and the output signal $f(\alpha)$ is transmitted to the connecting neuron. A number of activation functions (such as, rectifier, hyperbolic tangent, maxout, and sigmoidal) were studied in the previous study. However, hyperbolic tangent function is used due to the highest prediction accuracy achieved in detecting gene interactions.

$$f(\alpha) = \tanh(\alpha) = \frac{e^{2\alpha} - 1}{e^{2\alpha} + 1} \tag{1}$$

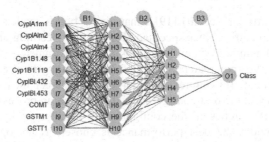

Fig. 2. A three-layered feedforward neural network.

The output error is calculated using cross entropy objective function given by:

$$E = - \sum_I (\log y_j * \hat{y}_j + \log(1 - y_j) * (1 - \hat{y}_j)) \tag{2}$$

where, the sum is the overall training inputs I. y_j is the actual value and $\hat{y}_j$ is the predicted value of the training example j. Backpropagation algorithm along with parallelized stochastic gradient decent (SGD) algorithm is used to optimize the weights to predict the output more accurately. The weights and biases are updated as follows to reduce the output error:

$$w_{jk} \leftarrow w_{jk} - \eta \frac{\partial E}{\partial w_{jk}} \tag{3}$$

$$B_{jk} \leftarrow B_{jk} - \eta \frac{\partial E}{\partial B_{jk}} \tag{4}$$

where, η is learning rate. The value of the learning rate is selected carefully as too large η values will never reach an optimal point, and too small values will take longer to converge. The weight value is decreased if $\frac{\partial E}{\partial w_{jk}}$ is positive, and the weight value is increased if $\frac{\partial E}{\partial w_{jk}}$ is negative.

2.2 Optimizing Hyperparameters

The objective of the deep learning algorithm is to find a function g that minimizes classification error. It produces the function g through the optimization of training criteria with respect to hyperparameters. Traditionally, a number of models are trained manually with various combinations of hyperparameters. The performances of all the models are compared to find the best model. This kind of manual search becomes tedious when the desired values of the network increases. Reproducing the results is one of the major drawbacks of manual search. Choosing set of configurations to the proposed method is a critical step in the hyperparameter optimization. Additionally, achieving the optimal hyperparameters is more complex when dealing with multidimensional data. Hence, automatic and reproducible approaches for tuning hyperparameters are required. The most widely used strategies such as, grid, and random search are evaluated in this study.

Grid search (Cartesian search) [19] exhaustively builds models for every combination of hyper-parametric values specified. That is, deep learning algorithm is trained accordingly with a number of configurations of hyperparameters. Bounds and steps between values of hyperparameters are specified to form a grid of configurations. The search begins at limited grid with relatively large steps between the parameter values by making the grid finer at the best configuration. This searching process further continues on a new grid till it searches all the configurations. Finally, the hyperparameter configuration that provides the best performance is chosen as the optimal value. Grid search is expensive as it searches exhaustively for all the configurations. Consider n hyperparameters each with 5 values, which make 5^n configurations in total. Hence, number of configurations in grid search is represented as G elements below:

$$G = \prod_{s=1}^{S} |H^s| \qquad (5)$$

The product over S sets leads grid search to suffer from issues of dimensionality as the number of values exponentially increases with the number of hyperparameters. Hence, grid search is feasible for a small number of configurations. However, grid search can compute in parallel to improve the computational power. In random grid search, the grid of hyperparameters is searched randomly. Adding new configurations or ignoring failed configurations are feasible in random search. It is simple, and as effective as full grid search. The number of trials is much less than grid with comparable performance.

2.3   Real World Data Application

Breast cancer is a complex disease that may be caused due to a number of unknown etiological aberrations. The data comprise of 410 samples obtained according to the requirements of the Institutional Review Board of Vanderbilt University Medical School [5]. The study is based on 207 white women with sporadic primary invasive breast cancer patients and 204 controls were treated at Vanderbilt University Medical Centre. The DNA of all the samples was isolated by using a DNA extraction kit (Gentra).The samples were used to amplify the desired gene segments using polymerase chain reaction (PCR) and were then analyzed. The genetic variants in five genes (COMT, CYP1A1, CYP1B1, GSTM1 and GSTT1) affected the metabolism of estrogens, which could increase the risk of breast cancer. The polymorphisms in these genes are summarized and reported in an earlier study [5]. The dataset considered 10 SNPs (Cyp1A1m1, Cyp1A1m2, Cyp1A1m4, Cyp1B1-48, Cyp1B1-119, Cyp1B1-432, Cyp1B1-453, COMT, GSTM1, and GSTT1) in five genes for the analysis. Since DNA is duplicated in each cell, three genotypes are formed and are numerically represented as zero for AA, one for Aa, and two for aa. There are 19 missing values and these are represented numerically by three.

Hypertension is also a complex disease that may occur due to genetic and environmental factors acting together or independently. The data comprises of 443 samples obtained from an outpatient clinic of the National Taiwan University Hospital (from July 1995 through June 2002) [20]. The study is based on 313 hypertensives and 130 normotensives among Taiwanese. Genotyping of genetic variants is performed by direct sequencing using a dye-terminator cycle sequencing method [21]. The data comprise of eight SNPs in four genes. They are: rs4762, and rs699 at AGT, rs5050, rs5051, rs11568020, and rs5049 at AGT 5', rs4646994 at ACE, and rs5186 at AT1-R. Numerically common homozygous is represented by zero, one for heterozygous, and two for variant homozygous.

3 Results and Discussion

The objective of this study is to tune the hyperparameters for improving the predictive performance of the deep learning algorithms. Grid search and random grid search are performed by optimizing the hyperparameters that can have impact on the model accuracy. Several experiments were conducted over real datasets by evaluating the model metrics of each model from the grid. The deep neural network is trained and analyzed in R to detect two-way SNP interactions [22]. In grid search, all possible combinations of hyperparameters (such as, hidden layers, epochs, activation function, input drop ratio, epsilon, momentum, learning rate, annealing rate, L1 and L2 penalties') are tested with Cartesian grid or exhaustive search. The search is performed by specifying parameter values that would be common to all the models, and a map that specifies the parameter spaces to be travelled. In random grid search, hyperparameters and search criteria is defined to tune the models. Hyperparameter values are chosen randomly within the specified values without any repeats by building the models sequentially. Optimal combination of parameter values is identified to maximize the

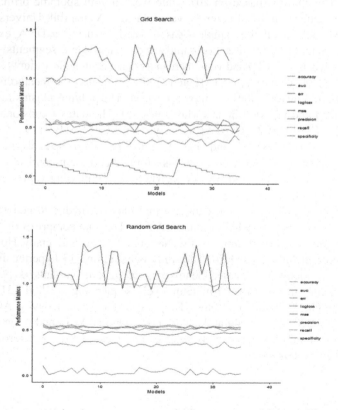

Fig. 3. Performance metrics (accuracy, auc, err, logloss, mse, precision, recall, and specificity) of DNN using (a) grid search (b) random grid search.

model accuracy. Results of each model are tested in the grid, and the best model with the highest accuracy or the predictive performance is selected.

Figure 3 illustrates the performance metrics of each model in terms of accuracy, auc, error, logloss, mean square error, precision, recall and specificity, using grid and random grid search on breast cancer data. The performance of the model is evaluated for the prediction accuracy on the test data. The auc of the best model identified from the grid search is 0.8242673 along with mse 0.1949181 on breast cancer data. The auc and mse of the best model identified for hypertension data are 0.6751302 and 0.266421 respectively. The search fairly performed well in large hyperparameter space. However, it was sensitive for few parameter combinations (peaked error functions). In random search, the hyperparameters are chosen randomly rather than a best guess. Mean square error (mse) and auc of each model during training, validation, and testing on breast cancer data is illustrated in Fig. 4. Auc and mse of the best model identified by random search are 0.8835072, and 0.1436914 respectively. The performances of all the models on hypertension data using random grid search is represented in Fig. 5. The corresponding auc and mse of the best model are 0.7014199, and 0.2801003. Internal 10-fold Cross validation is also performed on the entire data to confirm these parameters. It is observed that the best models identified by the random search have better validation aucs than the previous models identified by grid search. It is also observed that the variance between validation and testing is less than the variance observed in

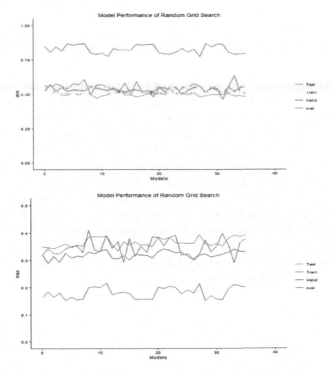

Fig. 4. Model performance of breast cancer data (a) auc vs models (b) mse vs models.

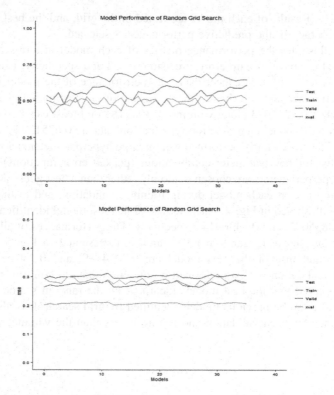

Fig. 5. Model performance of hypertension data (a) auc vs models (b) mse vs models.

grid search. Random search often performed well for more than four parameters by identifying best model in less time than performing exhaustive grid search. These findings confirmed that the random search worked well to find the models with the highest prediction accuracy and lowest mse for both real datasets. The optimal hyperparameters of the best model predicted by random search are used to detect SNP interactions. Top 30 SNP interactions responsible for breast cancer and hypertension are plotted in Fig. 6. The best two-way SNP interaction responsible for sporadic breast cancer identified by the tuned model is CypIA1m1 (presence of AA) and COMT (presence of aa). The interaction between rs699 (presence of AA) in AGT and rs5051 (presence of aa) in AGT 5' could be the most predominant cause of the hypertension. It is important to reduce the number of wrong predictions of cases or controls in reality. Hence, the learning algorithm is trained efficiently such that the model does not miss out any important interacting SNPs.

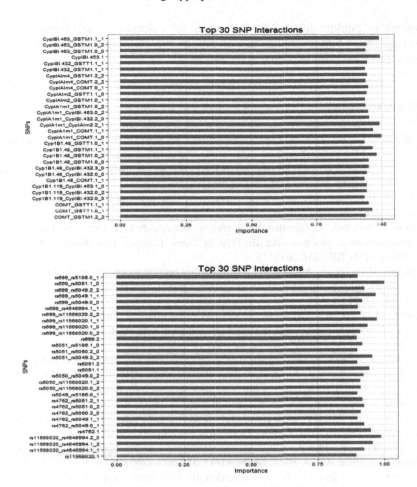

Fig. 6. Top 30 SNP interactions of (a) breast cancer data (b) hypertension data.

4 Conclusion

A brief study of previously proposed deep learning method to detect gene interactions is carried out in this paper. This study is extended to maximize the predictive performance of the models by tuning the hyperparameters using grid and random grid search approaches. Several experiments were performed on the real datasets by evaluating the model metrics of each model from the grid of large hyperparameter space. It is observed that the random search is more efficient than the grid search for optimizing the hyperparameters of deep learning algorithm. It worked well in predicting the best model with the highest prediction accuracy and lowest error. The optimal hyperparameters of the best model are used to improve the predictive performance of the deep learning algorithm for detecting gene interactions. The findings are confirmed by performing evaluations on the sporadic breast cancer and hypertension data. The top 30 interacting genes are illustrated on rank based method. Future studies will explore implementing random forest into the model for calculating the importance of the interacting genes. Further studies will investigate the performance of the model in the presence of genetic heterogeneity and phenocopy.

References

1. Padyukov, L.: Between the Lines of Genetic Code: Genetic Interactions in Understanding Disease and Complex Phenotypes. Academic Press, Cambridge (2013)
2. Gusareva, E.S., et al.: Genome-wide association interaction analysis for Alzheimer's disease. Neurobiol. Aging **35**(11), 2436–2443 (2014)
3. Cordell, H.J.: Detecting gene-gene interactions that underlie human diseases. Nat. Rev. Genet. **10**(6), 392–404 (2009)
4. Uppu, S., Krishna, A., Gopalan, R.: A review on methods for detecting SNP interactions in high-dimensional genomic data. IEEE/ACM Trans. Comput. Biol. Bioinf. **PP**(99) (2016). doi:10.1109/TCBB.2016.2635125
5. Ritchie, M.D., et al.: Multifactor-dimensionality reduction reveals high-order interactions among estrogen-metabolism genes in sporadic breast cancer. Am. J. Hum. Genet. **69**(1), 138–147 (2001)
6. Calle, M.L., et al.: MB-MDR: model-based multifactor dimensionality reduction for detecting interactions in high-dimensional genomic data. Stat. Med. **27**(30), 6532–6546 (2008)
7. Schwarz, D.F., König, I.R., Ziegler, A.: On safari to random jungle: a fast implementation of random forests for high-dimensional data. Bioinformatics **26**(14), 1752–1758 (2010)
8. Yang, C., et al.: SNPHarvester: a filtering-based approach for detecting epistatic interactions in genome-wide association studies. Bioinformatics **25**(4), 504–511 (2009)
9. Wan, X., et al.: BOOST: a fast approach to detecting gene-gene interactions in genome-wide case-control studies. Am. J. Hum. Genet. **87**(3), 325–340 (2010)
10. Purcell, S., et al.: PLINK: a tool set for whole-genome association and population-based linkage analyses. Am. J. Hum. Genet. **81**(3), 559–575 (2007)
11. Schwender, H., Ickstadt, K.: Identification of SNP interactions using logic regression. Biostatistics **9**(1), 187–198 (2008)
12. Zhang, Y., Liu, J.S.: Bayesian inference of epistatic interactions in case-control studies. Nat. Genet. **39**(9), 1167–1173 (2007)
13. Marvel, S., Motsinger-Reif, A.: Grammatical evolution support vector machines for predicting human genetic disease association. In: Proceedings of the 14th annual conference companion on Genetic and evolutionary computation. ACM (2012)
14. Motsinger, A.A., et al.: GPNN: Power studies and applications of a neural network method for detecting gene-gene interactions in studies of human disease. BMC Bioinformatics **7**(1), 39 (2006)
15. Bengio, Y., Goodfellow, I.J., Courville, A.: Deep Learning. An MIT Press book in preparation. Draft chapters available at http://www.iro.umontreal.ca/~bengioy/dlbook (2015)
16. LeCun, Y., Bengio, Y., Hinton, G.: Deep learning. Nature **521**(7553), 436–444 (2015)
17. Uppu, S., Krishna, A.: Improving strategy for discovering interacting genetic variants in association studies. In: Hirose, A., Ozawa, S., Doya, K., Ikeda, K., Lee, M., Liu, D. (eds.) ICONIP 2016. LNCS, vol. 9947, pp. 461–469. Springer, Cham (2016). doi:10.1007/978-3-319-46687-3_51

18. Uppu, S., Krishna, A., Raj, P.G.: A deep learning approach to detect SNP interactions. J. Softw. **11**(10), 960–975 (2016)
19. Bergstra, J., Bengio, Y.: Random search for hyper-parameter optimization. J. Mach. Learn. Res. **13**, 281–305 (2012)
20. Chiang, F.-T., et al.: Molecular variant M235T of the angiotensinogen gene is associated with essential hypertension in Taiwanese. J. Hypertens. **15**(6), 607–611 (1997)
21. Wu, S.-J., et al.: Three single-nucleotide polymorphisms of the angiotensinogen gene and susceptibility to hypertension: single locus genotype vs. haplotype analysis. Physiol. Genomics **17**(2), 79–86 (2004)
22. Aiello, S., Kraljevic, T., Maj, P.: h2o: R Interface for H2O. R package version, vol. 3 (2016)

Computational Efficacy of GPGPU-Accelerated Simulation for Various Neuron Models

Shun Okuno[1], Kazuhisa Fujita[1,2], and Yoshiki Kashimori[1(✉)]

[1] Department of Engineering Science, University of Electro-Communications,
Chofu, Tokyo 182-8585, Japan
chonmaru0kaite@gmail.com, kazu@spikingneuron.net,
kashi@pc.uec.ac.jp
[2] Tsuyama National College of Technology,
654-1 Numa, Tsuyama, Okayama 708-8506, Japan

Abstract. To understand the processing mechanism of sensory information in the brain, it is necessary to simulate a huge size of network that is represented by a complicated neuron model imitating actual neurons. However, such a simulation requires a very long computation time, failing to perform computer simulation with a realistic time scale. In order to solve the problem of computation time, we focus on the reduction of computation time by GPGPU, providing an efficient method for simulation of huge number of neurons. In this paper, we develop a computational architecture of GPGPU, by which computation of neurons is performed in parallel. Using this architecture, we show that the GPGPU method significantly reduces the computation time of neural network simulation. We also show that the simulations with single and double float precision give little significant difference in the results, independently of the neuron models used. These results suggest that the GPGPU computation with single float precision could be a most efficient method for simulation of a huge size of neural network.

Keywords: GPGPU · Accuracy evaluation · Spiking neural network

1 Introduction

Simulation of a huge size of neural network is necessary for understanding the neural mechanism of sensory processing. The network also needs to have a complicated neuron model reflecting actual neuronal structures. Spiking neural network (SNN) model is one of the important models of the cerebral cortex [1, 2]. The SNNs incorporate precise timing of spike structures leading to many interesting brain functions such as feature biding due to spike synchrony and spike timing-dependent plasticity. Notably, it is interesting to investigate the spike dynamics of the realistic brain constructed with a huge number of neurons, or tens to thousands millions of neurons. However, simulation of large scale-spiking networks requires a very long computing time, failing to perform computer simulation with a realistic time scale. For that reason, several computation methods have been developed to achieve high computing powers.

© Springer International Publishing AG 2017
D. Liu et al. (Eds.): ICONIP 2017, Part V, LNCS 10638, pp. 802–809, 2017.
https://doi.org/10.1007/978-3-319-70139-4_81

The SNN simulators have been traditionally simulated on large-scale clusters, super computers, and on dedicated hardware architectures. Alternatively, graphics processing units (GPU) with computer unified device architecture (CUDA) can provide a low-cost, programmable, and high-performance platform for simulation of SNNs. Fidjeland and Shanahan [3] developed a platform for simulators which achieves high performance on parallel commodity hardware in the form of GPUs. Nageswaran et al. [4] demonstrated an efficient, biologically, realistic large-scale simulator that runs on a single GPU. Pallipuram et al. [5] compared two GPGPU architectures and two GPGPU programing models using a two-layered network. Since CPUs and graphic boards keep evolving, it is important to study how to develop a cheaply, effective computing system using a GPGPU system.

In this study, we focus on the reduction of computing time by a GPGPU method that provides an efficient method for simulation of a huge number of spiking neurons. We develop a computational architecture of GPGPU, by which computation of neurons is performed in parallel. Large-scale SNNs consist of four types of spike neuron models, or leaky integrate-and-fire neuron model, Izhikevich neuron model, two-compartment model, and Hodgkin-Huxley model. Using this architecture, we show that the GPU method significantly reduces the computing time of neural network simulations. We also show that the simulations with single and double float precision give little significant difference in the results, independently of the neuron models used. These results suggest that the GPU computation with single float precision could be one of efficient methods for simulation of a huge size of SNNs.

2 Model

2.1 Leakey Integrate-and-Fire (LIF) Model

The LIF model is the most simplified model, which contains an integration process of input, and a spike generation process with a firing threshold [6]. The membrane potential, V, is determined by

$$\tau \frac{dV}{dt} = -V + I, \tag{1}$$

where τ is the time constant of V, and I is an input. A neuron emits a spike once the membrane potential reaches a threshold θ, and then it is rest to a resting state, $V = 0$. The parameters were set as follows: $\tau = 10\,\mathrm{ms}$, $\theta = 10\,\mathrm{mV}$, and $I = 15\,\mathrm{mV}$.

2.2 Izhikevich Neuron Model

The Izhikevich neuron model has been proposed to reproduce several types of spike patterns of a single neuron, such as tonic, burst, and chattering patterns [7]. The dynamics of a single neuron is described by the dynamics of a membrane potential and a recovery variable, which is a reduced form of Hodgkin-Huxley model. The membrane potential, V, and the recovery variable, U, are given by

$$\frac{dV}{dt} = 0.04V^2 + 5V + 140 - U + I, \tag{2}$$

$$\frac{dU}{dt} = a(bV - U), \tag{3}$$

where I is an input, and a and b are constant parameters. A neuron emits a spike for the following condition;

$$\text{if } V \geq \theta, \quad \text{then}$$
$$\begin{cases} V \leftarrow c \\ U \leftarrow U + d, \end{cases} \tag{4}$$

where θ is the firing threshold, and c and d are constant parameters. The parameter values were set as follows: $a = 0.02$, $b = 0.2$, $c = -65$ mV, $d = 8$, and $\theta = 20$ mV.

2.3 Two-Compartment Model

The two-compartment model has been proposed by Doiron et al. [8] in order to reproduce a bursting behavior of pyramidal cell in an electrosensory system. The model has two compartments of a soma and a dendrite. The membrane potential of a soma and a dendrite, V_s and V_d, are determined by

$$\begin{aligned} \frac{dV_s}{dt} &= I + g_{Na}\, m_{s,\infty}^2 (1 - n_s)(V_{Na} - V_s) + g_{Dr,s}\, n_s^2 (V_K - V_s) \\ &\quad + \frac{g_c}{\kappa}(V_d - V_s) + g_{L,s}(V_L - V_s), \end{aligned} \tag{5}$$

$$\begin{aligned} \frac{dV_d}{dt} &= g_{Na,d}\, m_{d,\infty}^2 h_d (V_{Na} - V_d) + g_{Dr,d}\, n_d^2 p_d (V_K - V_d) \\ &\quad + \frac{g_c}{(1 - \kappa)}(V_s - V_d) + g_{L,d}(V_L - V_d), \end{aligned} \tag{6}$$

where $g_{Na,s}$, $g_{Dr,s}$, and $g_{L,s}$ are the conductance of Na+, K+, and leak channels, respectively. $m_{s,\infty}$ and n_s, respectively, are the probabilities of opened states of Na+ and K+ channels. The conductance and probabilities of the dendrite have the similar definition to those of soma. κ is the ratio of the cross-sectional area of the soma to the dendrite, and g_c is the conductance of the connection between soma and dendrite. The details of the model and parameters are described in [8].

2.4 Hodgkin-Huxley Model

The Hodgkin-Huxley model is a neuron model on the basis of physiologically detailed behavior of ion channels of an action potential [9]. The membrane potential, V, is given by

$$C\frac{dV}{dt} = g_{Na}m^3h(V_{Na} - V) + g_K n^4(V_K - V) + g_L(V_L - V) + I, \qquad (7)$$

where g_{Na}, g_K, and g_L are the conductance of Na+, K+, and leak channels, respectively. C is the membrane capacitance. m, n, and h are the probabilities determining the opened and closed states of Na+ and K+ channels. The details of the model are described in [9].

3 Architecture and Flow in Computational Processing

3.1 Flows of Computational Processing

Figure 1A shows the processing flows of a single core and Open MP. The box of "calculate potential" denotes the core process of the flows. In the Open MP, the process is parallelized. The flow chart of the GPU computation is shown in Fig. 1B. The membrane potentials of neurons are calculated by a parallel processing of GPU. In this process, the data processed by GPU are fed back to the memory area of CPU. The data transfer between a GPU and a CPU has been reported to be a time-consuming process in GPU calculation.

(A)   (B)

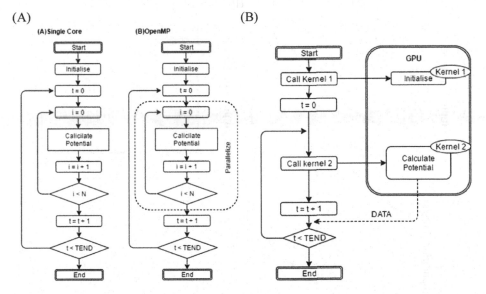

Fig. 1. (A) Flow charts of simulations with a single core (left) and Open MP (right). (B) Flow chart of simulation with a GPU.

3.2 Experimental Environment

The experimental environment is as follows.

CPU: Intel Xenon E5-2630 v.2, 2.60 GHz x 12
CPU memory: DDR3 16 GB

GPU: NVIDIA Geforce GTX1080
GPU memory: 8 GB
GPU memory band: 320 GB/sec

The GPU programming was developed by CUDA developmental environment (Ver. 7.5) on Linux.

4 Results

4.1 Computation Time of Different Neuron Models

Figure 2A–C show the computation time of four spiking networks, with a single core, Open MP, and a GPU, respectively. The four spiking networks used were constructed with the LIF model, the Izhikevich model, the two-compartment model, and the Hodgkin-Huxley model, respectively. These networks have no interconnection between each other. The simulations were performed with the neuron numbers of 2^n ($n = 8, 10, 12, 14, 16, 18$). In all computational architectures, the simple neuron models such as the LIF and the Izhikevich model showed the computation time several times

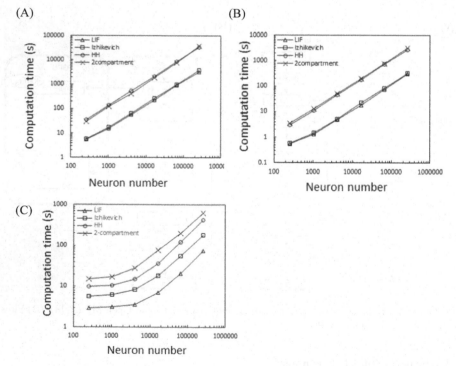

Fig. 2. Computation time of simulations of four spiking networks, with three architectures, or (A) a single core, (B) Open MP, and (C) a GPU. The neuron models used are the leaky integrate-and–fire (LIF) model, the Izhikevich model, the Hodgkin-Huxley (H-H) model, and the two-compartment model.

faster than more complicated models such as the two-compartment model and the Hodgkin-Huxley model. Furthermore, the performance of GPU showed the shortest computation time in the three computational architectures for the neuron numbers larger than 2^{12}, whereas the Open MP had the best performance for the neuron numbers less than 2^{12}. This indicates that a cheap graphic board, Geforce GTX 1080, accelerates the computational power of an expensive device of Xenon E5-2630 for a large-scale spiking network. Thus the GPU computational architecture provides a good cost performance in parallel computing of a large-scale network.

4.2   Influence of Data Transfer on Computation Time

Data transfer between graphics and memory is an important factor for determining the computation time of spiking network simulation. If the time of data transfer is larger enough to be neglected compared with the time of the core process of simulation, the computation time would be significantly affected by the data-transfer process. We have to estimate the rate of the data transfer time to the time of simulation process alone. Figure 3 shows the computation time of the four spiking networks without data transfer process. The performance showed the computation time shorter than the time of the GPU method shown in Fig. 2C. Particularly, the performance without data transfer is superior to that of Open MP for small numbers of neurons ($< 2^{12}$). This indicates that for the small numbers of neurons, the data-transfer process can be a time-consuming process compared with the core process of simulation. It is important for the efficient computation of GPU to solve how to improve the efficacy of data-transfer process.

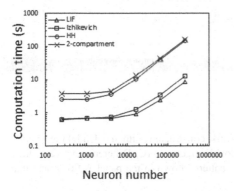

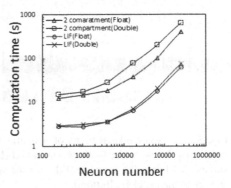

Fig. 3. Computation time of simulation of four spiking neural networks, without data-transfer process. The architecture used is a GPU.

Fig. 4. Computation time of simulations of two spiking networks, with single and double floating point. The architecture used is a GPU.

4.3   Comparison of Accuracy in Computation with Single and Double Precision

Computation with single precision floating point is a possible method of efficient computation. However, it depends on the content of computation. To examine the extent to which single precision provides accuracy enough to simulate spoking net-works, we compared the accuracy of the dynamics of spiking networks using single and double precision. Figure 4 shows the computation time with single and double precision for the LIF and two-compartment model. In both cases, the computation time was shorter in single precision than in double precision, indicating that single precision significantly reduces the computation time. Figure 5A and B, respectively, show the temporal courses of membrane potentials for a Hodgkin-Huxley neuron, each was calculated with single and double precision. The difference between the membrane potentials calculated with single and double precision is also shown in Fig. 5C. The

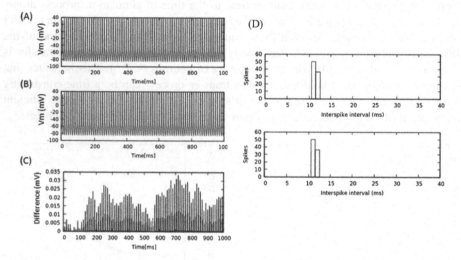

Fig. 5. (A), (B) Spike patterns of a neural network constructed with the Hodgking-Huxley neuron model, with (A) single precision and (B) double precision. (C) Difference between these spikes. (D) Interspike intervals (ISIs) for the spike pattern shown in the panel A (Top) and that shown in the panel B (Bottom).

two types of calculations yielded little significant difference in the spike patterns. The interspike intervals (ISIs) shown in Fig. 5D also show little difference for both precision. Together, GPU computation with single precision has a good balance between computational time and accuracy, and offers a powerful method for investigating the dynamics of a large-scale spiking network.

5   Conclusion

We have demonstrated the computation time and accuracy of the simulations of spiking neural networks, each consisting of one of four neuron models. Three results are shown in the present study. First, GPU computation has the best performance for simulation of large-scale networks, whereas Open MP shows the best performance for simulation of smaller size of networks. Second, data transfer process can be a key process for accelerating GPU computation. Third, computational performance with single precision provides accuracy enough to study the dynamical property of a large-scale spiking neural network. In summary, GPU architecture with single precision would offer a powerful architecture for simulation of a large scale spiking network.

Acknowledgement. This article was supported by Grant in aid for Scientific Research from the Japan Society for the promotion of Science (Number 15K07146).

References

1. Mass, W., Bishop, C.M.: Pulsed Neural Networks. MIT Press, Cambridge (1999)
2. Brette, R., et al.: Simulation of networks of spiking neurons: a review of tools and strategies. J. Comput. Neurosci. **23**(3), 349–398 (2007)
3. Fidjeland, A.K., Shanahan, M.P.: Accelerated simulation of spiking neural networks using GPUs. In: Proceedings of WCCI 2010 IEEE World Congress on Computational Intelligence, pp. 536–543. Barcelona, Spain (2010)
4. Nageswaran, J.M., et al.: A configurable simulation environment for the efficient simulation of large-scale spiking neural networks on graphical processors. Neural Netw. **22**(5–6), 791–800 (2009)
5. Pallipuram, V.K., Bhuiyan, M., Smith, M.C.: A comparative study of GPU programing models and architectures using neural networks. J. Supercomput. **61**, 673–718 (2012)
6. Tuckwell, H.C.: Introduction to Theoretical Neurobiology. Cambridge University Press, Cambridge (1988)
7. Izhikevich, E.M.: Simple model of spiking neurons. IEEE Trans. Neural Netw. **14**(6), 1569–1572 (2003)
8. Doiron, B., et al.: Ghostbursting: a novel neuronal burst mechanism. J. Comput. Neurosci. **12**, 5–25 (2002)
9. Hodgking, A.L., Huxley, A.F.: A quantitative description of membrane current and its application to conduction and excitation in nerve. J. Physiol. **117**(4), 500–544 (1952)

A Haptics Feedback Based-LSTM Predictive Model for Pericardiocentesis Therapy Using Public Introperative Data

Amin Khatami[1(✉)], Yonghang Tai[1], Abbas Khosravi[1], Lei Wei[1],
Mohsen Moradi Dalvand[1], Jun Peng[2], and Saeid Nahavandi[1]

[1] Institute for Intelligent Systems Research and Innovation, Deakin University,
Geelong, VIC 3216, Australia
{skhatami,yonghang,abbas.khosravi,Lei.Wei,mohsen.m.dalvand,
saeid.nahavandi}@deakin.edu.au
[2] Thoracic Surgery Department, Yunnan First People's Hospital, Kunming, China
389647518@qq.com

Abstract. Proposing a robust and fast real-time medical procedure, operating remotely is always a challenging task, due mainly to the effect of delay and dropping of the speed of networks, on operations. If a further stage of prediction is properly designed on remotely operated systems, many difficulties could be tackled. Hence, in this paper, an accurate predictive model, calculating haptics feedback in percutaneous heart biopsy is investigated. A one-layer Long Short-Term Memory based (LSTM-based) Recurrent Neural Network, which is a natural fit for understanding haptics time series data, is utilised. An offline learning procedure is proposed to build the model, followed by an online procedure to operate on new experiments, remotely fed to the system. Statistical analyses prove that the error variation of the model is significantly narrow, showing the robustness of the model. Moreover, regarding computational costs, it takes 0.7 ms to predict a time step further online, which is quick enough for real-time haptic interaction.

Keywords: Force prediction · Haptics predictive models · Real-time · LSTM

1 Introduction

Accurate and robust force feedback prediction is vital for haptics rendering, especially in complex virtual environments due to problems such as network delay and discontinuous forces. Although the development of hardware devices speeds up haptics interpretations, there is commonly a dispute between high requirements of simulation and calculating power of the real-time interaction devices. The importance of force feedback prediction is widely recognised especially when a surgical simulator does operations remotely. By monitoring the prediction of a

© Springer International Publishing AG 2017
D. Liu et al. (Eds.): ICONIP 2017, Part V, LNCS 10638, pp. 810–818, 2017.
https://doi.org/10.1007/978-3-319-70139-4_82

few steps further, an improvement in the performance is promising. This enables a user to estimate and obtain more real senses as realistically as possible [15].

A variety of research has been conducted, proposing linear predictive models for haptics rendering issues [3,10–14,17]. In more detail, however, although the methods of [3,11,12] improved the performance, the stability and smoothness of the force feedback is still not guaranteed [11]. Due to the development of deep approaches, recently, scientists are investigating the deep structural-based techniques, especially LSTMs for classifying haptics signals [4,5,16]. The LSTM models are chosen because they have recurrent architectures which are a natural fit for understanding haptics time series data [5].

Motivations: Inspired by the popularity of the linear approach and usefulness of the LSTMs in sequence classification commonly conducted on haptics data [5], we investigate the impact of the recurrent networks to predict the force feedback information downloaded from the public database from the thoracic department of Yunnan First People's Hospital which recorded the introperative force feedback through the pericardiocentesist therapy[1].

Contributions: There are several contributions of this study; (1) Proposing a robust *real-time* haptics feedback predictive model based on LSTM-based recurrent networks for percutaneous heart biospy. (2) Providing a prediction-based dataset for deep learning techniques based on the data collected and published by Yunnan First People's Hospital. (3) Conducting an augmentation procedure which generalises the model properly compared with well-known regularisation techniques in deep learning.

The study is organised as follows: Sect. 2 contains a brief explanation of the proposed LSTM model. Experimental results along with a discussion and comparison are conducted in Sect. 3, followed by a conclusion in Sect. 4.

2   Methodology

Deep learning techniques are commonly used in medical application, promising very good performance in terms of classification, prediction, and retrieval domains [1,7–9]. Among them, Recurrent Neural Networks, short RNN, are playing an important role in practice, especially for time series domains. Inspired by the usefulness of the RNNs and following up the applicability of a Long Short-Term Memory (LSTM) model which is a type of RNN and is commonly used in online applications [4,5], we are applying the LSTM to our real-time application for predicting haptics forces.

LSTM-based RNNs was originally proposed in [6]. These networks are able to learn sequences of observations which causes the model to be well-suited to time series applications. The main motivation behind proposing the LSTMs is to avoid vanishing gradients, which was an issue in traditional RNNs, by introducing a new structure of memory cell, as seen in Fig. 1. An LSTM model calculates a

---

[1] http://civ.ynnu.edu.cn/ChineseShow.aspx?ID=1.

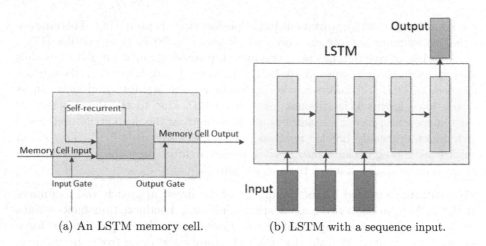

(a) An LSTM memory cell.   (b) LSTM with a sequence input.

Fig. 1. The black diagram of an LSTM.

mapping from a sequence of $x = (x_1, x_2, ..., x_T)$ as input to a sequence of $h = (h_1, h_2, ..., h_T)$ as output, by following the below equations iteratively from $t = 1$ to T:

$$f_t = \sigma(W_f.[h_t - 1, x_t] + b_f) \tag{1}$$

$$i_t = \sigma(W_i.[h_t - 1, x_t] + b_i) \tag{2}$$

$$\tilde{C}_t = tanh(W_C.[h_t - 1, x_t] + b_C) \tag{3}$$

$$C_t = f_t \odot C_{t-1} + i_t \odot \tilde{C}_t \tag{4}$$

$$o_t = \sigma(W_o.[h_{t-1}, x_t] + b_o) \tag{5}$$

$$h_t = o_t \odot tanh(C_t) \tag{6}$$

where the W terms are weight matrices. The b is bias vector, and σ and $tanh$ are activation functions. Also, C, i, f, and o are cell activation, input, forget, and output gate vectors, respectively.

Figure 2 illustrates the proposed real-time haptics feedback predictive model. As seen, after training the model offline, with a simulation manner, the remote procedure of haptics rendering feeds new experiments to the optimised network, using the tuned parameters to predict the next stage of forces captured by the sensors of the haptics instrument. As explained in next section, the online prediction is too quick, which is a must for a real-time application. The importance of the proposed model is highlighted once a delay is happening during an online simulation.

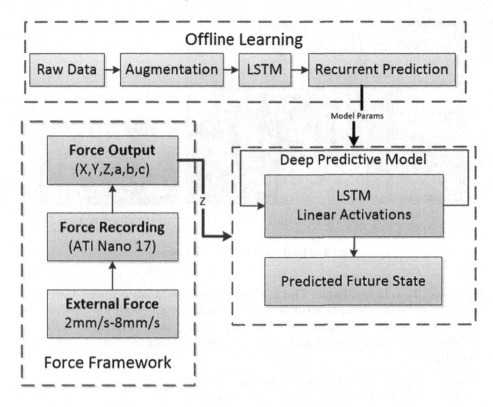

Fig. 2. The black diagram of our real-time haptic force feedback predictive model.

3   Experimental Results

This section demonstrates the empirical results of the proposed predictive model by defining seven strategies, conducted on the LSTM, followed by statistical analyses to report the proper model for the prediction model. The following subsections explain the dataset details and evaluation terms, followed by discussion.

Accessing publicly available Data: The authors have made contact with surgeons from Yunnan First People's Hospital and accessed the data from a publicly available website. According to the introduction of the dataset, all data and images are collected on a custom-built instrument and are based on animal organs purchased from a supermarket. Professional surgeons were invited to perform the procedures to ensure the credibility of the acquired data. The room setup was mimicking a surgical theatre but is not a real one. The authors have also requested additional details about the experiments from the surgeons, as explained in this paper.

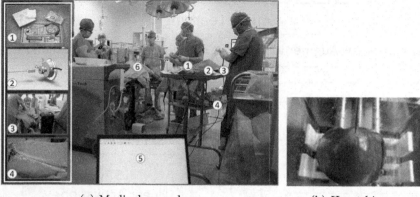

(a) Medical procedure.   (b) Heart biopsy.

Fig. 3. The data preparing procedure.

Haptics (Touch feedback) Data: Interoperated data are compared with the experimental data to verify whether the algorithm we proposed for the dynamic puncture has a high fitting degree with the modeling value. The detailed system implementation in the mimicked operation room is demonstrated in Fig. 3, as following:

- Step 1: Pericardiocentesis instruments
- Step 2: Force sensor + 3D Print Connector
- Step 3: Line Puncture Devices Integration
- Step 4: Data Conversion
- Step 5: Analysis Software
- Step 6: Mimicked surgical Environments

Normalization and Subsampling: As depicted in Fig. 4, the values contain negative and positive haptics feedbacks. We remove the negative values because

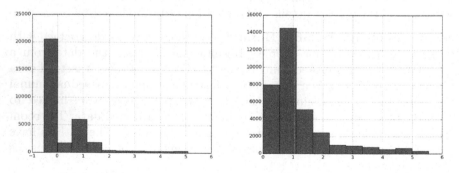

(a) The histogram of the original data   (b) The histogram of the augmented data

Fig. 4. The histogram of the datasets

the prediction of the forward percutaneous is investigated. We also normalise the raw data by subtracting the mean value of the data and creating the predictive set by considering X and Y as force feedbacks at given time of (t) and $(t + 1)$, respectively. An input with sequence length of 100 is fed to the network to predict one further step. It means that the prediction should be decided based on the previous 99 steps. Hence, by using a shifting windows, a dataset consisting of the samples with $(11330, 100)$ dimension is created. Note that 80% of the set is assigned for training and the rest for test.

Data Augmentation: Similar to [5], three different signals were experimented on the organ. We subsample each signal at different starting point. The signals are merged to enlarge the data set. After augmentation, the dataset, called data set 2, is increased to $(34020, 100)$.

Performance Metric: Mean Squared Error (MSE), Mean Absolute Error (MAE), and R Squared (R^2) are three criteria utilised to evaluate the performance. MSE provides the impact of how the magnitude of the error is. MAE, which is the absolute variation between the values of predictions and actual ones, provides the impact on how wrong the predictions are. R^2, which is called coefficient of determination as well, gives the impact on the excellence of fitting the predictions with the actual ones. Note that the closer value to 1, the higher fitting rate to the actual data.

Defining Scenarios: Seven cases studies are investigated by evaluating an LSTM network on the original and augmented sets. Different regularisation techniques, activations, and optimisers are evaluated as different case studies. We also compare the proposed LSTM network with the stacked one to highlight the impact of the depth analysis.

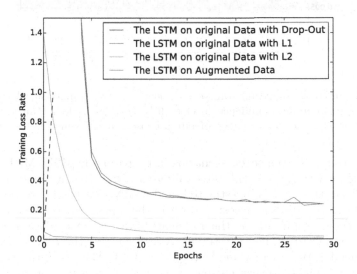

Fig. 5. The loss convergence speed during training.

Experiments and Discussion: The similar explored LSTM model conducted in [5] for a classification approach is followed in this study to provide a predictive model. The LSTM is chosen here due to the fact that these models have a recurrent architecture and are a natural fit for understanding haptic data [5]. The LSTM includes 10 recurrent units, followed by a fully connected layer with one output with *linear* activation function. The linear dense layer is utilised to combine all knowledge from previous layer into one single value, which is the 100th time step of the sequence. To avoid overfitting, after investigation, a 20% Drop-Out is applied to both the LSTM and the dense layers. For the learning approach, we use the standard loss function of MSE with 30 epochs. The 30 is chosen because at this step all explored networks are converged, as illustrated in Fig. 5. Regarding our proposed model, we can regularise the model by considering early stopping procedure at stage 3 which shows that the way of augmentation causes a quick convergence, compared with that of the other regularisation techniques.

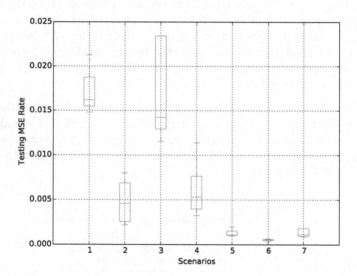

Fig. 6. The boxplot comparing the seven scenarios. As comparison, the H-test [2] obtains a corresponding confidence level of 95% (*p*-value is less than 0.05), which strongly illustrates the performance of our proposed model which is the scenario 6.

Table 1 compares the seven scenarios in terms of the MSE, MAE, and R^2. The results come from a 5-fold cross validation. Note that a range of penalties for the L1 and the L2 was investigated, and finally as a result, we consider the rate of 0.01 to penalise their costs. As seen, the magnitude of the error for the proposed model, evaluating on the augmented force feedback data, is too small, compared to the rest of the case studies. Also, regarding the R^2 criterion for our proposed model, the mean value is very close to 1. Moreover, the variation of results of the 5 folds are very narrow, as compared with that of the rest of the experiments.

Table 1. A comparison among the seven case studies, evaluated by MSE, MAE, and R^2 with a 5-fold cross validation procedure.

Scenarios	Strategies	MSE	MAE	R^2
		Mean ± std	Mean ± std	Mean ± std
1	LSTM + L1 (Original set)	0.0174 ± 0.003	0.064 ± 0.042	0.684 ± 0.422
2	LSTM + L2 (Original set)	0.005 ± 0.002	0.049 ± 0.017	0.995 ± 0.002
3	LSTM + DropOut (Original set)	0.022 ± 0.015	0.099 ± 0.027	0.980 ± 0.014
4	LSTM + L2 (Augmented set)	0.006 ± 0.003	0.218 ± 0.287	0.994 ± 0.003
5	LSTM + DropOut + adam (Augmented set)	0.001 ± 0.0004	5.312 ± 0.094	0.997 ± 0.001
6	**LSTM + DropOut + rmsprop (Augmented set)**	**0.0005 ± 0.0001**	**0.014 ± 0.002**	**0.999 ± 9.12E-5**
7	Stacked LSTM + DropOut + rmsprop (Augmented set)	0.002 ± 0.001	0.023 ± 0.005	0.9985 ± .0009

The boxplot illustrated in Fig. 6 highlights the aforementioned discussion. Moreover, statistical speaking, the H-test [2] rejects the null hypothesis of all cases studies, because the p-values are lower than 0.05 (95% confidence level).

As a comparison between scenarios 6 and 7, we found that a small change of the number of recurrent units in the LSTM architecture did not affect the final performance considerably, however stacking LSTMs reduced the prediction score by 4 times, in term of the MSE.

Regarding the computational costs, the average cost to predict one step further by looking at the 99 previous steps takes 0.7 μs, which enables the model utilising in a real-time application.

4   Conclusion

The role of force predictive models are vital especially for a real-time haptic rendering operations. The importance of these feedback predictive models is more emphasised when some issues like delay and dropping the speed of the net are happening during operations. Hence, the prediction of the force feedback for some further steps is very important. To achieve a robust predictive model on haptic data, inspired by the usefulness of the LSTM-based RNNs and considering their natural fit for understanding haptic time series signals, the LSTMs were investigated. The evaluation was performed on several experiments captured by 6DOF sensor during simulating a surgical simulation. The robust and accurate results with a significantly narrow variation during experiments along with quick and fast responding make the model suitable for real-time application.

In the future, the generalise version of our LSTM-based haptic prediction model will have the potential to be used for various surgical applications for surgeons to use as an assistive tool to predict various conditions and complications during operations.

References

1. Babaie, M., Kalra, S., Sriram, A., Mitcheltree, C., Zhu, S., Khatami, A., Rahnamayan, S., Tizhoosh, H.R.: Classification and retrieval of digital pathology scans: a new dataset. arXiv preprint (2017). arXiv:1705.07522

2. Breslow, N.: A generalized Kruskal-Wallis test for comparing K samples subject to unequal patterns of censorship. Biometrika **57**(3), 579–594 (1970)
3. Duriez, C., Andriot, C., Kheddar, A.: A multi-threaded approach for deformable/rigid contacts with haptic feedback. In: Proceedings of 12th International Symposium on Haptic Interfaces for Virtual Environment and Teleoperator Systems, 2004. HAPTICS 2004, pp. 272–279. IEEE (2004)
4. Gamboa, J.C.B.: Deep learning for time-series analysis. arXiv preprint (2017). arXiv:1701.01887
5. Gao, Y., Hendricks, L.A., Kuchenbecker, K.J., Darrell, T.: Deep learning for tactile understanding from visual and haptic data. In: 2016 IEEE International Conference on Robotics and Automation (ICRA), pp. 536–543. IEEE (2016)
6. Hochreiter, S., Schmidhuber, J.: Long short-term memory. Neural Comput. **9**(8), 1735–1780 (1997)
7. Khatami, A., Babaie, M., Khosravi, A., Tizhoosh, H., Salaken, S.M., Nahavandi, S.: A deep-structural medical image classification for a radon-based image retrieval. In: 2017 IEEE 30th Canadian Conference on Electrical and Computer Engineering (CCECE), pp. 1–4. IEEE (2017)
8. Khatami, A., Khosravi, A., Lim, C.P., Nahavandi, S.: A wavelet deep belief network-based classifier for medical images. In: Hirose, A., Ozawa, S., Doya, K., Ikeda, K., Lee, M., Liu, D. (eds.) ICONIP 2016. LNCS, vol. 9949, pp. 467–474. Springer, Cham (2016). doi:10.1007/978-3-319-46675-0_51
9. Khatami, A., Khosravi, A., Nguyen, T., Lim, C.P., Nahavandi, S.: Medical image analysis using wavelet transform and deep belief networks. Expert Syst. Appl. **86**, 190–198 (2017)
10. Khatami, A., Mirghasemi, S., Khosravi, A., Lim, C.P., Nahavandi, S.: A new PSO-based approach to fire flame detection using K-Medoids clustering. Expert Syst. Appl. **68**, 69–80 (2017)
11. Ortega, M., Redon, S., Coquillart, S.: A six degree-of-freedom god-object method for haptic display of rigid bodies with surface properties. IEEE Trans. Vis. Comput. Graph. **13**(3), 458–469 (2007)
12. Otaduy, M.A., Lin, M.C.: Stable and responsive six-degree-of-freedom haptic manipulation using implicit integration. In: First Joint Eurohaptics Conference and Symposium on Haptic Interfaces for Virtual Environment and Teleoperator Systems, 2005. World Haptics 2005, pp. 247–256. IEEE (2005)
13. Picinbono, G., Lombardo, J.C.: Extrapolation: a solution for force feedback. In: International Scientific Workshop on Virtual Reality and Prototyping, pp. 117–125 (1999)
14. Picinbono, G., Lombardo, J.C., Delingette, H., Ayache, N.: Improving realism of a surgery simulator: linear anisotropic elasticity, complex interactions and force extrapolation. J. Vis. Comput. Animat. **13**(3), 147–167 (2002)
15. Ruffaldi, E., Morris, D., Edmunds, T., Barbagli, F., Pai, D.K.: Standardized evaluation of haptic rendering systems. In: 2006 14th Symposium on Haptic Interfaces for Virtual Environment and Teleoperator Systems, pp. 225–232. IEEE (2006)
16. Sung, J., Salisbury, J.K., Saxena, A.: Learning to represent haptic feedback for partially-observable tasks. arXiv preprint (2017). arXiv:1705.06243
17. Wu, J., Song, A., Li, J.: A time series based solution for the difference rate sampling between haptic rendering and visual display. In: 2006 IEEE International Conference on Robotics and Biomimetics. ROBIO 2006, pp. 595–600. IEEE (2006)

Sleep Apnea Event Detection from Nasal Airflow Using Convolutional Neural Networks

Rim Haidar$^{(\boxtimes)}$, Irena Koprinska, and Bryn Jeffries

School of Information Technologies, The University of Sydney,
Sydney, Australia
rhai6781@uni.sydney.ed.au, {irena.koprinska,
bryn.jeffries}@sydney.edu.au

Abstract. Obstructive sleep apnea-hypopnea syndrome is a respiratory disorder characterized by abnormal breathing patterns during sleep. It causes problems during sleep, including loud snoring and frequent awaking. This study proposes a new approach for the detection of apnea-hypopnea events from the raw signal data of nasal airflow using convolutional neural networks. Convolutional neural networks are a prominent type of deep neural networks known for their ability to automatically learn features from high dimensional data without manual feature engineering. We demonstrate the applicability of this technique on a dataset of 24,480 samples (30 s long) extracted from nasal flow signals of 100 subjects in the MESA sleep study. The performance of the convolutional neural network model is compared with another approach that uses a support vector machine model with statistical features generated from the flow signal. Our results show that the convolutional neural network outperformed the support vector machine approach, achieving accuracy and F1-score of 75%.

Keywords: Sleep apnea detection · Convolutional neural networks · Support vector machines

1 Introduction

Obstructive sleep apnea-hypopnea syndrome is a respiratory disorder characterized by abnormal breathing patterns during sleep. These abnormal patterns, known as apneic events, include obstructive apnea events (complete or almost complete cessation of airflow) and hypopnea events (reduction of airflow) [1]. Sleep apnea causes loud snoring, nocturnal choking or gasping, frequent awakenings, disrupted sleep, insomnia and excessive daytime sleepiness [2]. It is estimated that it affects between 2 and 4% of the adult population [3].

Polysomnography (PSG) is the golden standard for the diagnoses of sleep apnea [4]. PSG is a process in which an individual is monitored overnight with a signal acquisition device that collects electrical signals from a variety of sensors and electrodes attached to the body. The PSG data consists of signals such as: respiratory signals (nasal airflow, thoracic abdominal movement), oxygen saturation, electroencephalography and electrocardiography. In addition to the sleep state and respiratory details, PSG provides detailed information about heart rate, body position and muscle contraction [4].

© Springer International Publishing AG 2017
D. Liu et al. (Eds.): ICONIP 2017, Part V, LNCS 10638, pp. 819–827, 2017.
https://doi.org/10.1007/978-3-319-70139-4_83

To diagnose sleep apnea events, the PSG signals are processed and analyzed by physicians and sleep experts, who rely on statistical techniques to facilitate the decision-making process [5]. However, due to the complexity and volume of the data (long PSG recordings), manually analyzing the data is very time and labor consuming, and expensive. Therefore, several methods and applications have been proposed to automatically detect the sleep apnea events.

A PSG signal is a time series representing a set of sequential measurements collected over an extended period. To automatically analyze PSG data, most approaches typically apply signal processing techniques such as wavelet and Fourier transforms, and then extract statistical or other manually engineered features, that are used by a classification algorithm. Without feature selection, these methods have problems dealing with data that is as complex, high-dimensional and noisy as PSG data. On the other hand, good feature selection is domain specific, and also expensive and time consuming when performed manually [6].

A possible alternative is to apply deep neural networks – a recently developed, powerful set of algorithms, which can automatically learn and extract features from the raw data without prior knowledge. Over the past few years Convolutional Neural Networks (CNNs), a type of deep neural networks, have gained a lot of interest showing excellent performance in various domains such as image classification [7], speech recognition [8] and pattern recognition [9].

The aim of this study is to assess the ability of CNNs to detect sleep apneic events from a high dimensional data, without feature engineering and compare their performance with a traditional approach involving feature engineering. In particular, we present an approach based on CNN and compare it with a typical approach used in previous studies, which applies wavelet transform and then extracts statistical features, utilized in conjunction with a Support Vector Machine (SVM) classifier.

2 Related Work

Several methods for detecting apnea-hypopnea events from respiratory signals have been proposed. Most of them apply signal processing techniques to preprocess the data and extract informative features, which are then used as inputs to classification algorithms. For instance, in [10, 11] the data was decomposed using wavelet transformation and features based on the wavelet coefficients were extracted and used as inputs to an SVM classifier. Other types of classifiers used include feed forward neural networks [12] and ensemble methods [13]. Features are typically extracted from the abdominal, chest, nasal or thoracic respiratory signals, and less frequently from oxygen saturation and other repository signals. In [26], features extracted from both airflow and blood oxygen saturation of 15 PSG records were used as inputs to a time delay neural network. Nasal air flow signal was utilized in conjunction with adaptive fuzzy logic algorithms to detect normal and abnormal events [14, 15]. In this paper, we extend this previous work by studying the performance of CNNs without feature selection and comparing it with a traditional state-of-the-art method – SVM, which uses a subset of manually selected wavelet features.

3 Data

3.1 Dataset

We use data from the MESA sleep study [16]. Between 2010 and 2011, 2057 participants were enrolled in a sleep examination that included full overnight unattended PSG, 7-day wrist-worn actigraphy and a sleep questionnaire. The full overnight PSG recordings with a duration of at least 8 h included: airflow (via oral/nasal thermistor and nasal pressure transducer), bilateral electrocochleography, electrocardiography, chin electromyography, thoracic and abdominal respiratory inductance plethysmography, electroencephalography (EEG), oxyhemoglobin saturation (finger pulse oximetry), leg movements and body position [16].

Registered polysomnologists have manually scored the respiratory events from the respiratory data (airflow, cannula, snore, abdominal, chest, and saturation). Obstructive apnea events were detected when the airflow signal was flat or nearly flat for at least 10 s. Hypopnea events were detected if there was at least 50% reduction in both thoracic and abdominal channels, or if there was a clear (greater than 50%) reduction on a good nasal pressure signal, or if the amplitude of the thorax and abdomen signals decreased by at least (approximately) 30% of the amplitude of the "baseline" and lasted for at least 10 s.

3.2 Data Preparation

For this study, we use only the nasal airflow signals. A sample of 100 subjects was randomly selected. The length of each record is at least 8 h with a sampling rate of 32 Hz. Each 30 s segment was labelled as *abnormal* if it included an obstructive apnea or a hypopnea event lasting for at least 10 s; if it didn't include such events, it was labelled as *normal*. Figure 1 shows examples of normal breathing and the two types of abnormal breathing events: hypopnea and obstructive apnea.

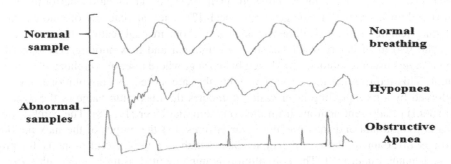

Fig. 1. Nasal airflow for normal breathing, hypopnea and obstructive apnea events

The dataset contains an equal number of examples from each class: 12,240 normal and 12,240 abnormal segments. The dimensionality of each sample is 960 (30 s × 32 Hz). All data preprocessing and preparation was performed in Matlab.

4   Method

Figure 2 summarizes our method. From a machine learning point of view, the task is to build a classifier that can accurately predict the two classes: normal and abnormal. We built two classifiers: (1) using CNN and all 960 original features, without feature selection, and (2) using SVM and 15 wavelet features, selected after the application of wavelet transformation and feature extraction.

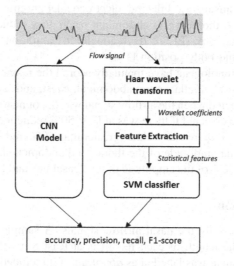

Fig. 2. Flow chart of the proposed method using CNN and SVM

4.1   CNN Model

CNNs are prominent examples of deep neural networks, initially developed for computer vision tasks. They have achieved an impressive classification performance in various domains including images and speech [7]. They are capable of automatically learning features from high dimensional data without manual feature engineering.

The two main layers in CNNs are convolutional and max-pooling. The convolutional layers use a technique called weight sharing, where a set of weights are set to be equal. This reduces the numbers of weights that are stored. A convolutional layer is followed by a max-pooling layer which computes the maximum value of the selected set (pool) of adjacent neurons from the convolutional layer [17, 18]. The combination of a convolutional and max-pooling layer ensures that the output of the max-pooling layer is invariant to shifts in the input data, which is a very useful property for processing natural data [18]. The convolution is implemented using a filter with specific kernel size which defines the number of nodes that share weights.

The architecture of our CNN model is shown in Fig. 3. It consists of three 1-D convolutional layers, each followed by a max pooling layer, and one fully connected layer with a soft-max activation function. There are two output nodes corresponding to the two classes and outputting the probability for each class (normal or abnormal). The new example is assigned to the class with the higher probability.

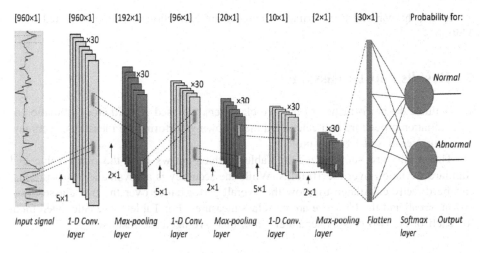

[960×1]   [960×1]   [192×1]   [96×1]   [20×1]   [10×1]   [2×1]   [30×1]   Probability for:

Input signal   1-D Conv.   Max-pooling   1-D Conv.   Max-pooling   1-D Conv.   Max-pooling   Flatten   Softmax   Output
              layer   layer   layer   layer   layer   layer   layer

Fig. 3. Architecture of the CNN model

Each convolutional layer consists of 30 filters with $[5 \times 1]$ kernel size, 5 strides and a Rectified Linear Unit (ReLU) activation function. The type of activation function was selected after evaluating the performance of several options (sigmoid, soft-max and ReLU). The best accuracy and fastest training time were obtained using the ReLU activation function. Each max pooling layer consists of $[2 \times 1]$ nodes.

The CNN was trained using the backpropagation algorithm and Adam optimizer [19], to optimize the categorical cross entropy objective function. To avoid overfitting, a dropout [20] was employed during the weight optimization at all layers. Dropout includes randomly selecting a fraction of neurons at each layer at each training epoch and not including them in the weight optimization.

The CNN was implemented using the Keras framework [21] with Tensorflow backend [22]. For the experimental evaluation, we used the following CNN parameters: batch size = 100, dropout rate = 0.5, maximum number of epochs 500.

4.2 Feature Extraction and SVM Model

SVM is a state-of-the-art machine learning algorithm [23, 24]. It uses an optimization method to find the maximum margin hyperplane separating two classes in a binary classification task.

To extract features for SVM, we followed the method proposed in [10]. We firstly applied a Haar wavelet packet transform [25] to the raw data of the nasal flow signal and then extracted the same 15 statistical features as in [10], calculated for the wavelet coefficients x, namely: mean value, variance, standard deviation, kurtosis, skewness and geometrical mean and mean absolute deviation for both x and x^2, and the log mean value for x^2.

For the experimental evaluation, we used SVM with RBF kernel and wavelet decomposition at level 4. These parameters were selected empirically.

SVM, the wavelet transformation and feature extraction were implemented using Matlab.

5   Results and Discussion

To evaluate the performance of the two classifiers, we used 10-fold cross validation as an evaluation procedure and calculated the following performance measures: accuracy, precision, recall and F1-measure.

The results are summarized in Table 1 which shows both the mean value and standard deviation over the 10 folds. While the accuracy is an evaluation measure that can be directly calculated to show the overall performance over the two classes, precision, recall and the F1 score are per-class measures. For Table 1, we calculated them for each of the two classes separately, and then averaged the results.

Table 1. Performance of CNN and SVM (10-fold cross validation) – mean value and standard deviation

	Accuracy [%]	Precision [%]	Recall [%]	F1 score [%]
CNN	74.70 ± 1.43	74.50 ± 1.49	74.70 ± 3.75	74.60 ± 2.38
SVM	72.00 ± 1.22	72.11 ± 1.27	72.00 ± 1.51	72.00 ± 1.25

We can see that CNN outperformed SVM on all performance measures – e.g. it achieved 75% accuracy while SVM achieved 72%. As a baseline, we can consider the performance of a classifier that simply predicts the majority class, which in our case will be 50% accuracy as the two classes are equally distributed. Thus, both CNN and SVM significantly outperform this baseline.

The big advantage of CNN over SVM is that it doesn't require feature engineering to construct and select useful features – it takes as an input the original 960 features and can automatically select the most informative ones. In comparison, the features used in SVM required the application of wavelet transform and then feature extraction and selection.

An accuracy of 75% is a very promising result especially given that we used only one type of signal – the nasal airflow. In future work, we plan to use other types of signals such as the thoracic, abdominal and chest flows.

The low standard deviations in Table 1 indicate low variability of the results across the 10 cross validation runs, for both CNN and SVM and all performance measures.

Figure 4 shows the precision, recall and F1 score for each class separately, for both CNN and SVM. We can see that there are some differences between CNN and SVM – CNN has higher precision for the abnormal class than the normal one, and SVM shows the opposite pattern. Similarly, the CNN has higher recall for the normal class than the abnormal one, and this pattern is the opposite for SVM. A further cost-sensitive evaluation can be conducted by considering the cost of misclassifying normal as abnormal events and vice versa.

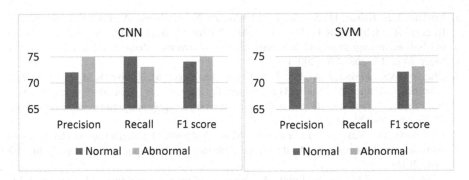

Fig. 4. Performance of CNN and SVM for each class separately (10-fold cross validation)

6   Conclusions

In this paper, we proposed a novel CNN model for detecting obstructive sleep apnea-hypopnea events, using data from a single respiratory channel: the nasal airflow. The CNN model learns the features automatically, without the need for feature extraction and selection. We compared its performance with a state-of-the-art SVM model using pre-selected statistical features, extracted after the application of wavelet transformation, which has shown good results in previous work. The evaluation was conducted using data for 100 people, consisting of 24,480 samples (50% normal and 50% abnormal) with a dimensionality of 960. Our results showed that CNN outperformed SVM on all performance measures, achieving an overall accuracy and F-score of 75% compared to 72% for SVM. The main advantage of CNN over SVM is the ability to learn informative features automatically from the high dimensional data, without the need for further signal processing and feature engineering.

Future work will include using more than one respiratory signal, e.g. using also the thoracic, abdominal and chest flows. We also plan to increase the number of patients – in this work we used data from only 100 patients as a pilot study, in order to investigate the potential of CNN. These extensions may further improve the accuracy and lead to the development of better methods and tools for sleep apnea detection based on CNNs.

Acknowledgments. The MESA sleep dataset was supported by the National Heart, Lung, and Blood Institute (NHLBI) at the National Institutes of Health. It is available through NHLBI National Sleep Research Resource at https://www.sleepdata.org/datasets/mesa. MESA Sleep was supported by contract NHLBI R01 L098433.

References

1. Mannarino, M.R., Filippo, F.D., Pirro, M.: Obstructive sleep apnea syndrome. Eur. J. Intern. Med. **23**, 586–593 (2012)
2. McNicholas, W.T.: Diagnosis of obstructive sleep apnea in adults. Proc. Am. Thoractic Soc. **5**, 154–160 (2008)

3. Epstein, L.J., Kristo, D., Strollo, P.J., Friedman, N., Malhotra, A., Patil, S.P., Ramar, K., Rogers, R., Schwab, R.J., Weaver, E.M., Weinstein, M.D.: Clinical guideline for the evaluation, management and long-term care of obstructive sleep apnea in adults. J. Clin. Sleep Med. **15**, 263–276 (2009)

4. Chesson, A.L., Ferber, R.A., Fry, J.M., Grigg-Damberger, M., Hartse, K.M., Hurwitz, T.D., Johnson, S., Littner, M., Kader, G.A., Rosen, G., Sangal, R.B., Schmidt-Nowara, W., Sher, A.: Practice parameters for the indications for polysomnography and related procedures. Sleep **20**, 406–422 (1997)

5. Karamanli, H., Yalcinoz, T., Yalcinoz, M.A., Yalcinoz, T.: Prediction model based on artificial neural networks for the diagnosis of obstructive sleep apnea. Sleep Breath. **20**, 509–514 (2016)

6. Längkvist, M., Karlsson, L., Loutfi, A.: A review of unsupervised feature learning and deep learning for time-series modeling. Pattern Recogn. Lett. **42**, 11–24 (2014)

7. Krizhevsky, A., Sutskever, I., Hinton, G.E.: ImageNet classification with deep convolutional neural networks. In: Advances in Neural Information Processing Systems, pp. 1097–1105 (2012)

8. Abdel-Hamid, O., Mohamed, A.-R., Jiang, H., Penn, G.: Applying convolutional neural networks concepts to hybrid NN-HMM model for speech recognition. In: IEEE International Conference on Acoustics, Speech and Signal Processing (ICASSP), pp. 4277–4280. IEEE (2012)

9. Wang, T., Wu, D.J., Coates, A., Ng, A.Y.: End-to-end text recognition with convolutional neural networks. In: 21st International Conference on Pattern Recognition (ICPR), pp. 3304–3308. IEEE (2012)

10. Maali, Y., Al-Jumaily, A.: Automated detecting sleep apnea syndrome: a novel system based on genetic SVM. In: 11th International Conference on Hybrid Intelligent Systems (HIS). IEEE (2011)

11. Maali, Y., Al-Jumaily, A.: Hierarchical parallel PSO-SVM based subject-independent sleep apnea classification. In: Huang, T., Zeng, Z., Li, C., Leung, C.S. (eds.) ICONIP 2012 Part IV. LNCS, vol. 7666, pp. 500–507. Springer, Heidelberg (2012). doi:10.1007/978-3-642-34478-7_61

12. Fontenla-Romero, O., Guijarro-Berdiñas, B., Alonso-Betanzos, A., Moret-Bonillo, V.: A new method for sleep apnea classification using wavelets and feedforward neural networks. Artif. Intell. Med. **34**, 65–76 (2005)

13. Avcı, C., Akbaş, A.: Sleep apnea classification based on respiration signals by using ensemble methods. Bio-Med. Mater. Eng. **26**, S1703–S1710 (2015)

14. Morsy, A.A., Al-Ashmouny, K.M.: Sleep apnea detection using an adaptive fuzzy logic based screening system. In: 27th Annual International Conference of the Engineering in Medicine and Biology Society (IEEE-EMBS), pp. 6124–6127. IEEE (2006)

15. Al-Ashmouny, K.M., Morsy, A.A., Loza, S.F.: Sleep apnea detection and classification using fuzzy logic: clinical evaluation. In: 27th International Conference of the Engineering in Medicine and Biology Society (IEEE-EMBS). IEEE (2006)

16. Dean, D.A., Goldberger, A.L., Mueller, R., Kim, M., Rueschman, M., Mobley, D., Sahoo, S. S., Jayapandian, C.P., Cui, L., Morrical, M.G., Surovec, S., Zhang, G.Q., Redline, S.: Scaling up scientific discovery in sleep medicine: the national sleep research resource. Sleep **5**, 1151–1164 (2016)

17. Le, Q.V.: A Tutorial on Deep Learning Part 2: Autoencoders, Convolutional Neural Networks and Recurrent Neural Networks. Google Brain (2015)

18. Hinton, G.E., Srivastava, N., Krizhevsky, A., Sutskever, I., Salakhutdinov, R.R.: Improving neural networks by preventing co-adaptation of feature detectors. arXiv preprint arXiv:1207.0580 (2012)

19. Kingma, D., Ba, J.: Adam: a method for stochastic optimization. arXiv preprint arXiv:1412. 6980 (2014)
20. Srivastava, N., Hinton, G., Krizhevsky, A., Sutskever, I., Salakhutdinov, R.: Dropout: a simple way to prevent neural networks from overfitting. J. Mach. Learn. Res. **15**, 1929–1958 (2014)
21. Tensorflow. https://www.tensorflow.org/. Accessed 08 Aug 2017
22. Keras - The Python Deep Learning Library. https://www.tensorflow.org/. Accessed 08 Aug 2016
23. Suykens, J.A., Vandewalle, J.: Least squares support vector machine classifiers. Neural Process. lett. **9**, 293–300 (1999)
24. Joachims, T.: Learning to Classify Text Using Support Vector Machines - Methods, Theory, and Algorithms. Kluwer Academic Publishers, Boston (2002)
25. Stanković, R.S., Falkowski, B.J.: The Haar wavelet transform: its status and achievements. Comput. Electr. Eng. **29**, 25–44 (2003)
26. Tian, J., Liu, J.: Apnea detection based on time delay neural network. In: 27th Annual International Conference of the Engineering in Medicine and Biology Society, IEEE-EMBS 2005, pp. 2571–2574. IEEE (2005)

A Deep Learning Method to Detect Web Attacks Using a Specially Designed CNN

Ming Zhang[✉], Boyi Xu, Shuai Bai, Shuaibing Lu, and Zhechao Lin

National Key Laboratory of Science and Technology on Information System
Security, Beijing Institute of System Engineering, Beijing, China
zhangming2013@alumni.sjtu.edu.cn

Abstract. With the increasing information sharing and other activities conducted on the World Wide Web, the Web has become the main venue for attackers to make troubles. The effective methods to detect Web attacks are critical and significant to guarantee the Web security. In recent years, many machine learning methods have been applied to detect Web attacks. We present a deep learning method to detect Web attacks by using a specially designed CNN. The method is based on analyzing the HTTP request packets, to which only some preprocessing is needed whereas the tedious feature extraction is done by the CNN itself. The experimental results on dataset HTTP DATASET CSIC 2010 show that the designed CNN has a good performance and the method achieves satisfactory results in detecting Web attacks, having a high detection rate while keeping a low false alarm rate.

Keywords: Web attacks · Deep learning · CNN

1 Introduction

The Internet has brought great convenience and happiness to people's life. The World Wide Web (abbreviated the Web) is the primary tool for billions of people to interact with the Internet and has made large contributions to the development of the Information Age. However, people are suffering threats and losses increasingly from the Internet. The Cyber-attacks make waves more and more frequently. What is worse, with the increasing information sharing and other activities conducted on the World Wide Web, the Web has become the main venue for attackers to engage in a range of cybercrimes. As early as 2007, Symantec Corporation had observed that instead of trying to penetrate networks with high-volume broadcast attacks, attackers have adopted stealthier, more focused techniques targeting computers through the World Wide Web, and the majority of effective malicious activities have become Web-based [1]. Another security company, Cenzic, reported in 2014 that 96% of the tested internet applications had vulnerabilities with a median of 14 per application, resulting in that hackers are increasingly focusing on and are succeeding with application layer attacks [2]. It is no doubt that Web security deserves enough attention.

There are a number of technical solutions to guarantee the Web security, including Web application security scanners, penetration testing, fuzzing tools used for input testing, Web application firewalls (WAF), Web intrusion detection systems (Web IDS)

© Springer International Publishing AG 2017
D. Liu et al. (Eds.): ICONIP 2017, Part V, LNCS 10638, pp. 828–836, 2017.
https://doi.org/10.1007/978-3-319-70139-4_84

and so on. For Web protection systems, having an effective method that can inspect Web traffic and detect attacks differing from normal behaviors is crucial and fundamental. There are two basic methods to detect Web attacks, the signature-based [3] and the anomaly-based [4]. The signature-based method builds the detection model from known attacks and any behavior having the corresponding attack signatures is identified as an attack. On the contrary, the anomaly-based method creates a profile from normal behaviors and any violation is identified as an attack. Obviously, both of the methods must have enough characterization and generalization ability of abnormal or normal behaviors, whereas it is difficult to do that in practice. With the popularity of machine learning, especially the rise of deep learning, it is possible to let machines learn features and patterns from data and then automatically distinguish different categories of things. In recent years, lots of machine learning methods have been applied to detect Web attacks.

In this paper, we present a method based on deep learning to detect Web attacks, strictly speaking, to detect server-side attacks. We describe a specially designed convolutional neural network and expound the steps and details to detect Web attacks.

The rest of the paper is organized as follows. Some related work is introduced in Sect. 2. The method based on deep learning to detect Web attacks is described in Sect. 3. Experimental results and discussions are presented in Sect. 4. Finally, Sect. 5 concludes the paper.

2 Related Work

Kruegel et al. have presented a multi-model approach to detect Web attacks in [5]. The approach analyzes HTTP requests and uses a number of different models built on different features, including attribute length, attribute character distribution, structural inference, invocation order and so on.

Ma et al. [6] have explored online learning approaches for detecting malicious Web sites using lexical and host-based features of the associated URLs. Their work is to protect users (or clients) from scams, whereas our study is on detecting server-side attacks.

Torrano et al. [7] have proposed an anomaly-based approach to detect intrusions in Web traffic. The approach relies on a XML file to classify the incoming requests as normal or anomalous.

Corona et al. [8] have presented a multiple classifier system to detect Web attacks by modeling legitimate requests. The system employs a set of predefined models, which are established on different message fields in HTTP requests and built on two basic models: the statistical distribution model and the Hidden Markov Model.

Zolotukhin et al. [9] have proposed an anomaly detection method for Web attacks through analysis of HTTP logs. The method employs the n-gram models to extract relevant features from three fields in HTTP logs, including Web resources, query attributes and user agents. Correspondingly, three machine learning algorithms are used, namely, Support Vector Data Description (SVDD), K-means, and Density-Based Spatial Clustering of Applications with Noise (DBSCAN).

Choras and Kozik [10] have proposed a machine learning approach to model normal behaviors of Web applications and to detect Cyber-attacks. The model is based on information obtained from HTTP requests and consists of patterns that are obtained using graph-based segmentation technique and dynamic programming.

Saxe and Berlin [11] have exposed a deep learning approach to a number of security detection problems including the malicious URLs detection. Similar to our work, they also use the Convolutional Neural Network, but the embedding approaches and the network architectures are different.

3   Method

3.1   Preprocessing HTTP Requests

The HTTP protocol is the foundation of data communication for the Web. The client submits a HTTP request message to the server and the server returns a HTTP response message to the client. If the server is attacked, that means it receives one or more malicious request messages. Based on this, the detection method is designed by inspecting the HTTP request packets (messages) to detect the server-side Web attacks. Because Web attacks detection belongs to the application-layer security solutions and it works after the packets are parsed, our detection method is also available for communications using HTTPS protocol.

An HTTP request message consists of a request line, several request header fields and an optional message body (for POST request messages). Figure 1 is an example of a GET request message obtained from the dataset HTTP DATASET CSIC 2010 (described in Sect. 4).

```
GET http://localhost:8080/iisstart.htm HTTP/1.1
User-Agent: Mozilla/5.0
Pragma: no-cache
Cache-control: no-cache
Accept: text/xml,application/xml;q=0.9,text/plain;q=0.8
Accept-Encoding: x-gzip, x-deflate, gzip, deflate
Accept-Charset: utf-8, utf-8;q=0.5, *;q=0.5
Accept-Language: en
Host: localhost:8080
Cookie: JSESSIONID=11F98280E08EE19274786F4EDDDC821F
Connection: close
```

Fig. 1. A GET request message example.

The request line is the first line of the HTTP request message and comprises three parts: the HTTP-method, the HTTP-url (URL for short) and the HTTP-version. Furthermore, we narrow the detected focus to the URL in an HTTP request message (for POST request messages, the URL is defined as the combination of the HTTP-url and the message body). There are two reasons for doing this. One is that the vast majority of

Web attacks are implemented by manipulating the URLs. The other is for computational convenience. But without loss of generality, our detection method can be applied to the whole HTTP request message.

Suppose a URL is as Fig. 2 shows.

```
http://localhost:8080/tienda1/publico/vaciar.jsp?B2=Vaciar
+carrito%27%3B+DROP+TABLE+usuarios%3B+SELECT+*+FROM+datos+
WHERE+nombre+LIKE
```

Fig. 2. A URL example.

The preprocessing is to segment the URL to a sequence of words (including the string containing some necessary special characters). The URL can be split into words by special characters like "/", "&", "=", "+", etc. If the URL is encoded, the decoding may be operated first. Theoretically, the preprocessing should reflect the difference between abnormal and normal URLs. But in practice, the preprocessing may be a continuously trying and optimizing process. For the above URL, the sequence of words after preprocessing can be as Fig. 3 shows (separated by commas).

```
http, localhost, 8080, tienda1, publico, vaciar.jsp, B2,
Vaciar, carrito, DROP, TABLE, usuarios, SELECT, *, FROM,
datos, WHERE, nombre, LIKE
```

Fig. 3. Sequence of words after preprocessing the URL.

3.2 Word Embedding

Word embedding is used to map the words to vectors, which are the inputs to the Convolutional Neural Network (CNN, described in the next subsection). There are many branches and research groups working on word embeddings. For example, a team at Google led by Tomas Mikolov created *word2vec*, a toolkit that can be used directly to generate vectors. Instead of generated by existing word embedding tools, the words' embedding vectors in our Web attacks detection method are learned in the training process. In fact, there is an embedding layer joint with the CNN, and the embedding vectors are optimized through back-propagations of the whole network. We believe that the embedding vectors generated by this way are more helpful to the detection of Web attacks, because they are task-specific and more reflective of their semantic meaning whereas the vectors generated by third-party tools are relatively static. We have also confirmed our supposition in experiments.

3.3 The Specially Designed CNN

The architecture of the CNN specially designed to detect Web attacks is as Fig. 4 shows. It is a feed-forward neural network and is formed by four distinct layers: the convolution layer, the max-pooling layer, the fully-connected layer and the Softmax layer. The input to the CNN is a matrix composed of vectors embedded from the word sequence which are obtained by preprocessing the URL. Suppose the length of the word sequence is l and the dimension of the embedding vector is k, so the size of the input matrix is $l \times k$. The details of each layer are as follows.

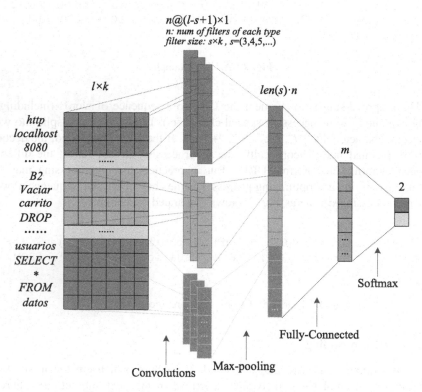

Fig. 4. Architecture of the specially designed CNN.

Convolution Layer

The convolution layer convolves the input matrix with different sizes of filters (or kernels). The filter sizes are set to $s \times k$, $s = (3, 4, 5, \ldots)$, with k equaling to the width of the input matrix and s taking different values. So the types of filters can be represented as $len(s)$. Every filter walks through the input matrix with stride being one and is convolved with a local area of size $s \times k$, thus producing a feature map (the convolved result) of size $(l - s + 1) \times 1$. For each type of filters, there are n filters of the same size to convolve with the input matrix, so the produced results are n feature maps, denoted as $n@(l - s + 1) \times 1$. Because there are $len(s)$ types of filters, the full output of the convolution layer has $len(s)$ clusters actually, with each cluster having n feature maps of size $(l - s + 1) \times 1$. The convolution layer exploits the spatially-local correlation and learns filters that activate when detecting some specific types of features at some local areas of the input. Different sizes of filters can extract more rich features by sliding on different local areas of the input.

Max-Pooling Layer

The max-pooling layer reduces the dimensionality of each feature map outputted by the convolution layer but retains the most important information. It takes the largest element from each feature map, and then concatenates them together to produce a vector. According to the types and numbers of filters in the convolution layer, we can get the length of the max-pooled vector is $len(s) \cdot n$. Note that though the inputs may have different lengths, but the results of the max-pooling layer are always of the same length.

Fully-Connected Layer

The fully-connected layer has m $(m < len(s) \cdot n)$ neurons, with each neuron connected to all the neurons in the max-pooling layer. The fully connected layer does high-level reasoning and extracts high-order features of the input. Because $m < len(s) \cdot n$, it also has the function of reducing dimensionality.

Softmax Layer

The CNN designed for detecting Web attacks is to recognize whether the HTTP request message is normal or abnormal, so the Softmax layer (also the output layer) has 2 neurons with each representing normal or abnormal. The neurons in the Softmax layer are also fully connected to the neurons in the previous fully-connected layer. The output values of the 2 neurons are set to lie between 0 and 1 and sum to 1. The predicted class of the URL is decided by which neuron outputs a larger value.

Some other tricks may be considered. For example, one can add dropout after the max-pooling layer or other layers to avoid overfitting at the training stage. The choice of the loss function is also critical. For our designed CNN, the cross-entropy loss is recommended.

4 Experiments and Results

To evaluate the effectiveness of the method for detecting Web attacks, we conducted experiments on the dataset HTTP DATASET CSIC 2010 [12].

4.1 Data Preparation

The HTTP DATASET CSIC 2010 dataset contains thousands of Web requests automatically generated by the Information Security Institute of CSIC (Spanish Research National Council), and has been widely used for testing the Web attacks detection systems. The dataset contains 36,000 normal requests and 24,668 abnormal requests. The abnormal requests include Web attacks such as SQL injection, buffer overflow, information gathering, files disclosure, CRLF injection, XSS, server side include, parameter tampering and so on.

We randomly selected about 70% of the dataset as training data, 5% as the validation data, the rest 25% as the test data. The data distribution is shown in Table 1.

Table 1. Experimental data distribution.

	Training	Validation	Test
Normal	25,200	1,800	9,000
Abnormal	17,268	1,233	6,167
Total	42,468	3,033	15,167

4.2 Model Parameters and Evaluating Criteria

Based on data characteristics and empirical experiences, we set parameters of the CNN as follows. For the input matrix, the height l depends on the length of the word sequence, and the width k, i.e. the dimension of the embedding vector is set to 128. In the convolution layer, for the filter size $s \times k$, let $s = (3, 4, 5, 6)$ and $k = 128$, so $len(s) = 4$. For each type of filters, the number n is set to 128. The neurons number m of the fully-connected layer is set to 256. We add dropout after the max-pooling layer at the training stage with the keeping probability being 0.5 and choose cross-entropy as the loss function.

To evaluate the effectiveness of the method for detecting Web attacks, we use three criteria: the *detection rate*, the *false alarm rate* and the *accuracy*. Following the notions usually used in machine learning methods, we use *TP*, *FP*, *TN* and *FN* to represent the number of true positives, false positives, true negatives and false negatives respectively. The *detection rate* (i.e. *true positive rate*) is defined as the proportion of the detected abnormal requests accounting for the total abnormal ones. The *false alarm rate* (i.e. *false positive rate*) is defined as the ratio between the number of normal requests wrongly categorized as abnormal and the total number of actual normal requests. An ideal Web attacks detection method must have both the high *detection rate* and the low *false alarm rate*. The *accuracy* is defined as the proportion of requests including normal and abnormal to be correctly classified. The formulas of the three criteria are presented below.

$$detection\ rate = \frac{TP}{TP + FN}. \tag{1}$$

$$false\ alarm\ rate = \frac{FP}{FP + TN}. \tag{2}$$

$$accuracy = \frac{TP + TN}{TP + FP + TN + FN}. \tag{3}$$

4.3 Results and Discussions

We trained the CNN for 10 epochs using batch training approach. The batch size is set as 64. We recorded the training accuracy and loss every one step and recorded the validation accuracy and loss every 100 steps. The trends of the metrics are presented in Fig. 5. Figure 5(a) shows the accuracy trends, where the orange curve represents the training accuracy and the dark cyan represents the validation accuracy. We can see that after about 4,000 steps (about 6 epochs) of training, both the training and validation accuracies have achieved above 95%. Figure 5(b) shows the loss trends, where the orange curve represents the training loss and the dark cyan represents the validation loss. Obviously, both the training and validation losses decrease rapidly towards 0. Such trends of accuracy and loss reflect the good performance of the CNN.

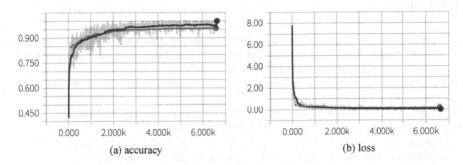

(a) accuracy   (b) loss

Fig. 5. Accuracy and loss in the training stage.

After 10 epochs of training, we run the trained CNN on test data to evaluate its ability of detecting Web attacks. As Table 2 shows, the *detection rate* is 93.35%, the *false alarm rate* is 1.37%, and the *test accuracy* is 96.49%. This demonstrates that with a certain amount of training, the CNN has achieved satisfactory results in detecting Web attacks, having a high detection rate while keeping a low false alarm rate.

Table 2. Evaluating results.

Detection rate	False alarm rate	Test accuracy
93.35%	1.37%	96.49%

5 Conclusion

A deep learning method to detect Web attacks is explored, which is based on a specially designed convolutional neural network. The method is able to detect various Web attacks through inspecting the HTTP request packets. Firstly, data preprocessing is studied, which chooses useful information from HTTP request packets and produce lots of word sequences. Secondly, the embedding approach used to map words to vectors is studied. The embedding vectors are learned in the training stage and not generated by the existing word embedding tools. Finally, a special CNN consisted of various layers is designed. It is able to extract features automatically, and then classify the HTTP request packets to normal or abnormal class. We conducted experiments on the dataset HTTP DATASET CSIC 2010 to evaluate the effectiveness of the method. The results show that the designed CNN can be trained easily and the detection method achieves a high detection rate with few false alarms in detecting Web attacks.

For reducing computational complexity, the method in this article only focuses on detecting Web attacks hidden in URLs. Future work will try modeling the whole HTTP request messages. Other embedding approaches and neural networks are also worth studying.

References

1. Symantec Internet Security Threat Report: Trends for July–December 2007. http://eval. symantec.com/mktginfo/enterprise/white_papers/b-whitepaper_exec_summary_internet_ security_threat_report_xiii_04-2008.en-us.pdf
2. Application Vulnerability Trends Report 2014. http://www.cenzic.com/downloads/Cenzic_ Vulnerability_Report_2014.pdf
3. Axelsson, S.: Research in intrusion-detection systems: a survey. Technical report 98–17, Department of Computer Engineering, Chalmers University of Technology (1998)
4. Garcia, T.P., Diaz, V.J., Macia, F.G., et al.: Anomaly-based network intrusion detection: techniques, systems and challenges. Comput. Secur. **28**(1), 18–28 (2009)
5. Kruegel, C., Vigna, G., Robertson, W.: A multi-model approach to the detection of web-based attacks. Comput. Netw. **48**(5), 717–738 (2005)
6. Ma, J., Saul, L.K., Savage, S., et al.: Identifying suspicious URLs: an application of large-scale online learning. In: Proceedings of 26th Annual International Conference on Machine Learning, pp. 681–688 (2009)
7. Torrano, G.Z., Perez, V.A., Maranon, G.A.: An anomaly-based approach for intrusion detection in web traffic. J. Inf. Assur. Secur. **5**(4), 446–454 (2010)
8. Corona, I., Tronci, R., Giacinto, G.: SuStorID: a multiple classifier system for the protection of web services. In: Proceedings of IEEE 21st International Conference on Pattern Recognition (ICPR), pp. 2375–2378 (2012)
9. Zolotukhin, M., Hamalainen, T., Kokkonen, T., et al.: Analysis of http requests for anomaly detection of web attacks. In: Proceedings of IEEE 12th International Conference on Dependable, Autonomic and Secure Computing (DASC), pp. 406–411 (2014)
10. Choras, M., Kozik, R.: Machine learning techniques applied to detect cyber attacks on web applications. Log. J. IGPL **23**(1), 45–56 (2015)
11. Saxe, J., Berlin, K.: eXpose: a character-level convolutional neural network with embeddings for detecting malicious URLs, file paths and registry keys. arXiv preprint arXiv:1702.08568 (2017)
12. HTTP DATASET CSIC 2010. http://www.isi.csic.es/dataset/

An Integrated Chaotic System with Application to Image Encryption

Jinwen He[1,2], Rushi Lan[2(✉)], Shouhua Wang[1], and Xiaonan Luo[2]

[1] School of Communication and Information Technology,
Guilin University of Electronic Technology, Guilin 541004, China
[2] Key Laboratory of Intelligent Processing of Computer Image and Graphics,
Guilin University of Electronic Technology, Guilin 541004, China
rslan2016@163.com

Abstract. Chaotic maps are widely applied in many applications. This paper proposes an integrated chaotic system (ICS) to improve the performance of some representative chaotic maps. ICS conducts cascade and nonlinear combination operations to three seed maps such that it has more complex chaotic behaviors and high security levels. A new image encryption algorithm is also developed using ICS. Simulation results on different types of images and security analysis demonstrate that the proposed approach has satisfactory properties in image encryption.

Keywords: Chaotic map · Image encryption · Integrated chaotic system

1 Introduction

Image encryption is an attractive research topic in the fields of information security, image processing and signal processing. So far, a large number of image encryption algorithms have been developed from different perspectives. Among these algorithms, the chaos-based encryption approaches are particularly popular [1,2]. The reason lies in that chaotic maps have many distinctive properties, such as unpredictability, ergodicity, and initial value sensitivity [3]. Owing to these excellent properties, chaos-based image encryption attracts increasing research interests and acquires sustained developments [4].

Existing chaotic maps are usually divided into two categories: one-dimension (1D) and high-dimension (HD). The 1D chaotic maps include Sine map, Tent map, Logistic map, Gaussian map [17], and Dyadic transform [5], etc. HD chaotic maps consider at least two variables [10,12]. Some new-generated 1D chaotic maps [11,16] and HD chaotic maps [9] are proposed with the rapid development of chaotic encryption technologies.

To 1D chaotic maps, their chaotic ranges are limited and easy to predict because of their simple structures. Although some chaotic maps have been proposed to broaden chaotic range [6], the complexity of the whole structure is still uncomplicate. Compared with the 1D chaotic maps, HD chaotic maps usually

© Springer International Publishing AG 2017
D. Liu et al. (Eds.): ICONIP 2017, Part V, LNCS 10638, pp. 837–847, 2017.
https://doi.org/10.1007/978-3-319-70139-4_85

have better chaotic behaviors and are hard to predict. Even though, their computation cost and implementation difficulty are very high [8,15,18]. Some 1D chaotic maps have complex structures such as cascade chaotic system [7] and nonlinear system [6]. The former one cascades existing chaotic maps to form a complicate chaotic map with limited chaotic range, while the latter one has a wider chaotic range with a lower security complexity.

In this paper, we propose an integrated chaotic system (ICS) for image encryption. ICS integrates three seed chaotic maps via cascade and nonlinear combination operations. Three example chaotic maps, generated by ICS, are also given and studied. Compared with some existing chaotic maps, the derived ones have better chaotic behaviors. In summary, the main contributions of this work are three-fold. First, we develop ICS with a complex structure to generate chaotic maps. Second, we further propose a novel image encryption based on ICS. Third, several experiments are carried out to evaluate the proposed methods from different aspects, and satisfactory performance has been achieved.

2   Proposed Chaotic System

This section presents the proposed ICS in detail. As aforementioned, ICS is an integrated version of some basic chaotic maps. Here we will first introduce these representative maps, and then give the definition and examples of ICS successively.

2.1   Used Control/Seed Maps

This part briefly reviews three representative chaotic maps, namely Sine map, Tent map, and Logistic map, which will be used as control/seed maps in proposed system. These basic maps are defined as follows [7]:

$$x_{n+1} = \mathbb{S}(x) = u\sin(\pi x_n), \tag{1}$$

$$x_{n+1} = \mathbb{T}(x) = \begin{cases} 2ux_n, & x_n < 0.5, \\ 2u(1 - x_n), & x_n \geq 0.5, \end{cases} \tag{2}$$

$$x_{n+1} = \mathbb{L}(x) = 4u(1 - x_n), \tag{3}$$

where the parameter $u \in [0, 1]$. Note that the outputs of all the above maps are normalized to [0, 1]. Figure 2(a)–(c) show the bifurcation diagrams of these maps. We can observe that their outputs are distributed in limited chaotic ranges.

2.2   Definition of ICS

To address the limitation of existing chaotic maps, we develop a novel chaotic system that consists of more complex structure to enlarge the chaotic ranges. Let $\mathbb{F}(x)$, $\mathbb{G}(x)$, and $\mathbb{H}(x)$ be three seed chaotic maps, which can be anyone of

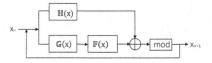

Fig. 1. Structure of proposed ICS.

Logistic map, Sine map, and Tent map. Mathematically, the proposed ICS is defined as follows:

$$x_{n+1} = (\mathbb{F}(\mathbb{G}(x_n)) + \mathbb{H}(x_n)) \mod 1, \tag{4}$$

where x_n is the iteration value, x_{n+1} is the output of the whole chaotic system. The structure of ICS is depicted in Fig. 1. From Eq. (4) and Fig. 1, we can observe that ICS integrates three seed maps by two operations. First, the cascade operator is applied to $\mathbb{F}(x)$ and $\mathbb{G}(x)$. After that, a nonlinear combination operator, including addition and modulo operations, is used to $\mathbb{F}(\mathbb{G}(x))$ and $\mathbb{H}(x)$. These two operators ensure ICS has better chaotic behaviors.

2.3 Examples of ICS

As shown in Eq. (4), setting $\mathbb{F}(x)$, $\mathbb{G}(x)$, and $\mathbb{H}(x)$ with different seed maps, ICS is able to produce several new chaotic maps. For convenience, we denote the used Logistic, Sine, and Tent maps by $\mathbb{L}$, $\mathbb{S}$, and $\mathbb{T}$ respectively. Three novel chaotic maps will be given here as examples of ICS. We name these maps by $\mathbb{LT}$-$\mathbb{S}$, $\mathbb{ST}$-$\mathbb{L}$, and $\mathbb{LT}$-$\mathbb{L}$ for short. Table 1 shows the definitions and detailed descriptions of three newly generated chaotic maps in detail. Figure 2(d)–(f) show the bifurcation diagrams of newly-generated chaotic maps too. It is apparently that the new-generated chaotic maps have wider chaotic ranges than their seed maps shown in Fig. 2(a)–(c).

Table 1. New chaotic maps generated by ICS.

New maps	Description	
LT − S	$x_{(n+1)} =$	$\begin{cases} (4u(2ux_n)(1-2ux_n) + (1-u)sin(\pi x_n)) \mod 1 & \text{for } x_n < 0.5 \\ (4u(2u(1-x_n))(1-2u(1-x_n)) + (1-u)sin(\pi x_n)) \mod 1 & \text{others} \end{cases}$
ST − L	$x_{(n+1)} =$	$\begin{cases} (usin(\pi 2ux_n) + (4-4u)x_n(1-x_n)) \mod 1 & \text{for } x_n < 0.5 \\ (usin(\pi 2u(1-x_n)) + (4-4u)x_n(1-x_n)) \mod 1 & \text{others} \end{cases}$
LT − L	$x_{(n+1)} =$	$\begin{cases} (4u2ux_n(1-2ux_n) + (4-4u)x_n(1-x_n)) \mod 1 & \text{for } x_n < 0.5 \\ (4u2u(1-x_n)(1-2u(1-x_n)) + (4-4u)x_n(1-x_n)) \mod 1 & \text{others} \end{cases}$

3   Image Encryption via ICS

This section introduces a new image encryption algorithm named ICS-IE using ICS. To this end, we first develop a ICS transform to change the locations of pixels in an image, and then introduce the detailed derivation of image encryption with ICS.

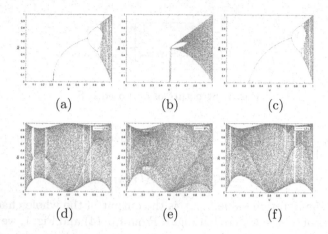

Fig. 2. Bifurcation diagrams of (a) Sine map, (b) Tent map, (c) Logistic map, (d) $\mathbb{LT}$-$\mathbb{S}$, (e) $\mathbb{ST}$-$\mathbb{L}$, and (f) $\mathbb{LT}$-$\mathbb{L}$ respectively.

3.1   ICS Transform

The aim of ICS transform is to increase the diffusion and confusion of an image by changing the locations of pixels. ICS includes two steps. The first step is to convert the chaotic sequence with the range of $[0, 1]$ into an integer sequence with a range of $[0, N]$. This transform can be represented as follows:

$$S_i = bin2dec(bitshift(dec2bin(\lfloor 1000X_i \rfloor + 2), -1)) \mod N, \qquad (5)$$

where $bin2dec$ is used to convert a binary number to decimal one, while $dec2bin$ is the reverse process. The term $bitshift(U, -1)$ (U is binary number) returns the value of U shifted to the left by one bit. X_i is a chaotic sequence with range of N generated by the ICS, and S_i is the iterate sequence with range of N. To improve the security level and unpredictability of ICS, a switch is further added to $\mathbb{F}(x)$ in Eq. (4). f_n is set as a control map to determine the switch. $\mathbb{F}_1(x)$ and $\mathbb{F}_2(x)$ are two choices. In this work, as an example, Tent map is chosen as the control map. $\mathbb{F}_1(x)$ and $\mathbb{F}_2(x)$ are set to Logistic map and Sine map. $\mathbb{G}(x)$ and $\mathbb{H}(x)$ are Tent map and Logistic map. So the definition of the example chaotic map is shown as follows:

$$x_{n+1} = \begin{cases} (4u(2ux_n)(1 - 2ux_n) + (4 - 4u)x_n(1 - x_n)) \mod 1 & f_n < 0.5, x_n < 0.5, \\ (4u(2ux_n)(1 - 2u(1 - x_n)) + (4 - 4u)x_n(1 - x_n)) \mod 1 & f_n < 0.5, x_n \geq 0.5, \\ (u\sin(\pi\, 2ux_n) + (4 - 4u)x_n(1 - x_n)) \mod 1 & f_n \geq 0.5, u\sin(\pi\, x_n) < 0.5, \\ (u\sin(\pi\, 2u(1 - x_n)) + (4 - 4u)x_n(1 - x_n)) \mod 1 & f_n \geq 0.5, u\sin(\pi\, x_n) \geq 0.5, \end{cases} \qquad (6)$$

The second step of ICS transform is to disturb the locations of pixels, which can be achieved by:

$$S = W_c^T I W_r^T, \qquad (7)$$

where $w_{c(i,j)} = \begin{cases} 1 & \text{for } (T_r(j), j) \\ 0 & others \end{cases}$ and $w_{r(k,l)} = \begin{cases} 1 & \text{for } (k, T_c(k)) \\ 0 & others \end{cases}$. W_r and W_c are the row and column matrices of the input image I, whose size is $M \times N$. T_r and T_c

are two sequences generated by Eq. (5) with length of M and N. In addition, i and j are in a range of $[0, N]$, while k and l are in a range of $[0, M]$. They are all integers. In the decryption process, the image I will be recovered by the following way:

$$I = (W_C^T)^{-1}I(W_r^T)^{-1}. \tag{8}$$

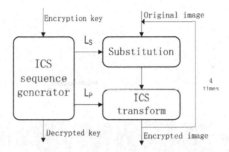

Fig. 3. The flowchat of ICS-IE algorithm.

3.2 The Derivation of ICS-IE

ICS-IE algorithm includes two major process: substitution and permutation. The flowchart and the pseudo code of the ICS-IE algorithm is shown in Fig. 3 and Algorithm 1. The encryption key k_e here is consist of f_0, x_0 and u. f_0 and x_0 are the initial values of f_n and x_n. u is a parameter in Eq. (6). f_0 should be updated for $L(L$ is set to 4 in this work) times according to the following formula:

$$f_0^i = \begin{cases} 1/2(f_0 + p), & \text{for } i = 1 \\ 1/2(f_0^{i-1} + x_0), & \text{for } i > 1 \end{cases} \tag{9}$$

where p is a random number with a range of $[0, 1]$, and x_0 is the initial value in the ICS process.

The substitution process encrypts the image in the follow manner:

$$E(m, n) = (\lfloor (X_s(k) \times F) \rfloor - I(m, n)) \mod F, \tag{10}$$

where $\lfloor . \rfloor$ is the floor function. $I(m, n)$ is the input image, and $E(m, n)$ is the output image of the substitution process. $X_s(k)$ is the ICS sequence for substitution. F is the maximum value of the input image. All parameters in the equation are integers. m and n respectively are with the range of $[0, M]$ and $[0, N]$. In the image decryption process, the output image $E(m, n)$ will be transformed to input the original one $I(m, n)$ by the following formula:

$$I(m, n) = (\lfloor (X_s(k) \times F) \rfloor - E(m, n)) \mod F. \tag{11}$$

In summary, the proposed ICS transform is used in permutation process, which can efficiently change the value and pixel positions of the image $E(m, n)$ after substitution process. The diffusion and confusion properties of encrypted images will get enhanced.

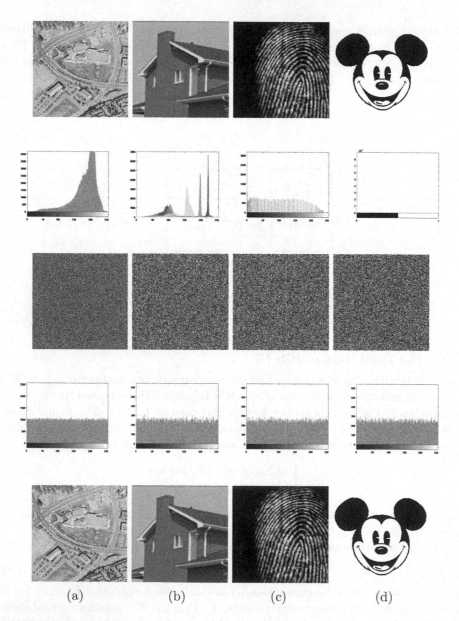

Fig. 4. Simulation results of different types of images. The first and second rows show original images and their histograms. The third and fourth rows show the ciphertext images and their histograms. The last row shows the decrypted image. (a) grayscale image. (b) color image. (c) biometrics image. (d) binary image.

Algorithm 1. ICS-IE

Input: Encryption key k_e, cycle times L, and original image with size of $M \times N$.

- Step 1: Initial the control parameter f_0, x_0, and u.
- Step 2: For $l = 1$ to L, do
    - (a) Generate the ICS sequence X_s with Eq. (6). The length of X_s is $L_s+L_p= M \times N + M + N$.
    - (b) $X_s \leftarrow (1{:}L_s)$ Conducts substitution operation with the Eq. (10).
    - (c) $X_s \leftarrow (L_p{:}L_s+L_p)$ Performs permutation operation with the ICS Transform.

Output: The encrypted image and decryption key.

---

4 Experiments

This section provides several experimental results to evaluate the proposed method. We first present the encryption results using different types of images, and then the security analysis of ICS-IE is given.

4.1 Encryption Results

In this part, we first show the encryption and decryption performance of the proposed Algorithm 1. Four types of images are considered here, including grayscale images, binary images, biometrics images, and color images. The encryption results are shown in Fig. 4(a)–(d). We can observe that all the encrypted images are noise-like ciphertext and unrecognized. These results indicate that ICS-IE algorithm is able to break the break the correlations of image pixels and achieves satisfactory performance in image encryption. As seen in Fig. 4, the histograms of encrypted images are uniform-distributed compared to the unevenly distributed original images.

4.2 Security Analysis

Security analysis is a significant aspect to evaluate an encryption algorithm. In this subsection, key sensitivity, differential attack, noise attack and data loss attack will be tested to show the properties of proposed ICS-IE. Several grayscale images are selected as test samples here.

Key Sensitivity Analysis. Key sensitivity test is to detect the variation of encryption result when there is a slight change (such as 10^{-14}) in the encryption keys. The key sensitivity simulation results are illustrated in Fig. 5. K_1 and K_2 are two encryption keys with the tiny difference of 10^{-14}. Figure 5(a) is the original image, encrypted respectively with K_1 and K_2 to form Fig. 5(b) and (c). Figure 5(d) shows their pixel-to-pixel differences. Figure 5(e) shows the decryption result of Fig. 5(b) with correct decryption key K_1. Figure 5(f) and (g) show the results obtained wrong keys and their pixel-to-pixel difference is shown in Fig. 5(e).

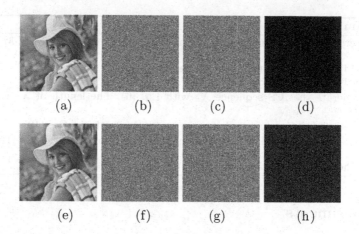

Fig. 5. Key sensitivity analysis. (a) Original image. (b) Encryption image C_1 with K_1. (c) Encryption image C_2 with K_2. (d) Difference between encryption images $|C_1-C_2|$. (e) Decryption image D_1. (f) Decryption image D_2 from C_1 with K_1. (g) Decryption image D_3 from C_1 with K_2. (h) Difference between two encryption images $|D_2-D_3|$.

These results indicate that ICS-IE algorithm has a high sensitivity to security keys in both encryption and decryption process. Only a tiny change of 10^{-14} can result in a significant difference in encryption/decryption results, which means the proposed ICS-IE algorithm has a large key space to defend the inimical deciphering.

Differential Attack. The unified average changed intensity ($UACI$) is usually used to test differential attack in image encryption. It measures the ability of resisting differential attack, and the definition is shown as follows:

$$UACI = \frac{1}{MN} \sum_{m=1}^{M} \sum_{n=1}^{N} [\frac{|E_1(m,n) - E_2(m,n)|}{L-1}] \times 100, \qquad (12)$$

E_1 and E_2 are two encrypted images with size of $M \times N$ and their original images only have one bit of pixel difference. L is the grayscale level of the image. For a 8-bit grayscale image, L will be 256. $UACI$ calculates the pixel changes between E_1 and E_2 by counting the average value of changed pixels.

In 8-bit grayscale image, the expected value of $UACI$ is 33.4635% [19]. Table 2 lists the $UACI$ values acquired by several different encryption schemes (10 experimental images are obtained from the SC-SIPI image database). The scores achieved by ICS-IE algorithm is extremely closed to the expected value. In conclusion, ICS-IE scheme has an excellent property of resisting the differential attacks.

Noise and Data Loss Attack. For an image encryption scheme, it should be flexible to resist the noise and data loss. Figure 6(b)–(c) show the different noise

Table 2. UACI scores of grayscale images for different image encryption algorithms.

Filename	Wu [14]	Liao [13]	Hua [7]	Bao [6]	ICS-IE
5.1.10	33.56	33.54	33.51	33.24	33.54
5.2.08	33.43	33.43	33.43	33.31	33.42
5.3.02	33.48	99.62	33.47	33.29	33.46
7.1.10	33.52	16.86	33.47	33.59	33.49
7.2.01	33.48	33.47	33.47	33.42	33.45
Boat.512	33.49	33.63	33.52	33.37	33.50
Elaine.512	33.44	33.44	33.38	33.37	33.49
Gray21.512	33.47	33.48	33.40	33.36	33.50
Numbers.512	33.54	33.45	33.42	33.77	33.49
Ruler.512	33.42	33.06	33.40	33.43	33.43
Mean	33.4826	38.5002	33.4463	33.4150	33.4770

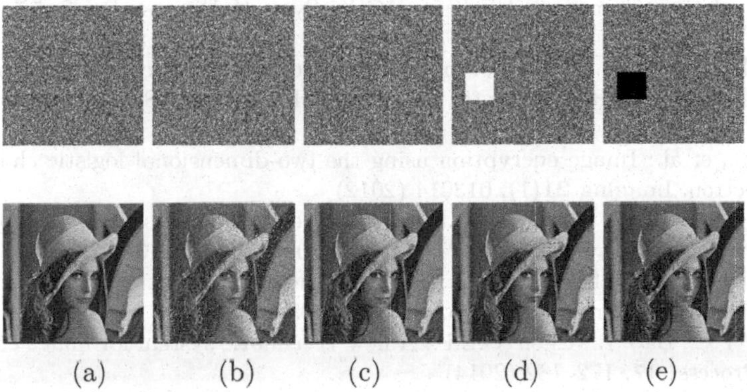

(a)   (b)   (c)   (d)   (e)

Fig. 6. The images on the first row are the original encrypted image, and its damaged versions by 1% gaussian noise, 5% Salt and Pepper noise, 50 * 50 data loss with white square, and 50 * 50 square data loss with black square respectively. (a)–(e) are decrypted results of corresponding encrypted images.

attack to image reconstruction. In the top row, the original pictures are added with 1% Gaussian noise and 5% Salt and Pepper noise, then they decrypted with ICS-IE algorithm and the decryption results are shown in the bottom row. As can be seen, the decrypted images can be recognized clearly. Therefore, ICS-IE can effectively resist the noise attack. Figure 6(d)–(e) also show the data loss attack to the image decryption. In the top row, Fig. 6(d) and (e) are individually cut with 50 × 50 size of square pixel in the encrypted image, the former is cut with white square, the latter cut with black square. As seen in the bottom row of Fig. 6(d) and (e), the reconstruction results are recognizable. Thus we can conclude that ICS-IE algorithm is able to effectively resist the data loss.

5 Conclusion

This paper presented ICS as a new system for image encryption. Integrating three seed maps, it is able to generate several complex chaotic maps, which have wider chaotic ranges and higher security levels. As a result, ICS can produce random and unpredictable chaotic sequences. Based on ICS, we also proposed a novel image encryption algorithm. Simulation results and security analysis have shown that ICS-IE algorithm achieved satisfactory performance in image encryption and can resist different attacks.

Acknowledgement. This work was supported in part by the National Natural Science Foundation of China (Nos. 61320106008 and 61571236).

References

1. Chen, S.L., Hwang, T.T., Lin, W.W.: Randomness enhancement using digitalized modified logistic map. IEEE Trans. Circ. Syst. II Express Briefs **57**, 996–1000 (2010)
2. Addabbo, T., et al.: A class of maximum-period nonlinear congruential generators derived from the Rényi chaotic map. IEEE Trans. Circ. Syst. I Regul. Pap. **54**, 816–828 (2007)
3. Wu, Y., et al.: Image encryption using the two-dimensional logistic chaotic map. J. Electron. Imaging **21**(1), 013014 (2012)
4. Zhou, Y.C., Bao, L., Chen, C.L.P.: Image encryption using a new parametric switching chaotic system. Sig. Process. **93**, 3039–3052 (2013)
5. Hilborn, R.C., et al.: Chaos and nonlinear dynamics: an introduction for scientists and engineers. Comput. Phys. **8**, 689–689 (1994)
6. Zhou, Y.C., Bao, L., Chen, C.L.P.: A new 1D chaotic system for image encryption. Sig. Process. **97**, 172–182 (2014)
7. Zhou, Y.C., et al.: Cascade chaotic system with applications. IEEE Trans. Cybern. **45**, 2001–2012 (2015)
8. Chen, H.K., Lee, C.I.: Anti-control of chaos in rigid body motion. Chaos Solitons Fractals **21**, 957–965 (2004)
9. Gao, T.G., Chen, Z.Q.: A new image encryption algorithm based on hyper-chaos. Phys. Lett. A **372**, 394–400 (2008)
10. Wu, Y., Hua, Z.Y., Zhou, Y.C.: n-Dimensional discrete Cat map generation using Laplace expansions. IEEE Trans. Cybern. **46**, 2622–2633 (2016)
11. Hua, Z.Y., Zhou, Y.C.: Image encryption using 2D logistic-adjusted-sine map. Inf. Sci. **339**, 237–253 (2016)
12. Hua, Z.Y., et al.: 2D Sine Logistic modulation map for image encryption. Inf. Sci. **297**, 80–94 (2015)
13. Liao, X.F., Lai, S.Y., Zhou, Q.: A novel image encryption algorithm based on self-adaptive wave transmission. Sig. Process. **90**, 2714–2722 (2010)
14. Wu, Y., Noonan, J.P., Agaian, S.: A wheel-switch chaotic system for image encryption. In: 2011 International Conference on System Science and Engineering (ICSSE), pp. 23–27. IEEE (2011)
15. Chen, G.R., Yu, X.H.: Chaos Control: Theory and Applications, vol. 292. Springer Science Business Media, Berlin (2003)

16. Hsiao, H.I., Lee, J.H.: Color image encryption using chaotic nonlinear adaptive filter. Sig. Process. **117**, 281–309 (2015)
17. Lian, K.Y., et al.: Synthesis of fuzzy model-based designs to synchronization and secure communications for chaotic systems. IEEE Trans. Syst. Man Cybern. Part B Cybern. **31**, 66–83 (2001)
18. Shen, C.W., et al.: A systematic methodology for constructing hyperchaotic systems with multiple positive Lyapunov exponents and circuit implementation. IEEE Trans. Circ. Syst. I Regul. Pap. **61**, 854–864 (2014)
19. Fu, C., et al.: A chaos-based digital image encryption scheme with an improved diffusion strategy. Opt. Express **20**, 2363–2378 (2012)

Fast, Automatic and Scalable Learning to Detect Android Malware

Mahmood Yousefi-Azar[1(✉)], Len Hamey[1], Vijay Varadharajan[1],
and Mark D. McDonnell[2]

[1] Department of Computing, Faculty of Science and Engineering,
Macquarie University, Sydney, NSW, Australia
mahmood.yousefiazar@hdr.mq.edu.au,
{len.hamey,vijay.varadharajan}@mq.edu.au
[2] Computational Learning Systems Laboratory,
School of Information Technology and Mathematical Sciences,
University of South Australia, Mawson Lakes, Australia
mark.mcdonnell@unisa.edu.au

Abstract. We propose a novel scheme for Android malware detection. The scheme has two extremely fast phases. First *term-frequency* simhashing (*tf*-simhashing) extracts a fixed sized vector for each binary file. The hashing algorithm embeds the frequency of n-grams of bytes into the output vector which can be reshaped into an image representation. In the second phase, we propose a convolutional extreme learning machine (CELM) learns to distinguish between hashes of malicious and clean files as a two class classification task. This scalable scheme is extremely fast in both learning and predicting. The results show that *tf*-simhashing in an image-shape representation together with CELM provides better performance than three non-parametric models and one state-of-the-art parametric model.

Keywords: Android malware detection · Convolutional extreme learning machine · *Term-frequency* simhashing · Static analysis

1 Introduction

Smart phone malware detection is an important application of cyber-security. In particular, Google's Android operating system is vulnerable to over 16 million malware programs [6].

Malware can be analyzed using static features or dynamic features. Static features are computationally cheaper and more suitable for malware detection on the user device. Commonly used features are byte n-grams that can represent multi-byte instructions and data. Larger n-grams are more specific for particular malware, but large conventional n-grams exponentially increase the dictionary space. In this paper, we propose *term-frequency* simhashing (*tf*-simhashing) features that are independent of the dimensionality of the underlying n-grams.

© Springer International Publishing AG 2017
D. Liu et al. (Eds.): ICONIP 2017, Part V, LNCS 10638, pp. 848–857, 2017.
https://doi.org/10.1007/978-3-319-70139-4_86

We also propose using a Convolutional Extreme Learning Machine (CELM) to classify the files as malware or benign in a two class classification. Together, *tf*-simhashing and CELM are scalable and fast. Experiments on a dataset of 23,169 files show that the speed and performance of our scheme is among the best for Android malware detection. In particular, the detection performance is comparable with ensemble models [13] but with much reduced computational overhead.

In summary, the contributions of this paper are the following:

- We introduce *tf*-simhashing, a fast algorithm which maps n-grams of each file into a fixed sized vector of continuous values. Previous use of simhashing requires extensive dynamic analysis [4].
- We introduce Convolutional ELM for malware detection. Previously, ELM was combined with principal component analysis (PCA) to detect Android malware [15]. However, this approach is computationally expensive and not suitable for mobile devices. Our proposed scheme does not require any computationally expensive transformation.
- Our approach utilizes a single classifier rather than an ensemble model [13, 15] while yielding competitive detection performance. This makes the scheme very appropriate for smart mobile phone platforms.
- Our method is scalable—both the feature extraction phase and classification process can be used for big data.

The paper is organized as follows. Section 2 presents related work, Sect. 3 discusses *tf*-simhashing, Sect. 4 discusses CELM, and Sect. 5 presents evaluation metrics, experimental methods, results and discussion.

2 Related Work

Image representations have been used for malware detection. Nataraj et al. [11] and Kancherla and Mukkamala [7] reshape the binary file into an image which is then processed to analyse malware with fairly good performance. Similarly, Han et al. [3] used an entropy-based image representation. Earlier, Han et al. [4] hashed a disassembled file into a color image and combined it with an execution trace of application programming interface (API) calls. Although this process is effective, the disassembly and dynamic analysis have high computational cost.

Malisa et al. [8] used locality-sensitive hashing (LSH) for screenshot analysis to detect Android malware. Unlike other hashes, LSH maps similar data to the same hash code with high probability, maximizing the probability of a collision for similar inputs [5]. Simhashing is an LSH algorithm that approximates the cosine similarity to find similar strings [2].

Previously, we showed that CELM is very effective for image recognition [14]. CELM is a parametric model that employs semi-supervised learning. We reshape the *tf*-simhash vector into an image to provide input to CELM.

Compared with the existing techniques, our approach does not require the computational overhead of disassembly or program tracing. For example, the state-of-the-art AMADROID system [9] involves extracting API calls and constructing a call graph, creating extremely large feature vectors. In contrast, *tf*-simhash provides a fixed-size feature vector independent of the length of the binary file. CELM provides fast learning in constant time given the constant input size.

3   Hashing Algorithm

Simhashing is based on Signed Random Projections (SRP). Each SRP is a random unit vector I representing a hyperplane through the origin. A hash bit is generated by determining on which side of the hyperplane the feature point V lies: $hash(V) = sign(V \cdot I)$.

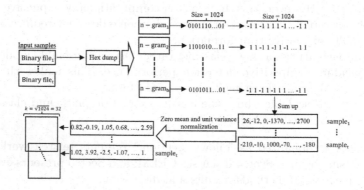

Fig. 1. A detailed schematic of *tf*-simhashing algorithm where input is arbitrary size binary file and output is a fixed size image. t is the number of files in the dataset. b is the vocabulary size. Numerical data are only examples to provide better comprehension.

For two inputs vectors V and U, the probability of identical hashes depends on the angle between them:

$$Pr[hash(V) = hash(U)] = 1 - \frac{\theta(V, U)}{\pi}. \tag{1}$$

tf-simhashing computes simhashes on n-grams and combines them by addition, yielding term frequency (*tf*) for each hash bit.

Figure 1 shows the proposed scheme to generate a fixed size image out of an arbitrary size binary file. Each n-gram of the binary file is simhashed to a fix size vector of −1 and 1. These vectors are then summed for all n-grams, producing a weighted tf for each hash bit. This *tf*-simhash representation provides two vectors that are close to each other when two files have many common n-grams.

The raw *tf*-simhash vectors have high variance which is corrected by normalisation. Each component $i \in (1, N_H)$ of the summed hash vector is linearly transformed to have zero mean and unit variance by $Z_i = \frac{X_i - \mu}{\sigma}$ where the hash size N_H is typically 1024, and μ and σ are the mean and standard deviation of that component across the entire training set. This one-dimensional representation can be used for classification. The results (see Sect. 5.1) show that reshaping this one-dimensional vector to two-dimensional matrix provides a more discriminative feature space.

4   Convolutional Extreme Learning Machine

Figure 2 presents the proposed CELM. The convolutional layer filters (weights) are learned with an unsupervised algorithm, applying K-means clustering to randomly selected patches of the tf-simhash images. For speed, we use the K-mean++ algorithm. After training of the convolutional layer, the training samples are passed through the feature representation block and the random projections layer. This provides supervised training data for the final layer.

We denote the size of the learnt two-dimensional filters as $W \times W$, the size of the pooling window as $Q \times Q$, the image dimension as $k \times k$, and the number of training samples as K. We store all training images in a matrix $\mathbf{X}$ of size $k^2 \times K$. The functions $g_1(.)$ and $g_2(.)$ are non-linear transformations applied term-wise to matrix inputs to produce matrix outputs of the same size. The symbol $*$ in the Fig. 2 represents convolution operation. Mathematically, the entire flow is $\mathbf{F} = g_2(\mathbf{W}_{\text{Pool}}g_1(\mathbf{W}_{\text{Filter}}\mathbf{X}))$ where $g_1(.)$ and $g_2(.)$ are applied term by term to all elements of their arguments. The matrices $\mathbf{W}_{\text{Filter}}$ and $\mathbf{W}_{\text{Pool}}$ are sparse Toeplitz matrices. In practice, we used 2D convolution technique.

As in [10,14], our non-linear hidden-unit functions $g_1(.)$ and $g_2(.)$ are those of *LP-pooling*, which is of the form $g_1(u) = u^p$ and $g_2(v) = v^{1/p}$. We use $p = 2$ so that $\mathbf{F} = \sqrt{(\mathbf{W}_{\text{Pool}}(\mathbf{W}_{\text{Filter}}\mathbf{X})^2}$.

The first block of the discriminative section (see Fig. 2) is a random projection layer. It results in M units in the output of the first block. The second block is the only trained layer (i.e. obtain $\mathbf{W}_{\text{out}}$ size $N \times M$ is the real-valued output weights matrix for the classifier stage) with N linear units corresponding to N classes.

In the following components of the discriminative block, the classifier is trained using the approach described by [10]. In this paper, we use a method where the weights, $\mathbf{W}_{\text{in}}$, are set by learning from the training set instead of randomly generating.

Given a choice of $\mathbf{W}_{\text{in}}$, the output weights matrix is determined as:

$$\mathbf{W}_{\text{out}} = \mathbf{Y}_{\text{label}}\mathbf{A}_{\text{train}}^{+}, \tag{2}$$

where $\mathbf{A}_{\text{train}}^{+}$ is the size $K \times M$ Moore-Penrose pseudo inverse corresponding to $\mathbf{A}_{\text{train}}$. This solution is equivalent to least squares regression applied to an overcomplete set of linear equations, with an N-dimensional target. In practice, the most efficient approach is to use Cholesky factorisation to solve the following set of $N \times M$ linear equations for the $N \times M$ unknown variables in $\mathbf{W}_{\text{out}}$:

$$\mathbf{Y}_{\text{label}}\mathbf{A}_{\text{train}}^{\top} = \mathbf{W}_{\text{out}}(\mathbf{A}_{\text{train}}\mathbf{A}_{\text{train}}^{\top} + c\mathbf{I}) \tag{3}$$

ELM Scheme

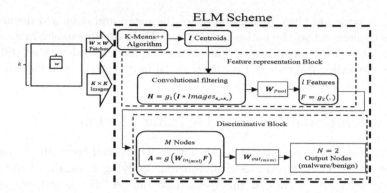

Fig. 2. CELM scheme [14].

5   Experiments and Results

We use a confusion matrix to evaluate the proposed model.

		Predicted	
		Benign	Malware
Actual	Benign	True Negative (TN)	False Positive (FP)
	Malware	False Negative (FN)	True Positive (TP)

Note that "positive" denotes malware. TN and TP are the counts of correct detection of benign and malware respectively. Conversely, FP and FN are the counts of incorrect malware detection and failed malware detection respectively.

The most common performance evaluators are Precision, Recall (Detection rate), f1-score, g-mean, Accuracy, and False positive rate (FPR) as follows:

$$\text{Precision} = \frac{TP}{TP+FP}, \text{Recall} = \frac{TP}{TP+FN}, \text{f1-score} = 2 * \frac{precision*recall}{precision+recall}$$

$$\text{g-mean} = \sqrt{precision * recall}, \text{Accuracy} = \frac{TN+TP}{TN+FP+FN+TP}, \text{FPR} = \frac{FP}{TN+FP}$$

We also used the area under the receiver operating characteristic curve (AUC) as an important evaluation metric. AUC is the probability that a classifier will rank a randomly chosen positive sample higher than a randomly chosen negative sample.

We collected 12089 clean files consisting of 3581 .apk from three different *Android* markets[1,2,3] together with 2962 images, 500 .docx, 4497 .pdf and 549 .rar and .jar files from the *contagiodump* website [4]. We included non-apk files to diversify the patterns to be recognised. Malware (VirusShare_Android_20130506.zip)

---

[1] http://www.appsmob.com/.

[2] https://f-droid.org/.

[3] http://www.mobomarket.net/.

[4] http://contagiodump.blogspot.com.au/2013/03/16800-clean-and-11960-malicious-files.html.

was collected from *virusshare* website[5], a freely available virus repository. This file consists of 11080 malicious Android applications. The package includes a wide range of malicious codes for Android platforms such as Android/Lotoor[6], Android/TrojanSMS.Boxer.AA[7], Android/TrojanSMS.FakeInst[8], Generic Fake Alert.a[9] and etc. Table 1 shows the statistics of the dataset. To avoid redundant data, the samples were checked with the MD5 algorithm.

For comparison, our baseline classifiers were Support Vector Machines (SVMs), Gradient Boost (XGBoost), K-Nearest Neighbors (K-NN) and Convolutional Neural Network (CNN). Because the first three models are non-parametric, the number of parameters depends on the training set. Both CNN and the proposed CELM method are parametric models with a fixed number of parameters.

SVM and K-NN were implemented in sklearn. The hyper-parameters for SVM and K-NN were optimised by grid search. For SVM, the best parameters were $C = 0.1$, $\gamma = 100$ and RBF kernel. For K-NN, 5 neighbours with uniform weighting were used. For XGBoost, we used the same hyper-parameters as the winner of the 2015 Kaggle Competition [12].

CNN was implemented in TensorFlow using two 7×7 convolutional layers with 16 and 32 features respectively. Each convolutional layer was followed by 2×2 average pooling layers. A fully connected layer of 512 hidden neurons connected to an output layer with one neuron for benign and a second for malware. ReLU activation function was used throughout. Training was performed with 25% dropout probability.

The CELM filter size was $W = 7 \times 7$, with pooling $Q = 8$ and downsampling factor $D = 5$. CELM was implemented in Matlab using the fast implementation of *MatConvNet*.

In our experiments, $N_H = 1024$ throughout. Thus, SVM, XGBoost and K-NN used 1024 element vectors whereas for CNN and CELM the input was reshaped to a 32×32 image. Our experiments used 2-gram throughout and 5-fold cross-validation for performance evaluation.

Table 1. The statistics of the dataset.

Type	Qty	Max. size (MB)	Min. size (KB)	Ave. size (MB)
Malware	11080	26.4	0.2	0.7
Benign	12089	98.8	0.4	2.1

---

[5] https://virusshare.com/.

[6] https://www.symantec.com/security_response/writeup.jsp?docid=2012-091922-4449-99.

[7] https://www.welivesecurity.com/wp-content/media_files/SMS_Trojan_Whitepaper.pdf.

[8] http://www.virusradar.com/en/Android_TrojanSMS.FakeInst.GM/description.

[9] https://www.mcafee.com/threat-intelligence/malware/default.aspx?id=143470.

5.1 Results and Discussion

Table 2 compares CELM with the baseline methods. We used 5-fold cross-validation to obtain the mean and standard deviation of each performance measure except FPR for which only the mean is reported.

The data representation is based on bigrams of the binary file. Results are presented using the *tf*-simhashing representation for all learning algorithms, and also using Information Gain (IG) to select features for XGBoost [12]. XGBoost with *tf*-simhashing is much more precise than IG-based feature selection but provides lower recall. For all other performance measures, XGBoost with *tf*-simhashing is superior to XGBoost with IG. The XGBoost (IG) baseline is the 2015 Kaggle competition winner—our experiments used their published code [12]. Other researchers have also used IG-based representation for malware detection [1].

Overall, CELM with *tf*-simhashing outperforms the baseline methods. SVM with *tf*-simhashing has the best precision and FPR, while CNN has the best recall. However, the measures that combine precision and recall favour CELM. In particular, CELM outperforms CNN which is the state of the art in image recognition. CELM has only one convolutional layer which may be more suitable for the non-natural images produced by *tf*-simhashing. The random projection layer following the pooling layer also contributes to the power of CELM.

We also compare CELM with the state-of-the-art MAMADROID results [9] using the 2013 data that is most similar to out dataset. The classification performances of CELM and MAMADROID are not significantly different. However, CELM is much faster than MAMADROID for both training and testing. In particular, CELM's high speed and low memory requirements make it suitable for implementation on a mobile phone.

Table 2. Performance results for baseline and CELM methods. Mean and standard deviation of 5-fold cross-validation is shown. MAMADROID results are taken from [9].

Model	Precision	Recall	f1-score	g-mean	Accuracy	AUC	FPR
SVM	**97.12%**	94.59%	95.83%	95.84%	96.07%	96.01%	**2.57%**
	(±0.35%)	(±0.31%)	(±0.14%)	(±0.33%)	(±0.14%)	(±0.13%)	
XGBoost	95.22%	95.97%	95.59%	95.59%	95.78%	95.84%	4.42%
	(±0.44%)	(±0.20%)	(±0.19%)	(±0.29%)	(±0.19%)	(±0.18%)	
XGBoost (IG [12])	77.21%	97.1%	86.02%	86.59%	84.91%	85.41%	26.30%
	(±0.47%)	(±0.25%)	(±0.29%)	(±0.34%)	(±0.36%)	(±0.35%)	
K-NN	95.39%	95.79%	95.59%	95.58%	95.77%	95.73%	4.27%
	(±0.31%)	(±0.13%)	(±0.12%)	(±0.20%)	(±0.12%)	(±0.11%)	
CNN	84.37%	**98.58%**	90.93%	91.20%	90.59%	91.45%	16.73%
	(±0.59%)	(±0.20%)	(±0.35%)	(±0.34%)	(±0.40%)	(±0.38%)	
CELM	96.38%	96.27%	**96.33%**	**96.48%**	**96.48%**	**96.47%**	3.32%
	(±0.44%)	(±0.35%)	(±0.22%)	(±0.22%)	(±0.22%)	(±0.20%)	
MAMADROID [9]	97.00%	95.00%	**96.00%**	**95.99%**	—	—	—
	(±0.44%)	(±0.35%)	(±0.22%)	(±0.22%)	(±0.22%)	(±0.20%)	

As a more detailed analysis, we present the confusion matrices for each *tf*-simhashing method in Table 3. Although CNN detected the most malware, it has the highest false positive rate which makes it inappropriate for real-world tasks. On the other hand, SVM has the lowest false positives but fails to detect the most malware files. The performance of CELM balances the false negative and false positive rates.

Table 3. Confusion matrices of various classifiers.

SVM	XGBoost	K-NN	CNN	CELM
$\begin{bmatrix} 11778 & 311 \\ 599 & 10481 \end{bmatrix}$	$\begin{bmatrix} 11555 & 534 \\ 447 & 10633 \end{bmatrix}$	$\begin{bmatrix} 11576 & 513 \\ 466 & 10614 \end{bmatrix}$	$\begin{bmatrix} 10066 & 2023 \\ 157 & 10923 \end{bmatrix}$	$\begin{bmatrix} \mathbf{11688} & \mathbf{401} \\ \mathbf{414} & \mathbf{10666} \end{bmatrix}$

Table 4. Average execution time for training 18,535 and testing 4,634 samples.

Model	SVM	XGBoost	K-NN	CNN	CELM
Train set	78.95s	42.77s	01.34s	668.04s	**9.41s**
Test set	16.23s	00.01s	16.19s	016.92s	**2.16s**

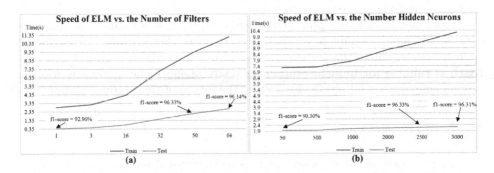

Fig. 3. Execution time (i.e. the second) for training and testing CELM with (a) increasing the number of filter of the convolutional layer (b) with increasing the number of hidden units.

Execution time is also important for end users. K-NN does not construct a model during training, which makes training fast but testing slow as shown in Table 4. XGBoost provides very fast testing but training is slow. The worst execution times are for CNN and SVM which have slow training and testing. In constrast, CELM provides balanced training and testing times which makes it suitable for regular updates as new malware is identified.

Figure 3 explores the time-performance trade-off for CELM architectures. Increasing the number of filters from 1 to 64 significantly increases the training and testing time, but also improves the f1-score from 92.96% to 96.14%. Increasing the number of hidden neurons from 50 to 3000 has a negligible effect on testing time, but improves the f1-score from 90.30% to 96.31%.

tf-simhashing is fast. Our Python implementation hashes an 80MB file within one millisecond on an Intel(R) Core(TM) i7-4790 CPU @ 3.60GHz.

6 Conclusion

In this paper, we propose an extremely fast scheme for Android malware detection. The scheme consists of two phases: a hashing algorithm for feature extraction and learning to detect malicious samples. Inspired by simhashing, we proposed *term-frequency* simhashing which hashes the frequency of n-grams to a fixed sized vector. This vector can be converted to an image-shape representation of the file. The whole feature extraction phase requires only static analysis and unlike the related work, does not require any pre-processing. Also, we propose to use a convolutional extreme learning machine (CELM) to generate predictions of our two class classification task. This parametric model complements our scalable scheme. Experiments show that with two phases of training (unsupervised filter learning of the convolutional layer and analytical supervised learning) the CELM model performs better than all four baseline models. Although the two-dimensional representation does not seem intuitive for our convolutional model, experiments show that the random projection layer of the CELM strongly improves generalization.

Experiments show the proposed scheme has high performance and requires a relatively small amount of computation. In particular, our approach achieves performance equal to the state-of-the-art MAMADROID system, but our approach is suitable for implementation on smart phones. The implementation of our scheme on an Android device will be the future of this study.

References

1. Aung, Z., Zaw, W.: Permission-based android malware detection. Int. J. Sci. Technol. Res. **2**(3), 228–234 (2013)
2. Charikar, M.S.: Similarity estimation techniques from rounding algorithms. In: Proceedings of the Thiry-Fourth Annual ACM Symposium on Theory of Computing, pp. 380–388. ACM (2002)
3. Han, K.S., Lim, J.H., Kang, B., Im, E.G.: Malware analysis using visualized images and entropy graphs. Int. J. Inf. Secur. **14**(1), 1–14 (2015)
4. Han, K.S., Kang, B.J., Im, E.G.: Malware analysis using visualized image matrices. Sci. World J. **2014**, 15 p. (2014). doi:10.1155/2014/132713. Article ID 132713
5. Indyk, P., Motwani, R.: Approximate nearest neighbors: towards removing the curse of dimensionality. In: Proceedings of the Thirtieth Annual ACM Symposium on Theory of Computing, pp. 604–613. ACM (1998)
6. AV-TEST Institute: https://www.av-test.org/en/news/news-single-view/current-risk-scenario-av-test-security-report-facts-at-a-glance/

7. Kancherla, K., Mukkamala, S.: Image visualization based malware detection. In: 2013 IEEE Symposium on Computational Intelligence in Cyber Security (CICS), pp. 40–44. IEEE (2013)
8. Malisa, L., Kostiainen, K., Och, M., Capkun, S.: Mobile application impersonation detection using dynamic user interface extraction. In: Askoxylakis, I., Ioannidis, S., Katsikas, S., Meadows, C. (eds.) ESORICS 2016. LNCS, vol. 9878, pp. 217–237. Springer, Cham (2016). doi:10.1007/978-3-319-45744-4_11
9. Mariconti, E., Onwuzurike, L., Andriotis, P., De Cristofaro, E., Ross, G., Stringhini, G.: Mamadroid: detecting android malware by building markov chains of behavioral models. arXiv preprint (2016). arXiv:1612.04433
10. McDonnell, M.D., Vladusich, T.: Enhanced image classification with a fast-learning shallow convolutional neural network. In: 2015 International Joint Conference on Neural Networks (IJCNN), pp. 1–7. IEEE (2015)
11. Nataraj, L., Karthikeyan, S., Jacob, G., Manjunath, B.: Malware images: visualization and automatic classification. In: Proceedings of the 8th International Symposium On Visualization For Cyber Security, p. 4. ACM (2011)
12. Wang, X., Liu, J., Chen, X.: First place team: say no to overfitting (2015)
13. Yerima, S.Y., Sezer, S., Muttik, I.: High accuracy android malware detection using ensemble learning. IET Inf. Secur. 9(6), 313–320 (2015)
14. Yousefi-Azar, M., McDonnell, M.D.: Semi-supervised convolutional extreme learning machine. In: International Joint Conference on Neural Networks (IJCNN 2017). IEEE (2017, accepted)
15. Zhang, W., Ren, H., Jiang, Q., Zhang, K.: Exploring feature extraction and ELM in malware detection for android devices. In: Hu, X., Xia, Y., Zhang, Y., Zhao, D. (eds.) ISNN 2015. LNCS, vol. 9377, pp. 489–498. Springer, Cham (2015). doi:10.1007/978-3-319-25393-0_54

Intrusion Detection Using Convolutional Neural Networks for Representation Learning

Zhipeng Li, Zheng Qin$^{(\boxtimes)}$, Kai Huang, Xiao Yang, and Shuxiong Ye

School of Software, Tsinghua University, Beijing 100084, China
{lizp14,huang-k15,ysx15}@mails.tsinghua.edu.cn,
qingzh@mail.tsinghua.edu.cn, yangxiao356@126.com

Abstract. The intrusion detection based on deep learning method has been widely attempted for representation learning. However, in various deep learning models for intrusion detection, there is rarely convolutional neural networks (CNN) model. In this work, we propose a image conversion method of NSL-KDD data. Convolutional neural networks automatically learn the features of graphic NSL-KDD transformation via the proposed graphic conversion technique. We evaluate the performance of the image conversion method by binary class classification experiments with NSL-KDD Test$^+$ and Test^{-21}. Different structures of CNN are testified for comparison. On the two NSL-KDD test datasets, CNN performed better than most standard classifier although the CNN did not improve state of the art completely. Results show that the CNN model is sensitive to image conversion of attack data and our proposed method can be used for intrusion detection.

Keywords: Intrusion detection · Convolutional neural networks · NSL-KDD · Representation learning

1 Introduction

It is well known that intrusion detection system (IDS) is a network or host system that can identify unsafe event and give an alarm. There are various categories of intrusion detection systems according to different classification criterions. IDS can be categorized as misuse detection [10] and anomaly detection according to the principle of attack recognition [4]. Misuse detection are able to detect attacks based on signatures of these attacks, however it can not identify novel attacks without manual rules. The problem of identifying novel attacks has become a main research focus in this field [17].

In recent years, deep learning has grown very fast and achieved good results in many scenarios [11]. Many scholars try to use deep learning technologies for anomaly detection. Various types of deep learning method for anomaly detection are proposed. Some of these techniques include: Self –Taught Learning, Deep Belief Networks, Auto Encoder, LSTM and so on [5–7,14,16]. These feature learning approaches and models have been successful to a certain extent and match or exceed state of the art techniques.

© Springer International Publishing AG 2017
D. Liu et al. (Eds.): ICONIP 2017, Part V, LNCS 10638, pp. 858–866, 2017.
https://doi.org/10.1007/978-3-319-70139-4_87

Convolutional Neural Networks (CNN) is a well-known deep learning model proposed for image classification [9]. Due to the good performance of CNN, a large number of applications based on the CNN model are proposed. Yandre M.G. Costa *et al.* use CNN trained with textural descriptors for music genre recognition [3]. Wang *et al.* propose a malware traffic classification method using convolutional neural networks [19]. Yoon Kim take experiments with convolutional neural networks trained for sentence level classification tasks [8]. However, there is no related work about intrusion detection application using CNN.

This paper presents an intrusion detection method using convolutional neural networks. The intrusion detection model adopts a novel representation learning method of graphic conversion. The method of transforming standard KDD'99 or NSL-KDD data form into image form is introduced in detail. The performance of the model is tested by several experiments with a popular NSL-KDD data set. The results show that our method has good performance on NSL-KDD Test$^+$ and Test^{-21}. Using convolutional neural networks for image conversion feature learning in intrusion detection is practicable.

The remainder of this paper is organized as follows: Sect. 2 presents the overview of the intrusion detection system architecture and the model function module in detail. Section 3 describes the experiment taken on NSL-KDD and the analysis of the results. Section 4 provides conclusion remarks of our work and future works.

2   Proposed Method

Intrusion detection system is a network security protection system deployed on network or host [12]. The practical input is raw network packet data. In this paper, we use visual conversion of the NSL-KDD format to evaluate the performance of convolutional neural networks in intrusion detection. There are a number of works and tools designed for transforming raw packet into NSL-KDD data formats. Each sample of the NSL-KDD dataset contains 41 features that contain integer or float features, symbolic features and binary features. We design a data-preprocessing module to convert various feature attributes into binary vectors. Then we convert the data into image form. Finally, CNN identifies the category of image conversion. The overall processing flow is shown in Fig. 1.

2.1   KDD'99 and NSL-KDD Data Format

In intrusion detection, the most famous dataset is KDD'99 dataset [13]. KDD'99 dataset is created by Stolfo *et al.* The dataset is an abstract description of DARPA'98 IDS evaluation program [15]. DARPA'98 program simulates a typical LAN of the US air force and records the raw data for 7 weeks. The 4 gigabytes of compressed raw data is labeled normal or various types of attack. The KDD'99 dataset is used as a standard dataset for evaluating IDS. Until recent years, many scholars point out some severe problems of KDD'99 dataset such as duplication of data, unbalanced distribution of data and so on. NSL-KDD appears as a dataset that overcomes the shortcomings of KDD'99 dataset [18].

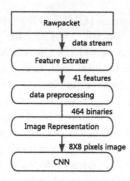

Fig. 1. An overview of IDS using CNN

NSL-KDD retains the data characteristics consistent with the KDD'99 dataset. Data are mainly categorized as five classes: Normal, Dos, Probe, U2R, R2L. At the same time, NSL-KDD feature attributes can be classified into three groups: basic features, traffic features, content features. The 41 features have numeric or symbolic values. Numeric features are integer or float type [2].

2.2   Preprocessing of the Experimental Data

Aim to identify the image conversion of NSL-KDD data with CNN, we design a method to convert NSL-KDD data format into the visual image type. To achieve this goal, we map various types of features into binary vector space and then transform the binary vector into image.

Symbolic Features. There are three symbolic datatypes in NSL-KDD data attributes: protocol_type, flag and service. We use one-hot encoder mapping these features into binary vectors as show in Fig. 2. For example, protocol_type has three values (tcp, udp, icmp), turns into binary vectors with three dimensions (100, 010, 100).

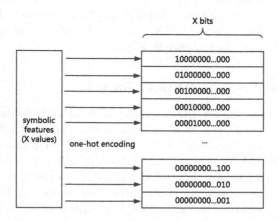

Fig. 2. One-hot encoding

Continuous Features. Continuous features include integer and float types of features. We use standard scaler to normalize the continuous data into the range [0, 1]. Standard scaler refers to scaling data to a specific interval. In this paper, the Min-Max normalization method is used. That is:

$$x_{new} = \frac{x - x_{min}}{x_{max} - x_{min}},\tag{1}$$

where x stands for numeric feature value, x_{min} stands for the minimal value of the feature, x_{max} stands for the max value, x_{new} stands for value after the normalization. After the normalization process, we discretize the scaled continuous value into 10 intervals. Then we use one-hot encoder encoding the order number of intervals into 10 binary vectors show in Fig. 3.

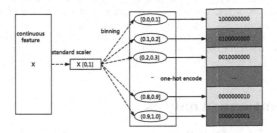

Fig. 3. Discretization and binarization on continuous features

After the preprocess, NSL-KDD data form turns into a binary vector with 464 dimensions. Then we turn each 8 bits into a grayscale pixel. The binary vector with 464 dimensions turns into a 8*8 grayscale image with vacant pixel padded by 0. Some typical data are shown in Table 1. The grayscale images transformed from the corresponding data are shown in Fig. 4. For better contrast, we display two different samples in each category.

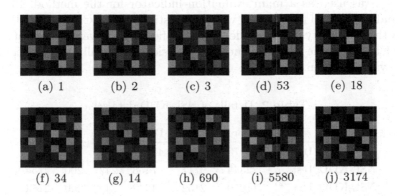

(a) 1   (b) 2   (c) 3   (d) 53   (e) 18

(f) 34   (g) 14   (h) 690   (i) 5580   (j) 3174

Fig. 4. Images of NSL-KDD samples

Table 1. Samples of NSL-KDD data

sample number	sample content	type
1	0, tcp, ftp_data, SF, 491, 0, 0, 0, 0, 0, 0, 0, 0, 0, 0, 0, 0, 0, 0, 0, 0, 0, 0, 2, 2, 0, 0, 0, 0, 1, 0, 0, 150, 25, 0.17, 0.03, 0.17, 0, 0, 0, 0.05, 0, normal	normal
2	0, udp, other, SF, 146, 0, 0, 0, 0, 0, 0, 0, 0, 0, 0, 0, 0, 0, 0, 0, 0, 0, 0, 13, 1, 0, 0, 0, 0, 0.08, 0.15, 0, 255, 1, 0, 0.6, 0.88, 0, 0, 0, 0, 0, normal	normal
3	0, tcp, private, S0, 0, 0, 0, 0, 0, 0, 0, 0, 0, 0, 0, 0, 0, 0, 0, 0, 0, 0, 123, 6, 1, 1, 0, 0, 0.05, 0.07, 0, 255, 26, 0.1, 0.05, 0, 0, 1, 1, 0, 0, neptune	dos
53	0, udp, private, SF, 28, 0, 0, 3, 0, 0, 0, 0, 0, 0, 0, 0, 0, 0, 0, 0, 0, 0, 2, 2, 0, 0, 0, 0, 1, 0, 0, 255, 2, 0.01, 0.02, 0.01, 0, 0, 0, 0.77, 0, teardrop	dos
18	0, icmp, eco_i, SF, 18, 0, 0, 0, 0, 0, 0, 0, 0, 0, 0, 0, 0, 0, 0, 0, 0, 0, 0, 1, 1, 0, 0, 0, 0, 1, 0, 0, 1, 16, 1, 0, 1, 1, 0, 0, 0, 0, ipsweep	probe
34	0, tcp, private, REJ, 0, 0, 0, 0, 0, 0, 0, 0, 0, 0, 0, 0, 0, 0, 0, 0, 0, 0, 0, 2, 1, 0, 0, 1, 1, 0.5, 1, 0, 255, 1, 0, 0.31, 0.28, 0, 0, 0, 0.29, 1, portsweep	probe
14	0, tcp, ftp_data, SF, 334, 0, 0, 0, 0, 0, 0, 1, 0, 0, 0, 0, 0, 0, 0, 0, 0, 0, 2, 2, 0, 0, 0, 0, 1, 0, 0, 2, 20, 1, 0, 1, 0.2, 0, 0, 0, 0, warezclient	r2l
690	0, tcp, telnet, RSTO, 125, 179, 0, 0, 0, 1, 1, 0, 0, 0, 0, 0, 0, 0, 0, 0, 0, 0, 1, 1, 0, 0, 1, 1, 1, 0, 0, 4, 4, 1, 0, 0.25, 0, 0.25, 0.25, 0.75, 0.75, guess_passwd	r2l
5580	0, tcp, ftp_data, SF, 0, 5696, 0, 0, 0, 0, 1, 0, 0, 0, 0, 0, 0, 0, 0, 0, 0, 0, 1, 1, 0, 0, 0, 0, 1, 0, 0, 1, 81, 1, 0, 1, 0.02, 0, 0, 0, 0, buffer_overflow	u2r
3174	98, tcp, telnet, SF, 621, 8356, 0, 0, 1, 1, 0, 1, 5, 1, 0, 14, 1, 0, 0, 0, 0, 0, 1, 1, 0, 0, 0, 0, 1, 0, 0, 255, 4, 0.02, 0.02, 0, 0, 0, 0, 0, 0, rootkit	u2r

3   Experiment and Analysis

3.1   Implementation Details

The experiment was taken on a Dell 7910 workstation with TITAN X Pascal. TensorFlow was adopted as a deep learning framework software [1]. We used ResNet 50 and GoogLeNet as CNN models for comparison. We used 100 epochs with 256 batch size to train ResNet 50 and 100 epochs with 64 batch size to train GoogLeNet. Both models used gradient descent optimizer as optimizer and cross entropy as cost function.

3.2   Performance Metric

Accuracy was used as a main evaluation indicator for the method. Precision, recall and F1 score of the method were also testified in the experiment. Precision reflects the sensitivity of the model. Recall reflects the coverage capacity of the model. F1 score is the harmonic mean of precision and recall. And test confusion matrix was also given.

Table 2. Details of NSL-KDD data set

Data set	Records	Normal	DoS	Probe	U2R	R2L
NSL-KDD Train	125973	67343	45927	11656	52	995
NSL-KDD Test$^+$	22544	9711	7458	2421	200	2754
NSL-KDD Test^{-21}	11850	2152	4342	2402	200	2754

3.3 Comparative Analysis

Two test datasets were used to evaluate the performance of the CNN model. Detail of the train set and test set is shown in Table 2. In order to test intrusion detection and identify the ability to discover new attacks, NSL-KDD divides data into training set and test set which makes the test more realistic. There are 17 additional attack types in the test set. NSL-KDD employs the 21 learned machines (7 learners, each trained 3 times) to label the records of the entire KDD train and test sets, which provides user with 21 predicated labels for each record. $Test^+$ is randomly sampled from the KDD'99 test sets and the $Test^{-21}$ contains all the data misclassified by all 21 learners [18].

We test the proposed method's performance for binary labeled class both on NSL-KDD $Test^+$ and $Test^{-21}$. Table 3 shows the accuracy, precision, recall and f1 score of binary labeled class on $Test^+$ and $Test^{-21}$. The results confusion matrices are show in Fig. 5. As results shown, the CNN based method can achieve a high recall score when test data set contains a large amount of attack data.

Table 3. Perfomance of binary labeled class

	Accuracy	Precision	Recall	F1 score
ResNet 50 NSL-KDD $Test^+$	79.14%	91.97%	69.41%	79.12%
ResNet 50 NSL-KDD $Test^{-21}$	81.57%	81.81%	99.63%	89.85%
GoogLeNet NSL-KDD $Test^+$	77.04%	91.66%	65.64%	76.50%
GoogLeNet NSL-KDD $Test^{-21}$	81.84%	81.84%	100%	90.01%

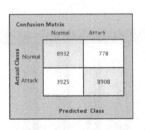

(a) NSL-KDD $Test^+$ (b) NSL-KDD $Test^{-21}$

(c) NSL-KDD $Test^+$ (d) NSL-KDD $Test^{-21}$

Fig. 5. Confusion matrix of binary test

We believe that the uneven distribution of Test^{-21} data leads the CNN tends to identify the data into attack class data. Although the percentage of anomaly traffic is very small in practice, the ability to recognize attack in some test or specific attack scenarios is meaningful for IDS.

According to the results shown in Table 4 and Fig. 6, our proposed method has the relatively good accuracy compared to other method. The accuracies of other method is measured by Tavallaee et al. [18].

Table 4. Comparison of different method

Classifier	Accuracy on Test$^+$	Accuracy on Test^{-21}
J48	81.05%	63.97%
Naive bayes	76.56%	55.77%
NB Tree	82.02%	66.16%
Random forest	80.67%	62.26%
Random tree	81.59%	58.51%
Multi-layer perceptron	77.41%	57.34%
SVM	69.52%	42.29%
Proposed method1 (ResNet50)	79.14%	81.57%
Proposed method2 (GoogLeNet)	77.04%	81.84%

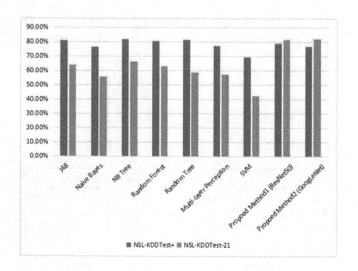

Fig. 6. Comparison of different method

4 Conclusion and Future Work

In this paper we have evaluated the Convolutional Neural Network for intrusion detection using image conversion of NSL-KDD dataset. We also introduced visual data conversion representation method specially designed for CNN. Performance of the method is tested by NSL-KDD Test$^+$ and Test^{-21}. By comparing with some standard classifier, the CNN does not improve state of the art, but CNN uses the image form of data as input without feature selection. That is a great advantage of deep learning methods. Results demonstrate that CNN can be used as anomaly detection classifier and our method of data conversion has good performance. As future work, we will explore a better image conversion representation techniques, since CNN are highly sensitive to the structural features of an image. How to retain the structural information of data at utmost is the main research target. In addition, we will consider use real raw packet as experimental data.

References

1. Abadi, M., Agarwal, A., Barham, P., Brevdo, E., Chen, Z., Citro, C., Corrado, G.S., Davis, A., Dean, J., Devin, M., et al.: Tensorflow: large-scale machine learning on heterogeneous distributed systems. arXiv preprint (2016). arXiv:1603.04467
2. Aggarwal, P., Sharma, S.K.: Analysis of KDD dataset attributes-class wise for intrusion detection. Procedia Comput. Sci. **57**, 842–851 (2015)
3. Costa, Y.M., Oliveira, L.S., Silla, C.N.: An evaluation of convolutional neural networks for music classification using spectrograms. Appl. Soft Comput. **52**, 28–38 (2017)
4. Denning, D.E.: An intrusion-detection model. IEEE Trans. softw. Eng. **2**, 222–232 (1987)
5. Erfani, S.M., Rajasegarar, S., Karunasekera, S., Leckie, C.: High-dimensional and large-scale anomaly detection using a linear one-class SVM with deep learning. Pattern Recogn. **58**, 121–134 (2016)
6. Gao, N., Gao, L., Gao, Q., Wang, H.: An intrusion detection model based on deep belief networks. In: 2014 Second International Conference on Advanced Cloud and Big Data (CBD), pp. 247–252. IEEE (2014)
7. Kim, J., Kim, J., Thu, H.L.T., Kim, H.: Long short term memory recurrent neural network classifier for intrusion detection. In: 2016 International Conference on Platform Technology and Service (PlatCon), pp. 1–5. IEEE (2016)
8. Kim, Y.: Convolutional neural networks for sentence classification. arXiv preprint (2014). arXiv:1408.5882
9. Krizhevsky, A., Sutskever, I., Hinton, G.E.: Imagenet classification with deep convolutional neural networks. In: Advances in Neural Information Processing Systems. pp. 1097–1105 (2012)
10. Kumar, S., Spafford, E.H.: A pattern matching model for misuse intrusion detection (1994)
11. LeCun, Y., Bengio, Y., Hinton, G.: Deep learning. Nature **521**(7553), 436–444 (2015)
12. Lee, W., Stolfo, S.J., Mok, K.W.: A data mining framework for building intrusion detection models. In: Proceedings of the 1999 IEEE Symposium on Security and Privacy, pp. 120–132. IEEE (1999)

13. Lee, W., Stolfo, S.J., et al.: Data mining approaches for intrusion detection. In: USENIX Security Symposium, pp. 79–93. San Antonio, TX (1998)
14. Li, Y., Ma, R., Jiao, R.: A hybrid malicious code detection method based on deep learning. Methods 9(5), (2015)
15. Lippmann, R., Cunningham, R.K., Fried, D.J., Graf, I., Kendall, K.R., Webster, S.E., Zissman, M.A.: Results of the DARPA 1998 offline intrusion detection evaluation. In: Recent Advances in Intrusion Detection, vol. 99, pp. 829–835 (1999)
16. Niyaz, Q., Sun, W., Javaid, A.Y., Alam, M.: A deep learning approach for network intrusion detection system. In: Proceedings of the 9th EAI International Conference on Bio-inspired Information and Communications Technologies (Formerly BIONETICS), BICT-15, vol. 15, pp. 21–26 (2015)
17. Özgür, A., Erdem, H.: A review of KDD99 dataset usage in intrusion detection and machine learning between 2010 and 2015. PeerJ PrePrints 4, e1954v1 (2016)
18. Tavallaee, M., Bagheri, E., Lu, W., Ghorbani, A.A.: A detailed analysis of the KDD cup 99 data set. In: IEEE Symposium on Computational Intelligence for Security and Defense Applications, CISDA 2009, pp. 1–6. IEEE (2009)
19. Wang, W., Zhu, M., Zeng, X., Ye, X., Sheng, Y.: Malware traffic classification using convolutional neural network for representation learning. In: 2017 International Conference on Information Networking (ICOIN), pp. 712–717. IEEE (2017)

Detect Malicious Attacks from Entire TCP Communication Process

Peng Fang, Liusheng Huang$^{(\boxtimes)}$, Xinyuan Zhang, Hongli Xu,
and Shaowei Wang

School of Computer Science and Technology,
University of Science and Technology of China, Anhui 230022, Hefei, China
{fape,dwz,wangsw}@mail.ustc.edu.cn, {lshuang,xuhongli}@ustc.edu.cn

Abstract. Malicious attack identification plays an essential role in network security monitoring. Current popular technologies are mainly to select a closely related set of attributes from a packet header for fingerprinting malicious attacks. Those methods are not effective enough because malicious attacks can be disguised as normal applications and we cannot observe their characteristics from only the packer's header. In this paper, we will employ the attributes generated from the entire TCP communication process to identify malicious attacks. A challenging point of our method is how to choose the right attributes from up to 248 properties of TCP flows for fingerprinting low proportion of malicious attacks. A wide variety of real-world viruses are analyzed as the malicious samples, such as extortion virus WannaCry. The experiment results demonstrate that the proposed method can not only fingerprint the viruses but also can accurately identify the types of virus.

Keywords: Fingerprint · Malicious attacks · TCP · Discriminators

1 Introduction

With the development of internet technology and the emergence of a variety of businesses, there is a growing demand for QoS (Quality of Service). The traffic of web browsing, e-mail, and other traditional services are surpassed by more and more new applications. Online shopping, instant messaging, social networking site business and mobile trading have become new leaders in the electronic business. Various applications bring conveniences to users, but they also increase the burden on network administrators. Mobile viruses, trojans and other malicious code spread through a variety of encryption protocols. They disguised as normal traffic to avoid firewall and security software monitoring. Traditional intrusion detection technologies cannot deal with all kinds of unknown application attacks. Traffic classification technologies have become a hot topic to handle this problem. Traditional classification schemes are difficult to identify malicious attacks with high accuracy. They always use packet's headers or destination ports to classify the traffic. The packet's headers often cannot get sufficient information, and

© Springer International Publishing AG 2017
D. Liu et al. (Eds.): ICONIP 2017, Part V, LNCS 10638, pp. 867–877, 2017.
https://doi.org/10.1007/978-3-319-70139-4_88

the ports are often deliberately altered. The average accuracy of those technologies for traffic classifications is 50–70% [1,2]. Now malicious attacks continue to evolve, they can be disguised as normal applications to communicate with hosts. As a result, there is no difference between malicious attacks and benign applications in packet's head. In this situation, some popular malicious attacks detection technologies using the attribute sets extracted from packet's headers are invalid.

To motivate the problem mentioned above, we propose a new method which employs the attributes extracted from entire TCP communication process. The number of the attributes is up to 248. The most challenging work is to choose the features from 248 kinds of properties that can represent the characteristic of various malicious attacks. Six different viruses are analyzed as the malicious samples. The viruses include WannaCry, Ponmocup, Phishing, CryptoWall, ZeroAccess and BlackHole. By carefully designing the feature vector which contains features from multi-domains and doing a series of experiments to decide the classifier, the F-measure of proposed method is 94.2%. The main contributions of this work are as follows:

(1) A novel method which employs the features extracted from entire TCP communication process is presented to fingerprint malicious attacks.
(2) We propose a heuristic dimension reduction method based on the threshold to select most related attribute sets from TCP flows. To choose the optimal number of attributes, various machine learning methods are used in experiments.
(3) We conduct experiments to evaluate the performance of our proposed method on real-world malicious traffic. The experiment results demonstrate that the proposed method outperforms existing algorithms. It can not only fingerprint the virus but also can accurately identify the type of virus.

2   Related Work

Traffic classification has become a research hotspot in recent years. In 2006, Bernaille et al. [3] classified different types of TCP-based applications using an unsupervised machine learning methods called simple K-means. The method mainly used the first few packets of the traffic flow. They are the result of negotiations between the two applications which are usually a pre-defined sequence of information and is distinct among applications. However, this method assumes that they can always capture the first few packets of the communication. The assumption is too harsh. In 2006, Nguyen and Armitage [4] presented a novel supervised method using only the most recent N packets of a flow. It is called a classification sliding window. Since the application classification requires real-time processing, the most recent N packets can ensure the timeliness of each classification and reduce the buffer space. However, this method is not suitable for malicious attacks detection. Crotti et al. [5] proposed a traffic classification method through simple statistical fingerprinting. This method is based on the three attributes captured in IP packets. Their size, inter-arrival time and arrival order are employed to fingerprint applications such as HTTP, SMTP and POP3. They consider the characteristics of a single packet rather than the entire communication process, and they do not concern the safety of the applications.

In 2006 Wright et al. [6] employed the packet size, timing, and directions to identify the common application protocols. The Hidden Markov Models (HMM) is used to build classifiers which is also used in [7–9]. They deal with the identification of the application protocols such as AIM, HTTP, HTTPS and SSH. They also showed that the number of the tunnels can be tracked when there is only a single application protocol leak sufficient information about the flows in the tunnel. Their purpose is not the identification of the malicious attacks.

Boukhtouta et al. [10] proposed a method to fingerprint the malicious traffic in 2014. They employed the 29 kinds of the discriminators captured from the packet header which are presented in [11]. Different machine learning methods are used in their works such as $C4.5$, Naive Bayesian, SVM and Boosted $J48$, Boosted Naive Bayesian classifier. Their work is built on the dataset. Since the malicious attacks are very good at disguising themselves, we cannot identify them using the packet header only. The machine learning methods used in their works cannot get good results in our research.

3   Framework of Application Fingerprinting

3.1   Modeling

Our geometric model is based on 248 different discriminators which are discussed in [12]. It is unreliable to classify traffic by employing client ports or server ports because the port can be deliberately modified. Actually, there are 246 kinds of discriminators employed in this paper. Our model is described as follows:

Let $X = (x_1, x_2, x_3, \ldots, x_n)$ be a collection of all attributes, where n is the number of the discriminators. In this paper, we fix $n = 246$. Let $f = (f_1, f_2, \ldots, f_k) \subseteq X$ be the fingerprint of identifying malicious attacks, where k is selected from threshold based dimension reduction method. The F-measure is employed to describe the accuracy rate.

Let f_{++} be the true positive, which corresponds to the number of positive samples correctly predicted by the classified model. Let f_{-+} be the false positive, which corresponds to the number of negative samples incorrectly predicted by the model. Let f_{+-} be the false negative, which corresponds to the number of positive samples incorrectly predicted by the model. The precision and recall is calculated as $p = \frac{f_{++}}{f_{++}+f_{-+}}$ and $r = \frac{f_{++}}{f_{++}+f_{+-}}$. Our goal is to maximum the F-measure as follows:

$$F_1 = \max_k \{\frac{2rp}{r+p} | (f_1, f_2, \ldots, f_k) \subseteq X\} \tag{1}$$

where $k = 1, 2, 3, \ldots, n$. If network administrators want to detect malicious attacks from traffic, it means that they want to detect as much correct information about the attacks as possible, and they may also hope that there are as few incorrect benign instances as possible. This is why we choose the F-measure as the metric to measure the effect. There are about 246 characteristics describing TCP flows. We employ heuristic attribute reduction to find the most relevant discriminators for malicious attacks. It is also essential to find the optimal machine learning methods by comparing experiment results. The procedure is designed as follows. First, we employ an algorithm to reassemble the TCP sessions. Next, 248 kinds of different discriminators are calculated from communication process. Then, we use a heuristic threshold-based dimension reduction method to select a most related attribute set from discriminators. Finally, we choose the optimal machine learning method to fingerprint malicious attacks.

3.2   Complete Reconstruction of TCP Session

TCP is a reliable transport protocol and transmits data through byte streams. If the byte stream it transmits exceeds the MTU, then the byte stream is divided into multiple small streams, and then sorted by serial number, indicating that those small streams are the same set of a session. Once the user receives the packet, the source and destination IP addresses, source and destination ports, and protocol are taken out to uniquely determine the packet. Because at the same time there may be a number of different sessions in progress, we set up a two-dimensional list to manage a wide range of TCP packets. The horizontal list represents the same TCP session, and each node refers to an IP packet. The vertical list represents multiple TCP sessions.

TCP session reorganization process is actually the insertion and deletion of the list. Whenever a packet is captured, it is checked whether the TCP session which the packet belongs to is already in the linked list. If the corresponding TCP session exists, insert it into the appropriate location according to the serial number. If the TCP session is not in the linked list, a new node is inserted into the end of the list. Finally, the serial number of each packet sequence is continuous, and the first packet is SYN packet, the last packet is FIN packet, then the message is considered complete. The focus and difficulty in this process lie in the problem of packet retransmission, out-of-order, time-out calculation *etc.* After TCP reorganization is completed, we will calculate the 248 discriminators from the complete sessions. We employ the statistical means to get different characteristics of the value, such as the average communication time, the average number of transmission packets, mean of control bytes in packets *etc.* The total 248 discriminators are presented in [12]. In addition, the TCP datagram in the network can be obtained by pulling the traffic in the equipment supporting the BGP protocol.

3.3   Heuristic Threshold Based Dimension Reduction

The information gain has a preference for the attributes with higher values. The information gain ratio has a preference for the attributes with lower values. To mitigate the impact of this preference, the heuristic method is employed in this section: we find out the attributes with the information gain values which are higher than the threshold, and then we employ information gain ratio to get the attributes with high values. Two parameters are needed in this algorithm: g-the threshold of selecting attributes through information gain, h-the threshold of selecting attributes through information gain ratio. Information entropy is the most commonly used measure of sample set purity. Assume that the proportion of the k-th sample in the current sample set D is $p_k(k = 1, 2, \ldots, y)$, the information entropy is defined as $Ent(D) = -\sum_{k=1}^{y} p_k \log_2 p_k$. Assume that the discrete attribute x_i has V attributes. If x_i is used to divide the sample set, it will generate V branch nodes, where the v-th branch node contains all the samples in D with the value x_{iv} in the attribute x_i, noted as D_v. The information gain is calculated as: $Gain(D, x_i) = Ent(D) - \sum_{v=1}^{V} \frac{|D_v|}{|D|} Ent(D_v)$. In general, the greater information gain means that the use of attribute x_i will obtain the greater purity. The attribute will be selected if the corresponding information gain is greater than the threshold. The information gain ratio will be calculated for the selected attributes. The information gain ratio is defined as: $Gain\_ratio(D, x_i) = \frac{Gain(D, x_i)}{IV(x_i)}$ where $IV(x_i)$ is the intrinsic value of x_i. The greater the number of the attribute x_i means the greater the $IV(x_i)$. It is defined as $IV(x_i) = -\sum_{v=1}^{V} \frac{|D_v|}{|D|} \log_2 \frac{|D_v|}{|D|}$. The attributes are selected eventually which are greater than the threshold. The heuristic threshold based dimension reduction method is evolved from $C4.5$ partitioning attribute method.

Actually, we will use different dimension reduction methods to select the most related attribute set such as information gain ratio and FCBF (Fast Correlation-Based Filter) [13]. The experiment results demonstrate that the heuristic threshold based dimension reduction method outperforms others.

4   Simulation Results and Analysis

4.1   Data and Tools

Since the traffic that contains the malicious attacks is very rare, in the first experiment, we employ the data used in [14]. The data sets are traced in several different periods from one site on the Internet. This site is a research-facility host to about 1000 users connected to the Internet via a full-duplex Gigabit Ethernet link. Because we want to fingerprint malicious attacks, we alter the data classes into two categories: ATTACK and BENIGN. The data structure is described in Table 1. The number of the attacks denote the number of TCP communication processes belong to the malicious attacks. WEKA [15] is employed in this paper to execute the machine learning algorithms such as Naive Bayes [16] and Random Forest. The compared results are presented in next section.

4.2   Machine Learning Method Selection

Due to the low proportion of the malicious attacks in original data sets, we will select the best performance machine learning algorithm through experiments. The datasets mentioned in Table 1 are employed to choose the optimal machine

Table 1. Data structure.

Datasets	Attack	Benign	Proportion of attacks	Duration(s)
01	122	24741	0.49%	1821.8
02	324	21961	1.45%	1784.1
03	134	19250	0.69%	1658.5
04	367	65881	0.55%	1664.5
05	446	64590	0.69%	1613.4

learning algorithm. Dataset 01 is used as the training samples and other datasets are used as the test samples. F-measure is employed in this paper. The results are presented in Table 2. The experiment is built on the 246 kinds of discriminators. As can be seen from Table 2, the performance of the Random Forest is the best. We choose the Random Forest as the machine learning method to fingerprint the malicious attacks.

Table 2. F-measure of different machine learning methods.

F-measure	02	03	04	05
Naive Bayes	0.033	0.016	0.495	0.469
C4.5	0.824	0.846	0.680	0.663
Random Forest	**0.890**	**0.904**	**0.851**	**0.807**

4.3 Compared with Existing Methods

In this section, we employ heuristic threshold method(Htm) to select optimal attributes. The Random Forest is used to fingerprint malicious attacks with the optimal attributes. We compare our method with BotFinder [17] and no attribute selection method(Nasm). The BotFinder employs 5 attributes as their feature space. The attributes include the average time interval, the average duration, the average number of source bytes and destination bytes, and the Fast Fourier Transformation. The following results demonstrate that the number of attributes cannot determine the implementation of results and our method outperforms existing methods. The intuitive comparisons are presented in Table 3.

We can find that the precisions of the other two methods are better than heuristic threshold method. Precision indicates that there is fewer benign traffic incorrectly classified as the malicious attacks traffic. This situation is led by the machine learning method. The Random Forest is the combined classifier. It combines n decision trees to build the model, so more attributes can produce more selections of the split. The recall indicates that there is few malicious traffic incorrectly classified as the benign traffic. If network administrators want to find the malicious traffic as much as possible, they may choose the higher recall even when there are some benign traffic included. The recall in Table 3 shows that our method is better than others. The F-measure of our method is better than others on two datasets. The other two methods account for one.

Table 3. Comparison of experimental results on different attribute sets.

Dataset	Value								
	Precision			Recall			F-measure		
	BotFinder	Nasm	Htm	BotFinder	Nasm	Htm	BotFinder	Nasm	Htm
02	**1**	1	0.978	**0.836**	0.802	**0.836**	**0.911**	0.890	0.902
03	**0.983**	0.974	0.960	0.881	0.843	**0.903**	0.929	0.904	**0.931**
04	0.863	1	0.942	0.755	0.741	**0.757**	0.805	**0.851**	0.840
05	0.960	0.975	**0.976**	0.706	0.688	**0.722**	0.814	0.807	**0.830**

4.4   Experiment on Real Virus Data

To demonstrate the practicability of the proposed method, we conduct experiments on real virus data. A computer virus is a set of computer instructions or program code that is programmed or inserted into a computer program to destroy computer functions or to destroy data, affect the use of the computer, and be self-replicating. Computer viruses are extremely dangerous. The virus is contagious, destructive, hidden, latent and triggering. We mainly selected the following several viruses to test: **Ponmocup, Phishing, BlackHole Exploit Kit, CryptoWall, WannaCry** and **ZeroAccess**.

Ponmocup is the world's most successful, oldest and largest botnet. And people have underestimated this unknown threat. Currently, it has infected 15 million devices and steals millions of dollars by plundered bank accounts. The Ponmocup botnet controls 2.4 million devices during the peak season of 2011, and now it controls about half the number of devices.

Phishing is an attack mode by sending a large number of spoofed spam messages that claim to come from a bank or other trusted institution, intended to lure recipients of sensitive information (such as username, password, account ID, ATM PIN, or credit card details).

BlackHole Exploit Kit is as of 2012 the most prevalent web threat, where 29% of all web threats detected by Sophos and 91% by AVG are due to this exploit kit. Its purpose is to deliver a malicious payload to a victim's computer.

WannaCry, CryptoWall is a new type of computer virus which mainly spreads through the mail, the program Trojan horse and hanging horse. The virus is extremely dangerous which will bring incalculable loss to the user. This virus uses a variety of encryption algorithms to encrypt the file, infected files generally can not be decrypted unless the private key of decryption is available.

ZeroAccess is a Trojan horse that uses an advanced rootkit to hide. It can also create a hidden file system, downloads more malware, and opens a back door on the compromised computer.

In order to make the experiment versatile and persuasive, we employ the data sets of real world intrusions which were exposed on the website [18]. The malware traffic was captured when the intrusion is in progress. We reassemble the benign and malicious TCP segments, and the assembled TCP sessions are described in Table 4. The duration represents the time required for malware to complete an attack. Different viruses have different purposes and communication features. Some viruses may cause infected people to upload large amounts of data over a short period of time while others may download large amounts of data. Most viruses work at late night and have relatively short duration. Those features may cause differences between multiple viruses and benign traffic. In order to mitigate the effects of different bandwidths, we removed all the features associated with the bandwidth rate during feature extraction. Our experiments are conducted on the basis of those differences. Because the time complexity of Nasm is very high, resulting in a lack of real-time performance. In the real data environment, we do not consider Nasm. We only compare our heuristic method with the BotFinder, the results are described in Fig. 1.

Table 4. Real world intrusions.

Datasets	Attribute		
	Session	Proportion	Duration(s)
Benign	8504	97.5%	1801.2
Ponmocup	39	0.48%	1563.1
Phishing	87	1.0%	588.5
BlackHole	9	0.10%	33.5
CryptoWall	35	0.40%	90.4
ZeroAccess	48	0.52%	388.4
WannaCry	195	0.2%	695.5

Our method can fingerprint BlackHole Exploit Kit with an accuracy of 100%. The phenomenon indicates that the virus has distinguishable features which lead to the ability to be accurately identified. The F-measure of the Phishing is 93.8% while CryptoWall is 97.1% and ZeroAccess is 98.9%. The accuracy of the BotFinder is much lower by comparison. The F-measure of the BlackHole is 0 which represents BotFinder cannot be qualified to recognize BlackHole. BotFinder is designed for recognizing bots, but as we can see that this method is invalid in some cases. The method proposed in this paper not only recognizes the viruses but also can accurately identify the types of virus. Our method has strong applicability when the intruder deliberately hides his own behavior. This is also the unique advantage of the presented method.

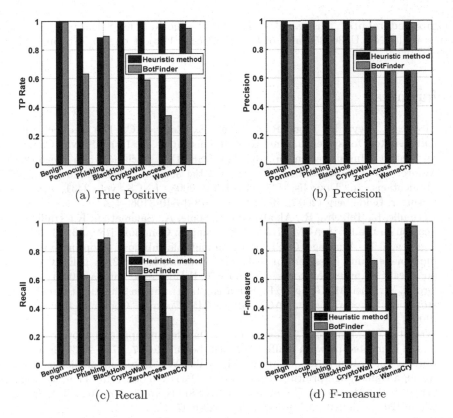

Fig. 1. The identification results of different viruses. Random Forest is employed in this experiment. The black pillars denote our method, the gray pillars represent the BotFinder.

5 Conclusion

The emergence of extortion virus, zombie virus, phishing and so on will cause immeasurable losses to users, especially the emerging WannaCry virus case. In this paper, we present a novel method to fingerprint the malicious attacks. To the best of knowledge, we are the first to propose the idea using 248 discriminators extracted from the entire TCP communication process to fingerprint malicious attacks. To reduce the complexity of the network traffic, we employ a heuristic threshold dimension reduction method which is also easily applied to other missions. In order to improve the accuracy of the fingerprinting, three popular machine learning methods are compared in our experiments. Random Forest is the most suitable learning method for fingerprinting malicious attacks. WannaCry, Ponmocup, Phishing, CryptoWall, ZeroAccess, and BlackHole are analyzed as the malicious samples. The experiment results demonstrate that the proposed method can not only fingerprint the viruses but also can accurately identify the types of virus.

Acknowledgments. This paper is supported by the National Science Foundation of China under No. U1301256 and 61472385, Special Project on IoT of China NDRC (2012 − 2766).

References

1. Moore, D., Keys, K., Koga, R., Lagache, E., Claffy, K.C.: The coralreef software suite as a tool for system and network administrators. In: 15th USENIX Conference on System Administration, pp. 133–144. USENIX Association, San Diego (2001)
2. Moore, A.W., Papagiannaki, K.: Toward the accurate identification of network applications. In: Dovrolis, C. (ed.) PAM 2005. LNCS, vol. 3431, pp. 41–54. Springer, Heidelberg (2005). doi:10.1007/978-3-540-31966-5_4
3. Bernaille, L., Teixeira, R., Akodkenou, I., Soule, A., Salamatian, K.: Traffic classification on the fly. J. ACM SIGCOMM Comput. Commun. Rev. **36**, 23–26 (2006)
4. Nguyen, T.T., Armitage, G.: Training on multiple sub-flows to optimise the use of machine learning classifiers in real-world ip networks. In: 31th IEEE Conference on Local Computer Networks, pp. 369–376. IEEE Press, Tampa (2006)
5. Crotti, M., Dusi, M., Gringoli, F., Salgarelli, L.: Traffic classification through simple statistical fingerprinting. J. ACM SIGCOMM Comput. Commun. Rev. **37**, 5–16 (2007)
6. Wright, C.V., Monrose, F., Masson, G.M.: On inferring application protocol behaviors in encrypted network traffic. J. Mach. Learn. Res. **7**, 2745–2769 (2006)
7. Zhioua, S., Jabeur, A.B., Langar, M., Ilahi, W.: Detecting malicious sessions through traffic fingerprinting using hidden Markov models. In: Tian, J., Jing, J., Srivatsa, M. (eds.) SecureComm 2014. LNICSSITE, vol. 152, pp. 623–631. Springer, Cham (2015). doi:10.1007/978-3-319-23829-6_47
8. Rabiner, L.R.: A tutorial on hidden Markov models and selected applications in speech recognition. J. Proc. IEEE **77**, 257–286 (1989)
9. Durbin, R., Eddy, S.R., Krogh, A., Mitchison, G.: Biological sequence analysis: probabilistic models of proteins and nucleic acids. J. Pro. Sci. **8**, 695–695 (1999)
10. Boukhtouta, A., Lakhdari, N.E., Mokhov, S.A., Debbabi, M.: Towards fingerprinting malicious traffic. J. Proc. Com. Sci. **19**, 548–555 (2013)
11. Alshammari, R.: Automatically generating robust signatures using a machine learning approach to unveil encrypted VOIP traffic without using port numbers, IP addresses and payload inspection. Doctoral Dissertation (2012)
12. Moore, A., Zuev, D., Crogan, M.: Discriminators for use in flow-based classification. Queen Mary and Westfield College, Department of Computer Science (2005)
13. Yu, L., Liu, H.: Feature selection for high-dimensional data: A fast correlation-based filter solution. In: 20th International Conference on Machine Learning, pp. 856–863. AAAI Press, Washington (2003)
14. Moore, A.W., Zuev, D.: Internet traffic classification using Bayesian analysis techniques. J. ACM SIGMETRICS Perform. Eval. Rev. **33**, 50–60 (2005)
15. Hall, M., Frank, E., Holmes, G., Pfahringer, B., Reutemann, P., Witten, I.H.: The WEKA data mining software: an Update. J. ACM SIGKDD Explor. Newsl. **11**, 10–18 (2009)
16. John, G.H., Langley, P.: Estimating continuous distributions in Bayesian classifiers. In: 11th Conference on Uncertainty in Artificial Intelligence, pp. 338–345. Morgan Kaufmann, Montreal (1995)

17. Tegeler, F., Fu, X., Vigna, G., Kruegel, C.: Botfinder: finding bots in network traffic without deep packet inspection. In: 8th International Conference on Emerging Networking Experiments and Technologies, pp. 349–360. ACM, Nice (2012)
18. Real World Intrusions Trace. https://www.netresec.com/?page=PcapFiles

Exploiting Cantor Expansion for Covert Channels over LTE-Advanced

Zhiqiang He, Liusheng Huang$^{(\boxtimes)}$, Wei Yang, and Zukui Wang

School of Computer Science and Technology,
University of Science and Technology of China, Hefei 230027, Anhui, China
{hezhq,zukwang}@mail.ustc.edu.cn, {lshuang,qubit}@ustc.edu.cn

Abstract. Worldwide, the Long Term Evolution Advanced technology has an unprecedented development and popularization in recent years. With the advantages of mobile communication technology, more and more researchers are focused on the security of mobile communication. Until then, some researches about covert channels over the 4th generation mobile communication technology had been proposed. Cantor Expansion is a permutation to a bijection of natural number, so it can be used as a coding scheme for a covert channel. In this paper, a novel class of covert channel based on Cantor Expansion (for decoding) and its inverse operation (for encoding) is proposed and designed for this mobile network. The description, analyses and evaluation of this covert channel will be present in the main part of this paper. Moreover, the peak value of camouflage capability can reach 1470 kbps. Nevertheless it doesnt affect the bandwidth of overt channel and it is difficult to be detected.

Keywords: LTE-Advanced · 4G · Information security · Covert channel · Cantor Expansion

1 Introduction

With the rapid development of the 4th generation mobile communication technology (4G) in recent years, the 3rd Generation Partnership Project (3GPP) organization are constantly improving its protocol specification [1–4], and it had gradually presented Long Term Evolution (LTE), LTE-Advanced and LTE-Advanced Pro. The data transmission rate of LTE-Advanced technology has been promoted to a new height of 220 Mbps from the original 21 Mbps, which has been widely used by many mobile communication suppliers all over the world. In October 2016, the Global Mobile Suppliers Association (GSA) announced the following data: [5] There were 771 operators invested in the LTE in 195 countries, and 6,504 LTE user devices had been announced (GSA October 10, 2016).

From 1973 to now, Lampson [6] first proposed the concept and gave the general communication model for covert channels. Afterwards, the related theory of covert channel has been improved so much in these years. Preliminary researches mainly focused on transferring hidden information through the Internet Protocol between computers. Along with the popularity of LTE-Advanced

© Springer International Publishing AG 2017
D. Liu et al. (Eds.): ICONIP 2017, Part V, LNCS 10638, pp. 878–887, 2017.
https://doi.org/10.1007/978-3-319-70139-4_89

mobile network, Mobile Internet has become the indispensable way for people to communicate with each other. Compared with the traditional mobile network, the LTE-Advanced network got high-quality advantages in the bandwidth, and it is quite seamless to the Internet. However, when it gives people the convenience for information communication, the security will cope with a severe test. Enemy personnel can use these new features to transfer the concealed information more easily than before. Covert channels technology can help to transmit a variety of private information through the normal communication channel, which is a great threat to the defense department and the government.

For the positive demand, covert channels over LTE-Advanced are of great importance to improve the survival and concealment of communication channels employed by commercial espionage. In the 4G era, the well-designed covert channel can ensure the real-time information and the security of the intelligence personnel. For the reverse demand, it is fatal if someone else is able to detect these covert channels and extract information from covert communication, so it is essential to preserve commercial secrets. All of above, the research of the covert channels over LTE-Advanced is imperative. In the future, researchers can find the detection methods through analyzing the creativity of covert channels, which are useful to protect the security of the mobile network.

The rest of this paper is organized as follows. In the next section, we review previous researches on the covert channels over LTE-Advanced. In Sect. 3, we describe the basement of Cantor Expansion (CE) and its inverse operation (ICE). In Sect. 4, our covert channel scheme will be introduced. In Sect. 5, we analyze our scheme with six scenarios. Finally, we make a conclusion in Sect. 6.

2   Related Work

In LTE and LTE-Advanced system, a lot of novel covert channel algorithms are proposed. To recapitulate briefly, they can be divided into three categories:

The first category is based on the wireless carrier, which mainly uses OFDM idle carrier for covert communication or embedded the secret information in the carrier modulation process. Zaid Hijaz put forward a method to construct covert communication based on spectrum structure of OFDM, which conducted covert channel construction of using idle subcarriers in OFDM channel [7]; Padma Priya and Praveen Kumar utilized 1/4 OFDM Cyclic Prefix (CP) as a symbol, and then the secret data is embedded in the CP to complete the construction of the covert channel [8]; Krzysztof Szczypiorski filled the embedding secret information in the physical layer of OFDM, which developed with the application of WiPad with high steganographic capacity [9]. While the data rate is 54 Mbps, the total capacity of WiPad data frame steganography can reach 1.65 Mbit/s.

The second category is based on the mobile application, and it primarily uses voice, mobile multimedia applications such as data flow as the carriers. W. Mazurczyk proposed TranSteg (Transcoding Steganography) for the use of information hiding re-encoding speech, which is especially suitable for use in IP mobile network [10]. W. Mazurczyk also designed a Lost Audio Packets Steganography (LACK) covert channel based on VOIP voice communication data-packet-sending-delay, to conceal information in the man-made transmission of lost data [11]. In the streaming media transmission module, Hideki Noda proposed a covert channel based on wavelet decomposition, which can carry out secret information transmission of video [12]. Whats more, Amr A. Hanafy embedded the secret information in the pixel information of video frames, in order to put forward a kind of covert channel based on video streaming media [13].

The last but not least one is based on LTE and LTE-Advanced communication protocol, which principally exploits packet-timing-order or storage redundancy of the LTE protocol stacks to conceal information. Fahimeh Rezaei created covert channels based on LTE-Advanced protocol. The secret information is embedded in the redundant reserved field of protocol [14]. The authors wrote [15] to propose a covert channel by using the padding field of Transport Block (TB) to hide data. Furthermore, the covert channel capacity has been analyzed with different Modulation and Coding Scheme (MCS) and Transport Block Size (TBS). The maximum camouflage speed can reach 1.162 Mb/s in exceptional circumstances. It is noteworthy that, the method we used to create this covert channel belongs to this category.

3   Cantor Expansion

Cantor Expansion is a permutation to a bijection of natural number, and it is usually used to construct the space compression of the hash table. The essence of CE is calculated for the order of the arrays for the whole permutations from the smallest to the largest, so it is reversible.

Definition: Assuming there are N numbers in all. The CE of X consists of coefficients $a_1, \cdots, a_N \in \mathbb{N}$ such that $1 \leq a_i \leq N, i \in \mathbb{N}$ and

$$X = a_1(N-1)! + a_2(N-2)! + \cdots + a_i(N-i)! + \cdots + a_{N-1}1! + a_N0! \quad (1)$$

In short, the serial number X is determined by the order of its various digital full permutations from the smallest to the largest (starting at number "0"), as we will show in Sect. 3.1.

Considering N permutation $\{1, 2, \cdots, N\}$, the amount of CE is $N!$, so it just needs a little space to store these arrays but not all the permutation. Just like the Eq. 1, it can be introduced to correspond inverse permutation. Thus, it is able to be implemented to the hash table space compression. As mentioned later, ICE will be exploited to encode messages (Algorithm 1) by the method of this covert channel, while CE will be used for decoding.

3.1   Examples of Cantor Expansion and Its Inverse Operation

Cantor Expansion is a wide array of natural numbers which can be used as a bijection, Hash function. Take the array $(2, 1, 4, 3)$ as an example, so $N = 4$ and its expansion can be calculated as below (from the equal sign to the tail end):

Algorithm 1. Encoding Algorithm (Based on ICE)

Require: The amount of PDCP PDUs used to code in each RLC PDU: $\mathbf{N}$
The minimum PDCP SN in this RLC PDU: SN_{min}
Hidden message: $\mathbf{X}$
Ensure: $0 < \mathbf{X} \le \mathbf{N}!$
1: $y = \mathbf{X} - 1$
2: **for** $k = 1 : \mathbf{N}$ **do**
3:   $Used_k = 0$
4: **end for**
5: **for** $i = 1 : \mathbf{N}$ **do**
6:   **if** $y = 0$ **then**
7:     break;
8:   **end if**
9:   $tmp = \left\lfloor \frac{y}{(\mathbf{N}-1-i)!} \right\rfloor$
10: **for** $j = 0 : tmp$ **do**
11:    **if** $Used_j = 1$ **then**
12:      $tmp + +$
13:    **end if**
14: **end for**
15: $a_i = tmp + 1$
16: $Used_{tmp} = 1$
17: $y = y \bmod (\mathbf{N} - i)!$
18: **end for**
19: The expansion for $\mathbf{X}$ will be $a_1(\mathbf{N} - 1)! + \cdots + a_i(\mathbf{N} - i)! + \cdots + a_{\mathbf{N}}0!$
20: **for** $k - 1 : \mathbf{N}$ **do**
21: $SN_k = SN_{min} + a_k - 1$
22: **end for**
23: **Output:** The SN array $SN_1, \cdots, SN_{\mathbf{N}}$

Algorithm 2. Decoding Algorithm (Based on CE)

Require: The SN array: $SN_1, \cdots, SN_{\mathbf{N}}$
1: Get the minimum of SN:
  $SN_{min} = \min \{SN_1, \cdots, SN_{\mathbf{N}}\} - 1$
2: $\mathbf{X} = 0$
3: **for** $k = 1 : \mathbf{N}$ **do**
4:   $a_k = SN_k - SN_{min}$
5: **end for**
6: **for** $i = 1 : \mathbf{N}$ **do**
7:   $tmp = 0$
8:   **for** j=i+1:($\mathbf{N}$) **do**
9:     **if** $a_i > a_j$ **then**
10:      $tmp + +$
11:    **end if**
12: **end for**
13: $\mathbf{X} = \mathbf{X} + tmp \times (N - i)!$
14: **end for**
15: **Output:** The hidden message: $\mathbf{X}+1$

The first element is "2", and then just 1 digit is smaller than it. Thus, $a_1 = 1$, and $a_1 \times (\mathbf{N} - 1)! = 6$; The second element is "1", no digit is smaller than it. Consequently, $a_2 = 0$, and $a_2 \times (\mathbf{N} - 2)! = 0$; The third element is "4", so there are 3 smaller digits including "1", "2" and "3", but digit "1" and "2" had already been used, so $a_3 = 1$, and $a_3 \times (\mathbf{N} - 3)! = 1$; Finally, the fourth element is "3", but there isn't unused smaller digit yet, so $a_4 = 0$, and $a_4 \times (\mathbf{N} - 4)! = 0$; All products are added: $6 + 0 + 1 + 0 = 7$. Hence, the array is in the eighth $(7+1)$ position of the permutations. Therefore, the serial number is $\mathbf{X} = 8$. It is not difficult to be checked by Eq. 1.

Correspondingly, the inverse operation of Cantor Expansion (ICE) is existing. And there is a case in the method of inverse operations. Assuming this is the 8th positions in the permutation of 4 digits: Since $8 - 1 = 7$; We can get that $7 = 1 \times 3! + 0$; Next $1 = 0 \times 2! + 1$; And then $0 = 1 \times 1! + 0$; Lastly, $0 = 0 \times 0!$. According to the above results: There is just $a_1 = 1$ digit ("1") smaller than the first element, so the first element is "2"; There is just $a_2 = 0$ unused digit smaller than the second element, hence the second element is "1"; There is just $a_3 = 1$ unused digit ("3") smaller than the third element, so the third element is "4"; Then the remaining fourth element is "3". Therefore, the 8th array is $(2, 1, 4, 3)$.

3.2 Time Complexity

CE and ICE can match the decimal information with permutation. By analyzing the complexity of them, we can conclude that, when the amount of order to participate is N, both the time complexity of CE and ICE are $\mathcal{O}(N)$. However, the time complexity of directly sorting and searching in the full array tables is $\mathcal{O}(N!)$. And the fastest speed of Binary-Search is just $\mathcal{O}(N \lg N)$. As a result, CE and ICE take significantly advantages in the processing for the rapid conversion in the encoding information compared with other methods. Moreover, they can get highly available automation of encoding and ensure the efficiency of covert communication in the package sorting of covert channel communication.

4   The Proposal of Covert Channel

The LTE-Advanced standard specifies a mobile communication network supplying data rates up to 220 Mbps. These high data rates will be able to sustain new applications and services such as VoIP, streaming multimedia, video conferencing or even a high-speed cellular modem. The protocol stack functions are composed of the Medium Access Control (MAC), Radio Link Control (RLC) and Packet Data Convergence Protocol (PDCP).

4.1   Packet Transmission over LTE-Advanced

With reviewing the protocol specification of LTE-Advanced [3,4], there are a certain amount of PDCP Protocol Data Units (PDU) in each RLC PDU. What's more, PDCP Sequence Number (SN) is an available field. From the perspective of User Equipment (UE), if there is a downlink radio bearer RLC entity in AM mode, when the UE is switched, UE and the load associated with the PDCP entity firstly received some PDCP PDUs from the source eNobeB. After switching, UE will start to receive PDCP PDUs from destination eNobeB. The front part of PDUs are sent from source eNodeB to destination eNodeB, and some of them are transferred from source eNodeB to UE, but this has not yet been confirmed. Therefore, the PDCP PDUs received by UE before and after may be out-of-order or repeated. PDCP SN is the tool to judge the disorder and repeat. For the general mode of operation (i.e. not switching), if it generates a random sequence (due to Automatic Repeat Request), packet order can be ranked by RLC according to the RLC SN. The RLC ensures in-order delivery of PDUs. Out-of-order packets can be delivered during the handover. The PDU SN carried by the RLC header is independent of the PDUs SN (i.e. PDCP SN). An RLC PDU is built from (one or more) PDCP PDU for downlink. Packet order

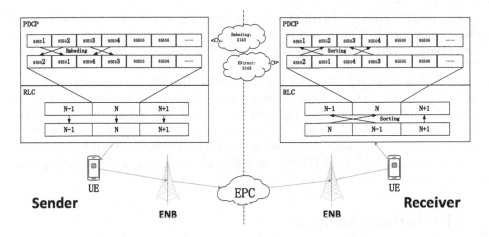

Fig. 1. An brief example of this covert channel

is corrected in the RLC by using SN in the RLC header. After then, when RLC PDU are submitted to the PDCP, PDCP PDUs are in the order of sequence. Therefore, this mechanism can be utilized to design coding scheme with CE and ICE.

4.2   Design of Covert Channel

Now, the following basic models are proposed. In most cases, people are intent on delivering information in relatively stable areas for keeping the signal smoothly. Thus, it can be assumed that the two UEs are going to communicate with each other are immobile, so it is not switching between any eNodeB. All the PDUs are transferred by stable eNodeB. Thus, if the special PDCP PDUs array in one RLC entity is transferred from source UE to destination UE, the PDCP PDUs array in that RLC entity is the same as before. Because the PDCP entity doesn't sort the array without switching between eNodeB. At the same time, this PDCP PDUs array can be employed as a carrier to encode and decode the messages based on ICE and CE. Let's cite an instance as presented in Fig. 1: If there are just four PDCP PDUs (SN=85501, 85502, 85503, 85504) in one RLC PDU of source UE, the array of them can be changed to 85502, 85501, 85504, 85503 in order to encode messages. This array is same while it is received by destination UE. Considering the last four digits of this array is (**2, 1, 4, 3**) so it has been transmitted successfully. Based on the detail in Sect. 3, serial number "**8**" is transferred successfully from source UE to destination UE.

Combining this theory with Sect. 3, a novel kind of method to transfer hidden message including encoding and decoding algorithm is presented (Algorithms 1 and 2 which refer to ICE and CE respectively). Accordingly, the transmission of hidden information can be expressed as follows. First of all, the **Sender** should discuss which serial number to start transmitting hidden data and the length of the permutation (**N**) with the **Receiver**. Afterwards, the **Sender** encode the hidden message by Algorithm 1 in his UE, and then communicate to the **Receiver**. Finally, the **Receiver** read the permutation of the selected PDCP PDUs and then decode it to obtain the hidden information. Last but not the least, in order to guarantee the communication quality of the covert channel, their UE must be always kept in their place without any movement, which is a normal and easy condition. Through these steps, the hidden message can be transferred successfully. To all appearances, this covert channel wouldn't affect the overt channel capacity. As a special timing covert channel in LTE-Advanced, it is based on the order of PDUs for encoding and it doesn't need another special byte to deliver hidden message.

5   Evaluation Analyses

As shown in Algorithms 1 and 2, the encoding and decoding algorithms are devised on account of ICE and CE respectively. According to the analyses, both of their computations are the same as CE and ICE, which had been explained in Sect. 3.2 as $\mathcal{O}(n)$. On the basis of the coding scheme of ICE and CE, the more different digits take participation in the full permutations, the more information can be encoded and decoded.

In the light of the above analyses, the amount of PDCP PDUs in each RLC PDU (N_{PDCP}) is a crucial component to evaluate camouflage capability of this covert channel. However, this amount is affected by many factors, such as Transport Block Size (TBS), IP Packet Size (IPS) and so on. For the sake of convenience, we define the following variables: The size of PDCP header is H_{PDCP} bits; The size of RLC header is H_{RLC} bits; The size of MAC header is H_{MAC} bits; And the size of padding field is S_{pad} bits. On the basis of [2–4], the relationship between them is very complicated as the equation below:

$$N_{PDCP} \times (H_{PDCP} + IPS) + H_{RLC} + H_{MAC} + S_{pad} = TBS \qquad (2)$$

Hence:

$$N_{PDCP} = \left\lfloor \frac{TBS - H_{RLC} - H_{MAC}}{H_{PDCP} + IPS} \right\rfloor \qquad (3)$$

On the whole, we can calculate the theoretically maximum value of N_{PDCP} based on the Eq. 3. It is easy to know, we should choose the maximum of TBS and the minimum of $H_{PDCP}, H_{RLC}, H_{MAC}, IPS$. According to the research [16], current IP packet sizes seem mostly bimodal at 40 B and 1500 B (About 40% and 20% of packets respectively). So we can assume that, IPS is 320 bits (40 B) for our theoretical analyses. Then the minimum of $H_{PDCP}, H_{RLC}, H_{MAC}$ is 8, 0 and 8 bits respectively according to [2–4]. The Transport Block (TB), delivered from the PHY to the MAC, and comprises data from the aforementioned radio subframe. It may contain multiple or partial packets, which is depended on scheduling and modulation. Besides, TBS is influenced by [1] the TBS Index (I_{TBS}) and the total number of allocated Physical Resource Block

(PRB) (N_{PRB}). I_{TBS} is affected by the modulation. And it is obvious that the more I_{TBS} is, the more the TBS is in the same channel bandwidth, which can serve to obtain the maximum of camouflage capability. So far, the UE layer supports the most common three modulation for QPSK (Quadrature Phase Shif Keying), 16-QAM (Quadrature Amplitude Modulation) and 64-QAM while the corresponding I_{TBS} are 9, 15 and 26. For the convenience of analyses, the six different scenarios of the most fashionable channel bandwidth (1.4 MHz, 3 MHz, 5 MHz, 10 MHz, 15 MHz, 20 MHz) are about to evaluate camouflage capability of this covert channel in this section. So we can get Table 1 from [1] with Eq. 3.

Table 1. Case study scenarios

Scenarios	Channel bandwidth	N_{PRB}	Modulation	I_{TBS}	TBS (bits)	N_{PDCP} (Maximum)
A	1.4 MHz	6	QPSK	9	936	2
			16-QAM	15	1800	5
			64-QAM	26	4392	13
B	3 MHz	15	QPSK	9	2344	7
			16-QAM	15	4584	13
			64-QAM	26	11064	33
C	5 MHz	25	QPSK	9	4008	12
			16-QAM	15	7736	23
			64-QAM	26	18336	55
D	10 MHz	50	QPSK	9	7992	24
			16-QAM	15	15264	46
			64-QAM	26	36696	149
E	15 MHz	75	QPSK	9	11832	36
			16-QAM	15	22920	69
			64-QAM	26	55056	167
F	20 MHz	100	QPSK	9	15840	48
			16-QAM	15	30576	93
			64-QAM	26	75376	229

Both the length of downlink and uplink wireless frame structures are 10 ms. Consequently, there are $1000 \times N_{PDCP}$ PDCP PDUs sent from source UE to destination UE per second, while there is only one in a TB and the speed of TB is 1000 per second. The next step is tantamount to analyse the coding scheme. The total number of permutations of N_{PDCP} is $N_{PDCP}!$. Thus, there are $\log_2 (N_{PDCP}!)$ bits hidden information can be sent in one TB (1 ms). Therefore, the camouflage capability per second can be calculated by the following Eq. 4.

$$\mathbf{C} = 1000 \times \log_2 \left(\left\lfloor \frac{TBS - H_{RLC} - H_{MAC}}{H_{PDCP} + IPS} \right\rfloor ! \right) = 1000 \sum_{i=1}^{N_{PDCP}} \log_2 i \quad (4)$$

As presented in Fig. 2, the concealed information capacity can be reach from 1 kbps to 1470 kbps in these scenarios mentioned above. This result is the ideal

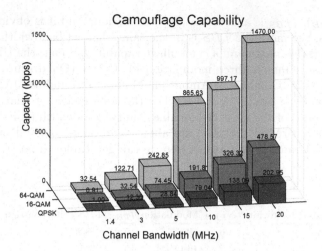

Fig. 2. The maximum of camouflage capability

peak value in theory. But there isn't appropriate simulator to achieve this covert channel yet. However, this method is a novel kind of timing covert channel. In summary, one of the best advantages of this covert channel is its stronger anti-detection ability with large capability, because it doesn't affect the bandwidth of overt channel.

6   Conclusion and Future Work

In this paper, we exploit Cantor Expansion to construct covert channels over LTE-Advanced, which is a novel class of timing channel. Intuitively, this method is distinguished from the habitual covert channel. It is based on the distinctive feature of the relationship of RLC PDU and PDCP PDU. It can work well without switching between UEs, which is reasonable to the stability of communications. There isn't additional bits transfer for the entire transmission process. Compared with other covert channels as described in [7–9,14,15] based on LTE or LTE-Advanced communication protocol, this covert channel has wonderful performance both at camouflage capability and resistance of detection.

After this paper, we are going to concentrate on setting up the real environment of Software Defined Radio (SDR) device. In order to evaluate this covert channel better, some experiments will be carried out in a realistic environment in the future.

Acknowledgments. This work was supported by the National Natural Science Foundation of China (No. 61572456), and the Natural Science Foundation of Jiangsu Province of China (No. BK20151241).

References

1. 3GPP: 3rd Generation Partnership Project: Technical Specification Group Radio Access Network, Evolved Universal Terrestrial Radio (E-UTRA) Physical Layer Procedures (Release 14). Technical Specification 36.213 (2016)
2. 3GPP: 3rd Generation Partnership Project: Technical Specification Group Radio Access Network, Evolved Universal Terrestrial Radio Access (E-UTRAN), Medium Access Control (MAC) Protocol Specifications (Release 14). Technical Specification 36.321 (2016)
3. 3GPP: 3rd Generation Partnership Project: Technical Specification Group Radio Access Network, Evolved Universal Terrestrial Radio Access (E-UTRAN), Radio Link Control (RLC) Protocol Specification (Release 13). Technical Specification 36.322 (2016)
4. 3GPP: 3rd Generation Partnership Project: Technical Specification Group Radio Access Network, Evolved Universal Terrestrial Radio Access (E-UTRAN), Packet Data Convergence Protocol (PDCP) Specifications (Release 14). Technical Specification 36.323 (2016)
5. GSA: Evolution to LTE report: 4G Market and Technology Update. Global Mobile Suppliers Association, 26th October (2016)
6. Lampson, B.W.: A note on the confinement problem. Commun. ACM 16(10), 613–615 (1973)
7. Hijaz, Z., Frost, V.S.: Exploiting OFDM systems for covert communication. In: Military Communications Conference 2010, pp. 2149–2155. IEEE MILCOM (2010)
8. Kumar, P.P., Amirtharajan, R., Thenmozhi, K., Rayappan, J.B.B.: Steg-OFDM blend for highly secure multi-user communication. In: Wireless Communication, Vehicular Technology, Information Theory and Aerospace & Electronic Systems Technology, pp. 1–5. IEEE Wireless VITAE (2011)
9. Szczypiorski, K., Mazurczyk, W.: Hiding data in OFDM symbols of IEEE 802.11 networks. In: Multimedia Information Networking and Security, pp. 835–840. IEEE Wireless MINES (2010)
10. Mazurczyk, W., Szaga, P., Szczypiorski, K.: Using transcoding for hidden communication in IP telephony. Multimed. Tools Appl. 70(3), 2139–2165 (2014)
11. Mazurczyk, W., Lubacz, J.: LACKa VoIP steganographic method. Telecommun. Syst. 45(2–3), 153–163 (2010)
12. Noda, H., Furuta, T., Niimi, M., Kawaguchi, E.: Video steganography based on bit-plane decomposition of wavelet-transformed video. In: Electronic Imaging 2004, pp. 345–353. International Society for Optics and Photonics (2004)
13. Hanafy, A.A., Salama, G.I., Mohasseb, Y.Z.: A secure covert communication model based on video steganography. In: Military Communications Conference 2008, pp. 1–6. IEEE MILCOM (2008)
14. Rezaei, F., Hempel, M., Peng, D., Qian, Y., Sharif, H.: A secure covert communication model based on video steganography. In: Wireless Communications and Networking Conference 2013, pp. 1903–1908. IEEE WCNC (2013)
15. Janicki, A., Mazurczyk, W., Szczypiorski, K.: Steganalysis of transcoding steganography. Ann. Telecommun. (annales des télécommunications) 69(7–8), 449–460 (2014)
16. Sinha, R., Papadopoulos, C., Heidemann, J.: Internet packet size distributions: Some observations. USC/Information Sciences Institute, Technical report, ISI-TR-2007-643 (2007)

AI Web-Contents Analyzer for Monitoring Underground Marketplace

Yuki Kawaguchi[1], Akira Yamada[2], and Seiichi Ozawa[1(✉)]

[1] Graduate School of Engineering, Kobe University, Kobe, Japan
ozawasei@kobe-u.ac.jp
[2] KDDI Research, Inc., Fujimino, Japan

Abstract. It is well known that products for cyber-attacks such as exploits and malware codes are illegally traded on hidden web services called *Dark Web* that are not indexed by conventional search engines we usually use. In general, it is not easy to capture the whole picture of trade activities on Dark Web because special browsers and tools are needed to visit such dark market sites and forums. And they usually require us to make a registration and/or to pass a qualification test. However, to understand the trends of cyber-attacks, there is no doubt that Dark Web is one of the useful information sources. In this paper, we try to understand the sales trends of illegal products for cyber-attacks from the largest marketplace called *AlphaBay*, which is relatively easier to collect information without passing any qualification tests, To monitor business trades on Dark Web, we develop an AI web-contents analyzer, which consists of a Tor crawler to collect the product information and a topic analyzer to capture the trends of what people are interested in and popular products of cyber-attacks. For this purpose, we use a topic model called Latent Dirichlet Allocation (LDA) and we show that the topic analysis would be helpful for predicting new cyber-attacks.

Keywords: Cybersecurity · Topic model · Dark Web · Underground market

1 Introduction

A recent pandemic of a ransomware called *WannaCry* [1] in May 2017 is still fresh in our memory and the news on malware victims shook lots of people over the world. The number of such serious cyber-crimes are rapidly increasing and they are threatening to our daily life.

Who attacks? In reality, malware authors do not always plot cyber-attacks directly. The surprising fact is that quite a few malware authors create malwares to earn money for living. Then, attackers buy such malwares to attack for their own purposes: stealing information/money, interfering services, blackmailing, and so on. In addition, there are a few intermediate stages until a malware is finally circulated: finding a vulnerability, creating an exploit using the vulnerability and a malware as a final product. Therefore, it requires a certain period of

© Springer International Publishing AG 2017
D. Liu et al. (Eds.): ICONIP 2017, Part V, LNCS 10638, pp. 888–896, 2017.
https://doi.org/10.1007/978-3-319-70139-4_90

time to perform a cyber-attack since vulnerabilities are first found. In some cases, the above products for cyber-attacks are assumably traded on *Dark Web* where normal web browsers cannot access and the traffic anonymization technique of onion routing is used so as not to be tracked. It is well known that there are several hidden marketplaces on Dark Web, and the largest one is called *AlphaBay* where lots of illegal products such as malwares, drugs, and weapons. Therefore, if we analyze trades on such underground markets, it would help to understand the trends of cyber-crimes and hopefully to predict new cyber-attacks.

For this purpose, there have been proposed several approaches to collect and analyze information on Dark Web [3,4]. Nunes et al. [3] collected merchandise information from multiple marketplaces and community forums. They developed a binary classifier that can identify whether the site contents are related to cyber-attacks or not, and showed that the proposed classifier was useful for detecting new attacks. Moore and Thomas [4] developed a Tor crawler to collect Dark Web contents automatically, and they classify Tor websites into 12 categories such as arms, drugs, extremism, and hacking by support vector machine. Since these approaches use classifier models, class labels are needed for training and labeling costs are generally expensive.

In this paper, we develop a different type of a Dark Web analyzer that automatically corrects and analyzes the merchandise information of products in the malware category at the largest marketplace called *AlphaBay*. The proposed system extracts various merchandise information such as product name/description, price, on-sale date, sales numbers, and vendor/trust levels. Then, the topics of product descriptions using Latent Dirichlet Allocation (LDA) [5] to classify products into several subcategories of malware products. This system allows us to monitor the trends of popular products at AlphaBay.

Section 2 explains the developed crawler and the collected information on merchandise information. In Sect. 3, we present a new web-mining system to study the trading trends at AlphaBay and we also present a method to analyze the descriptions of cyber-attack products using LDA. Finally, we show our conclusions and future work in Sect. 4.

2 AI Web-Contents Analyzer for Monitoring Marketplace

2.1 Monitoring Information on Market Activities

To monitor the market activities on Dark Web, let us target AlphaBay marketplace [6] in this paper, which is known as the largest shopping site in the Tor hidden service [7]. Figure 3 shows a search result on the AlphaBay marketplace, and we can see several botnets & malware products such as RAT for sale.

As seen in Fig. 3, each column has some information on a product such as title, item number, category, vendor name, views, and price. Therefore, to monitor the market activities, we need to extract product information from web contents by scraping and parsing HTML contents. We extract the following product information in Table 1. Here, *vendor level* and *trust level* are the scales for the reliability of a vendor and a seller, respectively (Fig. 1).

Fig. 1. A search result at AlphaBay [6]

2.2 AI Web-Contents Analyzer

The developed AI web-contents analyzer consists of the following two components: Tor crawler and topic analyzer. Figure 2 illustrates a system configuration. The developed crawler can automatically connect to the Tor network, move to *Botnets & Malware* category on AlphaBay, and collect all web pages of products. Then, a parser extracts the product information in Table 1 from the collected HTML contents, and all of the information are stored in a JSON file. The Tor crawler is developed using a JavaScript library called *Nightmare.js* providing useful functions to manipulate a browser. We also implement an automatic login authentication that requires an operator to input ID, passcode and captcha at first, and pass the authentication without giving ID, passcode, and captcha afterwards. This automatic login is achieved by saving cookies until its timeout comes.

In the topic analyzer, a topic model called Latent Dirichlet Allocation (LDA) [5] is used to analyze product descriptions. LDA can separate product descriptions into several categories (topics). Since a topic consists of several frequent words to represent a group of related documents, an operator can easily know a

Table 1. Extracted product information.

(1) Product name	(2) Page views	(3) Product description
(4) Price (USD and BTC)	(5) User comments	(6) Vendor name
(7) Categories	(8) Sales of products	(9) Product release date
(10) Users' feedbacks	(11) Vendor level	(12) Trust level

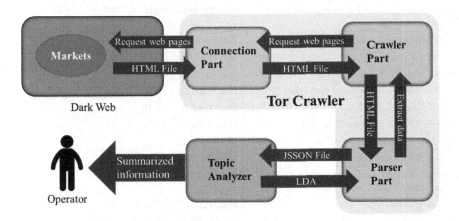

Fig. 2. Architecture of proposed AI web-contents analyzer.

product category by looking at a set of frequent words. The benefit to use LDA is that the learning is conducted in an unsupervised fashion. Therefore, no class label is needed to build the proposed AI web-contents analyzer.

3   Analysis of AlphaBay Trading

In this section, we study the trading activities at the AlphaBay marketplace. First, we analyze web contents collected by the developed AI web-contents analyzer and discuss the sales trends of popular products.

3.1   Sales Trends of Popular Products

The collection of web contents was carried out during the two periods: (1) June 7th - 13th, 2017 and (2) June 14th - 20th, 2017. We collected 1500 pages of botnets & malware products per day and sort them in order of popularity and the new arrival. Let us first compare the top 10 products between the two successive terms.

Table 2 shows the lists of the top 10 sales products. All of the top-sales products were sold more than 10 per week and the highest one, *Hack*, was sold more than 1100. However, the sales number for *Hack* in Table 2 (a) is suspicious because the product ID and the vendor name are exactly the same as those

Table 2. Top 10 sales products at AlphaBay.

Product name	Release date	Price	Sales/Week	Trust level
(a) Top 10 sales (6/7–6/13)				
Hack	May 26 2017	$0.00	1120	4
Jabber spammer	May 10 2017	$10.00	125	3
Super cheap fullz	Apr 17 2015	$0.88	32	7
Always encrypt and stay safe in cyberspace PGP	Oct 29, 2015	$0.00	27	6
Wifi automated cracker	Jan 16 2016	$0.98	25	7
Bitcoin stealer	Nov 4 2016	$0.00	20	4
Super bluetooth hack phones	Jan 14 2016	$0.99	18	7
FREE 230517 update Aegiscrypter 9.5	Dec 8 2016	$0.00	16	4
BlackShades RAT 5.5.1	Jul 11 2015	$1.10	15	7
FREE 230517 Update Zeus 3.0.1	May 8 2017	$0.00	14	4
(b) Top 10 sales (6/14–6/20)				
2017 Ultimate bank cashout guide	May 26 2017	$0.00	36	4
Always encrypt and stay safe in cyberspace PGP	Oct 29 2015	$0.00	18	6
SGCorp coinGrab	Jun 8 2017	$25.00	15	3
BlackShades RAT 5.5.1	Jul 11 2015	$1.10	14	7
FREE 230517 update loki 1.6	May 8 2017	$0.00	14	4
FREE 230517 update cyborg v3.9.2	Dec 8 2016	$0.00	14	4
Rent botnet flat rate	Oct 25 2016	$50.00	13	5
Super cheap fullz	Apr 17 2015	$0.88	12	7
Bitcoin stealer	Nov 4 2016	$0.00	12	4
Free SpyNote	Mar 5 2017	$0.00	10	7

for *2017 Ultimate Bank Cashout Guide*; therefore, there is a possibility that a vendor only changed the name and description of the product for some reason. It is assumed that the vendor operates some kind of cheating acts to increase sales and pretend as a popular item, or the vendor's account was hacked and the product information was rewritten. Even if the top sales numbers are suspicious, the results in Table 2 implies that the underground trade is more active than we expected.

As seen from Table 2, we found that there are many free and inexpensive products on AlphaBay. In addition, the release dates of many products are before Year 2016. We first assumed that malwares using the latest vulnerability are actively traded on such underground markets. In reality, however, it seems that people who do shopping at AlphaBay tend to choose trustworthy conventional products.

Next, let us consider why quite a few vendors are selling products for free. The hypothesis is that vendors want to raise their trust level by providing free products. Figure 3 illustrates a distribution of free and non-free products for different trust levels. We investigated 166 vendors, and it turned out that 73 vendors provided products for free. As seen from Fig. 3, except for the trust level 3, the number of free-product vendors are decreasing as the trust level is rising. Even for vendor level 3, the percentage of free-product vendors is not low.

There could be several reasons why vendors provide products for free. A primary reason might be to raise a trust level. However, it is also likely that a vendor is plotting to distribute malware instead of actual products.

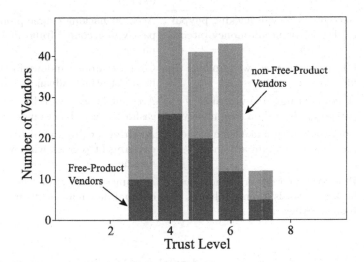

Fig. 3. The relation between the number of free-product vendors and vendor levels.

3.2   Evaluation of Topic Analyzer

Latent Dirichlet Allocation (LDA) [5] is a graphical model that can find out a hidden structure of documents (i.e., topics) and has often been used for a text mining purpose. LDA provides topic information to documents in an unsupervised way; thus, no class label is needed in learning. It means that LDA is also useful to analyze Dark Web contents whose class labels are not always given to a wide variety of illegal products at underground markets. In LDA, topics are estimated by finding co-occurrence relationships among words over document. For example, if *bot* and *windows* simultaneously occur over different product descriptions, and this pair often appears in different products, such products would be identified as a botnet toolkit working on the Windows environment.

Another reason to use LDA is that an obtained topic is represented as a set of typical words in a similar document category. Therefore, if a new type of products appears, it results in a change in topic distributions; that is, we can follow a trend of malware products in an unsupervised fashion.

To evaluate the topic analyzer, we apply LDA to the descriptions of the top 100 sales products which were sold during the two periods: June 7th to 13th and June 14th to 20th, 2017. Table 3 shows the result of topic analysis when the number of topics is set to 5.

Table 3. Topics of product descriptions.

Topic #	6/7–6/13	6/14–6/20
1	Account login vpn hacking paypal bank pass premium money bitcoin	Account hacking course premium pass vpn accounts fraud money phone
2	File bot http windows php anti use files user network	Com password windows file files http bot attack advanced computer
3	Bitcoin carding free victim cc paypal guide cashout btc money	Account login paypal bitcoin cashout btc cc bank guide carding
4	Com tools http make time money pass bank premium account	**Antidetect** program software use **carding** http crack windows fraud internet
5	Password recovery crack advanced hack passwords forensic phone tools hacking	Accounts phone want crack support android new make software hack

As seen from Table 3, words 'antidetect' and 'carding' are newly appeared in Topic 4 during June 14th to 20th. This fact means that a malware product that is undetectable and performs illegal financial transactions was well sold at that time. Actually, there is one malware called *SGCorp CoinGrab* matched the above features in the top ten list in Table 2. Therefore, one can say that our topic analyzer can detect new products that are gathering high attention.

3.3 Sales Trends of Ransomware

WannaCry is a notorious ransomware that encrypts user data in a hard disk for blackmailing. Victim users are usually required a large amount of money to decrypt their data back [1]. WannaCry was used for a worldwide cyber-attack in May 2017. After WannaCry, various malwares such as *UIWIX*, *EternalRocks*, and *Adylkuzz* that use the same exploit called *Eternalblue* were spread over the world. In this section, we investigate the influence of the WannaCry outbreak to the sales trends for other ransomwares on AlphaBay.

First, let us study the change of users' interests in ransomware by examining the number of page views around May 12th, 2017 when WannaCry was first identified. Table 4 shows a list of ransomwares appeared around the WannaCry outbreak. As seen from Fig. 4, a significant increase in page views is identified for Stampado 2 May 12th, while the products other than Stampado 2 are not clearly increased. Stampado is not a family of WannaCry which was first sold in

Table 4. Sale products of ransomwares at AlphaBay

Name	Release date	Price
Stampado 2	Jul 12, 2016	$39.00
Blackmail bitcoin ransomware (With source code)	May 1, 2017	$0.00
BEST blackmail bitcoin ransomware 2017 (With source code)	May 7, 2017	$0.00
BTC ransomware	May 9, 2017	$0.00
Load for your Botnet - ransomware - Bank Bot - clickBot	May 9, 2017	$20.00
The underground collection	May 11, 2017	$0.00
Philadelphia ransomware	May 15, 2017	$29.00

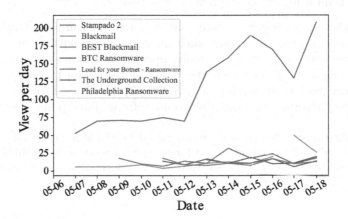

Fig. 4. The number of views per day

mid-2016. BTC Ransomware and Philadelphia Ransomware are not WannaCry families as well, but they were put on a store shelf at AlphaBay around the outbreak of WannaCry. Therefore, it is conjectured that there could be several vendors who try to sell old ransomwares, taking advantage of an outbreak of WannaCry.

From the above observations, it is concluded that monitoring the trends of malware sales at the underground marketplace could be very helpful for predicting new cyber-attacks.

4   Conclusions

In this paper, we developed an AI web-contents analyzer for the study of descriptions of underground marketplaces such as AlphaBay. The developed analyzer consists of the two components: Tor Crawler and topic analyzer that can automatically collect underground products such as malwares, extract important product information, and categorize collected web contents into several groups. Through this web contents analysis, we can monitor the sales trends of illegal products at AlphaBay and it might enable us to predict new cyber-attacks from the sales trends.

To evaluate our analyzer, the sales product information at AlphaBay was analyzed around May 12th, 2017 and from June 7th to 20th 2017. As a result, we can find the following sales trends:

- Customers at AlphaBay tend to choose trustworthy conventional products. Therefore, it is assumed that many vendors would provide their products for free to raise their vendor levels.
- The developed topic analyzer can detect new products with high attention.
- Monitoring the trends of malware sales at the underground marketplace could be very helpful for predicting new cyberattacks.

Although the developed AI web-contents analyzer can collect cyber-attack information from Dark Web, only such information would not be sufficient to predict cyber-attacks precisely. Usually, open information on the surface web (e.g., SNS, blogs, and security reports) would be helpful for this purpose. Therefore, our final goal is to develop an *OSINT* (Open Source INTeligence) which can provide cyber-attack information to users from global point of views.

Acknowledgement. This research was achieved by the Ministry of Education, Science, Sports and Culture, Grant-in-Aid for Scientific Research (B) 16H02874 and the Commissioned Research of National Institute of Information and Communications Technology (NICT), Japan.

References

1. Goodin, D.: NSA backdoor detected on 55,000 windows boxes can now be remotely removed. https://arstechnica.com/security/2017/04/nsa-backdoor-detected-on-55000-windows-boxes-can-now-be-remotely-removed/ ARSTechnica
2. McCoy, D., Bauer, K., Grunwald, D., Kohno, T., Sicker, D.: Shining light in dark places: understanding the tor network. In: Borisov, N., Goldberg, I. (eds.) PETS 2008. LNCS, vol. 5134, pp. 63–76. Springer, Heidelberg (2008). doi:10.1007/978-3-540-70630-4_5
3. Nunes, E., et al.: Darknet and deepnet mining for proactive cybersecurity threat intelligence. In: IEEE Conference on Intelligence and Security Informatics, pp. 7–12 (2016)
4. Moore, D., Thomas, R.: Cryptopolitik and the Darknet. Survival **58**(1), 7–38 (2016)
5. Blei, D.M., Ng, A.Y., Jordan, M.I.: Latent dirichlet allocation. J. Mach. Learn. Res. **3**, 993–1022 (2003)
6. Van Buskirk, J., Naicker, S., Bruno, R.B., Breen, C., Roxburgh, A.: Drugs and the internet (2016)
7. Steven, N.: Buying drugs online remains easy, 2 years after FBI killed Silk Road, 2 October 2015. https://www.usnews.com/news/articles/2015/10/02/buying-drugs-online-remains-easy-2-years-after-fbi-killed-silk-road

Towards an Affective Computational Model for Machine Consciousness

Rohitash Chandra[(⊠)]

Centre for Translational Data Science, The University of Sydney,
Sydney, NSW 2006, Australia
rohitash.chandra@sydney.edu.au

Abstract. In the past, computational models for machine consciousness have been proposed with varying degrees of challenges for implementation. Affective computing focuses on the development of systems that can simulate, recognize, and process human affects which refer to the experience of feeling or emotion. The affective attributes are important factors for the future of machine consciousness with the rise of technologies that can assist humans and also build trustworthy relationships between humans and artificial systems. In this paper, an affective computational model for machine consciousnesses with a system of management of the major features. Real-world scenarios are presented to further illustrate the functionality of the model and provide a road-map for computational implementation.

Keywords: Affective computing · Machine consciousness · Learning algorithms · Neuro-psychology

1 Introduction

Throughout modern digital history, there have been a number of developments in areas of artificial intelligence that mimic aspects or attributes of cognition and consciousness. These developments have been made with the hope to replicate and automate some of the tasks that are undertaken by humans given the industrial demand and constraints of humans on carrying out demanding tasks in limited time. In an attempt to empirically study consciousness, Tononi proposed the information integrated theory of consciousness to quantify the amount of integrated information an entity possesses which determines its level of consciousness [32]. The theory depends exclusively on the ability of a system to integrate information, regardless of having a strong sense of self, language, emotion, body, or an environment. David Chalmers highlighted the explanatory gap in defining consciousness and indicated that the hard problem of consciousness emerge from attempts that try to explain it in purely physical terms [5]. Integrated information theory is based on phenomenological axioms which *begins with consciousness* and indicates that complex systems with some feedback states could have varying levels of consciousness [23]. Howsoever, this does not support the concept of conscious experience. Chalmers argued that the science of consciousness

© Springer International Publishing AG 2017
D. Liu et al. (Eds.): ICONIP 2017, Part V, LNCS 10638, pp. 897–907, 2017.
https://doi.org/10.1007/978-3-319-70139-4_91

must integrate third-person data about behavior and brain processes with first-person data about conscious experience [7]. Some examples include comparing conscious and unconscious processes, investigating the contents of consciousness [24], finding neural correlates of consciousness [6], and connecting consciousness with physical processes.

The field of affective computing focuses on the development of systems that can simulate, recognize, and process human affects which essentially is the experience of feeling or emotion [26,27]. Affective computing could provide better communication between humans and artificial systems that can lead to elements of trust and connectivity with artificial systems [31]. The motivation to have affective models in artificial consciousness would be towards the future of mobile technologies and robotic systems that guide in everyday human activities. For instance, a robotic system which is part of the household kitchen could further feature communication that builds and connectivity from features of affective computing [33]. In the near future, there will also be a growing demand for sex robots, therapeutic and nursing robots which would need affective computing features [1,29]. Moreover, the emergence of smart toys and robotic pets could be helpful in raising children and also assist the elderly [29]. Although mobile application-based support and learning systems have been successfully deployed, they are often criticized for having less physical interactions [22]. In such areas, affects in robots could lead to further help such as stress management and counselling. Hence, the affective attributes are important factors for the future of machine consciousness with the rise of technologies that can assist humans and also build trustworthy relationships with them.

In this paper, an affective computational model for machine consciousnesses is presented that features an algorithm for management of the major aspects of consciousness that range from information processing to critical thinking. Real-world problem scenarios is presented to further illustrate the functionality of the model and a road-map for software-based implementation has been also discussed.

The rest of the paper is organised as follows. Section 2 provides related work and Sect. 3 presents the affective computational model for machine consciousness. Section 4 provides a discussion while Sect. 5 concludes the paper with further research directions.

2   Related Work

The field of natural language processing aims to make computer systems understand and manipulate natural languages to perform the desired tasks [9]. It has been one of the major attributes of cognition and consciousness [15]. One of the major breakthroughs that used natural language processing for cognitive computing has been the design of Watson, which is a system capable of answering questions posed in natural language developed by Ferrucci [10]. Watson won the game of Jeopardy against human players [19]. It had access to 200 million pages of structured and unstructured content including the full text of Wikipedia.

Moreover, IBM Watson was not connected to the Internet during the game. There are a number of applications of Watson technology that includes various forms of search that have semantic properties [11]. Furthermore, it can help in developing breakthrough research in medical and life sciences with a further focus on Big Data challenges. Hence, it was shown that Watson can accelerate the identification of novel drug candidates and novel drug targets by harnessing the potential of big data [8].

With such a breakthrough for development of Watson for cognitive computing, there remains deep philosophical questions from perspective of natural and artificial consciousness [18]. Koch evaluated Watson's level of consciousnesses from perspective of integrated information theory of consciousnesses [17] that views the level of consciousness based on complexity and how integrated the forms of information are in the system. Watson's capabilities motivated to further study the philosophy, theory, and future of artificial intelligence based upon Leibniz's computational formal logic that inspired a 'scorecard' approach to assessing cognitive systems [3]. Metacognition refers to a higher order thinking skills that includes knowledge about when and how to use particular strategies for learning or for problem solving [12]. In relation to metacognition, Watson relied on a skill very similar to human self-knowledge as it not only came up with answers but also generated a confidence rating for them. Therefore, Watson possessed elements of metacognition similar to the human counterparts in the game of Jeopardy [13]. More recently, *AlphaGo* was developed by *Google* to play the board game *Go* which became the first program to beat a professional human player without handicaps on a full-sized 19 × 19 board [14]. Although AlphaGo has been very successful, one can argue that it demonstrated a very constrained aspect of human intelligence that may not necessarily display consciousness.

3 An Affective Computational Model

3.1 Problem Scenarios

The details of the affective model is presented with problem scenarios that are intended to demonstrate its effectiveness. Depending on the experience, there is an expression which would be involuntarily stored as either long or short-term memory depending on the nature of the experience. Moreover, there is also conceptual understanding of implications to the observer and how it changes their long and short-time goals. The output in terms of action or expression could also be either voluntary or involuntary. In some situations, one reacts without controlling their emotions while in others, one does not react. A conscious decision is made depending on the type of personality, depth of knowledge (machine learning models) from past experience (audio, visual and other data). Figure 1 shows a general view of state-based information processing based on experience which acts as input or action while the response acts as the reaction given by the behavior or expression. Figure 2 shows an over overview of the affective model of consciousness that is inter-related with Fig. 1. The states in Fig. 2 shown in blue represent the metaphysical while those in black are the physical states.

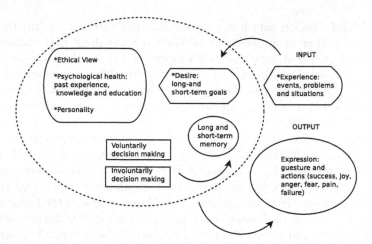

Fig. 1. Response after processing the features that contribute to a system that replicates elements of consciousness

The accounts of situations is further presented that require problem-solving skills feature different states of consciousness. The description of the scenario and how it will be tackled by the proposed affective model is presented with three distinct scenarios as follows.

Scenario 1: *Ramon is traveling on a flight from India to Japan and has a connecting flight from Shanghai, China. His flight lands in Shanghai and he is required to make it to the connecting flight gate. Ramon's boarding pass has gate information missing and since his flight landed about and hour late, he needs to rush to the connecting gate. Ramon is not sure if he will pass through the immigration authority. His major goal is to reach a connecting flight gate. In doing so, he is required to gather information about his gate and whether he will go through the immigration processing counter. He encounters a series of emotions which includes fear of losing the connecting flight and hence exhibits a number of actions that show his emotive psycho-physical states which include sweating, exaggerating while speaking and even shivering due to fear.*

In order for Ramon to successfully make it to the connecting flight on time, he will undergo a series of states in consciousnesses which is described in detail with state references from Fig. 2 as follows.

1. Exit flight and find the way to transfer desk.
   (a) Search for information regarding "transfers and arrivals" through vision recognition system (State 2 and then State 6).
   (b) Process information and make decision to move to the area of "transfers" (State 2 and State 5).
2. Since information that no baggage needs to be collected was already given, check boarding pass for baggage tag sticker.
   (a) Process visual information by checking boarding pass (State 2 and 6)

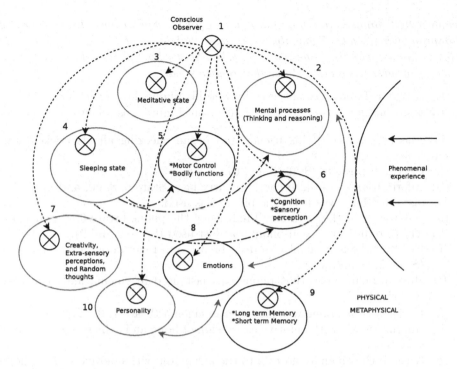

Fig. 2. The conscious observer is defined as the root of consciousness which is also referred to as "qualia". It can enter different states while also having the property to exist within two states, i.e. it can self-replicate as a process, gather knowledge and update long and short-term memories. The blue states are metaphysical and black states are physical. (Color figure online)

3. Confirm with the officer at transfer desk if there is a need to go through the Immigration Counter.
   (a) Find and walk to transfer desk (State 2, 6, and 5)
   (b) Communicate with the officer at transfer desk (State 2 and 6)
   (c) Fear and emotions during communication (State 2, 5, 8, and 10)
4. Information was given by the officer that there is a need to go through immigration booth, hence, prepare boarding pass and passport.
   (a) Rush to the immigration processing section (State 5 and 6).
   (b) Wait in queue and go through a number of emotions such as fear of losing flight and also sweat (State 5, 6, 8, and 10).
5. After immigration processing, find gate information and move to gate and board connecting flight.
   (a) Rush to the gate. In the process breath heavily and also sweat (State 2, 5, and 6).
   (b) Wait at the gate with some random thoughts and then board when called (State 7, 8, 6, 2 and 5).

Scenario 2: *Thomas is in a mall in Singapore for his regular Saturday movies and shopping with friends. Suddenly, he realizes that he can't locate his phone. He tries to brainstorm about the moments when he used his phone. He goes through a series of intense emotive states that includes fear.*

In order for Thomas to successfully find his phone, he will undergo a series of states in consciousnesses with reference from Figure 2 as follows.

1. Thomas first informed his friends and began checking all his pockets and carry bag.
   (a) Check all pockets (State 5 and 6).
   (b) Inform friends and also check in carry bag (State 6, 8, 10, and 5)
2. Brainstorm where was last time phone was used.
   (a) Ask friends if they can recall him using the phone (State 6, 8, and 10).
   (b) Try to remember when phone was last used (State 2 and 9).
   (c) Finally, take a moment of a deep breath and relax in order to remember (State 2, 9, and 3).
3. Recalled from memory that phone was last used in cinema and then rush there to check.
   (a) Recalled that phone was last used in cinema (State 1, 3, 7, and 9).
   (b) Inform friends with emotive expression of hope and achievement (State 6 and 8).
   (c) Rush to the cinema and talk to the attendant with emotive state of hope and fear (State 5, 6 and 8).
   (d) Attendant locates the phone and informs (State 6).
   (e) Emotive state of joy and achievement (State 8).

3.2   Artificial Qualia Manager

The affective computational model has the potential to replicate elements of human consciousness. It can exhibit characteristics with human touch with emotive states through synergy with affective computing. There is a need for management of components in the affective model which would help the property of conscious experience. Hence, there is a need for a manager for qualia. This could be seen as a root algorithm that manages the states with features that can assign the states based on the goal and the needs (instincts) and qualities (such as personality and knowledge).

The artificial qualia manager could be developed with the underlying principle in the case of a security guard officer that monitors a number of video feedbacks from security cameras and uses radio communication with other security guards. The officer follows a channel of communication strategies if any risks or security impeachment occurs. The artificial qualia manager would be monitoring and managing the states to assign jobs for reaching the goal through automated reasoning in machine consciousness as given in Algorithm 1.

In Algorithm 1, the goal and data from audio and visual inputs are used to determine and effectively manage the sequence of states of affective model of consciousness presented in Fig. 2. Once the goal is reached, a series of states can be

---

Algorithm 1. Artificial Qualia Manager

Data: Data from sensory perception (video, audio, and sensor data)
Result: States for consciousness
Initialization (knowledge and personality)

Φ is the list of states
Σ is the final gaol to reach
Ω is the list of actions required to reach the goal

while *alive* **do**
  traverse-states(Σ, Φ])

    while *goal Σ not reached* **do**
      if *challenge* **then**
        1. Nominate a state
        2. Attend to challenge (injury, pain, emotion)
        3. Store short-term and long-term memory

      end
      if *goal Σ reached (success)* **then**
        1. Output through expression (action, gesture, emotion)
        2. Store short-term and long-term memory

      end
      if *goal Σ not reached (failure)* **then**
        1. Output through expression (action, gesture, emotion)
        2. Store short-term and long-term memory

      end
    end
    1. Generate random thoughts based on problem and emotion
    2. Automated reasoning and planning for states needed for future goal(s)
    3. Address the requirements to revisit failed goals

end

---

used for expression which can include a set of emotions. Note that audio and visual data needs to undergo through processing with machine learning tools which would then output some information. For instance, if the goal is regarding finding date information for a boarding pass, then the task would be to be first to translate this higher level task into a sequence of lower level tasks that would execute machine learning components. After these components are triggered, they would return information which will be used by the algorithm to make further decision of states needed to reach the goal. This is illustrated in Fig. 3. There needs to a be a property of states for tasks based on their importance. For instance, priority is given to emergency situations while trying to fill a goal. While fulfilling a goal, priority would be given to aspects such as safety and security. The goal could be similar to those given in Scenario 1 and Scenario 2 where Ramon boards connecting flight and Thomas locates his phone, respectively.

3.3 Implementation Strategies

Multi-task learning is motivated by cognitive behavior where the underlying knowledge from one task is helpful to one or several other tasks. Hence, multi-task learning employs sharing of fundamental knowledge across tasks [4, 25]. The

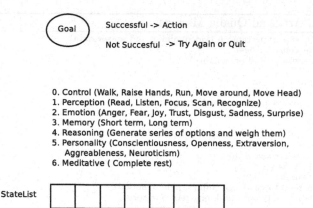

Fig. 3. States in affective computational for the artificial qualia manager

affective computational model can feature multi-task learning for replicating sensory perception through recognition task that includes vision, sensory input for touch and smell and auditory tasks such as speech verification, speech recognition, and speaker verification. Shared knowledge representation would further be used for recognition of objects, faces or facial expression where visual and auditory signals would be used in conjunction to make a decision.

In the identification of objects, its is important learn through the experience of different senses that can be seen as a modular input to biological neural system [16]. Modular learning would help in decision making in cases where one of the signals is not available [30]. For instance, a humanoid robot is required to recognize someone in the dark when no visual signal is available, it would be able to make a decision based on the auditory signal. Ensemble learning could take advantage of several machine learning models which can also include deep learning for visual or auditory based recognition systems [28]. Ensemble learning can also be used to address multi-label learning where instances have multiple labels which is different from multi-class learning [34]. Furthermore, language models that feature neural networks [2,21] could be used in conjunction with speech recognition methods [20].

4   Discussion

It is important to highlight the potential of animal consciousness as it can motivate models for consciousness that fill the gaps in models for human consciousness. In simulation or the need to implant certain level of consciousnesses to robotic systems, it would be reasonable to begin with animal level where certain tasks can be achieved. For instance, a robotic system that can replicate cognitive abilities and level of consciousness for rats can be used for some tasks such as burrowing holes, navigation in unconstrained areas for feedback of videos or information, in disasters such as earthquakes and exploration of remote places, and evacuation sites.

Deep learning, data science and analytics can further help in contribution towards certain or very limited areas of machine consciousness. This is primary to artificially replicate areas of sensory input such as artificial speech recognition and artificial vision or perception. Howsoever, with such advancements in replication of sensory perceptions, one encounters further challenges in developing software systems that manage aspects of perception that lead to a conscious state. With the rise of technologies such as Internet of Things (IoT), sensors could be used to replicate biological attributes such as pain, emotions, feeling of strength and tiredness. However, modelling these attributes and attaining same behaviour in humans may not necessarily mean that the affective model would address hard problem of consciousness. However, at least the model would be seen to replicate conscious experience to a certain degree that could be similar to humans and other animals. Such an affective model, with future implementations could give rise to household robotic pets that would have or could develop emotional relationship with humans. The concern would in giving autonomous control or decision making through simulated aspects of emotion and human behaviour. Humans are well known to be poor decision makers when in emotional states which also resort to level of aggression and violence. Therefore, simulation of affective states need to take into account of safety and security for the future robots that assist humans.

5 Conclusions and Future Work

The paper presented the notion of using affects in computational model for machine consciousness in order to give a human-like expression or behavior for artificial systems. The challenges lie in further refining specific features such as personality and creativity. Howsoever, the proposed model can be a baseline and motivate the coming decade of simulation and implementation of machine consciousness for artificial systems such as humanoid robots. The simulation for affective model of consciousness with the features of artificial qualia manager can also be implemented with the use of robotics hardware. In their absence, simulation can also be implemented through collection of audiovisual data and definition of certain goals. The affective model is general and does not only apply to humanoid robots, but can be implemented in service application areas of software systems and technology.

Future research could concentrate in simulation of the proposed model and development of areas such as artificial personality in machine consciousness. The incorporation of technologies such as IoT, semantic web, cognitive computing and machine learning could guide in simulation of aspects of machine consciousness.

References

1. Bendel, O.: Surgical, therapeutic, nursing and sex robots in machine and information ethics. In: van Rysewyk, S.P., Pontier, M. (eds.) Machine Medical Ethics. ISCASE, vol. 74, pp. 17–32. Springer, Cham (2015). doi:10.1007/978-3-319-08108-3_2

2. Bengio, Y., Ducharme, R., Vincent, P., Jauvin, C.: A neural probabilistic language model. J. Mach. Learn. Res. **3**(February), 1137–1155 (2003)

3. Bringsjord, S., Govindarajulu, N.S.: Leibniz's art of infallibility, Watson, and the philosophy, theory, and future of AI. In: Müller, V.C. (ed.) Fundamental Issues of Artificial Intelligence. SL, vol. 376, pp. 183–200. Springer, Cham (2016). doi:10.1007/978-3-319-26485-1_12

4. Caruana, R.: Multitask learning. Mach. Learn. **28**(1), 41–75 (1997)

5. Chalmers, D.J.: Facing up to the problem of consciousness. J. Conscious. Stud. **2**(3), 200–19 (1995)

6. Chalmers, D.J.: What is a neural correlate of consciousness. In: Neural Correlates of Consciousness: Empirical and Conceptual Questions, pp. 17–40 (2000)

7. Chalmers, D.J.: How can we construct a science of consciousness? Ann. New York Acad. Sci. **1303**(1), 25–35 (2013)

8. Chen, Y., Argentinis, J.E., Weber, G.: IBM Watson: how cognitive computing can be applied to big data challenges in life sciences research. Clin. Ther. **38**(4), 688–701 (2016)

9. Chowdhury, G.G.: Natural language processing. Ann. Rev. Inf. Sci. Technol. **37**(1), 51–89 (2003)

10. Ferrucci, D.: Build Watson: an overview of deepqa for the jeopardy! challenge. In: Proceedings of the 19th International Conference On Parallel Architectures And Compilation Techniques, pp. 1–2. ACM (2010)

11. Ferrucci, D., Levas, A., Bagchi, S., Gondek, D., Mueller, E.T.: Watson: beyond jeopardy!. Artif. Intell. **199**, 93–105 (2013)

12. Flavell, J.H.: Metacognition and cognitive monitoring: a new area of cognitive-developmental inquiry. Am. Psychol. **34**(10), 906 (1979)

13. Fleming, S.M.: Metacognition in mammals and machines. Sci. Am. Mind **25**(5), 35–35 (2014)

14. Gibney, E.: Google AI algorithm masters ancient game of go. Nature **529**, 445–446 (2016)

15. Halliday, M.A.K., Matthiessen, C.M., Yang, X.: Construing experience through meaning: A Language-Based Approach to Cognition. MIT Press, Cambridge (1999)

16. Johnson, M.K.: A multiple-entry, modular memory system. In: Psychology of Learning and Motivation, vol. 17, pp. 81–123. Academic Press (1983)

17. Koch, C.: A theory of consciousness. Sci. Am. Mind **20**(4), 16–19 (2009)

18. Koch, C., Tononi, G.: A test for consciousness. Sci. Am. **304**(6), 44–47 (2011)

19. Markoff, J.: Computer wins on jeopardy!: trivial, its not. New York Times, 16 (2011)

20. Mesnil, G., He, X., Deng, L., Bengio, Y.: Investigation of recurrent-neural-network architectures and learning methods for spoken language understanding. In: Interspeech, pp. 3771–3775 (2013)

21. Mikolov, T., Karafiát, M., Burget, L., Cernocký, J., Khudanpur, S.: Recurrent neural network based language model. In: Interspeech, vol. 2, p. 3 (2010)

22. Mira, J.J., Navarro, I., Botella, F., Borrás, F., Nuño-Solinís, R., Orozco, D., Iglesias-Alonso, F., Pérez-Pérez, P., Lorenzo, S., Toro, N.: A Spanish pillbox app

for elderly patients taking multiple medications: randomized controlled trial. J. Med. Internet Res. **16**(4), e99 (2014)

23. Oizumi, M., Albantakis, L., Tononi, G.: From the phenomenology to the mechanisms of consciousness: integrated information theory 3.0. PLoS Comput. Biol. **10**(5), e1003588 (2014)

24. Overgaard, M., Overgaard, R.: Neural correlates of contents and levels of consciousness. Front. psychol. **1**, 1 (2010)

25. Pan, S.J., Yang, Q.: A survey on transfer learning. IEEE Trans. Knowl. Data Eng. **22**(10), 1345–1359 (2010)

26. Picard, R.W.: Affective computing: from laughter to IEEE. IEEE Trans. Affect. Comput. **1**(1), 11–17 (2010)

27. Picard, R.W., Picard, R.: Affective Computing, vol. 252. MIT press, Cambridge (1997)

28. Schmidhuber, J.: Deep learning in neural networks: an overview. Neural Netw. **61**, 85–117 (2015)

29. Sharkey, A., Sharkey, N.: Granny and the robots: ethical issues in robot care for the elderly. Eth. Inf. Technol. **14**(1), 27–40 (2012)

30. Solomatine, D.P., Siek, M.B.: Modular learning models in forecasting natural phenomena. Neural Netw. **19**(2), 215–224 (2006)

31. Tao, J., Tan, T.: Affective computing: a review. In: Tao, J., Tan, T., Picard, R.W. (eds.) ACII 2005. LNCS, vol. 3784, pp. 981–995. Springer, Heidelberg (2005). doi:10.1007/11573548_125

32. Tononi, G.: An information integration theory of consciousness. BMC Neurosci. **5**(1), 1 (2004)

33. Yamazaki, K., Watanabe, Y., Nagahama, K., Okada, K., Inaba, M.: Recognition and manipulation integration for a daily assistive robot working on kitchen environments. In: 2010 IEEE International Conference on Robotics and Biomimetics (ROBIO), pp. 196–201. IEEE (2010)

34. Zhang, M.L., Zhou, Z.H.: A review on multi-label learning algorithms. IEEE Trans. Knowl. Data Eng. **26**(8), 1819–1837 (2014)

Measuring Self-monitoring Using Facebook Online Data Based on Snyder's Psychological Theories

Ying Liu[1], Yongfeng Huang[2](✉), and Xuanmei Qin[2]

[1] School of Electronic Engineering,
Beijing University of Posts and Telecommunications, Beijing, China
liuyingchina@bupt.edu.cn
[2] Tsinghua National Laboratory for Information Science and Technology,
Department of Electronic Engineering, Tsinghua University, Beijing, China
yfhuang@mail.tsinghua.edu.cn, qinxuanmei@gmail.com

Abstract. Measuring psychological concept self-monitoring (SM) is useful for understanding how people employ impression management strategies in their social interactions. Recently, researchers have attempted to utilize the online user data to measure users' SM value. However, in earlier researches, self-monitoring individuals' specific behavioral and psychological characteristics haven' t been sufficiently considered in the process of features extraction. In this paper, motivated by psychologist Snyder's SM psychological theories, we propose to extract the behavior character of self-monitoring individuals in social network at the macro-level to measure SM. Besides, some other SM relevant features, situational factors, implicit topic words in status updates and demographics are also extracted. Furthermore, a new SM measuring method is presented by exploiting various kinds of users' online data. The experimental results on a benchmark dataset show that all these features are effective and our SM measuring method can outperform many baseline methods.

Keywords: Self-monitoring · Social media · Community · Psycho-informatics

1 Introduction

Self-monitoring (SM) is an important psychological concept introduced by Mark Snyder in the 1970s. High self-monitoring individuals are always shown as "social chameleons", whereas low self-monitoring individuals are introverted relatively [1]. Measuring SM enables us to understand how people control their expressive behaviors and self-presentations in social interactions and gain further insights into the phenomenon of consuming behavior, close relationships and behavior at workplace [2,3].

Traditional SM measuring methods mainly rely on psychometric questionnaires, which have two shortcomings. Firstly, participants' self-reports can be

© Springer International Publishing AG 2017
D. Liu et al. (Eds.): ICONIP 2017, Part V, LNCS 10638, pp. 908–918, 2017.
https://doi.org/10.1007/978-3-319-70139-4_92

influenced by the social desirability and the state of mind as reporting. Secondly, employing psychometric questionnaires at high frequency would be prohibitively expensive and time-wasting. Recently, as Internet users have created large amount of online data, researchers attempt to exploit online data to measure SM. For example, Youyou et al. Demonstrated that Facebook Likes are positively relevant to SM. Their measurement results are even higher than the evaluation by users' friends [4]. He et al. Also found some textual features of the status updates extracted by text analysis tool LIWC in different SM-level groups [5]. However, the specified features extracted by LIWC may not precisely correspond to the implicit topics on Facebook users' status updates. Moreover, from the network perspective, recent researches linked self-monitoring individuals' characteristics to egocentric network and triads [3,6–9], which only focused on the social network at the micro-level without considering the organizational form on a larger perspective.

Psychologist Snyder has proposed a series of theories about self-monitoring individuals' psychosocial characteristics [1,10,11]. Snyder proposed that high self-monitoring individuals are more active in interpersonal relationships and sensitive to situational factors. In linguistic expression, high self-monitoring individuals may also monitor their self-presentation and expressive style. In earlier researches, social activity and situational factors have not been discussed to measure SM. Self-monitoring individuals' linguistic expression features also have not been mined deeply on Facebook status updates.

In this paper, motivated by Snyder's psychological theories, a series of new features are extracted from Facebook online data to measure SM. Firstly, we attempt to cross-fertilize psychological and community-level organizational domain to extract social network activity and situational features from friendship dyads data. Secondly, some implicit topics are extracted from status updates which supplement self-monitoring individuals' linguistic features. Thirdly, demographics are also attempted to measure SM. In the end, a new SM measuring method is proposed in our research. The experimental results show that all kinds of our features are effective and our method can outperform many baseline methods.

2   Methods

2.1   Community-Level Social Network Activity Features Extraction from Friendship Data

Inspired by Snyder's theory that the consequences of self-monitoring can be reflected in the aspects of social interaction and interpersonal relationships [11], we assume that features relevant to social activity may perform well to measure SM.

In our research, we attempt to use the concept of community in complex networks to denote psychologist's description of social worlds and then extract social activity features from the partitioned communities. To detect the community structure on Facebook social network, the widespread infomap algorithm is applied in our work [12]. The extraction methods of the i-th Facebook user's community-level social network activity features are shown as follows.

First, the number of three kinds of connections, first degree connections (NF), second degree connections (NS) and third degree connections (NT) are discussed in our SM measuring method. Specially, first degree connections are the Facebook friends whom $user_i$ is directly connected with. Second degree connections are the Facebook users who are connected to $user_i$'s first degree connections. And third degree connections are the Facebook users who are connected to $user_i$'s second degree connections. Due to the number of connections can indicate users' social activity to a certain extent, NF, NS and NT are extracted as $user_i$'s possible valid SM features.

Furthermore, the number of first degree connections within $user_i$'s community (NFW), the number of first degree connections outside $user_i$'s community (NFO), the number of $user_i$'s associated communities (NAC), and the entropy of communities which $user_i$'s first degree connections belong to (ENF) are also extracted. For example, in Fig. 1, as $user_i$'s first degree connections. $user_j$ and $user_{j+1}$ are within $user_i$'s community. $user_{j+2}$, $user_{j+3}$ and $user_{j+4}$ are outside $user_i$'s community. And $user_i$ is associated with community A, community B and community C. As these four features can reflect the width of social relationships and the social activity of Facebook users, they may validly measure SM. We use an unweighted and undirected graph $G = (V; E)$ to represent the Facebook social network that is being analyzed, which has $|V| = N$ vertices (or Facebook users) and $|E| = M$ edges (or Facebook friendship dyads). Let C_i denote $user_i$'s community and A denote the relationship between any two of the Facebook users. Among them, NFW of $user_i$ is given by

$$NFW_i = \sum_{j=1}^{N_i} A_{ij}, \ A_{ij} = \begin{cases} 1, \ if \ A_{ij} \in E, C_iC_j = 1 \\ 0, \ otherwise \end{cases}, \tag{1}$$

where $C_iC_j = 1$ means $user_i$ and $user_j$ are in the same community. N_i means the total number of users in $user_i$'s community. The NFO of $user_i$ is given by

$$NFO_i = \sum_{j=1}^{N-N_i} A_{ij}, \ A_{ij} = \begin{cases} 1, \ if \ A_{ij} \in E, C_iC_j = 0 \\ 0, \ otherwise \end{cases}. \tag{2}$$

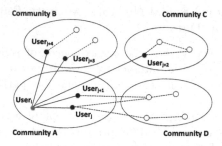

Fig. 1. An illustrative example of community-level social activity features extraction.

The number of $user_i$'s associated communities (NAC) is the sum of unduplicated communities which $user_i$'s first degree connections belong to. The NAC is given by

$$NAC_i = \sum_{j=1}^{N} A_{ij}, \ A_{ij} = \begin{cases} 1, & if \ A_{ij} \in E, j = 1 \ or \ A_{ij} \in E, C_j C_{j-1}...C_1 = 0, j > 1 \\ 0, & otherwise \end{cases}$$

$$(3)$$

The ENF of $user_i$ is given by

$$ENF_i = \sum_{i=1}^{NAC_i} (-p_i * \ln(p_i)) , \tag{4}$$

$$p_i = \frac{\sum\limits_{j'=1}^{N} A_{ij'}, \ A_{ij'} = \begin{cases} 1, & if \ A_{ij'} \in E, v_{j'} \in C_k \\ 0, & otherwise \end{cases}}{\sum\limits_{j=1}^{N} A_{ij}, \ A_{ij} = \begin{cases} 1, & if \ A_{ij} \in E \\ 0, & otherwise \end{cases}}, \forall k = 1, 2, ..., NAC_i , \tag{5}$$

where $v_{j'}$ means $vertex_{j'}$ (or $user_{j'}$).

2.2 Situational Factors Extraction from Friendship Data

Snyder proposed that self-monitoring individuals differ in the sensitiveness to situational factors. To measure self-monitoring individuals' actions, one should seek the characteristics of their situations [10]. Motivated by this, it is assumed that situational factors on Facebook, such as, friends and community environment may be informative in measuring SM.

Friends. Due to homogeneity in social interaction, first degree connections' community-level social activity features may be related to user's SM. Among them, average degree of $user_i$'s first degree connections (ADF) is given by

$$ADF_i = \frac{\sum\limits_{z=1}^{N} \sum\limits_{j=1}^{Nf_i} A_{jz}, \ A_{jz} = \begin{cases} 1, & if \ A_{jz} \in E, v_j \in Vf_i \\ 0, & otherwise \end{cases}}{Nf_i} , \tag{6}$$

where $Vf_i = \{friend_1, friend_2, ..., friend_{N_i}\}$ is the set of $user_i$'s first degree connections. Nf_i is the number of $user_i$'s first degree connections. And average entropy of $user_i$'s first degree connections (AEF) is given by

$$AEF_i = \frac{\sum\limits_{j=1}^{Nf_i} ENF_j}{Nf_i} , \tag{7}$$

where ENF_j is as computed in formula Eq. 4. Nf_i is the number of $user_i$'s first degree connections.

What's more, the difference between $user_i$'s NF and ADF (DDF) and the difference between $user_i$'s ENF and AEF (DEF) are also taken into account to measure SM.

Community Environment. Considering the influence of the environment, users' community-level social network activity has relativity to his community. Average degree of $user_i$'s community (CAD) is given by,

$$CAD_i = 2 * \frac{N_i}{M_i} \; , \tag{8}$$

where N_i is the total number of nodes (or Facebook users) and M_i is the total number of edges (or Facebook friendship dyads) in $user_i$'s community.

Therefore, $user_i$'s relative degree (RD) in the community is given by

$$RD_i = \frac{\sum\limits_{j=1}^{N} A_{ij}, \; A_{ij} = \begin{cases} 1, \; if \; A_{ij} \in E \\ 0, \; otherwise \end{cases}}{CAD_i} \; . \tag{9}$$

2.3 Topic Words Frequency Features Extraction from Facebook Status Updates

Snyder also proposed that high self-monitoring individuals may monitor their self-presentation and expressive behavior when they realize that their expressions are lacking or socially inappropriate [1]. Considering the different language expression styles between high and low self-monitoring individuals, topic words' frequencies are chosen as possible valid features.

In our SM measuring method, Latent Dirichlet allocation (LDA) algorithm is applied to extract topics from Facebook status updates [13]. Moreover, as the number of users who fill out the SM psychometric questionnaire is limited, the topics extracted only in their status updates may not be effective and obvious. The dataset is expanded by a large of extra unlabeled status updates to help improve LDA model's topic-extraction result.

2.4 Demographic Features Extraction from Facebook Profile Data

Recent years, psychological studies have widely discussed the relationship between demographics and personality. As SM is relevant to personality, there may also exist deep association between demographics and SM.

Demographics features are categorical. To eliminate the influence of the size of value, One-Hot Encoding is applied to process them. For example, to feature gender, its initial value is 1 or 0, One-Hot Encoding changes the initial value into two vectors $[0, 1]$ and $[1, 0]$ to represent two different genders.

2.5 Our SM Measuring Method

In our SM measuring method, various kinds of Facebook online data are mined to extract self-monitoring individuals' behavior features based on Snyder's psychological theories. Some other valid features in earlier researches, such as Facebook Likes [4], LIWC textual features [5], smiley frequencies [5] and micro-network structure features [3, 6–8] are also considered in our method. Then the extracted SM relevant features are as input characteristics for the classifier to distinguish high and low self-monitoring individuals. The process details of our method are shown in Fig. 2.

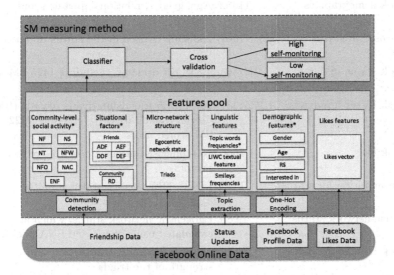

Fig. 2. Our SM measuring method based on Snyder's psychological theories. *Means the features proposed in our research.

3 Experiments

3.1 Dataset

The basis of our research relies on the dataset shared by Mypersonality Project (http://mypersonality.org/wiki). Considering the rationality in classifying high and low self-monitoring individuals, we follow the approach of Snyder by only including users with extremely high and low SM-scale scores, which are defined as scores are above 15 or below 9 [1]. Finally, 11,652 Facebook users' online data are used in the current research to measure SM. Among them, 5,770 (49.5%) are extremely low self-monitoring individuals and 5,882 (50.5%) are extremely high self-monitoring individuals. Detailed description of Facebook users' online data is shown in Table 1.

3.2 Performance Evaluation

Newly Extracted Features' Performance and Effectiveness. Eight parallel experiments are conducted to verify newly extracted features' effectiveness as shown in Table 2. In each experiment, when combining the new features with

Table 1. Detailed description of the dataset in our experiments. N_users means number of users.

Dataset	N_users	Description
Self-monitoring score	11652	25-items SM scale score
Facebook demographics	11652	Age, gender, relationship status and interested in
Smileys used in status updates	1007	293 smileys used in the status updates
Facebook status updates	103,404	4,848,777 updates, including 1475 SM-scored users
Facebook friendship dyads	4,935,337	23,577,255 friendship dyads records (friend1 userid, friend2 userid), including 2322 SM-scored users
Facebook likes	1666	100-dimensional Facebook likes reduced vectors
LIWC annotation for status updates	1474	LIWC 2015 tags for status updates aggregated on a user level
Egocentric network status	988	Egocentric network status, which are calculated using the R and SNA package
Facebook friendship triads	1174	Number of triads, transitivity, and tie strength of the triads

earlier valid features based on the same sample to measure SM, the measurement accuracies can be seen improved, which demonstrates new features' effectiveness. The size of each analysis sample is about 1000. Specifically, each kind of new features' performance and effectiveness is shown as follows.

When detecting communities, 4,935,337 Facebook users' approximately 23 million friendship dyads are input into infomap algorithm. As output, five kinds of community partition results, from coarse to fine grain, with every user's only community number return to us. The second one is chosen as our final result, in which approximately 5 million users are partitioned into 310 communities. The effectiveness of community-level social activity features is shown in Tables 2(a) and (b). And the effectiveness of situational factor features is shown in Tables 2(c) and (d). We can also see that the effectiveness of situational factor features is weaker than social activity features.

Results from Tables 2(e) and (f) show that topic word frequency features are effective and the implicit topics extracted by LDA perform better than textual features extracted by LIWC when measuring SM. The influence of parameter the number of high frequency words and parameter the number of extracted topics on SM measuring accuracy is shown in Fig. 3(a). It can be seen that the 50 topics extracted from 2000 high frequency words perform best to measure SM. Moreover, the topics extracted by LDA are fairly distinct as shown in Table 3. For example, topic1, 2, 12, 16 are related to politics, religion, family and swear

Table 2. SM classification accuracies of eight parallel experiments. Support Vector Machine (SVM) is chosen as the classifier. CS means community-level social activity features. SF means situational factor features. TP means topic features. DM means demographic features. LW means textual features extracted by LIWC. LK means Facebook Likes features.

Features	Accuracy	Features	Accuracy	Features	Accuracy	Features	Accuracy
CS	0.5776	CS	0.5453	SF	0.5647	SF	0.5561
LW	0.5776	LK	0.6401	LW	0.5776	LK	0.6401
CS+LW	**0.6294**	CS+LK	**0.6653**	SF+LW	**0.5813**	SF+LK	**0.6506**
(a)		(b)		(c)		(d)	
Features	Accuracy	Features	Accuracy	Features	Accuracy	Features	Accuracy
TP	**0.5946**	LK	0.6032	DM	0.5672	DM	0.5671
LW	0.5783	LW+LK	0.6633	LW	0.5783	LK	0.6419
TP+LW	**0.6297**	TP +LK	**0.6835**	DM+LW	**0.5858**	DM+LK	**0.6748**
(e)		(f)		(g)		(h)	

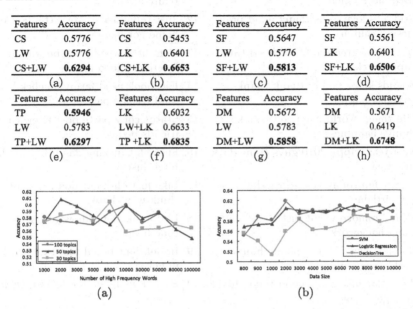

(a)   (b)

Fig. 3. (a) The influence of parameter *number of high frequency words* and parameter *number of extracted topics* on SM measuring accuracy. (b) The influence of parameter *data size* of demographic features on SM measuring accuracy under different classifiers.

words. And it can be found that topic family, leisure and religion act effective to measure SM in both methods of LDA and LIWC [5].

The effectiveness of demographic features is shown in Tables 2(g) and (h). And the influence of parameter data size of demographic features on SM measuring accuracy is shown in Fig. 3(b). As the increase of data size based on different classifiers, the measuring accuracy of demographic features approaches to 0.6 approximately.

Our SM Measuring Method's Performance. In this subsection, we compare our SM measuring method with a series of baseline valid methods. The methods to be compared are: (1) Facebook likes [4]. Youyou et al. chose Facebook Likes as online data to measure SM. (2) LIWC words and smileys frequencies [5]. He et al. Used existing tool LIWC to extract textual features from Facebook status updates. Also, smileys were extracted as unstructured textual features in their research. (3) Egocentric network status [3, 6–8]. Another important part

Table 3. Top words in top 20 valid topics for measuring SM.

Top words		Top words	
1	World Obama American president government men	11	Tonight going night wants tomorrow really loves day
2	God lord love jesus thank good pray blessed church	12	Just today kids baby little house home got like mom
3	Day today great good weekend time happy week	13	Com http www youtube watch facebook video ref
4	Friends know family thank thanks birthday happy	14	Facebook need farmville click help send friends page
5	Life things world live people make change happiness	15	Status copy paste post know cancer friends people
6	New year happy 2010 guys come days christmas	16	Fuck shit fucking damn like ass hell bitch just man
7	Happy fun today going love day wait got birthday	17	Like just oh know don really people think want hate
8	School tomorrow got just today day going homework	18	Ll don like love know just ve let time heart
9	Got just new like car phone hours work going time	19	Im lol dont bored just thats xx like ill got
10	Work day just going good today night time got home	20	En har er die og het kan nu ek van

Table 4. Comparison of SM classification results among various methods. LW means LIWC textual features. SM means smileys. EN means egocentric network status.

Method	Accuracy	Precision	Recall	F1
[5]Likes	0.6447	0.6512	0.7889	0.7125
Ours	0.6968	0.6799	0.8573	0.7468

(a)

Method	Accuracy	Precision	Recall	F1
[10]LW+SM	0.5854	0.6153	0.7337	0.6826
Ours	0.6733	0.6402	0.8779	0.7401

(b)

Method	Accuracy	Precision	Recall	F1
[6-9]EN	0.5636	0.5485	0.8785	0.6737
Ours	0.6301	0.6126	0.8854	0.7386

(c)

of social network features relevant to self-monitoring individuals' organizational behavior concentrated on brokerage, network centrality and network size.

Three parallel experiments are conducted to verify the performance of our SM measuring method as shown in Table 4. The size of analysis sample in the three parallel experiments is 1666, 1007 and 998 respectively. In each experiment, based on the same analysis sample, one kind of baseline valid method and our method are both adopted to measure SM. KNN-imputed method ($k = 10$) is adopted to impute the missing features. Support Vector Machine (SVM) is chosen as the classifier. Each experiment is repeated 10 times independently and average results are reported in Table 4. We can see that our measuring method performs better compared with the baseline methods.

4 Conclusion

Motivated by Snyder's self-monitoring psychological theories, in this paper we propose to measure SM by extracting and fusing different kinds of online user information. Moreover, to improve the measuring accuracy of SM from Facebook online data, a new SM measuring method is presented in our research. The experimental results show that each kind of our proposed features is effective and our method can significantly outperform existing methods.

Acknowledgement. We are grateful to David Stillwell and Michal Kosinski for sharing myPersonality project data with us. The research is supported by the National Key Research and Development Program of China (No. 2016YFB0800402); the National Natural Science Foundation of China (No. U1536201).

References

1. Snyder, M.: Self-monitoring of expressive behavior. J. Pers. Soc. Psychol. **30**(4), 526–537 (1974)
2. Kim, D.H., Seely, N.K., Jung, J.H.: Do you prefer, pinterest or Instagram? The role of image-sharing SNSs and self-monitoring in enhancing ad effectiveness. J. Comput. Hum. Behav. **70**, 535–543 (2017)
3. Wang, S., Hu, Q., Dong, B.: Managing personal networks: an examination of how high self-monitoring individuals achieve better job performance. J. Vocat. Behav. **91**, 180–188 (2015)
4. Youyou, W., Kosinski, M., Stillwell, D.: Computer-based personality judgments are more accurate than those made by humans. Proc. Natl. Acad. Sci. USA **112**(4), 1036–1040 (2015)
5. He, Q., Glas, C.A.W., Kosinski, M., Stillwell, D.J., Veldkamp, B.P.: Predicting self-monitoring skills using textual posts on Facebook. J. Comput. Hum. Behav. **33**, 69–78 (2014)
6. Fang, R., Landis, B., Zhang, Z., Anderson, M.H., Shaw, J.D., Kilduff, M.: Integrating personality and social networks: a meta-analysis of personality, network position, and work outcomes in organizations. J. Organ. Sci. **26**(4), 1243–1260 (2015)
7. Oh, H., Kilduff, M.: The ripple effect of personality on social structure: self-monitoring origins of network brokerage. J. Appl. Psychol. **93**(5), 1155–1164 (2008)
8. Sasovova, Z., Mehra, A., Borgatti, S.P., Schippers, M.C.: Network churn: the effects of self-monitoring personality on brokerage dynamics. J. Adm. Sci. Quart. **55**(4), 639–670 (2010)
9. Kalish, Y., Robins, G.: Psychological predispositions and network structure: the relationship between individual predispositions, structural holes and network closure. J. Soc. Netw. **28**(1), 56–84 (2006)
10. Snyder, M.: Self-monitoring processes. J. Adv. Exp. Soc. Psychol. **12**, 85–128 (1979)

11. Snyder, M., Gangestad, S., Simpson, J.A.: Choosing friends as activity partners: the role of self-monitoring. J. Pers. Soc. Psychol. **45**(5), 1061–1072 (1983)
12. Rosvall, M., Bergstrom, C.T.: Maps of random walks on complex networks reveal community structure. Proc. Natl. Acad. Sci. USA. **105**(4), 1118–1123 (2008)
13. Blei, D.M., Ng, A.Y., Jordan, M.I.: Latent dirichlet allocation. J. Mach. Learn. Res. **3**(January), 993–1022 (2003)

Coevolution of Cooperation and Complex Networks via Indirect Reciprocity

Aizhi Liu[1,2], Lei Wang[1,2], Yanling Zhang[1,2], and Changyin Sun[1,2(✉)]

[1] School of Automation and Electrical Engineering,
University of Science and Technology Beijing, Beijing 100083, China
ustb_laz@xs.ustb.edu.cn, wanglei413@126.com, yanlzhang@ustb.edu.cn
[2] School of Automation, Southeast University, Nanjing 210096, China
cysun@seu.edu.cn

Abstract. Most previous research on indirect reciprocity was in well-mixed population. Distinguishing the interacting network from learning network provides a chance to study indirect reciprocity in networks. Unlike previous research, we propose a coevolution model of cooperation and complex networks via indirect reciprocity, where an individual can interact globally but update strategy locally. Based on this model, we describe the simulation results of coevolution, including the effects of rewiring mechanism on the evolution of cooperation, and how the evolution of cooperation affects networks restructure. Results show that rewiring mechanism favors the evolution of cooperation and the evolution of cooperation can restructure social networks. To understand and explain the results in detail, we graphically depict the snapshots of coevolution process. These findings facilitate us to further understand the evolution of cooperation and the restructure of complex networks.

Keywords: Coevolution · Cooperation · Complex networks · Indirect reciprocity · Rewiring mechanism

1 Introduction

Although Cooperation is widespread in the real world, how natural selection favors cooperation in fierce competition is still a mystery for humans [1]. Evolutionary game theory provides a powerful framework to illustrate the evolution of cooperation [2]. Evolutionary game theory proposes that many mechanisms strongly favor the persistence and emergence of cooperation behavior [3–5]. Indirect reciprocity is one of the fundamental mechanisms that sustain mutual cooperation [3,4,6–10]. It is particularly worth mentioning that it can explain the cooperation between strangers based on reputation [7]. Thanks to reputation, helpful actions are reciprocated not by the recipient as in direct reciprocity, but by other third parties [8]. After the ground breaking research on indirect reciprocity [9,10], reputation assessment attracts many researchers' attention [11–13]. Recently action rules [14] and the spread of reputation [15] become two key research directions of indirect reciprocity.

© Springer International Publishing AG 2017
D. Liu et al. (Eds.): ICONIP 2017, Part V, LNCS 10638, pp. 919–926, 2017.
https://doi.org/10.1007/978-3-319-70139-4_93

However, most previous research on indirect reciprocity was in mixed-well population. Indeed, complex networks exist among individuals in the real world. In various networks, individuals can interact, communicate, share information, and learn from each other. Why did most previous research rarely consider indirect reciprocity in networks? One reason is that most previous research assumes interacting and learning in the same network and thus can not rule out direct reciprocity. But interaction and learning networks are often separated in real life. For example, we can interact with anyone on Internet, but we often compare with our neighbors and imitate them. Distinguishing interacting network from learning networks provides a chance to study indirect reciprocity in networks. Some recent research discussed the independent structures of interaction and learning neighborhood. One research based on prisoner's dilemma game in lattice network shows that interaction network has a critical effect on the evolution of cooperation, but learning network has a weaker impact on the evolution of cooperation [16]. Another study, which was also based on independent structures of interaction and learning neighborhood in complex networks, found that using coalitions to share information and rewiring to change neighbors can promote cooperation through indirect reciprocity [17]. However, their research did not allow disconnected nodes and individuals could reconnect to the one with the highest image score with no restriction. But in a diverse society, there are isolated individuals during evolution, request and consent are needed to establish a new link.

According to donation game and rewiring mechanism, we propose a coevolution model of cooperation and complex networks via indirect reciprocity, where an individual can interact globally but update strategy locally. Here we allow isolated nodes in our network. And a new link is established when both request and consent exist. Based on this model, we first compared the results of static and dynamic complex networks, and found that rewiring mechanism favors the evolution of cooperation whereas static complex networks inhibit the evolution of cooperation. Then we describe the effects of the evolution of cooperation on network restructuring. To better understand and explain the results, we graphically depict the snapshots of the coevolution process. Finally, we end this paper with some concluding remarks.

2   Model

Here we consider a population of n individuals. Initially, all individuals are placed in a complex network (Scale-Free Network $S_n^{\bar{k}, -\lambda}$) owning at least one neighbor, S represents Scale-Free Network, $\bar{k}$ is the average degree, and λ is power exponent. Unlike previous research, each edge of network serves as the strategy learning channel. To model real-world interactions over social networks, each individual can interact with any other one from the whole population. In a generation, m random pairs of individuals, of which one is the potential donor and the other is the recipient, are chosen to play donation games from the entire population. On average, an individual will be chosen $2m/n$ times, and $2m/(n(n-1)) < 1$ ensures that each pair is expected to meet less than once, making direct reciprocity cannot work here.

Donation Game. Each individual has a strategy of whether to help or not according to the image score of the recipient. The strategy is given by a number s: a donor with this strategy provides help if, and only if, the image score r of the potential recipient is as least s. For example, the strategy of donor i is given by s_i and the image level of recipient j by r_j. If $s_i \leq r_j$, the donor will help the recipient at a cost c to himself, in which case the recipient receives a benefit of value b (with $b > c$). Otherwise, the donor decides not to help, then both of them receive zero pay-off. Here the strategy value s ranges from -5 to $+6$. The strategy $s = -5$ represents unconditional cooperators, whereas the strategy $s = 6$ represents defectors. In the first generation, population starts with a random distribution of strategies.

Reputation Assessment Rule: Image Scoring. When the donor i decides to help, his image score increases by one unit; if the donor does not cooperate, it decreases by one unit. This reputation assessment rule is called image scoring which is a classic rule. At the beginning of each generation, the image score of all agents are reset to 0. Here the image score r ranges from -5 to $+5$. If an individual's image score is over $+5$ or below -5, it will be set to be the upper or lower bound.

Strategy Update Rule: Imitating the Best and Asynchronous Update. Here we adopt asynchronous update that a random individual is chosen to update strategy at the end of each generation. Then the chosen individual will compare his pay-off to his neighbors' and imitate the strategy of the one with the highest pay-off. If the chosen individual owns the highest pay-off, the strategy will remain the same. This update rule is called imitating the best. However, if an isolated node is chosen, he will update a random strategy because there is no neighbor to imitate.

Rewiring. Real-world the relationships are not static, and individuals can change their links as they want. By rewiring individuals can modify their neighborhood if they are not satisfied with their neighbors. After the strategy update, an individual will be randomly chosen to rewire. First, the chosen individual x decides to break a link or not according to the break link probability. Here we define the break link probability p_x for individual x, which depends on the proportion of "bad neighbors". Here we call those neighbors whose image scores are smaller than 0 as the "bad neighbors". If the total number of individual x's neighbors is A_x and the number of his "bad neighbors" is B_x. The break link probability p_x for individual x can be expressed by $p_x = B_x/A_x$. You can see that for a chosen individual, the more "bad neighbors" the larger probability to break a link. Once an individual decides to break a link, he will dismiss an edge with his "bad neighbors" randomly. After the break, the individual has an opportunity to request a new link to a random individual whose image score is larger than 0, excluding his present neighbors. It can be accepted or not depending on the requester's image score: if his image score is larger than 0, the request is accepted and a new link is established, otherwise it will be refused and there is no new link. The rewiring process is much closer to actual experiences that we keep away from the bad neighbors and wish to meet new friends with a good reputation. This process requires the new friend with a good reputation to share the same idea. So a new link is established when both request and consent are met. Here we allow isolated nodes in our network. Once an isolated node is chosen to rewire, he will request for a new link directly.

3   Results

In this section, we will report the Monte Carlo simulation results of our model in detail. We start form a Scale-Free Network $(S_n^{\bar{k},-\lambda})$, where population size $n = 100$, the average degree $\bar{k} = 4$, and the power exponent $\lambda = 1.36$. In each generation, $m = 125$ pairs individuals are randomly chosen from the population to play the donation game. In the following section, we mainly focus on the coevolution of cooperation and complex networks via indirect reciprocity.

First, by comparing the results in static network and dynamic network, we show the effects of rewiring mechanism on the evolution of cooperation. As shown in Fig. 1(a), there is a gradual decrease in cooperation level in structural population without rewiring mechanism, and it reduces to 0 at last. Some strategies

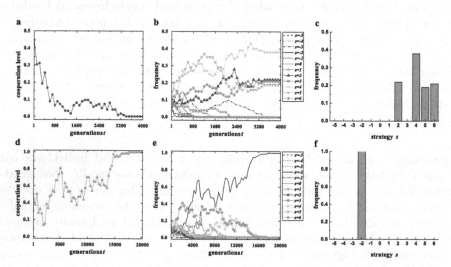

Fig. 1. The effects of rewiring mechanism on the evolution of cooperation. (a–c) respectively show the cooperation level and the frequency of strategies during evolution, and the final distribution of strategies in static networks without rewiring mechanism. (d–f) respectively show that rewiring mechanism leads to an increase in cooperation level, an decrease in strategy variety, and finial dominant strategy $s = -2$.

survive and others disappear in fierce competition in the course of evolution. In the end, the distribution of strategies reaches a stable state that strategy $s = 2$, $s = 4$, $s = 5$ and $s = 6$ survive and dominate the whole population as shown in Fig. 1(b)(c). However, cooperation level is a fluctuant increase process in structural population with rewiring mechanism, and reaches 1 at last as shown in Fig. 1(d). Meanwhile, we can see that those high strategies gradually disappear during the process and finally $s = -2$ dominates the whole population in Fig. 1(e)(f). By comparing results above, we conclude that static networks inhibit the evolution of cooperation and rewiring mechanism can promote the evolution of cooperation, as our model suggests.

Subsequently, we show the effects of the evolution of cooperation on networks restructuring. As shown in Fig. 2(a), the network average degree decreases first and then increases after reaching the minimum, and finally reaches 3.34. Comparing the final degree distribution in Fig. 2(c) with the initial state in Fig. 2(b), we can find that the topology evolves from the initial state of $S_{100}^{4,-1.358}$ to $S_{100}^{3.34,-1.957}$ at the end. Importantly, the final average degree is reduced compared with the initial state, indicating that the average neighbors of individuals in this population become fewer. But the power exponent of network's topology is increasing, which means a few nodes are connected more while most nodes have less connections. From these results we draw a conclusion that the evolution of cooperation can restructure the social networks.

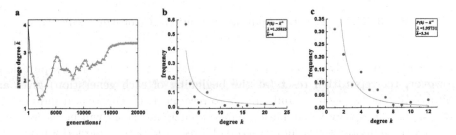

Fig. 2. The effects of the evolution of cooperation on networks restructuring. (a) The average degree changes during the coevolution with rewiring mechanism, (b) the initial network degree distribution, (c) the final network degree distribution.

Finally, we adopt programming language R to visualize the coevolution process and give some typical snapshots in Fig. 3 to explain the results in detail. As shown in Fig. 3, each node represents an individual, each edge serves as the strategy learning channel, and the color on nodes represents its strategy (the lower strategy, the darker color). The first snapshot is the initial distribution of strategies and topology of complex network $S_{100}^{4,-1.358}$. The last snapshot is the final state when the evolution is stable at the 20000th generation, and the light blue color represents the strategy $s = -2$ which dominates the whole population. From the second ($t = 5000$) and the third ($t = 10000$) snapshots, we can see that some individuals with high strategy (i.e., those who often refuse to

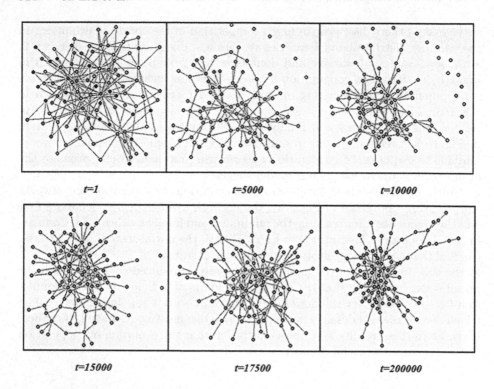

Fig. 3. The typical stages during coevolution. (Color figure online)

help others) gradually become isolated by others, due to their bad reputation. However, the reputation reset (at the beginning of each generation) gives an opportunity for the isolated individuals to reconnect. If the isolated individuals are not chosen to play the donation game, their image score will be 0. So when the isolated individual with image score of 0 requests or is requested a new link, the connection will be established. Therefore, the isolated nodes can reconnect into the networks as shown in the second ($t = 15000$) and the third ($t = 17500$) snapshots. Then these individuals can imitate the strategy of others with the highest pay-off but low strategy in strategy-updating process. If they still choose to defect because of keeping the high strategy, they will be isolated again.

The above results show that, reputation-based rewiring mechanism plays an important role in promoting the evolution of cooperation during the coevolution process. By dismissing link with a bad neighbor and building a new link with a good individual, the individuals with bad reputation get isolated and those who have a good reputation gather together in networks. Applying the imitating the best strategy rule, individuals with a good reputation can learn from each other, and strategies of isolated individuals cannot spread. Only when isolated individuals have a good image, they can reconnect into the networks. However, they can not own a good image and high payoff at the same time. According to strategy update rule, these strategies of individuals with low pay-off cannot spread. Thus reputation-based rewiring mechanism can favor the emergence of cooperation effectively.

During the coevolution process, since individuals with high strategy and bad reputation are gradually isolated, the network average degree decreases at the early generations. Afterwards, some individuals reconnect into networks, so the network average degree increases but no more than the initial state. However, since some individuals with high strategy and bad reputation are isolated once again, the network average degree decreases at the same time. After a long fluctuation process, the network average degree reaches a stable value and is lower than the initial one. Some individuals with good reputation can attract many connections, but most individuals own limited links. That's why the power exponent of the complex network is increasing during the coevolution process.

4 Conclusions

In this paper, we propose a coevolution model of cooperation and complex networks via indirect reciprocity, where an individual can interact globally but update strategy locally. Unlike previous research, we allow isolated nodes in our networks and new link is established once request and consent are met. Simulation results show that rewiring mechanism promotes the evolution of cooperation, and the evolution of cooperation can restructure social networks. To understand and explain the results in detail, we give more intuitive explanation by snapshot adopting complex networks visualization technology. These findings facilitate us to further understand the evolution of cooperation and the restructuring of complex networks.

Acknowledgments. This work was supported by the Fundamental Research Funds for the Central Universities (No. FRF-TP-15-116A1), the China Postdoctoral Science Foundation (No. 2015M580989), and the National Natural Science Foundation of China (Nos. 61603036, 61520106009 and 61533008).

References

1. Rand, D.G., Nowak, M.A.: Human cooperation. Trends Cogn. Sci. **17**, 413–425 (2013)
2. Smith, J.M.: Evolution and the Theory Games. Cambridge University Press, Cambridgeshire (1982)
3. Nowak, M.A.: Five rules for the evolution of cooperation. Science **314**, 1560–1563 (2006)
4. Zaggl, M.A.: Eleven mechanisms for the evolution of cooperation. J. Inst. Econ. **10**, 197–230 (2014)
5. Liu, A.Z., Zhang, Y.L., Chen, X.J., Sun, C.Y.: Randomness and diversity matter in the maintenance of the public resources. EPL **117**, 58002 (2017)
6. Nowak, M.A., Sigmund, K.: Evolution of indirect reciprocity. Nature **437**, 1291–1298 (2005)
7. Wedekind, C.: Give and ye shall be recognized. Science **280**, 2070–2071 (1998)

8. Swakman, V., Molleman, L., Ule, A., Egas, M.: Reputation-based cooperation: empirical evidence for behavioral strategies. Evol. Hum. Behav. **37**, 230–235 (2015)
9. Nowak, M.A., Sigmund, K.: Evolution of indirect reciprocity by image scoring. Nature **393**, 573–577 (1998)
10. Nowak, M.A., Sigmund, K.: The dynamics of indirect reciprocity. J. Theor. Biol. **194**, 561–574 (1998)
11. Sigmund, K.: Moral assessment in indirect reciprocity. J. Theor. Biol. **299**, 25–30 (2012)
12. Ohtsuki, H., Iwasa, Y.: How should we define goodness? Reputation dynamics in indirect reciprocity. J. Theor. Biol. **231**, 107–120 (2004)
13. Ohtsuki, H., Iwasa, Y.: The leading eight: social norms that can maintain cooperation by indirect reciprocity. J. Theor. Biol. **239**, 435–444 (2006)
14. Whitaker, R.M., Colombo, G.B., Allen, S.M., Dunbar, R.I.M.: A dominant social comparison heuristic unites alternative mechanisms for the evolution of indirect reciprocity. Sci. Rep. **6**, 31459 (2016)
15. Seki, M., Nakamaru, M.: A model for gossip-mediated evolution of altruism with various types of false information by speakers and assessment by listeners. J. Theor. Biol. **407**, 90–105 (2016)
16. Tian, L.L., Li, M.C., Wang, Z.: Cooperation enhanced by indirect reciprocity in spatial prisoners dilemma games for social P2P systems. Physica A. **462**, 1252–1260 (2016)
17. Peleteiro, A., Burguillo, J.C., Chong, S.Y.: Exploring indirect reciprocity in complex networks using coalitions and rewiring. In: 13th International Conference on Autonomous Agents and Multiagent Systems, Paris, pp. 669–676. AAMAS Press (2014)

Erratum to: A Brain Network Inspired Algorithm: Pre-trained Extreme Learning Machine

Yongshan Zhang[1], Jia Wu [2], Zhihua Cai[1(✉)], and Siwei Jiang[1]

[1] Department of Computer Science, China University of Geosciences,
Wuhan 430074, China
{yszhang, zhcai}@cug.edu.cn
[2] Department of Computing, Faculty of Science and Engineering,
Macquarie University, Sydney, NSW 2109, Australia
jia.wu@mq.edu.au

Erratum to:
Chapter "A Brain Network Inspired Algorithm: Pre-trained
Extreme Learning Machine" in: D. Liu et al. (Eds.):
Neural Information Processing, Part V, LNCS 10638,
https://doi.org/10.1007/978-3-319-70139-4_2

The original version of this chapter contained an error. The title of the paper was incorrect in the original publication. The title should be read as follows: A Brain Network Inspired Algorithm: Pre-trained Extreme Learning Machine.

The updated online version of this chapter can be found at
https://doi.org/10.1007/978-3-319-70139-4_2

Author Index